Lecture Notes in Physics

Edited by H. Araki, Kyoto, J. Ehlers, München, K. Hepp, Zürich
R. Kippenhahn, München, H. A. Weidenmüller, Heidelberg
J. Wess, Karlsruhe and J. Zittartz, Köln

Managing Editor: W. Beiglböck

273

Models and Methods in Few-Body Physics

Proceedings of the 8th Autumn School on
the Models and Methods in Few-Body Physics
Held in Lisboa, Portugal, October 13–18, 1986

Edited by
L. S. Ferreira, A. C. Fonseca and L. Streit

Springer-Verlag
Berlin Heidelberg New York London Paris Tokyo

Editors

L. S. Ferreira
Departamento de Física, Universidade de Coimbra
P-3000 Coimbra, Portugal

A. C. Fonseca
Centro Física Nuclear, Av. Gama Pinto, 2
P-1699 Lisboa, Portugal

L. Streit
Fakultät für Physik, Universität Bielefeld
D-4800 Bielefeld 1, Germany

ISBN 3-540-17647-0 Springer-Verlag Berlin Heidelberg New York
ISBN 0-387-17647-0 Springer-Verlag New York Berlin Heidelberg

Printing: Druckhaus Beltz, Hemsbach/Bergstr.;
Bookbinding: J. Schäffer GmbH & Co. KG., Grünstadt
2153/3140-543210

To the memory of our
colleague and friend

Harald Zingl

PREFACE

The Autumn School in Physics at the Instituto de Investigação em Física e Matemática (Complexo II do INIC) was first organized in 1979 under the auspices of the Instituto Nacional de Investigação Científica (INIC), the Junta Nacional de Investigação Científica e Tecnológica (JNICT) and the Fundação Calouste Gulbenkian (FCG). In Portugal, a country where there is little tradition of strong investment of public funds in the development of science and technology, it has not always been an easy task to organize this School in the past. Much of the credit goes to the organizers of the early Schools who had to cope with underfunding and lack of both adequate secretarial infrastructures and of understanding and support from government agencies. Fortunately over the years the situation has improved and for the first time this year the funds assigned to the organization of the School doubled those of the previous year. Although the present budget of the Autumn School is still rather small compared to those of other prestigious Schools that regularly take place in Europe and North America, we are particularly indebted to INIC, JNICT and FCG for recognizing the importance of the Autumn School for the physics community in Portugal. Their renewed interest brought about the possibility of inviting a large body of speakers this year whose expertise spans all major developments that have taken place in the area of Few-Body Physics since the last Trieste School in 1978. All the speakers have devoted a great deal of effort and care to the preparation of their manuscripts and we expect this series of lectures to be of value to students and researchers in few-body physics and in the related areas of nuclear chemistry and particle physics. The main purpose of these lectures is to present in a pedagogical way the major methods that are presently used to obtain accurate solutions for dynamical systems involving few particles or constituents. The underlying unity of presentation lies in transparency and completeness, which is sometimes achieved through simple examples that illustrate complicated details of the theory. For this and for the prompt submission of manuscripts we thank each individual speaker.

Last but not least we would like to express our gratitude to the Luso-American Foundation for making it possible, at very short notice, to fly one of the speakers from Australia, and to the staff at Complexo II whose help in the organization of this School was invaluable. Particular recognition goes to Fátima Loureiro for her excellent secretarial help, to Maria João Garrido for being a perfect treasurer, and to António Silva for patiently correcting the final manuscripts and for typing some of the lectures.

Lisboa, January 1987 The editors

8th AUTUMN SCHOOL

Scientific Committee:

L.S. Ferreira, Univ. Coimbra
A.C. Fonseca, CFNUL-INIC
L. Streit, Univ. Bielefeld

Organizing Committee:

A.C. Fonseca, Chairman
A. Barroso
J. Dias de Deus
L.S. Ferreira
R.V. Mendes
L. Streit

Sponsors:

The Instituto Nacional de Investigação Científica
The Junta Nacional de Investigação Científica e Tecnológica
The Fundação Calouste Gulbenkian
The Fundação Luso-Americana para o Desenvolvimento
The Foreign Ministeries of Austria, Belgium, France
and the Netherlands through the Cultural Exchange Program
between Portugal and each of the above countries
The Banco Português do Atlântico.

CONTENTS

PART I - NON RELATIVISTIC FADDEEV-LIKE EQUATIONS FOR THREE- AND FOUR-PARTICLES

PART II - INTEGRAL AND DIFFERENTIAL METHODS FOR THE SOLUTION OF THE SCHRÖDINGER EQUATION

PART III - RELATIVISTIC FEW-BODY EQUATIONS AND BAG MODELS

PART IV - ANTINUCLEON PHYSICS

PART V - FEW-BODY SYSTEMS WITH CHARGED PARTICLES AND CALCULATION OF ELECTROMAGNETIC OBSERVABLES

PART VI - OTHER APPLICATIONS OF SCATTERING THEORY METHODS

PART I

NON RELATIVISTIC FADDEEV-LIKE EQUATIONS FOR THREE- AND FOUR-PARTICLES

FEW-BODY EQUATIONS AND THEIR SOLUTIONS IN MOMENTUM SPACE

W. Glöckle
Institut für Theoretische Physik II
Ruhr-Universität Bochum
D 463o Bochum 1, West-Germany

Contents:

1. Two- and three-body scattering

Let V be a two-body interaction by which two particles are scattered from initial momentum states $|\vec{p}_1\rangle|\vec{p}_2\rangle$ to final momentum states $|\vec{p}_1{}'\rangle|\vec{p}_2{}'\rangle$. We would like to know the transition amplitude for that process. In lowest order in V it is clearly

$$\langle \vec{p}_1{}'\vec{p}_2{}'|t^{(1)}|\vec{p}_1\vec{p}_2\rangle = \langle \vec{p}_1{}'\vec{p}_2{}'|V|\vec{p}_1\vec{p}_2\rangle \tag{1.1}$$

The interaction, however, may act a second time. Between two interactions the particles have to propagate freely and one gets

$$\langle \vec{p}_1{}'\vec{p}_2{}'|t^{(2)}|\vec{p}_1\vec{p}_2\rangle = \langle \vec{p}_1{}'\vec{p}_2{}'|V g_0 V|\vec{p}_1\vec{p}_2\rangle \tag{1.2}$$

We shall see in section 3.3 that the free propagator g_0 is given by

$$g_0 = \lim_{\varepsilon\to 0} \frac{1}{e + i\varepsilon - h_0} \tag{1.3}$$

where e is the available energy $(p_1^2/2m_1 + p_2^2/2m_2)$ and h_0 the operator of kinetic energy. If V acts two times then also three, four, ... times, always with free propagations in between and we arrive at the complete picture

$$\langle \vec{p}_1{}' \vec{p}_2{}' | t | \vec{p}_1 \vec{p}_2 \rangle = \langle \vec{p}_1{}' \vec{p}_2{}' | V + V g_0 V + V g_0 V g_0 V + \cdots | \vec{p}_1 \vec{p}_2 \rangle \tag{1.4}$$

This is the Born- or Neumann-series. One way to sum it up is by rewriting it into an integral equation. Indeed, we may present the infinite series for the t-operator as

$$t = V + V g_0 (V + V g_0 V + V g_0 V g_0 V + \cdots \quad) \tag{1.5}$$

and recover t in the bracket again. Therefore t obeys

$$t = V + V g_0 t \tag{1.6}$$

which is called a Lippmann-Schwinger (LS) equation [1]. We shall describe in section 3.3 representations of that integral equation and their solution.

Three-body scattering is richer in asymptotic configurations than 2-body scattering. We show in Fig. 1 the three possible configurations with two fragments, one a pair bound by a two-body interaction and the other one an elementary particle. It is obviously convenient to denote these configurations or channels by the number of the unbound particle, as indicated in Fig. 1.

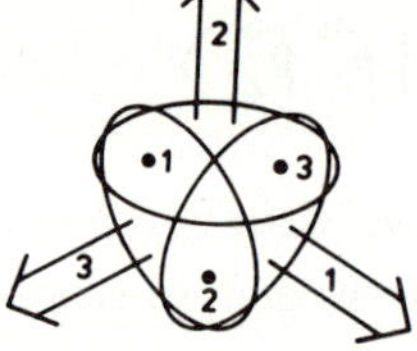

Fig. 1: The three two-body fragmentation channels in a three-body system

Of course, the bound pairs can be in their ground state or in excited states. If the energy is high enough, there is a fourth asymptotic configuration, the break-up channel, where all particles are free.

Each of the four channels can initiate a scattering process whicn in general will have a probability to end in any of the four channels again. In the usual experimental set up (see however the end of section 3.4) the initial state consists of two fragments. As long as they are well separated in space and the two fragments do not interact that state is governed by a simple channel Hamiltonian. Let us regard the configuration 1 in which particles 2 and 3 interact and form a bound state φ and particle 1 is free. The channel Hamiltonian is

$$H_1 \equiv h_0^{(1)} + \left(h_0^{(2)} + h_0^{(3)} + V_{23} \right) \equiv H_0 + V_{23} \tag{1.7}$$

and the eigenstate describing the initial configuration is

$$|\phi_1\rangle = |\vec{p}_1\rangle |\vec{p}_{23}\rangle |\varphi_{23}\rangle \tag{1.8}$$

In analogy to the notation used to distinguish the channels in Fig. 1 we write

$$V_1 \equiv V_{23} \quad , \quad V_2 \equiv V_{31} \quad , \quad V_3 \equiv V_{12} \tag{1.9}$$

Thus the channel states with two fragments obey

$$H_i \phi_i \equiv (H_0 + V_i)\phi_i = E \phi_i \tag{1.10}$$

where E is the total (asymptotically available) energy. It is the sum of kinetic energies for the two fragments and the (negative) bound state energy ε_i of the pair. In the break-up configuration for well separated noninteracting particles the channel Hamiltonian is simply H_0 and its eigenstate

$$|\phi_0\rangle = |\vec{p}_1 \, \vec{p}_2 \, \vec{p}_3\rangle \tag{1.11}$$

The Schrödinger equation (1.10) for the channel states ϕ_i can be rewritten into the integral form

$$\phi_i = \frac{1}{E - H_0} V_i \phi_i \tag{1.12}$$

The free resolvent operator

$$G_0(z) \equiv \frac{1}{z - H_0} \tag{1.13}$$

is apparently singular if z lies in the spectrum of H_0, which extends from E = 0 to infinity. As is well known, however, the limits $z = E + i\varepsilon$, $\varepsilon \to 0$, exist and they are different for $E > 0$. In other words, $G_0(z)$ is defined on the complex energy plane cut along the positive real axis.

In (1.12) no precaution or prescription is necessary for $E > 0$ since V_i does not act on particle i nor on the center of mass motion of the bound pair. We decompose H_o in the case of $i = 1$ as

$$H_o = h_o^{(1)} + h_{o,CM}^{(23)} + h_{o,rel}^{(23)} \tag{1.14}$$

The first two parts act directly on the momentum eigenstates in ϕ_1 and (1.12) reduces simply to $(\varphi_1 \equiv \varphi_{23})$

$$\varphi_1 = \frac{1}{\varepsilon_1 - h_{o,rel}^{(23)}} V_1 \varphi_1 \tag{1.15}$$

which is well defined since ε_1 is negative.

That homogeneous integral equation (1.15) for a bound state reveals that V_1 acts an infinite number of times. Indeed, any number of interactions and free propagations in between is implied by (1.15)

$$\varphi_1 = g_o(\varepsilon_1) V_1 \, g_o(\varepsilon_1) V_1 \cdots\cdots \varphi_1 \tag{1.16}$$

This property will be important below.

Let us now regard all possible processes contributing to the transitions from the initial configuration 1 to all final configurations 1,2,3 and the break-up configuration, which we denote by O. That central problem for three-particle scattering is depicted in Fig. 2.

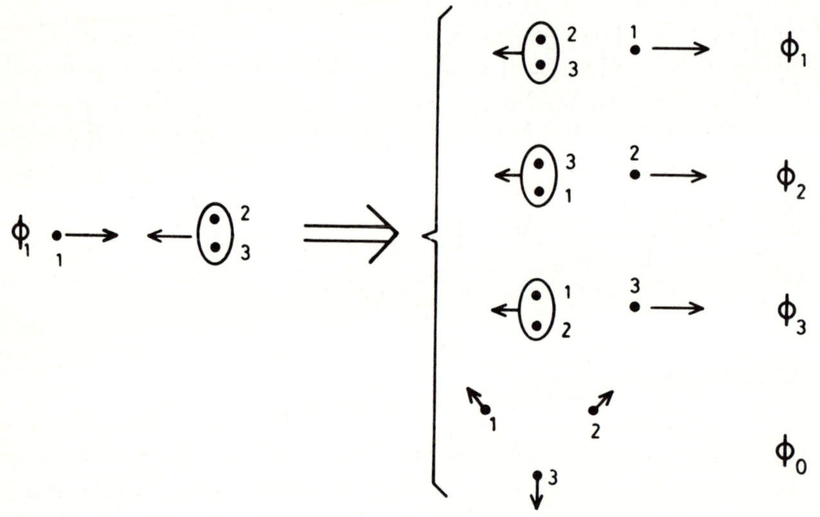

Fig. 2: Possible scattering processes initiated in channel 1

We denote the four transition amplitudes by

$$(\phi_k | U_{k1} | \phi_1) \quad , k = 0,1,2,3 \tag{1.17}$$

and begin with transitions to the direct (k = 1) and rearrangement configurations (k = 2,3). In order not to overload the presentation we neglect three-body forces here and refer to section 3.4 for their inclusion. In lowest order in the pair interactions the initial particle 1 interacts once with the constituents 2 and 3 of the bound fragment:

$$(\phi_k | U_{k1}^{(1)} | \phi_1) = (\phi_k | V_2 + V_3 | \phi_1) \quad k = 1,2,3 \tag{1.18}$$

For k = 2 and 3 we may use equivalent forms. One has

$$(\phi_2 | V_2 | \phi_1) = (\phi_2 | V_1 | \phi_1) \tag{1.19}$$

$$(\phi_3 | V_3 | \phi_1) = (\phi_3 | V_1 | \phi_1) \tag{1.20}$$

This is an immediate consequence of (1.10). Therefore we may write as well

$$(\phi_1 | U_{11}^{(1)} | \phi_1) = (\phi_1 | V_2 + V_3 | \phi_1) \tag{1.21}$$

$$(\phi_2 | U_{21}^{(1)} | \phi_1) = (\phi_2 | V_3 + V_1 | \phi_1) \tag{1.22}$$

$$(\phi_3 | U_{31}^{(1)} | \phi_1) = (\phi_3 | V_1 + V_2 | \phi_1) \tag{1.23}$$

Here the interactions of the free particle with the constituents of the bound pair in the final state occur .

The pair interactions may act a second time with a free propagation in between and we expect

$$(\phi_k | U_{k1}^{(2)} | \phi_1) \overset{?}{=} (\phi_k | (V_1 + V_2 + V_3) G_0 V_2 + (V_1 + V_2 + V_3) G_0 V_3 | \phi_1) \tag{1.24}$$

As we shall show in section 3.4 the free propagator for three particles is

$$G_0 = \lim_{\varepsilon \to 0} \frac{1}{E + i\varepsilon - H_0} \tag{1.25}$$

Let us regard

$$(\phi_1 | V_1 G_0 V_2 + V_1 G_0 V_3 | \phi_1) \tag{1.26}$$

a contribution to (1.24) for $k = 1$. From the discussion following eqn (1.15) we know that the effect of the pair interaction V_1 followed by a free propagation is already included in ϕ_1 to all orders. In other words simply using (1.12) that contribution (1.26) shrinks to

$$(\phi_1 | V_2 + V_3 | \phi_1) \tag{1.27}$$

which is the first order expression (1.21). Similarily, the action of V_2G_o or V_3G_o onto ϕ_2 or ϕ_3 from the left just reproduces the channel states and the corresponding contributions in (1.24) just reduce to the first order expressions. Consequently, the real second order contributions are

$$(\phi_1 | U_{11}^{(2)} | \phi_1) = (\phi_1 | (V_2 + V_3) G_o (V_2 + V_3) | \phi_1) \tag{1.28}$$

$$(\phi_2 | U_{21}^{(2)} | \phi_1) = (\phi_2 | (V_3 + V_1) G_o (V_2 + V_3) | \phi_1) \tag{1.29}$$

$$(\phi_3 | U_{31}^{(2)} | \phi_1) = (\phi_3 | (V_1 + V_2) G_o (V_2 + V_3) | \phi_1) \tag{1.30}$$

We recognize that the interactions to the left (the last ones read from the right) are with the free <u>final</u> particle.

Let us consider the third order processes

$$(\phi_k | U_{k1}^{(3)} | \phi_1) \overset{?}{=} (\phi_k | (V_1 + V_2 + V_3) * \\ * \; G_o (V_1 + V_2 + V_3) G_o (V_2 + V_3) | \phi_1) \tag{1.31}$$

We pick out the suspicious candidate

$$(\phi_k | V_k G_o (V_1 + V_2 + V_3) G_o (V_2 + V_3) | \phi_1) \tag{1.32}$$

and see immediately that it shrinks to

$$(\phi_k | (V_1 + V_2 + V_3) G_o (V_2 + V_3) | \phi_1) \tag{1.33}$$

which we discussed already in second order. Therefore, the proper third order contributions are

$$(\phi_1 | U_{11}^{(3)} | \phi_1) = (\phi_1 | (V_2 + V_3) G_o (V_1 + V_2 + V_3) G_o (V_2 + V_3) | \phi_1) \tag{1.34}$$

$$(\phi_2|U_{21}^{(3)}|\phi_1) = (\phi_2|(V_3+V_1)\,G_0\,(V_1+V_2+V_3)\,G_0(V_2+V_3)|\phi_1) \qquad (1.35)$$

$$(\phi_3|U_{31}^{(3)}|\phi_1) = (\phi_3|(V_1+V_2)\,G_0\,(V_1+V_2+V_3)\,G_0(V_2+V_3)|\phi_1) \qquad (1.36)$$

Their structure is: interactions of the initial free particle, free propagation followed by interactions of all three particles, again free propagation and finally interactions of the final free particle. Clearly this will also be the general sequence of processes in all higher orders: aside from the restrictions to the extrem right and left, where the interactions are only with the free particle, the interactions in between are with all particles.

In (1.21), (1.28), (1.34), U_{11} ends (to the left) either by V_2 or V_3. Correspondingly we decompose U_{11} into two parts and regard the one ending by V_2:

$$U_{11}^{(V_2)} = V_2 + V_2\,G_0\,(V_2+V_3) + V_2 G_0\,(V_1+V_2+V_3)\,G_0(V_2+V_3)$$
$$+ V_2\,G_0\,\sum_i V_i\;G_0\,\sum_j V_j\;G_0\,(V_2+V_3) + \cdots \qquad (1.37)$$

Then we modify the terms containing only V_2 as

$$V_2\,G_0 G_0^{-1} + V_2\,G_0 V_2\,G_0 G_0^{-1} + V_2\,G_0 V_2\,G_0\,V_2\,G_0\,G_0^{-1} + \cdots \qquad (1.38)$$

Since U_{11} is to be applied onto ϕ_1 we can replace G_0^{-1} by V_1 and get

$$U_{11}^{(V_2)} = V_2\,G_0\,(V_1+V_3) + V_2\,G_0\,V_2\,G_0\,(V_1+V_3)$$
$$+ V_2\,G_0(V_1+V_3)\,G_0\,(V_2+V_3) + V_2\,G_0(V_1+V_3)\,G_0 \,* \qquad (1.39)$$
$$*\,\sum_j V_j\;G_0\,(V_2+V_3) + \cdots$$
$$= V_2\,G_0\,\left[(V_1+V_3) + (V_1+V_3)\,G_0\,(V_2+V_3) + (V_1+V_3)\,G_0\sum_j V_j\;G_0\,(V_2+V_3) + \cdots\right]$$
$$+ V_2\,G_0\,V_2\,G_0\,\left[(V_1+V_3) + \cdots\right]$$
$$+ \cdots$$

Comparing with (1.22), (1.29), (1.35) we discover the first few terms of U_{21} in the brackets. It needs some contemplation to convince oneself that this is true for all higher orders and we find

$$U_{11}^{(V_2)} = V_2 G_0 U_{21} + V_2 G_0 V_2 G_0 U_{21} + \cdots$$
$$= (V_2 + V_2 G_0 V_2 + V_2 G_0 V_2 G_0 + \cdots) G_0 U_{21} \tag{1.4o}$$

The remaining infinite sequence in (1.4o) describes pure two-body processes with particle 2 being a spectator. It is the sequence of the type already encountered in (1.5) and we denote it by T_2:

$$T_2 \equiv V_2 + V_2 G_0 V_2 + V_2 G_0 V_2 G_0 V_2 + \cdots \tag{1.41}$$

As in (1.6) we can sum it into the integral equation

$$T_2 = V_2 + V_2 G_0 T_2 \tag{1.42}$$

Proceeding in the same manner with the second part $U_{11}^{(V_3)}$ which ends to the left by V_3 and introducing the T_3-operator driven by V_3 we find altogether

$$U_{11} = T_2 G_0 U_{21} + T_3 G_0 U_{31} \tag{1.43}$$

Encouraged by that result one can expect that

$$U_{21} = \; ? \; + T_3 G_0 U_{31} + T_1 G_0 U_{11} \tag{1.44}$$

$$U_{31} = \; ? \; + T_1 G_0 U_{11} + T_2 G_0 U_{21} \tag{1.45}$$

where the question marks indicate possible driving terms. This would be a closed system of integral equations for the three transition operators U_{k1}, $k=1,2,3$. A glance at U_{21} in (1.22), (1.29), (1.35) reveals that it decomposes as expected into two parts ending by V_3 and V_1, respectively. The discussion of the corresponding infinite sequences of processes is the same as for U_{11} and we indeed find (1.44). There is one term, however, the single term V_1 which plays a specific role since channel 1 is singled out as the initial one. As we saw V_1 acting on ϕ_1 is the same as G_0^{-1} acting on ϕ_1. Therefore, the question marks in (1.44) and (1.45) have to be replaced by G_0^{-1}. Summarizing (1.43)-(1.45) we have gotten the Alt-Grassberger-Sandhas (AGS) equations [2]:

$$U_{k1} = \bar{\delta}_{k1}\, G_0^{-1} + \sum_{l \neq k} T_l\, G_0\, U_{l1} \tag{1.46}$$

Clearly the index 1 can be replaced by 2 or 3 if the initial channel is chosen differently.

Using algebraic manipulations the AGS equations can be derived quite easily and rigorously [3]. We shall give an example in section 3. 4. The physical insight, however, may be more transparent by deriving (1.46) via the Born series as done above.

If we iterate (1.46) we find the multiple scattering series

$$U_{k1} = \bar{\delta}_{k1} G_0^{-1} + \sum_{\substack{l \neq k \\ l \neq 1}} T_l + \sum_{l' \neq k} T_{l'}\, G_0 \sum_{\substack{l \neq l' \\ l \neq 1}} T_l + \cdots \tag{1.47}$$

which describe a sequence of two-body scattering processes between <u>different</u> pairs. Since the pair interactions are summed up to infinite orders into two-body T-operators T_k, consecutive T_k's have to be different. As a consequence the kernel of (1.46)

$$K \equiv \begin{pmatrix} 0 & T_2 & T_3 \\ T_1 & 0 & T_3 \\ T_1 & T_2 & 0 \end{pmatrix} G_0 \tag{1.48}$$

which has the typical Faddeev structure [4] with zeros along the diagonal gets connected after one iteration. In other words, after one iteration only terms of the type $T_k\, G_0\, T_l\, G_0$ with $k \neq l$ occur which connect <u>all</u> three particles by interactions. This makes the kernel well behaved for numerical treatments [5].

Finally let us regard the transition operator into the break-up configuration. In the first few orders the transition amplitude will be

$$\begin{aligned}(\phi_0 | U_{01} | \phi_1) = (\phi_0 | &(V_2+V_3) + (V_1+V_2+V_3)\, G_0 (V_2+V_3) \\ &+ (V_1+V_2+V_3)\, G_0 (V_1+V_2+V_3)\, G_0 (V_2+V_3) + \cdots | \phi_1)\end{aligned} \tag{1.49}$$

Aside from a missing single V_1 term the three series of processes ending by V_1 or V_2 or V_3 to the left, respectively, have the same structure. Let us regard for example $U_{01}^{(V_2)}$ which by definition ends with V_2:

$$U_{01}^{(V_2)} \equiv V_2 + V_2 G_0 (V_2+V_3) + V_2 G_0 (V_1+V_2+V_3) G_0 (V_2+V_3)$$

$$+ V_2 G_0 (V_1+V_2+V_3) G_0 (V_1+V_2+V_3) G_0 (V_2+V_3) + \cdots \tag{1.50}$$

Clearly it is identical to (1.37) and therefore

$$U_{01}^{(V_2)} = T_2 G_0 U_{21} \tag{1.51}$$

$$U_{01}^{(V_3)} = T_3 G_0 U_{31} \tag{1.52}$$

$$U_{01}^{(V_1)} = T_1 G_0 U_{11} - V_1 \tag{1.53}$$

For on-shell scattering $(\phi_0 | V_1 | \phi_1)$ does not contribute and we find

$$U_{01} = T_1 G_0 U_{11} + T_2 G_0 U_{21} + T_3 G_0 U_{31} \tag{1.54}$$

The determination of the break-up operator U_{01} is therefore reduced to quadrature once the operators for transitions between two-body fragmentation channels are determined.

2. Two- and three-body bound states

We encountered already in (1.15) the homogeneous eigenvalue equation

$$|\varphi\rangle = \frac{1}{\varepsilon - h_0} V |\varphi\rangle \equiv g_0(\varepsilon) V |\varphi\rangle \tag{2.1}$$

for the two-body bound state φ and its energy $\varepsilon < 0$. Here h_0 is the kinetic energy of relative motion and V the two-body interaction. Since that homogeneous equation has a nontrivial solution only for the discrete negative bound state energies the Lippmann Schwinger equation (1.6) has a unique solution for $e > 0$.

Three particles interact in general by 2- and 3-body forces. The Schrödinger equation reads

$$(H_0 + V_1 + V_2 + V_3 + V_4) \Psi = E \Psi \tag{2.2}$$

where we denoted the proper three-body force by V_4. In integral form the homogeneous eigenvalue problem for the bound state is

$$\Psi = \frac{1}{E - H_0} (V_1 + V_2 + V_3 + V_4) \Psi \equiv G_0 \sum_{i=1}^{4} V_i \Psi \qquad (2.3)$$

Again that homogeneous equation implies an infinite sequence of interactions and propagations:

$$\Psi = G_0 \sum V_i \, G_0 \sum V_j \cdots \quad \Psi \qquad (2.4)$$

Among the sequential processes there are disconnected ones with only one type of pair interactions (and therefore a spectator particle) and connected ones with interactions between different pairs. Like in the AGS-equations we may also try to sum up the pair interactions within one pair to infinite order. To that aim we split Ψ into 4 parts according to the last interaction to the left in (2.3) or (2.4):

$$\Psi = \sum_{i=1}^{4} \psi_i \qquad (2.4a)$$

with

$$\psi_i \equiv G_0 V_i \Psi \qquad (2.5)$$

Now

$$\psi_i = G_0 V_i \sum_{j=1}^{4} \psi_j \qquad (2.6)$$

and the term with $j = i$ is clearly responsible for disconnected processes interacting only by V_i. We separate it and bring it to the left hand side

$$(1 - G_0 V_i)\, \psi_i = G_0 V_i \sum_{j \neq i} \psi_j \qquad (2.7)$$

For i = 1,2,3 the inversion of ($1 - G_0 V_i$) is essentially a two-body problem, since the degree of freedom of the third particle occurs only in the kinetic energy leading just to an energy shift in the pure two-body problem. Using the T-operator defind in (1.41) or (1.42) it is easy to see that

$$(1 + G_0 T_i)(1 - G_0 V_i) = 1 \qquad (2.8)$$

and consequently

$$\psi_i = G_0 T_i \sum_{j \neq i} \psi_j \qquad (2.9)$$

This is also true for $i = 4$ if we define a three-body T-operator T_4 based on the action of the three-body force V_4 alone

$$T_4 = V_4 + V_4 G_0 T_4 = V_4 + T_4 G_0 V_4 \tag{2.10}$$

The set (2.9) for i,j = 1, ...4 are the Faddeev equations [4]. The four components ψ_i sum up to the bound state. The kernel of (2.9) with $V_4 = 0$

$$K \equiv G_0 \begin{pmatrix} 0 & T_1 & T_1 \\ T_2 & 0 & T_2 \\ T_3 & T_3 & 0 \end{pmatrix} \tag{2.11}$$

gets connected after one iteration and is closely related in structure to the kernel (1.48) of the AGS-equations. It is easy to see that the two kernels have the same spectrum of discrete eigenvalues. As a consequence the AGS-equations define the transition operators uniquely, since the corresponding homogeneous set has nontrivial solutions only when (2.9) has a solution, namely at the three-body bound state energies.

Identical particles

One has to impose either symmetry (bosons) or antisymmetry (fermions) onto Ψ under all pair-exchanges P_{ij}. Then from (2.5) it is evident that the Faddeev components ψ_i, i = 1,2,3 are identical in form and turn into each other under cyclical permutations

$$\psi_2 = P_{12} P_{23} \psi_1 \tag{2.12}$$

$$\psi_3 = P_{13} P_{23} \psi_1 \tag{2.13}$$

As a consequence only one equation is necessary to determine $\psi \equiv \psi_1$ and a second one if one includes a three-body force:

$$\psi = G_0 T (P\psi + \psi_4) \tag{2.14}$$

$$\psi_4 = G_0 T_4 (1+P)\psi \tag{2.15}$$

Here P are the two cyclical permutations ($P = P_{12}P_{23} + P_{13}P_{23}$) and T the two-body T-operator. The bound state is

$$\Psi = (1+P)\psi + \psi_4 \tag{2.16}$$

and has the required symmetry property if the Faddeev component ψ is either symmetric (bosons) or antisymmetric (fermions) under exchange of the special pair interacting through T. The component ψ_4 has clearly the same symmetry property as Ψ itself.

3. Numerical methods in momentum space

3.1 Two-body bound states

Bound states have definite angular momenta and parity. For spinless bosons, like noble gas atoms, it is therefore adequate to choose basis states $|plm\rangle$ to a given magnitude of relative momentum p, orbital angular momentum l and magnetic quantum number m. If we take the complete set of momentum states $|\vec{p}\rangle$ to be normalised as

$$\langle \vec{p}\,' | \vec{p} \rangle = \delta^3(\vec{p} - \vec{p}\,') \tag{3.1}$$

we define the basis states $|plm\rangle$ by

$$\langle \vec{p}\,' | plm \rangle = \frac{\delta(p'-p)}{pp'} Y_{lm}(\hat{p}\,') \tag{3.2}$$

As a consequence one has

$$\langle p'l'm' | plm \rangle = \frac{\delta(p-p')}{pp'} \delta_{ll'} \delta_{mm'} \tag{3.3}$$

and

$$\sum_{lm} \int dp\, p^2 \, |plm\rangle \langle plm| = 1 \tag{3.4}$$

Often the potential is given in configuration space. Let $\vec{r} = \vec{x}_1 - \vec{x}_2$ and $|\vec{r}\rangle$, $|rlm\rangle$ be complete sets of states defined in analogy to (3.1) - (3.4). For equal mass particles the conjugate momentum is $\vec{p} = \frac{1}{2}(\vec{k}_1 - \vec{k}_2)$. One has

$$\langle \vec{r} | \vec{p} \rangle = \frac{1}{(2\pi)^{3/2}} e^{i\vec{p}\cdot\vec{r}} \tag{3.5}$$

or using (3.2)

$$\langle rlm | pl'm' \rangle = \delta_{ll'} \delta_{mm'} \sqrt{\frac{2}{\pi}}\, j_l(pr) \tag{3.6}$$

These relations lead then to the partial-wave representation of V in momentum space

$$\langle plm|V|p'lm\rangle \equiv V_l(pp') = \frac{2}{\pi}\int_0^\infty dr\, r^2\, j_l(pr)\, V(r)\, j_l(p'r) \tag{3.7}$$

In the case of a Yukawa or Gaussian interaction, for instance, that integral can be performed analytically. For interactions between atoms numerical evaluation is usually required. Rewriting the integral (3.7) to a cosin- or sin-transform the Filon method [6)] is quite useful. A simpler way may be to use

$$\begin{aligned}\langle plm|V|p'lm\rangle &= \int d\vec{q} \int d\vec{q}\,'\, \langle plm|\vec{q}\rangle \langle \vec{q}|V|\vec{q}\,'\rangle\, * \\ *\, \langle \vec{q}\,'|p'lm\rangle &= \int d\hat{q}\, Y^{*}_{lm}(\hat{q}) \int d\hat{q}'\, Y_{lm}(\hat{q}')\, \langle p\hat{q}|V|p'\hat{q}'\rangle\end{aligned} \tag{3.8}$$

Since for a local potential

$$\langle \vec{q}|V|\vec{q}\,'\rangle \equiv \tilde{V}(|\vec{q}-\vec{q}\,'|) = \frac{4\pi}{(2\pi)^3}\int dr\, r^2\, j_0(|\vec{q}-\vec{q}\,'|r)\, V(r) \tag{3.9}$$

and $\langle plm|V|p'lm\rangle$ is independent of m we easily get

$$V_l(pp') = 2\pi \int_{-1}^{1} dx\, P_l(x)\, \tilde{V}\left(\sqrt{p^2+p'^2-2pp'x}\right) \tag{3.1o}$$

As examples we show in Fig. 3 the potential HFDHE2 between two ^{4}He-atoms in the state l = 0 and in Fig. 4 the potential between two hydrogen atoms in the attractive singlet state for l = 3. The enhancements for p = p' are caused by the repulsive cores. The attraction in the hydrogen case is reflected by the negative parts.

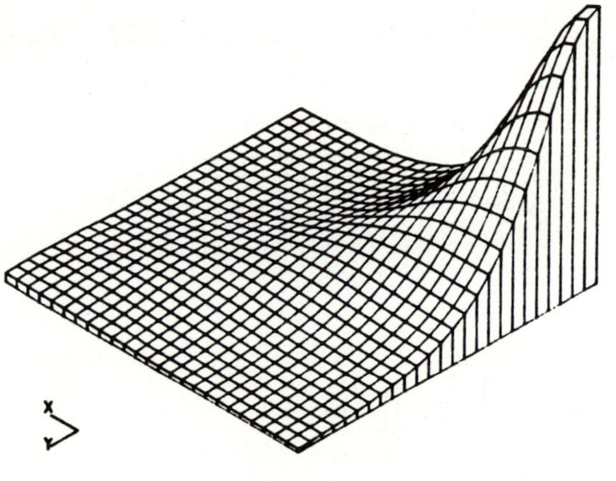

Fig. 3;
The ^{4}He-^{4}He intermolecular potential HFDHE2 in momentum space and projected to the state l = o.

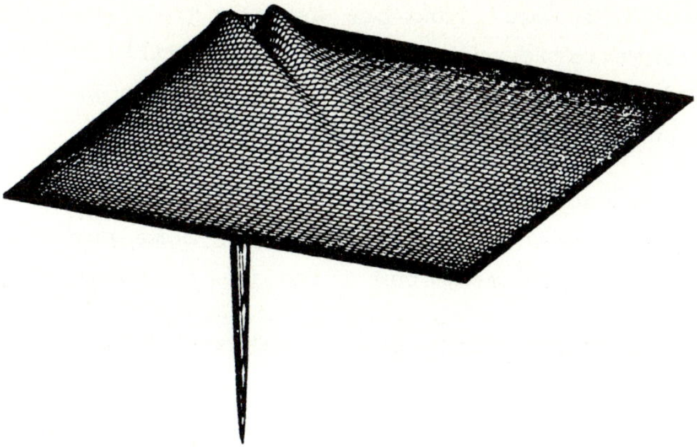

Fig. 4: The H-H singlet potential in momentum space and projected to the state l = 3.

The momentum space representation of the two-body bound state eigenvalue problem is

$$\varphi_l(p) \equiv \langle plm|\varphi\rangle = \frac{1}{\varepsilon - p^2/2\mu}\int_0^\infty dp' p'^2 V_l(pp')\varphi_l(p') \qquad (3.11)$$

(μ reduced mass). One may map the infinite or a sufficiently large p-interval onto $[-1,1]$ and use a Gaussian-Legendre quadrature approximation. The resulting homogeneous algebraic system of equations can be solved by standard routines. One varies the energy ε in such a manner that one eigenvalue of the kernel gets 1. As an example we show in Fig. 5 the dimer for two ^{4}He-atoms.

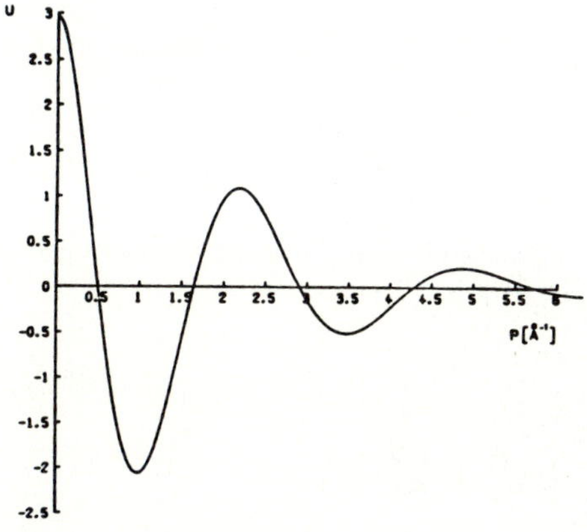

Fig. 5: The two-body bound state for two ^{4}He-atoms

Relativistic two-body equations are as easy to solve in momentum space [7] as nonrelativistic ones. The essential difference is the replacement of $p^2/2\mu$ by $\left(2\sqrt{m^2+p^2} - 2m\right)$ which in configuration space cannot be handled in such a trivial manner.

For fermions like nucleons one has to include spin into the basis states: $|p(ls)jm\rangle$. Here the orbital angular momentum l and total spin s is coupled to the angular momentum j and its magnetic quantum number m.

For two nucleons the Reid soft core potential [8] is often used as a reference potential. It has the structure

$$V = V_{centr}(r) + V_{ls}(r)\,\vec{l}\cdot\vec{s} + V_T(r)\left(3\,\vec{\sigma}_1\cdot\hat{r}\;\vec{\sigma}_2\cdot\hat{r} - \vec{\sigma}_1\cdot\vec{\sigma}_2\right) \tag{3.12}$$

where the radial functions $V(r)$ are state dependent and expressed as superpositions of Yukawas. The angular and spin-dependent parts with respect to our basis states are well known and the radial integrals of the type (3.7) can be carried through analytically. In the two-nucleon bound state with j = **1** a 2x2 potential matrix $V_{ll'}(pp')$ results with l,l' = o,2. Consequently (3.11) is replaced by two coupled equations.

One-boson exchange potentials for two nucleons [9] are given naturally in momentum space and their partial wave representation is straightforward though quite tedious. A popular two-nucleon potential based on dispersion relations for the 2π-exchange is the Paris potential [10].

We show in Fig. 6 the s- and d-wave parts of the deuteron wavefunction in momentum space for the three NN potentials mentioned.

3.2 Three-body bound states

It is usage to describe the relative motions of three particles by Jacobi momenta

$$\vec{p}_1 = \frac{1}{2}\left(\vec{k}_2 - \vec{k}_3\right) \tag{3.13}$$

$$\vec{q}_1 = \frac{2}{3}\left(\vec{k}_1 - \frac{1}{2}\left(\vec{k}_2 + \vec{k}_3\right)\right) \tag{3.14}$$

and cyclical permutations thereof (equal mass particles). Then the kinetic energy is

$$H_0 = p_i^2/2\mu + q_i^2/2M \quad , \quad i = 1,2,3 \tag{3.15}$$

where $\mu = \frac{1}{2} m$, $M = \frac{2}{3} m$ and m is the particle mass. Let $|\vec{k}_1 \vec{k}_2 \vec{k}_3\rangle \equiv |\vec{k}_1\rangle |\vec{k}_2\rangle |\vec{k}_3\rangle$ be momentum eigenstates for three free particles, normalised as in (3.1), then one defines momentum states $|\vec{p}_1 \vec{q}_1 \vec{P}\rangle$ by

$$\langle \vec{k}_1 \vec{k}_2 \vec{k}_3 | \vec{p}_1 \vec{q}_1 \vec{P}\rangle = \delta^3\left(\vec{p}_1 - \tfrac{1}{2}(\vec{k}_2 - \vec{k}_3)\right)\delta^3\left(\vec{q}_1 - \tfrac{2}{3}\left(\vec{k}_1 - \tfrac{1}{2}(\vec{k}_2 + \vec{k}_3)\right)\right)\delta^3\left(\vec{P} - \vec{k}_1 - \vec{k}_2 - \vec{k}_3\right) \qquad (3.16)$$

As a consequence

$$\langle \vec{p}_1 \vec{q}_1 | \vec{p}_1{}' \vec{q}_1{}' \rangle = \delta^3(\vec{p}_1 - \vec{p}_1{}')\,\delta^3(\vec{q}_1 - \vec{q}_1{}') \qquad (3.17)$$

and

$$\int d\vec{p}_1\, d\vec{q}_1\, |\vec{p}_1 \vec{q}_1\rangle\langle \vec{p}_1 \vec{q}_1| = 1 \qquad (3.18)$$

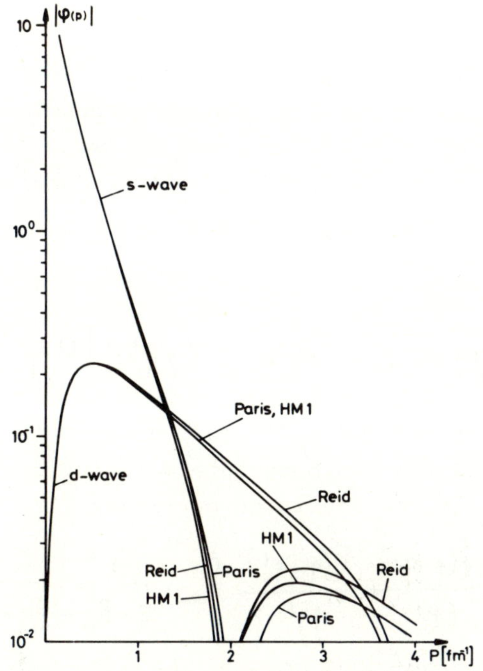

Fig. 6: The deuteron wavefunctions in momentum space for the Reid[8]-, OBEP[9] and Paris-potential[10].

a) Three bosons

We first regard the Faddeev equation for a bound state of three bosons. We can assume that only low angular momenta are important and use therefore a partial-wave representation. Let $\vec{\ell}$ and $\vec{\lambda}$ be the angular momenta related to $\vec{p} \equiv \vec{p}_1$ and $\vec{q} \equiv \vec{q}_1$. They are coupled to the conserved total orbital angular momentum $\vec{L}$:

$$\vec{\ell} + \vec{\lambda} = \vec{L} \tag{3.19}$$

Generalising (3.2) one introduces the complete set of partial wave basis states $|pq(\ell\lambda)LM\rangle$ by

$$\langle \vec{p}'\vec{q}'|pq(\ell\lambda)LM\rangle = \frac{\delta(p'-p)}{pp'}\frac{\delta(q'-q)}{qq'}\mathcal{Y}^{LM}_{\ell\lambda}(\hat{p}'\hat{q}') \tag{3.20}$$

To start with we neglect a three-body force. Then the Faddeev equation (2.14) reads

$$\psi = G_0 T P \psi \tag{3.21}$$

In order that $\Psi = (1+P)\psi$ is totally symmetric, the Faddeev component has to be symmetric within the two-body subsystem described by $\vec{p}$; consequently only basis states with l = even are allowed. In the following basis states with the appropriate symmetry will be denoted by $|pq\alpha\rangle$ for short. As a first step we write (3.21) as

$$\begin{aligned}\psi_\alpha(pq) &\equiv \langle pq\alpha|\psi\rangle \\ &= \sum_{\alpha'}\int dp'p'^2\int dq'q'^2\langle pq\alpha|G_0T|p'q'\alpha'\rangle\langle p'q'\alpha'|P\psi\rangle\end{aligned} \tag{3.22}$$

The two-body T-operator is defined by (1.42). Since $G_0(E)$ is diagonal in all quantum numbers and the two-body interaction V conserves the quantum numbers q and λ of the spectator particle:

$$\langle pq\alpha|G_0(E)|p'q'\alpha'\rangle = \frac{\delta(p-p')}{pp'}\frac{\delta(q-q')}{qq'}\delta_{\alpha\alpha'}\frac{1}{E-\frac{p^2}{m}-\frac{3}{4m}q^2} \tag{3.23}$$

$$\langle pq\alpha|V|p'q'\alpha'\rangle = \frac{\delta(q-q')}{qq'}\delta_{\lambda\lambda'}\delta_{\ell\ell'}V_\ell(pp') \tag{3.24}$$

one concludes that T must have the representation

$$\langle pq\alpha | T(E) | p'q'\alpha' \rangle \equiv \frac{\delta(q-q')}{qq'} \delta_{\lambda\lambda'} \delta_{\ell\ell'} t_\ell \left(p, p', E - \frac{3}{4m} q^2 \right) \tag{3.25}$$

Therefore the L.S. eqn. (1.42) reduces to a pure two-body problem, however taken at an off-shell energy $z = E - \frac{3}{4m} q^2$:

$$t_\ell (p,p',z) = V_\ell (p,p') + \int_0^\infty dp'' p''^2 V_\ell (p,p'') \frac{1}{z - p''^2/m} t_\ell (p'',p',z) \tag{3.26}$$

The three-body binding energy E is negative, thus z is negative which leads to a well behaved kernel in (3.26). As with (3.11) that equation can be trivially solved by standard routines. We show in Fig. 7 the solution of (3.26) for l = o and the HFDHE2-potential between two ^{4}He-atoms at a negative energy not far from the two-body binding energy ε .

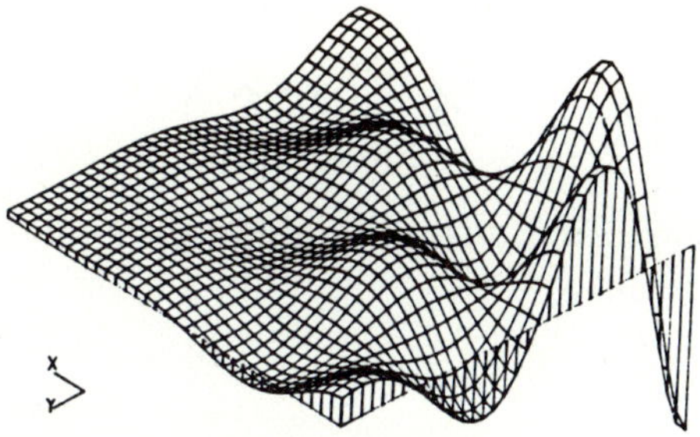

Fig. 7: The two-body t-matrix in the state l=o for the HFDHE2 potential between two ^{4}He-atoms.

From the formal solution of t expressed in terms of the resolvent operator to the two-body Hamiltonian it is evident that t will reflect the p-dependence of the bound state wavefunction (if z is close to ε). Comparing Fig. 5 with Fig. 7 this is manifest and explains the structure.

Inserting (3.25) into (3.22) and another complete set of basis states between P and ψ leads to

$$\psi_\alpha (pq) = \frac{1}{E - p^2/m - \frac{3}{4m} q^2} \int_0^\infty dp' p'^2 \sum_{\alpha''} \int_0^\infty dp'' p''^2 \int_0^\infty dq'' q''^2$$
$$t_\ell \left(p, p', E - \frac{3}{4m} q^2 \right) \langle p'q\alpha | P | p''q''\alpha'' \rangle \psi_{\alpha''} (p''q'') \tag{3.27}$$

We face now the central problem for three particles, namely to live with three two-body subsystems. This is manifest in the matrixelement for the cyclical permutation operators:

$$\langle pq\alpha | P_{12} P_{23} | p'q'\alpha' \rangle \equiv {}_1\langle pq\alpha | P_{12} P_{23} | p'q'\alpha' \rangle_1 \tag{3.28}$$

For notational clarity subscripts 1 are introduced to indicate that q is the momentum of particle 1 in the total momentum zero frame and p the relative momentum in the 23-subsystem. The cyclical permutation $P_{12}P_{23}$ replaces that choice by a state with subscript 2:

$$\langle pq\alpha | P_{12} P_{23} | p'q'\alpha' \rangle = {}_1\langle pq\alpha | p'q'\alpha' \rangle_2 \tag{3.29}$$

Thus in the ket the quantum numbers refer to the subsystem (31) and spectator particle 2, respectively. The calculation of that matrix-element is straightforward and we indicate only the first steps:

$$\begin{aligned} {}_1\langle pq\alpha | p'q'\alpha' \rangle_2 &= \int d\vec{p}_1 \, d\vec{q}_1 \int d\vec{p}_2 \, d\vec{q}_2 \, \langle pq\alpha | \vec{p}_1 \vec{q}_1 \rangle_1 \, {}_1\langle \vec{p}_1 \vec{q}_1 | \vec{p}_2 \vec{q}_2 \rangle_2 \, * \\ * \, {}_2\langle \vec{p}_2 \vec{q}_2 | p'q'\alpha' \rangle_2 &= \int d\hat{p}_1 \, d\hat{q}_1 \int d\hat{p}_2 \, d\hat{q}_2 \\ & \mathcal{Y}^{LM*}_{\ell\lambda}(\hat{p}_1 \hat{q}_1) \, {}_1\langle p\hat{p}_1 \, q\hat{q}_1 | p'\hat{p}_2 \, q'\hat{q}_2 \rangle_2 \, \mathcal{Y}^{LM}_{\ell'\lambda'}(\hat{p}_2 \hat{q}_2) \end{aligned} \tag{3.3o}$$

The connection between different choices of Jacobi momenta follows easily from (3.13)-(3.14); for instance

$$\vec{p}_1 = -\frac{1}{2}\vec{p}_2 + \frac{3}{4}\vec{q}_2 \tag{3.31}$$

$$\vec{q}_1 = -\vec{p}_2 - \frac{1}{2}\vec{q}_2 \tag{3.32}$$

Therefore

$${}_1\langle \vec{p}_1 \vec{q}_1 | \vec{p}_2 \vec{q}_2 \rangle_2 = \delta^3\left(\vec{p}_1 + \frac{1}{2}\vec{p}_2 - \frac{3}{4}\vec{q}_2\right) \delta^3\left(\vec{q}_1 + \vec{p}_2 + \frac{1}{2}\vec{q}_2\right) \tag{3.33}$$

As we shall see below it is advantageous to use the following equivalent form

$${}_1\langle \vec{p}_1 \vec{q}_1 | \vec{p}_2 \vec{q}_2 \rangle_2 = \delta^3\left(\vec{p}_1 - \frac{1}{2}\vec{q}_1 - \vec{q}_2\right) \delta^3\left(\vec{p}_2 + \vec{q}_1 + \frac{1}{2}\vec{q}_2\right) \tag{3.34}$$

Inserting (3.34) into (3.3o) leads to

$$ {}_1\langle pq\alpha | p'q'\alpha' \rangle_2 = \int d\hat{q}_1 \int d\hat{q}_2 \, \mathcal{Y}^{LM*}_{\ell\lambda}\left(\widehat{\tfrac{1}{2}q\hat{q}_1 + q'\hat{q}_2}, \hat{q}_1 \right) * $$

$$ * \frac{\delta(p-\pi_1)}{p^2} \frac{\delta(p'-\pi_2)}{p'^2} \mathcal{Y}^{LM}_{\ell'\lambda'}\left(\widehat{-q\hat{q}_1 - \tfrac{1}{2}q'\hat{q}_2}, \hat{q}_2 \right) \qquad (3.35) $$

where

$$ \pi_1 = \left| \tfrac{1}{2}q\hat{q}_1 + q'\hat{q}_2 \right| \qquad (3.36) $$

$$ \pi_2 = \left| q\hat{q}_1 + \tfrac{1}{2}q'\hat{q}_2 \right| \qquad (3.37) $$

In the simplest case $\ell = \lambda = \ell' = \lambda' = L = 0$ this trivially reduces to $(x \equiv \hat{q}_1 \cdot \hat{q}_2)$

$$ {}_1\langle pq\alpha | p'q'\alpha' \rangle_2 = \frac{1}{2} \int_{-1}^{1} dx \, \frac{\delta(p-\pi_1)}{p^2} \frac{\delta(p'-\pi_2)}{p'^2} \qquad (3.38) $$

In the general case one finds [11)]

$$ {}_1\langle pq\alpha | p'q'\alpha' \rangle_2 = \sum_k g_k \sum_{\ell_1+\ell_2=\ell} \sum_{\ell_1'+\ell_2'=\ell'} q^{\ell_2+\ell_2'} q'^{\ell_1+\ell_1'} (-)^{\ell'} $$

$$ \sqrt{\hat{\ell}\hat{\lambda}\hat{\ell}'\hat{\lambda}'} \, \hat{k} \left(\tfrac{1}{2}\right)^{\ell_2+\ell_1'+1} \sqrt{\frac{(2\ell+1)!}{(2\ell_1)!\,(2\ell_2)!}} \sqrt{\frac{(2\ell'+1)!}{(2\ell_1')!\,(2\ell_2')!}} \qquad (3.39) $$

$$ \sum_{ff'} \begin{Bmatrix} \ell_1 & \ell_2 & \ell \\ \lambda & L & f \end{Bmatrix} C(\ell_2 \lambda f, 00) \begin{Bmatrix} \ell_2' & \ell_1' & \ell' \\ \lambda' & L & f' \end{Bmatrix} C(\ell_1' \lambda' f', 00) $$

$$ \begin{Bmatrix} f & \ell_1 & L \\ f' & \ell_2' & k \end{Bmatrix} C(k \ell_1 f', 00) \, C(k \ell_2' f, 00) $$

where $\hat{m} \equiv (2m+1)$ and

$$ g_k \equiv \int_{-1}^{1} dx \, P_k(x) \, \frac{\delta(p-\pi_1)}{p^{\ell+2}} \frac{\delta(p'-\pi_2)}{p'^{\ell'+2}} \qquad (3.4o) $$

The second cyclical permutation is

$$P_{13} P_{23} = P_{23} P_{12} P_{23} P_{23} \tag{3.41}$$

Since the basis states $|pq\alpha\rangle$ are symmetric under (23)-exchanges, P_{23} acting to the right or left can be replaced by 1 and we find altogether

$$\langle pq\alpha | P | p'q'\alpha' \rangle = \int_{-1}^{1} dx \, \frac{\delta(p-\pi_1)}{p^{l+2}} \frac{\delta(p'-\pi_2)}{p'^{l'+2}} \, G_{\alpha\alpha'}(q,q',x) \tag{3.42}$$

where $G_{\alpha\alpha'}(qq'x)$ is given by (3.38) or generally by (3.39) multiplied by 2.

The Faddeev equation (3.27) now reads

$$\begin{aligned} \psi_\alpha(pq) = {} & \frac{1}{E - p^2/m - 3/4m\, q^2} \sum_{\alpha'} \int_0^\infty dq' q'^2 \int_{-1}^{1} dx \\ & t_l(p, \pi_1, E - 3/4m\, q^2)/p^l \; G_{\alpha\alpha'}(q,q',x) \, \psi_{\alpha'}(\pi_2, q') \end{aligned} \tag{3.43}$$

This is a set of coupled integral equations in two variables.

Whereas the solution of the two-body L.S. eqn.(3.26) requires high momenta if V has a repulsive core like for two atoms or nucleons it is a well established experience that the Faddeev component $\psi_\alpha(pq)$ drops very fast in the q-variable. This is intuitively clear since q describes the relative motion of a particle with respect to the center of mass of a pair. That third particle therefore feels a folding potential which is far less violent than V and does not induce high momenta. Our specific choice for evaluating the permutation operator (3.34) lead to the argument π_2 in ψ which because of (3.37) can be estimated as

$$\pi_2 \le \frac{3}{2} q_{max} \tag{3.44}$$

Here q_{max} is the relatively low maximal q-value necessary to describe $\psi_\alpha(pq)$. Therefore one does not need the p-dependence of $\psi_\alpha(pq)$ beyond $\frac{3}{2} q_{max}$ in solving (3.43). That long range p-dependence can be determined by quadrature in a second step. This is an important practical point to keep the number of unknowns small in solving (3.43).

A numerical solution of (3.43) requires a discretisation of the q'- and x-integral by N_q and N_x quadrature points, respectively. Obviously this leads to $N_q^2 \cdot N_x$ different π_2-values which is uncomfortably large. A way out is interpolation. It has always the form

$$f(x) \rightarrow \sum_i S_i(x)\, f(x_i) \tag{3.45}$$

where $S_i(x)$ are known functions and $f(x_i)$ are the (known or unknown) function values at an appropriately chosen set of grid points. We developped such a procedure using Spline functions [12]. In that manner the unknown Faddeev components under the integral in (3.43) are approximated by

$$\psi_{\alpha'}(\pi_2, q') = \sum_k S_k(\pi_2)\, \psi_{\alpha'}(p_k, q') \tag{3.46}$$

Thus the skew argument π_2 occurs now in known Spline functions. Of course the set of p_k -values has to be chosen sufficiently dense to guarantee the desired quality of interpolation.

Denoting by ω_ℓ, q_ℓ and ω_s, q_s the weights and positions, respectively, for the quadrature rules in the q'- and x-integrals the discretised representation of (3.43) is

$$\begin{aligned} \psi_\alpha(p_i q_j) &= \frac{1}{E - p_i^2/m - 3/4m\, q_j^2} \sum_{\alpha'} \sum_{\ell k} \Big[\omega_\ell q_\ell^2 \sum_s \omega_s \\ &\quad t_\ell(p_i, \pi_1, E - 3/4m\, q_j^2)\, G_{\alpha\alpha'}(q_j, q_\ell, x_s)\, S_k(\pi_2) \Big] \psi_{\alpha'}(p_k q_\ell) \\ &\equiv \sum_{\alpha' \ell k} K^{ij,k\ell}_{\alpha\alpha'}\, \psi_{\alpha'}(p_k q_\ell) \end{aligned} \tag{3.47}$$

We find an algebraic set of homogeneous equations [11]. The number of unknowns is in obvious notation

$$N = N_\alpha \cdot N_p \cdot N_q \tag{3.48}$$

Often a fairly good first orientation for the binding energy is to assume that the interaction acts only in the state l = o which for the ground state L = o encompasses also λ = o. Then (3.43) turns into

$$\psi(pq) = \frac{1}{E - p^2/m - 3/4m\, q^2} \int_0^\infty dq' q'^2 \int_{-1}^{1} dx\, t(p, \pi_1, E - 3/4m\, q^2)\, \psi(\pi_2, q') \tag{3.49}$$

or in discretised form into

$$\psi(p_i q_j) = \frac{1}{E - p_i^2/m - \frac{3}{4m} q_j^2} \sum_{lk} \Big[\omega_l q_l^2 \sum_s \omega_s \; t(p_i, \pi_1, E - \tfrac{3}{4m} q_j^2)\, S_k(\pi_2) \Big]\, \psi(p_k, q_l) \tag{3.5o}$$

We show in Fig. 8 the solution of (3.5o) for three ^{4}He-atoms interacting by the HFDHE2-potential, which corresponds to a trimer ground state energy of -.08 K . Allowing also d-wave interactions leads to two coupled equations and changes the energy to -.11 K . The second Faddeev component is shown in Fig. 9.

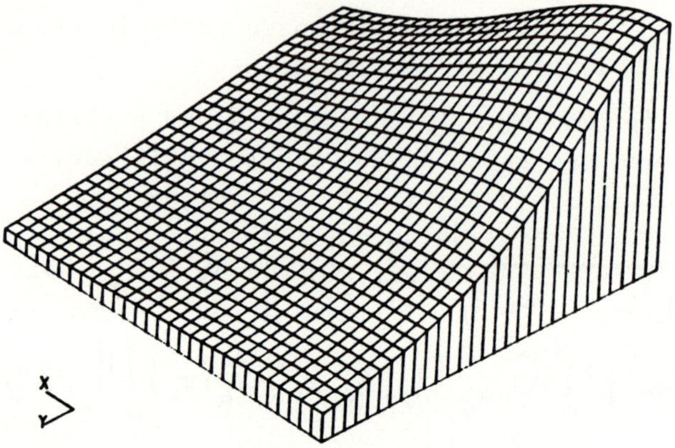

Fig. 8: The Faddeev component $(l = \lambda = 0)$ for three ^{4}He-atoms

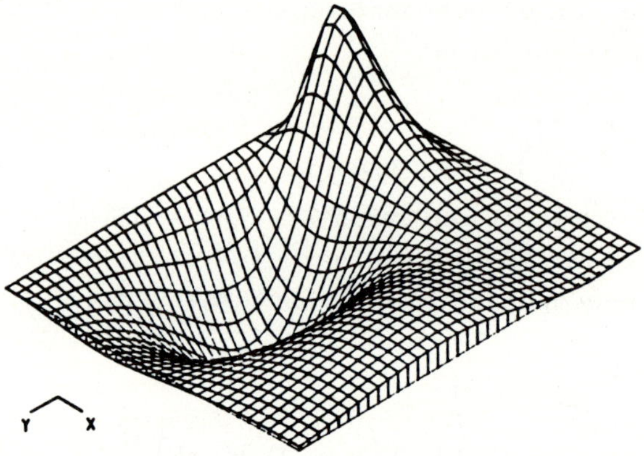

Fig. 9: The Faddeev component $(l = \lambda = 2)$ for three ^{4}He-atoms

A simple test case for (3.49) borrowed from nuclear physics is the Malfliet-Tjon potential V [13])

$$V(r) = \left(V_R\, e^{-\mu_R r} - V_A\, e^{-\mu_A r}\right)\frac{1}{r} \qquad (3.51)$$

which leads to a binding energy of -7.54 MeV using $q_{max} = 3fm^{-1}$; $N_q = 12$, $N_p = 18$ grid points for Spline interpolation and $N_x = 14$.

($V_R = 1438.4812$ MeV fm, $V_A = 570.316$ MeV fm, $\mu_R = 3.110\,fm^{-1}$, $\mu_A = 1.550\,fm^{-1}$, $m = 938.903$ MeV)

Let us regard now some techniques to solve the eigenvalue problem (3.47). It is of the type

$$\eta_\nu \psi_\nu = K \psi_\nu \qquad (3.52)$$

where the interest is in the eigenstate with $\eta_\nu = 1$. Since the once iterated kernel is connected and therefore of the Hilbert-Schmidt type the eigenvalues form a discrete set and cummulate at $\eta_\nu = 0$. The kernel K depends on the energy through the free propagator and the two-body T-operator. One varies the energy such that one eigenvalue is 1. If $\eta_\nu = 1$ is the largest eigenvalue in magnitude a very simple algorithm popularised by Malfliet-Tjon [13]) can be used. Putting

$$\psi^{(n+1)} = K \psi^{(n)} \qquad (3.53)$$

with an arbitrary starting function $\psi(o)$ leads to the general result

$$\lim_{n\to\infty} \psi^{(n+1)} / \psi^{(n)} = \eta_{max} \qquad (3.54)$$

where η_{max} is the largest eigenvalue in magnitude. If this is the physical one, $\eta_\nu = 1$, that algorithm is ideal to implement on the computer. For interactions with repulsive cores negative eigenvalues occur, which are related to "bound states" in the sign reversed potentials. Since the repulsive cores are typically quite strong, these "bound states" are usually stronger bound than the real ones. Consequently η_{max} is negative and larger in magnitude than 1. In this case the iterative procedure converges to the unphysical eigenstate and eigenvalue. One way to project out the unphysical state in the iterative procedure is described in [14]). An even simpler device is to replace the kernel K by K+Z, where Z is a positive number [15]). Then the new eigenvalues will be $\eta_\nu + Z$ and the physical one, $1 + Z$, can be made the largest one in magnitude. If in addition to the ground state also excited states are of interest then the method of inverse iteration is very useful [16]. Let η^* be an approximation to the searched eigenvalue η of K. Then (3.52) can be written as

$$(K-\eta^*)\psi = (\eta-\eta^*)\psi \tag{3.55}$$

or

$$(K-\eta^*)^{-1}\psi = \tilde{\eta}\,\psi \tag{3.56}$$

with

$$\tilde{\eta} = \frac{1}{\eta-\eta^*} \tag{3.57}$$

If $|\eta-\eta^*| < |\eta'-\eta^*|$ for all eigenvalues $\eta' \neq \eta$ then obviously $\tilde{\eta}$ is the largest eigenvalue in magnitude of the kernel $(K-\eta^*)^{-1}$. It can be determined by the iteration scheme

$$\psi^{(n)} = (K-\eta^*)^{-1}\psi^{(n-1)} \tag{3.58}$$

Instead of constructing the inverse matrix one solves the linear system

$$(K-\eta^*)\psi^{(n)} = \psi^{(n-1)} \tag{3.59}$$

Clearly we put $\eta^*=1$ and determine for each energy $\tilde{\eta}$ and therefore $\eta = 1/\tilde{\eta} + \eta^*$. If E is close to the searched for eigenvalue $\tilde{\eta}$ gets quite large (typically 1000 or more) and η approaches 1 with an adjustable accuracy.

As an example we show in table 1 the variation of the ground state and first excited state for three ^{4}He-atoms as a function of an overall strength parameter g of the interaction. For the physical value g = 1 the ^{4}He trimer supports a ground state and one excited state. For g = 1.4 the two-body binding energy, shown in column 1, overruns the position of the excited state, which moves then into a second unphysical sheet. This is a behaviour typical for an Efimov state. If the interaction is reduced by $\approx$ 2% a second excited state emerges, which also belongs to the family of Efimov states.

Table 1: Bound states for 2 and 3 ^{4}He-atoms for different potential strengths.

g	ε	ground state	1. excit. state	2. excit. state
1.4	-.31	-1.5		
1.3	-.1857	-1.o	-.1858	
1.2	-.o9o1	-.6	-.o9o6	
1.1	-.o276	-.3	-.o289	
1.o	-.ooo8	-.1	-.oo16	
o.98	-.1(-5)	-.o8	-.ooo2	-.2(-5)

b) Three fermions

We consider three nucleons. The basis states have to be enriched by spin and isospin degrees of freedom:

$$|pq\alpha\rangle \equiv |pq(ls)j(\lambda\tfrac{1}{2})J(jJ)\mathcal{J}M(t\tfrac{1}{2})TM_T\rangle \tag{3.6o}$$

Here $(ls)j$ denotes the coupling of orbital angular momentum l and total spin s to total angular momentum j for the two-body subsystem and $(\lambda\frac{1}{2})J$ the corresponding coupling for the third particle. Finally j and J are coupled to the total angular momentum $\mathcal{J}$ and its magnetic quantum number M. The isospin of the subsystem is t and TM_T are the total isospin quantum numbers. Since the Faddeev component has to be antisymmetric under exchange of the two nucleons described by the quantum numbers l,s,t only combinations of l,s,t are allowed such that

$$(-)^{l+s+t} = -1 \tag{3.61}$$

If we neglect the three-nucleon force the Faddeev equation (2.14) has the representation

$$\psi_\alpha(pq) = \frac{1}{E - p^2/m - 3/4m\, q^2} \sum_{\alpha'} \int dp'\, p'^2\, t^{jt}_{ls,l's}(pp'E - 3/4m\, q^2) \tag{3.62}$$

$$\sum_{\alpha''} \int dp''\, p''^2 \int dq''\, q''^2 \langle p'q\alpha'|P|p''q''\alpha''\rangle\, \psi_{\alpha''}(p''q'')$$

Here the two-body t-matrix is defined by the LS equation

$$t^{jt}_{ls,l's}(pp'z) = V^{jt}_{ls,l's}(pp') + \sum_{l''} \int dp''\, p''^2\, V^{jt}_{ls,l''s}(pp'') \frac{1}{z - p''^2/m}\, t^{jt}_{l''s,l's}(p''p'z) \quad (3.63)$$

If the tensor force is acting like in the states $^3S_1 - {}^3D_1\, , {}^3P_2 - {}^3F_2, \cdots$ (3.63) consist of two coupled equations otherwise (3.63) is uncoupled like in the states $^1S_0, {}^3P_1, {}^1P_1, \cdots$. The evaluation of the matrixelement of P includes now spin and isospin. A change of representation to LS-coupling is convenient. It results the same form as (3.42) with a rather lengthy geometrical coeffìcient $G_{\alpha\alpha'}(qq'x)$ [11]. With (3.42) inserted into (3.62) we end up with

$$\psi_\alpha(pq) = \frac{1}{E - p^2/m - 3/4m\, q^2} \sum_{\alpha'} \sum_{\alpha''} \int_0^\infty dq''\, q''^2 \int_{-1}^{1} dx\; t^{jt}_{ls,l's}(p, \pi_1, E - 3/4m\, q^2)\, G_{\alpha'\alpha''}(qq''x)\, \psi_{\alpha''}(\pi_2, q'') \quad (3.64)$$

This set of coupled equations can be treated as described above. We show in table 2 the first 18 combinations of quantum numbers if one keeps the two-nucleon force different from zero for $0 \leq j \leq 2$.

Table 2: Angular momenta in the triton; t is the subsystem isospin.

No.	l	s	j	t	λ	J
1	o	o	o	1	o	1/2
2	o	1	1	o	o	1/2
3	2	1	1	o	o	1/2
4	o	1	1	o	2	3/2
5	2	1	1	o	2	3/2
6	1	o	1	o	1	1/2
7	1	o	1	o	1	3/2
8	1	1	o	1	1	1/2
9	1	1	1	1	1	1/2
1o	1	1	1	1	1	3/2
11	1	1	2	1	1	3/2
12	3	1	2	1	1	3/2
13	1	1	2	1	3	5/2
14	3	1	2	1	3	5/2
15	2	o	2	1	2	3/2
16	2	o	2	1	2	5/2
17	2	1	2	o	2	3/2
18	2	1	2	o	2	5/2

Numerical experience [14),17),18)] tells that a cut-off value of q_{max} = $(4-6)fm^{-1}$ is sufficient. Typical quadrature and interpolation points are N_q = 12, N_x = 6, and N_p = 24, respectively. In that case the number of unknowns $N = N_\alpha \cdot N_q \cdot N_p$ are 1440 or 5184 for a five or 18 channel calculation, respectively. We show in table 3 results from various groups based on the Reid- and Paris-potential.

Table 3: Triton binding energies based on the two-nucleon potentials of Reid and the Paris group.

α	Reid-potential [8)]	Paris-potential [10)]
5	-7.0 [19)] -7.02 [20)18)14)] -7.04 [17)] -7.03 [21)]	-7.303 [18)] -7.48 [21)]
18	-7.232 [18)] -7.23 [17)22)] -7.24 [21)]	-7.38 [18)] -7.56 [21)] -7.33 [25)]
34	-7.35 [22)]	-7.64 [23)]

Note the very good agreement between calculations carried through in momentum space and configuration space. In the latter case the Faddeev equations are partial integro-differential equations [19)].

The discrepancy to the experimental number of E = -8.48 MeV is a challenge for our understanding of nuclear dynamics.

c) A perturbative scheme for a three-body force

In the convential picture of boson exchanges between nucleons the potential energy in the triton is ≈ -50. MeV in comparison to the kinetic energy of ≈ 43. MeV [18)]. An acceptable contribution of a three-nucleon force would be of the order of -1. to -2. MeV. One can therefore expect that this effect can be treated in low order perturbation theory. This can be formulated directly with the coupled set of Faddeev equations [24)25)]. In this case $G_0(E)$, $t(E)$, $t_4(E)$ have to be expanded around the energy $E^{(0)}$, which is the triton binding energy without three-body force V_4. Then like in usual perturbation theory the Faddeev components ψ and ψ_4 are decomposed into a sum of terms corresponding to different powers of V_4. Though quite straightforward a possibly simpler way is to develop perturbation theory on the level of the

Schrödinger equation and solve the various orders by a Faddeev decomposition: The Schrödinger equation

$$\left(H_0 + \sum_{i=1}^{3} V_i + V_4\right)\Psi = E\Psi \tag{3.65}$$

is equivalent to

$$\left(H_0 + \sum_{i=1}^{3} V_i - E^{(0)}\right)\Psi^{(0)} = 0 \tag{3.66}$$

$$\left(H_0 + \sum_{i=1}^{3} V_i - E^{(0)}\right)\Psi^{(1)} = -V_4\Psi^{(0)} + E^{(1)}\Psi^{(0)} \tag{3.67}$$

$$\left(H_0 + \sum_{i=1}^{3} V_i - E^{(0)}\right)\Psi^{(2)} = -V_4\Psi^{(1)} + E^{(1)}\Psi^{(1)} + E^{(2)}\Psi^{(0)} \tag{3.68}$$

where we put

$$\Psi = \Psi^{(0)} + \Psi^{(1)} + \Psi^{(2)} + \cdots \tag{3.69}$$

and

$$E = E^{(0)} + E^{(1)} + E^{(2)} + \cdots \tag{3.70}$$

Projections by $\langle\Psi^{(0)}|$ from the left lead to the well known first and second order energy shifts caused by V_4. We saw in section 2 how the first equation (3.66) is turned into a Faddeev form:

$$\Psi^{(0)} = (1+P)\,\psi^{(0)} \tag{3.71}$$

$$\psi^{(0)} = G_0 T P\,\psi^{(0)} \tag{3.72}$$

We can repeat that technique now for the second equation (3.67), which we first rewrite into

$$\Psi^{(1)} = G_0 \sum V_i \Psi^{(1)} + G_0\left(V_4 - E^{(1)}\right)\Psi^{(0)} \tag{3.73}$$

This suggests the Faddeev decomposition

$$\Psi^{(1)} = (1+P)\,\psi^{(1)} + G_0 V_4 \Psi^{(0)} \tag{3.74}$$

with

$$\psi^{(1)} = G_0 V_1 \Psi^{(1)} - E^{(1)} G_0 \psi^{(0)} \tag{3.75}$$

Inserting (3.74) on the right hand side of (3.75), separating $\psi^{(1)}$ and inverting $(1-G_0V_1)$ as in (2.7) leads to

$$\begin{aligned}\psi^{(1)} &= G_0TP\,\psi^{(1)} + G_0T\,G_0V_4\,\Psi^{(0)} - E^{(1)}(G_0+G_0TG_0)\,\psi^{(0)} \\ &\equiv G_0TP\,\psi^{(1)} + \phi^{(1)}\end{aligned} \tag{3.76}$$

Therefore the perturbed Faddeev component $\psi^{(1)}$ is determined by an inhomogeneous equation where the kernel K is the same as in the homogeneous problem (3.72) for the unperturbed component $\psi^{(0)}$. This is consistent only if the driving term is orthogonal to the left hand eigenvector $\tilde{\psi}^{(0)}$ of K:

$$\tilde{\psi}^{(0)}\,K = \tilde{\psi}^{(0)} \tag{3.77}$$

Clearly one has

$$\tilde{\psi}^{(0)} = \psi^{(0)}\,PTP \tag{3.78}$$

and the necessary and sufficient condition for a solution of (3.76) is

$$\langle \tilde{\psi}^{(0)} | \phi^{(1)} \rangle = 0 \tag{3.79}$$

Using (3.78) and (3.77) this is seen to be just

$$\langle \psi^{(0)} | V_4 | \Psi^{(0)} \rangle = E^{(1)} \langle \psi^{(0)} | \Psi^{(0)} \rangle \tag{3.80}$$

Since $\psi^{(0)}$ and $V_4\,\Psi^{(0)}$ are invariant under cyclical permutations the result (3.80) can also be put into the standard form

$$\langle \Psi^{(0)} | V_4 | \Psi^{(0)} \rangle = E^{(1)} \langle \Psi^{(0)} | \Psi^{(0)} \rangle \tag{3.81}$$

for the energy shift in first order. We present the result up to now in the form

$$\psi^{(1)} = K\,\psi^{(1)} + \phi^{(1)} \tag{3.82}$$

$$\phi^{(1)} = \phi_1^{(1)} + \phi_2^{(1)} \tag{3.83}$$

$$\phi_1^{(1)} = -E^{(1)}(G_0 + G_0TG_0)\,\psi^{(0)} \tag{3.84}$$

$$\phi_2^{(1)} = G_0 T \tilde{\phi}_2^{(1)} \tag{3.85}$$

$$\tilde{\phi}_2^{(1)} = G_0 V_4 \Psi^{(0)} \tag{3.86}$$

Let us now regard the second order equation (3.68), which we write in integral form

$$\Psi^{(2)} = G_0 \sum_{i=1}^{3} V_i \Psi^{(2)} + G_0 (V_4 - E^{(1)}) \Psi^{(1)} - E^{(2)} G_0 \Psi^{(0)} \tag{3.87}$$

Having the decomposition of $\Psi^{(0)}$ and $\Psi^{(1)}$ in mind it is suggestive to put

$$\Psi^{(2)} = (1+P)\psi^{(2)} + G_0 V_4 \Psi^{(1)} - E^{(1)} G_0^2 V_4 \Psi^{(0)} \tag{3.88}$$

with

$$\psi^{(2)} = G_0 V_1 \Psi^{(2)} - E^{(1)} G_0 \psi^{(1)} - E^{(2)} G_0 \psi^{(0)} \tag{3.89}$$

Then inserting (3.88) back into (3.89) and inverting the disconnected parts as usual we find

$$\psi^{(2)} = G_0 T P \psi^{(2)} + G_0 T (G_0 V_4 \Psi^{(1)} - E^{(1)} G_0^2 V_4 \Psi^{(0)}) - E^{(1)} (G_0 + G_0 T G_0) \psi^{(1)} - E^{(2)} (G_0 + G_0 T G_0) \psi^{(0)} \tag{3.9o}$$

Let us add (3.76) to (3.9o):

$$\psi^{(1)} + \psi^{(2)} = K (\psi^{(1)} + \psi^{(2)}) - E^{(1)} (G_0 + G_0 T G_0)(\psi^{(0)} + \psi^{(1)}) - E^{(2)} (G_0 + G_0 T G_0) \psi^{(0)} + G_0 T (G_0 V_4 (\Psi^{(0)} + \Psi^{(1)}) - E^{(1)} G_0 \tilde{\phi}_2^{(1)}) \tag{3.91}$$

In the third term we may replace $\psi^{(0)}$ by $\psi^{(0)} + \psi^{(1)}$ and in the last term $E^{(1)}$ by $E^{(1)} + E^{(2)}$ introducing errors of next order in V_4. We end up with

$$\psi^{(1)} + \psi^{(2)} = K (\psi^{(1)} + \psi^{(2)}) + \phi^{(2)} \tag{3.92}$$

$$\phi^{(2)} = \phi_1^{(2)} + \phi_2^{(2)} \tag{3.93}$$

$$\phi_1^{(2)} = -\left(E^{(1)} + E^{(2)}\right)\left(G_0 + G_0 T G_0\right)\left(\psi^{(0)} + \psi^{(1)}\right) \tag{3.94}$$

$$\phi_2^{(2)} = G_0 T \tilde{\phi}_2^{(2)} \tag{3.95}$$

$$\tilde{\phi}_2^{(2)} = G_0 V_4 \left(\Psi^{(0)} + \Psi^{(1)}\right) - \left(E^{(1)} + E^{(2)}\right) G_0 \tilde{\phi}_2^{(1)} \tag{3.96}$$

Obviously the driving term has the same structure as in (3.83) and this remains true for all higher orders - which is ideal for a recursive numerical evaluation. The orthogonality requirement

$$\langle \tilde{\psi}^{(0)} | \phi^{(2)} \rangle = 0 \tag{3.97}$$

leads to

$$\langle \psi^{(0)} | V_4 | \Psi^{(0)} + \Psi^{(1)} \rangle = \left(E^{(1)} + E^{(2)}\right) \langle \psi^{(0)} | \Psi^{(0)} + \Psi^{(1)} \rangle \tag{3.98}$$

This is the well known energy shift of second order. Again the common structure with the expression (3.8o) is ideal for the numerical performance.

A final remark concerns the way one can solve the inhomogeneous equations (3.82), (3.92). In a three-nucleon system the largest eigenvalue in magnitude of K is a negative one, $\eta^{(-)} < -1$. We transform the inhomogeneous equations

$$\psi' = \phi + K \psi' \tag{3.99}$$

into

$$\psi' = \phi + \left(K - \eta^{(-)}\right) \psi' + \eta^{(-)} \psi' \tag{3.1oo}$$

or

$$\psi' = \phi' + K' \psi' \tag{3.1o1}$$

with

$$\phi' = \frac{1}{1 - \eta^{(-)}} \phi \tag{3.1o2}$$

$$K' = \frac{1}{1 - \eta^{(-)}} \left(K - \eta^{(-)}\right) \tag{3.1o3}$$

The admixture of $\psi^{(0)}$ in ψ' is irrelevant (note $\langle \tilde{\psi}^{(0)} | \phi \rangle = 0$) and we can project it out explicitely:

$$\Lambda \equiv 1 - \frac{| \psi^{(0)} \rangle \langle \tilde{\psi}^{(0)} |}{\langle \tilde{\psi}^{(0)} | \psi^{(0)} \rangle} \tag{3.104}$$

Then one solves

$$\Lambda \psi' = \phi + K' \Lambda \psi' \tag{3.105}$$

which obviously has a convergent Neumann series.

d) The 2π -exchange three-nucleon force in the triton

In the conventional meson exchange between nucleons two-boson exchange three-nucleon forces naturally arise. Among them the 2π -exchange [26)27)] is of longest range:

$V_4 =$ 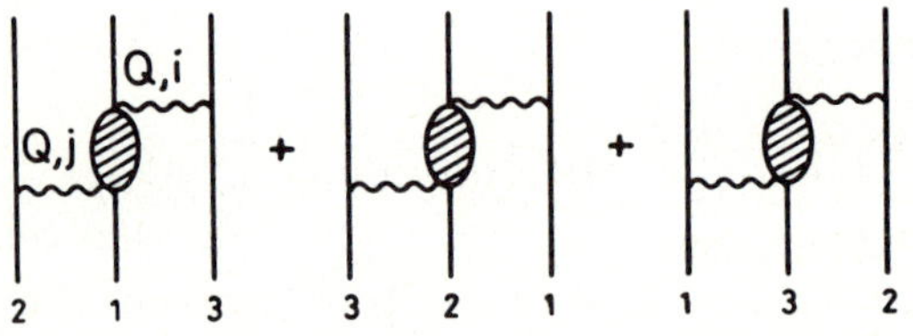 (3.106)

The shaded oval denotes the off-shell πN -amplitude without the part describing the propagation of a nucleon (the forward propagating part of the spin 1/2 Feynman propagator). That part would clearly lead to a sequence of OPE-potentials between different pairs, which is not a three-nucleon force. In a nonrelativistic reduction adequate to the Schrödinger equation the first diagram is

$$V_4^{(1)} = \frac{1}{(2\pi)^6}\, g\, \frac{\vec{\sigma}_2 \cdot \vec{Q}}{2m}\, \frac{H(\vec{Q}^2)}{\vec{Q}^2 + \mu^2}\, \tau_2^j\, T^{ij}\, \tau_3^i\, \frac{H(\vec{Q}'^2)}{\vec{Q}'^2 + \mu^2}\, \frac{\vec{\sigma}_3 \cdot \vec{Q}'}{2m}\, g \tag{3.107}$$

The off-shell πN -amplitude in a low momentum expansion is

$$\begin{aligned} T^{ij}(\vec{Q}, \vec{Q}') = {} & \delta_{ij} \left(a + b\, \vec{Q}\cdot\vec{Q}' + c\,(\vec{Q}^2 + \vec{Q}'^2) \right) \\ & + i\, \varepsilon_{ijk}\, \tau_1^k\, i\vec{\sigma}_1 \cdot (\vec{Q} \times \vec{Q}')\, (d_3 + d_4) \end{aligned} \tag{3.108}$$

where the constants a, b, c, d_3, d_4 can be correlated with observables and are therefore known. The πNN form factor squared is usually parametrised as

$$H(\vec{Q}^2) = \left(\frac{\Lambda^2 - \mu^2}{\Lambda^2 + \vec{Q}^2} \right)^2 \tag{3.109}$$

Once $V_4^{(1)}$ is known the total force (3.106) is also known:

$$V_4 = V_4^{(1)} + P_{12} P_{23} V_4^{(1)} P_{13} P_{23} + P_{13} P_{23} V_4^{(1)} P_{12} P_{23} \tag{3.110}$$

In (3.81), (3.86), (3.96), (3.98) occurs the combination

$$V_4 (1+P) = (1+P) V_4^{(1)} (1+P) \tag{3.111}$$

Therefore it is sufficient to know the matrixelements of $V_4^{(1)}$ with respect to our basis states (3.60):

$$\langle pq\alpha | V_4^{(1)} | p'q'\alpha' \rangle \tag{3.112}$$

This poses a technical problem since the Jacobi momenta (3.13)-(3.14) are not the natural variables for the pion momenta $\vec{Q}$ and $\vec{Q}'$ occuring in $V_4^{(1)}$. One has

$$\vec{Q} = \vec{p} - \vec{p}' - \tfrac{1}{2}(\vec{q} - \vec{q}') \tag{3.113}$$

$$\vec{Q}' = \vec{p} - \vec{p}' + \tfrac{1}{2}(\vec{q} - \vec{q}') \tag{3.114}$$

We want to illustrate the problem with a simplified expression

$$W_1 \equiv \frac{1}{\vec{Q}^2 + \mu^2} \, \frac{1}{\vec{Q}'^2 + \mu^2} \tag{3.115}$$

In LS-coupling the relevant matrixelement is

$$\langle pq(\ell\lambda)LM | W_1 | p'q'(\ell'\lambda')LM \rangle \tag{3.116}$$

$$= \int d\hat{p}\, d\hat{q}\, d\hat{p}'\, d\hat{q}'\, \mathcal{Y}^{LM*}_{\ell\lambda}(\hat{p}\hat{q})\, \frac{1}{(\vec{p}-\vec{p}\,'-\frac{1}{2}(\vec{q}-\vec{q}\,'))^2+\mu^2}\, *$$

$$\frac{1}{(\vec{p}-\vec{p}\,'+\frac{1}{2}(\vec{q}-\vec{q}\,'))^2+\mu^2}\, \mathcal{Y}^{LM}_{\ell'\lambda'}(\hat{p}'\hat{q}')$$

This can either be calculated directly using the QRN method [28)29)] or by a partial wave decomposition [3o)], which we sketch briefly. We expand W_1 into a sum of Legendre polynomials depending on the angle

$$x_1 \equiv \widehat{\vec{p}-\vec{p}\,'} \cdot \widehat{\vec{q}-\vec{q}\,'} \tag{3.117}$$

This is

$$W_1 = \sum_{\ell_1} \frac{2\ell_1+1}{2} P_{\ell_1}(x_1)\, g_\ell\left(|\vec{p}-\vec{p}\,'|, |\vec{q}-\vec{q}\,'|\right) \tag{3.118}$$

with

$$P_\ell(x_1) = \frac{4\pi}{\sqrt{\hat{\ell}_1}} (-)^{\ell_1}\, \mathcal{Y}^{00}_{\ell_1\ell_1}\left(\widehat{\vec{p}-\vec{p}\,'}, \widehat{\vec{q}-\vec{q}\,'}\right) \tag{3.119}$$

To proceed further we use

$$Y_{\ell m}(\widehat{\vec{a}+\vec{b}}) = \sum_{\ell_1+\ell_2=\ell} \frac{a^{\ell_1} b^{\ell_2}}{|\vec{a}+\vec{b}|^\ell} \sqrt{\frac{4\pi(2\ell+1)!}{(2\ell_1+1)!(2\ell_2+1)!}}\, \mathcal{Y}^{\ell m}_{\ell_1\ell_2}(\hat{a}\hat{b}) \tag{3.12o}$$

It remains to expand

$$\frac{g_{\ell_1}(|\vec{p}-\vec{p}\,'|, |\vec{q}-\vec{q}\,'|)}{|\vec{p}-\vec{p}\,'|^{\ell_1}\, |\vec{q}-\vec{q}\,'|^{\ell_1}} = \sum_{\ell_2\ell_3} \frac{2\ell_2+1}{2}\, \frac{2\ell_3+1}{2} P_{\ell_2}(x_2) P_{\ell_3}(x_3) H_{\ell_1\ell_2\ell_3} \tag{3.121}$$

where

$$H_{\ell_1\ell_2\ell_3} = \int dx_1 \int dx_2 \int dx_3\, P_{\ell_1}(x_1) P_{\ell_2}(x_2) P_{\ell_3}(x_3)\, * \tag{3.122}$$

$$\frac{1}{A_1^{\ell_2} A_2^{\ell_3}\left(A_1^2+\frac{1}{4}A_2^2-A_1A_2x_1+\mu^2\right)\left(A_1^2+\frac{1}{4}A_2^2+A_1A_2x_1+\mu^2\right)}$$

with

$$A_1 = \sqrt{p^2 + p'^2 - 2pp'x_2} \qquad (3.123)$$

$$A_2 = \sqrt{q^2 + q'^2 - 2qq'x_3} \qquad (3.124)$$

We end up with angular integrals of the form

$$X \equiv \int d\hat{p}\, d\hat{q}\, d\hat{p}'\, d\hat{q}'\; \mathcal{Y}^{LM^*}_{\ell\lambda}(\hat{p}\hat{q}) \left\{ \mathcal{Y}^{\ell_1}_{\lambda_1\lambda_2}(\hat{p}\hat{p}')\, \mathcal{Y}^{\ell_1}_{\lambda_3\lambda_4}(\hat{q}\hat{q}') \right\}^{0*} \qquad (3.125)$$

$$\mathcal{Y}^{00}_{\ell_2\ell_2}(\hat{p}\hat{p}')\, \mathcal{Y}^{00}_{\ell_3\ell_3}(\hat{q}\hat{q}')\; \mathcal{Y}^{LM}_{\ell'\lambda'}(\hat{p}'\hat{q}')$$

The curly bracket indicates zero angular momentum coupling. The evaluation of X requires various recouplings and the use of

$$\mathcal{Y}^{\ell m}_{\ell_1\ell_2}(\hat{a}\hat{a}) = \sqrt{\frac{\hat{\ell}_1\,\hat{\ell}_2}{4\pi\,\hat{\ell}}}\; C(\ell_1\ell_2\ell, 00)\; Y_{\ell m}(\hat{a}) \qquad (3.126)$$

One ends up with

$$X = \frac{1}{(4\pi)^2}(-)^{\ell+\ell'+\ell_1+L} \sqrt{\hat{\ell}_1\hat{\ell}_2\hat{\ell}_3\hat{\lambda}_1\hat{\lambda}_2\hat{\lambda}_3\hat{\lambda}_4} \qquad (3.127)$$

$$(\lambda_1\ell_2\ell, 00)(\lambda_2\ell_2\ell', 00)(\lambda_3\ell_3\lambda, 00)(\lambda_4\ell_3\lambda', 00)$$

$$\begin{Bmatrix} \lambda_2 & \lambda_1 & \ell_1 \\ \ell & \ell' & \ell_2 \end{Bmatrix} \begin{Bmatrix} \lambda_4 & \lambda_3 & \ell_1 \\ \lambda & \lambda' & \ell_3 \end{Bmatrix} \begin{Bmatrix} \ell & \ell' & \ell_1 \\ \lambda' & \lambda & L \end{Bmatrix}$$

and altogether with

$$\langle pq(\ell\lambda)LM | W_1 | p'q'(\ell'\lambda')LM \rangle$$

$$= \frac{(4\pi)^2}{8}(-)^{\ell+\lambda'+L} \sum_{\ell_1\ell_2\ell_3} \hat{\ell}_1\,\hat{\ell}_2\,\hat{\ell}_3\; H_{\ell_1\ell_2\ell_3}\, * \qquad (3.128)$$

$$\sum_{\lambda_1+\lambda_2=l_1} p^{\lambda_1} p'^{\lambda_2} \sqrt{\frac{(2l_1+1)!}{(2\lambda_1)!\,(2\lambda_2)!}} \sum_{\lambda_3+\lambda_4=l_1} q^{\lambda_3} q'^{\lambda_4}$$

$$\sqrt{\frac{(2l_1+1)!}{(2\lambda_3)!\,(2\lambda_4)!}} \,(\lambda_1 l_2 l, 00)(\lambda_2 l_2 l', 00)(\lambda_3 l_3 \lambda, 00)$$

$$(\lambda_4 l_3 \lambda', 00) \begin{Bmatrix} \lambda_2 & \lambda_1 & l_1 \\ l & l' & l_2 \end{Bmatrix} \begin{Bmatrix} \lambda_4 & \lambda_3 & l_1 \\ \lambda & \lambda' & l_3 \end{Bmatrix} \begin{Bmatrix} l & l' & l_1 \\ \lambda' & \lambda & L \end{Bmatrix}$$

The x_1 -integral in (3.122) can be carried through analytically and one is left with a two-fold integral to be done numerically. The actual expression (3.1o7)-(3.1o8) for $V_4^{(1)}$ has additional angular and spin-dependence, which leads to quite involved forms for its partial wave representation in momentum space [3o)25)].

What has been achieved up to now along this line? The homogeneous and inhomogeneous Faddeev equations for the unperturbed ($V_4 = o$) and perturbed ($V_4 \neq o$) Faddeev components have been solved including the 18 angular momentum combinations of Table 2. The Reid- and Paris-potentials were used. The results [25)] are shown in table 4.

Table 4: Energy-shifts $E^{(i)}$ of various orders in the triton caused by the 2π -exchange three-nucleon force

i	$E^{(i)}$	$\sum_{j=1}^{i} E^{(j)}$	$E^{(0)} + \sum_{j=1}^{i} E^{(i)}$	
o	o	o	-7.24	
1	-o.99	-o.99	-8.23	
2	-o.69	-1.68	-8.92	Reid-
3	-o.1o	-1.78	-9.o2	potential
4	-o.o6	-1.84	-9.o8	
5	-o.o1	-1.84	-9.o8	
o	o	o	-7.33	
1	-o.72	-o.72	-8.o5	
2	-o.93	-1.66	-8.99	Paris-
3	-o.o7	-1.73	-9.o6	potential
4	-o.13	-1.86	-9.19	
5	-o.oo	-1.86	-9.19	

Notice the strong second order energy shift caused by the first order change of the wavefunction. The resulting binding energy of -9.08 MeV (based on the Reid potential) agrees fairly well with the results gained by the Los Alamos group [22] (configuration space calculation) and Sendai group [23] (configuration and momentum space calculation) once the different values for the cut-off parameter Λ are taken into account. These independent calculations are absolutely necessary to test the codes.

The energy shifts (3.81), (3.98) ... shown in table 4 result from 18 x 18 contributions

$$\Delta E = \sum_{\alpha\alpha'=1}^{18} \langle \Psi_\alpha | V_4 | \Psi_{\alpha'} \rangle \equiv \sum_{\alpha\alpha'} \Delta E^{(\alpha\alpha')} \qquad (3.129)$$

where Ψ_α are partial wave projected states. It is well established that in the triton the first two or three angular momentum contributions of table 2 contribute about 90% to the norm of the total state. Therefore one expects that a much smaller number of terms is essentially responsible for Δ E, namely the ones which couple to the first 2 or 3 channels only:

$$\Delta \bar{E} = \sum_{\alpha\alpha'=1}^{3} \Delta E^{(\alpha\alpha')} + \left(\sum_{\alpha=1}^{3} \sum_{\alpha'=4}^{18} + \sum_{\alpha=4}^{18} \sum_{\alpha'=1}^{3} \right) \Delta E^{(\alpha\alpha')} \qquad (3.130)$$

This is indeed the case as shown in table 5 for the energy shift in first order and all orders included [25].

Table 5: The main contribution to the energy shift Δ E caused by the 2π-exchange three-nucleon force in the triton, as explained in the text.

Potential		$\Delta \bar{E}$	rest	ΔE
Reid	first order	-0.975	-0.025	-1.00
Reid	all orders	-1.78	-0.06	-1.84
Paris	all orders	-1.81	-0.05	-1.86

The reduction in the number of terms simplifies the calculation appreciably.

Back to physics. Table 4 shows an overbinding of the triton. We mention a few possible reasons. In setting up the form (3.108) for the off-shell πN-amplitude the contribution of an excited nucleon, the delta resonance, in the intermediate state was chosen in a static approximation. Since $m_\Delta - m_N \approx 300$ MeV only, the Δ should be allowed to propagate like the nucleon, which will reduce the amplitude and therefore the effect of the 2π-exchange three-nucleon force [31]. First estimates in nucleon matter for the $\pi\rho$- exchange three-nucleon force [32] show repulsion. This is encouraging and these mechanisms should be included in studying the triton. There is reason to believe that in the conventional picture of meson exchanges including three-nucleon forces the properties of the triton can be understood.

Relativistic effects are most naturally treated in momentum space. The techniques described can directly be used to solve relativistic few-body equations [7]. Such studies have been undertaken in the instant form of relativistic dynamics and in the light front form. In the models used relativistic effects in the triton binding energy were extremely small.

3.3 Two-body scattering

The LS equation for a two-body scattering state

$$|\varphi_{\vec{p}}^{(+)}\rangle = |\vec{p}\rangle + \frac{1}{p^2/2\mu + i\varepsilon - h_0} V |\varphi_{\vec{p}}^{(+)}\rangle \tag{3.131}$$

together with the well known expression for the scattering amplitude

$$f = \langle \vec{p}\,' | V | \varphi_{\vec{p}}^{(+)} \rangle \equiv \langle \vec{p}\,' | t | \vec{p} \rangle$$

leads immediately to the LS equation (1.6) for the t-operator. In a partial-wave representation $|p(ls)jm\rangle$ introduced in section 3.1 and suitable for two nucleons it has the form

$$t^{j}_{ls,l's}(p,p') = V^{j}_{ls,l's}(p,p') + \sum_{l''} \int dp''\, p''^2\, V^{j}_{ls,l''s}(p,p'') \tag{3.132}$$
$$* \frac{1}{p^2/2\mu + i\varepsilon - p''^2/2\mu}\, t^{j}_{l''s,l's}(p'',p')$$

This equation can either be solved directly or better replaced by a real k -matrix equation based on the principal-value prescription. Then a well known algebraic relation connects the on-shell (p=p') t- and k-matrices. The unitary S-matrix is then determined by

$$S^{j}_{ls,l's}(p) = \delta_{ll'} - 4i\mu p\, t^{j}_{ls,l's}(p,p) \tag{3.133}$$

which has a well known parametrisation in terms of phases and mixing parameters (in the case of coupled equations). These basic relations are developped for instance in [11].

For higher energies many partial waves may contribute. While the individual contributions oscillate strongly in the scattering angle their sum, the full scattering amplitude, is much smoother. Therefore a direct solution of (1.6) without angular momentum decomposition is adequate [33]. Let us regard two-body scattering from a Yukawa potential

$$V(\vec{p},\vec{p}\,') = \frac{\lambda}{(\vec{p}-\vec{p}\,')^2 + \gamma^2} \tag{3.134}$$

The LS equation (1.6) reads now

$$t(\vec{p},\vec{p}\,') = V(\vec{p},\vec{p}\,') + \int d\vec{p}\,''\, V(\vec{p},\vec{p}\,'')\, \frac{1}{p'^2/2\mu + i\varepsilon - p''^2/2\mu}\, t(\vec{p}\,'',\vec{p}\,') \tag{3.135}$$

Clearly t is a scalar and depends only on $|\vec{p}|, |\vec{p}\,'|$ and $x \equiv \hat{p}\cdot\hat{p}'$. If we identify the $\hat{p}'$ -direction with the z-axis the azimuthal angle φ'' of the $\vec{p}\,''$-integral occurs only in $V((\vec{p}-\vec{p}\,'')^2)$. We define

$$v(p,p'',x,x'') \equiv \int_0^{2\pi} d\varphi''\, V\Big(p^2 + p''^2 - 2pp''\big(xx'' + \sqrt{1-x^2}\,\sqrt{1-x''^2}\,\cos\varphi''\big)\Big) \tag{3.136}$$

and get

$$t(p,p',x) = \frac{1}{2\pi}\, v(p,p',x,1) + \int_0^\infty dp''\, p''^2 \int_{-1}^{1} dx''\, v(p,p'',x,x'')\, \frac{1}{p'^2/2\mu + i\varepsilon - p''^2/2\mu}\, t(p'',p',x'') \tag{3.137}$$

This is a two-dimensional integral equation for the half-shell t-matrix including its angular dependence. Again it is adviseable to relate t to a k-matrix, here defined by

$$k(p,p',x,x') = \frac{1}{2\pi} v(p,p',x,x') + \int_0^\infty dp''\, p''^2 \int_{-1}^{1} dx''\, v(p,p'',x,x'') \frac{1}{p'^2/2\mu - p''^2/2\mu}\, k(p'',p',x'',x') \tag{3.138}$$

Then it is straightforward to verify that

$$t(p,p',x) = k(p,p',x,1) + \int_{-1}^{1} dx''\, k(p,p',x,x'')\, t(p',p',x'') \tag{3.139}$$

For p=p' this is a one-dimensional integral equation for the on-shell t-matrix.

As an example we choose the Malfliet-Tjon potential III[13], which is a superposition of attractive and repulsive Yukawa potentials, typical for a two-nucleon interaction. In this case the φ-integral in (3.136) can be carried through analytically. The integrals in (3.138) require typically $N_p = 20$ and $N_x = 14$ quadrature points. We choose an energy $p^2/2\mu$ = 69o MeV (2μ = m, m = nucleonmass). In Figs. 1oa,b are shown the real part of $t(p,p,x)$ as a function of x in comparison with partial wave decompositions including 5, 1o and 2o orbital angular momenta. The advantage of solving (3.135) directly without angular momentum decomposition is obvious.

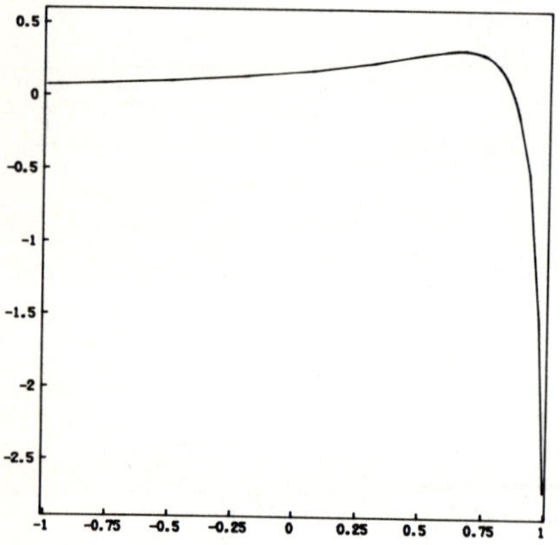

Fig. 1oa: The real part of t(ppx) against x

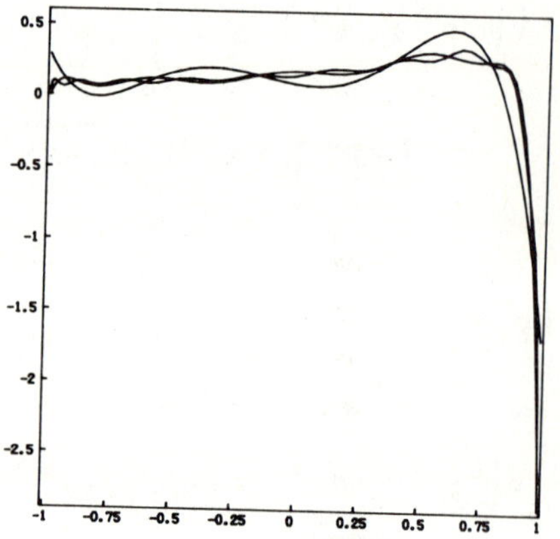

Fig. 1ob: The real part of t(ppx) against x in partial wave decompositions with l_{max} = 5,1o,2o.

3.4 Three-body scattering

We saw in section 2 how the infinite sequences of pair interactions between three particles can be summed up into a coupled set of three integral equations. Their solutions are channel to channel transition operators $U_{\alpha\beta}$. We want to include now a three-body force. Instead of using the same technique we shall derive the generalised AGS-equations [34] with the help of LS-equations [35][11]. In the three-body system there are four asymptotic configurations, which all can initiate a scattering process. Consequently there are four different types of scattering states. It is well known that the stationary scattering states can be gained with the help of the following limiting process (Möller-wave-operators)

$$\Psi_i^{(+)} = \lim_{\varepsilon\to 0} \frac{i\varepsilon}{E+i\varepsilon - H}\,\phi_i \qquad i = 0,1,2,3 \tag{3.14o}$$

Using the resolvent identity $(H \equiv H_0 + V_j + V^j)$

$$\frac{1}{z-H} = \frac{1}{z-H_j} + \frac{1}{z-H_j} V^j \frac{1}{z-H} \tag{3.141}$$

eqn. (3.14o) achieves the form of an integral equation

$$\Psi_i^{(+)} = \lim_{\varepsilon\to 0} \frac{i\varepsilon}{E+i\varepsilon - H_j}\,\phi_i + \frac{1}{E+io-H_j} V^j \Psi_i^{(+)} \tag{3.142}$$

Let us regard the initial channel 1 and choose j = 1,2,3. For j=1 the driving term is clearly just ϕ_1, whereas it vanishes for j=2 and 3. These relations are known as Lippmann identities [36]. In this manner we get the triad of LS equations [37]:

$$\Psi_1^{(+)} = \phi_1 + G_1(V_2+V_3+V_4)\,\Psi_1^{(+)} \tag{3.143}$$

$$\Psi_1^{(+)} = G_2(V_3+V_1+V_4)\,\Psi_1^{(+)} \tag{3.144}$$

$$\Psi_1^{(+)} = G_3(V_1+V_2+V_4)\,\Psi_1^{(+)} \tag{3.145}$$

The scattering state $\Psi_2^{(+)}$ $(\Psi_3^{(+)})$ can be treated in exactly the same manner; then the driving term $\phi_2(\phi_3)$ shows up in the second (third) equation. As a consequence $\Psi_2^{(+)}$ and $\Psi_3^{(+)}$ obey the homogeneous version of eqn (3.143). This is the well known defect of the single LS equation (3.143), not to define the solution uniquely [4]. Any linear combination $\Psi_1^{(+)} + \alpha\,\Psi_2^{(+)} + \beta\,\Psi_3^{(+)}$ is a solution of (3.143).

If however we require in addition that (3.144) and (3.145) are fulfilled these admixtures of $\Psi_2^{(+)}$ and $\Psi_3^{(+)}$ are excluded and the triad (3.143)-(3.145) defines $\Psi_1^{(+)}$ uniquely. The fourth possible solution of the Schrödinger equation, $\Psi_0^{(+)}$, obeys always inhomogeneous equations as can be seen from (3.142).

Transition amplitudes are known to have the general structure

$$T_{fi} = \langle \phi_f | V^f | \Psi_i^{(+)} \rangle \equiv \langle \phi_f | U_{fi} | \phi_i \rangle \tag{3.146}$$

The essential parts $U_{fi} | \phi_i \rangle \equiv V^f | \Psi_i^{(+)} \rangle$ of these matrixelements occur already on the right hand sides of (3.143)-(3.145). It just remains to derive integral equations. To that aim it is convenient to introduce a fourth auxiliary transition operator, which shows up naturally in a fourth LS equation

$$\Psi_1^{(+)} = G_4 \left(V_1 + V_2 + V_3 \right) \Psi_1^{(+)} \tag{3.147}$$

We define

$$U_{41} \phi_1 \equiv \left(V_1 + V_2 + V_3 \right) \Psi_1^{(+)} \tag{3.148}$$

Using (3.144) - (3.148) we read off that

$$U_{11} \phi_1 \equiv \left(V_2 + V_3 + V_4 \right) \Psi_1^{(+)} \tag{3.149}$$

can be written as

$$U_{11} \phi_1 = V_2 G_2 U_{21} \phi_1 + V_3 G_3 U_{31} \phi_1 + V_4 G_4 U_{41} \phi_1 \tag{3.150}$$

Similarily we find

$$U_{21} \phi_1 = V_3 G_3 U_{31} \phi_1 + V_1 \phi_1 + V_1 G_1 U_{11} \phi_1 + V_4 G_4 U_{41} \phi_1 \tag{3.151}$$

and corresponding expressions for U_{31} and U_{41} . As in (1.46) we replace $V_1 \phi_1$ by $G_0^{-1} \phi_1$ and using $V_j G_j \equiv T_j G_0$ we end up with the generalized AGS-equations

$$U_{i1} = \bar{\delta}_{i1} G_0^{-1} + \sum_{\substack{j \neq i \\ j=1}}^{4} T_j G_0 U_{j1} \qquad i = 1, \ldots, 4 \tag{3.152}$$

For identical particles the transition amplitude from a two-fragmentation channel to say channel 1 is

$$T_{1s} = \langle \phi_1 | V^1 | \Psi_s^{(+)} \rangle \tag{3.153}$$

where $\Psi_s^{(+)}$ is the properly symmetrised scattering state

$$\Psi_s^{(+)} = \Psi_1^{(+)} + \Psi_2^{(+)} + \Psi_3^{(+)} \tag{3.154}$$

Consequently

$$T_{1s} \equiv \sum_{j=1}^{3} \langle \phi_1 | U_{1j} | \phi_j \rangle \equiv \langle \phi_1 | \mathcal{U} | \phi_1 \rangle \tag{3.155}$$

The set (3.152) together with the corresponding ones for initial channels 2 and 3 lead now to the integral equation obeyed by the transition operator $\mathcal{U}$ and the auxiliary one, $\mathcal{U}_4$:

$$\mathcal{U} = P G_0^{-1} + P T G_0 \mathcal{U} + T_4 G_0 \mathcal{U}_4 \tag{3.156a}$$

$$\mathcal{U}_4 = (1+P) G_0^{-1} + (1+P) T G_0 \mathcal{U} \tag{3.156b}$$

or to

$$\mathcal{U} = X + X G_0 T G_0 \mathcal{U} \tag{3.157}$$

with

$$X = P G_0^{-1} + (1+P) T_4 \tag{3.158}$$

For reasons mentioned below it is advantageous to use a different operator $\mathcal{T}$ [38.] instead of $\mathcal{U}$:

$$\mathcal{U} \equiv X + X G_0 \mathcal{T} \tag{3.159}$$

Then from (3.158) follows the new set

$$\mathcal{T} = T P + T G_0 P \mathcal{T} + T G_0 \mathcal{T}_4 \tag{3.160}$$

$$\mathcal{T}_4 = T_4 (1+P) + T_4 G_0 (1+P) \mathcal{T} \tag{3.161}$$

In terms of $\mathcal{T}$ and $\mathcal{T}_4$ the elastic and break-up transition operators are

$$\mathcal{U} = P G_0^{-1} + P \mathcal{T} + \mathcal{T}_4 \tag{3.162}$$

$$\mathcal{U}_0 = (1+P)\mathcal{T} + \mathcal{T}_4 \tag{3.163}$$

This set lends itself to a systematic perturbational treatment of V_4:

$$\mathcal{T}^{(0)} = TP + T G_0 P \mathcal{T}^{(0)} \tag{3.164}$$

$$\mathcal{T}_4^{(0)} = 0 \tag{3.165}$$

$$\mathcal{T}^{(1)} = TP + T G_0 P \mathcal{T}^{(1)} + T G_0 \mathcal{T}_4^{(1)} \tag{3.166}$$

$$\mathcal{T}_4^{(1)} = V_4(1+P) + V_4(1+P) G_0 \mathcal{T}^{(0)} \tag{3.167}$$

$$\mathcal{T}^{(2)} = TP + T G_0 P \mathcal{T}^{(2)} + T G_0 \mathcal{T}_4^{(2)} \tag{3.168}$$

$$\mathcal{T}_4^{(2)} = V_4(1+P) + V_4(1+P) G_0 \mathcal{T}^{(1)} + V_4 G_0 \mathcal{T}_4^{(1)} \tag{3.169}$$

etc.

Numerical studies of that system are underway [39]. The complexity of the scattering problem caused by the free propagator can be seen in an one channel example of the unperturbed problem (3.164). Let us regard bosons with all angular momenta taken to be zero. Then the basis states are simply denoted by $|pq\rangle$ and we get

$$\langle pq|\mathcal{T}^{(0)}|\phi\rangle = \langle pq|TP|\phi\rangle + \langle pq|T G_0 P \mathcal{T}^{(0)}|\phi\rangle \tag{3.17o}$$

$$= \int dp'\, p'^2\, T(p,p',E-\tfrac{3}{4}q^2)\, \langle p'q|P|\phi\rangle$$

$$+ \int dp'\, p'^2\, T(p,p',E-\tfrac{3}{4}q^2)\, \frac{1}{E+i\varepsilon-p'^2-\tfrac{3}{4}q^2}\, \langle p'q|P\mathcal{T}^{(0)}|\phi\rangle$$

We use the partial wave representation (3.38) of the permutation operator P and the momentum space representation of the channel state ϕ

$$\langle pq|\phi\rangle = \varphi(p)\,\frac{\delta(q-q_0)}{q^2} \tag{3.171}$$

and get

$$\langle pq|\tau^{(0)}|\phi\rangle = \int_{-1}^{1}dx\, t(p,\pi_1^0,E-\tfrac{3}{4}q^2)\,\varphi(\pi_2^0) \tag{3.172}$$
$$+\int dq'q'^2\int_{-1}^{1}dx\, t(p,\pi_1,E-\tfrac{3}{4}q^2)\,\frac{1}{E+i\varepsilon-q^2-q'^2-qq'x}\,\langle\pi_2 q'|\tau^{(0)}|\phi\rangle$$

In π_1^0 and π_2^0 the q' is replaced by q^0 (see eqns (3.36),(3.37)).

There are two types of singularities. The two-body t-matrix $t(pp'z)$ has a pole at the two-body binding energy ε_b which leads to a pole of $\langle pq|\tau^{(0)}|\phi\rangle$ at q^0 defined by the on-shell relation

$$E = \frac{3}{4}q_0^2 + \varepsilon_b \tag{3.173}$$

That pole is easily handled by a subtraction technique. If we would have used the integral equation for $\mathcal{U}$ that pole would have been smeared out into an unpleasant logarithmic singularity. The second singularity comes from the free propagator. It occurs if

$$\bar{x} \equiv \frac{E-q^2-q'^2}{qq'} \tag{3.174}$$

lies between -1 and 1. This is the case if q and q' lie in the moon shaped area of Fig. 11.

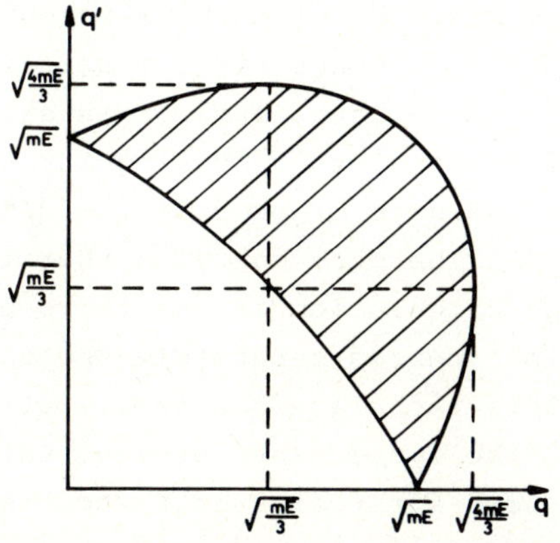

Fig. 11: The free propagator leads to logarithmic singularities along the boundary of the dashed area.

In this case one can replace the x-integral in (3.172), which is of the type

$$H = \int_{-1}^{1} dx\, f(x) \frac{1}{E+i\varepsilon - q^2 - q'^2 - qq'x} \qquad (3.175)$$

by

$$H = f(\bar{x}) \int_{-1}^{1} dx \frac{1}{E+i\varepsilon - q^2 - q'^2 - qq'x} + \int_{-1}^{1} dx \frac{f(x) - f(\bar{x})}{E+i\varepsilon - q^2 - q'^2 - qq'x} \qquad (3.176)$$

The second part is regular in q and q' and the first part can be integrated analytically and leads to the well known logarithmic singularities [40] along the boundary of the dashed area in Fig. 11.

It was Kloet and Tjon [41] who pioneered the solution of the Faddeev equation in momentum space for local potentials. They generated the Neumann series and summed it up by the Padé method. This appears to be the right strategy since for a realistic treatment of the three nucleon system, for instance, the Faddeev kernel will have a very large dimension like for the triton. Therefore an iterative procedure seems to be adequate where the matrix representation of the Faddeev kernel is always applied onto a column vector. The first term in the Born series is the driving term in (3.172) which is a nonsingular function in p and q. It is convenient to approximate its q-dependence by Splines which allows a simple evaluation of the singular q'-integral along the real axis. Also as in the bound state problem the x-integral is easily performed once the π_2 and π_1 dependence finds itself in Splines. In this manner one arrives at the second Born term. It is easy to see that the application of the singular kernel to a nonsingular function yields a q-dependence which has an infinite slope at a $q = \sqrt{4E/3}$, the maximal on-shell spectator momentum. The part carrying that infinite slope is known analytically. In the calculation of the third and higher order Born terms that part can be treated separately and only the rest is represented by Splines. This technique has been applied successfully to the πd system [42], to the n+d → n+n+p process using separable forces [43] and is presently used for realistic forces in the three-nucleon system [39] and for three hydrogen atoms [44].

Instead of using the Padé method a direct inversion is also possible if the large matrix to be inverted can be tamed by a method proposed by Brakhage [45]. This is presently done in the calculation of the recombination rate for three spin-polarised hydrogen atoms in a strong magnetic field to hydrogen molecules and free atoms. This is an example where the initial state consists of three colliding bodies.

Solving the Faddeev equations in the three-nucleon system for elastic and break-up reactions, including spin observables, will be an important test for the quality of our conventional meson exchange picture for two- and three-nucleon forces. Even if quark and gluon degrees of freedom will turn out to be important in the description of nuclear dynamics the technical know how for solving few-body equations will remain useful as long as the new dynamics can be formulated in terms of effective forces between nucleons.

Acknowledgement: I thank A. Bömelburg for carefully reading the manuscript.

References

1) B.A. Lippmann, J. Schwinger, Phys.Rev.79,469(1950)
2) E.O. Alt, P. Grassberger, W. Sandhas, Nucl.Phys.B2,167(1967)
3) W. Sandhas, Acta Physica Austriaca, Suppl.IX,57(1972); in few-body nuclear physics, Trieste lectures, IAEA, Vienna, 1978
4) L.D. Faddeev, Sov.Phys.JETP12,1014(1961)
5) L.D. Faddeev: Mathematical Aspects of the Three-Body Problem in Quantum Scattering Theory (Davey, N.Y. 1965)
6) M. Abramowitz, I.A. Stegun: Handbook of Mathematical Functions, Dover Publ.,N.Y., page 890
7) W. Glöckle, T.S.H. Lee, F. Coester, Phys.Rev.C33,709(1986); L. Müller, Nucl.Phys.A360,331(1981); Nuov.Cim.75A,39(1983)
8) R. Reid, Ann.Phys.50,411(1968)
9) K. Erkelenz, Phys.Rep.13C,191(1974); K. Holinde, Phys.Rep.68C,121(1981); R. Machleidt, K. Holinde, Ch. Elster, to appear in Phys.Rep. C
10) M. Lacombe, B. Loisseau, S.M. Richard, R. Vinh Mau, J. Cotê, P. Pirès, R. de Tourreil, Phys.Rev. C21,861(1980)
11) W. Glöckle, The Quantum Mechanical Few-Body Problem (Springer-Verlag, 1983)
12) W. Glöckle, G. Hasberg, A.R. Neghabian, Z.Phys.A305,217(1982)
13) R.A. Malfliet, J.A. Tjon, Nucl.Phys.A127,161(1969)
14) W. Glöckle, Nucl.Phys.A381,343(1982)
15) G.L. Payne, private communication
16) Th. Cornelius, W. Glöckle, to appear in J.Chem.Phys.(1986)
17) A. Bömelburg, Phys.Rev.C28,403(1983)
18) Ch. Hajduk, P.U. Sauer, Nucl.Phys.A369,321(1981)
19) A. Laverne, C. Gignoux, Nucl.Phys.A203,597(1973)
20) G.L. Payne, J.L. Friar, B.F. Gibson, Phys.Rev.C22,832(1980)
21) S. Ishikawa, T. Sasakawa, T. Sawada, T. Ueda, Phys.Rev.Lett.53, 1877(1984)

22) C.R. Chen, G.L. Payne, J.L. Friar, B.F. Gibson, Phys.Rev.Lett.55, 374(1985)
23) T. Sasakawa, S. Ishikawa, Few-Body Systems 1,3(1986)
24) A. Bömelburg, W. Glöckle, Phys.Rev.C28,2149(1983)
25) A. Bömelburg, Phys.Rev.C34,14(1986)
26) S.A. Coon, M.D. Scadron, P.C. McNamee, B.R. Barrett, D.W.E. Blatt, B.H.J. McKellar, Nucl.Phys.A371,242(1979)
27) H.T. Coelho, T.K. Das, M.R. Robilotta, Phys.Rev.C28,1812(1983)
28) H. Tanaka, H. Nagata, Prog.Theor.Phys.Suppl.56,121(1974)
29) A. Bömelburg, private communication
3o) S.A. Coon, W. Glöckle, Phys.Rev.C23,179o(1981)
31) Ch.Hajduk, P.U. Sauer, Shin Nan Yang, Nucl.Phys.A4o5,6o5(1983)
32) R.G. Ellis, S.A. Coon, B.H.J. McKellar, Nucl.Phys.A438,631(1985)
33) J. Holz, W. Glöckle, to be published
34) K.L. Kowalski, Phys.Rev.D7,18o6(1973); Nucl.Phys.A264,173(1976)
35) W. Glöckle, R. Brandenburg, Phys.Rev.C27,83(1983)
36) B.A. Lippmann, Phys.Rev.1o2, 264(1956); G. Bencze, C. Chandler, Phys.Lett.9oA,162(1982)
37) W. Glöckle, Nucl.Phys.A141,62o(197o), S.K. Adhikari, W. Glöckle, Phys.Rev.C21,54(198o); F.S. Levin, W. Sandhas, to appear in Phys.Rev.C(1986)
38) A. Bömelburg, W. Glöckle, W. Meier, in Few-Body Problems in Physics, Vol.II,ed.B. Zeitnitz, North Holland 1984, page 483
39) R. Brandenburg, H. Witala, private communications
4o) E.W. Schmid, H. Ziegelmann, The Quantum Mechanical Three-Body Problem (Pergamon Press, Oxford 1974)
41) W.M. Kloet, J.A. Tjon, Ann.Phys.79,4o7(1973); for extended work see C. Stolk, J.A. Tjon, Nucl.Phys.A295,384(1978)
42) A. Matsuyama, Phys.Lett.152B,42(1985)
A. Matsuyama, T.S.H. Lee, to be published
43) W. Meier, W. Glöckle, Phys.Lett.138B,329(1984) and
W. Meier, PHD Thesis (Bochum 1983) unpublished
44) L.P.H. de Goey, T.H.M. v.d.Berg, N. Mulders, H.T.C. Stoof, B.J. Verhaar, W. Glöckle, to appear in Phys.Rev.A and to be published
45) H. Brakhage, Numerische Mathematik 2,183(196o)

METHOD OF CONTINUED FRACTIONS APPLIED TO THREE-BODY CALCULATIONS

Tatuya Sasakawa
Department of Physics
Tohoku University
Sendai 980, Japan

Method of continued fractions for solving a local or a nonlocal potential is given for scattering and bound state problems. The method is used for calculations of three-body systems.

1. INTRODUCTION

This talk is devoted to the method of continued fractions (MCF) for solving scattering or bound state problems. The detailed account of the method was given in a series of articles [1,2,3]. The application to a three-nucleon bound state was made in the article [4].

In the past, a countless number of methods for calculations of scattering and bound states have been proposed. Now, we believe that MCF is best for various reasons: (1) The algorithm is very simple. (2) We enjoy always a very rapid convergence. (3) The method is applicable to local as well as non-local potentials, or even to three- (and more-) body problems on the same basis. (4) In the course of iterations, the kernels that vanish at large distances (non-singular kernels) are used. As a result, we don't need to calculate up to large distances. (5) The N-th step of this method is correct up to λ^{2N}, if we compare it with the perturbation expansion expressed in the power of the strength λ of the potential. (6) This method is closely related to the Schwinger variational principle [2]. (7) This method can be reduced to a kind of the Padé approximant [2]. (8) A great merit of MCF is its flexibility. In fact, each of the references [1],[2],[3] and [4] describes a different version. Owing to this flexible nature of MCF, we can find a version with which the computational time is minimized. By MCF, we can solve the 34-channel Faddeev equation for realistic two- and three-nucleon potentials in one minutes with SX1, a computer recently installed at Tohoku University.

2. POTENTIAL SCATTERING

2.1 Potential scattering (MCFV)

In this section, we demonstrate how to bring the scattering amplitude into the form of continued fractions.

For a given local or non-local potential V and the initial plane wave $|k\rangle$, we want to solve the Lippmann-Schwinger equation

$$|\phi\rangle = |k\rangle + G_o V|\phi\rangle \ . \tag{1}$$

We introduce a potential V_1 by

$$V_1 = V - \frac{V|k\rangle\langle k|V}{\langle k|V|k\rangle} \ . \tag{2}$$

We use Eq.(2) in Eq.(1) to obtain

$$\begin{aligned} |\phi\rangle &= |k\rangle + G_o\left(V_1 + \frac{V|k\rangle\langle k|V}{\langle k|V|k\rangle} \right)|\phi\rangle \\ &= \frac{1}{1-G_oV_1}|k\rangle + \frac{1}{1-G_oV_1}|k_1\rangle\frac{\langle k|V|\phi\rangle}{\langle k|V|k\rangle} \ , \end{aligned} \tag{3}$$

where the function $|k_1\rangle$ is defined by

$$|k_1\rangle = G_o V|k\rangle \ . \tag{4}$$

We note that the potential V_1 satisfies the orthogonality relation

$$V_1|k\rangle = \langle k|V_1 = 0 \ . \tag{5}$$

We define a function $|\phi_1\rangle$ by the Lippmann-Schwinger equation

$$|\phi_1\rangle = |k_1\rangle + G_o V_1|\phi_1\rangle \ . \tag{6}$$

Utilizing Eqs.(5) and (6), we express Eq.(3) in a form

$$|\phi\rangle = |k\rangle + |\phi_1\rangle\frac{\langle k|V|\phi\rangle}{\langle k|V|k\rangle} \ . \tag{7}$$

The scattering amplitude is given by $\langle k|V|\phi\rangle$. If we multiply Eq.(6) by $\langle k|V$, we obtain

$$\langle k|t|k\rangle \equiv \langle k|V|\phi\rangle = \frac{\langle k|V|k\rangle^2}{\langle k|V|k\rangle - \langle k|V|\phi_1\rangle} \ . \tag{8}$$

Our task is then to find an expression for calculating $<k|V|\phi_1>$. In analogy to Eq.(2), we define V_2 by

$$V_2 = V_1 - \frac{V_1|k_1><k_1|V_1}{<k_1|V_1|k_1>} \quad . \tag{9}$$

It is clear that the orthogonality relation (5) is generalized to

$$V_2|k_1> = <k_1|V_2 = <k|V_2 = V_2|k> = 0 \quad . \tag{10}$$

Using Eq.(9), we express Eq.(6) in the form

$$|\phi_1> = |k_1> + G_o(V_2 + \frac{V_1|k_1><k_1|V_1}{<k_1|V_1|k_1>})|\phi_1> \quad . \tag{11}$$

Taking similar steps that led Eq.(6) from Eq.(3), we express Eq.(11) in a form

$$|\phi_1> = |k_1> + |\phi_2> \frac{<k_1|V_1|\phi_1>}{<k_1|V_1|k_1>} \quad , \tag{12}$$

where $|\phi_2>$ is defined by

$$|\phi_2> = |k_2> + G_oV_2|\phi_2> \quad , \tag{13}$$

with

$$|k_2> = G_oV_1|k_1> \quad . \tag{14}$$

From Eq.(12), we obtain the following expression by a similar step that led Eq.(8)

$$<k_1|V_1|\phi_1> = \frac{<k_1|V_1|k_1>^2}{<k_1|V_1|k_1> - <k_1|V_1|\phi_2>} \quad . \tag{15}$$

Multiplying Eq.(12) by $<k|V$, we get

$$<k|V|\phi_1> = <k|V|k_1> + <k|V|\phi_2> \frac{<k_1|V_1|\phi_1>}{<k_1|V_1|k_1>} \quad . \tag{16}$$

The factor $<k|V|\phi_2>$ reads

$$<k|V|\phi_2> = <k|V|k_2> + <k|VG_oV_2|\phi_2> = <k_1|V_1|k_1> + <k_1|V_2|\phi_2> = <k_1|V_1|k_1> \quad . \tag{17}$$

In the last step, we used Eq.(10). We use Eq.(17) in Eq.(16), and Eq.(15) in Eq.(16) to obtain

$$<k|V|\phi_1> = <k|V|k_1> + <k_1|V_1|\phi_1>$$

$$= <k|V|k_1> + \frac{<k_1|V_1|k_1>^2}{<k_1|V_1|k_1> - <k_1|V_1|\phi_2>} \quad . \tag{18}$$

We can generalize Eq.(18). If we define τ_i by

$$\tau_i = <k_{i-1}|V_{i-1}|\phi_i> \qquad \text{for } i = 1, 2, 3, \ldots \tag{19}$$

we get the continued fractions for τ_i,

$$\tau_i = <k_{i-1}|V_{i-1}|k_i> + \frac{<k_i|V_i|k_i>^2}{<k_i|V_i|k_i> - \tau_{i+1}} \quad . \tag{20}$$

Thus the scattering amplitude (8) is expressed in terms of the continued fractions.

Various properties of this method as well as numerical examples were given in references [1] and [2]. Table 1 shows another example, where the phase shift (in radian) is given for the nucleon-nucleon scattering of ^1S_o state at E_{Lab} = 24 MeV by the Reid soft core potential [5]. In this calculation, the principal part of the Green function is used. Therefore, the amplitude and phase shift are related as

$$T_0^{(N)} = -\frac{1}{k}\tan(\delta^{(N)}) \quad . \tag{21}$$

Table 1. MCF applied to the two-nucleon Lippman-Schwinger equation of Reid soft core potential for ^1S_o state. E_{Lab} = 24 MeV.

N	$\delta^{(N)}$	N	$\delta^{(N)}$
1	0.415091	5	0.860626
2	0.604301	6	0.860656
3	0.851412	7	0.860652
4	0.860529	8	converge

2.2 Linear equation with symmetric kernel

If we multiply both sides of Eq.(1) by $V^{1/2}$, we get an equation

$$|\Phi> = |F_o> + K|\Phi> \ , \tag{22}$$

where

$$|\Phi> = V^{1/2}|\phi>, \ |F_o> = V^{1/2}|k> \text{ and } K = V^{1/2}G_oV^{1/2} \quad . \tag{23}$$

If G_o is real symmetric, as in the case of the principal value of a Green function, K is real symmetric. In this case, we define the following set of equations for i = 1,2, ...

$$|F_{i+1}\rangle = K_i|F_i\rangle \ , \tag{24}$$

$$K_{i+1} = K_i - \frac{K_i|F_i\rangle\langle F_i|K_i}{\langle F_i|K_i|F_i\rangle} \quad . \tag{25}$$

Then the vectors $|F_j\rangle$ satisfy the orthogonality relations

$$\langle F_j|K_i = K_i|F_j\rangle = 0, \ j = 0, 1, \ldots, i - 1. \tag{26}$$

For the vectors $|\Phi_i\rangle$ satisfying

$$|\Phi_i\rangle = |F_i\rangle + K_i|\Phi_i\rangle \tag{27}$$

we get the expression

$$|\Phi_i\rangle = |F_i\rangle + |\Phi_{i+1}\rangle \frac{\langle F_i|K_i|F_i\rangle}{\langle F_i|K_i|F_i\rangle - \langle F_i|K_i|\Phi_{i+1}\rangle} \ , \tag{28}$$

and the continued fractions

$$\langle F_i|K_i|\Phi_{i+1}\rangle = \langle F_i|K_i|F_{i+1}\rangle + \frac{\langle F_{i+1}|K_{i+1}|F_{i+1}\rangle^2}{\langle F_{i+1}|K_{i+1}|F_{i+1}\rangle - \langle F_{i+1}|K_{i+1}|\Phi_{i+2}\rangle} \quad . \tag{29}$$

The MCFG proposed in [2] is directly related to this form.

We note that this way of solving Eq.(22) is applicable not only to a scattering problem, but also to linear coupled equations, even including "ill-conditioned" equations. For instance, a famous example given by T.S. Wilson

$$\begin{aligned} 5x_1 + 7x_2 + 6x_3 + 5x_4 &= 23 \\ 7x_1 + 10x_2 + 8x_3 + 7x_4 &= 32 \\ 6x_1 + 8x_2 + 10x_3 + 9x_4 &= 33 \\ 5x_1 + 7x_2 + 9x_3 + 10x_4 &= 31 \end{aligned} \tag{30}$$

is known bv the fact that we can not get a correct answer ($x_1 = x_2 = x_3 = x_4 = 1$) by a single precision calculation for any usual method. However, if we apply the MCF demonstrated in this section, we get the correct answer, by a single precision calculation.

2.3 Bound state

In the case of the scattering problem, we have the initial vector $|k\rangle$, which plays an important role in the manipulation. On the other hand, the wave function of a bound state should satisfy a homogeneous equation

$$|\phi\rangle = G_o V|\phi\rangle \quad , \tag{31}$$

without the initial state. In this case, if we want to apply a similar MCF stated in secs. 2.1 and 2.2, we have to have inhomogeneous equations for $|\phi_i\rangle$, starting from the homogeneous equation (31). This is done as follows. Let $|f\rangle$ be a function which is regular at the origin and vanishes at large distances. Except for these requirements, the choice of this function is rather arbitrary.

In terms of this function, we define the potential V_1 by an equation which is similar to Eq.(2),

$$V_1 = V - \frac{V|f\rangle\langle f|V}{\langle f|V|f\rangle} \quad . \tag{32}$$

If we put this equation into Eq.(31), we can express the function $|\phi\rangle$ as

$$|\phi\rangle = |\phi_1\rangle\frac{\langle f|V|\phi\rangle}{\langle f|V|f\rangle} \quad , \tag{33}$$

where the function $|\phi_1\rangle$ is defined with the function $|f_1\rangle$ by

$$|f_1\rangle = G_o V|f\rangle \tag{34}$$

and

$$|\phi_1\rangle = \frac{1}{1 - G_o V_1}|f_1\rangle = |f_1\rangle + G_o V_1|\phi_1\rangle \quad . \tag{35}$$

If we multiply $\langle f|V$ from the left of Eq.(33), we obtain an equality

$$\langle f|V|\phi_1\rangle - \langle f|V|f\rangle = 0. \tag{36}$$

The solution of this equation determines the binding energy. Comparing Eq.(36) with Eq.(8), we realize that Eq.(36) is to find out the pole of the scattering amplitude.

Once we have an inhomogeneous equation (35), the calculation of $\langle f|V|\phi_1\rangle$ is done in the same manner as in the scattering problem described in sec. 2.1.

2.4 Notes

In sec.2.1, we can presume that V_i becomes smaller and smaller with increasing i because of the orthogonality equation (10), or more generally,

$$V_i|k_j\rangle = \langle k_j|V_i = 0, \text{ for } j = 0, 1, \ldots, i-1. \tag{37}$$

Therefore, we begin the calculation assuming that $V_i = 0$ for an assumed $i = N$; namely, we start the iteration of Eq.(20) from

$$\tau_{N-1} = \langle k_{N-1}|V_{N-1}|k_N\rangle \quad . \tag{38}$$

A similar remark applies also to the formula (29) in sec.2.2. We start from putting

$$\langle F_{N-1}|K_{N-1}|\Phi_N\rangle = \langle F_{N-1}|K_{N-1}|F_N\rangle \ . \tag{39}$$

In Appendix of [2], the method of calculating the amplitude and wave function is written in some detail.

3. THREE-BODY PROBLEM

3.1 Faddeev equation and MCF

Now we apply the MCF to the Faddeev equation. The solution of the three-body Schrödinger equation

$$(E - H_o - V_{12} - V_{23} - V_{31})\Psi = 0 \tag{40}$$

is expressed as a sum of three components

$$\Psi = \Phi(12,3) + \Phi(23,1) + \Phi(31,2) \quad . \tag{41}$$

The component $\Phi(12,3)$ represents the state in which a pair of particles 1 and 2 are interacting in the final state, while the particle 3 is standing as the spectator. With (41), we decompose (40) into

$$(E - H_o - V_{12})\Phi(12,3) = V_{12}(\Phi(23,1) + \Phi(31,2)) \tag{42}$$

and two other equations which are obtained from (42) by cyclic permutations of 1,2, and 3. Eq.(42) is called the Faddeev equation.

If three-body forces

$$W_{123} = W_{12,3} + W_{23,1} + W_{31,2} \tag{43}$$

are acting in the three-body system, we may extend the Faddeev equation (42) as

$$(E - H_o - V_{12})\Phi(12,3) = V_{12}[\Phi(23,1) + \Phi(31,2)]$$
$$+ W_{12,3}[\Phi(12,3) + \Phi(23,1) + \Phi(31,2)]. \tag{44}$$

Of course, there are other ways of extending (42) to accommodate three-nucleon potentials. In any way, if we solve the equation as exactly as possible, the way of accommodating the three-nucleon forces should not affect the calulated results. For the bound state, the Los Alamos-Iowa group [6] showed that if we perform the 34-channel calculations, the way of extending (42) to include (43) does not affect the result.

For simplicity, we use the following notations:

$$\Phi \equiv \Phi(12,3), \qquad \hat{Q}\Phi = \Phi(23,1) + \Phi(31,2), \qquad V \equiv V_{12}, \qquad U \equiv V\hat{Q} + W_{12,3}(1 + \hat{Q}). \tag{45}$$

Eq.(44) reads then

$$(E - H_o - V)\Phi = U\Phi \ . \tag{46}$$

Further, if we denote by G_o the Green function

$$G_o = \frac{1}{E - H_o - V} \ , \tag{47}$$

Eq.(46) is expressed in the Lippmann-Schwinger form; (1) for the scattering problem and (31) for the bound state problem.

3.2 n-d scattering

We calculated the n-d quartet scattering K-matrix $<F_o|U|\Phi>$ for the Reid soft core potential of ^1S_o, $^3S_1$ and $^3D_1$ states [5]. E_{Lab} is taken to be 2.5 MeV. For the relative angular momentum of ℓ = 0 and 2, the results are given in Table 2.

Table 2. MCF for the K-matrix elements of the nd doublet scattering. E_{Lab} = 2.5 MeV.

N	s-s	s-d	d-d
1	-0.502	-0.03	-0.058
2	35.099	0.527	-0.062
3	-2.250	0.014	-0.058
4	-0.760	-0.010	-0.058
5	-0.662	-0.014	-0.058
6	-0.724	-0.012	-0.058
7	-0.758	-0.011	-0.058
8	-0.749	-0.011	-0.058
9	-0.740	-0.011	-0.058
10	-0.739	-0.011	-0.058

Converge

3.3 Bound state

The method of sec. 2.3 is applied taking account of the non-symmetric nature of the kernel as stated in 3.1. The method was described in [4]. Choosing $|F_o>$ and $|f>$ as specified later, we introduce the potential U_i and functions $|F_i>$ by

$$U_{i+1} = U_i - \frac{U_i|F_i><f|U}{<f|U|F_i>} \ , \tag{48}$$

$$|F_{i+1}\rangle = G_o U_i |F_i\rangle \quad . \tag{49}$$

The wave function $|\Phi_i\rangle$ is defined by

$$|\Phi_i\rangle = |F_i\rangle + G_o U_i |\Phi_i\rangle \ . \tag{50}$$

The binding energy is obtained from the equation

$$\langle f|U|\Phi_1\rangle - \langle f|U|F_o\rangle = 0. \tag{51}$$

The matrix element $\langle f|U|\Phi_1\rangle$ is obtained by calculating the continued fractions

$$\langle f|U|\Phi_i\rangle = \frac{\langle f|U|F_i\rangle^2}{\langle f|U|F_i\rangle - \langle f|U|\Phi_{i+1}\rangle} \ , \tag{52}$$

starting from

$$|\Phi_N\rangle = |F_N\rangle \quad . \tag{53}$$

The function $|F_o\rangle$ and $|f\rangle$ are arbitrary. As to $|f\rangle$, we have chosen the following form. Let $\vec{x}$ be the relative coordinate between the interacting pair 1 and 2, and $\vec{y}$ be the coordinate of the spectator 3 relative to the center of mass of the pair 1 and 2. We designate by q(p) momentum of the interacting pair (the spectator). For a triton of the binding energy $|E|$, p and q satisfy the relationship

$$-|E| = (\hbar^2/M)q^2 + (3\hbar^2/4M)p^2, \tag{54}$$

where M is the nucleon mass. We take p as real, and q as pure imaginary; $q = i|q|$. Let $\phi_{1S}(|q|,x)$ [$\phi_{3S}(|q|,x)$ and $\phi_{3D}(|q|,x)$] be the normalized Sturm-Liouville function of the ${}^1S_o$[${}^3S_1$ and ${}^3D_1$] two-body state for a given energy $-(\hbar^2/M)q^2$, multiplied by $(\lambda_q/(1-\lambda_q))^{1/2}$, where λ_q is the eigenvalue. We let $u_o(py)$ stand for

$$u_o(py) = \sqrt{2/\pi}\ p \sin py/(py) \quad . \tag{55}$$

Denoting by χ the spin function of the spectator, we choose the function $|f\rangle$ to be given by

$$|f\rangle = \int_0^{P_M} dp\, u_o(py)\{\phi_{1S}(|q|,x)[{}^1S_o(\hat{x})\otimes\chi]_{J=1/2} + \phi_{3S}(|q|,x)[{}^3S_1(\hat{x})\otimes\chi]_{J=1/2}$$

$$+ \phi_{3D}(|q|,x)[{}^3D_1(\hat{x})\otimes\chi]_{J=1/2}\}, \tag{56}$$

where ${}^1S_o(\hat{x})$ etc. denote the spin-angular function of the interacting pair and J represents the total spin of triton. In (56), p_M is a cut-off momentum which is chosen judiciously. However, since the function $|f\rangle$ is a trial function, we need not to be nervous in choosing p_M.

As for $|F_o\rangle$, we have taken the following function,

$$|F_o\rangle = G_o V\hat{Q}|f\rangle \ . \tag{57}$$

For any perturbation method to converge very quickly, the starting function should be chosen so that it has an important character of the solution. This general requirement is fulfilled by the above choice of $|F_o\rangle$. Since the operator $G_o V\hat{Q}$ in (57) is a part of the kernel of $G_o U$, the function $|F_o\rangle$ and hence $|F_1\rangle$, which is the starting function of MCF, should be very similar in its behavior to the true wave function. Especially, due to the presence of the permutation operator $\hat{Q}$ in (57) and U in (45), the functions $|F_o\rangle$ and hence $|F_1\rangle$ have a node which is a characteristic of the Faddeev component for a soft core potential [7]. If we started the calculation of the continued fractions by such a function without a node as (56), the convergence would be very slow.

As denoted in the Introduction, the MCF is to some extent flexible. Here, we have chosen $|f\rangle$ to be given by (56),

$$|f\rangle = (56). \tag{58}$$

However, since $|f\rangle$ is arbitrary, some other choice is of course possible. For instance, taking (58) for $|f\rangle$ in equations

$$|\Phi\rangle = \frac{1}{1 - G_o U_1} G_o U |F_o\rangle \frac{\langle f|U|\Phi\rangle}{\langle f|U|F_o\rangle} \ , \tag{59}$$

and in (51), and

$$|f\rangle = |F_i\rangle \tag{60}$$

for $i \geq 1$ in (48), we made the order-by-order comparison of the left-hand side of (51) for RSC5 taking $|E| = 7.031$ MeV [4]. In the case (I)[(II)] of Table 3, the result of the choice (58) [(60)] is demonstrated. From this table, we see that the choice (60) is better than (58) for the order-by-order convergence. However, the choice (60) takes more time than the choice (58), because for (60) we have to calculate $\langle F_i|U|\Phi_j\rangle$, $N-1 \geq j \geq i+1$, thus one more loop than the choice (58) (although, in practice, the difference of the computational time is not significant for a very fast computer).

Table 3. Order-by-order comparison of convergence, used (58) for |f>(Case I) and (60) for |f>(Case II). Diff(m) denotes the value of the righthand side of (51) when the continued fraction (52) starts from m [=N in (53)]. This table illustrates the calculation for RSC5 (|E|=7.031 MeV).

	Diff(m)	
m	(I)	(II)
1	-0.02529	-0.22963
2	-0.00416	-0.12590
3	0.01555	-0.00512
4	0.00163	0.00465
5	0.00039	-0.00168
6	-0.00292	-0.00044
7	-0.00012	-0.00008
8	-0.00006	0.00002
9	-0.00068	-0.00004
10	-0.00001	0.00000
11	0.00000	0.00000

References

1. Horáček, J., Sasakawa, T.: Phys. Rev. A28, 2151 (1983)

2. Horáček, J., Sasakawa, T.: Phys. Rev. A30, 2274 (1985)

3. Horáček, J., Sasakawa, T.: Phys. Rev. C32, 70 (1985)

4. Sasakawa, T., Ishikawa, S.: Few-Body Systems 1, 3 (1986)

5. Ishikawa, S.: Proceedings of IUPAP International Conference Few Body XI (Sasakawa, T., et al. ed.) to appear. Amsterdam: North-Holland 1987

6. Chen, C. R., Payne, G. L., Friar, J. L., Gibson, B. F.: Phys. Rev. Lett. 55, 374 (1985)

7. Sasakawa, T., Okuno H., Sawada, T.: Phys. Rev. C23, 904 (1981)

CONFIGURATION-SPACE FADDEEV CALCULATIONS: NUMERICAL METHODS

G. L. Payne
Department of Physics and Astronomy
The University of Iowa
Iowa City, Iowa 52242

I. Introduction

The traditional approach of nuclear physics describes nuclei by means of a model in which nonrelativistic nucleons interact via two-body forces, and possibly three-body forces. It is only for the few-nucleon system that one can do accurate numerical tests of this model. It has long been possible to solve the two-nucleon problem for realistic two-body interactions, even on small computers. However, the solution of the three-nucleon problem is considerably more difficult, and it has only been during the past decade that the computational sophistication has improved to the point where any disagreement between the calculated values and the experimental results can be attributed to a failure of the models, and not to a lack of numerical accuracy. This is particularly true for the bound state of the trinucleon system, where many groups[1-5] can now calculate the bound-state energy and wave function for this system. The results of these various groups for the binding energy agree to within 10 keV, both for the case with only two-body interactions and for the case with two- and three-body interactions. Our ability to do theoretical calculations for the trinucleon scattering problem is not as advanced as for the bound state problem. It has only been during the past few years that accurate calculations with realistic forces have become feasible,[6-8] and no consensus between the various groups has been reached. Hopefully, within the next decade it will become possible to perform scattering calculations with the same precision that has been achieved for the bound-state system. In this paper we review one of the techniques used to solve the bound-state problem. Similar techniques can also be used for the scattering problem; however, the boundary conditions are considerably more difficult for the scattering problem.

There are several techniques for solving the nonrelativistic Schrödinger equation for a three-body system. One can solve the equation in momentum space,[2,4] configuration space,[1] or with a combination of the two.[3] In configuration space the Schrödinger equation can be written either as a differential or an integral equation.[9] In this paper we will review the solution of the differential equation in configuration space. For this case one has the choice of using a variational method for the Schrödinger equation,[5] or of using a standard numerical method to solve the Faddeev-Noyes equations.[10] There are several numerical advantages to solving the Faddeev-Noyes equations instead of the Schrödinger equation, these will be discussed in Section VI. The Faddeev-Noyes equations are a set of coupled elliptic partial differential equations, and there are many methods for solving these types of equations. These equations were first solved numerically by the Grenoble group,[11] who used the method of finite-differences. The Los Alamos-Iowa group[1] also solve these coupled differential equations, but choose to use a spline expansion method that provides an expansion of the wave function which is convenient for additional calculations. We review the spline expansion technique for the solution of the configuration-space Faddeev-Noyes equations.

In Section II we introduce the spline expansion technique by solving a two-body problem using this method. In Section III the spline expansion for the three-body problem is presented, and in Section IV the numerical methods used to solve the resulting large matrix equations are discussed. In Section V we present some numerical results to illustrate the convergence properties of the numerical techniques.

II. Two-Body Problem

In order to illustrate the use of spline expansions to solve bound-state eigenvalue problems, we first consider the two-body problem. There are many numerical techniques for solving the two-body Schrödinger equation for the bound state, and the orthogonal collocation method we use is a standard numerical method for solving differential equations of this form.[12] For simplicity, we consider the case of two spinless particles, each with mass M, interacting via a central potential, V(r). The Schrödinger equation,

$$- \frac{\hbar^2}{M} \nabla^2 \Psi(\vec{r}) + V(r)\Psi(\vec{r}) = E\Psi(\vec{r}) , \qquad (II.1)$$

can be solved by using the standard partial wave expansion

$$\Psi(\vec{r}) = \sum_{\ell} \sum_{m} \frac{u_\ell(r)}{r} Y_{\ell m}(\theta,\phi) , \qquad (II.2)$$

where we follow the usual practice and introduce the reduced wave function, $u_\ell(r)$. The primary reason for introducing the reduced wave function is that the boundary conditions for this function are easier to impose than the boundary conditions for the full wave function. The full wave function must be finite at the origin, and this implies that the reduced wave function is zero at the origin. The differential equation for the $u_\ell(r)$ is obtained by using the orthogonality properties of the spherical harmonics, $Y_{\ell m}$. Substituting the expansion in (II.2) into the Schrödinger equation, multiplying by $-M/\hbar^2$, and taking the inner product with $Y_{\ell m}(\theta,\phi)$ gives the ordinary differential equation

$$\left[\frac{d^2}{dr^2} - \frac{\ell(\ell+1)}{r^2} - \frac{M}{\hbar^2} V(r)\right] u_\ell(r) = \kappa^2 u_\ell(r) , \tag{II.3}$$

where

$$\kappa^2 = -\frac{M}{\hbar^2} E , \tag{II.4}$$

with $E < 0$ for a bound state.

Equation (II.3) is a second-order ordinary differential equation, and in order to obtain a unique solution one must specify the boundary conditions. As discussed above the appropriate boundary conditions for the bound-state problem are that the reduced wave function be zero at the origin and that it go to zero for large values of r. For a short range interaction the asymptotic form of the wave function is proportional to $e^{-\kappa r}$. We can simplify the numerical calculations by factoring out this asymptotic behavior. Thus, we define a new function $f_\ell(r)$ by assuming that the reduced wave function has the form

$$u_\ell(r) = f_\ell(r) e^{-\kappa r} . \tag{II.5}$$

Substituting this expression for $u_\ell(r)$ into Equation (II.3) yields the following differential equation for the unknown function $f_\ell(r)$:

$$\left[\frac{d^2}{dr^2} - \frac{\ell(\ell+1)}{r^2} - U(r)\right] f_\ell(r) = + 2\kappa \frac{d}{dr} f_\ell(r) , \tag{II.6}$$

where we have defined $U(r) = M V(r)/\hbar^2$. The boundary conditions for this new function are:

$$f_\ell(0) = 0 \ , \tag{II.7a}$$

and

$$\frac{d}{dr} f_\ell(r)\Big|_{r=R} = 0 \ , \tag{II.7b}$$

where R is the matching radius; that is, a value of r for which the wave function has the asymptotic form of a constant times $e^{-\kappa r}$.

Now we expand the function $f_\ell(r)$ in a complete set of basis functions; that is, we write

$$f_\ell(r) = \sum_{n=0}^{N+1} a_n s_n(r) \ , \tag{II.8}$$

where the choice of the basis functions is arbitrary. A basis set with many numerical advantages, which are discussed below, is the set of spline functions. A spline function is a function consisting of polynomial pieces on subintervals, joined together with certain smoothness conditions. These functions are defined by dividing the interval [0,R] into I subintervals defined by the breakpoints $r_0, r_1, \ldots, r_I$. The breakpoints can be chosen so that there are more breakpoints in the region where the function to be fitted has more structure, and there are fewer breakpoints in the region where the function is smooth. This is one of the advantages of the splines as a basis set; by a careful choice of breakpoints one can reduce the number of basis functions and still obtain a good approximation to the function to be fitted. For the bound state wave function $f_\ell(r)$, which approaches a constant for large values of r, the function will have more structure for small values of r; consequently, we can choose the breakpoints with a small separation for small values of r and a larger separation as r becomes larger. A simple method for doing this is to use a scale factor S_r and the relation

$$r_{i+1} - r_i = S_r(r_i - r_{i-1}) \ , \tag{II.9}$$

where for the interval [0,R], $r_0 = 0$ and $r_I = R$. The value of r_1 is chosen so that $r_I = R$; that is, we choose

$$r_1 = \left(\frac{S_r - 1}{S_r^{\,I} - 1}\right) R . \tag{II.10}$$

Given the breakpoints, the spline functions are defined as piecewise polynomials of degree k with a continuous m^{th} derivative. The degree of the polynomials is constrained by the value of m, in order for the mth derivative to be continuous the polynomial functions must be of degree m+1 or larger. Since we are using the spline expansion to solve a second-order differential equation, the spline functions must have continuous first derivatives. The splines could have a continuous higher derivative, but it has been shown[13] that an optimal choice for a second-order differential equation is a spline basis set with continuous first derivatives. For this case the splines must be of degree two or higher. For our expansion we choose to use cubic splines, that is, splines of degree three. For technical reasons, odd-degree splines behave better than even-degree splines. Given that the splines are piecewise polynomials of degree three with continuous first derivatives does not uniquely specify the spline function. For this case two common choices are the cubic B-splines of de Boor[13] and the cubic Hermite splines.[12] The cubic Hermite splines have a simpler analytic form, and it has been shown that they have better numerical stability than the B-splines.[14] Therefore, we use the cubic Hermite splines as our basis set.

The cubic Hermite spline basis set consists of the functions

$$\phi_i(r) = \left(\frac{r - r_{i-1}}{r_i - r_{i-1}}\right)^2 \left[3 - 2\left(\frac{r - r_{i-1}}{r_i - r_{i-1}}\right)\right] \quad ; \quad r_{i-1} \le r < r_i \tag{II.11a}$$

$$\phi_i(r) = \left(\frac{r_{i+1} - r}{r_{i+1} - r_i}\right)^2 \left[3 - 2\left(\frac{r_{i+1} - r}{r_{i+1} - r_i}\right)\right] \quad ; \quad r_i \le r < r_{i+1} \tag{II.11b}$$

and

$$\xi_i(r) = \left(\frac{r - r_{i-1}}{r_i - r_{i-1}}\right)^2 (r - r_i) \quad ; \quad r_{i-1} \le r < r_i \tag{II.12a}$$

$$\xi_i(r) = \left(\frac{r_{i+1} - r}{r_{i+1} - r_i}\right)^2 (r - r_i) \; ; \quad r_i \le r < r_{i+1} \tag{II.12b}$$

defined at each breakpoint. The functions ϕ_i and ξ_i are plotted in Figure 1. These function have a continuous first derivative, and they are nonzero only in the interval $[r_{i-1}, r_{i+1}]$. If the interval is divided into I subintervals, there are 2I+2 splines. An example for I=4 is shown in Figure 2.

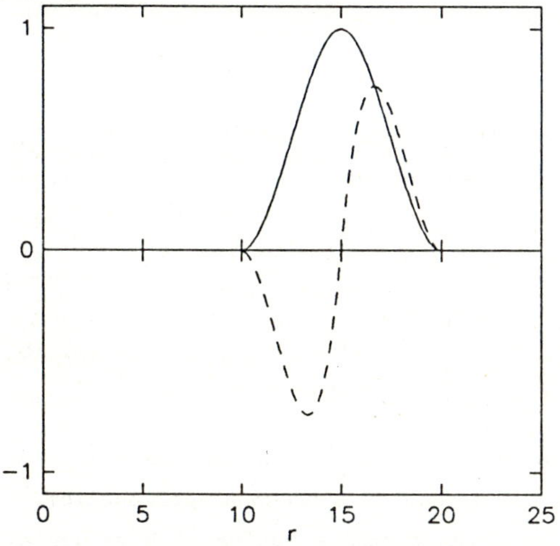

Figure 1. The functions ϕ_3 (solid line and ξ_3 (dashed line) for $I = 5$, $S_r = 1$, and $R = 25$.

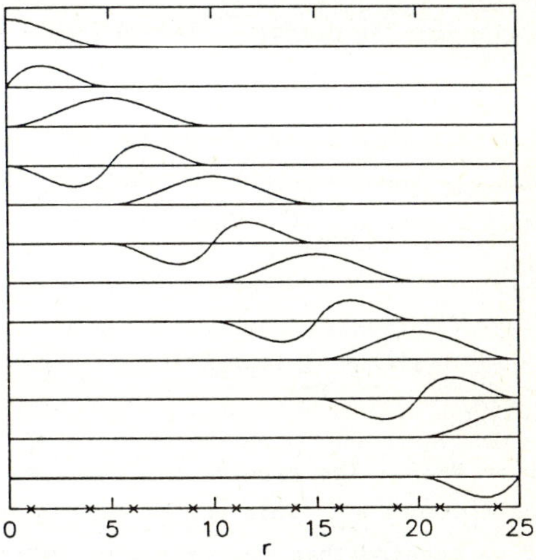

Figure 2. The cubic Hermite splines for $I = 5$, $S_r = 1$, and $R = 25$.

The 2(I+1) spline functions $s_n(r)$ are defined as

$$s_{2i}(r) = \phi_i(r) \; , \tag{II.13a}$$

and

$$s_{2i+1}(r) = \xi_i(r) \; , \tag{II.13b}$$

for $i=0,1,2,\ldots,I$. The Hermite spline functions are defined so that ϕ_i is unity at r_i, and ξ_i has a slope equal to unity at r_i. This normalization is convenient when the Hermite splines are used to fit an analytic function. This normalization is also convenient for imposing the boundary conditions on f_ℓ. For this basis one finds that

$$f_\ell(r_i) = a_{2i} , \tag{II.14a}$$

and

$$\frac{d}{dr} f_\ell(r)\Big|_{r=r_i} = a_{2i+1} . \tag{II.14b}$$

With this basis set it is easy to impose the boundary conditions on f_ℓ. The boundary condition (II.7a) implies that $a_0=0$, and the boundary condition (II.7b) implies that $a_{N+1}=0$. Thus, we simply replace the expansion in (II.8) by the expansion

$$f_\ell(r) = \sum_{n=1}^{N} a_n s_n(r) , \tag{II.15}$$

where N=2I. The problem has now been reduced to finding the N unknown expansion coefficients a_n. This can be done by using the method of orthogonal collocation.[12] If we require that the expansion satisfy the differential equation at N values of r, the procedure is a collocation method; the N discrete values r_m (m=1,2,3,...,N) are called the collocation points. Since there are N=2I unknowns, we can choose two collocation points on each interval. An optimal choice for these points are the two-point Gauss quadrature points.[13] If one chooses these points, the method is called the orthogonal collocation method.

Substituting the expansion (II.15) into (II.6) and setting r equal to r_m yields the N equations

$$\sum_{n=1}^{N} a_n\left[s_n''(r_m) - \frac{\ell(\ell+1)}{r_m^2} s_n(r_m) - U(r_m)s_n(r_m)\right]$$

$$= \kappa \sum_{n=1}^{N} a_n[\ 2s_n'(r_m)] , \tag{II.16}$$

where s_n'' is the second derivative of s_n. The set of equations (II.16) has the form

$$\sum_{n=1} A_{mn}a_n = \kappa \sum_{n=1}^{N} B_{mn}a_n \ , \qquad \text{(II.17)}$$

where

$$A_{mn} = s_n''(r_m) - \frac{\ell(\ell+1)}{r_m^2} s_n(r_m) - U(r_m)s_n(r_m) \ , \qquad \text{(II.18a)}$$

and

$$B_{mn} = - 2s_n(r_m) \ . \qquad \text{(II.18b)}$$

The equations in (II.17) can be written as the matrix equation

$$\underset{\approx}{A}\, \underset{\sim}{a} = \kappa\, \underset{\approx}{B}\, \underset{\sim}{a} \ . \qquad \text{(II.19)}$$

Since the Hermite splines are nonzero only on two subintervals, the matrices in (II.19) are banded. The resulting matrices for the basis shown in Figure 2 have the form shown in Figure 3.

$$\begin{bmatrix}
X & X & X & 0 & 0 & 0 & 0 & 0 & 0 & 0 \\
X & X & X & 0 & 0 & 0 & 0 & 0 & 0 & 0 \\
0 & X & X & X & X & 0 & 0 & 0 & 0 & 0 \\
0 & X & X & X & X & 0 & 0 & 0 & 0 & 0 \\
0 & 0 & 0 & X & X & X & X & 0 & 0 & 0 \\
0 & 0 & 0 & X & X & X & X & 0 & 0 & 0 \\
0 & 0 & 0 & 0 & 0 & X & X & X & X & 0 \\
0 & 0 & 0 & 0 & 0 & X & X & X & X & 0 \\
0 & 0 & 0 & 0 & 0 & 0 & 0 & X & X & X \\
0 & 0 & 0 & 0 & 0 & 0 & 0 & X & X & X
\end{bmatrix}$$

Figure 3. Nonzero elements of $\underset{\approx}{A}$ and $\underset{\approx}{B}$.

The banded structure of the matrix can be exploited to reduce the number of numerical operations required to solve the matrix equations. For the two-body case, the matrices are not very large and we do not exploit the banded structure. However, for the three-body case similar banded matrices are encountered, and we utilize the banded structure to considerably reduce the computer time required to solve the matrix equations. The algorithm for the three-body case will be presented in Section IV. For the two-body case the matrix equation in (II.19) is the generalized eigenvalue problem, and the eigenvalues and eigenvectors can be found by using the QZ algorithm of Moler and Stewart.[15] The numerical solution of (II.19) gives the eigenvalue κ, from which we can determine the bound-state energy by using (II.4), and the eigenvector consists of the spline expansion coefficients a_n. Thus, once we have determined the eigenvector, we can use Equations (II.5) and (II.15) to calculate the wave function for any value of r. This feature is particularly useful if we wish to use the wave function to calculate other properties of the bound state.

Since Equation (II.19) is an N×N matrix equation, there will be N eigenvalues. Also, the matrices are not symmetric and some of the eigenvalues may be complex. Obviously, all of the eigenvalues do not correspond to physical states of the system. Many of the eigenvalues are a result of using the boundary condition (II.7a) at a finite value of R. It is easy to identify the physical eigenstates, they should be independent of the value of the matching radius R for large enough values of R. That is, if we choose a value of R which corresponds to the asymptotic region of the wave function, the result should not depend upon the value of R as long as we use enough basis functions to obtain an accurate solution of the differential equation. To demonstrate this property we show in Table I the values of the eigenvalue corresponding to the ground states for various values of N and two values of R. For the two-body potential we have chosen the Malfliet-Tjon potential MT-III,

$$V(r) = V_1 \frac{e^{-\mu_1 r}}{r} - V_2 \frac{e^{-\mu_2 r}}{r} , \tag{II.20}$$

where $V_1 = 1438.72$ MeV-fm, $V_2 = 626.885$ MeV-fm, $\mu_1 = 3.11\ \mathrm{fm}^{-1}$, and $\mu_2 = 1.55\ \mathrm{fm}^{-1}$. As an additional check to insure that the eigenvector corresponds to the ground state one can plot the reduced wave function as a function of r. In Figure 4 we plot $f_\ell(r)$ and $u_\ell(r)$ for the ground state of the MT-III potential. From Figure 4, one can see that the wave function has the expected functional form.

Table I

Ground-state energy of the MT-III potential calculated using different values of the matching radius (R fm.) and different numbers of intervals (I).

I	R = 5	R = 10	R = 15
10	2.2251	2.2852	2.7114
12	2.2275	2.2202	2.3068
14	2.2282	2.2273	2.2197
16	2.2284	2.2294	2.2262
18	2.2285	2.2301	2.2291
20	2.2285	2.2303	2.2299
22	2.2286	2.2305	2.2303
24	2.2286	2.2305	2.2304
26	2.2286	2.2306	2.2305
28	2.2286	2.2306	2.2306
30	2.2286	2.2306	2.2306

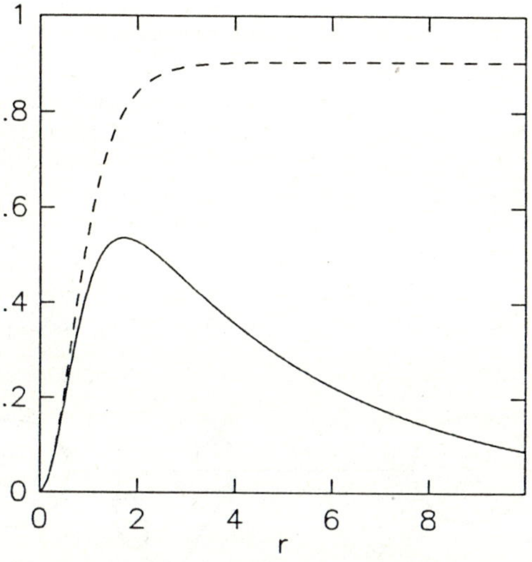

Figure 4. The reduced wave function $u_\ell(r)$ (solid line) and the function $f_\ell(r)$ (dashed line).

In the next section we generalize the procedures presented above to the three-body case. The basic methods are identical to those described above; the only difference is that the size of the numerical calculations is increased considerably.

III. Three-Body Problem: Bosons

To demonstrate how to use the spline techniques presented in Section II for solving the three-body problem using the Faddeev-Noyes equations, we first consider the case of three spinless bosons. In Section VI we extend the methods to include the case of three nucleons with spin and isospin.

For the case of three identical bosons each with mass M and coordinates $\vec{r}_1$, $\vec{r}_2$, and $\vec{r}_3$, the total wave function for the three-particle system can be expressed as the sum of the three Faddeev amplitudes

$$\begin{aligned} \Psi_T &= \Psi(\vec{x}_1,\vec{y}_1) + \Psi(\vec{x}_2,\vec{y}_2) + \Psi(\vec{x}_3,\vec{y}_3) \\ &= \Psi_1 + \Psi_2 + \Psi_3 \ . \end{aligned} \tag{III.1}$$

In Equation (III.1) we have used the Jacobi coordinates

$$\vec{x}_i = \vec{r}_j - \vec{r}_k \ , \tag{III.2a}$$

$$\vec{y}_i = \frac{1}{2}(\vec{r}_j + \vec{r}_k) - \vec{r}_i \ , \tag{III.2b}$$

where i,j,k imply cyclic permutation. The vectors $\vec{x}_1$ and $\vec{y}_1$ are shown in Figure 5.

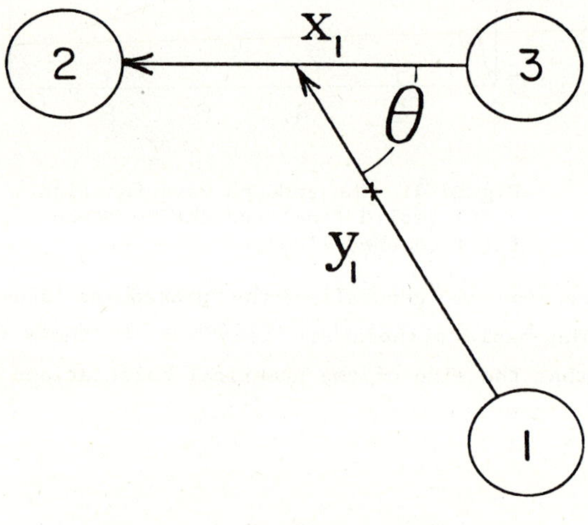

Figure 5. The Jacobi coordinates $\vec{x}_1$ and $\vec{y}_1$ for the three-body system.

For the case where the only interactions are the pairwise two-body potentials, $V(\vec{x}_i)$, the Hamiltonian for the system is

$$H = T + V(\vec{x}_1) + V(\vec{x}_2) + V(\vec{x}_3) \; ,$$

$$= T + V_1 + V_2 + V_3 \; , \qquad \text{(III.3)}$$

where T is the kinetic energy operator. In the center-of-mass coordinate system, the kinetic energy operator has the form

$$T = -\frac{\hbar^2}{M}\left(\nabla_x^2 + \frac{3}{4}\nabla_y^2\right) . \qquad \text{(III.4)}$$

The Faddeev-Noyes procedure for solving the three-body problem in configuration space consists of replacing the Schrödinger equation,

$$(H - E)\Psi_T = 0 \; , \qquad \text{(III.5)}$$

by the three coupled Faddeev-Noyes equations,

$$[T + V(\vec{x}_i) - E]\Psi(\vec{x}_i,\vec{y}_i) = - V(\vec{x}_i)[\Psi(\vec{x}_j,\vec{y}_j) + \Psi(\vec{x}_k,\vec{y}_k)] \; , \qquad \text{(III.6a)}$$

or

$$[T + V_i - E]\Psi_i = - V_i[\Psi_j + \Psi_k] \; , \qquad \text{(III.6b)}$$

with i, j, k cyclic. Adding the the three Faddeev-Noyes equations gives the original Schrödinger equation. For identical particles the three functions Ψ_i all have the same functional form, and it is only necessary to solve one of the three equations, which we choose to be i=1. To solve (III.6b), we follow the same procedures as for the two-body case. We first expand the Ψ_i in a complete set of orbital angular momentum states. For the angular momentum states we use the states formed by coupling the state vector for particles j and k with relative orbital angular momentum ℓ_α to the state vector for particle i with an angular momentum L_α

relative to the center of mass of particles j and k. That is, for a system with total orbital angular momentum L and z-component M, we use the partial wave expansion:

$$\Psi(\vec{x}_i,\vec{y}_i) = \sum_{\alpha} \frac{\phi_\alpha(x_i,y_i)}{x_i y_i} |\alpha_i\rangle , \tag{III.7}$$

where we have introduced the reduced channel wave function, and the angular basis states labeled by α are called channels. Each channel basis function is the orbital part of the Faddeev amplitude, that is

$$|\alpha_i\rangle = \sum_{m} \sum_{M} (\ell_\alpha \; m \; L_\alpha \; M \mid L \; M) Y_{\ell_\alpha m}(\hat{x}_i) Y_{L_\alpha M}(\hat{y}_i)$$

$$= [Y_{\ell_\alpha m}(\hat{x}_i) \otimes Y_{L_\alpha M}(\hat{y}_i)]_{LM} . \tag{III.8}$$

To obtain the differential equations for the reduced wave functions, we use the orthogonality properties of the $|\alpha_i\rangle$. Substituting the expansion (III.7) into (III.6) for i=1, multiplying by $-(M/\hbar^2)x_1y_1$, and taking the inner product with $|\alpha_1\rangle$ gives the set of coupled partial differential equations

$$\left[\frac{\partial^2}{\partial x_1^2} + \frac{3}{4}\frac{\partial^2}{\partial y_1^2} - \frac{\ell_\alpha(\ell_\alpha + 1)}{x_1^2} - \frac{3}{4}\frac{L_\alpha(L_\alpha + 1)}{y_1^2} - K^2\right]\phi_\alpha(x_1,y_1) - \sum_\beta v_{\alpha\beta}(x_1)\phi_\beta(x_1,y_1)$$

$$= \sum_\beta v_{\alpha\beta}(x_1)\left[\langle\beta_1|x_1y_1|\Psi(\vec{x}_2,\vec{y}_2)\rangle + \langle\beta_1|x_1y_1|\Psi(\vec{x}_3,\vec{y}_3)\rangle\right] , \tag{III.9}$$

where $K^2 = -M E/\hbar^2$, and $v_{\alpha\beta}(x_1) = M\langle\alpha_1|V(\vec{x}_1)|\beta_1\rangle/\hbar^2$. In addition, we have used the projection for the two-body interaction onto the complete set of basis functions $|\beta_1\rangle$.

The right-hand side of (III.9) can be simplified by using the relations:

$$\Psi(\vec{x}_2,\vec{y}_2) = P^- \Psi(\vec{x}_1,\vec{y}_1) , \tag{III.10a}$$

and

$$\Psi(\vec{x}_3,\vec{y}_3) = P^{+}\Psi(\vec{x}_1,\vec{y}_1) , \tag{III.10b}$$

where P^{+} and P^{-} are the cyclic permutation operators, that is, the result of P^{+} acting on a state with particle 1 coupled to the coupled pair (2,3) is

$$P^{+}|(2,3)1\rangle = |((1,2)3\rangle , \tag{III.11a}$$

and the result of P^{-} acting on the state is

$$P^{-}|(2,3)1\rangle = |(3,1)2\rangle . \tag{III.11b}$$

Now by using the relationship $P^{+} = P_{23}P^{-}P_{23}$, where P_{23} is the two particle exchange operator, one finds that

$$\begin{aligned}\langle\beta_1|\Psi(\vec{x}_3,\vec{y}_3)\rangle &= \langle\beta_1|P^{+}|\Psi(\vec{x}_1,\vec{y}_1)\rangle = \langle\beta_1|P_{23}P^{-}P_{23}|\Psi(\vec{x}_1,\vec{y}_1)\rangle \\ &= \langle\beta_1|P^{-}|\Psi(\vec{x}_1,\vec{y}_1)\rangle = \langle\beta_1|\Psi(\vec{x}_2,\vec{y}_2)\rangle ,\end{aligned} \tag{III.12}$$

where we have used the property that for identical bosons, the Faddeev amplitude Ψ_1 is symmetric under the exchange of particles 2 and 3. Using (III.12) in (III.9) and averaging over the values of M, that is, summing over M and dividing by $2L+1$, yields the set of equations

$$\left[\frac{\partial^2}{\partial x_1^2} + \frac{3}{4}\frac{\partial^2}{\partial y_1^2} - \frac{\ell_\alpha(\ell_\alpha + 1)}{x_1^2} - \frac{3}{4}\frac{L_\alpha(L_\alpha + 1)}{y_1^2} - K^2\right]\phi_\alpha(x_1,y_1) - \sum_\beta v_{\alpha\beta}(x_1)\phi_\beta(x_1,y_1)$$

$$= \frac{2}{2L+1}\sum_M\sum_\beta v_{\alpha\beta}(x_1)\langle\beta_1|x_1y_1|\Psi(\vec{x}_2,\vec{y}_2)\rangle . \tag{III.13}$$

The numerical calculations can be considerably simplified if we follow Noyes[10] and replace the variables x_i and y_i by the hyperspherical variables ρ and θ_i. These variables are defined by the relations:

$$x_i = \rho\cos\theta_i , \tag{III.14a}$$

and

$$y_i = \frac{\sqrt{3}}{2} \rho \sin\theta_i \ . \tag{III.14b}$$

The introduction of these variables leads to a banded matrix equation which requires less computer memory to store; in addition, the computer time to solve the equations is considerably reduced. After the change of variables, the coupled differential equations (III.13) become

$$(\Delta_\alpha - K^2)\phi_\alpha(\rho,\theta_1) - \sum_\beta v_{\alpha\beta}(\rho \cos\theta_1)\phi_\beta(\rho,\theta_1)$$

$$= \frac{2}{2L+1} \sum_M \sum_\beta v_{\alpha\beta}(\rho \cos\theta_1)\langle\beta_1|x_1y_1\Psi(\vec{x}_2,\vec{y}_2)\rangle \ , \tag{III.15}$$

where

$$\Delta_\alpha = \frac{\partial^2}{\partial\rho^2} + \frac{1}{\rho}\frac{\partial}{\partial\rho} + \frac{1}{\rho^2}\frac{\partial^2}{\partial\theta_1^2} - \frac{\ell_\alpha(\ell_\alpha + 1)}{\rho^2\cos^2\theta_1} - \frac{L_\alpha(L_\alpha + 1)}{\rho^2\sin^2\theta_1} \ . \tag{III.16}$$

There are several techniques for evaluating the matrix element on the right-hand side of (III.15). A common procedure is to use the method of Harper, Kim, and Tubis[16] which uses the addition theorem for the spherical harmonics[17] to express the amplitude $\Psi(\vec{x}_2,\vec{y}_2)$ as an infinite sum of the projections of $\Psi(\vec{x}_2,\vec{y}_2)$ onto the states $|\alpha_1\rangle$. This was the method used for our initial calculations for the trinucleon problem.[18] However, we have found that a method first suggested by Balian and Brezin[19] is more efficient for calculations which involve a large number of channels. The method of Balian and Brezin combined with the use of the bipolar harmonics[20] yields an algorithm which is very efficient on a vectorized computer such as the CRAY.

The method of Balian and Brezin consists of recognizing that the right-hand side of (III.15) is independent of the choice of the z-axis for our coordinate system. Consequently, we can choose the z-axis along $\hat{y}_1$, and since the integrand depends only on the angle between $\vec{x}_1$ and $\vec{y}_1$, we can use the relation

$$\int d\hat{x}_1 \int d\hat{y}_1 \rightarrow 8\pi^2 \int_{-1}^{1} d\mu \ , \tag{III.17}$$

where $\mu = \hat{x}_1 \cdot \hat{y}_1$. The numerical calculations are further simplified by choosing the coordinate system so that the vector $\hat{x}_1$ lies in the x-z plane. Then from the relations

$$\vec{x}_2 = -\frac{1}{2}\vec{x}_1 + \vec{y}_1 , \tag{III.18a}$$

and

$$\vec{y}_2 = -\frac{3}{4}\vec{x}_1 - \frac{1}{2}\vec{y}_1 , \tag{III.18b}$$

one can see that $\vec{x}_2$ and $\vec{y}_2$ also lie in the x-z plane. Therefore, all of the spherical harmonics will be real, and the integrand will be real. The numerical evaluation of the spherical harmonics is discussed in the appendix. Now the right-hand side of (III.15) has the form

$$\sum_\beta v_{\alpha\beta}(\rho \cos\theta_1) \sum_\gamma \int_{-1}^{1} d\mu \, \frac{x_1 y_1}{x_2 y_2}$$

$$\times \left[\frac{16\pi^2}{2L+1} \sum_M [Y_{\ell_\beta}(\hat{x}_1) \otimes Y_{L_\beta}(\hat{y}_1)]_{LM} \, \phi_\gamma(x_2, y_2) \, [Y_{\ell_\gamma}(\hat{x}_2) \otimes Y_{L_\gamma}(\hat{y}_2)]_{LM} \right] . \tag{III.19}$$

Using the relation

$$x_2 = \rho \cos\theta_2 = \left(\frac{x_1^2}{4} - x_1 y_1 \mu + y_1^2 \right)^{\frac{1}{2}} , \tag{III.20}$$

one can show for fixed values of x_1 and y_1, changing variables from μ to θ_2 gives

$$\int_{-1}^{1} \frac{x_1 y_1}{x_2 y_2} d\mu \rightarrow \frac{4}{\sqrt{3}} \int_{\theta^-}^{\theta^+} d\theta_2 , \tag{III.21}$$

where for equal mass particles $\theta^- = \left|\theta_1 - \frac{\pi}{3}\right|$ and $\theta^+ = \left| |\theta_1 - \frac{\pi}{6}| - \frac{\pi}{2} \right|$. After this change of variables, the right-hand side of (III.15) is

$$\sum_{\beta} v_{\alpha\beta}(\rho\ \cos\theta_1) \sum_{\beta} \int_{\theta^-}^{\theta^+} K_{\beta\gamma}(\theta_1,\theta_2)\phi_\gamma(\rho,\theta_2)d\theta_2\ , \tag{III.22}$$

where the kernel is given by

$$K_{\beta\gamma}(\theta_1,\theta_2) = \frac{16\pi^2}{2L+1} \sum_M [Y_{\ell_\beta}(\hat{x}_1) \otimes Y_{L_\beta}(\hat{y}_1)]_{LM}\ [Y_{\ell_\gamma}(\hat{x}_2) \otimes Y_{L_\gamma}(\hat{y}_2)]_{LM}\ \frac{4}{\sqrt{3}}\ . \tag{III.23}$$

The values of θ^+ and θ^- are shown in Figure 6.

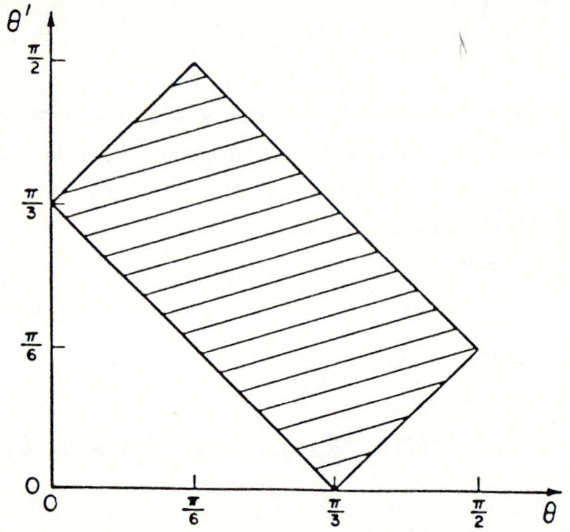

Figure 6. The domain of integration for the kernel $K_{\beta\gamma}(\theta,\theta')$.

To solve (III.15) we first follow the same procedure as for the two-body case and define a new function $F_\alpha(\rho,\theta_i)$ by writing the reduced wave function in the form:

$$\phi_\alpha(\rho,\theta_i) = F_\alpha(\rho,\theta_i)\ \rho\ e^{-K\rho}\ , \tag{III.24}$$

where we expect this new function to be a smoother function of ρ. Substituting (III.24) in (III.15) gives:

$$\left[\frac{\partial^2}{\partial\rho^2} + \frac{3}{\rho}\frac{\partial}{\partial\rho} + \frac{1}{\rho^2} - \frac{3K}{\rho} - 2K\frac{\partial}{\partial\rho} + \frac{1}{\rho^2}\frac{\partial^2}{\partial\theta^2} - \frac{\ell_\alpha(\ell_\alpha + 1)}{\rho^2\cos^2\theta} - \frac{L_\alpha(L_\alpha + 1)}{\rho^2\sin^2\theta}\right]F_\alpha(\rho,\theta)$$

$$- \sum_\beta v_{\alpha\beta}(\rho\cos\theta)F_\beta(\rho,\theta)$$

$$= \sum_\gamma v_{\alpha\gamma}(\rho\cos\theta)\sum_\beta \int_{\theta^-}^{\theta^+} K_{\gamma\beta}(\theta,\theta')F_\beta(\rho,\theta')d\theta' , \qquad \text{(III.25)}$$

where we have dropped the subscript on θ_1 and we have set $\theta_2 = \theta'$. Also, we have interchanged the subscripts β and γ.

This is a second-order elliptic partial differential equation, and in order to obtain a unique solution, it is necessary to specify the boundary conditions on a closed surface. The reduced wave function is zero when $x_1 = 0$ ($\theta = \frac{\pi}{2}$) and when $y_1 = 0$ ($\theta = 0$). Therefore $F_\alpha(\rho,\theta)$ has the boundary conditions:

$$F_\alpha(\rho,0) = 0 , \qquad \text{(III.26a)}$$

and

$$F_\alpha(\rho,\tfrac{\pi}{2}) = 0 . \qquad \text{(III.26b)}$$

Also, the reduced wave function goes as ρ^2 for small values of ρ; thus, the boundary condition for $\rho = 0$ is

$$F_\alpha(0,\theta) = 0 . \qquad \text{(III.27)}$$

Finally, for large values of ρ, the reduced wave function has the asymptotic form[11]

$$\phi_\alpha(\rho,\theta) = A(\theta)\frac{e^{-K\rho}}{\rho^{\frac{1}{2}}} . \qquad \text{(III.28)}$$

Consequently, $F_\alpha(\rho,\theta)$ must have the asymptotic form $A(\theta)\rho^{-\frac{3}{2}}$ and the boundary condition on F_α at $\rho = \rho_{max}$, where ρ_{max} is the matching radius, is the mixed boundary condition

$$\rho_{max} \left.\frac{\partial F_\alpha}{\partial \rho}\right|_{\rho=\rho_{max}} + \frac{3}{2} F_\alpha(\rho_{max},\theta) = 0 . \tag{III.29}$$

We now expand the unknown channel functions $F_\alpha(\rho,\theta)$. For our basis set we choose the bicubic Hermite splines; that is, we use the expansion

$$F_\alpha(\rho,\theta) = \sum_{m=0}^{M+1} \sum_{n=0}^{N+1} a_{\alpha mn}\, s_m(\rho) s_n(\theta) . \tag{III.30}$$

The boundary conditions (III.26a) and (III.26b) can be imposed by setting $a_{\alpha mn} = 0$ for $n = 0$ and for $n = N$. Similarly, the boundary condition (III.27) can be imposed by setting $a_{\alpha mn} = 0$ for $m = 0$. The boundary condition (III.29) can not be imposed in the same manner. However, by noting that the only ρ-splines with nonzero values or first derivatives at $\rho = \rho_{max}$ are s_M and s_{M+1}, one finds that substituting (III.30) into (III.29) yields

$$\rho_{max}\, a_{\alpha(M+1)n} + \frac{3}{2} a_{\alpha Mn} = 0 . \tag{III.31}$$

Therefore, one finds that $a_{\alpha(M+1)n} = - 3\, a_{\alpha Mn} / 2\rho_{max}$, and the last two terms in the sum over m have the form

$$a_{\alpha Mn} s_M(\rho) s_n(\theta) - \left(\frac{3}{2\rho_{max}}\right) a_{\alpha Mn} s_{M+1}(\rho) s_n(\theta)$$

$$= a_{\alpha Mn}\left[s_M(\rho) - \frac{3}{2\rho_{max}} s_{M+1}(\rho)\right] s_n(\theta) , \tag{III.32}$$

for all values of n. If we use the replacements

$$s_M(\rho) \rightarrow s_M(\rho) - \frac{3}{2\rho_{max}}\, s_{M+1}(\rho) , \tag{III.33a}$$

and

$$s_N(\theta) \to \xi_I(\theta) \qquad \text{(III.33b)}$$

the expansion in (III.30) becomes

$$F_\alpha(\rho,\theta) = \sum_{m=1}^{M} \sum_{n=1}^{N} a_{\alpha mn}\, s_m(\rho) s_n(\theta) \ . \qquad \text{(III.34)}$$

Following the same procedure we used for the two-body case, we define I ρ-breakpoints on the interval $[0,\rho_{max}]$ and J θ-breakpoints on the interval $[0,\frac{\pi}{2}]$. These breakpoints can be scaled in the same manner as we used for the two-body case. The ρ scale factor, S_ρ, is chosen to be greater than unity since the F_α will have more structure for small values of ρ. The θ scale factor, S_θ, is chosen to be less than unity since we want more θ-breakpoints near $\theta = \frac{\pi}{2}$ which correspond to small values of x_1. (Normally, the Faddeev amplitude has more structure in the region where the potential $V(x_1)$ is large.) Now we use the method of orthogonal collocation to determine the $N_c \times M \times N$ unknown expansion coefficients $a_{\alpha mn}$, where $M = 2I$, $N = 2J$, and N_c is the number of channels. The two Gauss quadrature points on each ρ-interval define the ρ collocation points ρ_p, and the two Gauss quadrature points on each θ interval define the θ collocation points θ_q. Substituting the expansion (III.34) into (III.25) and evaluating the resulting expression at the ρ and θ collocation points yields the set of equations:

$$\sum_{m=1}^{M} \sum_{n=1}^{N} a_{\alpha mn}\Bigg[s_m''(\rho_p)s_n(\theta_q) + \frac{3}{\rho_p} s_m'(\rho_p)s_n(\theta_q) + \frac{1}{\rho_p^2} s_m(\rho_p)s_n(\theta_q)$$

$$- \frac{3K}{\rho_p} s_m(\rho_p)s_n(\theta_q) - 2K\, s_m'(\rho_p)s_n(\theta_q) + \frac{1}{\rho_p^2} s_m(\rho_p)s_n''(\theta_q)$$

$$- \left(\frac{\ell_\alpha(\ell_\alpha + 1)}{\rho_p^2 \cos^2\theta_q} + \frac{L_\alpha(L_\alpha + 1)}{\rho_p^2 \sin^2\theta_q}\right) s_m(\rho_p)s_n(\theta_q)\Bigg] \qquad \text{(III.35)}$$

$$- \sum_{\beta=1}^{N_c} \sum_{m=1}^{M} \sum_{n=1}^{N} a_{\beta mn}\Big[v_{\alpha\beta}(\rho_p \cos\theta_q)\ s_m(r_p)s_n(\theta_q)\Big]$$

$$- \sum_{\beta=1}^{N_c} \sum_{m=1}^{M} \sum_{n=1}^{N} a_{\beta mn}\Big[\sum_{\gamma} v_{\alpha\gamma}(\rho_p \cos\theta_q)s_m(r_p)\int_{\theta^-}^{\theta^+} K_{\gamma\beta}(\theta_q,\theta')s_n(\theta')d\theta' \Big] \ .$$

The problem has now been reduced to solving the $N_c \times M \times N$ equations in (III.35) for the unknown $a_{\alpha mn}$ and the eigenvalue K. In the same manner as for the two-body case, the fact that the ρ-splines are nonzero only on two subintervals will produce a block diagonal matrix. However, the θ-splines appear inside an integral; consequently, at each θ-collocation point all of the θ-splines will yield nonzero values. Thus, for a single channel the matrix will have the form shown in Figure 3, where each X now represents an N×N matrix whose rows are labeled by q and whose columns are labeled by n. For more than one channel, the last term on the right-hand side of (III.35) will couple all of the channels, and the blocks will be $(N_c \times N) \times (N_c \times N)$ matrices. For more than a few channels, the computer time required to solve the matrix equation can become prohibitive. Therefore, we use a different algorithm to solve for the eigenvalues and eigenvectors.

To find one of the eigenvalues we note that (III.35) has the form:

$$\sum_{\beta=1}^{N_c} \sum_{m=1}^{M} \sum_{n=1}^{N} A_{\alpha pq,\beta mn}\, a_{\beta mn} = \sum_{\beta=1}^{N_c} \sum_{m=1}^{M} \sum_{n=1}^{N} B_{\alpha pq,\beta mn}\, a_{\beta mn} \ , \qquad \text{(III.36)}$$

which can be written as the matrix equation

$$\underset{\approx}{A}\, \underset{\sim}{a} = \underset{\approx}{B}\, \underset{\sim}{a} \ . \qquad \text{(III.37)}$$

This matrix equation can not be solved since the matrix $\underset{\approx}{A}$ contains the unknown eigenvalue K. In order to solve the equation, we introduce a new parameter λ and rewrite (III.37) in the form:

$$\underset{\approx}{A}\, \underset{\sim}{a} = \lambda \underset{\approx}{B}\, \underset{\sim}{a} \ . \qquad \text{(III.38)}$$

For a fixed value of K this becomes a generalized eigenvalue problem with eigenvalue λ. Thus, to find the bound-state energy, we search for the value of K for which the eigenvalue λ is unity. This procedure can be used to find any bound state, not just the ground state. The advantage of using this procedure is that now the matrix $\underset{\approx}{A}$ will be a block diagonal matrix where the size of these blocks is determined by the form of $v_{\alpha\beta}$. For a pure central potential

$$v_{\alpha\beta}(\rho \cos\theta) = U(\rho \cos\theta)\, \delta_{\alpha\beta} \ , \qquad \text{(III.39)}$$

where $U(x_1) = M\, V(x_1)/\hbar^2$. Consequently, the matrix $\underset{\approx}{A}$ will be a reduced matrix where each block corresponds to a channel. Thus, the inverse of this matrix can be evaluated for one channel at a time. The block corresponding to a particular channel will have the form shown in Figure 3, where each X will be an N×N matrix. This matrix can be inverted using a modification of the standard method for inverting tridiagonal matrices. For a realistic two-body potential with a tensor force, the potential couples at most two channels, and for this case the matrix $\underset{\approx}{A}$ will have a reduced form where the blocks correspond to either one or two channels. For the case with two channels, the blocks will be (2N)×(2N) matrices, and even for many channels the largest block will be of this size.

Since the matrix $\underset{\approx}{A}$ can be inverted in an efficient manner, the eigenvalues of (III.38) can be found by using the Lanczos algorithm, which is described in the next Section.

IV. Lanczos Algorithm

The generalized eigenvalue problem which we have to solve is

$$\underset{\approx}{A}\,\underset{\sim}{a} = \lambda \underset{\approx}{B}\,\underset{\sim}{a} , \tag{IV.1}$$

where $\underset{\approx}{A}$ and $\lambda\underset{\approx}{B}$ are nonsymmetric real matrices. The eigenvector $\underset{\sim}{a}$ contains the expansion coefficients for the reduced wave functions $F_\alpha(\rho,\theta)$, and we vary the value of K in $\underset{\approx}{A}$ until the eigenvalue λ is unity. Since we wish to find the eigenvalue closest to unity for a particular value of λ, we rewrite (IV.1) as

$$\underset{\approx}{A}^{-1}\,\underset{\approx}{B}\,\underset{\sim}{a} = \frac{1}{\lambda}\,\underset{\sim}{a} , \tag{IV.2}$$

or

$$\underset{\approx}{H}\,\underset{\sim}{a} = \Lambda\,\underset{\sim}{a} , \tag{IV.3}$$

so that $\Lambda = 1/\lambda \simeq 1$ becomes one of the large eigenvalues. If we have chosen the correct value of K, the value of Λ will be exactly equal to unity.

To solve (IV.3) we use the Lanczos[22] algorithm, as modified by Saad[23] for nonsymmetric matrices, to generate a small basis set which can be used to obtain an approximate solution using standard eigenvalue techniques. To generate our new basis set, we start with an initial vector $\underset{\sim}{a}_1$ and generate a basis set of vectors $\underset{\sim}{a}_i$ for $i=2,3,\ldots,\nu$. Since the matrix in (IV.3) is nonsymmetric, we must also generate the biorthogonal basis set b_i, starting with an initial vector b_1. The initial vectors are normalized so that $b_1^T a_1 = 1$.

Given the initial vectors, we then generate the vectors

$$a_2' = H\, a_1 - \alpha_1 a_1 \tag{IV.4a}$$

and

$$b_2' = H^T\, b_1 - \alpha_1 b_1 , \tag{IV.4b}$$

where α_1 is chosen so that b_2' is orthogonal to a_1. that is, we choose

$$\alpha_1 = b_1^T\, H\, a_1 . \tag{IV.5}$$

Note this also insures that a_2' is orthogonal to b_1.

In practice one never calculates the matrix H. The solutions to (IV.4) are found in two steps. First we find the vectors

$$c_1 = H\, a_1 \tag{IV.6a}$$

and

$$d_1 = H^T b_1 . \tag{IV.6b}$$

To find c_1 we rewrite (IV.4a) in the form

$$A\, c_1 = B\, a_1 = z_1 . \tag{IV.7}$$

Since the matrix A is a reduced matrix, we can solve (IV.7) one block at a time where a block corresponds to either one or two channels. Each of these blocks is itself a block diagonal matrix which can be solved using an algorithm for banded matrices. To solve (IV.6b) we first rewrite it in the form

$$\underset{\sim}{d}_1 = \underset{\approx}{B}^T(\underset{\approx}{A}^T)^{-1}\underset{\sim}{b}_1. \tag{IV.8}$$

Next, if we define the vector

$$\underset{\sim}{e}_1 = (\underset{\approx}{A}^T)^{-1}\underset{\sim}{b} \tag{IV.9}$$

and solve the related equation

$$\underset{\approx}{A}^T\underset{\sim}{e}_1 = \underset{\sim}{b}_1 \tag{IV.10}$$

for $\underset{\sim}{e}_1$, we can obtain $\underset{\sim}{d}_1$ from

$$\underset{\sim}{d}_1 = \underset{\approx}{B}^T\underset{\sim}{e}_1 \ . \tag{IV.11}$$

Given the vectors $\underset{\sim}{c}_1$ and $\underset{\sim}{d}_1$, we first find

$$\alpha_1 = \underset{\sim}{b}_1^T\underset{\sim}{c}_1 \tag{IV.5'}$$

and then

$$\underset{\sim}{a}_2' = \underset{\sim}{c}_1 - \alpha_1\underset{\sim}{a}_1 \tag{IV.4a'}$$

and

$$\underset{\sim}{b}_2' = \underset{\sim}{d}_1 - \alpha_1\underset{\sim}{b}_1 \ . \tag{IV.4b'}$$

To impose the normalization condition, we follow Saad[22] and choose

$$\underset{\sim}{a}_2 = \frac{\underset{\sim}{a}_2'}{|\underset{\sim}{b}_2'^T\underset{\sim}{a}_2'|^{\frac{1}{2}}} \tag{IV.12}$$

and

$$b_2 = \frac{b_2' |b_2'^T a_2'|^{\frac{1}{2}}}{(b_2'^T a_2')} . \qquad \text{(IV.13)}$$

Given the first two vectors, we can generate the remaining vectors by means of the recursion relations

$$a_{i+1}' = H\, a_i - \alpha_i a - \beta_i a_{i-1} \qquad \text{(IV.14a)}$$

and

$$b_{i+1}' = H^T b_i - \alpha_i b_i - \beta_i a_{i-1} . \qquad \text{(IV.14b)}$$

where

$$\alpha_i = b_i^T\, H\, a_i , \qquad \text{(IV.15)}$$

and

$$\gamma_i = |b_i'^T a_i'|^{\frac{1}{2}} \qquad \text{(IV.16a)}$$

$$\beta_i = b_i'^T a_i' / \gamma_i \qquad \text{(IV.16b)}$$

are the normalization factors that determine

$$a_i = a_i' / \gamma_i \qquad \text{(IV.17a)}$$

and

$$b_i = b_i' / \beta_i . \qquad \text{(IV.17b)}$$

Since the new basis vectors are generated by repeated application of the matrix $\underset{\approx}{H}$, the vectors should have large components of the eigenvectors with large eigenvalues. Thus, this procedure should give accurate results with a few basis vectors. Also, the convergence depends upon how close the initial vector is to the desired eigenvector. Normally, results accurate to five significant figures are obtained using 5-10 basis vectors.

Given the new basis set ($\nu \ll N_c \times M \times N$), we use the expansion

$$\underset{\sim}{a} = \sum_{\iota=1}^{\nu} \omega_i \underset{\sim}{a}_i \ . \tag{IV.18}$$

Substituting this expansion into (IV.3) and multiplying by $\underset{\sim}{b}_j^T$, we obtain

$$\sum_{i=1}^{\nu} \omega_i \underset{\sim}{b}_j^T \underset{\approx}{H} \underset{\sim}{a}_i = \Lambda \omega_j \ , \tag{IV.19}$$

or, using (IV.14a)

$$\sum_{i=1}^{\nu} \omega_i (\gamma_{i+1} \delta_{i+1,j} + \alpha_i \delta_{i,j} + \beta_{i-1} \delta_{i-1,j}) = \Lambda \omega_j \ . \tag{IV.20}$$

Thus, we now have to solve the matrix eigenvalue problem:

$$\begin{vmatrix} \alpha_1 & \beta_2 & & & & \\ \gamma_2 & \alpha_2 & \beta_3 & & & \\ & \gamma_3 & \alpha_3 & \beta_3 & & \\ & & \cdot & \cdot & \cdot & \\ & & & \cdot & \cdot & \beta_\nu \\ & & & & \gamma_\nu & \alpha_\nu \end{vmatrix} \begin{vmatrix} \omega_1 \\ \omega_2 \\ \omega_3 \\ \cdot \\ \cdot \\ \omega_\nu \end{vmatrix} = \Lambda \begin{vmatrix} \omega_1 \\ \omega_2 \\ \omega_3 \\ \cdot \\ \cdot \\ \omega_\nu \end{vmatrix} \tag{IV.21}$$

The convergence of this method is illustrated for a model problem in the next section.

V. Numerical Results

To illustrate the numerical properties of the configuration-space Faddeev calculations, we first consider the model problem of the MT-V potential. The Mt-V potential has the same form as the MT-III potential in (II.20), where for the MT-V: V_1 = 1438.4812 MeV-fm, V_2 = 570.3316 MeV-fm, μ_1 = 3.11 fm^{-1} and μ_2 = 1.55 fm^{-1}. Since this potential is an s-wave interaction, the Faddeev amplitudes will have only one channel with $\ell = L = 0$. For this case we can solve for the eigenvalue K directly. Rewriting (III.35) with all the terms multiplied by K on the right-side and the rest of the terms on the left-hand side gives a matrix equation of the form

$$\underset{\approx}{A}'\underset{\sim}{a} = K\,\underset{\approx}{B}'\underset{\sim}{a} \,, \tag{V.1}$$

where the matrices $\underset{\approx}{A}'$ and $\underset{\approx}{B}'$ have the same structure as $\underset{\approx}{A}$ and $\underset{\approx}{B}$ for the one channel case. For the multichannel case the matrix $\underset{\approx}{A}'$ will have nonzero matrix elements between all channels since the permuted terms are now in this matrix. Therefore, for more than one channel the matrix will not be a reduced matrix, and we have to invert the matrix for all of the channels. Consequently, this procedure will only be practical for a calculation with a few channels. However, for these cases we can use the Lanczos algorithm to solve the matrix eigenvalue equation.

For this case, where we solve for all of the channels at the same time, we can use the shifted equation

$$(\underset{\approx}{A}' - K_0\,\underset{\approx}{B}')\underset{\sim}{a} = (K - K_0)\underset{\approx}{B}'\underset{\sim}{a} \,, \tag{V.2}$$

to increase the rate of convergence for the algorithm. The inverse of the matrix on the left-hand side of (V.2) operating on the initial vector will produce a new vector with large components corresponding to eigenvectors of $\underset{\approx}{A}'$ with eigenvalues near K_0. Given an initial vector $\underset{\sim}{a}_1$, the vector can be expanded in the basis set of the unknown eigenvectors $\underset{\sim}{x}_j$ of the the matrix equation (V.1), i.e., we write

$$\underset{\sim}{a}_1 = \sum_j c_j\,\underset{\sim}{x}_j \,. \tag{V.3}$$

Then using (V.3) one finds that

$$(\underset{\approx}{A}' - K_0\,\underset{\approx}{B}')^{-n}\,\underset{\approx}{B}'\underset{\sim}{a}_1 = \sum_j c_j (K_j - K_0)^{-n}\,\underset{\sim}{x}_j \,. \tag{V.4}$$

Therefore, if $K_j - K_0$ is small, the Lanczos basis vectors will have a large component of the eigenvector corresponding to K_j, and a good approximation for this eigenvector can be obtained using a small number of the Lanczos basis vectors. To demonstrate this convergence, we show in Table II the results for the MT-V potential using three different values of K_0, where the value of K_0 is defined by $E_0 = \hbar^2 K_0^2/M$.

Table II

Bound-state energy (in MeV) for the MT-V potential calculated using Equation (V.2) with I = 15, J = 17, S_ρ = 1.3, S_θ = 0.8, and ρ_{max} = 20.0 fm.

ν	E_0 = 4.0	E_0 = 7.0	E_0 = 7.5	E_0 = 10.0
1	21.954	8.186	7.586	5.620
2	7.772	7.542	7.541	7.499
3	7.288	7.541	7.541	7.542
4	7.541	7.541	7.541	7.541
5	7.541	7.541	7.541	7.541

From Table II one can see that the rate of convergence is best for values of E_0 close to the eigenvalue.

In Table III we show the results for the eigenvalue λ of the modified Faddeev equation (III.38) calculated with E = 7.541 MeV for the same potential and spline parameters. Finally, to demonstrate the sensitivity of the value of λ to the energy, we list in Table IV the values of λ for a range of values of the energy.

Table III

Eigenvalue of the matrix equation (III.38) for the MT-V potential calculated using a value of E = 7.541 MeV and various numbers of Lanczos basis vectors, ν. (The spline parameters are the same as in Table II)

ν	λ	$\lambda - 1$
1	7.05238	6.05238
2	1.00437	4.36890×10^{-3}
3	1.16974	1.69736×10^{-1}
4	0.99034	-9.65434×10^{-3}
5	1.00039	3.90887×10^{-4}
6	1.00002	2.22921×10^{-5}
7	1.00000	-4.76837×10^{-7}

Table IV

Eigenvalue of the matrix equation (III.38) for the MT-V potential calculated using several values of the energy. (The spline parameters are the same as in Table II)

E	λ	$\lambda - 1$
7.0	0.97114	-2.88623×10^{-2}
7.2	0.98190	-1.80951×10^{-3}
7.4	0.99255	-7.44575×10^{-3}
7.6	1.00309	3.09169×10^{-3}
7.8	1.01352	1.35245×10^{-2}
8.0	1.02385	2.38492×10^{-2}

The MT-V potential has a short-range repulsive term, which should lead to small values of the Schrödinger wave function for small values of the interparticle separation. This not the case for the Faddeev amplitudes. In Figure 7 we plot Ψ_1 for the MT-V potential. (For an s-wave interaction Ψ_1 is independent of the angle θ between $\vec{x}_1$ and $\vec{y}_1$.) From Figure 7, we see that Ψ_1 is large and negative for small values of x_1. However, the total wave function has the expected behavior. This is seen in Figures 8 and 9, where we plot Ψ for two values of the angle between $\vec{x}_1$ and $\vec{y}_1$. Also, the total wave function is positive definite as expected for the ground state. In Figure 8 the value of θ is zero, which corresponds to the three particles in a collinear configuration, and the wave function has a small value for $x_1 \simeq 0$. (In Figures 7, 8, and 9 we have used x and y for x_1 and y_1.) In addition, Ψ is small when $x_1 \simeq 2y_1$ which corresponds to $x_2 \simeq 0$. In Figure 9, where $\vec{x}_1$ and $\vec{y}_1$ are perpendicular, the wave function has its largest value at $x_1 \simeq 1.0$ fm and $y_1 \simeq 0.866$ fm. These values of x_1 and y_1 correspond to the three particles forming an equilateral triangle with sides of length 1.0 fm. The MT-V potential has its minimum value at approximately 1.0 fm; thus, the particles have the largest probability of being found at positions where the two-body potential is most attractive.

VI. Conclusions

The coordinate-space Faddeev equations can be used to obtain accurate wave functions and binding energies for the three-particle bound-state system. The use of a spline basis set provides a convenient expansion of the Faddeev amplitudes which can be used to construct the total wave function. Thus, once the three-body Faddeev equations have been solved, the total wave function can be used to calculate other properties of the system, such as charge radii, asymptotic normalization constants, and charge densities. The use of the modified Lanczos method allows one to solve realistic models with many channels. Finally, the use of the bipolar harmonics, which are discussed in the appendix, can greatly reduce the computer time required for the numerical evaluation of the angular integrals.

Appendix: Bipolar Harmonics

There are many cases where the numerical values of the channel basis states are needed. For example, the numerical evaluation of the kernel in (III.35) requires the values of $|\alpha_1\rangle$ and $|\alpha_2\rangle$. Also, once the expansion coefficients for the Faddeev amplitudes have been found by solving (III.38), these coefficients can be used in (III.30) to evaluate $F_\alpha(\rho,\theta)$; then one can use (III.24) and (III.7) to evaluate the Faddeev amplitudes. Thus, to use the total wave function to calculate additional properties of the three-body system such as the charge radius[22] or the

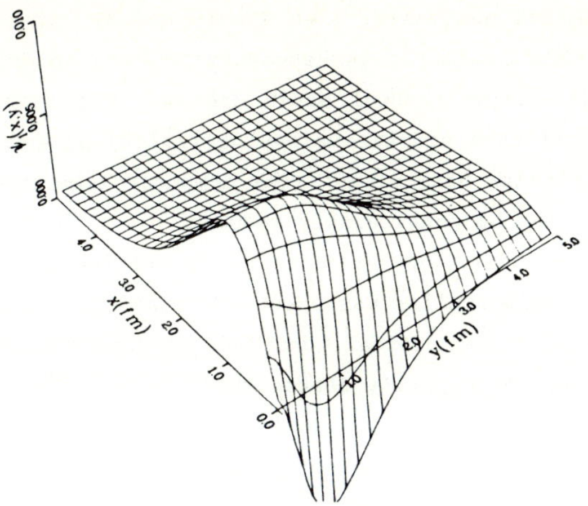

Figure 7. Faddeev amplitude Ψ_1 for the MT-V potential model ground state.

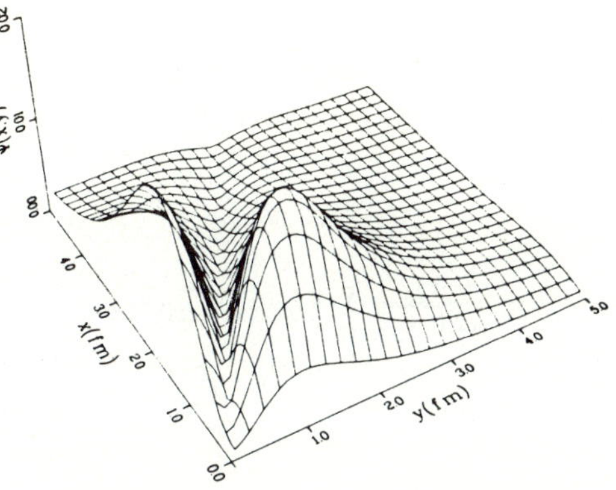

Figure 8. Complete Schrödinger wave function at fixed angle $\Theta = 0°$ for the MT-V potential model ground state.

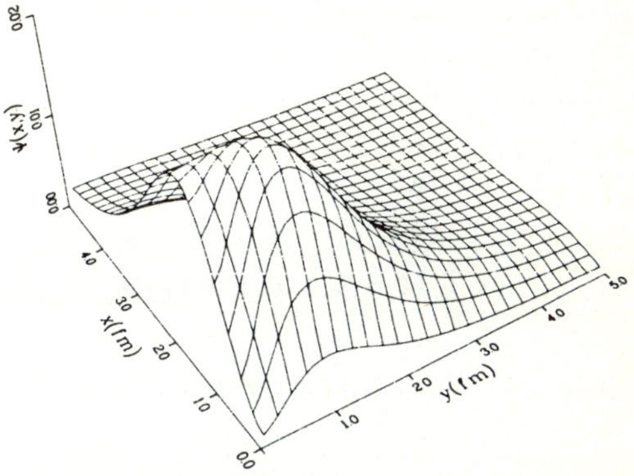

Figure 9. Complete Schrödinger wave function at fixed angle $\Theta = 90°$ for the MT-V potential model ground state.

asymptotic normalization constants,[23] we simply add the three Faddeev amplitudes. For a many channel calculation it is necessary to have an efficient method for the numerical evaluation of the channel basis states. The straightforward method of fixing $\hat{x}_i$ and $\hat{y}_i$, and then simply evaluating (III.8) is not an efficient method. Through the use of bipolar harmonics,[19] the computer time required to evaluate the channel basis states can be considerably reduced.

The bipolar harmonic expansion consists of writing the various channel basis states in terms of a much smaller basis set of "standard" orbital functions. Thus, we only need to evaluate the orbital functions to find the numerical values of the channel functions. For even-parity states there is one function for $L = 0$, one for $L = 1$, and two for $L = 2$. For the odd-parity states there are two functions for $L = 1$, and two for $L = 2$. The choice of these orbital functions is not unique, and we choose a set which is convenient for our numerical calculations. Once the orbital functions have been defined, we express the channel function as a sum of the functions, where the coefficients of the expansion are functions of $\mu_i = \hat{x}_i \cdot \hat{y}_i$. This decomposition of the channel functions not only facilitates the numerical evaluation of the Faddeev components, but also greatly simplifies the algebra involved in calculating matrix elements.In addition, it is useful for plotting the various parts of the wave functions.

The S-wave ($L = 0$) case is trivial since the functions are invariant under rotations. Thus, there is a single orbital function which we denote by $G = 1$. The coefficients for this case can be found by using

$$4\pi \sum_{m\,M} (\ell\ m\ L\ M \mid 0\ 0) Y_{\ell m}(\hat{x}_i) Y_{LM}(\hat{y}_i) = 4\pi\, \delta_{\ell L} \frac{(-1)^{-\ell}}{\sqrt{2\ell + 1}} \sum_m Y_{\ell m}(\hat{x}_i) Y^*_{\ell m}(\hat{y}_i)$$

$$= \delta_{\ell L} (-1)^{-\ell} \sqrt{2\ell + 1}\, P_\ell(\mu_i) \ . \tag{A.1}$$

Therefore, for S-waves we write

$$4\pi [Y_\ell(\hat{x}_i) \otimes Y_\ell(\hat{y}_i)]_0 = (-1)^{-1} \sqrt{2\ell + 1}\, P_\ell(\mu_i) \ . \tag{A.2}$$

For the P-waves ($L = 1$) the triangular condition for the angular momentum coupling limits the values of L to ℓ and $\ell \pm 1$. For the even parity-case the only possible value of L is $L = \ell$ since $\ell + L$ must be even. Therefore, we choose as the orbital function

$$H = 4\pi[Y_1(\hat{x}_i) \otimes Y_1(\hat{y}_i)]_1 , \tag{A.3}$$

and we write

$$4\pi[Y_\ell(\hat{x}_i) \otimes Y_\ell(\hat{y}_i)]_1 = B_\ell(\mu_i)H . \tag{A.4}$$

This definition of H differs by a factor of $x_i y_i$ from the definition used in reference 20. Since the coefficient B_ℓ depends only on the angle between $\hat{x}_i$ and $\hat{y}_i$, we can find the function B_ℓ by choosing a particular coordinate system. A convenient coordinate system is defined by letting $\hat{y}_i$ be along the z-axis and $\hat{x}_i$ in the x-z plane. For this choice of the coordinate system and for $M = 1$, (A.4) is

$$\sqrt{2\ell + 1}(\ell\ 1\ \ell\ 0 \mid 1\ 1)Y_{\ell 1}(\theta_x,0) = \sqrt{3}\ B_\ell(\mu_i)\ (1\ 1\ 1\ 0 \mid 1\ 1)Y_{11}(\theta_x,0) , \tag{A.5}$$

where we have used

$$\sqrt{4\pi}Y_{\ell m}(0,0) = \delta_{m0}\sqrt{2\ell + 1} , \tag{A.6}$$

and θ_x is the angle between $\hat{x}_i$ and $\hat{y}_i$. For the odd parity case $L = \ell \mp 1$, and for this case we choose the orbital functions

$$H' = \sqrt{4\pi}\ Y_1(\hat{x}_i) , \tag{A.7a}$$

and

$$H'' = \sqrt{4\pi}\ Y_1(\hat{y}_i) . \tag{A.7b}$$

Now we can write the odd-parity P-waves as

$$4\pi[Y_\ell(\hat{x}_i) \otimes Y_{\ell+1}(\hat{y}_i)]_1 = B'_\ell(\mu_i)H' + B''_\ell(\mu_i)H'' , \tag{A.8a}$$

and

$$4\pi[Y_{\ell+1}(\hat{x}_i) \otimes Y_\ell(\hat{y}_i)]_1 = B_\ell''(\mu_i)H' + B_\ell'(\mu_i)H'' \quad \text{(A.8b)}$$

where we have used

$$[Y_{\ell+1}(\hat{x}_i) \otimes Y_\ell(\hat{y}_i)]_1 = [Y_\ell(\hat{y}_i) \otimes Y_{\ell+1}(\hat{x}_i)]_1 \quad \text{(A.9)}$$

to derive (A.8b). The coefficients B_ℓ' and B_ℓ'' can be found by using the same procedure we used to determine B_ℓ. We find

$$B_\ell'(\mu) = \frac{(-1)^{\ell+1}}{\sqrt{2\ell + 1}} \frac{d}{d\mu} P_\ell , \quad \text{(A.10a)}$$

and

$$B_\ell''(\mu) = \frac{(-1)^{\ell}}{\sqrt{2\ell + 1}} \frac{d}{d\mu} P_{\ell+1} . \quad \text{(A.10b)}$$

The D-waves ($L = 2$) can have $L = \ell, \ell\pm2$ for the even-parity case, and, $L = \ell\pm1$ for the odd-parity case. For the even-parity orbital functions we use

$$I_i = \sqrt{4\pi}\, Y_2(\hat{x}_i) , \quad \text{(A.11a)}$$

$$J_i = \sqrt{4\pi}\, Y_2(\hat{y}_i) , \quad \text{(A.11b)}$$

and

$$K_i = \sqrt{2\ell + 1} \left(\frac{10}{3}\right)^{\frac{1}{2}} [Y_1(\hat{x}_i) \otimes Y_1(\hat{y}_i)]_2 . \quad \text{(A.11c)}$$

The even-parity D-waves can be written as

$$4\pi[Y_\ell(\hat{x}_i) \otimes Y_L(\hat{y}_i)]_2 = C_{\ell L}^I I_i + C_{\ell L}^J J_i + C_{\ell L}^K K_i . \quad \text{(A.12)}$$

Once again, we find the expansion coefficients by choosing a particular orientation of the coordinate system. Comparing the results of choosing $\hat{y}_i$ along the z-axis, and $\hat{x}_i$ along the z-axis one can show that

$$C^J_{\ell L}(\mu) = C^I_{L\ell}(\mu) \; , \tag{A.13a}$$

and

$$C^K_{\ell L}(\mu) = C^I_{L\ell}(\mu) \; . \tag{A.13b}$$

Using these symmetries, the expansion coefficients for $L = \ell$ can be found from

$$C^I_{\ell\ell}(\mu) = (-1)^\ell \left[\frac{4(2\ell + 1)}{\ell(\ell + 1)(2\ell + 3)(2\ell - 1)}\right]^{\frac{1}{2}} \frac{d^2}{d\mu^2} P_\ell \; , \tag{A.14a}$$

and

$$C^K_{\ell\ell}(\mu) = (-1)^{\ell+1} \left[\frac{(2\ell + 1)}{\ell(\ell + 1)(2\ell + 3)(2\ell - 1)}\right]^{\frac{1}{2}} \left[\frac{d}{d\mu} P_\ell + 2\mu \frac{d^2}{d\mu^2} P_\ell\right] . \tag{A.14b}$$

The coefficients for $L = \ell \pm 2$ are given by

$$C^I_{\ell\ell+2}(\mu) = (-1)^\ell \left[\frac{2}{3(\ell + 1)(\ell + 3)(2\ell + 3)}\right]^{\frac{1}{2}} \frac{d^2}{d\mu^2} P_\ell \; , \tag{A.15a}$$

$$C^I_{\ell+2\ell}(\mu) = (-1)^\ell \left[\frac{2}{3(\ell + 1)(\ell + 3)(2\ell + 3)}\right]^{\frac{1}{2}} \frac{d^2}{d\mu^2} P_{\ell+2} \; , \tag{A.15b}$$

and

$$C^K_{\ell\ell+2}(\mu) = (-1)^{\ell+1} \left[\frac{2}{3(\ell + 1)(\ell + 1)(\ell + 3)}\right]^{\frac{1}{2}} \left[(\ell + 2) \frac{d}{d\mu} P_\ell + \mu \frac{d^2}{d\mu^2} P_\ell\right] . \tag{A.15c}$$

For the odd-parity orbital functions we use

$$I_i' = 4\pi[Y_1(\hat{x}_i) \otimes Y_2(\hat{y}_i)]_2 , \qquad \text{(A.16a)}$$

and

$$J_i' = 4\pi[Y_2(\hat{x}_i) \otimes Y_1(\hat{y}_i)]_2 . \qquad \text{(A.16b)}$$

The odd-parity D-waves are then written in the form

$$4\pi[Y_\ell(\hat{x}_i) \otimes Y_L(\hat{y}_i)]_2 = D^I_{\ell L} I_i' + D^J_{\ell L} J_i' . \qquad \text{(A.17)}$$

The expansion coefficients for this case have the symmetry relation

$$D^J_{\ell L}(\mu) = D^I_{L\ell}(\mu) . \qquad \text{(A.18)}$$

Using the same technique which was used to determine the previous coefficients, we find that

$$D^I_{\ell\ell+1}(\mu) = (-1)^{\ell+1}\left[\frac{2}{3\ell(\ell+1)(\ell+2)}\right]^{\frac{1}{2}} \frac{d^2}{d\mu^2} P_{\ell+1} , \qquad \text{(A.19a)}$$

and

$$D^J_{\ell\ell+1}(\mu) = (-1)^{\ell}\left[\frac{2}{3\ell(\ell+1)(\ell+2)}\right]^{\frac{1}{2}} \frac{d^2}{d\mu^2} P_{\ell} , \qquad \text{(A.19b)}$$

Using the relations above, we can evaluate the channel functions for L = 0,1,2.

Acknowledgments

This work was supported, in part, by the U.S. Department of Energy.

References

1. C. R. Chen, G. L. Payne, J. L. Friar, and B. F. Gibson, Phys. Rev. C 31, 2266 (1985).
2. W. Glöckle, The Quantum Mechanical Three-Body Problem (Springer-Verlag, New York, 1983), and references therein.
3. S. Ishikawa, T. Sasakawa, T. Sawada, and T. Ueda, Phys. Rev. Lett. 53, 1877 (1984).
4. C. Hajduk and P. U. Sauer, Nucl. Phys. A322, 329 (1979).
5. J. Carlson, V. R. Pandharipande, R. B. Wiringa, Nucl. Phys. A401, 59 (1983).
6. T. Takemiya, Prog. Theor. Phys. 74, 301 (1985).
7. Y. Koike and Y. Taniguchi, Few Body Systems 1, 13 (1986).
8. R. Brandenburg, International Symposium on the Three-Body Force in the Three-Nucleon System (Springer-Verlag, Berlin), to be published.
9. G. L. Payne, W. H. Klink, W. N. Polyzou, J. L. Friar, and B. F. Gibson, Phys. Rev. C 30, 1132 (1984).
10. H. P. Noyes, in Three Body Problems in Nuclear and Particle Physics, edited by J.S.C. McKee and P. M. Rolph (North-Holland, Amsterdam, 1970).
11. P. Merkuriev, C. Gignoux, and A. Laverne, Ann. Phys. (NY) 99, 30 (1976).
12. P. M. Prenter, Splines and Variational Methods (Wiley, New York, 1975).
13. C. de Boor, A Practical Guide to Splines (Springer-Verlag, Berlin, 1978).
14. C. B. Moler and G. W. Stewart, SIAM J. Num. Anal. 10, 241 (1973).
15. E. P. Harper, Y. E. Kim, and A. Tubis, Phys. Rev. C 6, 126 (1972).
16. D. M. Brink and G. R. Satchler, Angular Momentum (Clarendon Press, Oxford, 1968).
17. G. L. Payne, J. L. Friar, B. F. Gibson, and I. R. Afnan, Phys. Rev. C 26, 1385 (1982).
18. B. Balian and E. Brezin, Nuovo Cim. 2, 403 (1969).
19. J. L. Friar, E. L. Tomusiak, B. F. Gibson, and G. L. Payne, Phys. Rev. C 32, 677 (1981).
20. C. Lanczos, J. Res. Nat. Bur. Stand. 49, 33 (1952).
21. Y. Saad, SIAM J. Num. Anal. 19, 485 (1982).
22. J. L. Friar, B. F. Gibson, E. L. Tomusiak, and G. L. Payne, Phys. Rev. C 24, 655 (1981).
23. J. L. Friar, B. F. Gibson, D. R. Lehman, and G. L. Payne, Phys. Rev. C 25, 1616 (1982).

SEPARABLE EXPANSION METHODS FOR THE TWO-BODY INTERACTION AND t-MATRIX

L. S. Ferreira
Dep. de Física
Universidade de Coimbra
3000 Coimbra, Portugal

Introduction

The Nucleon-Nucleon interaction is usually represented in a nonrelativistic framework by a potential model which obeys first principles and contains phenomenological fits to experimental data. The potential models are not unique. In spite of agreement they might show in the description of two-body data, substantial differences arrive in many-body calculations, where momentum is not necessarily conserved by each two-body pair but some momentum can be taken up by the surrounding nucleons thus requiring off-energy-shell matrix elements. The on-shell behaviour can be determined in the lower energy domain by the two-particle elastic scattering data and deuteron properties. If the properties of all few-particle systems were available the off-shell matrix elements of a potential could be completely determined. This is not the case, so uncertainties come in this context.

The use of a local potential stands from the sucess of its application to electromagnetic and gravitational forces. It cannot be fully justified in the present status of our Knowledge since the interaction pocesses relativistic components. We are still without a general theory where the Nucleon-Nucleon interaction is explicitly derived in a natural way.

The few phenomenological model available are not always suitable for applications. In spite of the recent developements in computational techniques which allowed great progress in practical calculations, as for example the exact solution of the Faddeev equations with a realistic potential, to do a systematic search and estimate partial effects, or to go beyond three body systems, the calculations became very heavy. In this sense the use of separable potentials has been very successful.

The first simplification that arises from a separable potential is to give an algebraic solution to the calculation of the two-body t-matrix, which becomes also separable. Different methods have been discussed in the literature to evaluate separable potentials or t-matrices for the two body system, and various model separable potentials were used in three body and nuclear matter calculations.

In these lectures we shall start by giving some generalities about the N-N interaction and structure of the realistic potential models used. Section II is devoted to separable interactions. The different methods to construct such interactions are discussed and the different models which appeared in the literature are reviewed. In section III as an example the Gamow separable expansion is applied to describe the on and off shell behaviour of the t-matrix for a realistic N-N interaction.

I. Potential Models for the Nucleon-Nucleon Interaction

Basically the interaction between two nucleons seems to have a short range repulsion responsable for the average separation of nucleons in nuclei. It is nearly charge independent, that is experiments of neutron-proton or proton-proton scattering give almost the same results. In order to explain the ground state properties of the simplest bound two-body system, the deuteron, in particular the magnetic dipole and electric quadrapole moments one has to consider a tensor force that couples the S and D partial waves giving 4% of D-state probability in the wave function. This force was expressed by Rarita and Schwinger[1)] through a potential of the form $V_T=f(r)S_{12}$ where $S_{12}=3(\vec{G}_1.\vec{r})(\vec{G}_2.\vec{r})-\vec{G}_1.\vec{G}_2$ is the usual tensor operator and f(r) a central functions. This potential tends to align the vector $\vec{r}$ along the direction of the total spin thus producing a distribution alongated along the direction of S giving a quadrupole moment.

To describe the hydrogen atom it is also necessary to introduce a spin orbit interaction $V_{LS}=g(r)\vec{G}.\vec{L}$ where L is the total angular momentum. The tensor and spin orbit forces prevent the orbital angular momentum of the two nucleons to be constant, thus bringing deviation from a central force situation.

The scattering data indicates that the most general form for the interaction should contain terms corresponding to the eight operators[2].

$$O_{ij} = 1,\ \vec{G}_i.\vec{G}_j,\ \vec{\tau}_i.\vec{\tau}_j,\ (\vec{G}_i.\vec{G}_j)(\vec{\tau}_i.\vec{\tau}_j),\ S_{ij},$$
$$S_{ij}(\vec{\tau}_i.\vec{\tau}_j),\ (\vec{L}.\vec{S})_{ij}(\vec{\tau}_i.\vec{\tau}_j)$$

where L is the relative angular momentum and S the total spin of the pair. Typically the NN potentials contain a strong short-range repulsion (SR), an intermediate range attraction and a long-range tail. This structure has been successfully described in terms of the exchange of various mesons. The long range is well understood in terms of the Yukawa theory where the π-meson, the linghtest particle which couples strongly to the nucleon, is exchanged between the two nucleons (OPE). Exchange of two mesons (TPE) gives a force of range $\hbar/2$ mc, thus the intermediate range is understood by the exchange of correlated pairs of scalar mesons. The short-range part is generally treated phenomenologically. The exchanges of the vector mesons ω,ρ was predicted by Breit in order to account for the spin-orbit interaction. It is now oldfashioned to interpret the short-range repulsion through vector meson exchange since at those energies the quark structure of the nucleons should manifest itself. There is still controversy on how to include the quarks in the nucleons, namely if the nucleon is made simply of three quarks which gives a quite large quark distribution, or if the quarks are compressed by the pressure of the pion-cloud exterior to the confinement region, and coupled to the interaction by an axial vector current. When the nucleons come close one has the interaction between two pion clouds.

The coupling constants can then be obtained taking account of the quark picture.

Some of the potential models are local. In other models to completely obtain the proper shape of phase shifts momentum dependent terms had to be introduced. The potential becomes non-local that is the matrix elements of the given interaction taken between free particle states, are a functional at two different radial positions: $\langle\vec{r}|V|\vec{r}'\rangle = (\vec{r},\vec{r}')$. In meson theoretical calculations of the N-N potential the inclusion of recoil of the source brings non-locality. Only for a point source we have a local character.

Various phenomenological potentials were constructed and improved in order to accomodate the new information brought by the experimental data. They have started from the basic operators indicated by the nucleon-nucleon data and tried to reproduce each N-N state with central functions of Yukawa type. This is the case of Hamada-Johnston[3)], Yale[4)], Tourreil-Sprung[5)], Malfliet-Tjon[6)] and Reid[7)] potentials. All of them reproduce the OPE potential for large r. The Hamada-Johnston and Yale potentials also contain terms quadratic in $\vec{L}$ and $\vec{L}.\vec{S}$. These models do not generally give good phase shifts for high partial wave states.

Boson exchange models were developed in the sixties, where the interaction between the two-nucleons is obtained from the superposition of Born terms given by simple exchange of various scalar and vector mesons. The choice of the form factors determined the models. The Bonn[8)] potential is in this category and has been applied not only to the properties of the two nucleon systems but also extensively to nuclear matter data. The Nijmegen[9)] potential developed from Regge-pole theory includes besides the traditional bosons the Pomeron and has non-local momentum dependent central parts.

The attraction in the intermediate range comes mainly from two-pion exchange processes, such as those where the nucleon resonances are excited in intermediate states. The Stony Brook[10)] potential incor porated this effect using a dispersion theory framework, but the short range repulsion is weaker than the one found in other models.

The Urbana[11)] potentials parameterise the N-N interaction with the eigth basic Bethe operators plus six additional terms chosen in order to simplify the many-body calculations, and express second order contributions of the UPE transition potentials that excit nucleons to isobar states. This potential has been extensively used in nuclear matter calculations and bound states.

Using π-N phase and the π-π interaction as input in dispersion relations the Paris group[12)] calculated the TPE contribution to the N-N force and extracted an equivalent potential which also include the π and ω exchange. The SR is purely phenomenological. The potential was not found to be convenient for practical use in many body calculations since it is energy dependent and has a dispersion integral and a sharp cutoff. A new parameterization[13)] was proposed containing a discrete sum of Yukawa type terms where the energy dependence is transformed in a p square dependence, and was applied to nuclear matter.

Perturbative and variational calculations show that realistic N-N potentials underbound the ^{3}H and ^{4}He and give too large radii, while infinite nuclear matter tends to be overbound, with too large saturation density. It was suggested that this discrepance might come from three body forces like for example the one derived from TPE processes.

With them, one gets the correct binding energy for ^{3}H but the same force will overbind ^{4}He and increase the deviations for nuclear matter. Another type of three body interaction can be constructed via a Δ(232) isobar, through a $\pi N\Delta$ coupling. It has been found by many authors that a strong saturation effect is observed in nuclear matter if the Δ degrees of freedom are included. In order to account for these coupling the Argonne V_{28}[14)] was derived. A simpler version without the three-nucleon potential, the Argonne V_{14}, is equivalent to the Urbana model and was used in calculations of light nuclei and nuclear matter. The Argonne and Paris potentials produce comparable results for the two body system, so differences in the nuclear structure calculations should be attributed to a different off-shell behaviour.

II. Separable Interactions

A separable potential can be expressed in terms of the product of two functions each one depending upon a different variable, that is $V(\vec{r},\vec{r}')=\lambda f(\vec{r})\, g(\vec{r}')$, where λ represents the strength of the interaction. The most general form for such potentials can be written as

$$V_N = \sum_{n,m=1}^{N} |f_n> \lambda_{nm} <f_m|$$

where N defines the rank of the expansion. This form might approximate a certain potential V or simply be a fitting of the N-N bound and scattering data. In the first case V_N becomes exact when N tends to infinity. In pratice one has to truncate the sum up to a certain optional rank.

Choosing a representation we can define the form factors of the interaction V_N. They will be in coordinate or momentum representation $<\vec{r}|f_m>=f_m(\vec{r})$ or $<\vec{k}|f_m>=g_m(\vec{k})$ respectively and related to each other by a Fourier transform. They are functions defined analytically or numerically and also depending in general on parameters to be adjusted to data. The quantities are the strength parameters of the interaction.

A separable interaction has various advantages. It provides an explicit solution for the two-body t-matrix obtained from matrix algebra without solution of any integral equation. Reduces the three-body Faddeev equations to a system of one dimensional equations saving enormous time in computation. Four body or nuclear matter calculations are also greatly simplified. They have however to fulfill certain requirements. They should reproduce the properties of the two-body system, namely, the deuteron properties and the scattering data. They should be of low rank. If they approximate an interaction V they should reproduce it in the proper energy domain on and off the energy shell. It should also be simple to construct V_N from V. They have also been applied to nuclear physics problems like for example to the description of resonances of spherical and non-spherical potentials[15] which are made separable through an expansion in an harmonic oscillator basis.

Many approaches to construct separable interactions and various separable potential models appeared in the literature and have been used in practical calculations for few-body system. They will be discussed next.

II 1. Methods to construct separable interactions

The two-body t-matrix is the solution of the partial wave Lippmann-Schwinger equation (LS),

$$T(s) = V + V\,G_o(s)T(s) \qquad (2.1)$$

where V is the two body potential and G_o the free particle Green's function. This study can be limited without loss of generality to the case of S waves.

Let $<\psi_n(s)>$ be the set of eigenvectors of the Kernel $V\,G_o(s)$ with eigenvalues λ_n^{-1} and $s<0$, then

$$V\,G_o(s)|\psi_n(s)> = \lambda_n^{-1}|\psi_n(s)> \qquad (2.2a)$$

Eq.(2.2a) corresponds to the homogeneous form of eq. (2.1)

$$\lambda_n V\,G_o(s)|\psi_n(s)> = |\psi_n(s)> \qquad (2.2b)$$

The functions $|\psi_n(s)>$ form a complete set normalized to $<\psi_n(s)|G_o(s)|\psi_m(s)> = -\delta_{nm}$. To prove the orthogonality relation the following matrix element can be considered,

$$<\psi_n|G_o(s)V\,G_o(s)|\psi_m> = \lambda_n^{-1}<\psi_n|G_o(s)^2\psi_m> =$$

$$= \lambda_m^{-1}<\psi_n|G_o(s)|\psi_m> \qquad (2.3)$$

where we have used eq. (2.2), and consequently the above normalization has to be fulfiled. The minus sign comes from the fact of having $s<0$. If we consider the adjoint equation to eq.(2)

$$<\xi_n(s)| = \mu_n(s)\ <\xi_n(s)|V\,G_o(s) \qquad (2.4)$$

taking the overlap $<\xi_n(s)|\psi_n(s)>$ we see that $\mu_n(s)=\lambda_n(s)$ and $|\xi_n(s)> \equiv G_o(s)|\psi_n(s)>$.

It can now be proved that the separable form for the potential

V, constructed with the function $|\psi_n(s)\rangle$,

$$V = \sum_{n=1}^{N} - \frac{|\psi_n(s)\rangle\langle\psi_n(s)|}{\lambda_n(s)} \tag{2.5}$$

is consistent with the eigenvalue eqs. (2.2) and (2.4)

$$|\psi_n(s)\rangle = -\lambda_n(s) \sum_{m=1}^{N} \frac{|\psi_m(s)\rangle\langle\psi_m(s)|}{\lambda_m(s)} G_o(s)|\psi_n(s)\rangle$$

$$= |\psi_m(s)\rangle\delta_{mn} \tag{2.6}$$

The separable representation of eq.(2.5) gives a separable solution for the two body t-matrix of eq.(2.1),

$$T(s) = \sum_{n,m=1}^{N} |\psi_n(s)\rangle\Delta_{nm}(s) \langle\psi_m(s)| \tag{2.7}$$

$$-|\Delta(s)^{-1}|_{nm} = \lambda_n\delta_{nm} + \langle\psi_n(s)|G_o(s)|\psi_m(s)\rangle$$

If in the definition of Δ we use the orthogonality condition of the vectors $|\psi_n\rangle$ the separable t-matrix becomes,

$$T_W(s) = \sum_{n=1}^{N} \frac{|\psi_n(s)\rangle\langle\psi_n(s)|}{1-\lambda_n(s)} \tag{2.8}$$

this is the Weinberg[16)] expansion (W). The method is cumbersome since one has to solve eq.(2.2) for each energy to obtain the eigenvectors. Also the convergence of the series depends on the nature of the potential, and it seems to be poor for large s.

It has been applied to a square well, Hulthén and Yukawa[17)] type of potentials, for two and three body problems. At $s>0$ it does not satisfy off-shell unitarity for N finite.

If s is fixed at some negative energy, s=-B, and eq.(2.2) is solved we can drop the energy dependence of $\langle\psi_n\rangle$ and λ_n. In this case the separable t-matrix has the form.

$$T_{UPE}(s) = \sum_{n,m} |\psi_n\rangle\Delta_{nm}(s) \langle\psi_m| \tag{2.9}$$

with. This is the Unitary Pole Expansion (UPE) given by Harms[18)], which satisfies off-shell unitarity and was applied to the Malfliet-Tjon and Reid and other potentials[19)] for the S state and then used as input in three body calculations.

If the two body system has a bound state, the energy B can be the binding energy and the vector $<\xi_n|$ of eq.(2.4) can be interpreted as the bound state of the potential $\lambda_n V$ so proportional to the original bound state. Taking only one term in the expansion of eq.(2.9) we get the Unitary Pole Approximation (UPA) of Lovelace[20] and Fuda[21],

$$\begin{aligned} T_{UPA}(s) &= |\psi_1> \Delta(s) <\psi_1| \\ \Delta(s) &= -[1+<\psi_1|G_o(s)|\psi_1>]^{-1} \end{aligned} \qquad (2.10)$$

which has a pole at the same position as the real matrix. In the neighbourhood of the pole the exact and UPA t-matrices agree. This also holds for the UPE t-matrix. The UPA has been extensively applied and its accuracy cheked for various potentials like for example the Yukawa and Reid soft core and to construct non central potentials. For details see the review of Levinger[22]. The UPE and UPA were also generalized to problems with higher number of particles[23], in particular to three and four-body systems.

If there are additional bound states and resonance poles, one can construct an energy dependent vector which is the superposition of the eigenstates corresponding to the various n bound states and resonances $|\psi_i>$ of the original Hamiltonian,

$$|\phi(s)>= \sum_{i=1}^{n} \alpha_i(s)|\psi_i> \qquad (2.11)$$

Bhatia and Walker[24] construct then a separable potential and t-matrix,

$$\begin{aligned} V_{Bw} &= \lambda V|\phi(s)><\bar{\phi}(s)|V \\ T_{Bw}(s) &= \frac{V|\phi(s)><\bar{\phi}(s)|V}{\lambda^{-1}-<\bar{\phi}(s)|VG_o(s)V|\phi(s)>} \end{aligned} \qquad (2.12)$$

where $<\bar{\phi}(s)|$ is a linear combination of the solution of the original Hamiltonian with Hermitean conjugate boundary conditions. The parameters $\alpha_i(s)$ are adjusted to the exact t-matrix and guarantee that $|\phi(s)>$ coincides with the exact vectors at the bound state or resonance poles. This t-matrix also satisfies unitarity and has the correct position of the various poles.

The use of the previous separable expansion in the three body problem made people aware of the need of a separable t-matrix that could describe well the half-off-shell and full off-shell behaviour of the exact one. The method proposed by Ernest, Shakin and Thaler[25)] gives a t-matrix which is exact on and half-shell at N chosen bound state and continuum energies. They define a separable potential with the condition that the eigenvectors of the corresponding Hamiltonian, are identical to the ones of the original Hamiltonian, at a certain energy.

Let $V_s=|g>\lambda<g|$ be the separable potential and V the real one. Each will satisfy a LS eq., with outgoing wave boundary conditions

$$|\phi^+_{k_s\,s}> =|k_s>+\lambda\ G^+_o(s)|g><g|\phi^+_{k_s\,s}> \qquad (2.13.a)$$

$$|\psi^+_{k_s\,s}> =|k_s>+ G^+_o(s)V|\psi^+_{k_s\,s}> \qquad (2.13.b)$$

with $k_s^2=s$. The first eq. can be solved immediately,

$$|\phi^+_{k_s\,s}>=|k_s> + \frac{\lambda\ G^+_o(s)|g><g|k_s>}{1-\lambda<g|G^+_o(s)|g>} \qquad (2.14)$$

For the two functions to be identical we must have from eqs.(2.13b) and (2.14), $G^+_o(s)V|\psi^+_{k_s\,s}> \alpha\, G^+_o(s)|g>$. Setting the multiplicative constant equal to one we see that $|g> \equiv V|\psi^+_{k_s\,s}>$ and since $|\phi^+_{k_s\,s}> \equiv |\psi^+_{k_s\,s}>$ eq.(2.14) gives immediately for λ, $\lambda=<\psi^+_{k_s\,s}|V|\psi^+_{k_s\,s}>^{-1}$. The vectors $|\psi_{k_s\,s}>$ are the bound or scattering states at energy s. The method can be generalized if the wave functions have to be reproduced at different energies s_n. Then

$$V_s = \sum_{ij} V|\psi_i><\psi_i|M|\psi_j><\psi_j|V \qquad (2.15)$$

where $|\psi_i>$ stands for $|\psi^+_{k_s\,s}>$ or $|\psi_{Bi}>$ in the case of scattering or bound states respectively, and the matrix M is defined by the relation

$$\delta_{im} = \sum_j<\psi_i|M|\psi_j><\psi_j|V|\psi_m> = \sum_j<\psi_i|V|\psi_j><\psi_i|M|\psi_m> \qquad (2.16)$$

By construction $V_s|\psi_i> = V|\psi_i>$, therefore the vectors are eigenvectors of the Hamiltonians and the two half-shell t-matrices are equal. The matrix M can be diagonalized by a unitary transformation U, giving

$$V_s = \sum_i |\hat{g}_i>\lambda_i<\hat{g}_i| \qquad (2.17)$$

with $|\hat{g}_i>\equiv V|\hat{\psi}_i>$, $|\hat{\psi}_i>\equiv \sum_j U_{ij}|\psi_j>$ and $\lambda_i\equiv<\hat{\psi}_i|M|\hat{\psi}_i>$. The equivalent of eq.(2.14) for the interaction of eq.(2.17) is

$$|\phi^+_{k_s s}>=|k_s>+ G^+_o(s) \sum_{ij} |\hat{g}_i>\Gamma_{ij}(s)<\hat{g}_j|k_s> \qquad (2.18)$$

with Γ_{ij} defined by $\sum_j \Gamma_{ij}(s)<\psi_j||V-VG^+_o(s)V|\psi_k>=\delta_{ik}$.

Finally the t-matrix will be,

$$T_{EST}(s) = \sum_{ij} |\hat{g}_i> \Gamma_{ij}(s) <\hat{g}_j| \qquad (2.19)$$

In the vicinity of the predetermined energies the off-shell t-matrix is also correctly reproduced. The method has been extensively used to describe simple potentials like a square well and even realistic interactions. We shall come to this point later. One of the disadvantages of the previous method is the choice of the energy value where the t-matrix is exact. If they are not properly chosen the resulting phase shifts can behave quite differently from the exact ones[26]. The method is also prone to zero-width resonances[27] coming from unphysical poles near the real energy axis of the matrix Γ.

Adhikari and Sloan[28] developed a more general scheme which contains as special cases the (EST), (UPE) and (W) approaches. They consider a rank N approximated potential V_N, with the property of being exact when it operates on any linear combination of selected functions $|u_n>$ namely,

$$V_N = \sum_{n,m=1}^{N} V|u_n> D_{nm} <v_m| \qquad (2.20)$$

where $(D^{-1})_{nm}=<v_n|u_m>$, n,m=1,...N and $<v_n|$ a second set of functions, whose choice is only making D non-singular. By construction $V_N|u_n>\equiv V|u_n>$, n=1,...N. They have chosen $|u_n>= G_o|f_n>$, n=1...N where $|f_n>$ are smooth functions of momentum that could describe well the momentum dependence of the t-matrix. The separable t-matrix becomes,

$$T_{AS}(s) = \sum_{n,m=1}^{N} V G_o|f_n> \Delta_{nm} <v_m| \qquad (2.21)$$

where $(\Delta^{-1})_{nm}=\langle v_n|(G_o-G_o V G_o)|f_m\rangle$, and $T_{AS}=V_N+V_N G_o T_{AS}$. There are now different possible choices for the vectors $\langle v_m|$ but they have to make V_N Hermitian so the t-matrix obeys the correct unitarity relation. If $|f_n\rangle=|\psi_n\rangle$ and $\langle v_m|=\langle\bar{\psi}_m|$ where $V\,G_o|\psi_n\rangle=\lambda^{-1}|\psi_n\rangle$ and $\langle\bar{\psi}_n|G_o V=\lambda_n^{-1}\langle\bar{\psi}_n$ with the orthogonality relation $\langle\bar{\psi}_m|G_o|\psi_n\rangle=\delta_{mn}\langle\bar{\psi}_n|G_o|\psi_n\rangle$ we obtain the Weinberg series. If there is a bound state at energy s=-B, chosing $|u_n\rangle=G_o(-B)|\psi_n(-B)\rangle$ and $\langle v_m|=\langle\psi_m(-B)|$ we get the UPE. Finally if $|u_n\rangle$ are chosen to be the eigenstate of the full Hamiltonian $H=H_o+V$ and $\langle v_m|$ are chosen to make the potential symmetric, $\langle v_m|=\langle u_m|V$ n=1,...N we obtain the EST expansion. The AS method was used[28] to obtain a separable approximation to the Malfliet-Tjon and Reid potentials and to obtain the virtual state of two nucleons[29]. It was also generalized to coupled channels[30] and illustrated for the simplified version of the $^3S_1-^3D_1$ Reid soft-core potential obtained by Pieper.

To finalize this review two more methods to obtain a separable t-matrix will be discussed. The Kowalski-Noyes[31] method and its extension to N-rank by Oryu[32]. In order to solve the LS equation, Kowalski and Noyes defined an off-shell function $\phi(p,k)=T_\ell(p,k;s)v_\ell(k,k)/T_\ell(k,k;s)$ and a modified potential $v_\ell(p,q)=v_\ell(p,q)-v_\ell(p,k)v_\ell(k,q)/v_\ell(K,K)$ and obtained a t-matrix,

$$T_\ell(p,q;s)=\phi_\ell(p,k)\,\frac{T_\ell(k,k;s)}{v_\ell^2(k,k)}\,\bar{\chi}(k,q)+\gamma_\ell(p,q;s) \qquad (2.22)$$

which has a rank one separable term plus a non-separable part. The quantity γ_ℓ satisfies an LS equation for the modified potential. This matrix has an unphysical pole on the first term where $T_\ell(k,k;s)=0$ as it was shown by Osborn[33]. Expanding the modified potential and recasting it in terms of a Weinberg-type series this term can be canceled, however the final t-matrix has infinit rank. There are other methods to avoid this singularities. One is given by Fuda[34] if the zero in one shell t-matrix is due to the presences of a hard core in the potential, the other is the rank three separable full off-shell t-matrix of Kowalski[35].

Oryu has extended the formalism to an n-rank separable part

plus a non-separable term. For details see Oryu's contribution to this school. The spurious singularity is also present but can be shifted to higher energies by increasing the rank of the separable expansion. However it cannot be guaranteed that these poles are removed except in the infinit rank limit.

For intermediate energies, Lim[36)] studdied a separable expansion method for the t-matrix which was tested for an Yukawa potential.

II. 2 Phenomenological Separable Potentials

The idea of using a separable potential was first introduced by Yamaguchi[37)] in the fifties and extended previously by numerous authors. They described only the ^1S_o and $^3S_1$-$^3D_1$ partial wave states with form factors which are Fourier transforms of Yukawa type interactions. Most of the potentials developed afterwards, had form factors of Yamaguchi type with parameters adjusted to fit the deuteron properties and low energy scattering data, like scattering lengths and effective range parameters. In Table I the development of such potentials since its first appearance until recent days is presented, with the type of fit to data used and type of calculation where they were applied.

The early calculation of Yamaguchi and Yamaguchi[37)] was only for n-p scattering and deuteron properties and to check the consistency of the data. The following potentials were immediately used in the three body bound state or scattering problems like for example Mitra[38)], Alt[39)], Doleschall[40)], Graz[41)], Haidenbauer-Plessas[42)], and Haidenbauer--Koike-Plessas[43)] potentials. Applications were also made to nuclear matter with the potentials of Tabakin[44)], Hammann[45)], Afnan-Clement--Serduke[46)], Kahana-Lee-Scott[47)] and potential of ref.(38). All the potentials in spite of describing well the two-body properties, can exhibit totally different aspects when incorporated in three body calculations. This is the case of the Tabakin potential which has a zero width resonance which produces an unphysically large value for the triton binding energy[48)]. Various parametrizations of the N-N interaction were also given by Morgan[49)] in order to include various partial wave states. Using the inverse scattering problem, Fiedeldey[50)] generated a family of potentials and studdied their off-shell behaviour.

The early potentials were of rank one with an attraction and a repulsion part to account for the change of sign in the ^1S_o and $^3S_1$ channels. Others were obtained from the (UPA) or (UPE) scheme for higher rank. The potentials also developed in order to accommodate higher partial waves required with the increase in sophistication of three body and nuclear matter calculations. The best fitting to the experimental data was done by the Graz group. However they had difficulties in reproducing accurately the scattering results in the coupled S-D channel. The most modern version of their separable potential, the Graz II [51)], also treats Coulomb distortion exactly allowing for a precise description of the p-p interaction in all partial waves up to $L\leq 2$.

More recently the separable potentials tried to reproduce an original realistic interaction via the (EST) and (AS) methods instead of a direct fitting of data. These last two methods as we have seen, allow a proper treatment of the off-shell behaviour of the N-N interaction giving a strong improvement in the quality of the resulting potential. The application of the EST method can give with low rank good results if the energies are chosen adequately, but one has to be careful with the energy dependence over the whole range to avoid unphysical poles. The Pieper[52)] potential was the first separable approach to a realistic interaction, derived from the (EST) method. Recent studies in the three--body problem emphasizes the importance of off-shell properties of the N-N force which go beyond the description of the Reid potential but are provided by meson-exchange ones. The recent separable parametrization of the Paris[42)] and Bonn[43)] from the (EST) method and Argonne V_{14}[53)] potentials, from the (AS) method allow to introduce such features. The first two representations have form factors constructed from rational functions depending on parameters which are adjusted to reproduce the off-shell behaviour of the exact potentials besides the choice of the reference energies imposed by the EST method. The phase shifts, and half-off-shell functions for $L\leq 4$ in the Paris version and ^1S_o and $^3S_1-^3D_1$ for the Bonn case are well reproduced. The same applies to deuteron properties. They have been applied to the three body problem and nuclear matter. For details concerning the three body calculation see the contribution of W. Plessas to this school. The comparison with the Paris potential is good for nuclear matter[54)] within the Brueckner scheme, where the involved domain of negative energies is smaller. For the Λ^{oo} approximation[55)] of the Green's function where negative energies are very important, the momentum distribution and mass operators are different. This probably shows the difference in behaviour at these energies of the exact and separable potentials.

The separable form for the Argonne V_{14} is an application of the general Gamow Separable Approximation (GSA)[26)] to construct a separable potential which will be discussed in the next section. It was initially applied to a square well and later to the Reid soft core potential, and has the advantage of having no choice of energies or free parameters.

III. The Gamow Separable Approximation (GSA)

In this section we present a separable representation of the Reid and Argonne potentials which follows from the general method of Adhikari and Sloan described in the section II and uses Gamow states to obtain the basis vectors for the expansion.

In the separable expansion of (AS) given by eq.(2.20), we have selected: $\langle v_m|=\langle u_m|V$, $n=1,\ldots N$ and $|u_m\rangle$ as the bound and Gamow states associated with the original potential V. From the properties of the AS expansion the first N bound and Gamow state of V_N coincide with those of V and the corresponding S-matrix has the first N poles at the correct positions. The method is free of spurious zero width resonances as it was analysed in detail in ref.26, and has no free parameters. Even the lowest energy resonances are generated at energies with a proper finite imaginary part, the higher energy resonances move away from the real axis even further.

The Gamow vectors are the solutions of the Schrodinger equation with purely outgoing wave boundary conditions, and correspond in general to complex energies. They were obtained numerically in configuration space integrating the Schrodinger equation from the origin up to a cut-off radius R, where the Gamow vector is matched to a pure outgoing free wave with complex momentum k ($Imk<0$). This is equivalent to assume that the potential is approximated by a potential of compact support, a reassonable approach provided R is sufficiently large. We have chosen $R=6.5$ fm.

As an example we show in table II the first few Gamow momenta for the Argonne V_{14} potential in the ^1S_o single channel. The first pole is the virtual state present in this channel of the N-N interaction. Changes in the value of R will make these poles to move in the complex k-plane, with the exception of the virtual state since it has physical meaning. However the final results for the t-matrix were stable when R was increased to 8.5 fm.

Results for the separable t-matrix of eq.(2.21) for the uncoupled S and P wave channels have shown excelent agreement with their exact counterparts for phase shifts, as well as for the off-shell terms, with a rank n=5 expansion. These results were obtained for the Reid and Argonne V_{14} potentials and are presented in ref.52. For brevity we show here only the on-shell results for the Argonne V_{14} interaction.

In order to apply the GSA coupled-channel cases we have first to obtain the Gamow states for the set of coupled Schrodinger radial equations.

$$\mu''_\alpha(r)+\left[k^2+k_\alpha^2 - \frac{\ell_\alpha(\ell_\alpha+1)}{r^2} - V_\alpha(r)\right]\mu_\alpha(r) = \sum_{\beta\neq\alpha} g_{\alpha\beta}(r)\mu_\beta(r) \qquad (3.1)$$

where the function $\mu_\alpha(r)$ is the radial wave function in the channel α with orbital angular momentum ℓ_α, the quantity k_α^2 is the shift in the relative energy due to threshold in each channel, and $g_{\alpha\beta}$ the coupling interaction.

Making the transformation,

$$\begin{aligned} \mu_\alpha(r) &= e^{ikr}\phi_\alpha(r) \quad r \leq R \\ &= e^{ikr}\phi_\alpha(R) \quad r > R \end{aligned} \tag{3.2}$$

with the boundary conditions $\phi_\alpha(0)=0$, $\phi'_\alpha(R)=0$ and expanding ϕ_α in a basis of functions $Q_m^\alpha=N_m^\alpha+j_{\ell_\alpha}(K_m^\alpha r)$ where N_m^α is a normalization constant and $j_\ell(x)$ are the spherical Bessel functions eq.(3.1) is reduced to an eigenvalue problem for a non-symmetric real matrix.

$$\sum_m |A_{nm}^\alpha + ikB_{nm}^\alpha| \, a_m^\alpha=\sum_m G_{nm}^{\alpha\beta} \, a_m^\beta \tag{3.3}$$

where A,B and G are matrices. They correspond to the matrix elements of the Schrodinger operator, first derivative and coupling interaction respectively taken between the basis functions. The quantities K_m^α are obtained from the boundary condition which reduces to $\frac{d}{dt}[r\, j_{\ell_\alpha}(K_m^\alpha r)]_{r=R}=0$ The transformation defined in eq.(3.2) is only exact if R is large enough. We should have instead of the exponential factor, the Hankel function which prevents us to recast eq.(3.1) in the form of an eigenvalue problem. Since this function behaves asymptotically like an exponential the only approximation involved is the cut-off of the centrifugal potential for r>R. For details see ref.(56).

It is interesting to see the similarities of this method and the calculation of G.Payne presented in this school.

The application of this method to the coupled $^3S_1-^3D_1$ channels of the Nucleon-Nucleon system gave as first eigenvalue the deuteron. We used both the Reid and Argonne V_{14} potentials. The momentum obtained was k=i 0.2315, in the case of Reid potential to be compared with the exact one given by Reid, k=i 0.2316. The agreement for the wave function was also excellent with a basis of 30 functions. Small deviations appeared beyond the cut-off radius R, where the wave function is assumed to decrease as an exponential. But this region is irrelevant for the ansatz considered in the AS method.

The first few eigenvalues, corresponding to the lower Gamow momenta are displayed in table III. With the eigenvactors associated to these momenta a separable potential can be constructed like in the single channel case and obtain the t-matrix from a straightforward generalization of AS to coupled channels. The phase parameters presented in table IV were obtained with the parametrization of the scattering matrix of Stapp for the Reid potential. The agreement with the results given by the exact potential is excellent for both S and D waves and the coupling parameter. The same applies for the off-shell t-matrix for the Reid and Argonne V_{14} potentials.

The Gamow separable approximation seems to be very successful in describing a realistic N-N interaction both on and off the energy shell for single and coupled channels situations. It should be noticed that the advantages of this method lies in the fact of being unambiguous and without free parameters.

REFERENCES

1) W.Rarita, J.S.Schwinger, Phys.Rev. 59 (1941) 436.
2) G.Breit, Reviews of Mod.Phys. 34 (1962) 766.
3) T.Hamada, I.D.Johnston, Nucl.Phys. 34 (1962) 382.
4) K.E.Lassila, M.H.Hull, M.Ruppel, F.A.McDonnald, G.Breit, Phys.Rev. 126 (1962) 881.
5) R. de Tourreil, D.W.L.Sprung, Nucl.Phys.A201 (1973) 193.
6) R.A.Malfliet, J.A.Tjon, Nucl.Phys.A127 (1969) 161.
7) R.V.Reid, Ann.Phys. 50 (1968) 411.
8) K.Holinde, R.Machleidt, Nucl.Phys. A247 (1975) 495.
9) M.M.Nagels, T.A.Rijken, J.J. de Swart, Phys.Rev. D17 (1978) 768.
10) A.D.Jackson, D.O.Riska, B.Verwest, Nucl.Phys.A249 (1975) 397.
11) I.E.Lagaris, V.K. Pandharipande, Nucl. Phys. A359 (1981) 331.
12) W.N. Cottingham, M.Lacombe, B.Loiseau, J.M.Richard, R.Vinh Mau, Phys Rev. D8 (1973) 800.
13) M.Lacombe, B.Loiseau, J.M.Richard, R.Vinh Mau, J.Coté, P.Pirés, R. de Tourreil, Phys. Rev. C21 (1980) 861.
14) R.B.Wiringa, R.A.Smith, T.L.Ainsworth, Phys.Rev. C (1984) 1207.
15) B.Gyarmatti, A.T. Kruppa, Phys.Rev. C34 (1986) 95.

16) S.Weinberg, Phys.Rev. 131 (1963) 440.

17) J.S.Levinger, A.H.Lu, R.Stagat, Phys. Rev. 179 (1969) 926.

18) E.Harms, Phys.Rev.C1 (1970) 1667.

19) E.O.Alt, W.Sandhas, Few-Body Problems in Nucl. and Particle Phys. (1975) ed. R.J. Slobodrian, B.Cujec, K. Ramavataram.

20) C.Lovelace, Phys.Rev. 135B (1964) 1225.

21) M.G.Fuda, Nucl.Phys.A116 (1968) 83.

22) J.S.Levinger, Springer Tracts in Modern Phys. (Springer, Berlin, 1974) vol 71 and references contained therein.

23) A.Casel, H.Haberzette, W.Sandhas, Phys.Rev.C25 (1982) 1738.

24) R.D.Bathia, J.F.Walker, Nucl.Phys. 192 (1972) 658.

25) D.J.Ernest, C.M.Shakin, R.M.Thaler, Phys.Rev.C8 (1973) 46.

26) M.Baldo, L.S.Ferreira, L.Streit, Phys.Rev.C32 (1985) 685.

27) J.Haidenbauer, W.Plessas, Phys.Rev. C27 (1983) 63.

28) S.K.Adhikari, I.H.Sloan, Phys.Rev. C11 (1975) 1133; Nucl.Phys. A241 (1975) 429;

29) S.K.Adhikari, A.C.Fonseca, L.Tomio, Phys.Rev. C27 (1983) 1826.

30) S.K.Adhikari, I.H.Sloan, Nucl.Phys. A251 (1975) 297.

31) K.L.Kowalski, Phys.Rev.Lett. 15 (1965) 798; H.P.Noyes, Phys.Rev. Lett. 15 (1965) 538.

32) S.Oryu, Prog. Theor. Phys. 52 (1974) 550.

33) T.A. Osborn, Nucl. Phys. A138 (1969) 305.

34) M.G.Fuda, Phys. Rev. 186 (1969) 1078.

35) K.L. Kowalski, Nucl. Phys. A190 (1972) 645.

36) T.K.Lim, J.Giannini, Phys. Rev. A18 (1978) 517.

37) Y.Yamaguchi, Phys. Rev. 95 (1954) 1628; Y.Yamaguchi, Y. Yamaguchi, Phys.Rev.95 (1954) 1635.

38) A.N.Mitra, Phys. Rev. 131 (1963) 1265.

39) E.O.Alt, B.L.G.Bakker, Z Physik A273 (1975) 37.

40) P.Doleschall, Nucl. Phys. A220 (1974) 491.

41) L.Crepinsek et al.Acta Phys.Austr.39 (1974) 345; ibid 42 (1975) 139; Oberhummer et al. ibid 42 (1975) 225.

42) J.Haidenbauer, W.Plessas, Phys.Rev.C30 (1984) 1822.

43) J.Haidenbauer, Y, Koike, W.Plessas, Phys.Rev.C33 (1986) 439.

44) F.Tabakin, Ann.Phys.30 (1964) 51; Phys.Rev.174 (1968) 1208.

45) T.F.Hammann, W. Ho-Kim, Nuovo Cim. 64B (1969) 356.

46) S.Kahana, H.C.Lee, C.K.Scott, Phys.Rev. 185 (1969) 1378.

47) I.R.Afnan, D.M.Clement, F.J.D. Serduke, Nucl.Phys.A170 (1971) 625.

48) G.Rupp, L.Streit, J.A.Tjon, Phys.Rev.C31 (1985) 2285.

49) T.R.Morgan, Phys.Rev. 175 (1968) 1260; ibid. 178 (1969) 1597.

50) H.Fiedeldey, Nucl.Phys. A189 (1972) 83.

51) W.Schweiger, W.Plessas, L.P.Kok, H.Van Haeringen, Phys.Rev. C27 (1983) 515; ibid. C28 (1983) 1414.

52) S.C.Pieper, Phys. Rev. C9 (1974) 883.

53) M.Baldo, L.S.Ferreira, L.Streit, T.Vertse, Phys.Rev.33 (1986) 1587.

54) H.Lampl, M.K.Weigel, Phys. Rev. 33 (1986)

55) R.D.Puff, Ann. Phys. 13 (1961) 317)

56) M.Baldo, L.S.Ferreira, L.Streit, to be published in Nucl.Phys.

TABLE I

Potential	Form-Factors f(p)	Fits to data	Applications
Yamaguchi (1954)	$\frac{1}{p^2 + \beta^2}$	a)	n-p scatt.
Mitra (1963)	$\frac{1}{p^2 + \beta^2} \quad ({}^3S_1, {}^1S_0)$	a) b)	c)
Nagvi (1964)	$\frac{p^a}{[p^2 + \beta^2]^b}$	a) b)	
Tabakin (1964)	$\frac{p^2}{[(p-\gamma)^2+\beta_1^2][(p+\gamma)^2+\beta_2^2]}$ $\frac{p^2}{p^2 + \beta^2}$	a) b) L≤2	c) d)
Morgan (1969)	$\frac{p^\ell}{(p^2+\beta^2)^{\alpha(\ell)}}$	b) L≤2	
Kahana, Lee, Scott (1969)	$\frac{p^a}{(p^2+\beta^2)^b}$	b) L≤2	d)
Hammann (1970)	$r^{\ell-1} \exp[-\alpha_{\ell SJT} r]$	a)	d)
Afnan (1971)	$\frac{p^\ell}{[p^2+\beta_\ell]^{(\ell+2)/2}}$	a) b)	c)
Doleschall (1974)	$p^L [1+\sum_{n=1}^{N} \gamma_N p^{2n}] \times [\prod_{n=0}^{N} (1+\beta_n p^2)]^{-1}$	a) b)	c)
Pieper (1974)	$\sum_{n=1}^{\infty} \frac{c_n}{[p^2+\beta_n^2]^\alpha} \quad \alpha=2,3$	EST to Reid	c)
Alt (1975)	$\sum_{i=1}^{4} \frac{c_i}{p^2+\beta_i^2}$	UPA to Malfliet--Tjon potential	c)
Graz (1982)	$\frac{p^{\alpha(\ell)}}{(p^2+\beta_\ell^2)^{\gamma(\ell)}}$ (n-p) $\times \exp\{2\xi \tan^{-1}[p/\beta_\ell]\}$ (p-p)	a) b)	c)
Heidenbauer Plessas (1984)	$\sum_{n=1}^{4} \frac{c_n p^{\alpha_n(\ell)}}{(p^2+\beta_n^2)^{\gamma_n(\ell)}}$	EST to Paris pot.	c) d)
Gamow Separable Approximation (1985)	Gamow states	AS to square well Reid, Argonne	c)
Heidenbauer, Koike, Plessas (1986)	$\sum_{n=1}^{4} \frac{c_n p^{\alpha_n}}{(p^2+\beta_n^2)^{\gamma_n}}$	EST to Bonn pot.	c)

TABLE II

Real K	Im K
0.0	0.053
0.472	0.714
1.081	0.801
1.646	0.849
2.194	0.881

TABLE III

Real K	Im K
0.0	0.2315
0.4123	-0.7797
0.8811	-0.2732
1.0360	-0.7911
1.4220	-0.3340
1.5640	-0.7940
1.9330	-0.3785

TABLE IV

K (fm^{-1})	$\delta(^3S_1$, radians)		$\delta(^3D_1$, radians)		$\rho_1 = \sin(2\varepsilon_1)$	
(LAB)	(GSA)	(EXACT)	(GSA)	(EXACT)	(GSA)	(EXACT)
.4	2.221	2.224	-0.009	-0.009	0.021	0.020
.6	1.331	1.335	-0.056	-0.063	0.074	0.072
.8	1.058	1.052	-0.117	-0.127	0.087	0.083
1.	0.831	0.833	-0.154	-0.190	0.118	0.106

Table Captions

TABLE I - Various separable potentials, corresponding form-factors, type of fit used in its construction and applications. a)-deuteron properties; b)-phase shifts; c)-3-body problem; d)-nuclear matter.

TABLE II - Gamow vector momenta in fm for the $^1S_0$ Argonne V_{14} potential, with cut-off radius R=6.5 fm.

TABLE III - Gamow vector momenta for the $^3S_1$-$^3D_1$ Reid soft core potential obtained from the diagonalization of eq.(3.3) with a basis set of 40 functions, in fm^{-1} with cut-off radius R=6.5 fm.

TABLE IV - Phase parameters calculated from the GSA to the Reid soft core potential compared with the corresponding exact values for different momenta.

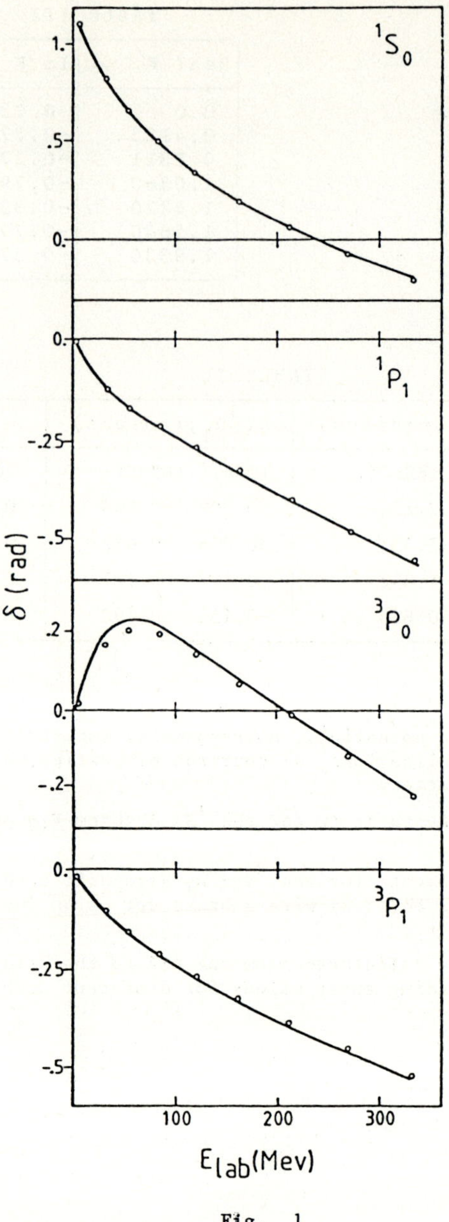

Fig. 1 .

Fig. 1 - T=1, S=0; T=0, S=0, and T=1, S=1 phase shifts for the Argonne V_{14} potential. The solid line represents the exact phases. The dots represent the rank five approximation.

SEPARABLE EXPANSION METHODS FOR THE THREE-BODY T-MATRIX

Shinsho Oryu

Department of Physics, Faculty of Science and Technology,
Science University of Tokyo, Noda, Chiba, 278 Japan

I. Introduction

Recent developments of the few-body problems are mainly owe to the separable expansion methods of the given potentials to avoid the complexity of many variables and the related analyticity of the equations. Not only 3- and 4-body equations but also N-body equations are given by the Fredholm (the 2-nd) type integral equations as well as the 2-body Lippmannn-Schwinger equation. Therefore, we start with the two-body problem as a prologue to study the three-body and four-body problems. In order to explain the given potentials by using the separable expansion method, we have some well-known techniques which are summerized skillfully by J. S. Levinger.[1]

We proposed a new method for making the generalized separable expansion (GSE) amplitude of the Lippmann-Schwinger (LS) equation with realistic potentials which can be given in the momentum space representation. We have two types of separable formalisms which were called the type-A and the type-B, respectively. The former one is introduced in the reference A,[2] and the latter one appears in the reference B[3] in which the relations with the old Bateman's method were discussed. Unfortunately, a clear method about "how to choose the parameters in the scattering problem " has never been proposed in the Bateman's method.[4] Moreover, it will be found that Bateman's formalism contains unphysical poles which are of the same kind as apeared in the Kowalski-Noyes equation.[5] Therefore, in the application of Bateman's method to the scattering problem , the kernel and parameters must be carefully investigated before used in few-body problems. Furthermore, it was proved in the reference C[6] that "the point-wise" Bateman's method in which the parameters are chosen on the optimal value on the mesh-points cannot satisfy the off-shell unitarity relation. Nevertheless, the Bateman's idea was useful to extend the Kowalski-Noyes equation up to the rank-N case in our work . We have proposed a new entire formalism based on Bateman's idea and some mathematically equivalent modifications were given with complete proofs in Ref. A.

Further developments in multi-channels cases were given in C and D.[7]

In this lecture, the idea of our methods will be briefly presented and some examples are illustrated. In section II, the idea of our separable expansion method is presented for type-A and type-B formalism . The unique parameter choice is proposed based on the discussion for the unitarity relation. In section III, the separable expansion formalisms of those types are generalized for the rank N case , in which the method to obtain parameters for the higher rank formalism is given. The convergence of separable amplitudes are shown for the Reid Soft Core ($^1S_0$ -state) potential as an example in section IV. Further developments of our separable expansion formalism to the three- and four-body systems are demonstrated in section V. The important properties of our separable expansion method are summerized, and the relation with the Kowalski-Noyes method is given in the last section.

II. FORMALISM

1) The type-A rank-1 formalism

First of all, let us start with the 2-body LS equation for a single channel case for simplicity. The partial wave expansion of this integral equation can be written in the form

$$t_\ell(p,p';z) = v_\ell(p,p') + \int_0^\infty dp''\, v_\ell(p,p'')G_0(p'';z)t_\ell(p'',p';z), \quad (2.1)$$

with

$$G_0(p;z) = \frac{1}{2\pi^2}\cdot\frac{p^2}{(z - p^2/2m)}. \quad \text{(m: the reduced mass)} \quad (2.2)$$

We define the reduced amplitude $t^{(2)}(p,p';z)$ and the reduced potential $v^{(2)}(p,p')$ as follows:

$$t^{(2)}(p,p') = \begin{vmatrix} t^{(1)}(k_1,k_1) & t^{(1)}(k_1,p') \\ t^{(1)}(p\ ,k_1) & t^{(1)}(p\ ,p') \end{vmatrix} \Big/ t^{(1)}(k_1,k_1) \quad (2.3)$$

and

$$v^{(2)}(p,p') = \begin{vmatrix} v^{(1)}(k_1,k_1) & v^{(1)}(k_1,p') \\ v^{(1)}(p\ ,k_1) & v^{(1)}(p\ ,p') \end{vmatrix} \Big/ v^{(1)}(k_1,k_1) \quad (2.4)$$

where we have suppressed the energy z and the partial wave index ℓ in the t-matrix $t_\ell^{(1)}(p,p';z) = t_\ell(p,p';z)$ and the potential $v_\ell^{(1)}(p,p') = v_\ell(p,p')$. It is easily seen that eqs. (2.3) and (2.4) satisfy the boundary conditions

$$t^{(2)}(p,k_1) = t^{(2)}(k_1,p') = t^{(2)}(k_1,k_1) = 0 \tag{2.5}$$

$$v^{(2)}(p,k_1) = v^{(2)}(k_1,p') = v^{(2)}(k_1,k_1) = 0, \tag{2.6}$$

where k_1 is a suitable parameter which has the same dimension as p and p'. Therefore, the original off-energy-shell t-matrix and the off-energy-shell potential can be respectively rewritten by using the definitions (2.3) and (2.4),

$$t(p,p') = \frac{t(p,k_1)t(k_1,p')}{t(k_1,k_1)} + t^{(2)}(p,p') \tag{2.7}$$

and

$$v(p,p') = \frac{v(p,k_1)v(k_1,p')}{v(k_1,k_1)} + v^{(2)}(p,p'). \tag{2.8}$$

where the first term of eq.(2.7) is the separable t-matrix and the second term is the non-separable one as well as the formalism for the potential descriptions in eq. (2.8). Furthermore, substituting eqs. (2.7) and (2.8) into eq. (2.1), it is led that the reduced amplitude (or the non-separable amplitude) $t^{(2)}(p,p')$ satisfies the following LS-type integral equation,

$$t^{(2)}(p,p') = v^{(2)}(p,p') + \int_0^\infty dp''v^{(2)}(p,p'')G_0(p'')t^{(2)}(p'',p'). \tag{2.9}$$

2) Unitarity Relation and Unique Parameter Choice

The unitarity relations for the expansion amplitudes are discussed in this paragraph. Generally, the given potentials are written by

$$v = v(\text{separable}) + v(\text{nonseparable}) = v_{sep} + v_{non} \tag{2.10}$$

and the t matrix is

$$t = t(\text{separable}) + t(\text{nonseparable}) = t_{sep} + t_{non} . \tag{2.11}$$

Since the total t matrix t satisfies the LS equation, t has to satisfy the unitarity condition:

$$t - t^+ = -2\pi i\, t^+\delta(z - H_0)t. \tag{2.12}$$

Substituting eq. (2.11) into this equation, we have

$$\begin{aligned}(t_{sep} - t^+_{sep}) + (t_{non} - t^+_{non}) &= -2\pi i(t^+_{sep} + t^+_{non})\delta(z - H_0)(t_{sep} + t_{non}) \\ &= -2\pi i t^+_{sep}\,\delta(z - H_0)t_{sep} - 2\pi i t^+_{sep}\,\delta(z - H_0)t_{non} \\ &\quad -2\pi i t^+_{non}\,\delta(z - H_0)t_{sep} - 2\pi i t^+_{non}\,\delta(z - H_0)t_{non}\,.\end{aligned} \tag{2.13}$$

Therefore, it is easily seen that the separable t matrix t_{sep} cannot satisfy the unitarity relation without the nonseparable term. However, we find that the following two cases allow us to obtain a unitarity relation for the separable t matrix without the nonseparable term.

(a) t_{non} is a real function, and the half on-(off)-shell t_{sep} is analytically exact.

(b) t_{non} and v_{non} satisfy the LS type of equation:

$$t_{non} = v_{non} + \int v_{non}\, G_0\, t_{non}\,, \tag{2.14}$$

and the half on-(off-) snell t_{sep} is analytically exact.

In case (a), we have the relations,

$$t_{non} - t^+_{non} = 0, \tag{2.15}$$

and

$$\delta(z - H_0)t_{non} = t^+_{non}\,\delta(z - H_0) = 0, \tag{2.16}$$

because the half off-shell t_{sep} becomes zero for exact half off-shell t_{sep}. Therefore eq. (2.13) becomes

$$t_{sep} - t^+_{sep} = -2\pi i t^+_{sep}\delta(z - H_0)t_{sep}. \tag{2.17}$$

In case (b), because of eq.(2.14) t_{non} satisfies

$$t_{non} - t^+_{non} = -2\pi i t^+_{non}\,\delta(z - H_0)t_{non}, \tag{2.18}$$

and for exact half off-shell t_{sep}, we obtain

$$\delta(z - H_0)t_{non} = t^+_{non}\,\delta(z - H_0) = 0 \,. \qquad (2.19)$$

By inserting eqs. (2.18) and (2.19) into eq. (2.13), we also obtain

$$t_{sep} - t^+_{sep} = -2\pi i t^+_{sep}\,\delta(z - H_0)t_{sep} \,. \qquad (2.20)$$

This is the unitary relation for the separable t matrix t_{sep}. It should be stressed that t_{sep} satisfies the relation without the nonseparable term. In any separable expansion method, there are only two cases (a) and (b) which will satisfy the off-shell unitarity relation for t_{sep}. Furthermore, we prefer case (b) rather than case (a), because in case (a) t_{non} is not always a real function, e.g., the three-body calculation by the contour deformation method demands the complex value of t_{non} which is no longer a real function.

In order to satisfy eq. (2.19), the unique parameter choice in our theory is

$$k_1 = k = \sqrt{2mz} \,, \qquad \text{(on-energy-shell momentum)} \qquad (2.21)$$

in which k_1 is continuous energy dependent function (value) for the positive energy region, and no longer a fixed mesh-point. Therefore, by means of eqs.(2.5) and (2.7) on-shell and half-off (on-) shell t matrices are exactly given by the separable terms. Furthermore, t_{non} satisfies a LS type equation as eq.(2.9), then eq.(2.18) is fulfilled. Consequently, our separable part of the t matrix satisfies the off-energy-shell unitarity relation under condition (b).

3) The type-B rank-1 formalism

Now, we introduce the new half off-shell functions

$$\varphi(p,k) = \frac{t(p,k)}{t(k,k)}\,v(k,k)$$
$$\chi(k,p') = \frac{t(k,p')}{t(k,k)}\,v(k,k) \qquad (2.22)$$

and define

$$t(k,k) = [v(k,k)]^2 / A(k,k). \qquad (2.23)$$

These definitions yield the following relations,

$$t(p,k) = \varphi(p,k)\,\frac{v(k,k)}{A(k,k)}\,,$$

$$\varphi(k,k) = v(k,k), \qquad \text{and} \qquad \chi(k,k) = v(k,k). \tag{2.24}$$

Substituting (2.22) - (2.24) into the half off-shell LS equation for t(p,k), we can obtain a new integral equation for the half off-shell function $\varphi(p,k)$, and the relation between A(k,k) and $\varphi(p,k)$:

$$\varphi(p,k) = v(p,k) + \int_0^\infty dp''\, v^{(2)}(p,p'')G_0(p'')\,\varphi(p'',k), \tag{2.25}$$

and

$$A(k,k) = v(k,k) - \int_0^\infty dp''\, v(p,p'')G_0(p'')\,\varphi(p,k). \tag{2.26}$$

By the same way, the conjugate equations of (2.25) and (2.26) are given by

$$\chi(k,p') = v(k,p') + \int_0^\infty dp''\,\chi(k,p'')G_0(p'')v^{(2)}(p'',k), \tag{2.25a}$$

and

$$A(k,k) = v(k,k) - \int_0^\infty dp''\, \chi(k,p'')G_0(p'')v(p'',k). \tag{2.26a}$$

It is easily seen that eqs.(2.25) and (2.25a) are nonsingular integral equations as well as eq.(2.9), because the pole of the Green's function is canceled with the reduced potential $v^{(2)}(p,p')$. Consequently, the off-shell t matrix is given by

$$t(p,p';z) = \frac{\varphi(p,k)\,\chi(k,p')}{A(k,k;z)} + t^{(2)}(p,p';z), \tag{2.27}$$

where the first term is the separable amplitude with the form factors $\varphi(p,k)$ and $\chi(k,p')$, and the nonseparable term satisfies eq.(2.9).

4) Bound State

It is well known that the bound states are represented by the poles of the N/D separation of the t matrix in the terminology of dispersion theory. Therefore, one can easily imagine that the separation is contained in the first term of the right-hand side of eq.(2.27). Although $A(k_1,k_1;z)$ is in general not identical to the D function, the condition that there be a bound state at $z = -E_B$ should be written as

$$A(k_1,k_1;-E_B) = 0. \tag{2.28}$$

Substituting this into eq.(2.26), we obtain

$$\begin{aligned} A(k_1,k_1;z) &= -\frac{(E_B + z)}{2\pi^2}\int_0^\infty dp''p''^2 \frac{v(k_1,p'')}{(E_B + p''^2/2m)} \\ &\times\left\{\frac{\varphi(p'',k_1;z)}{z - p''^2/2m} + \frac{\varphi(p'',k_1;-E_B) - \varphi(p'',k_1;z)}{E_B + z}\right\} \\ &= -(E_B + z)F(k_1,k_1;z,E_B), \end{aligned} \tag{2.29}$$

where $F(k_1,k_1;z;E_B)$ is a regular function at the bound-state energy $z = -E_B$. Thus the rank-one approximation for the t matrix is represented by

$$t(p,p';z) = \frac{\varphi(p,k_1;z)\,\chi(k_1,p';z)}{(z + E_B)F(k_1,k_1;z,E\ )} + t^{(2)}(p,p';z), \tag{2.30}$$

where the t matrix diverges at the bound-state energy $z = -E_B$ for any value of the fitting parameter k_1. However , it was pointed out in reference C that the parameter should be a positive value by the reason why the analyticity of the separable form factor can be satisfied:

$$k_1 \geq 0 \qquad (\text{for } z < 0\ ). \tag{2.31}$$

Otherwise the pinching singularity will occur for the choice of eq.(2.21).

III. Generalization

The modification of eq.(2.7) is given by

$$t(p,p') = -\frac{\begin{vmatrix} t(k_1,k_1) & t(k_1,p') \\ t(p\ ,k_1) & 0 \end{vmatrix}}{t(k_1,k_1)} + \frac{\begin{vmatrix} t(k_1,k_1) & t(k_1,p') \\ t(p\ ,k_1) & t(p\ ,p') \end{vmatrix}}{t(k_1,k_1)} .$$

This symmetry suggests a simple generalization of the formalism for the rank-N case:

$$t(p,p') = -\begin{vmatrix} t(k_1,k_1) & t(k_1,k_2) & \cdots & t(k_1,k_N) & t(k_1,p') \\ t(k_2,k_1) & t(k_2,k_2) & \cdots & t(k_2,k_N) & t(k_2,p') \\ \cdot & \cdot & \cdot & \cdot & \cdot \\ t(k_N,k_1) & t(k_N,k_2) & \cdots & t(k_N,k_N) & t(k_N,p') \\ t(p\ ,k_1) & t(p\ ,k_2) & \cdots & t(p\ ,k_N) & 0 \end{vmatrix} /\det[t(k_i,k_j)]$$

$$+ \begin{vmatrix} t(k_1,k_1) & t(k_1,k_2) & \cdots & t(k_1,k_N) & t(k_1,p') \\ t(k_2,k_1) & t(k_2,k_2) & \cdots & t(k_2,k_N) & t(k_2,p') \\ \cdot & \cdot & \cdot & \cdot & \cdot \\ t(k_N,k_1) & t(k_N,k_2) & \cdots & t(k_N,k_N) & t(k_N,p') \\ t(p\ ,k_1) & t(p\ ,k_2) & \cdots & t(p\ ,k_N) & t(p\ ,p') \end{vmatrix} /\det[t(k_i,k_j)], \quad (3.1)$$

or

$$t(p,p';z) = \sum_{i,j}^{N} \frac{\eta_{ij}(z)}{\eta(z)} t(p,k_j;z)t(k_i,p';z) + t^{(N+1)}(p,p';z), \quad (3.2)$$

where $\eta(z) = \det[t(k_i,k_j)]$ and $\eta_{ij}(z)$ is the i-j co-factor of $\eta(z)$. Here, it is proved that the reduced potential $v^{(N+1)}(p,p')$ and t-matrix $t^{(N+1)}(p,p')$ satisfy the LS-type equation as the generalization of eq.(2.9):

$$t^{(N+1)}(p,p') = v^{(N+1)}(p,p') + \int_0^\infty dp''\ v^{(N+1)}(p,p'')G_0(p'')t^{(N+1)}(p'',p). \quad (3.3)$$

On the other hand, the type-B formalism is easily introduced by using the generalization of eqs.(2.22)-(2.26):

$$t(p,p';z) = \sum_{i,j}^{N} \frac{\theta(z)_{ij}}{\theta(z)} \varphi(p,k_j;z)\chi(k_i,p';z) + t^{(N+1)}(p,p';z), \quad (3.4)$$

where the form factors satisfy nonsingular integral equations

$$\varphi(p,k_j;z) = v(p,k_j) + \int_0^\infty dp''v^{(N+1)}(p,p'')G_0(p'';z)\varphi(p'',k_j;z), \quad (3.5)$$

and

$$\chi(k_i,p';z) = v(k_i,p') + \int_0^\infty dp'' \chi(k_i,p'';z)G_0(p'';z)v^{(N+1)}(p'',p'), \tag{3.5a}$$

and

$$A(k_i,k_j;z) = v(k_i,k_j) - \int_0^\infty dp'' v(k_i,p'')G_0(p'') \varphi(p'',k_j;z), \tag{3.6}$$

$$= v(k_i,k_j) - \int_0^\infty dp'' \chi(k_i,p'';z)G_0(p'';z)v(p'',k_j). \tag{3.6a}$$

IV. Parameter Choice of Higher Ranks

These formalisms are similar to the Bateman's method as discussed in B, however, the essential difference will be shown in the choice of parameters k_1, k_2, ... , k_N. The unique k_1 value is already given in eqs.(2.21) and (2.31). These parameters, however, are not sufficient in the whole energy region; the parameters cause unphysical singularities of the term $t^{(2)}(p,p';z)$. For instance, the $^1S_0$ state of the nucleon-nucleon interaction at 125 MeV corresponds to a zero of the phase shift. In order to avoid such a difficulty and to get a good fit of the off-shell elements, we may determine the second fitting parameter k_2. Before such a procedure, the Osborn's theorem should be carefully taken into account in which it is mentioned that the local potential cannot be given by the separable expansion terms, and the norm of the nonseparable term diverges weakly.[8] It suggests that, even in our formalism which is mathematically correct for on- and half-on shell amplitudes, the nonseparable term cannot be neglected. Fortunately, we have a satisfactory explanation that since our separable amplitudes are practically used with the Green's function in the 3-body problems, then after the integrations over the intermediate variables the term $G_0t^{(N+1)}$ may be zero or very small as compared with values of the terms given by the separable expansion.

We proposed a method to determine the second fitting parameter k_2 by minimizing the norm $\bar{\chi}(k_1,k_2;z)$ over the entire energy region for fixed k_1. This norm is defined in the general rank case by

$$\bar{\chi}_N^2(k_1,k_2, \ldots ,k_N;z) = \sum_{\alpha,\beta=1}^{M} \int_0^\infty dp \int_0^\infty dp'$$

$$\times F^{\star}_{\alpha\beta}(p,p';k_1,k_2, \ldots ,k_N;z)F_{\beta\alpha}(p',p;k_1,k_2, \ldots ,k_N;z)$$

$$= \sum_{\alpha,\beta=1}^{M} \left\| t^{(N+1)}(p,p';z)G_0(p';z) \right\|^2 . \qquad (4.1)$$

Here α and β denote the channels. Using the rank-N nonseparable term of the t matrix given in eqs.(2.9) and (3.3), the integrand denotes

$$F_{\alpha\beta}(p,p';k_1,k_2, \ldots ,k_N;z) = t_{\alpha\beta}^{(N+1)}(p,p';z)G_0(p';z). \qquad (4.2)$$

In the same way, we can obtain a set of suitable parameters $k_3, k_4, \ldots,$ and k_N, step by step. For negative energies, one has to take a positive value parameter k_1, and also minimize the norm $\bar{\chi}_1^2(k_1,z)$ of eq.(4.1), because of the pinching singularity just mentioned above. Therefore, eq.(4.1) suggests that the higher rank's parameters are also energy dependent functions. This feature is clearly illustrated in Figs.1 and 2 for $z > 0$, and Figs.3, 4 and 5 for $z < 0$, in the case of the Reid Soft Core nucleon-nucleon potential for $^1S_0$ state. After the unique parameter choice: $k_1 = k$ (for $z > 0$), $\bar{\chi}_2^2(k_1,k_1;z)$ of eq.(4.1) is calculated.

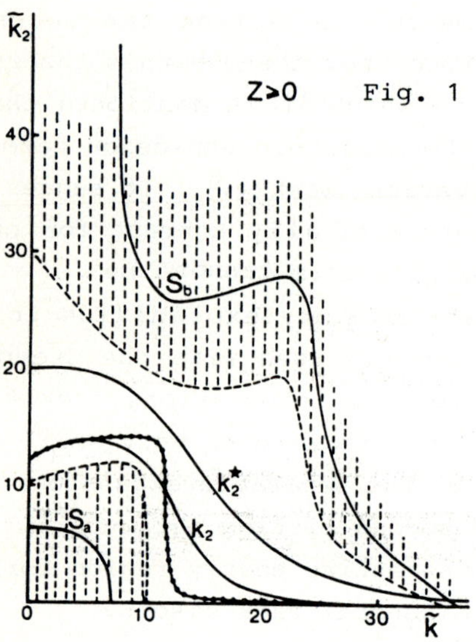

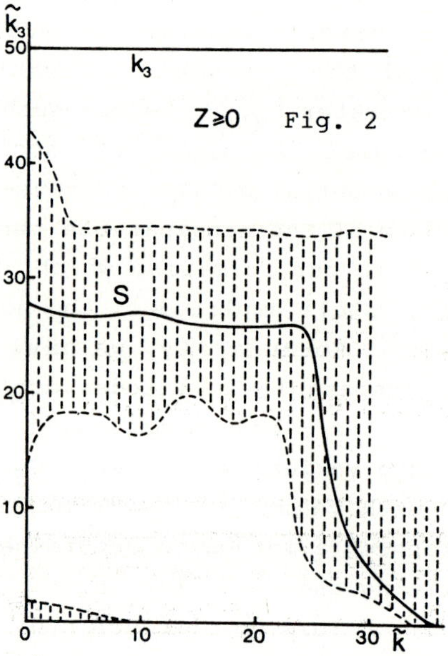

The quantity of $\bar{\chi}_2(k\ ,k_2;z)$-function is illustrated in the $k - k_2$ plane in Fig. 1. Here, $\tilde{k}$ and $\tilde{k}_2$ are denoted with the unit: @ = 0.2316 fm^{-1}. The solid lines S_a and S_b illustrate the singular or the very large value. The shadow area shows the large quantity of the function. Therefore, one can choose a good k_2 as a function of k, in which we gave three alternatives with two solid curves k_2 and k_2^*, and solid-dotted line which shows the minimum of $\bar{\chi}_2$-function. By the same way, the 3-rd parameter can be chosen with $\bar{\chi}_3^2(k_1, k_2, k_3;z)$ for the fixed values k_1, k_2. The solid line S in Fig. 2 denotes the most singular part of the function, in which the parameter k_3 is safely chosen as a constant for z < 1 GeV. For negative energies, Figs. 3, 4, and 5 illustrate each quantity $\bar{\chi}_1^2(k_1;z)$, $\bar{\chi}_2^2(k_1, k_2;z)$ and $\bar{\chi}_3^2(k_1, k_2, k_3; z)$, respectively. All the inscriptions are the same

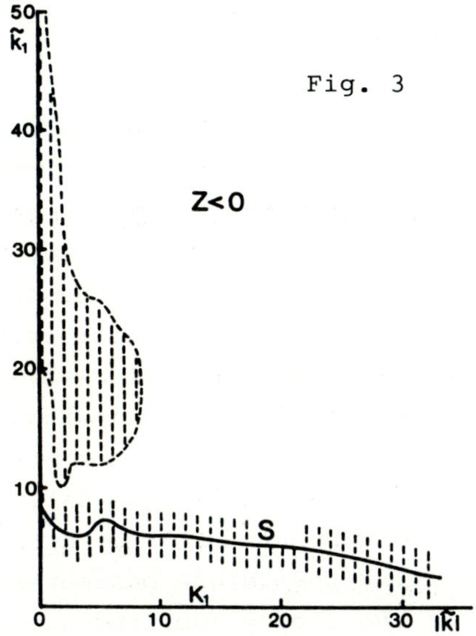

Fig. 3

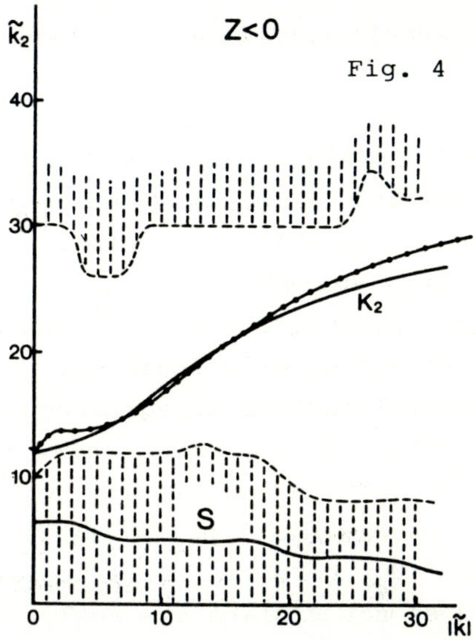

Fig. 4

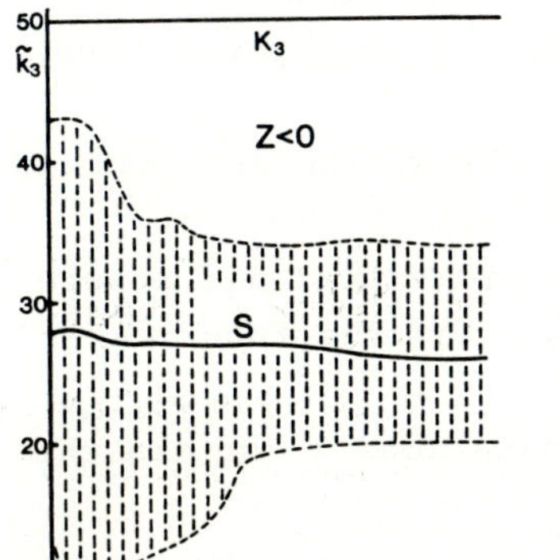

Fig. 5

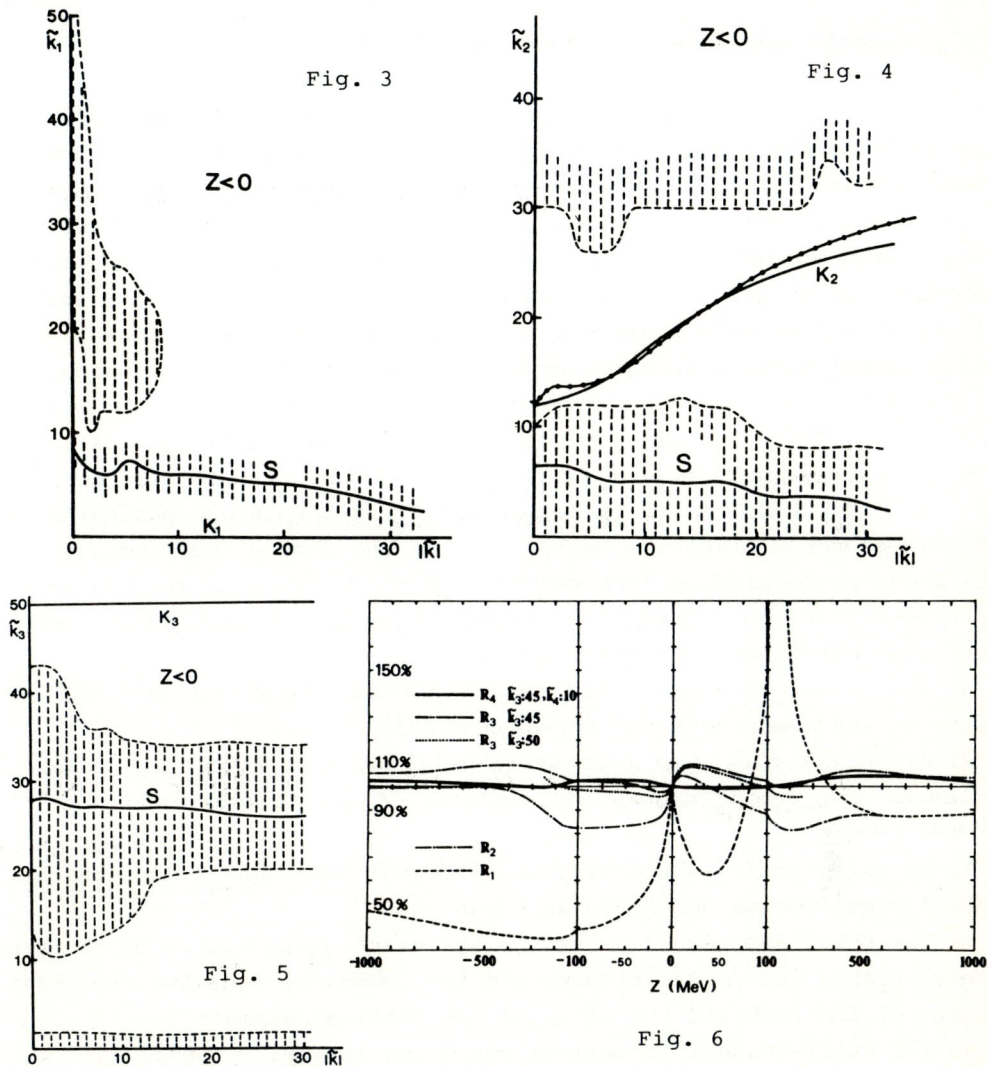

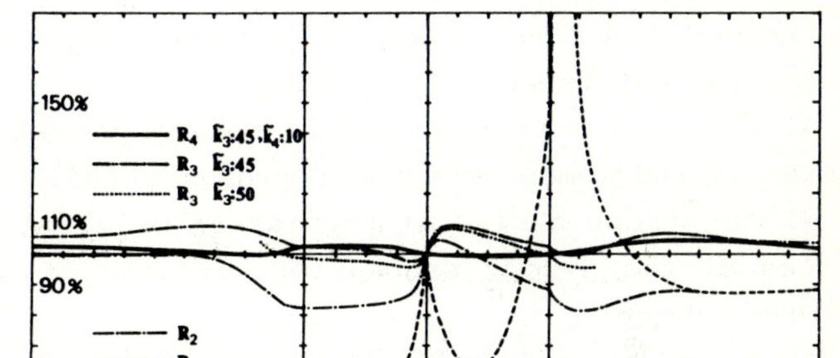

Fig. 6

as the positive energy cases. The convergence for the rank 1, 2, 3, and 4 cases is illustrated for the norm of the separable amplitudes by % in which they are treated by a ratio:

$$R_{sep}(N) = \left\{ \sum_{\alpha,\beta=1}^{M} \left\| t^{sep}_{\alpha\beta}(\text{rank-}N) G_0 \right\| \Big/ \sum_{\alpha,\beta=1}^{M} \left\| t_{\alpha\beta}(\text{total}) G_0 \right\| \right\} 100\ \% \tag{4.3}$$

Details of the parameter functions are given in C, in which the convergence was investigated for $0 < p,p' < 50$ @. It seems in Fig.6 that the rank-3 case is sufficient in the region -1000 MeV $\leq z \leq 1000$ MeV.

V. Separable Expansion of 3-Body Amplitude

It is known that all 3-body Faddeev equivalent equations like the Amado-Lovelace-Mitra (ALM) equation or the Alt-Glasberger-Sandhas (AGS) equation are Fredholm of 2-nd kind integral equations as well as the 2-body LS equation.[9] However, those equations have more complicated analytic structure which is constructed with the energy dependent potential, multi-channels and also 2-body propagator instead of the simple Green's function. Our separable expansion formalism is still useful in this integral equation, the reasons are following:

1) In the energy plane where the 3-body amplitude satisfies the unitarity and the analyticity conditions, our separable term satisfies the off-energy-shell unitarity relation without nonseparable residual term. Furthermore, it is possible to find the unique parameter set which does not violate the analyticity of the 3-body amplitudes.

2) Since our separable form factors and the parameters are the energy dependent functions, then the energy dependent 3-body potentials do not cause any trouble in the formalism.

3) The method is applicable to any potentials which are already given by the momentum space representation.

4) The unique choice of parameters will be completed at all possible poles of the 2-body propagator instead of the Green's function pole at the on-energy-shell momentum.

5) By using those parameters, the integral equations to obtain the form factors become non-singular equations.

6) The bound and resonance states are exactly reproduced by the rank one separable term. The binding and the resonance energies are independent of the rank and the value of the fitting parameters.

7) The multi-channel formalisms are given in Ref. C and D.

Therefore, our separable expansion amplitude is automaticaly described as given in Ref.D. As an example, we will show the role of the off-shell behavior of the 3-alpha amplitude in the 4-alpha binding energy, in which three-body channels are given by $(L_{ij}, L_{ij,k}) = (0,0)$, $(2,2)$ and $(4,4)$ with alpha-alpha orbital angular momenta L_{ij} and the spectator alpha-particle's orbital angular momenta $L_{ij,k}$ for the center of mass between two alphas. Fig.7 illustrated the contoure plots (p-p' plane) of the exact off-shell amplitude (solid lines) [fm^2 unit] for the ground 0_1^+ state of ^{12}C nucleus, and of the rank-1 separable amplitude for $k_1 = 2.5\ fm^{-1}$ (dotted lines) at $z = -5$ MeV. This amplitude is known as [3+1] sub-amplitude in the 4-alpha system. Furthermore, we have [2+2] sub-amplitude for the 4-body Faddeev-Yakubovsky equation.[10] The amplitude denotes $0_1^+(^8Be) + 0_1^+(^8Be)$ system. In Fig. 8, [2+2] amplitude is illustrated in total angular momentum $J^\pi = 0_1^+$ with the coupling: $(L_{ij} + L_{kl}) + L_{ij,kl} = J$, in which channels are defined by the set $(L_{ij}, L_{kl}, L_{ij,kl})$. Here, we adopted $(0,0,0)$, $(0,2,2)$ and $(0,4,4)$. All the inscriptions of Fig. 8 are the same as those of Fig. 7, in which the parameter $k_1 = 0.0 fm^{-1}$ is taken for $z = -5$ MeV. Both results show rather good fitting with the exact amplitudes, though 2 or 3 ranks may be still necessary for quantitative agreement.

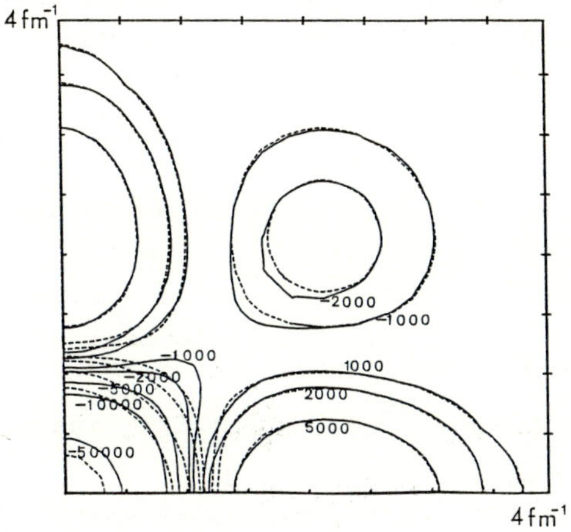

Fig. 7

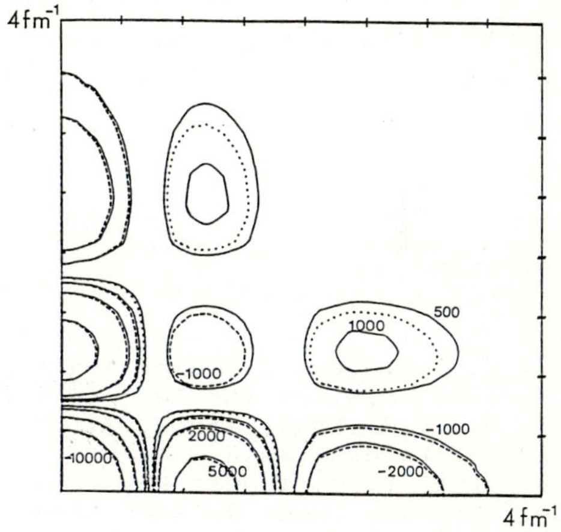

Fig. 8

VI. Conclusion and Discussions

The important features of our separable expansion formalism have been described by using several aspects such as the unitarity and the analyticity of the separable amplitudes. Then, our on- and half on-(off) shell amplitudes are mathematically exact. It should be stressed that all the separable expansion formalism which cannot represent on- and half-on (off) shell amplitudes exactly, cannot satisfy the off-shell unitarity relation without the non-separable term . Otherwise they are models for the realistic potentials. Even if one uses a separable expansion model, one has to take of the convergence of the norm for the non-separable terms with the Green's function by means of the Osborn's theorem.[8] Finally, we would like to show the relation between the Kowalski-Noyes equation and our formalism. The half-off-shell function defined by Noyes is $f(p,k) = t(p,k;z)/t(k,k;z)$ which is also given with our form factors by using eq.(2.22): $f(p,k) = \varphi(p,k_1)/v(k_1,k_1) = \varphi(p,k)/v(k,k)$. Substituting this into eq.(2.25), it satisfies the equation (6) of H. P. Noyes.[5] Let us multiply $G_0(p;z)$ on the both sides of eq.(2.9), and define $\Lambda(p,p') = v^{(2)}(p,p')G_0(p';z)$, and $R(p,p') = t^{(2)}(p,p')G_0(p';z)$, then they satisfy the integral equation for the non-separable terms given by K. L. Kowalski.[5] Therefore, the Kowalski-Noyes equation is equivalent with our rank-1 formalism for the positive energy case.

However, their equation is not correct for the negative energy case by means of the analyticity which was remarked in Ref. C.

Our formalism has several different features by type A, type B, the diagonal forms, and so on, however, these equations hold conspicuously good symmetric property. This fact will be an important advantage for the programing of the computer codes.

References

1. J. S. Levinger, Springer Tracts in Modern Physics 71- Nuclear Physics; The Two and Three-Body Problem.
2. S. Oryu, M. Araki and S. Satoh, Prog. Theor. Phys. Suppl. 61, 199 (1977): refered as A.
3. S. Oryu, Prog. Theor. Phys. 52, 550 (1974): referd as B.
4. H. Bateman, Proc. R. Soc. London A100,441 (1921); Messenger Math. 37, 179 (1908).
   V. F. Kharchenko, S. A. Strozhenko, and V. E. Kuzmichev, Nucl. Phys. A188, 609 (1972).
   B. Akhmadkhodzaev, V. B. Belyaev, J. Wrzecionko and A. L. Zubarev, Preprint JINR, E4-5763, Dubna, 1971.
5. K. L. Kowalski, Phys. Rev. Letters, 15, 798 (1965).
   H. P. Noyes, Phys. Rev. Letters, 15, 538 (1965).
6. S. Oryu, Phys. Rev. C27, 2500 (1983); refered as C.
7. S. Oryu, Prog. Theor. Phys. Suppl. 61, 180 (1977); referd as D.
8. T. A. Osborn, J. Math. Phys. 14, 373 (1973).
9. Walter Gloeckle, The Quantum Mechanical Few-Body Problem (Springer-Verlag 1983).
10. O. A. Yakubovsky, Sov. J. Nucl. Phys. 5, 937 (1967).

THREE-BODY PROBLEM WITH SEPARABLE-EXPANSION TECHNIQUES AND USE OF MODERN NUCLEON-NUCLEON FORCES

W. Plessas*)
Institute for Nuclear Physics
KFA Jülich
D-5170 Jülich, FRG

ABSTRACT

An outline is given of recent attempts towards introducing meson-theoretical nucleon-nucleon interactions into Faddeev calculations of the three-nucleon system. A method of separable expansion is explained that allows to reproduce the on-shell and off-shell behaviour of any given interaction, and its application is demonstrated for the Paris and Bonn nucleon-nucleon potentials. Corresponding results for nucleon-deuteron elastic scattering below E_N = 20 MeV are reported and discussed with special emphasis on off-shell effects of the nucleon-nucleon interaction. While our investigations already allow to discard unrealistic properties, as inherent in phenomenological separable potentials used before, no evidence is as yet found against the meson-theoretical forces considered here.

1. INTRODUCTION

Among quantum-mechanical few-body problems the three-body problem plays a decisive role. Beyond the two-body problem it provides for most particle systems the only one that is amenable to an accurate solution for both bound and scattering states under the explicit treatment of all degrees of freedom. By accurate we mean that the solution is either exact or is obtained via converged approximative methods; in the latter case it must be required that the method is mathematically convergent as such and that the approximation can be

---

*) On leave from Institute for Theoretical Physics, University of Graz, Universitätsplatz 5, A-8010 Graz, Austria

or has been driven so far as to yield a converged result in the sense that an ever diminishing remainder can be neglected at a certain step for the problem in question.

Often, especially in earlier investigations, model three-particle systems were considered, what was very useful in studying specific aspects, such as the structure and behaviour of the dynamical equations, the performance of particular techniques etc. However, it has always been an extremely difficult task to derive accurate solutions for three-particle systems under physically relevant circumstances (e.g., realistic interactions). Only over the last few years this has become possible for certain three-particle systems.

As a prominent example we have the three-nucleon (3-N) system, which has been studied intensively for now more than two decades. When the quantum theory of the three-body problem had become available, there started many attempts towards solving the 3-N system. Soon a wealth of insight was gained, which allowed a qualitative understanding of 3-N phenomena; this was achieved by the early seventies. At that time many simplifying assumptions had to be made in the calculations mainly with respect to the interaction between the nucleons. The N-N forces employed were essentially phenomenological, reproducing more or less the 2-N data, but without being derived from a theory of N-N dynamics. Consequently the conclusions that could be drawn from such kind of calculations were rather limited, sometimes even misleading. In particular, the most prominent question, namely, whether or not the then accepted theoretical models of the N-N interaction were able to explain 3-N phenomena could not be answered quantitatively.

As indicated above, it could be achieved only recently that our actual knowledge of the N-N interaction, as derived from basic dynamical concepts, can be subject to a check in the 3-N system not only at the bound-state but also in the scattering regime. This is of particular importance for estimating the domain of validity for concepts of the strong interaction between nucleons, such as meson exchange, QCD models etc. Indeed, it already turned out in several calculations of that type that the 3-N system provides some stringent tests for the (form of the) N-N potential.

To a large extent this achievement is due to the use of separable-expansion techniques in the solution of the two- and three-body equations. It is the purpose of the present paper to elaborate on this matter stressing the physical aspects in the modern treatment of the 3-N system. Of course, technical details are addressed too, but contrary to the oral presentation of the lecture at this school the physical and mathematical background as well as the formulation of two- and three-body systems with separable operators (kernels, potentials, ...) are left out here. These items are collected in a lecture I gave a year ago at a similar occasion[1]. The interested reader might find it worthwhile to study that article first and then to go ahead with the exposition below, which - somehow as a continuation - demonstrates the application of separable-expansion techniques to an up-to-date problem, namely, N-d scattering.

2. THE PROBLEM

A prominent reason for studying $A>2$ nuclear systems has always been to learn more about the fundamental N-N interaction. Evidence from the N-N system alone does not suffice to determine in general the interaction between two nucleons. N-N data constrain only the on-shell behaviour of the 2-N transition operator, i.e. the values of its matrix elements between improper states (eigenstates of the free relative-motion Hamiltonian)

$$\langle \vec{p}\,'|T(E_k \pm i0)|\vec{p}\rangle = T(\vec{p}\,',\vec{p};E_k \pm i0) \tag{2.1}$$

on the sphere (μ being the reduced mass):

$$\frac{p'^2}{2\mu} = \frac{p^2}{2\mu} = \frac{k^2}{2\mu} = E_k \; . \tag{2.2}$$

For the N-N system the on-shell elements of (2.1) are known - also with respect to their angular dependence - at a large number of energies, though not at all energies. Consequently the on-shell behaviour of the N-N interaction is comparatively rather well established; this can be seen through modern phase-shift analyses, whose results are quite reliable[2-4].

Still it happens that N-N potentials with fairly different characteristics (functional dependence on relative distance and momentum as well as spin, isospin, angular momentum, ...) can reproduce the whole N-N phenomenology to the same extent. Thus it has been hoped from the very beginning of few-nucleon physics that more complex systems would allow to discriminate between such differences in the 2-N potentials. Indeed, any system involving N-N as a subsystem can in principle provide information on the N-N transition operator off the energy shell;

the condition (2.2) needs no longer to be fulfilled. There are several prominent candidates of composite systems or reactions that might serve well for this purpose, like ^3H, ^3He and also N-N bremsstrahlung, γ-, e-, μ-, π-, N-induced reactions on the deuteron. Of course, there was also some controversy about whether or not constraints on the off-shell behaviour could be deduced from these phenomena[5].

With respect to the trinucleon most investigations were carried out with the N-N interaction in separable form. In the early seventies these forces were usually constructed from some ansatz with open parameters, whose values were determined by fitting N-N data[6]. Consequently the off-shell behaviour of the corresponding transition operators was completely arbitrary and often at variance with properties peculiar to meson-theoretical interaction models, which were already coming up at that time[7]. Since these separable forces were applicable in 3-N calculations, even in rather complex form, essentially all trinucleon observables could be calculated. As a result a qualitative understanding was achieved[8]. With regard to N-d scattering very elaborate calculations were done with the use of separable interactions leading to results qualitatively in agreement with experimental observables[9]. However, a quantitative reproduction of 3-N data was not reached. Also the question, whether fundamental dynamical models for the N-N interaction were capable of explaining the behaviour of the 3-N system, could not be answered.

At the same time evidence emerged from other sources that the off-shell properties of conventional separable forces might by no means be adequate. This was first observed through the deuteron wave function, which subsequently turned out to be an important quantity for gathering new insight into the off-shell behaviour of the N-N potential. In careful studies of the electromagnetic form factors of the deuteron[10] and the trinucleon[11] it was found that a deuteron wave function is preferred of the kind as suggested by meson-exchange models of the N-N interaction; that is with a certain dip at short distances in configuration space. More clearly this typical behaviour is seen in momentum space as in fig. 1. There the deuteron S- and D-state wave functions are demonstrated for a separable model constructed by Afnan, Clement, and Serduke[12] (ACS) and the Paris potential[13].

In this context the ACS 5.5 interaction serves as an example of purely phenomenological potentials fitted to N-N data[6,7] and the Paris potential can be considered as characteristic of meson-exchange N-N

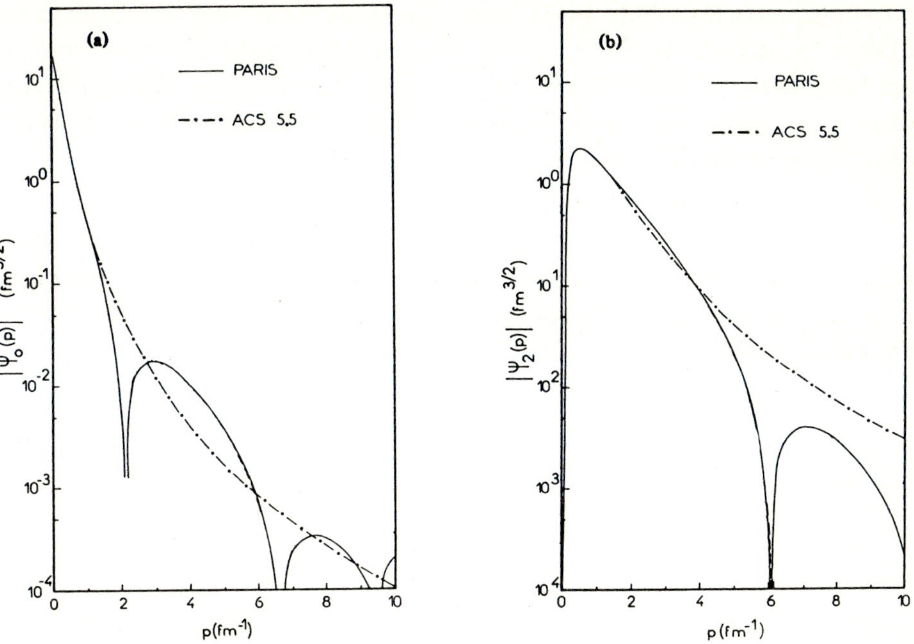

Fig. 1. a) S-state and b) D-state deuteron wave functions in momentum space for the Paris and ACS 5.5 potentials.

forces: while the former has momentum-space wave functions for both the S and D states, which never change sign, the latter shows an oscillatory behaviour with the first zeros at $p \approx 2$ fm^{-1} (S state) and $p \approx 6$ fm^{-1} (D state). Still both models reproduce the (static) properties of the deuteron, such as E_d, Q_d, p_D, and η_d, as well as the low-energy phase shifts in a similar way[6,12,13]. However, off-shell properties like in the ACS potential cause considerable effects in observables of e-d scattering, most strikingly in the tensor polarization p_{zz} (or T_{20}). A comparison of the corresponding results with now available experimental data allows to discard such an off-shell behaviour like in the ACS model[14]. On the other hand a Paris-type deuteron wave function leads to good agreement with these data and also with charge and magnetic form-factor measurements[15].

The same observation was made independently in studies of the π-d and N-d systems. For π-d scattering a strong correlation was found between the above-described details of the deuteron wave function, especially the zero in the momentum-space S-state wave function, and elastic scattering observables such as the differential cross section and again the tensor polarization[16]. Detailed investigations of the

N-d system finally also succeeded in locating several places with off-shell sensitivity[17-20]. All indications pointed into the same direction, namely, that a more realistic off-shell behaviour such as provided by meson-exchange models must be employed in studies of nuclear three-body systems.

In the following the construction of separable N-N potentials was refined so as to make allowance for realistic off-shell properties at least with regard to the most influential 2-N quantities. E.g., the Graz group constructed a separable model[21,22] (GRAZ-II), which provided not only a good fit to the whole N-N data but also had a deuteron wave function resembling the one of the Paris potential; also Doleschall made attempts in this direction[18,19].

The GRAZ-II interaction to some extent remedied the short-comings of earlier separable potentials, which paid no attention to the off-shell behaviour. It yielded e-d observables now in agreement with all available experimental data[23] and led to various improvements in the N-d system[24-26]. Nevertheless several problems remained. For instance, the backward n-d differential cross section, which is known to be effected by the details of the deuteron wave function, could still not be reproduced[26]. Consequently there was demand for further improvement on the N-N interaction to be employed in the 3-N system.

In the meantime considerable progress had been made in investigations of the trinucleon bound states with modern meson-theoretical N-N potentials. E.g., the Paris potential was used to calculate trinucleon binding energies and electromagnetic form factors[27,28], and also other interactions with realistic on-shell and off-shell properties were tested[29]. These studies revealed that existing experimental data on the trinucleon bound states cannot be explained by 2-N forces alone, even if the most elaborate models are employed[30]. Considerable effects are left possibly for 3-N forces. At present much effort is devoted to clarifying these problems[31].

In view of this development and the reasons discussed before, the application of meson-theoretical forces to the 3-N scattering problem was highly desired. Thereby it should be made clear, to which extent 3-N scattering observables can be explained by modern realistic models of the N-N interaction. Over the last few years investigations of this type were facilitated by the work of the Graz group.

3. SEPARABLE REPRESENTATION OF MESON-THEORETICAL N-N POTENTIALS

3.1 EST method

The task of introducing all essential features (on-shell and off-shell behaviour) of a certain N-N potential into 3-N calculations could be performed by separable expansions. It turned out that among the many separable-expansion procedures[1] a method proposed by Ernst, Shakin, and Thaler[32] (EST) was the most direct and effective one for this purpose. Of course, in order to avoid unreasonable results several cautions must be exerted in applying this separable-approximation scheme[33].

The EST method uses the wave functions of the given interaction V for constructing the separable form factors. This choice is very appropriate, because it allows the separable interaction $\tilde{V}$ to reproduce this wave function pointwise as a function of energy.

Let $|\psi_{E_1\ell}\rangle$ be the eigenstate of the problem V at some fixed energy E_1 in the partial wave ℓ, then the EST method leads to the rank-1 approximation

$$\tilde{V}_\ell = V_\ell|\psi_{E_1\ell}\rangle \lambda_\ell \langle\psi_{E_1\ell}| V_\ell \, . \tag{3.1a}$$

If λ_ℓ is determined from the condition

$$\lambda_\ell^{-1} = \langle\psi_{E_1\ell}|V_\ell|\psi_{E_1\ell}\rangle, \tag{3.2a}$$

it is guaranteed that the eigenstates for both problems V and $\tilde{V}$ coincide at the energy E_1, because

$$V_\ell|\psi_{E_1\ell}\rangle = \tilde{V}_\ell|\psi_{E_1\ell}\rangle. \tag{3.3}$$

Consequently the half-off-shell elements of the corresponding transition operators are also identical at E_1. The boundary conditions of the states $|\psi\rangle$ can be chosen freely. In actual calculations the choice of standing waves is preferable, because one can then deal with the real reaction matrix elements, which are more convenient in computer codes.

If the expansion (3.1) is extended to rank N, this amounts to an interpolation in energy for the representation of the eigenstates $|\psi_E\rangle$ over the whole (discrete and continuous) spectrum in a particular state ℓ. For

$$\tilde{V} = \sum_{i,j=1}^{N} V|\psi_{E_i}>\lambda_{ij}<\psi_{E_j}|V \tag{3.1b}$$

with

$$\sum_{j=1}^{N} \lambda_{ij}<\psi_{E_j}|V|\psi_{E_i}> = 0 \tag{3.2b}$$

the equation (3.3) holds at the N energies $E_1,\ldots,E_N$. Notice that the potential strengths (coupling parameters) λ_{ij} of the N separable terms are energy independent.

It is now a matter of the physical requirements of the system under consideration, which energies to select in order to keep the rank of the separable interaction $\tilde{V}$ as small as possible. Like in other separable-approximation schemes it is worthwhile to take those energies, where the transition operator becomes "large" or has poles (bound states or resonances). Thus the EST method comprises the possibilities of the unitary-pole expansion[34)] (UPE) as well as the so-called Gamow separable approximation[35)] (GSA); for the latter the energy must, of course, be allowed to become complex.

The EST method can easily be generalized to a multichannel problem. For the two-channel case of the N-N system with coupled partial waves the formalism is given in detail in the literature[32,33)], and it need not be repeated here. We just remark that for the application of the EST method only the wave functions or half-off-shell elements of the transition (reaction) matrix of the original interaction are required. According to (3.1) the form factors of the separable potential

$$\tilde{V}(p',p) = \sum_{i,j=1}^{N} g_i(p')\lambda_{ij}g_j(p) \tag{3.4}$$

are given by

$$g_i(p) = <\phi_E|V|\psi_{E_i}> , \quad E = \frac{p^2}{2\mu} . \tag{3.5}$$

Likewise the solution of the Lippmann-Schwinger equation

$$\tilde{T}(z) = \tilde{V} + \tilde{V}G_0(z)\tilde{T}(z) , \tag{3.6}$$

with $G_0(z) = (z-H_0)^{-1}$ being the free resolvent, can also be expressed by the half-off-shell T matrix elements corresponding to the original interaction V:

$$\mathcal{T}(p',p;z) = \sum_{i,j=1}^{N} g_i(p')D_{ij}(z)g_j(p) \tag{3.7a}$$

with

$$(D^{-1}(z))_{ij} = \langle\psi_{E_i}|[V-VG_o(z)V]|\psi_{E_j}\rangle \; . \tag{3.7b}$$

From eqs. (3.2b) and (3.7) it is evident that in the course of applying the EST method two N-dimensional matrices have to be inverted. Therefore one must avoid them to become singular (or "almost singular" in practical computer calculations). This is also to be considered in the selection of the interpolation energies E_i. In fact, former applications of the EST method sometimes did not pay proper attention to these aspects and generated unreasonable results[33]. We remark, however, that this is not an obstacle peculiar to the EST method; similar matrix inversions have to be performed in other separable-approximation schemes too[1].

3.2 Paris potential

As a first case for V to be represented in separable form we considered the Paris potential[13]. For the various partial waves we developed expansions of different ranks N, called PESTN [36,37]. It was found that already approximations of relatively low rank can reproduce essential on-shell and off-shell features of that interaction. For instance, the PEST1 potential in $^3S_1$-$^3D_1$ allows for an accurate separable parametrization of the Paris potential at the deuteron bound-state pole[36]. It may thus serve as a convenient form for use in particular problems like, e.g., elastic e-d scattering[14].

In ref. 36 an analytical representation of the numerical EST form factors (3.5) is provided. This is sometimes needed, like in the solution of few-nucleon (integral) equations with the contour-deformation technique. Above all the rational form factors of ref. 36 further allow for a rigorous treatment of the p-p system. Using the formulae derived before[22] by means of Coulombian asymptotic states[38], the Coulomb distortion of the hadronic interaction can be treated exactly to all orders of e^2 (fine-structure constant) for all PESTN potentials. This will certainly be important for future 3-N calculations, once long-range forces can be treated reliably in the formalism and the computer codes solving three-body integral equations.

For calculations of the 3-N problem the analytical form factors of ref. 36 are as good as the numerical ones. Of course, they show deviations, but only at off-shell momenta large enough to play no noticeable role in 3-N calculations. In Fig. 2 the half-off-shell behaviour is exemplified for the $^1S_0$ wave by means of the Noyes-Kowalski function

$$f(p,k) = \frac{T(p,k;E+i0)}{T(k,k;E+i0)} , \quad E = \frac{k^2}{2\mu} . \tag{3.8}$$

It is evident that the PEST1 interactions shown reproduce the Paris potential in the physically relevant domain, i.e. beyond the second zero of f(p,k) up to $p \approx 7\ fm^{-1}$, very accurately. Of course, the point E=0 was taken as the interpolation energy, but these rank-1 approximations remain acceptable up to $E_{lab} \approx 50$ MeV [36]. Towards higher energies another rank is necessary. The important ingredient, namely, that the $^1S_0$ half-off-shell function (and anlogously the momentum-space deuteron wave functions in $^3S_1$-$^3D_1$) change sign at $p \approx 2\ fm^{-1}$, is perfectly present in the PEST potentials. From that several improvements can be expected in applications to such problems as N-d and π-d scattering. Fig. 2 contains also a comparison between two different analytic representations of the rank-1 PEST approximation. PEST1 is the parametrization with 4 rational terms as in ref. 36 and PEST1-6 is the more ambitious one with 6 rational terms as in

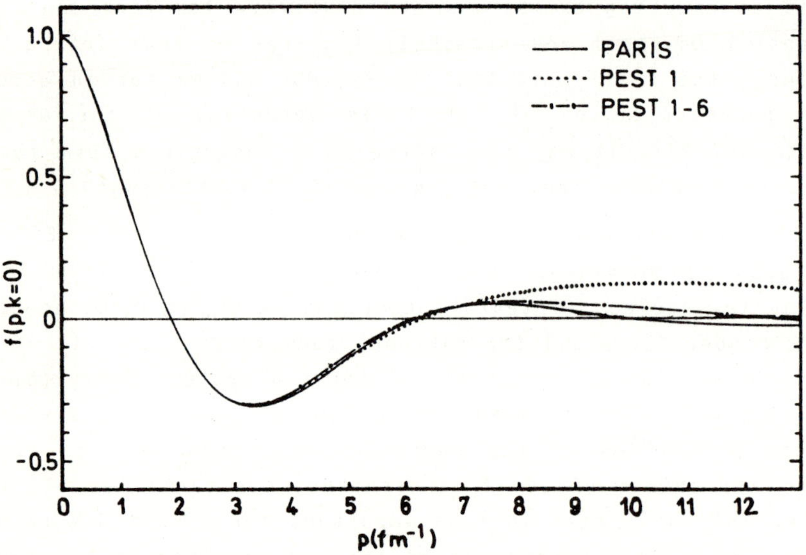

Fig. 2. Half-off-shell functions at E=0 for $^1S_0$(p-p) purely nuclear. See the text above for explanations.

ref. 39. It is seen that both parametrizations are practically the same until $p \approx 7.5\ fm^{-1}$, while they differ further out. However, as test calculations in the trinucleon have shown, differences of that kind do not play any noticeable role; therefore also parametrizations of the PEST1 type are sufficient.

The properties of the PEST potentials are discussed in detail in ref. 36 with respect to their on-shell, half-off-shell as well as fully off-shell behaviour in all relevant N-N partial waves. In general their performance is very satisfactory with the possible exception of the mixing parameter ε_1 at low energies for the PEST potentials up to rank 4. The situation is improved considerably by recent PEST approximations of rank 6 and 8 using two further interpolation points at negative energies[37)] (cf. fig. 3). In these cases also another analytic representation of the numerical form factors was employed using Gegenbauer polynomials[40)]; thus we denote them by PESTN-G. We will address these expansions again in the following chapter when discussing N-d scattering results.

Finally in this section we add a notice for prospective users of the PEST potentials. In the original parametrizations PEST3 and PEST2 for $^1S_0$ and $^3P_0$, respectively, we overlooked the occurrence of unphysical bound states at large negative energies ($E \lesssim -600$ MeV). This fault is removed in the second paper of ref. 36 by selecting somewhat different interpolation energies.

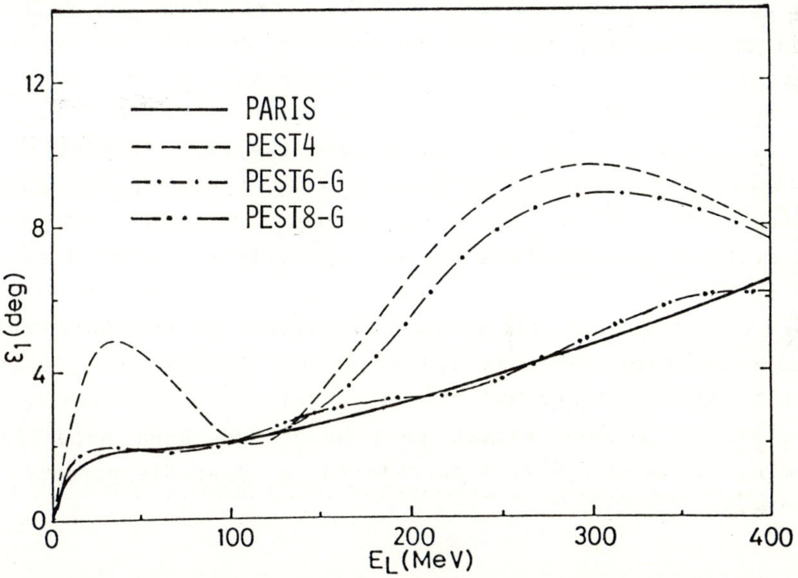

Fig. 3. Mixing parameter ε_1 for higher-rank PEST expansions of the Paris potential as a function of the laboratory kinetic energy.

3.3 Bonn potential

Along the lines discussed in the foregoing sections we recently also constructed a separable representation of the Bonn potential[41]. Like the model of the Paris group[13] the Bonn potential is also based on meson-exchange dynamics. But instead of $(\pi+2\pi+\omega)$ exchange[42] it relies on the one-boson exchange picture with the $\pi,\eta,\sigma,\delta,\rho$, and ω mesons[43], what leads to some differences in the off-shell behaviour. This is visible in the deuteron wave functions beyond $p \approx 3\ fm^{-1}$ in momentum space and correspondingly in the e-d scattering observables at increased momentum transfers[15]. It is thus interesting to compare the two interactions also in N-d scattering. In ref. 44 only the most important N-N partial waves are considered, namely, $^1S_0$ and $^3S_1$-$^3D_1$. The separable approximations for the case of the Bonn potential (i.e. the version with the internal code 080 presented in ref. 41) are called BESTN and their properties are discussed in detail in ref. 44.

A few clarifying remarks should be added here in view of the following comparison of the N-d results for Bonn and Paris. Firstly, in the $^1S_0$ state the Bonn potential is designed for the n-p system, while the Paris potential is for p-p. The latter one treats Coulomb effects and includes Coulomb distortion of the hadronic p-p interaction. Consequently its low-energy behaviour (effective-range parameters) is different from the one of the Bonn potential. In order to make a comparison reasonable we provided for both cases, Bonn and Paris, separable parametrizations reproducing the p-p as well as the n-p low-energy data[36,44]. Once the PEST(p-p) and BEST(n-p) separable parametrizations were obtained, this could be achieved by shifting the Bonn potential to the p-p on-shell point and the Paris potential to the n-p on-shell point through adjusting only the coupling parameters (potential strengths) λ_{ij}; thereby either off-shell behaviour was conserved. Uncertainties from the conceptually different treatment of $^1S_0$ can thus be avoided and the two cases can be applied on the same footing. Secondly, the low-energy behaviour of the Bonn potential under consideration (080) is deficient in $^3S_1$-$^3D_1$. It differs both from Paris and experiment and this is likely to be influential on N-d observables. In a more recent version of the Bonn potential this shortcoming is remedied[45]. A corresponding separable parametrization is under way.

4. N-d SCATTERING

4.1 Off-shell effects

The separable representations of the Bonn and Paris potentials immediately turned out to be very useful in studies of 3-N scattering. Already the relatively simple rank-1 approximations applied in the beginning led to valuable results. Of course, the first aim was to search for evidences of off-shell sensitivity in N-d observables. For this purpose calculations were performed with the PEST interaction and an on-shell-equivalent Yamaguchi potential Y(E) [39,46]. These two models clearly have very distinct off-shell properties (see fig. 4). A comparison of their results for specific N-d spin observables revealed their sensitivities on the N-N off-shell behaviour. Most remarkable in this respect are the spin-correlation parameter $C_{y,y}$ (vector-polarized deuteron on polarized nucleon) and the nucleon-to-nucleon spin transfers $K_y^{y'}$, $K_x^{x'}$, $K_z^{x'}$. From the examples in figs. 5 and 6 one can see that the differences between PEST1-6 and Y(E) are of considerable size and that the hitherto available data favour the realistic off-shell behaviour of PEST1-6 and thus of the Paris potential.

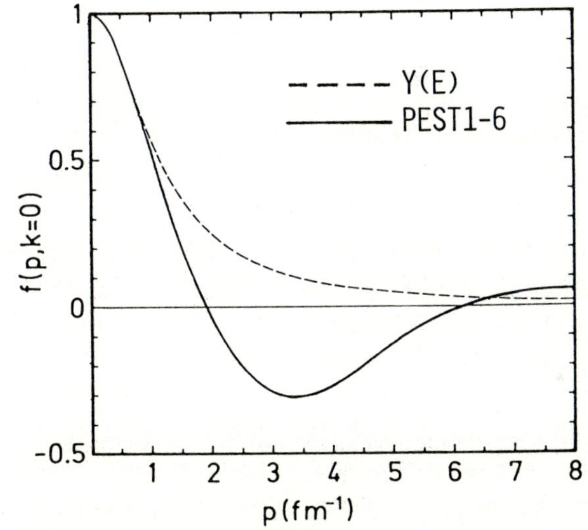

Fig. 4. Half-off-shell functions of the PEST1-6 potential and a phase-shift-equivalent Yamaguchi potential (ref. 39) in $^1S_0$.

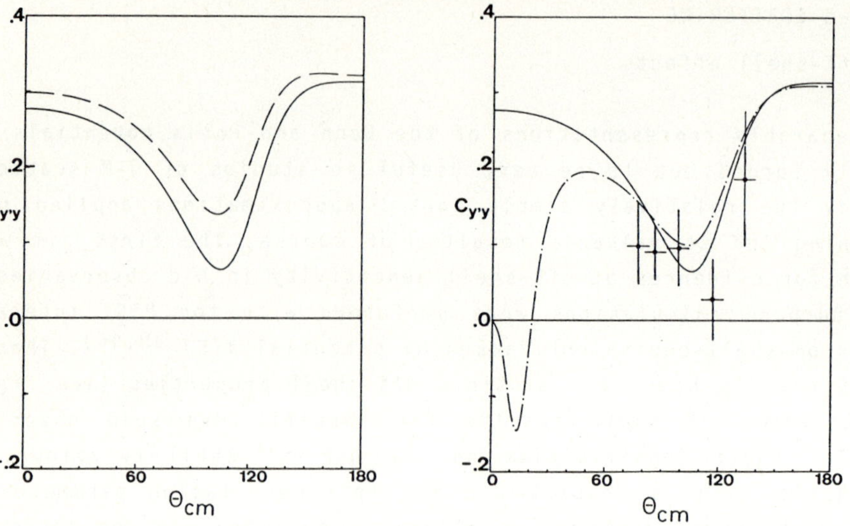

Fig. 5. Predictions for the N-d spin-correlation parameter $C_{y,y}$ at E_d = 10 MeV with PEST1-6 (solid line) and the phase-shift equivalent Y(E) (dashed line) in N-N S-waves ($^1S_0$+$^3S_1$) only. The graph at left shows the off-shell differences for the purely nuclear result (n-d scattering). At right the PEST1-6 result was Coulomb-corrected (ref. 46) in order to obtain the prediction for p-d scattering (dashed-dotted line). This is then compared to experiment (ref. 47). Coulomb corrections in the Y(E) calculation would amount to a similar shifting of the n-d result and thus lead to a prediction for p-d at variance with experimental data (cf. also fig. 6).

The reason, why these polarization observables are rather sensitive to the off-shell N-N interaction, lies in the fact that due to their spin structure they are mainly governed by N-N S-waves[46,49]. Because large (repulsive) centrifugal barriers are not so effective they allow to probe the N-N potential down to relatively small distances without much hindrance by peripheral forces. Indeed, the detailed studies of refs. 39 and 46 made clear that differences such as appearing in figs. 5 and 6 can be related to the features of the N-N interaction between $0.8 \lesssim r \lesssim 1.2$ fm. There the effect of 2π exchange (or alternatively ρ exchange) comes into play and changes the character of the N-N potential from the long-range π exchange. For instance, the isotriplet, spin-singlet central potential becomes repulsive in this region, what is in fact reflected by the zero in the half-off-shell function at $p \approx 2$ fm^{-1}. This behaviour is typical for most meson-exchange models and very well in agreement with the requirements in nuclear few-body systems (e-d, π-d, N-d, ...). On the other hand this range is also, where one may expect the interface to

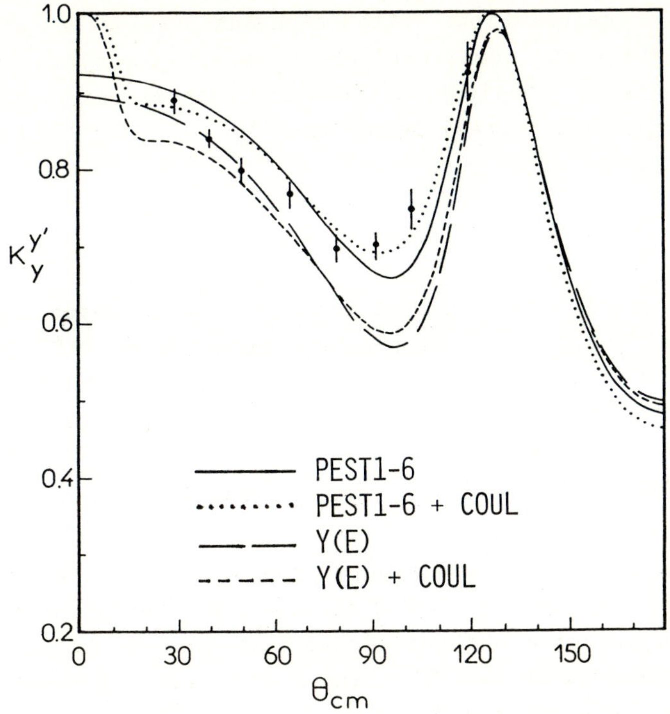

Fig. 6. $^2H(\vec{N},\vec{N})^2H$ spin-transfer coefficient $K_y^{y'}$ at E_N = 10 MeV for the same calculations as in fig. 5. The Coulomb-corrected Y(E) result (p-d) scattering falls clearly outside the experimental data of ref. 48.

the quark description of the N-N interaction. In this respect more insight is still needed and this makes the N-d spin observables addressed above so interesting - together with similar quantities from other few-body scattering systems, of course.

Searches for off-shell effects have recently been extended to N-N P waves[50)]. First calculations have shown that differences in the off-shell behaviour in P waves between Paris- and Yamaguchi-type are of considerable influence on certain N-d polarization observables, like the nucleon analyzing power A_y, the deuteron vector polarization T_{11} and, to a lesser extent, the tensor polarizations T_{20} and T_{21}; of course, these are the spin observables that receive dominant resp. large effects from N-N P waves anyway.

For the future it is desirable to study the N-d polarizations addressed above over a wider energy range both in theory and experiment. For the latter, measurements of n-d scattering are encouraged, because this would then allow to circumvent the yet not reliably established p-d calculations with rigorous inclusion of Coulomb effects. To avoid these complications is important, since it is conceivable that, e.g., Coulomb distortion changes or conceals the influence of the N-N off-shell behaviour.

4.2 Comparison of Paris and Bonn potentials

The next step in our study of N-d scattering with meson-theoretical N-N interactions was to include the more complicated PEST resp. BEST approximations of higher rank into 3-N Faddeev calculations. This could be performed by the code developed by Koike and Taniguchi (see, e.g., ref. 26).

First it was interesting to see, how differences between the Paris and Bonn potentials (this time, of course, for on-shell and off-shell quantities; cf. the tables and figures concerning $^1S_0$ and $^3S_1$-$^3D_1$ in refs. 15, 36 and 44) come up in N-d observables. Let us here consider as an example once again the nucleon-to-nucleon spin transfers. Fig. 7 shows the results with the EST rank-3 and rank-4 parametrizations of the Paris and Bonn potentials in $^1S_0$ and $^3S_1$-$^3D_1$, respectively. In this comparison the effect of N-N P and D waves as well as Coulomb effects are left out, since both of them are anyway small at this energy[46,49]. The results of PEST and BEST in fig. 7 are very similar with the only exception of the angular range around $\theta_{cm} \approx 90^0$. It is at this stage not possible to tell, whether it is an on-shell or off-shell effect. We only remark that the earlier influences of off-shell properties (with on-shell-equivalent interactions) appeared over the whole range of angles up to $\theta_{cm} \approx 130^0$ (see fig. 6.).

For the comparison of n-d differential cross sections we must add also the higher N-N partial wave, since their effect is not negligible even at E_n = 10 MeV [26]. In fig. 8 we demonstrate the result for the cases just considered in fig. 7, but with the (uncoupled) P and D waves supplemented by phenomenological separable potentials of Doleschall[51]. Also here differences among the theoretical predictions appear at all energies considered. They show up at forward and backward angles as well as in the cross section minima. The reasons are manifold: as already remarked the Paris and Bonn deuteron wave functions are different, both potentials have different A_S (asymptotic S-state normalization), p_D (D-state probability), and Q_d (quadrupole moment)[15]; also the on-shell behaviour in $^3S_1$-$^3D_1$ is distinct, especially in the low-energy limit[44]. Nevertheless both models may be

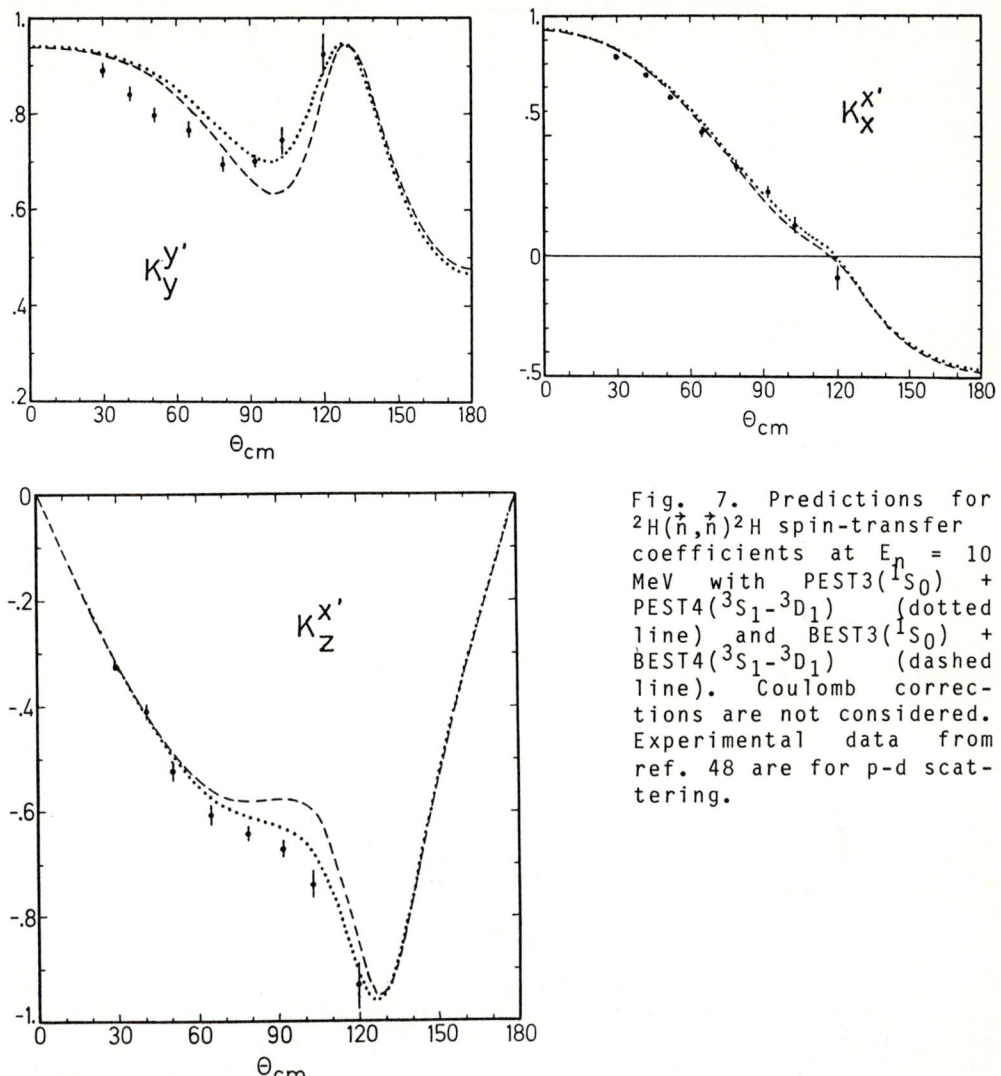

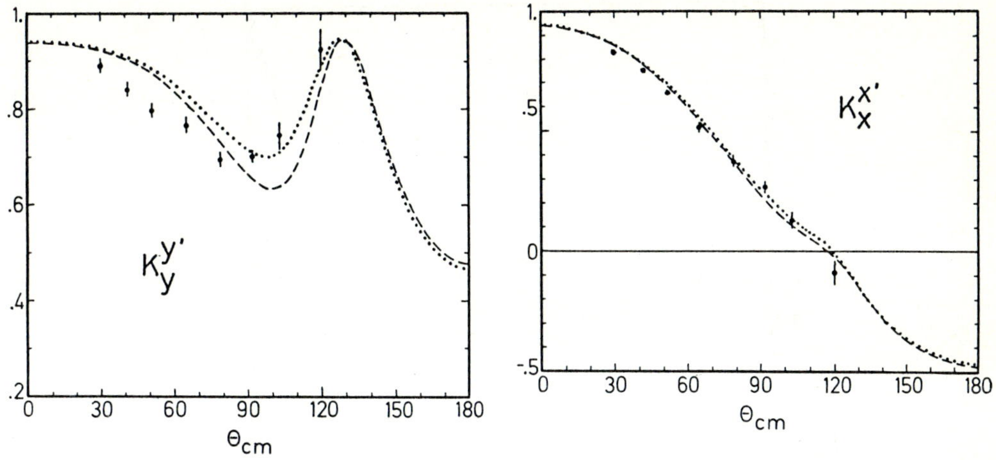

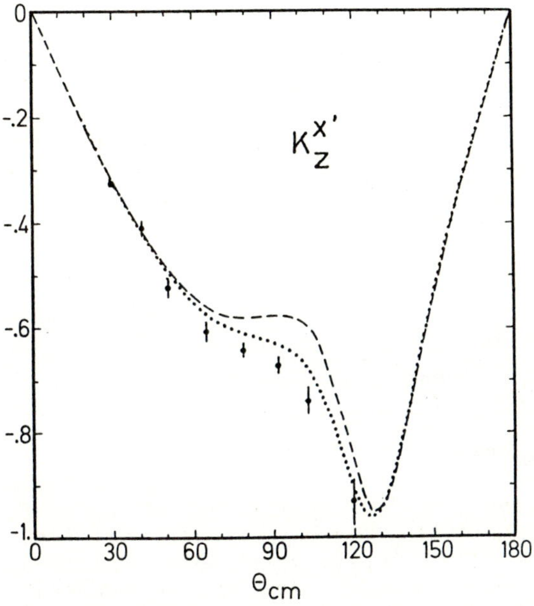

Fig. 7. Predictions for $^2H(\vec{n},\vec{n})^2H$ spin-transfer coefficients at E_n = 10 MeV with PEST3($^1S_0$) + PEST4($^3S_1$-$^3D_1$) (dotted line) and BEST3($^1S_0$) + BEST4($^3S_1$-$^3D_1$) (dashed line). Coulomb corrections are not considered. Experimental data from ref. 48 are for p-d scattering.

considered to be in fair agreement with experimental N-d data; at least they lead to remarkable improvements over earlier used phenomenological separable forces, which had the notorious difficulty to reproduce the height of the backward cross section. The increase for meson-theoretical models is likely to be caused by the different deuteron wave function and a more reasonable A_S. With regard to the Karlsruhe data[52)] (open circles in fig. 8) we remark that they might not be reliable in the backward domain. From the comparison to the Uppsala data[53)] as well as to more recent data from Karlsruhe[54)] (see also fig. 9 in the following section) they appear to be too high.

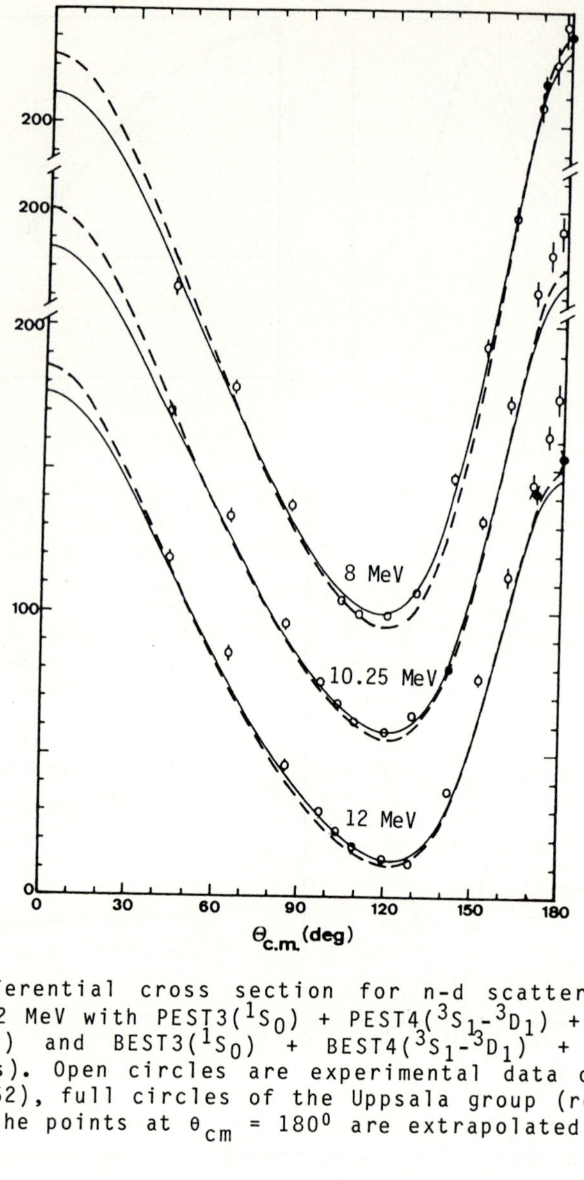

Fig. 8. Differential cross section for n-d scattering at E_n = 8, 10.25, and 12 MeV with PEST3($^1S_0$) + PEST4($^3S_1$-$^3D_1$) + Doleschall(P+D) (solid lines) and BEST3($^1S_0$) + BEST4($^3S_1$-$^3D_1$) + Doleschall(P+D) (dashed lines). Open circles are experimental data of the Karlsruhe group (ref. 52), full circles of the Uppsala group (ref. 53); for the latter case the points at θ_{cm} = 180^0 are extrapolated values.

The calculations reported up till now are still deficient in several respects. First of all the higher N-N partial waves are not supplied from the meson-theoretical models, furthermore the rank-4 approximation might not be sufficient in $^3S_1$-$^3D_1$ (cf. the mixing parameter ε_1 in fig. 3) and the same question can be asked with regard to $^1S_0$. However, the results obtained so far point into the right direction and we may go ahead with even more ambitious calculations.

4.3 Converged Paris results

As stated in the Introduction we can be sure that our separable-expansion procedure produces the final result only, if convergence at the 2-N and 3-N level is established. With regard to the Paris potential this matter has been investigated recently[37]. In particular it was shown that the $^1S_0$ and $^3S_1$-$^3D_1$ states are adequately treated by rank-3 and rank-6 PEST approximations, respectively. In this case the triton binding energy is reproduced with the value E_T = -7.31 MeV in a 5-channel calculation, what is in good agreement with the corresponding result obtained by a different method[27]. If in addition all N-N partial waves up to j=3 are included via the higher-rank PESTN parametrizations of ref. 36, convergence is also achieved for the n-d total and differential cross sections below E_n = 20 MeV and likewise for the vector-to-vector spin transfers of the reaction $^2H(\vec{N},\vec{d})N$ as well as the nucleon-to-nucleon spin transfers of $^2H(\vec{N},\vec{N})^2H$ at E_N = 10 MeV [40,55]. We demonstrate the pertinent result in fig. 9 for the differential cross section at several energies up to E_n = 20 MeV. These curves may now be regarded as the genuine predictions of the Paris potential[13]. The agreement with experiment is very satisfactory. The same is also true with respect to the other observables (spin transfers) mentioned above[40,55]. This is remarkable in two respects: Firstly, in view of the many ingredients needed in an N-N potential to reproduce the considered 3-N observables to such an accuracy, the performance of the Paris potential is simply convincing. On the other hand we must not forget that it fails in describing the triton. This is likely to be due, however, to 3-N forces[30]. If we assume this to be the reason for underbinding the triton, we can say that effects of the 3-N force are not visible in the above observables of elastic N-d scattering. In fact this is plausible and in agreement with our expectation: Because elastic N-d scattering occurs mainly in the quartet state[26], the three nucleons will not approach each other as closely as in the triton and consequently 3-N forces, typically of the range of 2π exchange, will not play a significant role. Thus we observe the corresponding observables to be governed essentially by 2-N forces. As such they are very well described by the Paris potential and we may take this fact as evidence that the concept of meson-exchange dynamics is appropriate for the long- and intermediate-range N-N interaction down to $r \approx 0.8$ fm, say (cf. the discussion in section 4.1).

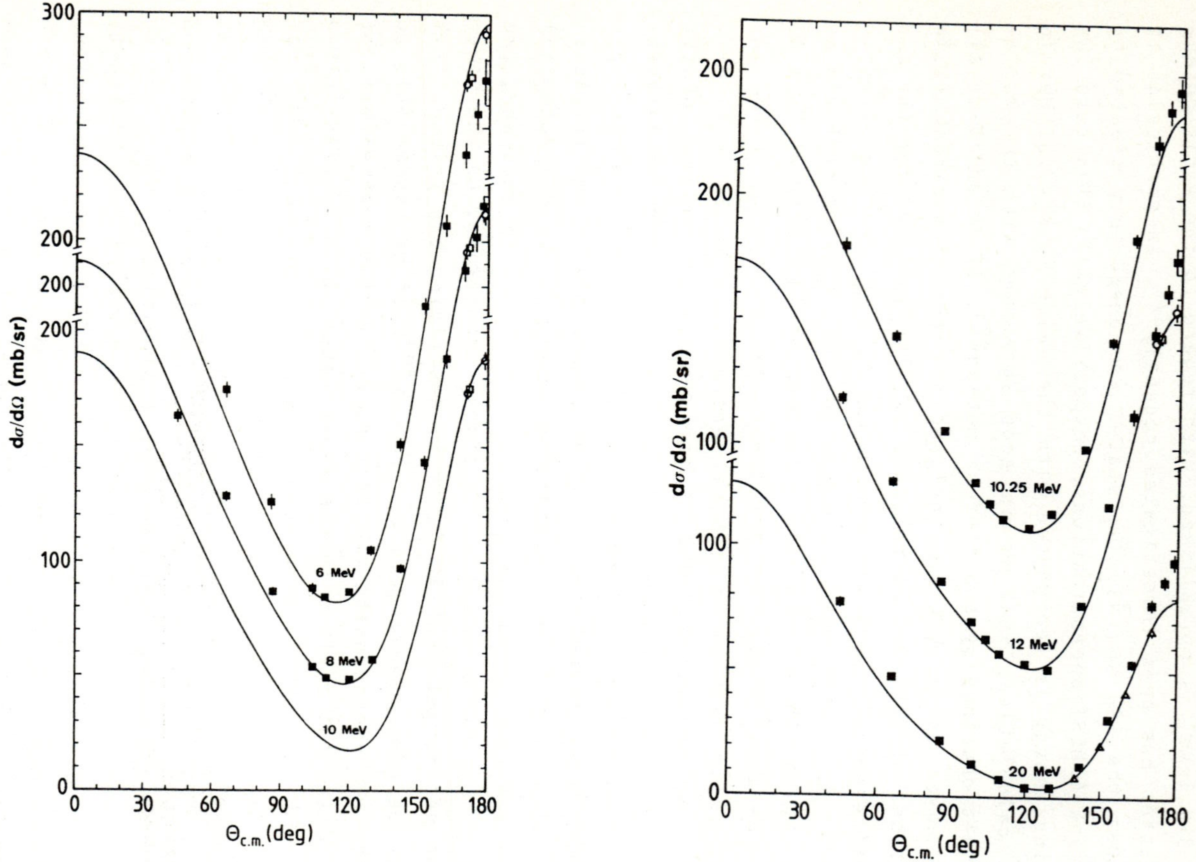

Fig. 9. Paris potential predictions for the n-d differential cross section as calculated with the separable representations PEST3-G($^1S_0$) + PEST6-G($^3S_1$-$^3D_1$) + PEST2($^1P_1$,$^3P_0$,$^3P_1$,$^1D_2$,$^3D_2$) + PEST3($^3P_2$-$^3F_2$) + PEST4($^3D_3$-$^3G_3$). Open circles and squares are Uppsala data (ref. 53), full squares are Karlsruhe data (ref. 52), and open triangles recent measurements of Hofmann (ref. 54). The interactions used here are described in detail in refs. 36, 37, and 40.

The results demonstrated above and in ref. 30 are the product of the presently most advanced Faddeev calculation of 3-N scattering. Never before separable interactions of such high ranks could be employed in so many N-N partial waves. Still there is further improvement needed and also possible. From the examination of other N-d spin observables we have the indication that the description of higher partial waves (P and D) is not sufficient in the PESTN parametrizations of ref. 36. Especially the neutron vector-analyzing power A_y, which is extremely sensitive to details in the N-N P-waves (on-shell as well as off-shell[50)]) is not yet established in a completely reliable manner from the Paris potential[31)].

Nevertheless we have at this stage several benchmark results, from where we can start out to examine specific details of the N-N interaction by means of the 3-N scattering system, e.g., the prominent question of whether or not quark degrees-of-freedom are visible (needed) in this domain of nuclear resp. particle physics. Once the technology for calculating these results with even the most advanced present-day models is now available, we can in future concentrate more on the physical questions and relate the observations we will make to our knowledge of the fundamental N-N interaction. In fact, this was the primary question, why physicists started out to investigate the 3-N problem.

ACKNOWLEDGMENT

The work described in this paper is the result of a long and fruitful collaboration with my colleagues J. Haidenbauer, Y. Koike, and H. Zankel. Their efforts were indispensable, and it is a pleasure that I can acknowledge them here. I am also grateful to L. Mathelitsch, W. Schweiger, and H. Zingl for many profitable discussions and their continuous interest. During several stages this work received financial support from Fonds zur Förderung der Wissenschaftlichen Forschung in Österreich under projects 5212 and 5733. Finally I also like to express my gratitude to the organizers for inviting me to this Autumn School, where I had the opportunity to present our approach to the 3-N problem in a more detailed talk.

REFERENCES

1. W. Plessas, in Few-Body Methods (Proceedings of the International Symposium on Few-Body Methods and Their Applications in Atomic, Molecular & Nuclear Physics, and Chemistry, Nanning, 1985), ed. by T.-K. Lim et al. (World Scientific, Singapore, 1986).

2. R. Dubois et al., Nucl. Phys. A377, 554 (1982).

3. R.A. Arndt et al., Phys. Rev. D 28, 97 (1983).

4. J. Bystricki, C. Lechanoine-Leluc, and F. Lehar, Saclay Report No. DPhPE 86-13 (1986).

5. See, e.g., D.D. Brayshaw, Phys. Rev. Lett. 32, 382 (1974); M.I. Haftel and E.L. Peterson, ibid. 33, 1229 (1974); D.D. Brayshaw, ibid 34, 1478 (1975); M.I. Haftel and E.L. Peterson, ibid. 34, 1480 (1975).

6. W. Plessas, L. Mathelitsch, F. Pauss, and H.F.K. Zingl, Nonlocal Separable Interactions in the Two-Nucleon System (Univ. of Graz Press, Graz, 1975).

7. W. Plessas, Acta Phys. Austriaca 54, 305 (1982).

8. See, e.g., the corresponding review resp. rapporteur talks in Few-Body Systems and Nuclear Forces (Lecture Notes in Physics, Vol. 87), ed. by H. Zingl et al. (Springer, Heidelberg, 1978).

9. P. Doleschall, Nucl. Phys. A201, 264 (1973); ibid. A220, 491 (1974).

10. L. Mathelitsch and H.F.K. Zingl, Nuovo Cim. 44A, 81 (1978).

11. M.I. Haftel, Phys. Rev. C 14, 698 (1976); M.I. Haftel and W.M. Kloet, ibid. 15, 404 (1977).

12. I.R. Afnan, D.M. Clement, and F.J.D. Serduke, Nucl. Phys. A170, 625 (1971).

13. M. Lacombe et al., Phys. Rev. C 21, 861 (1980).

14. B. Loiseau et al., Phys. Rev. C 32, 2165 (1985).

15. W. Plessas, K. Schwarz, and L. Mathelitsch, in Perspectives in Nuclear Physics at Intermediate Energies, ed. by S. Boffi et al. (World Scientific, Singapore, 1984).

16. G.H. Lamot, N. Giraud, and C. Fayard, Nuovo Cim. 57A, 445 (1980); N. Giraud, C. Fayard, and G.H. Lamot, Phys. Rev. C 21, 1959 (1980).

17. Y. Koike et al., Prog. Theor. Phys. 66, 1899 (1981).

18. F. Sperisen et al., Phys. Lett. 102B, 9 (1981).

19. P. Doleschall et al., Nucl. Phys. A380, 72 (1982).

20. R. Schmelzer et al., Phys. Lett. 120B, 297 (1983).

21. L. Mathelitsch, W. Plessas, and W. Schweiger, Phys. Rev. C 26, 65 (1982).

22. W. Schweiger et al., Phys. Rev. C 27, 515 (1983); ibid. 28, 1414 (1983).

23. K. Schwarz, W. Plessas, and L. Mathelitsch, Nuovo Cim. 76A, 321 (1983).

24. Y. Koike and Y. Taniguchi, Phys. Lett. 118B, 248 (1982).

25. K. Hatanaka et al., Nucl. Phys. A426, 77 (1984).

26. Y. Koike and Y. Taniguchi, Few-Body Systems 1, 13 (1986).

27. C. Hajduk and P.U. Sauer, Nucl. Phys. A369, 321 (1981); W. Strueve, C. Hajduk, and P.U. Sauer, ibid. A405, 620 (1983).

28. E. Hadjimichael, R. Bornais, and B. Goulard, Phys. Rev. Lett. 48, 583 (1982); E. Hadjimichael, B. Goulard, and R. Bornais, Phys. Rev. C 27, 831 (1983).

29. C.R. Chen et al., Phys. Rev. C 31, 2266 (1985).

30. J.L. Friar, B.F. Gibson, and G.L. Payne, Comm. Nucl. Part. Phys. 11, 51 (1983).

31. See, e.g., the corresponding review resp. rapporteur talks in the Proceedings of the 11th International Conference on Few-Body Systems in Particle and Nuclear Physics, Tokyo, 1986, to appear in Nucl. Phys.

32. D.J. Ernst, C.M. Shakin, and R.M. Thaler, Phys. Rev. C 8, 46 (1973).

33. J. Haidenbauer and W. Plessas, Phys. Rev. C 27, 63 (1983).

34. For a reference see, e.g., J.S. Levinger, in Nuclear Physics (Springer Tracts in Modern Physics, Vol. 71), ed. by G. Höhler (Springer, Heidelberg, 1974).

35. M. Baldo, L.S. Ferreira, and L. Streit, Phys. Rev. C 32, 685 (1985); M. Baldo et al., ibid. 33, 1587 (1986).

36. J. Haidenbauer and W. Plessas, Phys. Rev. C 30, 1822 (1984); ibid. 32, 1424 (1985).

37. J. Haidenbauer and Y. Koike, Phys. Rev. C 34, 1187 (1986).

38. H. van Haeringen, Charged-Particle Interactions (Coulomb Press, Leyden, 1985).

39. H. Zankel, W. Plessas, and J. Haidenbauer, Phys. Rev. C 28, 538 (1983).

40. Y. Koike, J. Haidenbauer, and W. Plessas, Preprint (1986).

41. R. Machleidt and K. Holinde, in Few-Body Problems in Physics, ed. by B. Zeitnitz (North-Holland, Amsterdam, 1984), p. 79.

42. W.N. Cottingham et al., Phys. Rev. D 8, 800 (1973); R. Vinh Mau, in Mesons in Nuclei I, ed. by M. Rho and D. Wilkinson (North-Holland, Amsterdam, 1979), p. 151.

43. K. Erkelenz, Phys. Rep. 13C, 191 (1974); K. Holinde and R. Machleidt, Nucl. Phys. A247, 495 (1975).

44. J. Haidenbauer, Y. Koike, and W. Plessas, Phys. Rev. C 33, 439 (1986).

45. R. Machleidt, K. Holinde, and C. Elster, Phys. Rep., to appear.

46. H. Zankel and W. Plessas, Z. Phys. A317, 45 (1984).

47. R. Schmelzer et al., in Few-Body Problems in Physics, ed. by B. Zeitnitz (North-Holland, Amsterdam, 1984).

48. F. Sperisen et al., Phys. Lett. 102B, 9 (1981).

49. Y. Koike et al., in Perspectives in Nuclear Physics at Intermediate Energies II, ed. by S. Boffi et al. (World Scientific, Singapore, 1985).

50. J. Haidenbauer et al., in Few-Body Approaches to Nuclear Reactions in Tandem and Cyclotron Energy Regions, ed. by T. Sawada et al. (World Scientific, Singapore, to appear).

51. F.D. Correll et al., Phys. Rev. C 23, 960 (1981).

52. P. Schwarz et al., Nucl. Phys. A398, 1 (1983).

53. G. Janson et al., in Few-Body Problems in Physics, ed. by B. Zeitnitz (North-Holland, Amsterdam, 1984), p. 529;
G. Janson, private communication.

54. K. Hofmann, Thesis (Univ. Karlsruhe, 1985).

55. Y. Koike, J. Haidenbauer, and W. Plessas, in Proceedings of the 11th International Conference on Few-Body Systems in Particle and Nuclear Physics, ed. by T. Sasakawa et al. (Tohoku University Report, Sendai, 1986), pp. 346, 348.

FOUR-BODY EQUATIONS IN MOMENTUM SPACE

A.C. Fonseca
Centro de Física Nuclear, Av.Gama Pinto 2
1699 Lisbon. Portugal

1. INTRODUCTION

Although one can find in the literature[1] many generalizations of the Faddeev equations to four and more particles, it was Yakubovsky[2] who first discovered how to set up a system of coupled integral equations for the wave function components, which has a unique correspondence to the Schrödinger equation and whose Kernel becomes connected after a finite number of iterations. Independently Alt, Grassberger and Sandhas (AGS)[3] formulated an equivalent set of equations for the t-matrix components, which they then solved in the context of the quasi-particle method and the k-matrix approximation[4]. Of all integral equation formulations these are the only ones that have the right-left off-shell properties that allow for a reduction in the dimensionality of the equations each time the subamplitudes are expressed in a separable form. For this reason they have been extensively used in most calculations that have been performed over the past fifteen years.

In the present series of two lectures we start by showing how to derive the AGS equations from first principles of multichannel quantum scattering theory and proceed to explain how to use them in a practical calculation. Although most of this work is now published we find it useful for future use by students in this field to present here many of the important results using a common notation. In Section 2. we derive the AGS equations and in Section 3. we make use of a recently published review article[5] to sketch the important equations leading to two-variable or one-variable integral equations with or without use of the convolution method. In Section 4. we apply them to four identical bosons and in Section 5. show a few results.

2. FOUR-BODY EQUATIONS

Lets assume that we have four particles interacting by short ranged pairwise potentials v_i where i denotes any pair. There are six pairs numbered from one to six and let V be

$$V = \sum_i v_i . \tag{2.1}$$

We can write the four-particle Lippmann-Schwinger equation

$$T = V + V G_o T \ , \qquad (2.2)$$

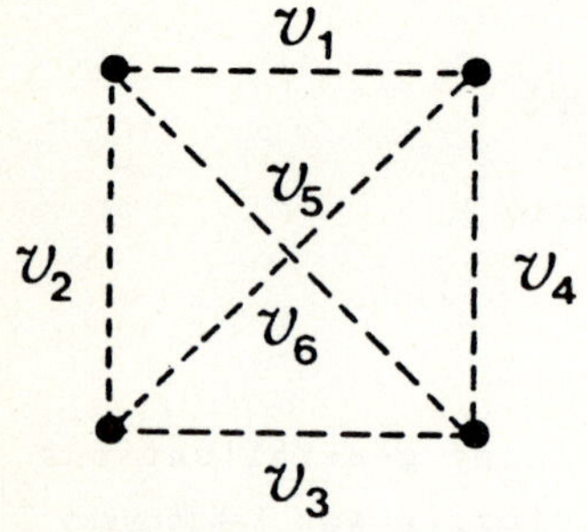

Fig.1 Four-Particles with pair potentials

where G_o is the four free particle Green's function and T is the t-matrix for the scattering of four-free particles into four-free particles. Although T is not an operator whose on-shell matrix elements are of physical interest, equation (2.2) is useful to demonstrate the disconnectedness of the Kernel and how to proceed to remedy such problem. From Eq.(2.2) we can write the Born series

$$T = V + V G_o V + V G_o V G_o V + \dots \ , \qquad (2.3)$$

whose graphical representation is shown in Fig.2. In all orders of iteration one finds diagrams that connect only one or two pairs and whose matrix elements lead to δ-functions in the initial and final momentum

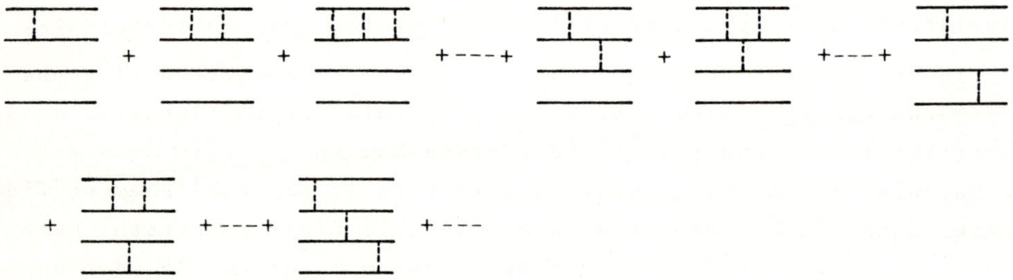

Fig.2 Graphical representation of Born series given in Eq.(2.3). The dashed line denotes v_i while G_o is represented by paralel lines in intermediate states

of each free particle or cluster†. These dangerous singularities make the Kernel a noncompact operator and Eq.(2.2) impossible to solve by standard numerical methods [6]. We now proceed to develop connected four-particle equations.

† A "Cluster" is a group of interacting particles

2.1 For the T Operator

One first step in that direction is to follow Faddeev's prescription by dividing T in six components T_i where i runs over all pairs (or three-cluster partitions†)[7]

$$T = \sum_i T_i, \tag{2.4}$$

where

$$T_i = v_i + v_i G_o T. \tag{2.5}$$

The operator T_i is the sum of all graphs that end on the left with v_i. Substituting (2.4) on the right side of (2.5)

$$T_i = v_i + v_i G_o \sum_j T_j , \tag{2.6}$$

we note that j may by identical to i. Therefore if one brings to the left the term j=i one gets

$$(1-v_i G_o) T_i = v_i + v_i G_o \sum_{j \neq i} T_j , \tag{2.7}$$

together with an alternation rule. Since v_i is only the pair interaction we know how to invert (2.7) by using standard two-body identities such as

$$(1+v_i G_i)(1-v_i G_o) = 1, \tag{2.8}$$

$$v_i G_i = t_i G_o , \tag{2.9}$$

which lead to the two-body Lippmann-Schwinger equation

$$t_i = v_i + v_i G_o t_i , \tag{2.10a}$$

$$t_i = v_i + t_i G_o v_i , \tag{2.10b}$$

embedded in four-particle space. Therefore multiplying on the left by $(1+v_i G_i)$ and using (2.9) together with (2.10b) we get

$$T_i = t_i + t_i G_o \sum_{j \neq i} T_j , \tag{2.11}$$

which is the equivalent of the Faddeév equations for a system of four particles. This is a 6×6 matrix equation whose Born series may be written as

$$T_i = t_i + t_i G_o \sum_{j \neq i} t_j + t_i G_o \sum_{j \neq i} t_j G_o \sum_{k \neq j} t_k + \ldots \tag{2.12}$$

As shown diagramatically in Fig.3 there are still disconnected diagrams in all orders of the expansion which again leads to the non-compactness of the Kernel.

---

† An "A-cluster partition" is a group of A non-interacting clusters. With four particles we can form six three-cluster partitions which are (12)34, (13)24, (14)23, (23)14, (24)13 and (34)12. In (12)34 particles (12) interact with each other while 3 and 4 are free.

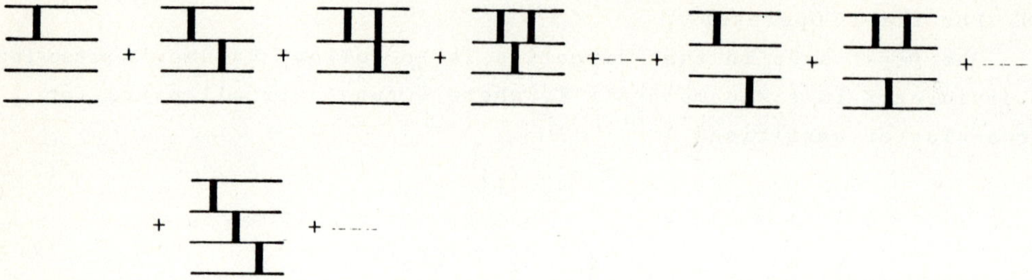

Fig.3 Graphical representation of the Born series given in Eq.(2.12). The solid vertical lines denote t_i.

Since the Faddeev decomposition is unsufficient to bring about connectedness for four particles, Yakubovsky generalized the method and, making use of the concept associated with chains of partitions, proposed the decomposition of the T_i's into a new set of components, which depend on three-cluster and two-cluster indices T_i^ρ given by

$$T_i^\rho = t_i \, G_o \sum_{\substack{j \neq i \\ j \subset \rho}} T_j \, , \tag{2.13}$$

where i and j are pairs (or three-cluster partitions) internal to the two-cluster partition ρ. With four particles one can form seven two--cluster partitions which may be of the (3)+1 kind or of the (2)+(2) kind. Those of (3)+1 kind are (123)4, (234)1, (341)2 and (412)3 while those of (2)+(2) kind are (12) (34), (13)(24) and (14)(23). In a chain of partition connectivety is increased from the initial four-cluster partition 1 2 3 4 where all four-particles are free to the final one -cluster partition (1234) where they all interact, through intermediate three-cluster and two-cluster partitions. A particular exemple may be 1234 → (12)34 → (123)4 → (1234) or 1234 → (23)14 → (23)(14) → (1234). In Table I we show all 18 different ways in which this can be done.

Returning to Eq.(2.11) one can consider a specific exemple such as i = (12). Then

$$\begin{aligned} T_{(12)} = t_{(12)} &+ t_{(12)} G_o T_{(13)} + t_{(12)} G_o T_{(23)} \\ &+ t_{(12)} G_o T_{(14)} + t_{(12)} G_o T_{(24)} \\ &+ t_{(12)} G_o T_{(34)} \, . \end{aligned} \tag{2.14}$$

Taking note of Eq.(2.13) one can easily rewrite (2.14) as

Table I. The 18 different chains of partition for a system of four particles

One-Cluster	Two-Cluster	Three-Cluster	Four-Cluster
	(123) 4	(12) 3 4	
		(23) 1 4	
		(31) 2 4	
	(234) 1	(23) 4 1	
		(34) 2 1	
		(42) 3 1	
	(341) 2	(34) 1 2	
		(41) 3 2	
(1234)		(13) 4 2	1 2 3 4
	(412) 3	(41) 2 3	
		(12) 4 3	
		(24) 1 3	
	(12) (34)	(12) 3 4	
		(34) 1 2	
	(13) (24)	(13) 2 4	
		(24) 1 3	
	(14) (23)	(14) 2 3	
		(23) 1 4	

$$T_{(12)} = t_{(12)} + T_{(12)}^{(123)4} + T_{(12)}^{(124)3} + T_{(12)}^{(12)(34)} , \qquad (2.15)$$

or in general

$$T_i = t_i + \sum_{\sigma \supset i} T_i^{\sigma} . \qquad (2.16)$$

Therefore each T_i gives rise to three T_i^{ρ}'s leading to a total of $6\times3=18$, 12 of which are of (3)+1 type and 6 of (2)+(2) type. One may now proceed by substituting (2.16) back in (2.13) to obtain a coupled set of equations for the T_i^{ρ}'s

$$T_i^{\rho} = t_i \, G_o \sum_{\substack{j \neq i \\ (j \subset \rho)}} (t_j + \sum_{\sigma \supset j} T_j^{\sigma}), \qquad (2.17)$$

where σ on the right may be identical to ρ. As it was done before we bring to the left the $\sigma=\rho$ term and gain an alternation rule on the summa tion over two-cluster partitions σ that contain j

$$T_i^{\rho} - t_i G_o \sum_{\substack{j \neq i \\ (j \subset \rho)}} T_j^{\rho} = t_i G_o \sum_{\substack{j \neq i \\ (j \subset \rho)}} t_j + t_i G_o \sum_{\substack{j \neq i \\ j \subset \rho}} \sum_{\sigma \supset j} \bar{\delta}_{\sigma\rho} \, T_j^{\sigma} . \qquad (2.18)$$

For a better understanding of Eq.(2.18) it may be useful to write down a specific example such as for $\rho=(123)4$

$$\begin{bmatrix} T_{(12)}^{(123)4} \\ T_{(13)}^{(123)4} \\ T_{(23)}^{(123)4} \end{bmatrix} - \begin{bmatrix} 0 & t_{(12)} & t_{(12)} \\ t_{(13)} & 0 & t_{(13)} \\ t_{(23)} & t_{(23)} & 0 \end{bmatrix} G_o \begin{bmatrix} T_{(12)}^{(123)4} \\ T_{(13)}^{(123)4} \\ T_{(23)}^{(123)4} \end{bmatrix} = \begin{bmatrix} 0 & t_{(12)} & t_{(12)} \\ t_{(13)} & 0 & t_{(13)} \\ t_{(23)} & t_{(23)} & 0 \end{bmatrix} \begin{bmatrix} t_{(12)} \\ t_{(13)} \\ t_{(23)} \end{bmatrix} +$$

$$+ \begin{bmatrix} 0 & t_{(12)} & t_{(12)} \\ t_{(13)} & 0 & t_{(13)} \\ t_{(23)} & t_{(23)} & 0 \end{bmatrix} G_o \begin{bmatrix} \sum_{\sigma\supset(12)} \bar{\delta}_{\sigma\rho} T^{\sigma}_{(12)} \\ \sum_{\sigma\supset(13)} \bar{\delta}_{\sigma\rho} T^{\sigma}_{(13)} \\ \sum_{\sigma\supset(23)} \bar{\delta}_{\sigma\rho} T^{\sigma}_{(23)} \end{bmatrix} \tag{2.19}$$

Therefore for $\rho=(123)4$ we have on the left side the Faddeev Kernel for subsystem ρ, which is a 3×3 matrix operator. If we chose $\rho=(12)(34)$ we get instead a 2×2 equation involving pairs (12) and (34) internal to (12)(34). In an matrix operator notation we can write (2.18) as

$$[\mathbb{1} - \mathbb{V}^{\rho}G_o]T^{\rho} = \mathbb{V}^{\rho}G_o t + \mathbb{V}^{\rho}G_o R, \tag{2.20}$$

where

$$[\mathbb{V}^{\rho}]_{ij} = t_i\,\bar{\delta}_{ij}\ , \tag{2.21}$$

for $i,j\subset\rho$ and zero otherwise. As in (2.7) we now look for an operator $\mathbb{K}^{\rho}$ such that

$$[\mathbb{K}^{\rho}]_{ij} = K^{\rho}_{ij}\ , \tag{2.22}$$

and

$$(\mathbb{1} + \mathbb{K}^{\rho}G_o)(\mathbb{1} - \mathbb{V}^{\rho}G_o) = \mathbb{1}\ , \tag{2.23}$$

or

$$\mathbb{K}^{\rho} = \mathbb{V}^{\rho} + \mathbb{K}^{\rho}G_o\,\mathbb{V}^{\rho}\ . \tag{2.24}$$

With the help of (2.21) and (2.22), Eq.(2.24) becomes

$$K^{\rho}_{ij} = t_i\,\bar{\delta}_{ij} + \sum_{k\neq j} K^{\rho}_{ik}\,G_o\,t_k\ , \tag{2.25}$$

which shows that K^{ρ}_{ij} is the sum of all graphs that end with an interaction in pair i and start with an interaction in any pair k other than j. Multiplying (2.20) on the left by $(\mathbb{1}+\mathbb{K}^{\rho}G_o)$ and using (2.24) together with (2.22) we get

$$T^{\rho}_i = \sum_{j\subset\rho} K^{\rho}_{ij}G_o t_j + \sum_{j\subset\rho} K^{\rho}_{ij}G_o \sum_{\sigma\supset j} \bar{\delta}_{\sigma\rho}\,T^{\sigma}_j\ , \tag{2.26}$$

which is the four-body equation we are looking for. From (2.26) we can write the Born series

$$T^{\rho}_i = \sum_{j\subset\rho} K^{\rho}_{ij}G_o t_j + \sum_{j\subset\rho} K^{\rho}_{ij}G_o \sum_{\sigma\supset j} \bar{\delta}_{\sigma\rho} \sum_{k\subset\sigma} K^{\sigma}_{jk}G_o t_k + \ldots \tag{2.27}$$

Taking the lowest term in $K^{\rho}_{ij}=t_i\bar{\delta}_{ij}$ we get

$$T^{\rho}_i = \sum_{j\subset\rho} t_i\bar{\delta}_{ij}G_o t_j + \sum_{j\subset\rho} t_i\bar{\delta}_{ij}G_o \sum_{\sigma\supset j} \bar{\delta}_{\sigma\rho} \sum_{k\subset\sigma} t_j\bar{\delta}_{jk}G_o t_k + \ldots \tag{2.28}$$

Because of the alternation rules in all three-and two-cluster partitions together with the need for σ to contain j, the second term is already connected as shown in Fig.4.

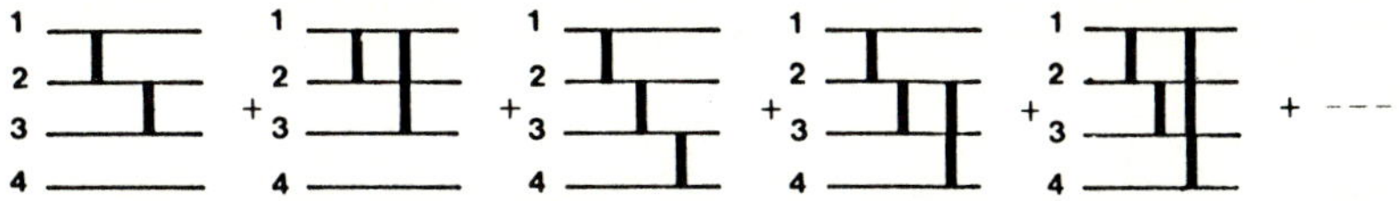

Fig.4 Graphical representation of Eq.(2.28) for ρ=(123)4 and σ=(234)1 or (23)(41).

Therefore the Kernel in Eq.(2.26) becomes connected after a finite number of iterations and the proof for compactness may follow .

Instead of using the subsystem operator $\mathbb{K}^{\rho}$ in (2.26) it is more advantageous for future manipulations to use the two-cluster subamplitudes $\mathbb{U}^{\rho}$ which satisfy the AGS equation [8]

$$U^{\rho}_{ij} = G_0^{-1} \bar{\delta}_{ij} + \sum_{k\neq i} t_k G_o U^{\rho}_{kj} \quad . \tag{2.29}$$

Multiplying Eq.(2.29) by $t_i G_o$ and comparing with Eq.(2.25) one easily relates $\mathbb{K}^{\rho}$ with $\mathbb{U}^{\rho}$

$$K^{\rho}_{ij} = t_i \; G_o \; U^{\rho}_{ij} \quad . \tag{2.30}$$

In addition to the operators $\mathbb{U}^{\rho}$ and $\mathbb{K}^{\rho}$ one can define a third one $\mathbb{M}^{\rho}$ which satisfies the equation

$$M^{\rho}_{ij} = t_i \; \delta_{ij} + \sum_{k\neq i} t_i \; G_o \; M_{kj} \quad , \tag{2.31}$$

and consits of the sum of all diagrams that end with an interaction in pair i and start with an interaction in pair j. Therefore

$$K^{\rho}_{ij} = \sum_{k\neq j} M^{\rho}_{ik} \quad . \tag{2.32}$$

Defining

$$[\; \mathbb{T}^{\rho}]_{ij} = t_i \; \delta_{ij} \quad , \tag{2.33}$$

we may rewrite Eq.(2.31) as

$$(\mathbb{1}-\mathbb{V}^{\rho} G_o) \; \mathbb{M}^{\rho} = \mathbb{T} \quad , \tag{2.34}$$

or using (2.23)

$$\mathbb{M}^{\rho} = (\mathbb{1}+\mathbb{K}^{\rho} G_o) \; \mathbb{T} \quad , \tag{2.35}$$

which reads

$$M^{\rho}_{ij} = t_i \delta_{ij} + K^{\rho}_{ij} G_o t_j \quad . \tag{2.36}$$

Using (2.32) and (2.36) in (2.26) we get

$$T_i^\rho = \sum_{j\subset\rho} M_{ij}^\rho - t_i + \sum_{j\subset\rho} \sum_{k\neq j} M_{ik}^\rho \, G_o \sum_{\sigma\supset j} \bar{\delta}_{\sigma\rho} \, T_j^\sigma \quad , \tag{2.37}$$

which is an Yakubovsky like equation for the T_i^ρ's. If instead one uses (2.30) back in (2.26) we get the corresponding AGS equation

$$T_i^\rho = \sum_{j\subset\rho} t_i G_o U_{ij}^\rho G_o t_j + \sum_{j\subset\rho} t_i G_o U_{ij}^\rho G_o \sum_{\sigma\supset j} \bar{\delta}_{\sigma\rho} \, T_j^\sigma \, . \tag{2.38}$$

which clearly shows the underlying structure of subamplitudes that is essential to any practical calculation. Finally puting together (2.4) and (2.16) we get

$$T = \sum_i t_i + \sum_i \sum_{\rho\supset i} T_i^\rho \, . \tag{2.39}$$

2.2 For the Green's Function G

Once we have obtained an equation for T that satisfies the requirements of connectivity, one can easily get a similar equation for the total Green's function G

$$G = G_o + G_o \, T \, G_o \, . \tag{2.40}$$

Substituting (2.39) in (2.40) one gets

$$G = G_o + \sum_i G_o t_i G_o + \sum_i \sum_{\rho\supset i} G_o T_i^\rho G_o \, . \tag{2.41}$$

Defining

$$G_i^\rho = G_o T_i^\rho G_o \quad , \tag{2.42}$$

Eq.(2.41) may be written as

$$G = G_o + \sum_i (G_i - G_o) + \sum_i \sum_{\rho\supset i} G_i^\rho \quad , \tag{2.43}$$

where G_i is the pair resolvent operator embedded in four-body space $G_i^{-1}(z) = G_o^{-1}(z) - v_i$. The G_i^ρ satisfy the equation

$$G_i^\rho = \sum_{j\subset\rho} G_o t_i G_o U_{ij}^\rho G_o t_j G_o + \sum_{j\subset\rho} G_o t_i G_o U_{ij}^\rho \sum_{\sigma\supset j} \bar{\delta}_{\sigma\rho} \, G_j^\sigma \tag{2.44}$$

which is obtained from (2.38) by multiplying from the left and right by G_o and using the definition (2.42).

2.3 For the Wave Function Ψ

By definition[6,9,10] the four-body scattering wave function $|\Psi^a\rangle$ is given by

$$|\Psi^a\rangle = \lim_{\varepsilon\to o} i\varepsilon \; G(E+i\varepsilon) |\psi^a\rangle \quad , \tag{2.45}$$

where ψ^a the incoming asymptotic wave function in some appropriate channel partition a. Assuming that the initial state a is the two-cluster state ρ_o

$$|\psi^{\rho_o}(\vec{k}_{\rho_o})> = |\chi^{\rho_o}> \, |\phi(\vec{k}_{\rho_o})> \quad , \tag{2.46}$$

where $|\chi^{\rho_o}>$ is the bound state wave function for the clusters in channel ρ_o and ϕ is the plane wave of momentum $\vec{k}_{\rho_o}$ between clusters. The wave function $|\chi^{\rho}>$ may be written as a sum of components

$$|\chi^{\rho}> = \sum_i |\chi_i^{\rho}> \tag{2.47}$$

where i runs over all pairs internal to partition ρ and the components $|\chi_i^{\rho}>$ satisfy the equation

$$|\chi_i^{\rho}> = \sum_k \bar{\delta}_{ki} \, G_o t_i |\chi_k^{\rho}> \tag{2.48}$$

whose Kernel is the same as of Eqs.(2.25), (2.29) and (2.31). For example if ρ=(123)4 is a physically realizable two-body channel (which means that system (123) has a bound state),then $|\chi_i^{\rho}>$ is the ith component of the three-body Faddeev[7] wave function for cluster (123).

Substituting (2.43) in (2.45) we get

$$|\Psi^{\rho_o}> = \sum_i \sum_{\rho\supset i} |\Psi_i^{\rho,\rho_o}> \quad , \tag{2.49}$$

because, by definition of wave function component,

$$|\Psi_i^{\rho,\rho_o}> = \lim_{\varepsilon\to o} \; i\varepsilon \; G_i^{\rho}(E+i\varepsilon)|\psi^{\rho_o}> \quad , \tag{2.50}$$

and the terms in G_o and G_i do not contribute

$$\lim_{\varepsilon\to o} i\varepsilon \; G_o(E+i\varepsilon)|\psi^{\rho_o}> = 0 \quad , \tag{2.51}$$

$$\lim_{\varepsilon\to o} i\varepsilon \; G_i(E+i\varepsilon)|\psi^{\rho_o}> = 0 \quad , \tag{2.52}$$

for lack of ψ^{ρ_o} being an eigenfunction of H_o or H_i=H_o+v_i. In order to get an equation for the wave function components we apply $|\psi^{\rho_o}>$ to the right of Eq.(2.44) and use (2.50) on both sides leading to

$$\begin{aligned} |\Psi_i^{\rho,\rho_o}> = \lim_{\varepsilon\to o} i\varepsilon & \sum_{j\subset\rho} G_o t_i G_o U_{ij}^{\rho} G_o t_j G_o |\psi^{\rho_o}> \\ & + \sum_{j\subset\rho} G_o t_i G_o U_{ij}^{\rho} \sum_{\sigma\supset j} \bar{\delta}_{\sigma\rho} \, |\Psi_j^{\sigma,\rho_o}> \, . \end{aligned} \tag{2.53}$$

Since, as shown in Appendix A,

$$U^{\rho}_{ij} = -\delta_{ij} G_i^{-1} + G_i^{-1} G_\rho G_j^{-1} , \quad (2.54)$$

where $G_\rho^{-1}(z) = z - H_o - V_\rho$ and $V_\rho = \sum_{i\subset\rho} v_i$, we get with the help of (2.9) and the well known Lippmann-Schwinger equation for the resolvent operators [11] G_i and G_ρ

$$G_i = G_o + G_o v_i G_i , \quad (2.55)$$

$$G_\rho = G_o + G_o V_\rho G_\rho , \quad (2.56)$$

$$\sum_{j\subset\rho} G_o t_i G_o U^{\rho}_{ij} G_o t_j G_o = -G_i + G_o + G_o v_i G_\rho \quad . \quad (2.57)$$

Using again (2.51) and (2.52) together with

$$|\psi_i^{\rho_o}> = G_o v_i \, |\psi^{\rho_o}> , \quad (2.58)$$

which is the usual defining equation for the wave function component in subsystem ρ_o we obtain

$$\lim_{\varepsilon\to o} i\varepsilon \sum_{j\subset\rho} G_o t_i G_o U^{\rho}_{ij} G_o t_j G_o |\psi^{\rho_o}> = \delta_{\rho,\rho_o} |\psi_i^{\rho_o}> , \quad (2.59)$$

where

$$|\psi_i^{\rho_o}> = |\chi_i^{\rho_o}>|\phi(\vec{k}_{\rho_o})> . \quad (2.60)$$

The four-body AGS equation for the wave function components then reads

$$|\Psi_i^{\rho,\rho_o}> = \delta_{\rho,\rho_o} |\psi_i^{\rho_o}> + \sum_{j\subset\rho} G_o t_i G_o U^{\rho}_{ij} \sum_{\sigma\supset j} \bar{\delta}_{\sigma\rho} |\Psi_j^{\sigma,\rho_o}>, \quad (2.61)$$

whose homogeneous version is the equation to be used to calculate four-body bound states and corresponding wave functions.

At this stage it may be useful to note that Eq.(2.61) could have been derived from the four-particle Schrödinger equation

$$G_o^{-1} \Psi = V \Psi \quad (2.62)$$

or

$$\Psi = G_o V \Psi \quad (2.63)$$

using the standard decomposition in wave function components. Defining

$$\Psi_i = G_o v_i \Psi \quad (2.64)$$

together with

$$\Psi = \sum_i \Psi_i \quad (2.65)$$

we get

$$\Psi_i = G_o t_i \sum_{j\neq i} \Psi_j , \quad (2.66)$$

after substituting (2.65) in (2.64) and proceeding as we did before for the T_i's. Next defining

$$\psi_i^{\rho} = G_o t_i \sum_{\substack{j \neq i \\ j \subset \rho}} \Psi_j \ , \tag{2.67}$$

and following similar steps as for the T_i^{ρ}'s one obtains (2.61) after adding $|\psi^{\rho_o}> \delta_{\rho,\rho_o}$ which is an homogeneous solution of $(\mathbf{1}-G_o \mathbf{V}^{\rho_o})$ that satisfies the appropriate asymptotic boundary conditions for an initial state in partition ρ_o. Therefore using (2.65) and (2.66) together with (2.67) we recover (2.49).

As first pointed out by Alt, Grassberger and Sandhas Eq.(2.61) may be writting in a matrix operator form by defining

$$\mathcal{G}_{o\,ij}^{\sigma\rho} = G_o t_i G_o U_{ij}^{\sigma} G_o t_j G_o \ \delta_{\sigma\rho} \ , \tag{2.68}$$

for $i,j \subset \rho$ and

$$\mathcal{V}_{ij}^{\sigma\rho} = (G_o t_i G_o)^{-1} \ \bar{\delta}_{\sigma\rho} \ \delta_{ij} \ , \tag{2.69}$$

leading to

$$|\Psi^{\rho_o}> = |\psi^{\rho_o}> + \mathcal{G}_o \mathcal{V} |\Psi^{\rho_o}> \ . \tag{2.70}$$

As it will be shown in Appendix A one of the great advantages of the AGS matrix approach [3] is that one can manipulate the operators $\mathcal{G}_o$ and $\mathcal{V}$ in a formal way, as if they were two-body operators, to generate new equations whose physical significance may be understood a posteriori.

2.4 For the AGS Operator $\mathcal{U}$

For exemple using the analogy with two-body scattering theory we can define an operator $\mathcal{U}$ such that

$$\mathcal{U}|\psi^{\rho_o}> = \mathcal{V}|\Psi^{\rho_o}> \tag{2.71}$$

where $\mathcal{U}$ satisfies equation

$$\mathcal{U} = \mathcal{V} + \mathcal{V}\mathcal{G}_o\mathcal{U} \ . \tag{2.72}$$

Using (2.68) and (2.69) one may explicitly write down

$$\mathcal{U}_{ij}^{\sigma\rho} = (G_o t_i G_o)^{-1} \ \bar{\delta}_{\sigma\rho} \ \delta_{ij} + \sum_{\alpha,k} \bar{\delta}_{\sigma\alpha} \ U_{ik}^{\alpha} \ G_o t_k G_o \mathcal{U}_{kj}^{\alpha\rho} \ , \tag{2.73}$$

which is the AGS equation for $\mathcal{U}$. The matrix elements of $\mathcal{U}$ between asymptotic states can be shown to give the appropriate two-cluster to two-cluster transition amplitudes. By using the defining equation (2.71) the matrix elements of $\mathcal{U}$ may be written as

$$<\psi^{\rho_o}|\mathcal{U}|\psi^{\sigma_o}> = <\psi^{\rho_o}|\mathcal{V}|\Psi^{\sigma_o}> , \tag{2.74}$$

which reads

$$<\psi^{\sigma_o}|\mathcal{V}|\Psi^{\rho_o}> = \sum_{\rho\sigma} \sum_{\substack{i\subset\sigma \\ j\subset\rho}} <\psi_i^{\sigma,\sigma_o}| \mathcal{V}_{ij}^{\sigma\rho}|\Psi_j^{\rho,\rho_o}> , \tag{2.75}$$

where

$$|\psi_i^{\sigma,\sigma_o}> = \delta_{\sigma\sigma_o}|\psi_i^{\sigma_o}> . \tag{2.76}$$

Using (2.69) and (2.76) together with (2.67) and (2.64) one may rewrite the right side of (2.75) as

$$\sum_{\rho} \sum_{\substack{i\subset\sigma_o \\ j\subset\rho}} <\psi_i^{\sigma_o}|(G_o t_i G_o)^{-1} \bar{\delta}_{\sigma_o\rho}\ \delta_{ij}\ G_o t_j \sum_{\substack{k\neq j \\ k\subset\rho}} G_o v_k|\Psi^{\rho_o}> , \tag{2.77}$$

which then becomes

$$\sum_{i\subset\sigma_o} <\psi_i^{\sigma_o}| \sum_{\rho\supset i} \bar{\delta}_{\sigma_o\rho} \sum_{\substack{k\neq i \\ k\subset\rho}} v_k|\Psi^{\rho_o}> . \tag{2.78}$$

One can easily verify that

$$\sum_{\rho\supset i} \bar{\delta}_{\sigma_o\rho} \sum_{\substack{k\neq i \\ k\subset\rho}} v_k = \sum_{k\not\subset\sigma_o} v_k , \tag{2.79}$$

whichever i may be. Therefore using (2.47) for the sum in i

$$<\vec{K}_{\sigma_o}|T^{\sigma_o\rho_o}|\vec{K}_{\rho_o}>=<\psi^{\sigma_o}|\mathcal{U}|\psi^{\rho_o}> = <\psi^{\sigma_o}|V^{\sigma_o}|\Psi^{\rho_o}> \tag{2.80}$$

where

$$V^{\sigma_o} = \sum_{k\not\subset\sigma_o} v_k , \tag{2.81}$$

is the t-matrix element for the transition between channel ρ_o and σ_o according to the definitions of multichannel quantum scattering theory[11].

Finally we would like to mention that Eqs.(2.37), (2.38), (2.44), (2.61) and (2.73) are three-vector variable integral equations which after triple partial wave decomposition reduce to three continuous variables and sums over relative orbital angular momenta coupled to total angular momentum $\mathcal{J}$ [12] (assuming the particles carry no intrinsic spin). The Jacobian coordinates of interest are shown in Fig.5 for (123)4 and (12)(34) two-cluster partitions. The momenta $\vec{p}_{1,2}$, $\vec{p}_{3,12}$, $\vec{p}_{4,123}$ and $\vec{p}_{12,34}$ are respectively the momenta between particles 1 and 2, particle 3 and pair (12), particle 4 and the three-particle system

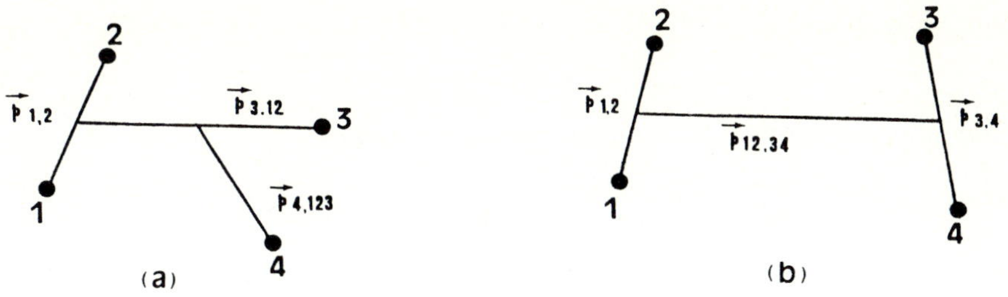

Fig.5 Standard Jacobian coordinates used in the four-body problem

(123) and pairs (12) and (34). The Jacobian coordinates of type (a) in Fig.5 are more suitable to the representation of (3)+1 subamplitudes while those of type (b) are useful to the representation of (2)+(2) subamplitudes. Further details are given in Appendix B where we also define $\vec{p}_i$, $\vec{p}_{\sigma,i}$ and $\vec{p}_\sigma$.

3. CALCULATION METHODS

We now proceed to illustrate in the simplest way the necessary steps one may follow to carry out a four-body calculation. Nevertheless because this was done with great care and sufficient detail in ref.[5] we abreviate this Section by sketching only the main steps and the final results. For simplicity we assume that all particles are distinguishable spinless bosons interacting by one term s-wave separable potentials. Although this may look like a restriction at the moment, the method may easily be generalized to fermions and any number of separable terms in any partial wave as a means to represent a realistic two-body potential. As shown in other contributions to this School[13,14] there are very accurate separable expansion methods for the two-body interaction which may prove to be of use in the four-body problem due to the difficulty of solving numerically at this time three continuous variable integral equations.

Assuming that the two-body interaction for pair k is

$$v_k = \lambda_k |f_k\rangle\langle f_k| \quad , \tag{3.1}$$

where λ_k is the potential strengh and $|f_k\rangle$ the two-body form factor, we use Eq.(2.10) for the two-body t-matrix to get

$$t_k = |f_k\rangle\tau_k(E_k)\langle f_k| \quad , \tag{3.2}$$

$$\tau_k^{-1} = \lambda_k^{-1} - \langle f_k|g_o(E_k)|f_k\rangle \ . \tag{3.3}$$

The pair energy is E_k and $g_o^{-1}(E_k) = (E_k - T_k)$ is the free resolvent for pair k where T_k is the relative kinetic energy. If pair k has a bound state at $E_k=-\varepsilon_k$ then

$$\lambda_k^{-1} = \langle f_k | g_o(-\varepsilon_k) | f_k \rangle \;, \tag{3.4}$$

and the two-body bound state wave function reads

$$|\phi_k\rangle = g_o(-\varepsilon_k)|\bar{f}_k\rangle \;, \tag{3.5}$$

where $|\bar{f}_k\rangle = N_k|f_k\rangle$ is the normalized form factor such that $\langle\phi_k|\phi_k\rangle=1$.

3.1 Two-Variable Four-Body Equation ([2V])

Starting with the AGS equation (2.73), multiplying on the left and right by $G_o t_i G_o$ and $G_o t_j G_o$, and introducing (3.2) everywhere one obtains

$$\tilde{\mathcal{P}}_{ij}^{\sigma\rho} = \bar{\delta}_{\sigma\rho}\;\delta_{ij}\;\tau_j + \sum_{\alpha k}\bar{\delta}_{\sigma\alpha}\;\tau_i\;X_{ik}^{\alpha}\;\tilde{\mathcal{P}}_{kj}^{\alpha\rho} \;, \tag{3.6}$$

where

$$\tilde{\mathcal{P}}_{ij}^{\sigma\rho} = \tau_i\;\langle f_i|G_o\mathcal{U}_{ij}^{\sigma\rho}G_o|f_j\rangle\tau_j \;, \tag{3.7}$$

$$X_{ij}^{\alpha} = \bar{\delta}_{ij}\;B_{ij}^{\alpha} + \sum_k \bar{\delta}_{ki}\;B_{ik}^{\alpha}\tau_k X_{kj}^{\alpha} \;, \tag{3.8}$$

$$B_{ij}^{\alpha} = \langle f_i|G_o|f_j\rangle \;. \tag{3.9}$$

The operator $\tilde{\mathcal{P}}$ involves an implicit integration over the relative momenta $\vec{p}_i$ and $\vec{p}_j$ of pairs i and j. Therefore Eq.(3.6) is, after partial wave decomposition, a two-continuous-variable integral equation ([2V]) in the momenta $\vec{p}_{\sigma,i}$ and $\vec{p}_\sigma$ for σ of (3)+1 type and $\vec{p}_j$ and $\vec{p}_\sigma$ for σ of (2)+(2) type where $\vec{p}_j$ is the momentum of the pair other than i which is internal to σ. The equation for X^α is the subsystem amplitude in partition α which, for α of (3)+1 type, corresponds to the Mitra-Amado-Lovelace[15,16,17] three-body equation where B is the particle exchange Born term. As shown in Appendix C the bound state equation for clusters in partition σ becomes

$$|\bar{g}_i^\sigma\rangle = \sum_k \bar{\delta}_{ik}\;B_{ik}^\sigma\;\tau_k\;|\bar{g}_k^\sigma\rangle \;, \tag{3.10}$$

and the wave function component is given by

$$|\chi_i^\sigma\rangle = G_o|f_i\rangle\tau_i\;|\bar{g}_i^\sigma\rangle \;, \tag{3.11}$$

where $|\bar{g}_i^\sigma\rangle$ is considered normalized such that $\langle\chi^\sigma|\chi^\sigma\rangle=1$. Both (3.8) and (3.10) are single-variable integral equations after partial wave decomposition whose solution may be readily obtained by standard discretization methods[18]. Further details may also be found in Section 3. of ref.[5].

The four-body matrix elements of interest are given by

$$\sum_{ij} \langle\psi_i^{\sigma}|U_{ij}^{\sigma\rho}|\psi_j^{\rho}\rangle = \sum_{ij} \langle\phi(\vec{k}_{\sigma})|\langle\bar{g}_i^{\sigma}|\tilde{\mathcal{P}}_{ij}^{\sigma\rho}|\bar{g}_j^{\rho}\rangle|\phi(\vec{k}_{\rho})\rangle \quad , \tag{3.12}$$

according to (2.60) and (3.11). The solution of Eq.(3.6) is as difficult as solving the Faddeev equation for three particles interacting with local potentials. Therefore all that has been discussed in a previous lecture[19] may be used here. The only additional difficulty arises from the number of coupled equations which increases from three to eighteen for a single term separable interaction such as 3.1 or 18×N for N terms in the separable expansion of the underlying two-body t-matrix. Exact calculations with realistic interactions should use the [2V] integral equation approach, particularly in bound state calculations of the four-nucleon system. Although in the scattering region one may still use them, it turns out that for simple separable potential calculations it may be faster computerwise to use any of the methods presented below.

3.2 One-Variable Four-Body Equation ([1V])

If an appropriate separable representation can be found for all two-cluster subamplitudes X^{α}, further reduction in the dimensionality of the equations takes place at the expense of a larger number of coupled equations. As discussed in Appendix C of ref.[5] one can write X^{α} as an operator of rank N

$$X_{ik}^{\alpha} = \sum_{mn}^{N} |g_m^{\alpha i}(E_{\alpha})\rangle D_{mn}^{\alpha}(E_{\alpha})\langle g_n^{\alpha k}(E_{\alpha})| \quad , \tag{3.13}$$

where the g's are generalized subsystem form factors, the D's generalized propagators and E_{α} the internal energy of the two-cluster partition α given by $E_{\alpha}=E-\frac{1}{2\mu_{\alpha}}\vec{p}_{\alpha}^{\,2}$. The form of Eq.(3.13) is general enough to accomodate most separable expansion methods used such as the Hilbert Schmidt Expansion (HSE)[20], the Generalized Unitary Pole Expansion (GUPE)[21], the Energy Dependent Pole Expansion (EDPE)[22], the Generalized Energy Dependence Expansions (GEDE1 and GEDE2) [23] and others[14].

Going back to (3.6) and introducing (3.13) together with

$$\mathcal{Z}_{rs}^{\sigma i,\rho j} = \langle g_r^{\sigma i}|\tilde{\mathcal{P}}_{ij}^{\sigma\rho}|g_s^{\rho j}\rangle \quad , \tag{3.14}$$

and

$$V_{rs}^{\sigma i,\rho i} = \langle g_r^{\sigma i}|\tau_i|g_s^{\rho i}\rangle \quad , \tag{3.15}$$

we get

$$\mathcal{Z}_{rs}^{\sigma i,\rho j} = \bar{\delta}_{\sigma\rho}\,\delta_{ij}\,V_{rs}^{\sigma i,\rho i} + \sum_{\alpha k}\sum_{mn}^{N} \bar{\delta}_{\sigma\alpha}\,V_{rm}^{\sigma i,\alpha i}\,D_{mn}^{\alpha}\,\mathcal{Z}_{ns}^{\alpha k,\rho j} \quad , \tag{3.16}$$

which is an effective two-body equation. If both σ and ρ are 3+1 states, $V^{\sigma i,\rho i}$ involves the exchange of the pair i that is common to both σ and ρ. On the contrary if σ is of 3+1 kind and ρ of 2+2 kind, the effective potential $V^{\sigma i,\rho i}$ corresponds to the exchange of a single particle, the one needed to form the second pair in partition ρ other than i.

The operator $\mathcal{Z}$, as given by (3.14), involves an implicit integration over the variables $\vec{p}_{\sigma,i}$ if σ is of (3)+1 kind and $\vec{p}_j$ (the pair other than i) if σ is of (2)+(2) kind. Therefore $\mathcal{Z}$ depends only on the relative momenta $\vec{p}_\sigma$ between clusters and after partial wave decomposition equation (3.16) reduces to a single-variable integral equation.

As mentioned above this method is very useful for simple separable potential calculations. Nevertheless because the (2)+(2) subamplitudes are unphysical and the resulting separable expansions may converge slowly, it turns out that one can rewrite the AGS equations in a way that the contribution of the (2)+(2) subsystem is included by the convolution method.

3.3 One Variable Four-Body Equations with Convolution ([1V+C])

Going back to Eq.(2.73) we note that because of $\bar{\delta}_{\sigma\alpha}$ the summation over α excludes $\alpha=\sigma$ but constrains α to contain pair i that is internal to σ. Therefore if σ is a partition of (3)+1 type α may be of both (3)+1 and (2)+(2) type. Nevertheless if σ is a (2)+(2) partition, α can only be of (3)+1 kind because no two different (2)+(2) two-cluster partition can share a pair that is internal to both of them. Using the symbols σ,ρ, α to denote only the (3)+1 partitions and the letter a to denote (2)+(2) partitions, we can split (2.73) into two equations.

$$\mathcal{U}_{ij}^{\sigma\rho} = \bar{\delta}_{\sigma\rho}\,\delta_{ij}(G_o t_i G_o)^{-1} + \sum_{\alpha k} \bar{\delta}_{\sigma\alpha}\, U_{ik}^{\alpha} G_o t_k G_o \mathcal{U}_{kj}^{\alpha\rho}$$

$$+ \sum_{ak} U_{ik}^{a}\, G_o t_k G_o \mathcal{U}_{kj}^{a\rho}\ , \tag{3.17}$$

and

$$\mathcal{U}_{ij}^{a\rho} = \delta_{ij}(G_o t_i G_o)^{-1} + \sum_{\alpha k} U_{ik}^{\alpha} G_o t_k G_o \mathcal{U}_{kj}^{\alpha\rho}\ . \tag{3.18}$$

The first one relates 3+1 states with 3+1 and 2+2 states while the second one relates only 2+2 with 3+1. Therefore substituting, (3.18) in (3.17) and defining

$$\tilde{\mathcal{U}}_{ij}^{\sigma\rho} = G_o t_i G_o \mathcal{U}_{ij}^{\sigma\rho} G_o t_j G_o\ , \tag{3.19}$$

together with

we get[24]

$$\tilde{\mathcal{V}}^{\sigma\rho}_{ij} = \bar{\delta}_{\sigma\rho}\delta_{ij}(G_o t_j G_o) + \sum_{a} G_o t_i G_o U^{a}_{ij} G_o t_j G_o , \tag{3.20}$$

$$\tilde{\mathcal{U}}^{\sigma\rho}_{ij} = \tilde{\mathcal{V}}^{\sigma\rho}_{ij} + \sum_{\alpha}\sum_{k\ell}\tilde{\mathcal{V}}^{\sigma\alpha}_{ik} U^{\alpha}_{k\ell}\tilde{\mathcal{U}}^{\alpha\rho}_{\ell j} , \tag{3.21}$$

which involves only the (3)+1 subamplitudes explicitly and an effective potential that holds the contribution of the (2)+(2) subsystem. Once i and j are fixed the summation over a is restricted to a single (2)+(2) partition.

If at this stage we use the operator identity (2.54) in (3.20) together with (2.9) we get

$$\tilde{\mathcal{V}}^{\sigma\rho}_{ij}(z) = \bar{\delta}_{\sigma\rho}\,\delta_{ij}\, G_o(z)t_j(z)G_o(z) + G_o(z)\left[v_i G_a(z) v_i - \delta_{ij} v_i G_o(z) t_i(z)\right] G_o(z). \tag{3.22}$$

Since $G_a(z)$ is the resolvent for two non interacting pairs it may be expressed in terms of the convolution of the two-body subsystem resolvents $G_i(z)$ and $G_j(z)$ in the form of[25]

$$G_a(z) = \frac{1}{2\pi i}\oint dz'\; G_i(z')G_j(z-z'), \tag{3.23}$$

where the contour integration goes counterclokwise around the singularities of $G_i(z)$.

If we now assume for simplicity that the two-body t-matrix is given by (3.2) and that the (3)+1 subamplitudes may be expressed in a separable form through (3.13), Eq.(3.21) becomes

$$\mathcal{Z}^{\sigma i,\rho j}_{rs} = \mathcal{V}^{\sigma i,\rho j}_{rs} + \sum_{\alpha k\ell}\sum_{mn}^{N}\mathcal{V}^{\sigma i,\alpha k}_{rm} D^{\alpha}_{mn}\mathcal{Z}^{\alpha\ell,\rho j}_{ns} , \tag{3.24}$$

where

$$\begin{aligned}\mathcal{V}^{\sigma i,\rho j}_{rs} &= \bar{\delta}_{\sigma\rho}\,\delta_{ij}\,\langle g^{\sigma i}_r|\tau_i|g^{\rho j}_s\rangle \\ &+ \frac{1}{2\pi i}\oint dz'\langle g^{\sigma i}_r|\tau_i(z')\langle f_i|G_o(z')G_o(z-z')|f_j\rangle\tau_j(z-z')|g^{\rho j}_s\rangle \\ &- \delta_{ij}\langle g^{\sigma i}_r|\left[1-\lambda_i\tau_i^{-1}(z)\right]\tau_i(z)|g^{\rho j}_s\rangle ,\end{aligned} \tag{3.25}$$

and $\mathcal{Z}$ is defined as in (3.14). While the first term in (3.25) already appeared in (3.16) the extra terms are new and correspond to a box-diagram where two non interacting pairs propagate between an initial and a final 3+1 state.

Although the convolution method is independent of separable representations for subsystem amplitudes and may be used in the context of three-vector variable formulations (Eq.3.21 with (3.22) and (3.23)), its advantage resides in the formulation of one-vector-variable integral equations where the contribution of (2)+(2) subsystem may be treated exactly at the expense of an extra integral over the singularities of a resolvent. These integrals are usually done using dispersion techniques together with subtraction methods[26].

4. FOUR IDENTICAL BOSONS

Having written general expressions for the four-body equations in operator form we now give explicit expressions for the relevant equations for a system of four identical bosons of mass M (2M=1).

4.1 Bound State Equation

Starting with Eq.(2.61) we first drop the inhomogeneous term and the subscript ρ_o that specifies the incoming channel state, to get a set of eighteen homogeneous coupled equations

$$|\Psi_i^\rho> = \sum_{\sigma k} G_o t_i G_o U_{ik}^\rho \bar{\delta}_{\sigma\rho} |\Psi_k^\sigma> \quad . \tag{4.1}$$

For i=(12) and ρ=(123)4, (4.1) reads

$$|\Psi_{(12)}^{(123)4}>=G_o t_{(12)} G_o \{U_{(12)(12)}^{(123)4} \left[|\Psi_{(12}^{(124)3}> + |\Psi_{(12)}^{(12)(34)}>\right]$$
$$+U_{(12)(13)}^{(123)4} \left[|\Psi_{(13)}^{(134)2} +|\Psi_{(13)}^{(13)(24)}>\right]+U_{(12)(23)}^{(123)4} \left[|\Psi_{(23)}^{(234)1}>+|\Psi_{(23)}^{(23)(14)}>\right]\} \quad . \tag{4.2}$$

Since all particles are identical one cannot distinguish one (3)+1 two-cluster partition from another and likewise for the (2)+(2) partitions. Therefore the functional form of the $|\Psi_i^\sigma>$ is the same for all σ of the same kind ((3)+1) or (2)+(2)) and independent of i because all pairs are alike. Consequently the number of independent components reduces down to two : one for partitions of (3)+1 type that we denote Ψ^{31} and a second one for partitions of (2)+(2) type that we denote Ψ^{22}. Furthermore matrix elements of the operator U_{ik}^σ are not observable but only $\sum_i U_{ik}^\sigma$ (or $\sum_k U_{ik}^\sigma$) because for a given pair in the initial state one cannot distinguish one pair from another in the final state. Therefore, as shown in Appendix C, there are only two independent subamplitudes U^{31} and U^{22} that satisfy formal equations

$$U^{31} = 2\, G_o^{-1} + 2\, t\, G_o\, U^{31} \quad , \tag{4.3}$$

and

$$U^{22} = G_o^{-1} + t\, G_o\, U^{22} \quad , \tag{4.4}$$

where t is the pair t-matrix. The resulting four-body equation for $|\Psi^{13}>$ becomes

$$|\Psi^{31}> = G_o t G_o U^{31}|\Psi^{31}> + G_o t G_o U^{31}|\Psi^{22}> . \quad (4.5)$$

Likewise if we start with any ρ of (2)+(2) kind and proceed as in (4.2) we get an equation for $|\Psi^{22}>$

$$|\Psi^{22}> = 2 G_o t G_o U^{22} |\Psi^{31}> . \quad (4.6)$$

The equation (4.5) and (4.6) are formal equations for a system of four identical bosons, which replace the original set of eighteen coupled equations.

As in the three-body problem[19] the solution of Eqs.(4.5) and (4.6) involves the projection into a complete set of momentum states which are expressed in the Jacoby variables of Fig.5. Denoting the orbital angular momenta of the pair, third particle to pair and fourth particle to remaining three-body system as ℓ, L and $\mathcal{L}$ respectively, we may generalize Eqs.(3.2) and (3.4) of ref.[19] to get

$$<\vec{p}'\vec{P}'\vec{\mathcal{P}}'|p\ P\ \mathcal{P}(\ell L)(L_3\mathcal{L})\mathcal{J}\mathcal{M}_j> = \frac{\delta(p'-p)}{p'p}\ \frac{\delta(P'-P)}{P'P}\ \frac{\delta(\mathcal{P}'-\mathcal{P})}{\mathcal{P}'\mathcal{P}}$$

$$\times\ \mathcal{Y}^{\mathcal{J}\mathcal{M}_j}_{\ell L\ L_3\mathcal{L}}(\hat{p}',\hat{P}',\hat{\mathcal{P}}') , \quad (4.7)$$

and

$$\sum_{\substack{\ell L L_3\\ \mathcal{L}\mathcal{J}\mathcal{M}_j}} \int dp\ p^2 \int dP\ P^2 \int d\mathcal{P}\mathcal{P}^2\ |p\ P\ \mathcal{P}(\ell L)(L_3\mathcal{L})\mathcal{J}\mathcal{M}_j>$$

$$\times\ <p\ P\ \mathcal{P}(\ell L)(L_3\mathcal{L})\mathcal{J}\mathcal{M}_j| = 1. \quad (4.8)$$

which are useful to represent (3)+1 components. For the (2)+(2) components one may define similar states in the variables p,q and $\mathcal{Q}$ where $\mathcal{Q}$ is the relative momentum between pairs. Next we proceed as in the three-body problem to obtain for a given $\mathcal{J}$ coupled equations in many channels. For $\mathcal{J}=0^+$ and $\ell=L=\mathcal{L}=0$ one gets a set of two coupled equations in the variables p,P and $\mathcal{P}$ or p,q and $\mathcal{Q}$ depending on the whether we have a (3)+1 or a (2)+(2) component. A note of care should be added concerning Eqs.(4.5) and (4.6) which are formal equations written in operator form without specifying alternation rules in the two-cluster partitions. In actual calculations, where one introduces complete sets (Eq.4.8) in intermediate states, one has to use Jacoby variables according to the alternation rules set up by the original equation. Therefore one should start with (4.1), introduce all necessary partitions of unity in the appropriate Jacoby coordinates and make the necessary transformations to end up with operators U^{31} and U^{22} defined as in

Appendix C together with single component states $|\Psi^{13}>$ and $|\Psi^{22}>$.

If the two-body t-matrix is separable in momentum space

$$t = |f> \tau <f| , \tag{4.9}$$

we substitute (4.9) in (4.5) and (4.6) to obtain

$$|\Psi^{31}> = G_o|f>\tau<f|G_oU^{31}|\Psi^{31}> + G_o|f>\tau<f|G_oU^{31}|\Psi^{22}> ,$$

$$|\Psi^{22}> = 2\ G_o|f>\tau<f|G_oU^{22}|\Psi^{31}> , \tag{4.10}$$

which again factorizes in the pair momentum by defining

$$|\Psi> = G_o|f>\ |R> . \tag{4.11}$$

Substituting (4.11) in (4.10) and using the equivalent of Eq.(C3) for identical bosons we get

$$|R^{31}> = \tau X^{31}|R^{31}> + \tau X^{31}|R^{22}> ,$$

$$|R^{22}> = 2\ \tau\ X^{22}|R^{31}>, \tag{4.12}$$

which are formal equations whose momentum space representation reads [27]

$$R^{31}(\vec{P},\vec{\mathcal{P}}) = \tau(E-\tfrac{3}{2}\vec{P}^2 - \tfrac{4}{3}\vec{\mathcal{P}}^2) \int \frac{d^3K}{(2\pi)^3}\{<\vec{P}|X^{31}(E-\tfrac{4}{3}\vec{\mathcal{P}}^2)|\vec{K}+\tfrac{1}{3}\vec{\mathcal{P}}>$$

$$\times R^{31}(\vec{\mathcal{P}}+\tfrac{1}{3}\vec{K},\vec{K}) + <\vec{P}|X^{31}(E-\tfrac{4}{3}\vec{\mathcal{P}}^2)|\vec{K}-\tfrac{2}{3}\vec{\mathcal{P}}>$$

$$\times R^{22}(\vec{\mathcal{P}}-\tfrac{1}{2}\vec{K},\vec{K})\} , \tag{4.13}$$

$$R^{22}(\vec{q},\vec{\mathcal{Q}}) = 2\ \tau(E-2\vec{q}^2-\vec{\mathcal{Q}}^2) \int \frac{d^3K}{(2\pi)^3}<\vec{q}|X^{22}(E-\vec{\mathcal{Q}}^2)|\vec{K}+\tfrac{1}{2}\vec{\mathcal{Q}}>$$

$$\times\ R^{31}(-\tfrac{2}{3}\vec{K}-\vec{\mathcal{Q}},\vec{K}) , \tag{4.14}$$

which,as mentioned before,are two-vector variable integral equations. The matrix element of X^{31} and X^{22} satisfy the equations shown in Appendix C.

For a multiterm separable expansion for t of the type

$$t = \sum_{mn}^{N} |f_m> \tau_{mn} <f_n| \tag{4.15}$$

we proceed to substitute (4.15) in (4.5) and (4.6), and generalize (4.11) to get

$$|R_m^{31}> = \sum_{nr}^{N} \{\tau_{mn}\ X_{nr}^{31}\ |R_r^{31}> + \tau_{mn}\ X_{nr}^{31}\ |R_r^{22}>\}$$

$$|R_m^{22}> = \sum_{nr}^{N} 2\ \tau_{mn}\ X_{nr}^{22}\ |R_r^{31}> . \tag{4.16}$$

As shown in Section 3.2 further reduction in the dimensionality of the equations may be obtained by expanding X^{31} and X^{22} in a separable form [28 - 32].

4.2 Scattering Operators and Equations

As we mentioned before, in the scattering region it may be faster computer wise to use one-dimension integral equations for the operator $\mathcal{Z}$, particularly if simple separable potentials (or expansions) with few terms are used for the two-body t-matrix and the energy is not too high. What is considered too high may change from one problem to another but a reasonable order of magnitude is two or three times the energy corresponding to the four-body breackup threshold.

Taking note of (2.80), (3.12) and (3.14)

$$<\vec{p}_\sigma \mid T^{\sigma\rho} \mid \vec{p}_\rho> = \sum_{ij} \bar{\mathcal{Z}}^{\sigma i,\rho j}_{m,n} \quad , \tag{4.17}$$

where m and n denote a specific bound state in partition σ and ρ respectively and $\bar{\mathcal{Z}} = N_\sigma N_\rho \mathcal{Z}$ where N_σ is the normalization factor that relates the normalized form factor $|\bar{g}^{\sigma i}_m>$ with $|g^{\sigma i}_m>$ such that $|\bar{g}^{\sigma i}_m>=N_\sigma |g^{\sigma i}_m>$. Since in (3.16) the summation in k only involves $\mathcal{Z}$, not V or D, for identical particles we can sum in i and j because there is no pair dependence due to the identity of all three-cluster partitions. A new equation that only couples two-cluster partitions may be defined

$$\mathcal{Z}^{\sigma\rho}_{rs} = \bar{\delta}_{\sigma\rho} V^{\sigma\rho}_{rs} + \sum_\alpha \sum_{mn}^{N} \bar{\delta}_{\sigma\alpha} V^{\sigma\alpha}_{rm} D^\alpha_{mn} \mathcal{Z}^{\alpha\rho}_{ns} \quad , \tag{4.18}$$

where $V^{\sigma\rho}$ involves the exchange of a pair if both σ and ρ are of (3)+1 type (the pair shared by both σ and ρ) and the exchange of a particle if σ and ρ are of different type. Furthermore, since one cannot distinguish 3+1 (or 2+2) states that differ from each other by the permutation of two particles, the matrix element that corresponds to a physically measurable quantity for the reaction 3+1 → 3+1 (3+1 → 2+2) is $\mathcal{Z}^{11}$ ($\mathcal{Z}^{21}$) defined as

$$\mathcal{Z}^{11} = \frac{1}{n_{31}} \sum_{\sigma\rho} \mathcal{Z}^{\sigma\rho} \qquad \sigma,\rho \supset (3)+1 \quad , \tag{4.19}$$

$$\mathcal{Z}^{21} = \frac{1}{n_{31}} \sum_{\sigma\rho} \mathcal{Z}^{\sigma\rho} \qquad \begin{array}{l}\sigma \supset (2)+(2) \\ \rho \supset (3)+1\end{array} \quad , \tag{4.20}$$

because the final result does not depend on the initially chosen (3)+1 partition ρ. The number of (3)+1 partitions is $n_{31}=4$ whereas the number of (2)+(2) partitions is $n_{22}=3$. Therefore using (4.19) and (4.20) in (4.18) we get

$$\begin{aligned}\mathcal{Z}^{11}_{rs} &= 3\, V^{11}_{rs} + \sum_{mn}^{N} 3\, V^{11}_{rm} D^1_{mn} \mathcal{Z}^{11}_{ns} + \sum_{mn}^{N} 4\, V^{12}_{rm} D^2_{mn} \mathcal{Z}^{21}_{ns} \quad , \\ \mathcal{Z}^{21}_{rs} &= 3\, V^{21}_{rs} + \sum_{mn}^{N} 3\, V^{21}_{rm} D^1_{mn} \mathcal{Z}^{11}_{ns} \quad .\end{aligned} \tag{4.21}$$

where D^1 and D^2 are the propagators for (3)+1 and (2)+(2) states. Likewise one may define

$$\mathcal{Z}^{12} = \frac{1}{n_{22}} \sum_{\sigma\rho} \mathcal{Z}^{\sigma\rho} \qquad \sigma \supset (3)+1, \quad \rho \supset (2)+(2) \tag{4.22}$$

and

$$\mathcal{Z}^{22} = \frac{1}{n_{22}} \sum_{\sigma\rho} \mathcal{Z}^{\sigma\rho} \qquad \sigma,\rho \supset (2)+(2) \tag{4.23}$$

as the operators for the reactions initiated by a 2+2 state. The corresponding set of equations is

$$\mathcal{Z}^{12}_{rs} = 4\, V^{12}_{rs} + \sum_{mn}^{N} 3\, V^{11}_{rm} D^1_{mn} \mathcal{Z}^{12}_{ns} + \sum_{mn}^{N} 4\, V^{12}_{rm} D^2_{mn} \mathcal{Z}^{22}_{ns} ,$$

$$\mathcal{Z}^{22}_{rs} = \sum_{mn}^{N} 3\, V^{21}_{rm} D^1_{mn} \mathcal{Z}^{12}_{ns} \quad . \tag{4.24}$$

If one takes advantage of the factors of $\sqrt{3}$ and $\sqrt{2}$ used to normalized three-body and (2)+(2) bound state clusters for systems of identical particles (see Appendix C) and symmetrize with respect to the exchange of identical pairs in the 2+2 state we get for (4.21) a system of two coupled equations whose momentum space representation reads [28]

$$\langle \vec{P}'|\mathcal{Z}^{11}_{rs}(E)|\vec{P}\rangle = \langle \vec{P}'|V^{11}_{rs}(E)|\vec{P}\rangle + \sum_{mn}^{N} \int \frac{d^3P''}{(2\pi)^3} \langle \vec{P}'|V^{11}_{rm}(E)|\vec{P}''\rangle$$

$$\times D^1_{mn}(E-\tfrac{4}{3}\vec{P}''^2) \langle \vec{P}''|\mathcal{Z}^{11}_{ns}(E)|\vec{P}\rangle$$

$$+ \sum_{mn}^{N} \int \frac{d^3Q}{(2\pi)^3} \langle \vec{P}'|\bar{V}^{12}_{rm}(E)|\vec{Q}\rangle D^2_{mn}(E-\vec{Q}^2) \langle \vec{Q}|\mathcal{Z}^{21}_{ns}(E)|\vec{P}\rangle , \tag{4.25}$$

$$\langle \vec{Q}|\mathcal{Z}^{21}_{rs}(E)|\vec{P}\rangle = \langle \vec{Q}|V^{21}_{rs}(E)|\vec{P}\rangle + \sum_{mn}^{N} \int \frac{d^3P''}{(2\pi)^3} \langle \vec{Q}|V^{21}_{rm}(E)|\vec{P}''\rangle$$

$$\times D^1_{mn}(E-\tfrac{4}{3}\vec{P}''^2) \langle \vec{P}''|\mathcal{Z}^{11}_{ns}(E)|\vec{P}\rangle , \tag{4.26}$$

where

$$\langle \vec{P}'|V^{11}_{rs}(E)|\vec{P}\rangle = g^1_r(E-\tfrac{4}{3}\vec{P}'^2;\vec{P}+\tfrac{1}{3}\vec{P}')\, \tau(E-\vec{P}'^2-(\vec{P}+\vec{P}')^2/2-\vec{P}^2)$$

$$\times g^1_s(E-\tfrac{4}{3}\vec{P}^2;\vec{P}'+\tfrac{1}{3}\vec{P}), \tag{4.27}$$

$$\langle \vec{Q}|V^{21}_{rs}(E)|\vec{P}\rangle = \langle \vec{Q}|\bar{V}^{21}_{rs}(E)|\vec{P}\rangle + \langle -\vec{Q}|\bar{V}^{21}_{rs}(E)|\vec{P}\rangle, \tag{4.28}$$

$$\langle \vec{Q}|\bar{V}^{21}_{rs}(E)|\vec{P}\rangle = g^2_r(E-\vec{Q}^2;\vec{P}+\tfrac{1}{2}\vec{Q})\, \tau(E-\frac{\vec{Q}^2}{2}-(\vec{Q}+\vec{P})^2-\vec{P}^2)$$

$$\times g^1_s(E-\tfrac{4}{3}\vec{P}^2;\vec{Q}+\tfrac{2}{3}\vec{P}) \tag{4.29}$$

Some authors [33] prefer to deal with a single equation by substituting (4.26) in (4.25). The corresponding equation for 3+1 → 3+1 amplitude is

$$\langle\vec{P}'|\mathcal{Z}^{11}_{rs}(E)|\vec{P}\rangle = \langle\vec{P}'|V^{ef}_{rs}(E)|\vec{P}\rangle + \sum_{mn}^{N}\int\frac{d^3P''}{(2\pi)^3}\langle\vec{P}'|V^{ef}_{rm}(E)|\vec{P}''\rangle \times D^1_{mn}(E-\tfrac{4}{3}\vec{P}'')\ \langle\vec{P}''|\mathcal{Z}^{11}_{rs}(E)|\vec{P}\rangle , \tag{4.30}$$

which has an effective potential

$$\langle\vec{P}'|V^{ef}_{rs}(E)|\vec{P}\rangle = \langle\vec{P}'|V^{11}_{rs}(E)|\vec{P}\rangle + \sum_{mn}^{N}\int\frac{d^3Q}{(2\pi)^3}\langle\vec{P}'|\bar{V}^{12}_{rm}(E)|\vec{Q}\rangle\ D^2_{mn}(E-\vec{Q}^2)\langle\vec{Q}|V^{21}_{ns}(E)|\vec{P}\rangle . \tag{4.31}$$

If we follow instead an approach based an Eq.(3.24) where V is given by (3.25) and proceed to introduce particle identity we obtain an effective equation for the 3+1 → 3+1 amplitude similar to (4.30) where the effective potential V^{ef} is obtained by convolution of τ's [34,35]

$$\langle\vec{P}'|V^{ef}_{rs}(E)|\vec{P}\rangle = \langle\vec{P}'|V^{11}_{rs}(E)|\vec{P}\rangle + \sum_{\mu=1}^{2} 2\int\frac{d^3Q}{(2\pi)^3}\langle\vec{P}'|\bar{B}^{12}_{r}(E)|\pm\vec{Q}\rangle\ G_\mu(E;\vec{P}',\vec{Q},\vec{P})\ \langle\vec{Q}|\bar{B}^{21}_{s}(E)|\vec{P}\rangle. \tag{4.32}$$

The + sign goes with μ=1 and the - sign with μ=2. The factor of two results from the identity of pairs in the intermediate state. The first term in (4.32) is given by (4.27) and

$$\langle\vec{Q}|\bar{B}^{21}_{r}(E)|\vec{P}\rangle = \frac{f(\vec{P}+\frac{1}{2}\vec{Q})\ g^1_s(E-\frac{4}{3}\vec{P}^2;\vec{Q}+\frac{2}{3}\vec{P})}{E+\varepsilon_2-\frac{1}{2}\vec{Q}^2-(\vec{Q}+\vec{P}')^2-\vec{P}'^2} , \tag{4.33}$$

which coincides with (4.29) on-shell if (C12) in Appendix C is used in (4.29) for $E=2\varepsilon_2+Q^2$. Assuming there is a two-body bound state with energy ε_2 the generalized propagators G_1 and G_2 are given by

$$G_1(E;\vec{P}',\vec{Q},\vec{P})=\tau(Y-\varepsilon_2) - \frac{(Y-U)(Y-U')}{\pi}\int_0^\infty dx\ \frac{\mathrm{Im}[\tau(x)]\tau(Y-2\varepsilon_2-x)}{(Y-U-\varepsilon_2-x)(Y-U'-\varepsilon_2-x)} , \tag{4.34}$$

$$G_2(E;\vec{P}',\vec{Q},\vec{P})=\frac{(Y-U'')}{-U''}\tau(Y-\varepsilon_2)+\frac{(Y-U)(Y-U'')}{Y-U-U''}\tau(U''-\varepsilon_2)\tau(Y-U''-\varepsilon_2) - \frac{(Y-U)(Y-U'')}{\pi}\int_0^\infty dx\ \frac{\mathrm{Im}[\tau(x)]\tau(Y-2\varepsilon_2-x)}{(x+\varepsilon_2-U'')(Y-U-\varepsilon_2-x)} \tag{4.35}$$

where

$$Y = E + 2\varepsilon_2 - Q^2$$

$$Y-U = E+\varepsilon_2-\tfrac{1}{2}\vec{Q}^2-(\vec{P}+\vec{Q})^2-\vec{P}^2 ,$$

$$Y-U' = E+\varepsilon_2-\tfrac{1}{2}\vec{Q}^2-(\vec{P}'+\vec{Q})^2-\vec{P}'^2 ,$$

$$Y-U'' = E+\varepsilon_2-\tfrac{1}{2}\vec{Q}^2-(\vec{P}'-\vec{Q})^2-\vec{P}'^2 . \tag{4.36}$$

This effective potential is depicted in Fig.6 where G_1 is the first box diagram and G_2 the second and third box diagrams which have a complete analog to time ordered field theory graphs in terms of two-body and three-body form factors and pair propagators.

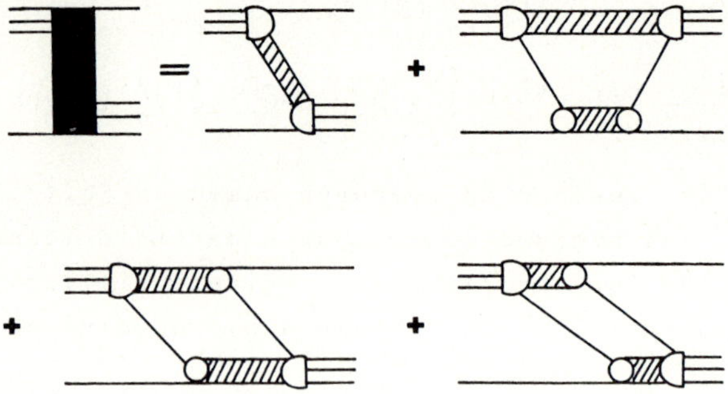

Fig.6 Graphical representation of the effective potential given by Eq.(4.32).

Both effective potentials (4.31) and (4.32) are equivalent and the choice between one or the other has to be decided by actual practice. Since the convolution method treats the (2)+(2) system exactly, avoids separable expansions of (2)+(2) subamplitudes leading to fewer number of coupled equations in the four-body sector (if (4.25) and (4.26) are used) and simplified convergence procedure. Furthermore in the convolution method the analytical structure of the (2)+(2) subamplitudes is explicitly shown in the propagators G_1 and G_2 which may prove advantageous to scattering calculations. This also allows the testing of very simple approximations [35] to the pair propagators that may prove usefull to the study of reaction models in more complex systems.

The integrals in (4.34) and (4.35) may be evaluated by Gauss quadrature together with subtraction methods after decomposition in partial fractions relative to each term in the denominator of the integrand.

5. APPLICATIONS AND CONCLUSIONS

Although over ninety percent of the calculations involving the solution of four-body integral equations address themselves to the four-nucleon system, there has been some work on photonuclear reactions in ^4He [36], bound state calculations of Hypernuclei ($^4_\Lambda H$ and $^4_\Lambda He$ [37] as well as $^4_C He$ [38]) and bound state as well as scattering calculations involving systems of four atoms [39-41]. Since one can find in the literature [5,42 - 45] very complete review articles on these subjects we only discuss here a few results concerning the four-nucleon system. This is by no means a review on the subject but a very short summary of what has been done and where new progress may be achieved.

Nucleons are spin 1/2 isospin 1/2 particles and therefore the equations of Section 4. have to be generalized to include spin and isospin degrees of freedom. Without entering into too much detail we outline in Appendix D the most important modifications, and refer to the appropriate literature for extra information. Most work involves only ^1S_o and $^3S_1-^3D_1$ in the t_{oo} approximation[5] where only the $\ell=0$ component of the nucleon-nucleon (N-N) triplet t-matrix is included. In scattering calculations the N-N interactions that have been used are simple one-term separable potentials[35] or few term separable representations of Malfliet-Tjon I-III (MT I-III)[46]. In bound state calculations converged results exist for both MT I-III and Reid potentials. To the exception of two calculations[46,47], all other work only includes L=0 (3)+1 subamplitudes. Only recently[48] was the importance of the p-wave N-N channels studied in the reaction $n\,^3H \rightarrow n\,^3H$ below four-body breackup threshold. The p-wave N-N channels were included in an approximate way and their effect found negligible.

In general one can say that there is good agreement between different bound state calculations using different sets of equations or expansion methods. This is shown in Table II for the ground state of ^4He and in Table III for the excited state where the calculation using the Bateman method is in desagreement with the other two.

Table II ^4He binding energy in MeV for different potentials between pairs and integral equation methods. Only L=0 (3)+1 subamplitudes were included

V_{NN}	[2V]	[1V] EDPE	[1V] GUPE	[1V] HS	[1V] Bateman	[1V+C] EDPE
Y1	45.7 [27]			45.7 [29]	45.2 [49]	45.6 [35]
Y2	42.4 [27]	42.3 [30]	41.7 [30]			42.3 [35]
MT I-III		30.4 [30]	27.3 [30]	29.6 [31]		
RSC				19.5* [32]		

* Tensor-force included through t_{oo} (see ref. [5]).

Table III - Excited 0^+ state in ^{4}He for potential Y1 in MeV. Only L=0 (3)+1 subamplitudes were included

V_{NN}	[1V] Bateman	[1V] HS	[1V+C] EDPE
Y1	10.88 [49]	11.69 [29]	11.63 [35]

In these calculations only the L=0 (3)+1 subamplitudes were included. Nevertheless the effect of the L=1 (3)+1 subamplitudes on ^{4}He states has been studied[48] and found negligible (less than 0.2% effect).

Threshold scattering results, have been performed by three groups[46,47,49] but only two may be compared with each other. Unfortunately by comparing columns two and three in Table IV one finds considerable discrepancy and a third independent calculation is very much needed. Nevertheless all the work performed

Table IV - Scattering lengh $a_{\mathcal{SF}}$ in fm for different potentials and integral equation methods.

V_{NN} Ref.	Y1 [47]	Y1 [47]	Y1 [49]	E [49]	MT I-III L1 [46]	MT I-III L2 [46]	MT I-III L4 [46]
L's	[L=0,1]		[L=0]	(3)+1	Subamplitudes		
Method	[1V+C]	(EDPE)	[1V]	(Bateman)	[1V]	(HS)	
a_{01}	3.24	3.26	3.77	3.89	3.74	4.14	4.09
a_{11}	3.00	3.06	3.13	3.22	3.32	3.61	3.61
a_{00}	9.79	9.75	12.34	14.95	-53.9	-14.8	-25.6
a_{10}	2.55	2.64	3.03	3.09	2.35	2.44	2.65

so far seems to indicate that $a_{01} > a_{11} > a_{10}$ independently of the potential chosen and that, to the exception of a_{00}, all other scattering lengh depend very tittle on the N-N interaction as expected by spin-isospin considerations. The values for a_{00} depend strongly on V_{NN} through the existence of a second 0^+ bound state. In general separable N-N potentials lead to a 0^+ excited state while local potentials with short range repulsion don't.

As the energy increases behond the first scattering threshold the contribution of the p-wave (3)+1 subamplitudes increases and is responsible for remarkable changes in the phase shifts and differential cross sections. There are only two full calculation in this energy region[46-47], and a few using k-matrix approximation[50]. The most recent one of the full calculations[47] involves the solution of a sixteen channel [1V+C] equation for Y4 potential which has 4% d-state probability. In the t_{oo} approximation the resulting ^3H binding energy is ε_3=8.66 MeV and ε_α=33.9 MeV together with ε_α^*=8.9 MeV. With this same potential, whose parameters are given in ref.[5], (same for Y1 and Y2) we also get a_{01}=3.89fm, a_{11}=3.53fm, a_{10}=2.91fm and a_{00}=12.34fm which to the exception of a_{00} are in good agreement with experiment where a_{01}^{exp}= 3.91± 0.12fm[42] and a_{11}^{exp}=3.60 ± 0.10fm. Typical results for the differential cross sections are shown in Figs 7. and 8. for $p\,^3He \rightarrow p\,^3He$ and $dd \rightarrow p\,^3H$ respectively. Although the results are very encouraging for $p\,^3He$ elastic scattering further basic improvements are still needed for the reactions initiated by dd, possibly adding the d-wave NN tensor components.

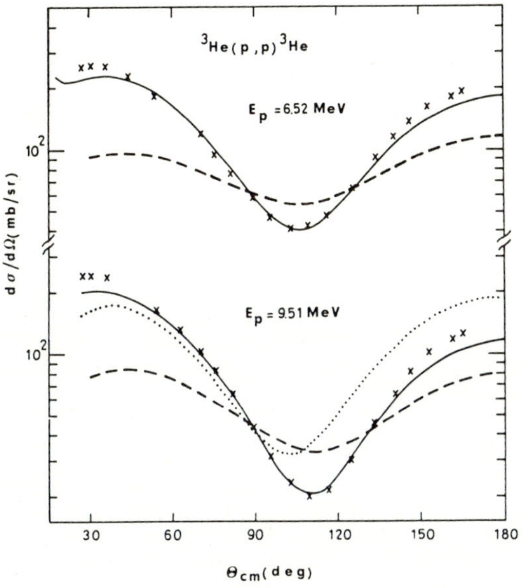

Fig.7 Angular distribution for $^3He(p,p)^3He$ at different proton laboratory energies. The solid line includes both L=0 and L=1 (3)+1 subamplitudes while the dashed line corresponds to including the L=0 subamplitudes alone. As for the dotted line it includes the L=1 subamplitudes in first order perturbation. The crosses are experimental points from ref. [51].

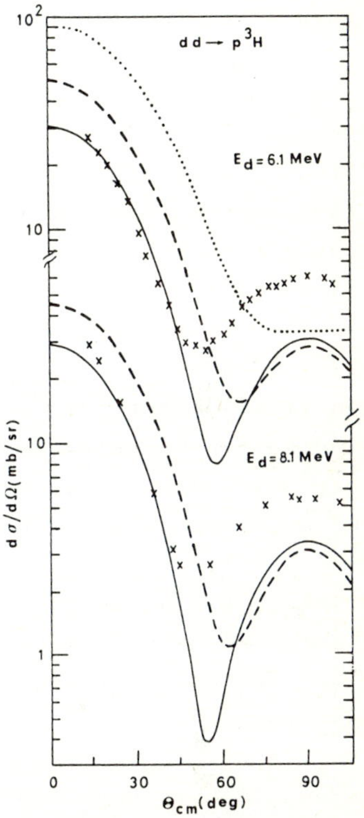

Fig.8 Same as in Fig.7 for $^2H(d,p)^3H$. The crosses are experimental points from ref. [52].

Therefore, as mentioned above, four-nucleon calculations, though the simplest one can perform with four interacting particles, are still at a rudimentary level compared to three-nucleon calculations particularly concerning the sofistication of the potentials that have been used. As mentioned in Appendix D four-body calculations are one order of magnitude more difficult to perform than the corresponding three-nucleon calculations, and the number of channels involved quickly rises into the few tens (even hundreds) as new N-N channels and two-cluster subamplitudes are added, not to mention those resulting from the separable representation of the operators. Nevertheless with the help of powerful separable techniques one is now on the threshold for new progress. Bound state calculations using two-variable integral equations with realistic potentials in $^3S_1-^3D_1$ and ^1S_o channels and all dominant (3)+1 and (2)+(2) subamplitudes are within reach of present day computers. This may provide reasonably accurate wave functions with which one may study assymptotic normalization constants, photo-nuclear processes in ^4He and electromagnetic observables. Since nucleons are strongly confined in ^4He, this nucleus may provide the most stringent test for the nuclear force either of the two-or three-body nature. Therefore progress in this domain is necessary though competition from other methods is "fierce". In the scattering region the integral equation approach stands as the only source for exact microscopic calculations. There progress may come slowlier because one may be able to get reasonable agreement with experiment even when simple potentials are used between pairs. This is due to strong spin-isospin blocking in many of the channels and also to strong constraints of analyticity and unitarity [53] on the amplitudes. Nevertheless one should quickly reach a point where all calculations include L=0 and L=1 subamplitudes as well as the small tensor components of the N-N force. Realistic potentials may also be included but only at the expense of great numerical effort if more than one or two terms are considered. As for the solution of three-variable integral equations, though technically possible within a few years with large and faster computers, it may turn out to be unnecessary except for potentials whose separable representation is unreliable or inaccurate. Due to the rich structure of subamplitudes, the effect of the two-body t-matrix is integrated first to calculate the two-cluster subamplitudes which are subsequently integrated through to get four-body amplitudes. Therefore one may not be sensitive to details that a good separable expansion is not able to represent accurately.

APPENDIX A - **Alt, Grassberger and Sandhas Matrix Equations.**

From multichannel quantum scattering theory[11] we know that the t-matrix T_{ba} for the transition from an initial state in channel a to a final state in channel b is given by

$$T^{+}_{ba} = V^{b} + V^{b} G V^{a} , \tag{A1}$$

$$T^{-}_{ba} = V^{a} + V^{b} G V^{a} , \tag{A2}$$

where G is the full resolvent and

$$V^{b} = \sum_{i \not\subset b} v_{i} , \tag{A3}$$

is the sum of all interactions that are external to b. The post form of the transition operator is given by (A1) while the prior form is given by (A2). Defining

$$V_{b} = \sum_{i \subset b} v_{i} , \tag{A4}$$

one may write

$$G_{b} = H_{o} - V_{b} , \tag{A5}$$

and because $V=V_b+V^b$

$$V^{b} = G_{b}^{-1} - G^{-1} . \tag{A6}$$

Subtituting (A6) in (A1) and (A2) we get

$$T^{+}_{ba} = -G_{a}^{-1} + G_{b}^{-1} G G_{a}^{-1} , \tag{A7}$$

$$T^{-}_{ba} = -G_{b}^{-1} + G_{b}^{-1} G G_{a}^{-1} . \tag{A8}$$

Another operator U_{ba} that is on-shell equivalent to (A7) and (A8) may be defined as

$$U_{ba} = T^{+}_{ba} + \bar{\delta}_{ab} G_{a}^{-1} = T^{-}_{ba} + \bar{\delta}_{ab} G_{b}^{-1} ,$$

which, unlike (A1) or (A2), is left-right symmetric

$$U_{ba} = -\delta_{ab} G_{a}^{-1} + G_{b}^{-1} G G_{a}^{-1} , \tag{A9}$$

or

$$G = \delta_{ab} G_{b} + G_{b} U_{ba} G_{a} . \tag{A10}$$

The operator U_{ba} is the AGS operator for which a connected Kernel matrix equation was presented in Section 2. . Here we rederive the same equations using AGS approach[3,9] which is not only elegant but also easily generalizable to any number of particles.

Starting from the Lippmann-Schwinger equation for the full resolvent

$$G = G_o + G_o \sum_i v_i G \ , \qquad (A11)$$

we define the operator M_{ij}

$$M_{ij} = \delta_{ij} v_i + v_i G v_j \ , \qquad (A12)$$

which is the sum of all terms that start with an interaction in pair j and end with an interaction in pair i. Substituting (A11) in (A12) we get

$$M_{ij} = \delta_{ij} v_i + v_i G_o v_j + v_i G_o \sum_k v_k G v_j \ , \qquad (A13)$$

which by using again (A12) in (A13) leads to

$$M_{ij} = \delta_{ij} v_i + v_i G_o \sum_k M_{kj} \ . \qquad (A14)$$

Bringing the term k=i to left side and inverting with the help of (2.8), (2.9) and (2.10b) we get

$$M_{ij} = \delta_{ij} t_i + t_i G_o \sum_k \bar{\delta}_{ki} M_{kj} \ , \qquad (A15)$$

which is similar to (2.31). The alternation rule is specified by $\bar{\delta}_{ij}=1-\delta_{ij}$. In AGS matrix notation this reads

$$\mathbb{G} = \mathbb{G}_o + \mathbb{G}_o \mathbb{V} \mathbb{G} \ , \qquad (A16)$$

where

$$[\mathbb{G}_o]_{ij} = G_o t_i G_o \delta_{ij} \ , \qquad (A17)$$

$$[\mathbb{V}]_{ij} = \bar{\delta}_{ij} G_o^{-1} \ , \qquad (A18)$$

$$[\mathbb{G}]_{ij} = G_o M_{ij} G_o \ . \qquad (A19)$$

For three particles (A15) is already connected and one may proceed to write an equation for U using the operators $\mathbb{G}_o$ and $\mathbb{V}$ as if they where two-body operators to obtain

$$\mathbb{U} = \mathbb{V} + \mathbb{V} \mathbb{G}_o \mathbb{U} \ , \qquad (A20)$$

which reads like (2.29) if one uses (A17) and (A18) in (A16).
For four particles (A15) is not connected and one proceeds to define an operator $\mathfrak{M}^{\sigma\rho}$ at the expense of (A16). Using the analogy with (A11) where the total V is distributed over all three-cluster partitions (or pairs) we now distribute $\mathbb{V}$ over all two-cluster partitions

$$\mathbb{V} = \sum_\alpha \mathbb{V}^\alpha \ , \qquad (A21)$$

where

$$[\mathbb{V}^\alpha]_{ij} = \bar{\delta}_{ij} G_o^{-1} \quad i,j \subset \alpha, \qquad (A22)$$
$$\phantom{[\mathbb{V}^\alpha]_{ij}} = 0 \quad i,j \not\subset \alpha.$$

The essential point is to prove that

$$\sum_{\alpha \supset i,j} \bar{\delta}_{ij} G_o^{-1} = \bar{\delta}_{ij} G_o^{-1} . \tag{A23}$$

In the four-body problem this is obvious because given two different pairs i and j there is only one α that contains both. Therefore one writes.

$$\mathbb{G} = \mathbb{G}_o + \mathbb{G}_o \sum_\alpha V^\alpha \mathbb{G} , \tag{A24}$$

and in analogy with (A12) one defines

$$\mathcal{M}^{\sigma\rho} = \delta_{\sigma\rho} \mathbb{V}^\sigma + \mathbb{V}^\sigma \mathbb{G} \mathbb{V}^\rho . \tag{A25}$$

Again substituting (A24) in (A25) and using once more (A25) in the new equation to redefine some of the terms we get

$$\mathcal{M}^{\sigma\rho} = \delta_{\sigma\rho} \mathbb{V}^\sigma + \mathbb{V}^\sigma \mathbb{G}_o \sum_\alpha \mathcal{M}^{\alpha\rho} \tag{A26}$$

where α may be equal to σ. Bringing the term $\alpha=\sigma$ to the left side and inverting one obtains a new equation

$$\mathcal{M}^{\sigma\rho} = \delta_{\sigma\rho} \mathbb{U}^\sigma + \mathbb{U}^\sigma \mathbb{G}_o \sum_\alpha \bar{\delta}_{\sigma\alpha} \mathcal{M}^{\alpha\rho} , \tag{A27}$$

where $\mathbb{U}^\sigma$ satisfies (A20)

$$\mathbb{U}^\sigma = \mathbb{V}^\sigma + \mathbb{V}^\sigma \mathbb{G}_o \mathbb{U}^\sigma , \tag{A28}$$

and

$$(\mathbb{1} + \mathbb{U}^\sigma \mathbb{G}_o)(\mathbb{1} - \mathbb{V}^\sigma \mathbb{G}_o) = \mathbb{1} . \tag{A29}$$

The operator $\mathbb{U}^\sigma$ is the same as defined in (A20) for pairs internal to σ and reads as (2.29). Again by analogy with (A17), (A18) and (A19) one may define

$$[\mathcal{G}_o]^{\sigma\rho} = \delta_{\sigma\rho} \mathbb{G}_o \mathbb{U}^\sigma \mathbb{G}_o , \tag{A30}$$

$$[\mathcal{V}]^{\sigma\rho} = \bar{\delta}_{\sigma\rho} \mathbb{G}_o^{-1} , \tag{A31}$$

$$[\mathcal{G}]^{\sigma\rho} = \mathbb{G}_o \mathcal{M}^{\sigma\rho} \mathbb{G}_o , \tag{A32}$$

that leads to

$$\mathcal{G} = \mathcal{G}_o + \mathcal{G}_o \mathcal{V} \mathcal{G} , \tag{A33}$$

and

$$\mathcal{U} = \mathcal{V} + \mathcal{V} \mathcal{G}_o \mathcal{U} , \tag{A34}$$

which is the four-body equivalent of (A20). Using (A30) and (A31) we may write

$$\mathcal{U}^{\sigma\rho} = \bar{\delta}_{\sigma\rho}\,\mathbb{G}_o^{-1} + \sum_{\alpha} \bar{\delta}_{\sigma\alpha}\mathbb{U}^{\alpha}\,\mathbb{G}_o\,\mathcal{U}^{\alpha\rho}. \tag{A35}$$

Using (A17) one obtain AGS equation (2.73)

$$\mathcal{U}_{ij}^{\sigma\rho} = \bar{\delta}_{\sigma\rho}\delta_{ij}(G_o t_i G_o)^{-1} + \sum_{\alpha k} \bar{\delta}_{\sigma\alpha}U_{ik}^{\alpha}G_o t_k G_o \mathcal{U}_{kj}^{\alpha\rho}, \tag{A36}$$

which is connected for a system of four particles. The method is easily generalizable to a system of N interacting particles.

APPENDIX B - Relative Momenta in the Four-Body Problem

Using the indices 1,2,3, and 4 to denote the particles, we define the relevant relative momenta of the four-body problem as mentioned before in Section 2. They are

$$\vec{p}_{1,2} = \frac{m_2}{M_{12}}\vec{p}_1 - \frac{m_1}{M_{12}}\vec{p}_2, \tag{B1}$$

$$\vec{p}_{3,12} = \frac{M_{12}}{M_{123}}\vec{p}_3 - \frac{m_3}{M_{123}}\vec{p}_{12}, \tag{B2}$$

$$\vec{p}_{4,123} = \frac{M_{123}}{M_{1234}}\vec{p}_4 - \frac{m_4}{M_{1234}}\vec{p}_{123}. \tag{B3}$$

$$\vec{p}_{12,34} = \frac{M_{34}}{M_{1234}}\vec{p}_{12} - \frac{M_{12}}{M_{1234}}\vec{p}_{34}. \tag{B4}$$

where $M_{12}=m_1+m_2$, $M_{123}=m_1+m_2+m_3$, $M_{1234}=m_1+m_2+m_3+m_4$, $\vec{P}_{12}=\vec{p}_1+\vec{p}_2$ and $\vec{P}_{123}=\vec{p}_1+\vec{p}_2+\vec{p}_3$. The momenta $\vec{p}_1,\vec{p}_2,\vec{p}_3$ and $\vec{p}_4$ are the momenta of each individual particle. In the Center of Mass where $\sum_{i=1}^{4} \vec{p}_i=0$ we have $\vec{P}_{12}=-\vec{P}_{34}$ and $\vec{p}_4=-\vec{P}_{123}$. Depending on the choice of Jacobian coordinates the total kinetic energy in the Center of Mass is given by ($\hbar=1$)

$$T_{CM} = \frac{1}{2\mu_{1,2}}\vec{p}_{1,2}^{\,2} + \frac{1}{2\mu_{3,12}}\vec{p}_{3,12}^{\,2} + \frac{1}{2\mu_{4,123}}\vec{p}_{4,123}^{\,2}, \tag{B5}$$

or

$$T_{CM} = \frac{1}{2\mu_{1,2}}\vec{p}_{1,2}^{\,2} + \frac{1}{2\mu_{3,4}}\vec{p}_{3,4}^{\,2} + \frac{1}{2\mu_{12,34}}\vec{p}_{12,34}^{\,2} \tag{B6}$$

where

$$\mu_{1,2} = \frac{m_1 m_2}{M_{12}} \tag{B7}$$

$$\mu_{3,12} = \frac{m_3 M_{12}}{M_{123}} \tag{B8}$$

$$\mu_{4,123} = \frac{m_4 M_{123}}{M_{1234}} \tag{B9}$$

$$\mu_{12,34} = \frac{M_{12}M_{34}}{M_{1234}} \tag{B10}$$

are the appropriate reduced masses. Other sets of relative momenta and reduced masses may be obtained by permutation of the indices 1,2,3,4. Also for convenience of notation the momenta defined in (B1), (B2) and (B3) may be denoted as $\vec{p}_i, \vec{p}_{\rho,i}$ and $\vec{p}_\rho$ in acordance with four-body conventions where i is a pair and ρ a two-body partition. The definition of $\vec{p}_{\rho,i}$ is only applicable to partitions of (3)+1 type. If E is the four-body energy, $E_\rho=E-\vec{p}_\rho^{\,2}/2\mu_\rho$ is the internal energy of partition ρ and $E_i=E_\rho-\vec{p}_{\rho,i}^{\,2}/2\mu_{\rho,i}$ is the energy for pair i inside a (3)+1 partion. Inside a (2)+(2) partition $E_i=E_\rho-\vec{p}_j^{\,2}/2\mu_j$ where j is the second pair.

APPENDIX C - Two-Cluster Subamplitudes

As shown in Section 2. the two-cluster subamplitudes satisfy the equation

$$U^\rho_{ij} = \bar{\delta}_{ij}G_o^{-1} + \sum_k \bar{\delta}_{ik}\, t_k G_o U^\rho_{kj} \; . \tag{C1}$$

If ρ is of the (3)+1 type then U^ρ is the AGS three-body t-matrix operator[8] embedded in four particle space. Equation (C1) is a matrix equation where i,j and k run over the pairs internal to ρ[(12), (13) and (23) if ρ=(123)4]. On the other hand if ρ is of (2)+(2) type then U^ρ is the AGS t-matrix operator for the scattering of two noninteracting pairs. Although the total four-body energy is conserved the energy of each pair is not. Therefore these amplitudes describe unphysical processes such as a bound pair being raised into the continuum while the other pair changes its energy state by the correspondent amount. Although such processes cannot take place on-shell they may occur off-shell in four-particle space. Again U^ρ is a matrix equation where i,j and k run over all pairs internal to ρ[if ρ=(12)(34) these are (12) and (34)].

If we now we introduce (3.2) for t_k together with left and right multiplication by $\langle f_i|G_o$ and $G_o|f_j\rangle$

$$\langle f_i|G_o\, U^\rho_{ij}\, G_o|f_j\rangle = \bar{\delta}_{ij}\, \langle f_i|G_o|f_j\rangle$$
$$+ \sum_k \bar{\delta}_{ik}\langle f_i|G_o|f_k\rangle\tau_k\langle f_k|G_o U^\rho_{kj} G_o|f_j\rangle. \tag{C2}$$

Since $\langle f_i|G_o\, U^\rho_{ij}\, G_o|f_j\rangle$ involves an implicit integration over the relative momentum in pairs i and j we define a new operator

$$X^\rho_{ij} = \langle f_i|G_o\, U^\rho_{ij}\, G_o|t_j\rangle \quad , \tag{C3}$$

which together with

$$B^\rho_{ij} = \langle f_i | G_o | f_j \rangle . \tag{C4}$$

leads to a new AGS equation

$$X^\rho_{ij} = \bar{\delta}_{ij} B^\rho_{ij} + \sum_k \bar{\delta}_{ik} B^\rho_{ik} \tau_k X^\rho_{kj} . \tag{C5}$$

Equation (C5) is solved at the energy $E_\rho = E - \frac{1}{2\mu_\rho} \vec{p}_\rho^{\,2}$ where E is the four--body Center of Mass energy, and $\vec{p}_\rho$ and μ_ρ are the relative momentum and reduced mass in channel ρ given by Eqs.(B3) or (B4) and (B9) or (B10) respectively. After partial wave decomposition this is a one variable integral equation that can be readily solved in a computer by standard numerical methods.

The homogeneous equation for the components of the bound state wave function in partition ρ also gets modified by the choice of a separable two-body t-matrix. Puting together (2.48) and (3.2) we get

$$|\chi^\rho_i\rangle = \sum_k \bar{\delta}_{ki} G_o |f_i\rangle \tau_i \langle f_i | \chi^\rho_k\rangle . \tag{C6}$$

which factorizes with respect to the relative momentum in pair i. Therefore defining

$$|\chi^\rho_i\rangle = G_o |f_i\rangle \tau_i |\bar{g}_i\rangle , \tag{C7}$$

leads to

$$|\bar{g}^\rho_i\rangle = \sum_k \bar{\delta}_{ki} \langle f_i | G_o | f_k\rangle \tau_k |\bar{g}^\rho_k\rangle , \tag{C8}$$

which is the equation for the bound state form factor that has a solution at $E_\rho = -\varepsilon_\rho$. If ρ is a partition of (3)+1 type ε_ρ is the three-body subsystem binding energy and (C8) is the equation for the three-body form factor mentioned in Section 3.1 that depends in the relative momentum between pair i and the third particle internal to ρ. The form factor $|\bar{g}^\rho_i\rangle$ is considered normalized such that $\langle\chi^\rho|\chi^\rho\rangle = 1$.

Although for (2)+(2) channel partitions one could in principle still solve equation (C8) for $E_\rho = -\varepsilon_\rho = -\varepsilon_i - \varepsilon_j$, this is not necessary because we know that the total bound state wave function for the two pairs is

$$|\chi^\rho\rangle = |\chi^\rho_i\rangle + |\chi^\rho_j\rangle = |\phi_i\rangle |\phi_j\rangle , \tag{C9}$$

where $|\phi_i\rangle$ and $|\phi_j\rangle$ are the wave functions for pair i and j given by (3.5).

Using the well know Faddeev relation (2.58)[7]

$$|\chi^\rho_i\rangle = G_o v_i |\chi^\rho\rangle , \tag{C10}$$

together with (C9), (3.1), (3.5) and (3.4) we get

$$|\chi_i^\rho\rangle = G_o(-\varepsilon_\rho)|\bar{f}_i\rangle g_o(-\varepsilon_j)|\bar{f}_j\rangle \ . \tag{C11}$$

Comparing (C7) with (C11) the normalized (2)+(2) form factor becomes.

$$|\bar{g}_i^\rho\rangle = N_i \tau_i^{-1}(E_i) g_o(-\varepsilon_j)|\bar{f}_j\rangle \ , \tag{C12}$$

that depends on the relative momentum in pair j.

For identical bosons of mass M all pairs are identical and as mentioned in Section 4.1 all (3)+1 (or (2)+(2)) partitions alike. Therefore all four (3)+1 subamplitudes are identical and the same for the three (2)+(2) subamplitudes. Returning to Eq.(C1) we note that individual matrix elements of U_{ij}^ρ are not observable because, given an initial state j, we cannot distinguish one final state i from another due to the identity of all pairs. Furthermore the result should not depend on which pair we have in the initial state. Consequently the operator whose matrix elements is of physical interest may be defined formally as

$$U^\rho = \frac{1}{n^\rho} \sum_{ij} U_{ij}^\rho \ , \tag{C13}$$

where n^ρ is the number of pairs internal to ρ (three for (3)+1 and two for (2)+(2) subsystems). Using (C13) in (C1) we get

$$U^\rho = \frac{1}{n^\rho} \sum_{ij} \bar{\delta}_{ij} \ G_o^{-1} + \frac{1}{n^\rho} \sum_{ij} \sum_{k\neq i} t_k G_o U_{kj}^\rho \ . \tag{C14}$$

Since t_k is the same for all k we formally get

$$U^\rho = (n^\rho-1)G_o^{-1} + (n^\rho-1)\, t\, G_o U^\rho \ . \tag{C15}$$

Denoting the two independent subamplitudes as U^{31} and U^{22} we may write

$$U^{31} = 2\, G_o^{-1} + 2\, t\, G_o\, U^{31} \ , \tag{C16}$$

and

$$U^{22} = G_o^{-1} + t\, G_o\, U^{22} \ . \tag{C17}$$

Again one should note that (C15), (C16) and (C17) are formal equations written in operator form without concern for alternation rules which have to be taken into account when matrix elements are to be calculated.

If we now assume as in (3.2) that t is separable in momentum space one may define operators X^{31} and X^{22} as in (C3). The resulting equation for X^{31} reads ($2M=\hbar=1$)

$$\langle\vec{P}'|X^{31}(E_{31})|\vec{P}\rangle = \langle\vec{P}'|B^{31}(E_{31})|\vec{P}\rangle + \int\frac{d^3P''}{(2\pi)^3}\langle P'|B^{31}(E_{31})|P''\rangle$$
$$\tau(E_{31}-\tfrac{3}{2}\vec{P}''^2)\langle\vec{P}''|X^{31}(E_{31})|\vec{P}\rangle, \tag{C18}$$

where

$$\tau^{-1}(E_2) = \lambda^{-1} - \int d^3p \,\frac{f^2(\vec{p})}{(E_2-2\vec{p}^2)}\,, \tag{C19}$$

and

$$\langle\vec{P}'|B^{31}(E_{31})|\vec{P}\rangle = 2\,\frac{f(\,\vec{P}'+\frac{1}{2}\vec{P}\,)\ f(\,\vec{P}+\frac{1}{2}\vec{P}'\,)}{E_{31}-\vec{P}'^2-(\vec{P}+\vec{P}')^2-\vec{P}^2}\,. \tag{C20}$$

These are Aaron. Amado and Yam[53] equations for three identical bosons. The equation for X^{22} may be obtained from (C18) by changing $31\to 22$, $\vec{P}\to\vec{q}$ and $\frac{3}{2}\to 2$ together with

$$\langle\vec{q}'|B^{22}(E_{22})|\vec{q}\rangle = \frac{f(\,\vec{q}\,)\ f(\,\vec{q}'\,)}{E_{22}-2\,\vec{q}'^2-2\,\vec{q}^2}\,. \tag{C21}$$

The three-boson bound state wave function becomes

$$\langle\vec{p}\vec{P}|\chi\rangle = \sum_i\frac{1}{\sqrt{3}}\,G_o\,(-\varepsilon_3-\tfrac{3}{2}\,\vec{P}_i^2-2\vec{p}_i^2)\ f(\,\vec{p}_i\,)\,\tau(-\varepsilon_3-\tfrac{3}{2}\vec{P}_i^2)\bar{g}(\vec{P}_i), \tag{C22}$$

where the sum on i denotes the different Jacoby coordinates of the three-body problem and ε_3 the three-body binding energy. The $\sqrt{3}$ factor is only usedwhen all particles are identical. The bound state form factor $\bar{g}$ is obtained from the solution of

$$\bar{g}(\vec{P}) = \int\frac{d^3P'}{(2\pi)^3}\langle\vec{P}|B^{31}(-\varepsilon_3)|\vec{P}'\rangle\tau(-\varepsilon_3-\tfrac{3}{2}\vec{P}'^2)\bar{g}(\vec{P}')\,. \tag{C23}$$

APPENDIX D - Subamplitudes of the Four-Nucleon Problem.

In its simplest form the four-nucleon problem already involves a large number of subamplitudes. This requires the use of specific language denominators that we outline here without going into too much detail[43]. Assuming that the two-body nucleon-nucleon interaction is restricted to the s-wave ($\ell=0$) channels $^3S_1$ (spin triplet) and ^1S_o (spin singlet) we proceed to the three-body sector where the spin (isospin) of a pair s (i) couples with the spin (isospin) of a nucleon to total spin S (isospin I). Since there are no tensor or spin orbit forces the total spin S, isospin I and three-body orbital angular momentum L are good quantum numbers. While S and I can only take the values $\frac{1}{2}$ (doublet) and $\frac{3}{2}$ (quartet) there is no limit on L except from practical considerations involving the three-body center of mass energy and the range of the effective exchange interaction for particle-pair scattering. At low energies the dominant partial waves are L=0 and L=1. Denoting as d a triplet pair and as ϕ a singlet pair, we show in Table V the relevant three-body sub amplitudes and the corresponding (2)+1 channels they are associated with. The spin doublet subamplitudes (which for L=0 carries the quantum numbers of the triton) requires the solution of two coupled equations, while the spin quartet and isospin quartet subamplitudes only require the solution of a single equation

Table V – Three-body subamplitudes

	(S,I)	(2)+1 Channels	L
Spin Douplet	$(\frac{1}{2},\frac{1}{2})$	d+N ϕ+N	0,1,...
Spin Quartet	$(\frac{3}{2},\frac{1}{2})$	d+N	0,1,...
Isospin Quartet	$(\frac{1}{2},\frac{3}{2})$	ϕ+N	0,1,...

In the presence of s-wave triplet and singlet pairs alone, the (2)+(2) subamplitudes involve d+d, $\phi+\phi$ and d+ϕ pairs, depending on the total four-body spin $\mathcal{S}$ and isospin $\mathcal{F}$. This is shown in Table VI where for each $\mathcal{S}$ and $\mathcal{F}$ we list all two-cluster subamplitudes required for the solution of the four-nucleon problem. If only the L=0 (3)+1 subamplitudes are included then $\mathcal{S}$, $\mathcal{F}$ and the total four-body angular momentum $\mathcal{L}$ are concerved leading to the simplest four-nucleon calculation one may perform. Using [1V+C] and a single term separable expansion in the triplet and singlet nucleon-nucleon channels we get, for $\mathcal{S} = \mathcal{F} = 0$, $n^{o}_{\frac{1}{2}\frac{1}{2}}$ coupled equations, where n^{L}_{SI} is the number of separable terms for the $(\frac{1}{2},\frac{1}{2})$ (3)+1 subamplitude. For $\mathcal{S} = \mathcal{F} = 1$ we get a higher number of coupled equations, usualy eight or nine for $n^{o}_{SI} = 3$. We note that, compared with [1V] , the number of coupled equations is reduced at the expense of extra integrals in the calculation of pair-pair propagators [(4.34) and (4.35)] for d+d, d+ϕ and $\phi+\phi$.

In the presence of spin and isospin quantum numbers the equations (4.25), (4.26) and (4.30) are to be generalized [43]. Although the structure remains identical each driving term [Eq.(4.27) and (4.29)] is affected by a spin-isospin coupling coefficient [12,43] that depends on $\mathcal{S}$, $\mathcal{F}$ and the quantum numbers of the initial and final states.

Table VI — Two-cluster subamplitudes of the four-nucleon problem for different total spin $\mathcal{S}$ and isospin $\mathcal{T}$.

		ISOSPIN $\mathcal{T}$		
		0	1	2
SPIN $\mathcal{S}$	0	$(\frac{1}{2},\frac{1}{2})$ (d+d) $(\phi+\phi)$	$(\frac{1}{2},\frac{1}{2})$ $(\frac{1}{2},\frac{3}{2})$ $(\phi+\phi)$	$(\frac{1}{2},\frac{3}{2})$ $(\phi+\phi)$
	1	$(\frac{1}{2},\frac{1}{2})$ $(\frac{3}{2},\frac{1}{2})$ (d+d)	$(\frac{1}{2},\frac{1}{2})$ $(\frac{3}{2},\frac{1}{2})$ $(\frac{1}{2},\frac{3}{2})$ $(\phi$+d)	$(\frac{1}{2},\frac{3}{2})$
	2	$(\frac{3}{2},\frac{1}{2})$ (d+d)	$(\frac{3}{2},\frac{1}{2})$	

The full addition of $L>0$ (3)+1 subamplitudes (seemingly L=1) brings the four-nucleon problem to the level of difficulty that is well above what is found in the three-nucleon problem when both s-and p-channels are included in the nucleon-nucleon force. For this reason the effect of L=1 (3)+1 subamplitudes has so far only been studied approximately by retaining in all orders the coupling between L=0 and L=1 channels [46,47]. In the presence of s-wave two-body forces alone (or truncated tensor force effects) the resulting number of coupled equations rises to a maximum of 18 and $\mathcal{S}$, $\mathcal{T}$ and $\mathcal{L}$ remain good quantum numbers.

If instead one follows the [2V] approach [Eq.(4.13) and (4.14)] one gets in $\mathcal{S}=\mathcal{T}=0$ only three coupled equations (four for $\mathcal{S}=\mathcal{T}=1$) at the expense of solving a two-variable integral equation.

REFERENCES

1) K.L.Kowalski in Lecture Notes in Physics 87, 393(1978); Nucl.Phys. A414, 465(1984); F.S.Levin, Nucl.Phys. A353, 143(1981).

2) O.A.Yakubovsky, Yad.Fiz. 5, 1312(1967) [Sov.J.Nucl.Phys.5,937(1967)].

3) P.Grassberger and W.Sandhas, Nucl.Phys.B2, 181(1967); E.O.Alt, P.Grassberger and W.Sandhas, Joint Institute for Nuclear Research, Report nº E4-6688(1972).

4) E.O.Alt, P.Grassberger and W.Sandhas, Phys.Rev.C1, 85(1970).

5) A.C.Fonseca in Few-Body Methods: Principles and Applications, edited by T.K.Lim, C.G.Bao, D.P.Hou and H.S.Huber [World Scientific, Singapore 1986].

6) E.W.Schmid and H.Ziegelmann, The Quantum Mechanical Three-Body Problem [Friedr. Vieweg + Sohn, Braunschweig 1974].

7) L.D.Faddeev, Z.Eksp.Teor.Fiz.39, 1459(1960) [Sov.Phys. JETP 12,1014(1961)]

8) E.O.Alt, P.Grassberger and W.Sandhas,Nucl.Phys.B2, 167(1967).

9) W.Sandhas, in Few Body Nuclear Physics [ICTP, Publication IAEA-SMR--45, 1978].

10) W.Glöckle, The Quantum Mechanical Few-Body Problem [Springer-Verlag, Heidelberg 1983].

11) L.S.Rodberg and R.M.Thaler, The Quantum Theory of Scattering [Academic Press, New York 1967].

12) V.F.Kharchenko and S.A.Shadchin, Yad.Fiz.22, 632 (1975); Preprint ITP-74-107 E, Kiev 1974.

13) L.S.Ferreira, contribution to this School.

14) S.Oryu, contribution to this School.

15) A.N.Mitra, Nucl.Phys. 32, 529(1962).

16) R.D.Amado, Phys.Rev. 132, 485(1963).

17) C.Lovelace, Phys.Rev.135, B1225(1964).

18) W.Plessas, contribution to this School.

19) W.Glöckle, contribution to this School.

20) S.Weinberg, Phys.Rev. 133, B232(1964).

21) A.Casel, H.Haberzettl and W.Sandhas, Phys.Rev.C25, 1728(1982).

22) S.Sofianos, N.J.McGurk and and H.Fiedeldey, Nucl.Phys.A318,295(1979).

23) A.C.Fonseca, H.Haberzettl and E.Cravo, Phys.Rev.C27, 939(1983).

24) H.Haberzettl and W.Sandhas, Phys.Rev. C24, 359(1981).

25) L.Bianchi and L.Favella, Nuovo Cimento 34, 1825(1964).

26) A.C.Fonseca and P.E.Schanley, Phys.Rev. D13, 2255(1976); Phys.Rev. C14, 1343(1976).

27) B.F.Gibson and D.R.Lehman, Phys.Rev. C14, 685(1976); C15, 2257(1977); C18, 1042(1978).

28) V.F.Kharchenko and V.E.Kuzmichev, Phys.Lett. 42B, 328(1972); Nucl. Phys. A183, 606(1972).

29) I.M.Narodetskii, E.S.Galpern and V.N.Lyakhovitsky, Phys.Lett.46B, 51(1973); I.M.Narodetskii, Nucl.Phys.A221, 191(1974).

30) S.Sofianos, H.Fiedeldey, H.Haberzettl and W.Sandhas, Phys.Rev. C26, 228(1982).

31) J.A.Tjon, Phys.Lett. 56B, 217(1975).

32) J.A.Tjon, Phys.Rev.Lett. 40, 1239(1978).

33) S.Sofianos, H.Fiedeldey and H.Haberzettl, Phys.Rev.C22, 1772(1980).

34) H.Haberzettl and S.A.Sofianos, Phys.Rev.C27, 2411(1983).

35) A.C.Fonseca, Phys.Rev.C30, 35(1984).

36) W.Böttger, A.Gasel and W.Sandhas, Phys.Lett.92B, 11(1980).

37) B.F.Gibson and D.R.Lehman, Phys.Lett.83B, 289(1971); Nucl.Phys.A329, 308(1979); Phys.Rev.C23, 404(1981).

38) B.F.Gibson, C.B.Dover, G.Bhamathi and D.R.Lehman, Phys.Rev.C27, 2085(1983).

39) J.A.Tjon, Phys.Rev.A21, 1334(1980).

40) S.Nakaichi, T.K.Lim, Y.Akaishi and H.Tanaka, Phys.Rev.A26, 32(1982).

41) A.C.Fonseca and T.K.Lim, Phys.Rev.Lett.55, 1285(1985).

42) J.A.Tjon in Proceedings of Eighth International Conference on Few-Body Systems and Nuclear Forces II, Graz 1978, edited by H.Zingl, M.Haftel and H.Zankel (Springer, Berlin, 1978); in Proceedings of the Ninth International Conference on the Few-Body Problem, Eugene, 1980, edited by F.Levin [Nucl.Phys.A353, (1981)] (North-Holland, Amsterdam 1981).

43) I.M.Narodetskii, Riv.Nuovo Cimento 4, 1(1981).

44) A.C.Fonseca, in Proceedings of the Tenth International IUPAP Conference on Few Body Problems in Physics, Karlsruhe, Germany 1983, edited by B.Zeitnitz [Nucl.Phys.A416, (1984)] (North-Holland, Amsterdam 1984).

45) H.Fiedeldey in Proceedings of the Eleventh International Conference on the Few-Body Problem, Tokyo-Sendai, 1986, in press.

46) J.A.Tjon, Phys.Lett. 63B, 391(1976).

47) A.C.Fonseca, Few-Body Systems 1, 69(1986).

48) A.C.Fonseca, in Proceedings of the Eleventh International Conference on the Few-Body Problem, Tokyo-Sendai, 1986,in press.

49) V.F.Kharchenko and V.P.Levashev, Nucl.Phys.A343, 249(1980).

50) S.A.Sofianos, H.Fiedeldey and W.Sandhas, Phys.Rev.C32, 400(1985) and references therein.

51) T.B.Clegg, A.C.L.Barnard, J.B.Swint and J.L.Weil, Nucl.Phys. 50, 621(1964).

52) J.E.Brolley, T.M.Putman and L.Rosen, Phys.Rev.107, 820(1957).

53) S.K.Adhikari and R.D.Amado, Phys.Rev.C15, 498(1977); S.K.Adhikari, Phys.Rev.C17, 903(1978).

54) R.Aaron, R.D.Amado and Y.Y.Yam, Phys.Rev.136, B650(1964).

Variational Operator Padé Approximants and Applications to the Nucleon-Nucleon Scattering

J. Fleischer *

Fakultät für Physik der Universität Bielefeld

Universitätsstr. 25

D-4800 Bielefeld 1, W. Germany

Abstract:

A review of the operator Padé approach to the summation of perturbation theory of quantum field theory is given. It is shown that the method yields solutions of the Schrödinger- and the Bethe-Salpeter-equation as special cases. Applications are given for the Nucleon-Nucleon scattering in terms of the Bethe-Salpeter equation in ladder approximation and a renormalizable gauge field theory including π-, ρ- and ω-exchange.

1. Introduction

"Padé approximants" (PA's) were introduced by H. E. Padé (1863 - 1953) in his thesis "Sur la représentation approchée d'une fonction par des fractions rationelles", presented at the Sorbonne on June 21st, 1892 - with Hermite being advisor. The idea is to start from a (formal) Taylor series expansion and construct the rational approximants as shown below.

Rational fractions can be rewritten as continued fractions and of course from this point of view the subject has a long history, dating back even to Euclid, for a review of which the interested reader is referred to Ref. 1.

---

* Report on work in collaboration with M. Pindor

In the physics literature PA's were also quite fashionable in the late 60s and the early 70s, where it was hoped that their application would help to obtain satisfactory results from renormalizable quantum field theories (QFT) in strong interactions (see e.g. Ref. 2, vol. 14 and references therein). In this report I hope to convince you that there finally exists an approximation method allowing to extract valuable information even from a one-loop approach in QFT of strong interactions. This method is in fact a quite involved procedure to evaluate the Schwinger variational principle[3] to calculate on-shell matrix elements of a so called [0/1] operator PA and which we therefore like to call "Variational Operator Padê Approximation". The variational parameters here are off-shell momenta and thus the method makes use of the full off-shell Green functions.

Of course, the situation has changed since the advent of Quantum Chromodynamics (QCD), the majority of physicists believing nowadays that the fundamental constituents of strongly interacting matter are in fact "quarks", the forces between them mediated by "gluons" and the observed hadrons being in fact bound states of quarks. There is overwhelming evidence for this picture- but nevertheless it is extremely difficult to derive low energy results from the QCD-Lagrangian (for an interesting development to obtain low energy results from QCD see e.g. Ref. 4). Therefore, the "old-fashioned" point of view of starting from a QFT given by a Lagrangian containing the observable hadrons may still have its own right as an "effective theory" which one might hope to derive from QCD. Renormalizability of the effective theory, however, does not seem to be of fundamental importance anymore since the effective theory of a renormalizable theory (QCD) need not necessarily be renormalizable again - but nevertheless in a nonrenormalizable theory one has to take into account arbitrary cutoffs if one proceeds with the calculation, and this remains an unsatisfactory situation as long as these cannot be derived from QCD as well.

In this report I shall demonstrate what I consider the "best" (though elaborate) approximation one can possibly achieve by going one order beyond the Born term. Everyone can then make his own use of the method. After having the method well established by reproducing results obtained from the Schrödinger- and the Bethe-Salpeter-equation (BSE) in ladder approximation some first results in terms of a renormalizable "effective" Lagrangian of the Nucleon-Nucleon (NN) interaction will be presented for the ^1S_o partial wave.

2. The Padé approximation as rational approximant

Given a function in terms of its formal power series expansion

$$f(z) = a_o + a_1 z + a_2 z^2 + \dots \tag{1}$$

an [M/N] PA to this function is given as the rational fraction

$$[M/N]_f = \frac{P_M(z)}{Q_N(z)} \tag{2}$$

where P_M and Q_N are polynomials of degree M and N, respectively, the coefficients of which are determined by the requirement

$$f(z) - \frac{P_M(z)}{Q_N(z)} = O(z^{M+N+1}). \tag{3}$$

For the general solution of these equations see Refs. 2 and 5. Here explicit expressions are only given for the lowest order approximants:

$$[0/1]_f = \frac{a_o^2}{a_o - a_1 z} \tag{4^a}$$

$$[1/1]_f = a_o + \frac{a_1^2 z}{a_1 - a_2 z} \tag{4^b}$$

As a famous example recall the series of Stieltjes, their general definition being [2,5]:

$$f(z) = \int_o^\infty \frac{d\varphi(u)}{1 + zu} = \sum_{n=o}^\infty f_n (-z)^n \tag{5^a}$$

$$f_n = \int_o^\infty u^n \, d\varphi(u), \tag{5^b}$$

where $\varphi(u)$ is a bounded nondecreasing function ($d\varphi \geq o$). Choosing, e.g., $\varphi(u) = -e^{-u}$, $d\varphi(u) = e^{-u}du$ gives

$$f_o(z) = \sum_{n=o}^\infty (-1)^n n! z^n, \tag{6}$$

which is clearly a divergent series. At $x = 1$ the value of $f_o(z)$ obtained from its integral representation is 0.5963.

The lowest approximants yield

$$[0/1]_{f_o} = \frac{1}{2} = 0.5000 \ , \ [1/1]_{f_o} = \frac{2}{3} = 0.6667$$

$$[1,2]_{f_o} = \frac{4}{7} = 0.5714 \ , \ [2,2]_{f_o} = \frac{8}{13} = 0.6154$$

The rate of convergence of the PA's at $x = 1$ and $x = \infty$ is essentially the same, namely for the diagonal approximants [M/M] the error is of the order $1/(M+1)$. Many other interesting proporties of the PA's for Stieltjes functions can be proven, for which the reader is referred to the above references. In perturbation theory of a QFT the number of Feynman graphs of the nth order roughly grows like n! (depending of course on the model) and this is what makes series like (6) so intriguing.

The physical problem we want to investigate in this article is the NN-scattering in the low energy regime (up to 250 MeV). The simplest model to study is the scattering of particles via a central potential V (r). The phase shifts can then be found from the K-matrix:

$$\langle p, l \mid K(E) \mid p', l\rangle = \frac{1}{2mp} \tan\delta_l (E) \tag{7}$$

with $p = p' = \hat{p}$ and $E = \frac{\hat{p}^2}{2m}$. K is the solution of the Lippmann-Schwinger equation

$$K(E) = V + V\, G_o^P(E)\, K(E) \tag{8}$$

with G_o^P the principal-value Green function

$$G_o^P(E) = \frac{1}{2} [(E - H_o + i\epsilon)^{-1} + (E - H_o - i\epsilon)^{-1}]. \tag{9}$$

The reason why the K-matrix is used when working with PA's is simply that the poles of the latter occur where the phase shift is going through 90°, while working with the S-matrix the situation would be more complicated.

Since we are mainly interested in a full relativistic treatment of the NN-scattering, we write in formal analogy to equ. (8) the BSE (in momentum space) as

$$\Phi = G + G\, S\, \Phi\ . \tag{10}$$

Here G is the kernel, which in a ladder approximation is calculated from a superposition of one-boson exchanges: π-, η-, ϵ-, δ-, ρ- and ω-mesons being exchanged.

Projecting into partial waves, one has to sandwich the above equation between Dirac particle states:

$$\Phi \rightarrow \Psi \, \Phi \, \Psi \quad \text{and} \quad G \rightarrow \Psi \, G \, \Psi \, , \tag{11}$$

where Ψ is meant to be a two-particle Dirac wave function of given momentum, helicity and "energy-spin", the later describing the possible couplings of positive and negative energy states, i. e.

$$| + \rangle = U^{(1)} \, U^{(2)} \qquad , \; | - \rangle = W^{(1)} \, W^{(2)}$$

$$| e \rangle = (U^{(1)}W^{(2)} + W^{(1)}U^{(2)})/\sqrt{2} \, , \; | o \rangle = (U^{(1)}W^{(2)} - W^{(1)}U^{(2)})/\sqrt{2} \tag{12}$$

are the basic states in the "energy-spin space", where U and W correspond to spinors of positive and negative energy, the upper indices $^{(1)}, ^{(2)}$ referring to nucleon 1 and 2, respectively [6]. For J = o four intermediate states couple [6,7]. They are

$$^1S_o^{\,+} \, , \; ^1S_o^{\,-} \, , \; ^3P_o^{\,e} \, , \; ^3P_o^{\,o}$$

$$^3P_o^{\,+} \, , \; ^3P_o^{\,-} \, , \; ^1S_o^{\,e} \, , \; ^1S_o^{\,o} \tag{13}$$

for the ^1S_o and the ^3P_o partial wave, respectively. The upper r.h. index refers to the energy-spin. In general (for J > o) eight states couple. Finally the coupled integral equation explicitely reads

$$\Phi(p,p_o,\alpha;\, p',p_o',\alpha') = G(p,p_o,\alpha;p',p_o',\alpha')$$

$$- \frac{i}{2\pi^2} \int dq dq_o \sum_{\beta,\gamma} G(p,p_o,\alpha;q,q_o,\beta) S(q,q_o,\beta,\gamma) \Phi(q,q_o,\gamma;p',p_o',\alpha'). \tag{14}$$

p, p' are the moduli of the outgoing and incoming three-momenta; p_o, p_o' are the corresponding relative energies in the CMS. The on-shell transition element is $\Phi(\hat{p}, o, 1; \hat{p}, o, 1)$ with $\hat{p} = \sqrt{E^2 - m^2}$ the on-shell c.m. momentum. For later use we have already written down the complete off-shell equation.

The two-nucleon propagator is independent of spin indices and has for J = o the general form[7]

$$S = \begin{pmatrix} S_{++} & & & \\ & S_{--} & & \\ & & S_{ee} & S_{eo} \\ & & S_{oe} & S_{oo} \end{pmatrix} \qquad (15)$$

with, e. g.

$$S_{\substack{++ \\ --}} = \frac{1}{(E \mp E(q) \pm i\epsilon)^2 - q_o^2} \qquad (16)$$

Explicit expressions for the various contributions to the kernel are given in Refs. 7 - 9. The BSE is finally solved by iteration and the obtained perturbation series summed by PA's. Examples, demonstrating their convergence are given in tables 2 and 3 of Ref. 7 for 100 MeV. For the ^3P_o the [1/1] (two-loop) is fairly accurate (10 % error) and the [1/2] is already excellent. For the ^1S_o, however, very high orders (at least up to the 10th) are necessary to obtain stable results. This is so, because the order of magnitude of the nth order contribution in that case is $\sim(-5.24)^n$.

In this context the PA's serve as a technical means to solve the BSE. The size of the (n x n) matrices involved in this problem (e.g. the kernel G in (14)) is of the order of $n \sim 1\,000$ after discretization of the integration momenta. In principle also matrix inversion could do the job. For the inversion of such large matrices, however, the computertime grows like $\sim n^3$ while using the Padê method it grows only like n^2. This clearly shows that only by the use of PA's one has a practicable means to solve the BSE.

In the next Sect. we consider the operator Padê method. Here we take the point of view that the Born term and its direct box graph corrections (but fully off-shell) are the ingredients to produce the ladder series. The method allows the generalization to take into account the full one-loop corrections of a given Lagrangian (renormalizable or not) and thus the new method has its own meaning beyond being a technical procedure to solve the BSE.

3. The Operator Padé Approximants

Next we consider a linear symmetric operator (K-matrix) in a Hilbert space H, with the formal expansion

$$K = K_o + K_1 + K_2 + \ldots, \tag{17}$$

where K_o, K_1, ... are again operators in H. For (17) the lowest order operator approximants in analogy to (4a,b) are

$$[0/1] = K_o \frac{1}{K_o - K_1} K_o \tag{18^a}$$

$$[1/1] = K_o + K_1 \frac{1}{K_1 - K_2} K_1 \tag{18^b}$$

Here we are merely interested in (18a), i.e. in the following we consider $K_{OPA} \equiv [0/1]$ only. Amazingly, however, due to the special form of the [1/1], the next order approximation has to be treated in exactly the same manner as the [0/1] except for the replacement $K_o \to K_1$ and $K_1 \to K_2$*. For the general formulation of higher orders see Refs. 10.

The meaning of (18a) becomes clear now by considering the BSE (10), which can be formally iterated

$$\Phi = G + GSG + GSGSG + \ldots , \tag{19}$$

and since this is a geometric series, it can be summed to yield

$$\Phi = G \frac{1}{G - GSG} G \tag{20}$$

in the form of (18a), i.e. as stated at the end of the last Sect., only $K_o = G$ (Born term) and $K_1 = GSG$ (direct box graph) are needed to formally solve the BSE.

What one finally wants are one-shell matrix elements of K_{OPA}. To calculate them, we make use of the Schwinger variational principle[3]. This principle was first evaluated by Cini and Fubini[11] by means of trial functions while later Nuttal[12] and independently Bessis and Pusterla[13] showed the equivalence of this approach to Padé approximants.

---

* The notation is not quite unique. Considering in QFT the oth term as zero, denoting the Born term by K_1, what we call here [0/1] can then be called [1/1].

The Schwinger variational principle states that the unique stationary value of the functional

$$R_{\alpha\beta}(\Psi,\Psi') = \langle\Psi'|K_o|\beta\rangle + \langle\alpha|K_o|\Psi\rangle - \langle\Psi'|K_o - K_1|\Psi\rangle \quad (21)$$

with respect to any variations of $|\Psi\rangle$ and $|\Psi'\rangle$ is given by the matrix element of K_{OPA}:

$$K^{\alpha\beta}_{OPA} = \langle\alpha|K_{OPA}|\beta\rangle, \quad (22)$$

which we want to calculate.

Varying, e.g., $R_{\alpha\beta}(\Psi,\Psi')$ with respect to Ψ' yields

$$\delta_{\Psi'} R_{\alpha\beta}(\Psi,\Psi') = \langle\delta\Psi'|\{K_o\,|\,\beta\rangle - (K_o - K_1)|\Psi\rangle\} = o \quad (23)$$

$$\text{i.e.} \quad |\,\Psi\rangle_{St} = \frac{1}{K_o - K_1} K_o\,|\beta\rangle \quad (24^a)$$

and similarly from the variation with respect to Ψ:

$$|\Psi'\rangle_{St} = \frac{1}{K_o - K_1} K_o\,|\alpha\rangle. \quad (24^b)$$

Inserting these into $R_{\alpha\beta}(\Psi,\Psi')$ gives $K^{\alpha\beta}_{OPA}$. Since in this report only J = o is considered, in the following $|\beta\rangle = |\alpha\rangle$ is assumed for the on-shell states.

In actual calculations one proceeds in two steps:

i. one confines oneselves to a finite-dimensional subspace of H, with basis states $|\varphi_1\rangle = |\alpha\rangle, |\varphi_2\rangle, \ldots, |\varphi_L\rangle$, spanning $H_L = P_L H$ ($P_L^2 = P_L$). It is the essence of the above mentioned development[14], that $|\Psi\rangle$ and $|\Psi'\rangle$ varied over H_L only, yield

$$\underset{|\Psi\rangle,\,|\Psi'\rangle \in H_L}{\text{stat. value}} \; R_{\alpha\alpha}(\Psi,\Psi') = \langle\alpha|K_o P_L \frac{1}{P_L(K_o - K_1)P_L} P_L K_o|\alpha\rangle \equiv R^L_{\alpha\alpha}(\varphi_2,\ldots,\varphi_L), \quad (25)$$

which is a "matrix-Padé" evaluated on H_L.

ii. the next step is to fix the $|\varphi_i\rangle$'s properly. The problem is that $R^L_{\alpha\alpha}(\varphi_2,\ldots,\varphi_L)$ may have many stationary points and one has to find out the right one. How to do this has been shown in Ref. 15. The therein given proof for the proper stationarity condition is somewhat lengthy. Investigating at first the quantity:

$$\Delta = |\,K^{\alpha\alpha}_{OPA} - R^L_{\alpha\alpha}\,|, \quad (26)$$

$\Delta = o$ would obvously be the solution of the problem.

Some rewriting yields

$$\Delta = |\langle\alpha| K_o \frac{1}{K_o-K_1} K_o|\alpha\rangle - \langle\alpha| K_o P_L \frac{1}{P_L(K_o-K_1)P_L} P_L K_o |\alpha\rangle|$$

$$= |\langle\alpha| K_o \frac{1}{K_o-K_1} [1 - (K_o-K_1)P_L \frac{1}{P_L(K_o-K_1)P_L} P_L] K_o |\alpha\rangle|$$

$$\equiv |\langle\varphi_\alpha|\gamma_\alpha\rangle| \quad \text{with} \tag{27}$$

$$\langle\varphi_\alpha| = \langle\alpha| K_o \frac{1}{K_o-K_1} , \tag{28^a}$$

$$|\gamma_\alpha\rangle = [1 - (K_o-K_1) P_L \frac{1}{P_L(K_o-K_1)P_L} P_L] K_o |\alpha\rangle , \quad \text{and} \tag{28^b}$$

$$P_L |\gamma_\alpha\rangle = o. \tag{29}$$

In fact it has been shown in Ref. 15, that the stationary point $R^L_{\alpha\alpha}$, achieved when $\langle\gamma_\alpha| = o$ coincides with the stationary point of the Schwinger functional. In other words: the $\langle\varphi_i|$'s $\in H$ have to determined such as to yield $\langle\gamma_\alpha| = o$. According to (27) this is apparently a sufficient condition for finding the proper value of $K^{\alpha\alpha}_{OPA}$.

Since we cannot achieve $\langle\gamma_\alpha| = o$ in any practical calculation, the best we can do is to minimize the norm of this vektor ($||\gamma||$) as a function of the states $|\varphi_2\rangle, \ldots, |\varphi_L\rangle$, which will be taken as off-shell states, using the off-shell momenta as variational parameters.

It is important to notice at this point that in the early trials to explore the variational principle as well in potential theory[16] as in the framework of the BSE[9,17], just the off-shell momenta in $R^L_{\alpha\beta}(\varphi_i)$ were varied and it was only looked for stationary values of this quantity. In many cases this gave surprisingly good results, though they were not always unique or (for strong potentials) would not even exist (see below). One has to notice that in the derivation of $\langle\gamma_\alpha| = o$ in Ref. 15 the states are assumed to vary over the <u>full H,</u> while the above mentioned procedure yields only a variation over certain subsets of H (trajectories) and can thus yield stationary points, which may be different from those obtained for $\langle\gamma_\alpha| = o$.

In potential theory[15], potentials were chosen in the form

$$V(r) = V_1\Theta(r_1 - r) + V_2\Theta(r_2 - r), \tag{30}$$

where V_1 and V_2 are constants and Θ is the step function.

The operator PA is

$$K_{OPA}(E) = V \frac{1}{V - VG_o^P V} V \tag{31}$$

and results obtained by minimizing $||\gamma||^2$ are shown in figs. 1 and 2 [15].

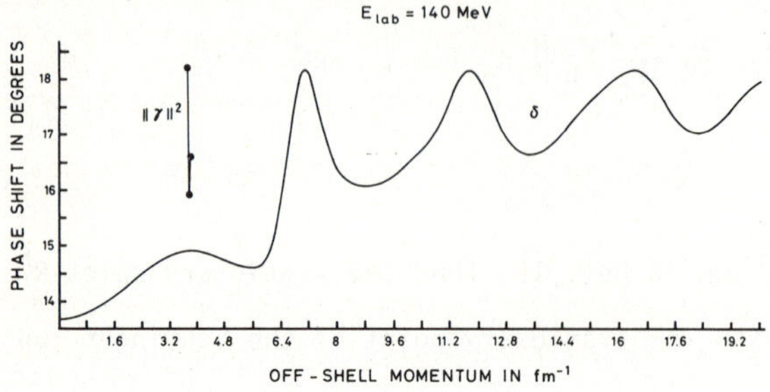

FIG. 1. Dependence of the phaseshift δ on the off-shell momentum at E_{lab}=140 MeV. For the potential [see (30)] the parameters are as in Ref. 16 (V_1=170.05 MeV, V_2=-34.01 MeV, r_1=0.68 fm, and r_2=1.92 fm). Only three points (see dots) of the $||\gamma||^2$ curve (in arbitrary normalization) appear in this figure and are connected by straight lines. The minimum of $||\gamma||^2$ is obtained at the off-shell momentum k_m=3.758 fm^{-1}. The value of the phase shift at this point is 14.9156° to be compared with the exact value of 14.9174° and the value at the first maximum of 14.9167°.

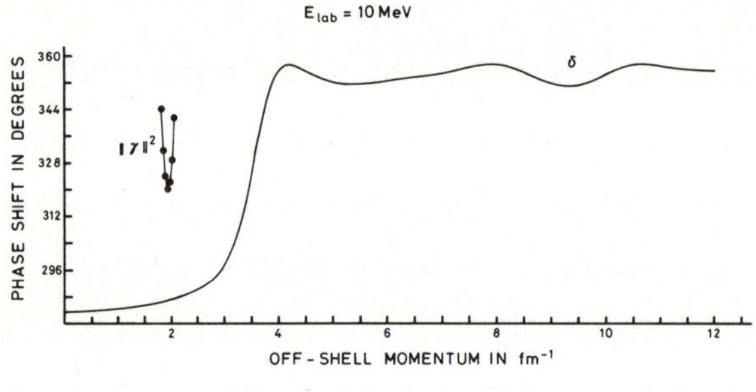

FIG. 2. Same as Fig. 1 for V_1= 1000 MeV, V_2= -100 MeV, r_1= 1 fm, r_2= 2 fm, and E_{lab}= 10 MeV. For these values of the parameters there exist two bound states, i.e., $\delta(E_{lab}=0)$= 360°. The minimum of $||\gamma||^2$ is obtained at k_m= 1.928 fm^{-1}. The value of the phase shift at this point is 287.05°, to be compared with the exact value of 289.34°. For two off-shell momenta the phase shift at the minimum of $||\gamma||^2$ is 289.26°.

For the BSE a similar calculation was performed, making use of the work of Ref. 9, in which case for the πNN-coupling pseudovector coupling had been chosen. For the calculation of the operator PA's, the full off-shell equation as given in equ. (14) has to be evaluated, the matrices in the "matrix-Padés" (see also (25)) being labelled by discretized off-shell momenta and spin indices $(p_i, p_{oi}, \alpha; p_k', p_{ok}', \alpha')$. For the calculation of $||\gamma||^2$, the calculation of a two-dimensional integral is necessary:

$$||\gamma||^2 = \int d\, q_o d\, q\, \gamma^2\, (q_o,\, q;\, p_{oi},\, p_i) \qquad (32)$$

but as variational parameters $-(p_{oi}, p_i)$ - only the modulus of the external momentum was chosen, while the relative energy was fixed at a small value: $p_{oi} = \frac{\hat{p}}{8}$. Results[15] are shown in tables I and II for the ^1S_o- and ^3P_o - partial waves and show in fact excellent agreement with the method of summing the iterated series, as described in Sect. 2.

Table I. ^1S_o NN phase shift for various energies E calculated from one-boson exchange in terms of the Bethe-Salpeter equation (δ_{BSE}) in comparison with the value of the phase shift δ_m, obtained at the off-shell momentum p_1 (in units of the nucleon mass), where $||\gamma||^2$ takes its minimal value $\gamma^{L=2}$. Since this off-shell momentum is surprisingly energy independent, we have also calculated the OPA at some energies (indicated by an asterisk) without performing a search for a minimum of $||\gamma||^2$. p_o is the on-shell momentum.

E (MeV)	$p_o = \hat{p}$	δ_{BSE}	δ_m	$\gamma^{L=2}$	p_1
10 *	0.073	58.59	57.38	2.403	1.24
25 *	0.115	49.37	49.55	0.777	1.24
50	0.163	39.13	39.47	0.327	1.24
100	0.231	24.95	25.38	0.157	1.24
150 *	0.283	14.74	14.92	0.114	1.24
200	0.326	6.47	6.45	0.098	1.24
250	0.365	-0.51	-0.75	0.092	1.25

Table II. Same as Table I for ^3P_o. For energies where the phase shift is negative (200 and 250 MeV), it is necessary to use two off-shell momenta as variational parameters (L=3) in order to obtain a good approximation. With the same off-shell momenta also the low-energy phase shifts (10 and 25 MeV) are reproduced properly without searching for a minimum of $||\gamma||^2$.

E (MeV)	$p_o = \hat{p}$	δ_{BSE}	δ_m	$\gamma_m^{L=2}$	p_1	$\gamma_m^{L=3}$	p_1	p_2
10*	0.073	4.10	4.06			0.002	0.633	1.44
25*	0.115	8.89	8.80			0.005	0.633	1.44
50	0.163	11.32	11.31	0.011	0.949			
100	0.231	8.33	8.32	0.018	0.859			
150	0.283	2.65	2.46	0.025	0.804			
200	0.326	-3.39	-6.63	0.024	1.14			
200	0.326	-3.39	-3.70			0.011	0.633	1.44
250	0.365	-9.27	-12.62	0.024	1.10			
250	0.365	-9.27	-9.64			0.013	0.639	1.44

4. Application of Operator Padé Approximants to a Renormalizable Quantum Fied Theory

A renormalizable field theory for the low energy NN-interaction has been worked out in Ref. 18. It takes into account the main contribution to the NN force: π-, ρ- and ω-exchange. Renormalizability then implies that this must be a Yang Mills model with spontaneous symmetry breaking. Isospin invariance and nucleon number conservation lead to the choice of an $SU(2)_L \times SU(2)_G \times U(1)_L \times U(1)_G$ symmetric Lagrangian, L and G standing for local and global, respectively. For any details see the above reference. We only mention that the unphysical particles "would be Goldstone bosons" and "Fadeev-Popov" ghosts have masses (in the 't Hooft gauge)

$$m = \sqrt{\xi_V}\, m_V \qquad (33^a)$$

and the vector-meson propagator is

$$\Delta^{(V)}_{\mu\nu}(k) = -\frac{g_{\mu\nu}}{k^2-m_V^2} - \frac{k_\mu k_\nu}{m_V^2}\left(\frac{1}{k^2-m_V^2} - \frac{1}{k^2-\xi_V m_V^2}\right) ; \; V = \rho,\omega.$$

If the gauge parameters for ρ- and ω-mesons are taken as $\xi_V = 1$, in this (Feynman-) gauge the unphysical particles have the same masses as the physical vector-mesons.

Starting from the full Lagrangian, instead of taking into account only the direct box graphs, according to the above outlined philosophy, one now has to calculate all one-loop Feynman-diagrams contributing to the off-shell NN-scattering amplitudes. The phase shifts are then calculated from the operator PA's as described in Sect. 3.

At this point a comment is in order concerning the off-shell behaviour of Green functions in connection with the formation of the operator PA's. First of all, in the usual formalism of gauge field theories, only the on-shell elements are considered as gauge invariant. Thus in principle we have to expect a dependence of our off-shell amplitudes on the gauge parameter, a problem which is ignored for the time being by the choice of $\xi_V = 1$ ($V = \rho,\omega$). In the "singlet formalism"[19], however, it is possible also to find a gauge invariant formulation of the off-shell Green functions at the price of introducing a bad high energy behaviour.

Another off-shell ambiguity is apparently the choice of the off-shell spinors in obtaining the partial wave amplitudes (see Sect. 2). We could, e.g., multiply the spinors U(p) and W(p) (see(12)) by a factor

$$f_p = \left(\frac{E + m}{E(p) + m}\right)^{\lambda} \tag{34}$$

without changing their on-shell values ($E(\hat{p}) = E$ i.e. $f_{\hat{p}} = 1$). Luckily such an extra off-shell factor drops out in the calculation of $R^L_{\alpha\beta}$ (see (25)), so that once the discretized off-shell momenta (variational parameters) are fixed, the phase shifts are independent of the choice of f_p. What depends, however, on its choice is the vector $\langle\gamma_\alpha|$ (see (28b)), i.e. $||\gamma||^2 = \langle\gamma_\alpha|\gamma_\alpha\rangle_q$ is multiplied by f_q^2, resulting for $\lambda > o$ in a damping of the large off-shell contributions in the integration of (32).

It is natural to require that the obtained results for the phase shifts should be stable against changes of λ (i.e. the location of the minimum of $||\gamma||^2$ is required to be stable in this case) and / or against introducing more off-shell momenta as variational parameters. Finally with such a procedure even the effect of the bad high energy behaviour of the Green functions in the singlet formalism may be eliminated.

Finally first results are represented for the ^1S_o partial wave[20]. This wave is of particular interest because as demonstrated in Sect. 2, very high order on-shell PA's would be necessary to yield acceptable results. Moreover, the variational principle always worked best for this wave[9,17], so that one can expect to obtain results most easily in this case.

One simplification, however, was necessary in order to get results with a realizable amount of computertime: in the calculation of $||\gamma||^2$ (32) not only the external relative energy was chosen to be $p_{oi} = o$ but also instead of performing an integration over q_o, simply $q_o = o$ was assumed as well. The integration over q is difficult in so far as due to the property (29) at each variational momentum p_i the integrand vanishes. Therefore up to 32 Gaussian integration points were used in the q-variable.

The coupling constants were chosen as

$$\frac{g^2_{\pi NN}}{4\pi} = 14.2 \; , \quad \frac{g^2_{\rho NN}}{4\pi} = 2.3 \quad \text{and} \quad \frac{g^2_{\omega NN}}{4\pi} = 15 \; . \tag{35}$$

and our results are presented in table III.

Table III. ^1S_o phase shifts for various energies. λ is the parameter defined in (34). γ_o^λ the corresponding Born term $||\gamma||^2$. $\gamma_m^{L=i}$ are the minima of $||\gamma||^2$ obtained with $p_1 = \hat{p}$ and i-1 off-shell variational parameters p_i. δ^i are the corresponding phase shifts. The p_i's were always chosen as $o < p_i < 3.5$ nucleon masses.

E_{Lab} (MeV)	λ	γ_o^λ	$\gamma_m^{L=3}$	δ^3	$\gamma_m^{L=4}$	δ^4	$\gamma_m^{L=5}$	δ^5
50.	0	1964.	261.1	47.2	254.8	29.7	96.8	40.5
	1	1073.	69.0	47.2	20.7	38.7		
	2	806.	35.1	40.3	5.2	40.0		
100.	0	1953.	96.9	39.2	95.6	28.0	22.9	29.5
	1	1058.	16.3	49.8	9.1	30.5		
	2	792.	8.7	49.8	2.5	31.6		
200.	0	1953.	39.7	12.9				
	1	1049.	5.3	14.1	3.5	13.2		
	2	785.	3.2	11.2	1.7	13.2		

First of all we see that the underlined results for the phase shifts at the given three energies agree within relatively small errors and are stable as requested as well against changes of λ as against changing the number of variational points. Moreover the results are in agreement with experimental values for the ^1S_o phase shift, which corresponding to Ref. 21 (selecting the np data, table VI of that Ref.) are

$$\delta_{^1S_o}(50) = 43.16^\circ \pm 0.98^\circ;\ \delta_{^1S_o}(100) = 28.97^\circ \pm 1.55^\circ \text{ and } \delta_{^1S_o}(200) = 10.22^\circ \pm 2.04^\circ.$$

The pp-data may be reproduced by choosing a somewhat smaller ωNN-coupling, which in fact has been chosen as relatively large (see (35)). It is remarkable, that this is the only free parameter in the present calculation. Finally it should also be mentioned that the computertime used for one energy is of the order of 4 hours on the CDC 7600. As a consequence, "fitting" the ^1S_o wave with this coupling constant is not an easy matter.

So far the deficiency of the above calculation are the empty entries in table III. Trials to fill them yielded completely unstable results. Though smaller $||\gamma||^2$ were obtained, the phase shifts changed drastically. Similarly for the ^3P_o, in which case no stable results at all were obtained by now. One can assume that this instability is related to the fact that the q_o-integration was not yet taken into account properly, which one would expect to smoothen and stabilize the calculations. At least for the $^3S_1$ and pseudoscaler πNN-coupling encouraging results were found, by taking into account the crossed box graph in a similar calculation[9] - though no search for a minimum of $||\gamma||^2$ was performed. In that case, however, it was absolutely necessary to take into account relative energies. Thus further investigations have to be performed in that direction.

References

1. C. Brezinski, in *Padé Approximation and its Applications Amsterdam 1980*, Springer Lecture Notes in Mathematics, Vol. 888, edited by M. G. de Bruin and H. van Rossum (Springer, Berlin and New York, 1981).
2. G. A. Baker, Jr. and P. Graves-Morris, *Encyclopedia of Mathematics and its Applications*, Vol. 13-14 (Addison-Wesley Publishing Company, London and Amsterdam, 1981).
3. S. R. Singh and A. D. Stauffer, Nuovo Cimento 22 B, 139 (1974).
4. J. Gasser and H. Leutwyler, Phys. Rep. 87 (1982) 77.
5. G. A. Baker, Jr., *Essentials of Padé Approximants* (Academic Press, New York and London, 1975).
6. J.J. Kubis, Phys. Rev. D 6, 547 (1977); J. Fleischer, Journal of Computational Physics 12, 112 (1973).
7. J. Fleischer and J. A. Tjon, Nucl. Phys. B 84, 375 (1975).
8. J. Fleischer and J. A. Tjon, Phys. Rev. D 15, 2537 (1977).
9. J. Fleischer and J. A. Tjon, Phys. Rev. D 21, 87 (1980).

10. J. D. Bessis, in "Padé Approximants", edited by P. R. Graves-Morris (Institute of Physics, London and Bristol, 1973);
M. Pindor and G. Turchetti, Nuovo Cimento A 71, 171 (1982).
11. M. Cini and S. Fubini, Nuovo Cimento 10, 1695 (1953) and N. C. 11, 142 (1954).
12. J. Nuttal, Phys. Lett. 23, 492 (1966) and Phys. Rev. 157, 1312 (1967)
13. D. Bessis and M. Pusterla, Nuovo Cimento 54, 243 (1968).
14. J. Nuttal, in *The Padé Approximants in Theoretical Physics*, edited by G. A. Baker, Jr., and J. L. Gammel (Academic, London and New York, 1970), pp. 219 - 230.
15. J. Fleischer and M. Pindor, Phys. Rev. D 24, 1978 (1981).
16. L. P. Benofy, J. L. Gammel and P. Mery, Phys. Rev. D 13, 3111 (1976);
L. P. Benofy and J. L. Gammel, in *Padé and Rational Approximation*, edited by E. B. Saff and R. S. Varga (Academic, London, 1977).
17. J. Fleischer and J. A. Tjon, in *Padé and Rational Approximation*, edited by E. S. Saff and R. S. Varga (Academic, London, 1977).
18. K. Fabricius and J. Fleischer, Phys. Rev. D 19, 353 (1979);
K. Fabricius and J. Fleischer, in *Few Body Systems and Nuclear Forces I*, Springer Lecture Notes in Physics, Vol. 82, edited by H. Zingl, M. Haftel and H. Zankel (Springer, Berlin and New York, 1978).
19. F. Jegerlehner and J. Fleischer, Phys. Lett. 151 B, 65 (1985); F. Jegerlehner and J. Fleischer, Acta Physica Polonica B 17, 35 (1986).
20. J. Fleischer and M. Pindor, to be published (supported by the Deutsche Forschungsgemeinschaft).
21. M. H. Mac Gregor, R. A. Arndt and R. M. Wright, Phys. Rev. 182, 1714 (1969).

PART II

INTEGRAL AND DIFFERENTIAL METHODS FOR THE SOLUTION OF THE SCHRÖDINGER EQUATION

VARIATIONAL METHODS FOR THE FEW-BODY BOUND STATE IN A HARMONIC OSCILLATOR BASIS

Sidney A. Coon
Department of Physics
University of Arizona
Tucson, Arizona 85721, USA

Oyanarte Portilho
Departamento de Fisica
Universidade de Brasilia
70.910 Brasilia-DF, Brazil

Abstract

The method of expanding the trial variational function into a complete set of harmonic oscillator functions is reviewed. The formalism for the two-body and three-body bound state is given. The treatment of two equal-mass bosons and a third boson of different mass is elaborated. Model calculations of quarkonium are compared with exact results and with integral equation methods. The convergence of a model 3α system is studied in detail, and the role of Jastrow-type correlations is examined. Phenomenological studies reviewed include realistic models of quarkonium, alpha-particle models of this carbon nucleus, and alpha-particle models of singly and doubly strange hypernuclei. The latter studies employed the best available phenomenological $\alpha\alpha$ interactions.

1. Introduction

"The choice of a numerical method for a given problem depends on the nature of the problem, and on the nature of our interest in it; and in particular, on whether we are interested in the results for themselves, or as a practice run for some more complicated calculations which we have in mind to do next ... To choose the relevant tool for the particular job in hand, one must compare like with like, and the popularity of variational methods then resides in the fact that for a given job they come out of such a comparision very well." These remarks open the definitive study of "Variational Techniques in the Nuclear Three-Body Problem" by Delves.[1] This lecture can be viewed as an appendix to this article in that we review our experience with two- ,three- and four-body bound states calculated variationally in coordinate space with a trial wave function expanded in terms of harmonic oscillator states. In our published papers, we were primarily concerned with investigations of the Hamiltonian and/or the structure of the wave function, and implicitly assumed that our choice of tools was appropriate. In this lecture, we try to examine the issues raised by Delves: "we should also consider the ease of setting up the equations, the stability of the solution against numerical (round-off) errors, and the provision of error estimates or error bounds for the approximate solution." As we give some details on the formalism and numerical results, we will address some of these questions, mostly by comparision with exact results or with results obtained by other popular numerical methods used in the few-body problem. Our choice of problems was determined primarily by the rather limited computational facilities we used.

We begin with a discussion of a variational approach to the bound state. The trial function takes the form of a linear expansion

$$\psi_t = \sum_{\nu=1}^{N} a_\nu \phi_\nu \quad . \tag{1}$$

The a_ν are parameters to be varied and the ϕ_ν is a set of known functions. The ϕ_ν may also contain parameters ϵ_μ, which will be varied. Then the functional to be minimized $E_v = \langle\psi_t|H|\psi_t\rangle/\langle\psi_t|\psi_t\rangle$ takes the form

$$E_v = \mathbf{a}^\dagger H_N \mathbf{a}/\mathbf{a}^\dagger N_N \mathbf{a} \quad , \tag{2}$$

where H_N and N_N are the NxN Hamiltonian and normalization matrices in the representation $[\phi_\nu]$ and $\mathbf{a}$ is the vector of coefficients a_ν. From (2) we obtain the defining equation for $\mathbf{a}$ and E_v:

$$(H_N - E_v N_N)\mathbf{a} = 0 \quad . \tag{3}$$

This is an eigenvalue problem, so for fixed ϵ_μ (and normalization matrix equal to the identity), one diagonalizes the Hamiltonian matrix to obtain the vector of coefficients $\mathbf{a}$. The minimum of E_v with respect to the parameter vector $\mathbf{a}$ always exists, since a finite eigenvalue problem of the form (3) is guaranteed to have N real eigenvectors $E_\nu(N)$ and N independent eigenvectors. One obtains from (3) not only an upper bound on the lowest eigenvalue E_1 but also on the higher eigenvalues $E_2, \ldots, E_N$.[1]

Next we summarize the convergence properties of a linear trial function.[1] Provided the set of expansion functions is suitably complete, one will eventually obtain convergence with increasing N to the exact value. Moreover, if the set is constructed systematically, then in general one can expect the convergence to be smooth. Given such a theory of convergence one can observe the results for increasing N to estimate the accuracy attained in a calculation. With a linear trial function, the expectation value W(N) of any bounded operator W will converge provided that the energy converges; and one can again estimate the accuracy obtained by watching the numerical convergence of W(N) with increasing N. Examples of this procedure will be displayed later on.

The advantage of a harmonic oscillator basis is that it is relatively straightforward to construct a complete set of three- or four-body functions of appropriate angular momentum and symmetry.[2-4] This means that we can avoid a Eular angle expansion of the trial function and work directly in a Cartesian space where the algebra is more simple. If the potentials are spin and isospin independent, the variational analysis can be done entirely in coordinate space. For simple potentials, the needed integrals can be done analytically so that one can include a large number of states before running against computer limitations.

The perceived disadvantage of these calculations is the slow rate of convergence as terms are added to the expansion. The convergence rate for a "non-smooth" potential like the Yukawa potential is of order $N^{(-2)}$, where N is the number of excitations of the three-dimensional harmonic oscillator (see Fig. 8 of Ref. 1). This corresponds to something like $N^{(-1)}$ for the eigenfunction, which is extremely slow convergence. However, in the end it may be more practical to calculate many terms easily in a slowly converging scheme than to solve painfully the coupled differential equations and numerically evaluate integrals for the elements of another expansion that converges faster. In the latter case, the trial function is often chosen to take advantage of a feature of the Hamiltonian, such as short-range correlations. These "ad hoc" trial functions are difficult to improve systematically, and it is difficult to estimate the accuracy of a calculation given only a single spot upper bound, even if accompanied by a spot lower bound.

The harmonic oscillator expansions do converge to the exact eigenvalues, but there remains some suspicion about the quality of the approximate eigenfunction. That is, it is often felt that the energy has converged, but it is more difficult to judge other physical observables or geometric properties of the system studied.

These common prejudices about this method will be reconsidered as we proceed to discuss selected problems in two- and three-body bound-state systems.

2. The two-body bound state

A. Formalism

The intrinsic Hamiltonian H_I for two particles with masses m_1 and m_2 and interacting through the potential $V(\mathbf{r})$ is

$$H_I = \frac{p'^2}{2\mu} + V(r') \quad , \tag{4}$$

where $\mathbf{r}' = \mathbf{r}_1' - \mathbf{r}_2'$ is the relative coordinate, μ is the reduced mass, and $\mathbf{p}' = -i\hbar\nabla'$. Primed variables are not dimensionless.

If these particles are interacting through the harmonic oscillator potential with frequency ω, we have

$$H_{HO} = \frac{p'^2}{2\mu} + \tfrac{1}{2}\mu\omega^2 r'^2 \tag{5}$$

and therefore

$$H_I = H_{HO} - \tfrac{1}{2}\mu\omega^2 r'^2 + V(r') \quad . \tag{6}$$

Using the dimensionless coordinate and momentum given by

$$r = \sqrt{\frac{\mu\omega}{\hbar}}\, r' \tag{7}$$

and

$$p = \frac{1}{\sqrt{\mu\hbar\omega}}\, p' \tag{8}$$

the above Hamiltonians become

$$H_I = H_{HO} - \tfrac{1}{2}\epsilon r^2 + V(r) \tag{9}$$

and

$$H_{HO} = \tfrac{1}{2}\epsilon(p^2 + r^2) \tag{10}$$

where we have introduced the nonlinear variational parameter

$$\epsilon = \hbar\omega \quad . \tag{11}$$

The matrix elements of H_I are[2]

$$\langle n'\ell|H_I|n\ell\rangle = \frac{\epsilon}{2}\Big[(2n+\ell+\tfrac{3}{2})\,\delta_{n'n} + \sqrt{n(n+\ell+\tfrac{1}{2})}\;\delta_{n',n-1} + \sqrt{(n+1)(n+\ell+\tfrac{3}{2})}\;\delta_{n',n+1}\Big] + \langle n'\ell|V(r)|n\ell\rangle \quad , \tag{12}$$

where $|n\ell\rangle$ is the eigenfunction of H_{HO}. Each pair of quantum numbers of the harmonic oscillator is characterized by the symbol ν. Each eigenfunction of H_I is expanded in a basis of eigenfunctions of H_{HO} with well defined spin, orbital, and total angular momentum

$$\Psi = \sum_{\nu} a_\nu |\nu S;\, JM\rangle$$

$$|\nu S;\, JM\rangle = \sum_{m} (\ell S m,\, M-m|JM)\; |n\ell m\rangle\; |S,\, M-m\rangle \quad . \tag{13}$$

By diagonalising the Hamiltonian matrix, whose elements are calculated with respect to the above basis, we obtain the spectrum (eigenvalues) and the coefficients a_ν (eigenvectors). The parameter ϵ is chosen so as to minimise the lowest eigenvalue. The summation in (13) is truncated, allowing ν to run from 1 to ν_{max} = 28, i.e., the Hamiltonian matrix is 28x28. Convergence can be examined by diagonalising submatrices of H_I and observing the change in the spectra with the basis dimension. The value of ϵ is always chosen to minimize the lowest eigenvalue for a given truncation; as ν_{max} increases, the dependence on ϵ becomes small.

B. Quarkonium: Model Studies

The interpretation of resonances of the J/ψ family and of the Υ family as bound states of $c\text{-}\bar{c}$ and $b\text{-}\bar{b}$ interacting by nonrelativistic quark-antiquark potentials has been successful phenomenologically, and inspired new investigations of solutions of the two-body Schrödinger equation.[5] The experimental data consist of the mass spectrum and decay properties such as leptonic, hadronic, and electric dipole transition widths. The mass spectrum corresponds to the eigenvalue spectrum of (6) and the leptonic and hadronic widths are proportional to $|\psi(0)|^2$, which corresponds to the probability of the quark and antiquark annihilation. For this lecture, we are not concerned with the validity of this model nor with relativistic corrections to the simple formulae for decay widths. Instead we ask, "how easily and how well can the equations be solved?"

We begin by considering separately the potentials $v_i(r') = \alpha_i r'^i$ (i = -1,1,2), which represent the Coulomb, linear-confining and quadratic-confining potentials, respectively. These problems have analytic solutions which are shown in Table I for the $\ell = 0$ case. The results of the variational calculation described above truncated at $\nu_{max} = 28$ also appear in Table I. The least accurate results seem to occur for the Coulomb potential over values of the principal quantum number n>1. In this case the eigenvalues accumulate at zero and are therefore difficult to reproduce accurately. For n=4 the eigenvalue is the wrong sign. The wave functions at the origin are especially bad, off by almost 50% even for the lowest eigenvalue. The convergence appears reasonable for the lowest energy as ν is increased: ν_{max} = (10,20,24,28) corresponds to E_ν = (0.2386, 0.2449, 0.2460, 0.2467), which is close to the exact answer E = 0.2500. We shall see in Sec. 3B, however, that this sequence does not represent very good convergence. The convergence of the n = 1 eigenfunction to its final value is represented by the comparision; ν_{max} = (10,28) yields $|\psi(0)|^2$ = (0.1517, 0.2322). The dimension cannot be increased beyond ν_{max} = 28 because of numerical instabilities. We close this alarming paragraph with the reassurance that this is the worst performance of the method in our experience (at least that which we will publish), so perhaps it is better to get it over with early in the lecture.

Table I. Comparison of ℓ=0 solutions of the two-body problem for the power law potentials $v(r) = \alpha_i r^i$. Energies are in units of $\alpha_i^{2/(2+i)} m^{-i/(2+i)}$ and the square of the wave function in units of $(m\alpha)^{3/(2+i)}$. $\hbar$=c=1. Labels exact, 28x28, and EV refer to the analytic solution, the variational solution for ν=28 in Eq. (13), and solutions of the singular integral equations obtained with 20 splines by Eyre and Vary (Ref. 8).

Potential	n	En			$\|\psi(0)\|^2$		
		exact	28x28	EV	exact	28x28	EV
Coulomb	1	-0.25	-0.2467	-0.250	0.03979	0.02322	0.03974
i = -1	2	-0.0625	-0.0621	-0.060	0.00497	0.00293	0.00492
	3	-0.0278	-0.0244	-0.021	0.00147	0.00131	0.00141
	4	-0.0156	+0.0101	-0.010	0.00062	0.00212	0.00066
Linear	1	2.33811	2.33811	2.35	0.7958	0.7943	0.105
i = +1	2	4.08795	4.08795	4.10	0.7958	0.7942	0.096
	3	5.52056	5.52054	5.51	0.7958	0.7942	0.088
	4	6.78671	6.78673	6.80	0.7958	0.7941	0.081
Harmonic	1	3.0	3.000	3.09	0.17959	0.17957	0.277
oscillator	2	7.0	7.000	7.05	0.26938	0.26971	0.329
i = +2	3	11.0	11.001	11.11	0.33673	0.33466	0.358
	4	15.0	15.035	15.93	0.39285	0.41356	0.381

On the other hand, the agreement with the exact results for the linear confining potential is excellent, as is the convergence (not shown). This should not be surprising, for it has long been known for this H_I that a truncation of (13) to a single term with variational improvement of epsilon yields an approximate eigenvalue 0.3% in error in the n = 1 state and only 8% high for the n=4 state.[6]. Even the n = 1 wave function at the origin is good to 8%. This method (truncation of (13) to one term) has been used by a group here in Lisboa[7] to study quarkonia. Each energy level was minimized independently so that the wave functions need not be orthogonal, and there is no reason to expect the method to be accurate in computing transition matrix elements (which they didn't, anyway). The wave functions obtained by diagonalizing the full H_I are, however, orthogonal.

We also display in Table I results of a method for treating the singularity structure of these potentials in the momentum space two-body Lippmann-Schwinger (LS) equation. Eyre and Vary[8] propose a regularization (screening) procedure which changes the long range behavior of the potential. The long range effects, such as confinement, are then treated as a perturbation. The eigenfunction is approximated by a sum of cubic B-splines with coefficients determined by the Galerkin method. Their results with 20 splines are comparable to the variational results with a basis of 28 terms. In particular, the eigenfunctions of the Coulomb case are much better, and the energies for all three potentials a little worse. The momentum space calculations illustrate the motivations behind the choice of a numerical method, as outlined by Delves above. Presumably one is not really interested in yet another method of working with a linear confining potential--the answer is already known. The work of Eyre and Vary demonstrates a method for treating singular potentials in integral equations that is also applicable to more general non-local interactions as will arise in applications of the Bethe-Salpeter equation to problems in QCD. (On the other hand, the harmonic oscillator expansion is not limited to local potentials--see Fig. 8 of Ref. 1.)

Table II. Comparison of ground-state energies in MeV for the Coulomb-plus-linear potential $v(r) = -ar^{-1} + br$, with a = 0.49 and b = 0.17 GeV^2. The masses of the charm and bottom quarks are m_c = 1.35 GeV and m_b = 4.77 GeV. These values fit spin-averaged data (K. J. Miller and M. G. Olsson, Phys. Rev. D25 (1982) 2382). The energies labeled exact are taken from Table II of L. Durand et al., Phys. Rev. D28 (1983) 607. Other notations as in Table I.

State	E(exact)	E(28x28)	E(EV)	E(SV)
Charmonium				
1S	364	365	368	364
1P	772	772	775	772
1D	1060	1060	1062	1060
b-quarkonium				
1S	- 98	- 98	- 97	
1P	349	350	353	
1D	585	585	588	

Table II contains results for the popular "Coulomb plus linear confinement" flavor-independent potential with parameters fitted to spin-averaged quarkonium data.[9] The provenance of the "exact" results is unknown so they can't be judged. It is possible that they were obtained by a discretization procedure which needed the solution of a 100x100 matrix.[9] The final column shows results obtained in momentum space by evaluating the limit in which the screening goes to zero. This yields an analytic expression for the confinement potential in the LS equation which is then solved by 31 splines and the Galerkin method.[10] The results with the screening method of Ref. 8 were obtained with 40 splines. Clearly the analytic regularization of the momentum space singularities is superior to the numerical regularization. In the end, all four numerical methods are in essential numerical agreement for the eigenvalues.

C. Quarkonium: Phenomenology

Portilho and Shokranian have employed the variational formalism of Section 1.A to study potential models of quarkonium in a series of papers on charmonium,[11] b-quarkonium and toponium,[12] and a study of bound states of three identical quarks[13] in the best confining potential found in the earlier papers. The potentials studied were phenomenological with a square-root or linear confining term, a Coulombic term proportional to the strong coupling constant, and spin-dependent terms of the Fermi-Breit type. The spin-dependent parts are handled from the beginning via equations (10) and (11), so the wave functions obtained describe fully each state characterized by the quantum numbers (ℓSJ). The matrix elements of such a general potential are given in equation (15) of Refs. 11 and 12. From these equations one can pick off the analytic matrix elements needed to calculate the two special cases (i = -1,1) of Table I. Their χ^2 test of the mass spectrum (Fig. 1) indicates that both charmonium and b-quarkonium are best described by the square root potential rather than the linear one. Lepton transition-width ratios (to minimize corrections to the simple nonrelativistic formula) for bottomium compare well with experimental values. A tensor force based on the linear and square root potentials for charmonium did not make any significant change in the mass spectra.

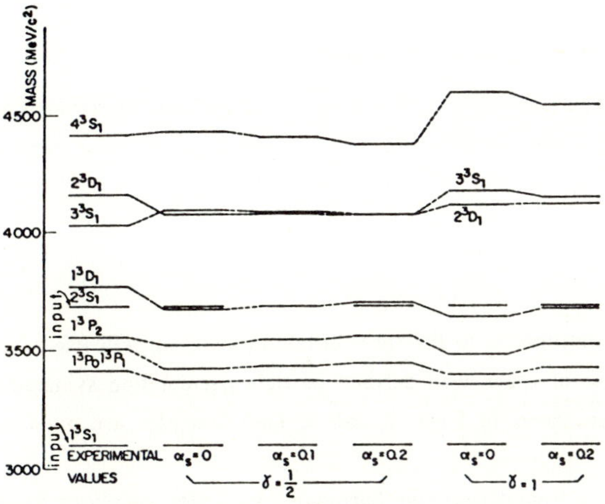

Fig. 1. Calculated and experimental values for the mass spectrum of the J/ψ family. The strength of the Coulombic part of H_I is given by α_s, and γ=(1,$\frac{1}{2}$) corresponds to a linear or square root confining potential.

How does one judge the applicability of the harmonic oscillator expansion to the non-relativistic quarkonium problem? It can easily handle spin-dependent and even tensor forces. It is not taxing numerically for slow computers. Based on Tables 1 and 2 , the eigenvalues are reliable and suitable for parameter fixing of phenomenological potentials. One might worry about the decay widths evaluated from the wavefunctions at the origin. Near the origin the $1/r$ singularity surely dominates the potential and, it is the Coulomb wave functions that the harmonic oscillator expansion has the most trouble mocking up.

3. The three-body bound state - particles of equal mass

A. Formalism

The construction of the trial wave function and evaluation of the Hamiltonian matrix elements needed for the three-body problem is fully described in Ref. 2. In this section we merely sketch the formalism for equal mass particles. We will defer the solution of a three-boson problem, requested by the organizers of the school, to the case of two equal masses and a third boson of a different mass (Section 4).

The trial wave function is given by the expansion

$$\Psi(\epsilon, \{3\}; JM) = \sum_{\nu} \mathbf{a}_{\nu}(\epsilon, \{3\}; JM) \mid \nu, \epsilon, \{3\}; JM\rangle \quad , \tag{14}$$

where ν stands for the set of oscillator quantum numbers ($n_I \ell_I$, n_{II} and ℓ_{II}) which are restricted by the cutoff condition

$$2n_I + \ell_I + 2n_{II} + \ell_I \leq N \quad , \tag{15}$$

and N is the maximum number of quanta defining the approximation. This summation is further restricted by the parity of the trial wave function: $(-)^{(\ell_I + \ell_{II})} = +$ or $-$. If the trial function is to be symmetric under the exchange of all three particles, then the summation is even further restricted by the condition

$$2n_I + \ell_I - 2n_{II} - \ell_{II} = 0 \text{ (modulus 3)} \quad . \tag{16}$$

This symmetry is labeled {3} in (14) and is appropriate to the three-α bound state we will use as an example. The construction of trial functions of other well-defined particle permutation symmetry, needed for the three-nucleon problem, is discussed in Refs. 2 and 3, and examples are given in Ref. 14.

The functions of (14) take the form of a summation over harmonic oscillator functions based on the two Jacobi relative coordinates defined below. This means that the basis states are translationally invariant. As the variational principle is applied to the intrinsic Hamiltonian (the total Hamiltonian minus the center of mass kinetic energy) with a translationally invariant trial function, the variational energy cannot contain any spurious effects of center-of-mass motion. In addition, the basis states have a well-defined orbital angular momentum which one chooses to be that of the system studied. Consider the ground state of a three-alpha system with J = L because S = 0. An explicit form of this $J^P = L^+$ function for a given ν would be[15]

$$|n_I\ell_I, n_{II}\ell_{II}, \epsilon, \{3\}; JM\rangle = [2/(1 + \delta_{n_I n_{II}}\, \delta_{\ell_I \ell_{II}})]^{1/2} \sum_{\substack{n_a\ell_a \\ n_b\ell_b}} (-)^{n_a-\ell_a}\, |n_a 2\ell_a, n_b\ell_b, \epsilon, JM\rangle$$

$$\times\ \langle n_a 2\ell_a, n_b\ell_b, J | n_I\ell_I, n_{II}\ell_{II}, J\rangle \quad . \tag{17}$$

The summations in (17) are restricted by the Moshinsky bracket conditions

$$2(n_a+n_b+\ell_a) + \ell_b = 2n_I + \ell_I + 2n_{II} + \ell_{II} \tag{18}$$

besides the triangular relations

$$|2\ell_a - \ell_b| \leq J = L \leq 2\ell_a + \ell_b \quad . \tag{19}$$

Finally, cutoff conditions (15) and (18) imply

$$2(n_a + n_b + \ell_a) + \ell_b \leq N \quad . \tag{20}$$

The indices a and b label harmonic oscillator states in the relative Jacobi coordinates

$$\mathbf{r}_a = \sqrt{1/2}\,(\mathbf{r}_1 - \mathbf{r}_2) \quad ,$$

$$\mathbf{r}_b = \sqrt{1/6}\,(\mathbf{r}_1 + \mathbf{r}_2 - 2\mathbf{r}_3) \quad , \tag{21}$$

where $\mathbf{r}_1$, $\mathbf{r}_2$, and $\mathbf{r}_3$ are the lab coordinates of the three-α particles. The coordinates and momenta are given in units of $(\hbar/m_\alpha\omega)^{1/2}$ and $(m_\alpha\hbar\omega)^{1/2}$, respectively, unless otherwise stated.

The matrix elements of the intrinsic Hamiltonian in this basis can be expressed as a sum over Talmi integrals and the B coefficients defined and tabulated in Ref.16. The coordinate $\mathbf{r}_a$ is useful in evaluating two-body potentials and $\mathbf{r}_b$ is also needed in case three-body potentials are present.[17,18] A special advantage of the method is that no difficulties arise from the ℓ-dependence of the potentials as the matrix elements are taken between states of definite ℓ_a. Explicit formulae for the Talmi integrals, many of which can be done analytically, are given in Ref. 2, and Refs. 15-19 for Gaussian, Yukawa, and Coulomb two-body potentials and two examples of three-body potentials with Gaussian form.

The variational calculation then proceeds as in the two-body case. The matrices to be diagonalized are larger, and one must sum over many more Talmi integrals than needed for the two-body problem. For a completely symmetric spin zero wave function, one finds that N = (8, 12, 16, 20, 24, 28) quanta corresponds to (11, 23, 41, 67, 102, and 147) basis states or matrices of this size to be diagonalized. Although these matrices are small enough for a modest computer, one still must be careful about numerical accuracy as N increases. We have had difficulties for basis sizes exceeding 150 and sometimes have found that the matrix needed for convergence simply would not fit into our computer. These practical difficulties can be overcome by careful attention to detail and access to large fast computers. (Strayer and Sauer[15] used a basis of 4654 states over ten years ago, apparently achieving a successful diagonalization.)

B. Three Equal Mass Bosons: Model Studies

Some years ago the Hokkaido group applied their ATMS (Amalgamation of Two-Body correlations in the the Mutiple Scattering process) method to the simple model problem[20] of three alpha particles in the ground state, in which the alpha particles interact pairwise via the phenomenological potential of Ali and Bodmer.[21] This potential has a strong inner repulsive core and an outside weak attraction which are features common to the nucleon-nucleon force (and not shared by the model quarkonium studies of Sec. 2). This model problem has been used as a test case for a proposed variational method[22] based upon the expansion of the trial wave function onto a complete set of correlated basis states obtained by multiplying the basis function of (17), for example, by a scalar symmetric correlation factor of the Jastrow form

$$F = \prod_{i<j} f(|\mathbf{r}_i - \mathbf{r}_j|) \quad . \tag{22}$$

If the correlation factor F = 1, then the regular formalism just discussed in Sec. 2.A is recovered. It is suggested that introduction of F improves the convergence of the ground state energy as well as the tail of the one-body and two-body distribution functions.[23]

For this lecture, we have recomputed this model problem in order to compare with the already existing results. The model is this: three particles interacting via the AB d_0 potential acting in all partial waves, and the Coulomb interaction is not included. The model thus differs from phenomemology because the true AB potentials differ in each partial wave and include the Coulomb interaction. The potential used is the S-wave

$$V(r) = V_R \exp[-(\mu_R r)^2] - V_A \exp[-(\mu_A r)^2] \quad , \tag{23}$$

where V_R = 500 MeV, V_A=130 Mev, μ_R = 0.7 fm-1, and μ_A = 0.475 fm-1. The dependence of the energy and radius on N as obtained in three separate calulations is displayed in Table 3. The radius of the system obtained from the wave function is that of point particles and is not, therefore, the charge radius.

Table III. Convergence of the model 3α system described in the text. An S-wave αα potential minus Coulomb is used in all partial waves. The number of quanta in the approximation is N. The Jastrow correlation factor F is employed in the right-hand columns, which are from Ref. 23 and a private communication.

Present calculation			(Ciofi degli Atti and Simula)				
			F = 1			F ≠ 1	
N	E (MeV)	radius (fm)	E (MeV)	radius (fm)	N	E (MeV)	radius (fm)
0			153.70	1.41	0	5.617	1.88
4			- 1.614	2.23	2	-3.193	2.33
8	-3.674	2.388	- 3.696	2.46	4	-4.883	2.43
12	-4.891	2.363	- 4.658	2.45	6	-5.136	2.42
16	-5.055	2.421	- 5.057	2.44	8	-5.173	2.43
20	-5.092	2.420	- 5.126	2.44	10	-5.176	2.43
24	-5.118	2.428	- 5.139	2.44	12	-5.177	2.43
28	-5.126	2.432	- 5.156	2.44			
32			- 5.169	2.43			
∞	-5.137	2.438	(- 5.143)			-5.177	
						ATMS (Ref. 20)	
						-5.18	2.43

With the aid of Table III we can give an example of an extrapolation of the result to N = ∞ symbol which will allow us to produce some estimate of the exact solution. By comparision of the extrapolated value with the best result obtained directly, we can give some measure of the expected accuracy of the result. The theoretical expectation for the Nth term of any systematic expansion is[1]

$$E_N = E + AN^{-Q} + O(N^{-(Q+1)}) \quad . \tag{24}$$

One should not fit the entire column of Table III to this form because convergence often has not really set in at small N, and because it neglects terms of higher order in 1/N. Indeed, during the fit one should verify that the convergence rates obtained do in fact appear to follow the functional form used. (They do not in the Coulomb example of Table I.) By differencing (23) we obtain

$$- \Delta E_N \equiv E_N - E_N+1 = AQN^{-(Q+1)} + O(N_{-}^{(Q+2)}) \quad ,$$

and hence to leading order

$$\ell n(-\Delta E_N) = \ell n\ AQ - (Q+1)\ell n\ N \quad . \tag{25}$$

Thus a logarithmic plot of $-\Delta E_N$ should be linear with slope -(Q+1) and intercept $\ell n(AQ)$. We estimate A and Q by a least-squares fit to those differences which seem to be linear on such a plot. Then we simply substitute in (24) to estimate an extrapolated value. Clearly this differencing procedure magnifies inaccuracies in the numerically computed eigenvalues but, on the other hand, if the difference plot is really linear, one should have some confidence in the extrapolated value.

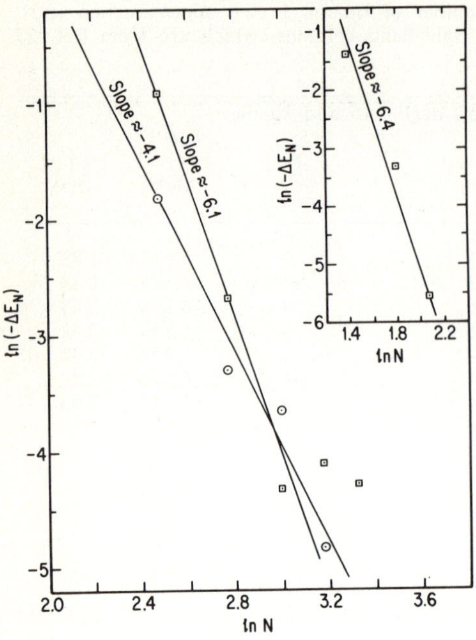

Fig. 2. The differences $\ell n(-\Delta E_N)$ versus $\ell n(N)$ obtained from Table 3. The circles represent the present results and squares those of Ref. 23. The inset shows the results corresponding to a Jastrow correlation factor in the trial wave functions.

Applying this procedure to the numbers of Table III yields Figure 2. We immediately notice that the energy differences from the present calculation (circles) are linear and that the corresponding harmonic oscillator expansion of the Rome group yields energy differences (squares) which seem to indicate nonconvergence above N=20 or 24. If we throw away the N=24, 28, and 32 eigenvalues of the Rome group, and fit the sequence (N = 12, 16, 20), we find the line indicated with slope of about -6 or Q~5. This is about the same as our convergence rate of Q~3. These differences in convergence rates from this difference method are probably not significant, and the extrapolated values would have a similar error which we do not attempt to estimate. The convergence rate for the Jastrow correlated basis is indicated by the slope of the line on the inset of Figure 1. *The convergence is in reality the same as that of the two calculations without a Jastrow factor!* The significant fact is not that the convergence rate is changed by the Jastrow correlations, but that convergence sets in at a much smaller value of N. Therefore much less work is required to get the answer.

Some of the work in such variational calculations can be reduced by taking advantage of the defined symmetry of the trial wave function. The trial function of the Rome group was symmetric under permutation of two of the particles but not all three. Therefore their trial wavefunction truncated at N=32 contained 525 basis states. (One can suggest that diagonalizing these large matrices led to the apparent numerical instability of Table III and Fig. 2, but this is pure speculation.) Imposing the Jastrow correlation factor on the trial wavefunction implies a converged result at N = 12 or 50 basis states in their calculation. Imposing the proper symmetry on the trial wave function without a Jastrow correlation factor yields a converged result at N=24 (102 basis states) or N=28 (147 basis states). The reduction in work afforded by the Jastrow factor is significant but not overpowering when compared to taking advantage of all the symmetries in this problem.

The same extrapolation procedure can be used for expection values of bounded operators such as the rms matter radius. It clearly jitters around for small N in our results, but one can be optimistic and extract a convergence rate of Q~3 from the sequence (20, 24, 28). This is comparable to the rate Q~3 for the energy. Not enough information is available for the other results of Table III to analyze convergence. For some problems, such as potentials with repulsive cores, operators such as the rms radius are so much smoother than the Hamiltonian that they are both numerically less troublesome and less sensitive to details of the wave fuction. Thus one can trust some aspects of the variational wave function. More troublesome operators such as the charge form factor at high momentum transfer will be discussed in Section 3.C.

It appears from Table III that the two systematic harmonic oscillator expansions yield about the same extrapolated ground state energy of -5.14 MeV, but the two variational calculations which attempt to take into account short-range two-body correlations would instead like an energy of -5.18 MeV. In fact, the converged result of the systematic linear expansion lies above both the upper and the lower bound of the ATMS solution[20] which are -5.18 MeV and -5.3 MeV, respectively. This is a distressing discrepancy which lies outside reasonable errors from the extrapolation procedure. It would be desirable for the practitioners of other methods discussed at this school to have a look at this model problem. Certainly at present one knows neither what the correct answer is nor which method is superior.

C. Three Equal Mass Particles: Phenomemology

One can immediately think of three-body bound states that could be compared with experimental data: Can the A = 3 nuclei be described by three interacting structureless nucleons? Can ^{12}C be described by three interacting structureless α particles? Can the baryons with spin 3/2 be described by three identical heavy structureless quarks with a nonrelativistic treatment? The method of HO expansions has acquired a certain notoriety from the attempts to answer the first question[14,19,23] with the complex nucleon-nucleon potentials termed "realistic." The experimental data is sparse to compare with the predictions of the mass spectrum obtained with the HO method by Portilho and Shokranian.[13] This leaves the middle phenomenological question which we address in this section.

The description of ^{12}C as three structureless α particles has a long history, which is partly summarized in our work on the subject.[17,18] In this game, the input is, *or should be*, the best $\alpha\alpha$ potential which describes the phase shifts and the near zero resonance assumed to correspond to the ground state of ^{8}Be. The energy spectrum and properties of the calculated 3α system are then compared to the experimental data on ^{12}C. The variational approach in an HO basis has many advantages for this type of study. The energies of the ground state *and low-lying states* are bounded by the eigenvalues which come out of the diagonalization of the Hamiltonian matrix. The 0^+ and 2^+ states are cleanly separated and can be studied separately. The elastic form factor and inelastic form factors associated with the transition $J_i \rightarrow J_j$ can be analytically evaluated as bilinear forms over the a_ν. A new type of generalized B coefficient has been developed for the inelastic form factors and the reduced electromagnetic transition widths.[24] Thus a variety of properties of the states of ^{12}C can be calculated reliably from the variational wave function and compared with experiment. Finally, the method easily handles the state dependence of realistic local $\alpha\alpha$ potentials, in contrast to the state-independent potentials required by hyperspherical expansion techniques actually used and evidently required also by the ATMS method.[23]

The Gaussian potentials Ali and Bodmer (AB) fit to the scattering data were dependent on the relative angular momentum. The short-range repulsion decreased with increasing ℓ as one would expect from an exclusion effect and the long range attraction was constructed from the $\ell = 4$ phase shifts. A newer potential with the same philosophy was fit by Chien and Brown (CB) to phase shifts at higher energies than those available for the fit of the AB potentials. It was obtained by double-folding an attractive Yukawa NN interaction over the α density and fitting the result plus the ℓ-dependent repulsive term to their accurate scattering data.[25] Both potentials include the Coulomb interaction.

All of our comparisions with experiment were made with the AB potential labeled (d_0', d_2, d_4). Its parameters are given most conveniently in Table 1 of Ref. 15. From the sequence N = (18, 20, 22, 24) and E_N = (-2.052, -2.079, -2.099, -2.108) MeV, we can extrapolate to E = -2.125 MeV as the prediction of this potential. This is in rather bad agreement with the 7.274 MeV binding energy of ^{12}C with respect to dissociation into three alphas. However, we shall see that other properties of ^{12}C are nicely matched by this model. The studies of Refs. 17 and 18 attempted to find 3α forces which would fix the binding energy discrepency without disrupting other positive features of the 3α model. Figures 3 and 4 (taken from Ref. 18) show the degree of success of this program. In addition, the reduced transition probability for the transition $O_2^+ \rightarrow O_1^+$ dropped by a factor of four to a value with the experimental error bars when a three-body force was added to the Hamiltonian.[18] The calculations were done in the N = 16 quanta approximation. As noted in Ref. 17, the height of the first maximum of the elastic form factor is decreased by 5% as N changes from 10 to 12 quanta, a smaller change than that of the lowest eigenvalue. It appears not so hard to calculate form factors resulting from realistic potentials as one might think.

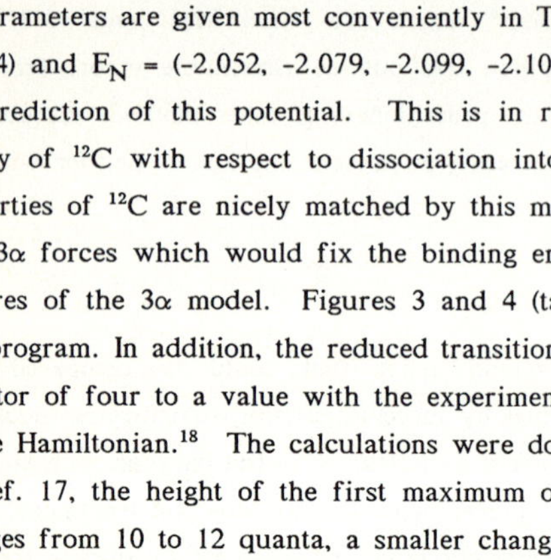

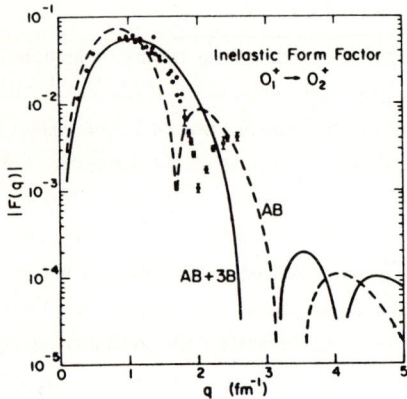

Fig. 3. The $O_1^+ \rightarrow O_2^+$ inelastic form factor of ^{12}C. The line AB is the prediction of the 3α system with the Ali-Bodmer potential described in the text. The effect of a three-alpha force is shown by the line AB + 3B.

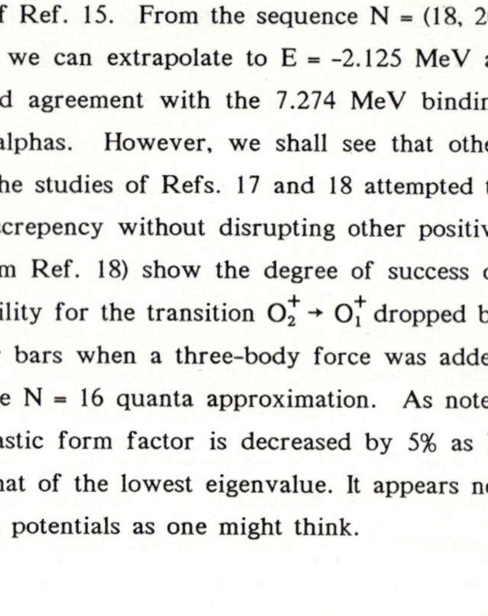

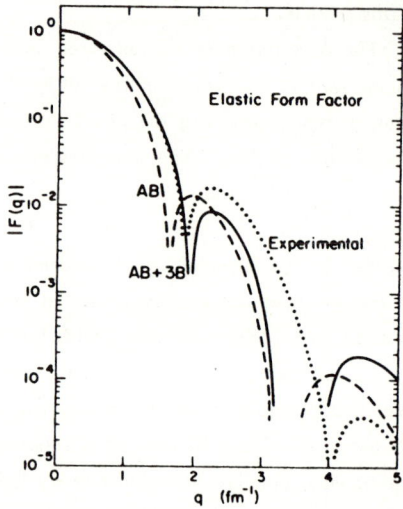

Fig. 4. Elastic form factor of ^{12}C.

One can then use the variational wave functions to obtain the geometrical arrangement of the three-α particles. We found that the ground state was most of the time in an equilateral triangle configuration and the first 0^+ excited state is a linear chain of alpha particles (Fig 5). But these configurations are not rigid; 50% of the maximum probability occurs at triangle deformation of ~25 degrees and linear chain deformation of ~40^0 as can be inferred from Figs. 6 and 7. These findings have implications for calculations of pion-carbon scattering or proton-carbon scattering which have assumed a rigid 3α structure.[26] These calculations were later repeated by others with the same results.[27]

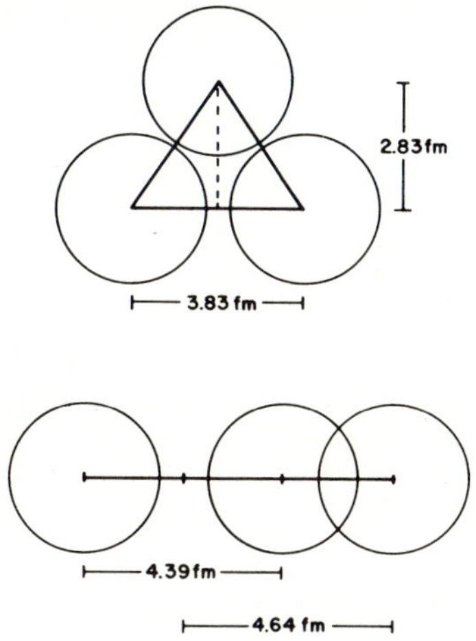

Fig. 5. Most probable geometrical configurations of the ground state and O_2^+ state, deduced from Figs. 6 and 7. The size of the α particles is also shown.

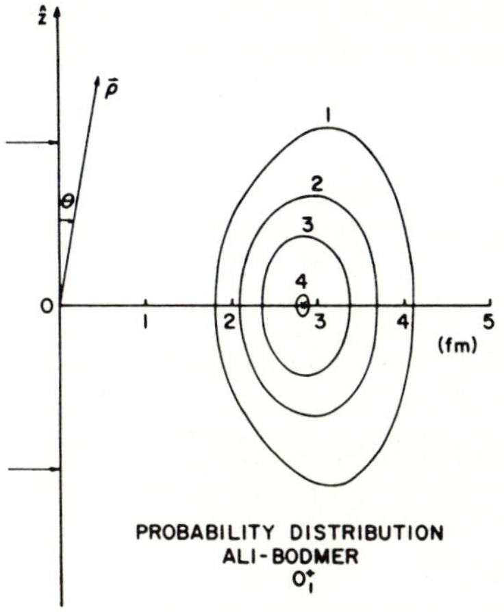

Fig. 6. Polar plot of the equi-probability curves for the ground-state using the Ali-Bodmer ($d_0'd_2d_4$) potential. The numbers beside the curves are expressed in an arbitrary scale, e.g., the curve labeled 4 is four times more probable than the curve labeled 1. The cross shows the most probable position of the third α particle and the arrows expectation value of $\mathbf{r}_{12}$.

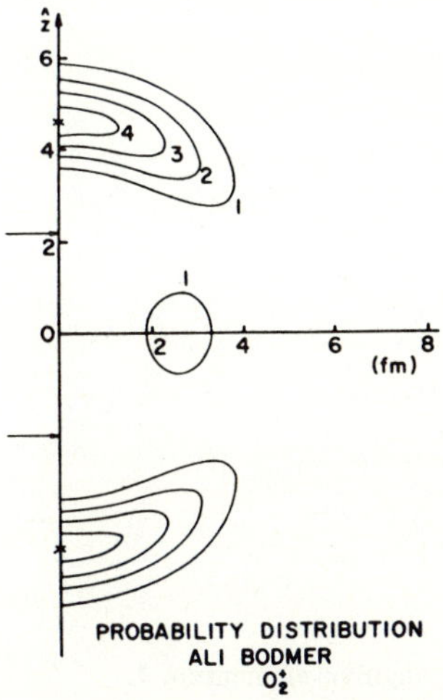

Fig. 7. The same as in Fig. 5, for the O_2^+ excited state.

4. The three-body bound state - two particles of equal mass

A. Formalism

We start with the following Jacobi coordinates for three particles with masses m_1, m_2, and m_3

$$\mathbf{r}_a = \sqrt{\frac{m_2 m_3}{m(m_2+m_3)}}\,(\mathbf{r}_2 - \mathbf{r}_3)$$

$$\mathbf{r}_b = \sqrt{\frac{m_1(m_2+m_3)}{mM}}\left[\mathbf{r}_1 - \frac{m_2\mathbf{r}_2+m_3\mathbf{r}_3}{m_2+m_3}\right] \tag{26}$$

$$\mathbf{r}_c = \frac{1}{M}(m_1\mathbf{r}_1 + m_2\mathbf{r}_2 + m_3\mathbf{r}_3) \quad ,$$

where

$$m = \frac{m_1m_2+m_1m_3+m_2m_3}{M} \tag{27}$$

is the reduced mass analogue for three particles and

$$M = m_1 + m_2 + m_3 \tag{28}$$

is the total mass.

The total kinetic energy operator may then be written as

$$T = -\frac{\hbar^2}{2m}(\nabla_a^2 + \nabla_b^2) - \frac{\hbar^2}{2M}\nabla_c^2 \quad . \tag{29}$$

On the other hand, the Hamiltonian for two oscillators in the coordinates $\mathbf{r}_a$ and $\mathbf{r}_b$, with common mass m and frequency ω, is

$$H_{HO} = -\frac{\hbar^2}{2m}(\nabla_a^2 + \nabla_b^2) + \frac{m\omega^2}{2}(r_a^2 + r_b^2) \tag{30}$$

with known eigenfunctions

$$\psi(\mathbf{r}_a, \mathbf{r}_b) = \phi_{n_a\ell_a m_a}(\mathbf{r}_a)\,\phi_{n_b\ell_b m_b}(\mathbf{r}_b) \tag{31}$$

and eigenvalues

$$E = \hbar\omega(2n_a + \ell_a + 2n_b + \ell_b + 3) \quad . \tag{32}$$

If we take the following linear combination, with a well-defined angular momentum **L**

$$|n_a\ell_a, n_b\ell_b, LM\rangle = \sum_{m_a} (\ell_a\ell_b m_a, M-m_a | LM)\, \phi_{n_a\ell_a m_a}(\mathbf{r}_a)\, \phi_{n_b\ell_b, M-m_a}(\mathbf{r}_b)\ , \tag{33}$$

we still have (32) as eigenvalues.

We can relate H_{HO} to the intrinsic Hamiltonian for three particles, interacting via potentials $V(r_{ij})$

$$H_I = -\frac{\hbar^2}{2m}(\nabla_a^2 + \nabla_b^2) + \sum_{i<j} V(r_{ij})$$

$$= H_{HO} - \frac{m\omega^2}{2}(r_a^2 + r_b^2) + \sum_{i<j} V(r_{ij})\ . \tag{34}$$

We expand the eigenfunctions of H_I in a basis formed by the harmonic oscillator wave functions given in Eq. (33)

$$\Psi = \sum_\nu a_\nu\, |\nu, LM\rangle\ , \tag{35}$$

where $\nu = (n_a\ell_a, n_b\ell_b)$ goes up to a certain maximum truncation value N which characterizes the approximation used. The coefficients a_ν and the frequency ω are taken as variational parameters.

We will apply this formalism to the hypernuclei ${}^9_\Lambda Be$ and ${}^6_{\Lambda\Lambda}He$ in the $\alpha\alpha\Lambda$ and $\alpha\Lambda\Lambda$ models, respectively. We suppose in the following that the equal particles ($\alpha\alpha$ in ${}^9_\Lambda Be$ and $\Lambda\Lambda$ in ${}^6_{\Lambda\Lambda}He$) are located at the coordinates $\mathbf{r}_2$ and $\mathbf{r}_3$, while the different particle is at $\mathbf{r}_1$. It is easy to verify that the condition that the wave function shall be symmetric under permutation between the two α's in ${}^9Be$ imposes ℓ_a to be even. On the other hand, for ${}^6_{\Lambda\Lambda}He$, $\Psi = |radial\rangle\, |spin\rangle$ shall be antisymmetric under permutation between the two Λ's (spin = $\frac{1}{2}$). We consider only singlet spin states, which are antisymmetric, and for this reason we shall again have symmetric radial wave functions and only even values for ℓ_a.

We need the following matrix elements

$$\langle\nu, LM|H_I|\nu', LM\rangle = \langle\nu, LM|H_{HO} - \frac{m\omega^2}{2}(r_a^2 + r_b^2) + V^{(1)}(r_{23}) + 2V^{(2)}(r_{12,13})\,|\nu', LM\rangle\ , \tag{36}$$

where the following interactions were considered:

(a) $\alpha\alpha$ interaction - Ali-Bodmer (d_0',d_2,d_4) potential[21] plus Coulomb interaction;

(b) $\alpha\Lambda$ interaction - Tang-Herndon (A and B) potentials[28]

- Bando potential;[29]

(c) ΛΛ interaction - Dalitz potential.[29,30]

The necessary matrix elements of each term in (36) are

$$\langle \nu,LM|m\omega^2 r_a^2|\nu',LM\rangle = \delta_{n_b n_b'}\ \delta_{\ell_a \ell_a'}\ \delta_{\ell_b \ell_b'}\ \epsilon \sum_{p=\ell_a}^{n_a+n_a'+\ell_a} B(n_a\ell_a,n_a'\ell_a',p)(p+\tfrac{3}{2}) \tag{37}$$

$$\langle \nu,LM|m\omega^2 r_b^2|\nu',LM\rangle = \delta_{n_a n_a'}\ \delta_{\ell_a \ell_a'}\ \delta_{\ell_b \ell_b'}\ \epsilon \sum_{p=\ell_b}^{n_b+n_b'+\ell_a} B(n_b\ell_b,n_b'\ell_b',p)(p+\tfrac{3}{2})\ . \tag{38}$$

As all the above potentials, except the Coulomb one, are of the Gaussian type, we also need the following matrix element of $V^{(1)}$ for the equal particles.

$$\langle \nu,LM|e^{-\tau r_a^2}|\nu',LM\rangle = \delta_{n_b n_b'}\ \delta_{\ell_a \ell_a'}\ \delta_{\ell_b \ell_b'} \sum_{p=\ell_a}^{n_a+n_a'+\ell_a} B(n_a\ell_a,n_a'\ell_a',p)(1+\tau')^{-p-\frac{3}{2}}\ , \tag{39}$$

and for $V^{(2)}(r_{12,13})$

$$\langle \nu,LM|e^{-\xi r_{12,13}^2}|\nu',LM\rangle = \langle \nu,LM|e^{-tr_a^2-ur_b^2\pm v\mathbf{r}_a\cdot\mathbf{r}_b}|\nu',LM\rangle$$

$$= \tfrac{1}{2}\sqrt{\pi(2\ell_a+1)(2\ell_b+1)(2\ell_a'+1)(2\ell_b'+1)}\ (-1)^L$$

$$\times \sum_{\ell(\text{even})} (\ell_a\ell_a'\ 00|\ell 0)(\ell_b\ell_b' 00|\ell 0)\ W(\ell_a\ell_a'\ \ell_b\ell_b';\ \ell L)\ \frac{(v/2)^\ell}{\Gamma(\ell+\frac{3}{2})}$$

$$\times \sum_p B(n_a\ell_a',n_a'\ \ell_a'\ ,p)\ \frac{\Gamma(p+\ell/2+\frac{3}{2})}{\Gamma(p+\frac{3}{2})(1+t)^{(p+\ell/2+\frac{3}{2})}}$$

$$\times \sum_q B(n_b\ell_b',n_b'\ \ell_b'\ ,q)\ \frac{\Gamma(q+\ell/2+\frac{3}{2})}{\Gamma(q+\frac{3}{2})(1+u)^{(q+\ell/2+\frac{3}{2})}}$$

$$\times F_{21}\left(p+\ell/2+\tfrac{3}{2},\ q+\ell/2+\tfrac{3}{2},\ \ell+\tfrac{3}{2}\ ;\ \frac{v^2}{4(1+t)(1+u)}\right) \tag{40}$$

and finally the Coulomb force takes the form

$$\langle \nu,LM \left| \frac{4e^2}{r_{23}} \right| \nu',LM\rangle = 2\alpha \sqrt{2m_\alpha c^2 \epsilon}\; \delta_{n_b n_b'} \; \delta_{\ell_b \ell_b'} \; \delta_{\ell_a \ell_a'} \times \sum_{p=\ell_a}^{n_a+n_a'+\ell_a} B(n_a\ell_a, n_b'\ell_b', p) \frac{p!}{\Gamma(p+\frac{3}{2})} \tag{41}$$

In these equations, α is the fine structure constant, the coefficients B were defined in Ref. 2 and

$$\epsilon = \hbar\omega \tag{42}$$

$$\tau' = \frac{2\hbar^2}{m_2\epsilon}\tau \tag{43}$$

$$t = \frac{\hbar^2}{2m_2\epsilon}\xi \tag{44}$$

$$u = \frac{\hbar^2(m_1+2m_2)\xi}{2m_1m_2\epsilon} \tag{45}$$

$$v = 2\sqrt{tu} \quad . \tag{46}$$

Once the matrix elements have been calculated, we look for a value of ϵ, which minimizes the lowest eigenvalue that comes from the matrix diagonalization. The coefficients a_ν of the expansion (35) are columns of the eigenvector matrix. With Ψ we can find, for instance, the expectation values for $\alpha\alpha$ and $\Lambda\Lambda$ distances, and for Λ-$\alpha\alpha$ and α-$\Lambda\Lambda$ center-of-mass distances

$$\langle r_{23}^2\rangle^{1/2} = \hbar \left\{ \frac{2}{m_2\epsilon} \sum_{\nu\nu'} a_\nu a_{\nu'} \; \delta_{n_b n_b'} \; \delta_{\ell_a \ell_a'} \; \delta_{\ell_b \ell_b'} \sum_p B(n_a\ell_a',\, n_a'\ell_a', p)(p+\tfrac{3}{2}) \right\}^{1/2} , \tag{47}$$

$$\langle r_{12}^2\rangle^{1/2} = \hbar \left\{ \frac{m_1+2m_2}{2m_1m_2\epsilon} \sum_{\nu\nu'} a_\nu a_{\nu'} \; \delta_{n_a n_a'} \; \delta_{\ell_a \ell_a'} \; \delta_{\ell_b \ell_b'} \sum_p B(n_b\ell_b',\, n_b'\ell_b', p)(p+\tfrac{3}{2}) \right\}^{1/2} . \tag{48}$$

B. The three-body bound state - two particles of equal mass : Model studies

We tested our codes based on this new formalism by recalculating the model of Bando *et al.*[29] for the spin-singlet state of ${}^{6}_{\Lambda\Lambda}He$. With $\Lambda\Lambda$ and $\Lambda\alpha$ potentials of Eqs. (2.5) and (2.6) of Ref. 29, we obtain the same binding energy for the 0^+ ground state. Our wave function is only slightly less extended than theirs. Let the rms distance between the like particles be denoted by $R_{\Lambda\Lambda}$ and the rms distance between the unlike particle and the center of mass of the like particles be $R_{\alpha\text{-}\Lambda\Lambda}$. Then Table IV displays the results for this $\alpha + \Lambda + \Lambda$ model.

Table IV. Properties of ${}_{\Lambda\Lambda}^{6}He$ as a $\alpha + \Lambda + \Lambda$ bound state.

	E(MeV)	$R_{\Lambda\Lambda}$(fm)	R_{α}-$\Lambda\Lambda$(fm)
Our results	-10.8	2.43	1.56
Ref. 29	-10.8	2.52	1.60

The excellent agreement with the experimental binding energy of -10±0.8 MeV is probably coincidental, as both potentials are single attractive Gaussians (the $\Lambda\alpha$ potential is basically the A potential of Ref. 28).

We present one more model problem that has been solved by the ATMS method. The model is ${}_{\Lambda}^{9}Be$ as an $\alpha\alpha\Lambda$ system with the $\alpha\alpha$ potential chosen to be the AB d_0 potential of (23) acting in all states and the $\Lambda\alpha$ potential, the "Isle" potential[31]

$$V_{\Lambda\alpha}(r) = V_R \exp[-(r/b_R)^2] - V_A \exp[-(r/b_A)^2] \quad , \tag{49}$$

where V_R = 450.43 MeV, V_A = 404.88 MeV, b_R = 1.25 fm, and b_A = 1.41 fm. The Coulomb force is neglected in this model. The Hokkaido group has kindly furnished details of their results with this model.[32] The variational wave function is assumed to be of the Jastrow type, earlier considered by Bodmer and Ali.[33] It takes the form

$$\psi = f_{\alpha\alpha}(r_{23})\, f_{\Lambda\alpha}(r_{12})\, f_{\Lambda\alpha}(r_{13}) \quad , \tag{50}$$

and the f_{ij}'s are determined by the Eular-Lagrange equations discussed in Ref. 20. We applied the formalism of Section 4.A to this problem approximating, as did the Hokkaido group, the mass of the α particle by four nucleon masses. Table V shows the results.

Table V. Properties of ${}_{\Lambda}^{9}Be$ as a α+α+Λ bound state.

	$E_{\alpha\alpha}$(MeV)	$E_{\Lambda\alpha\alpha}$(MeV)	B_{Λ}(MeV)
Our results	-1.37	- 9.17	7.80
Ref. 32	-1.37	-10.09 (Temple lower bound -10.41)	8.72

In this case, as in that of Table III, the harmonic oscillator expansion converges to a value which lies above the upper- and lower-bound of a version of the ATMS method.

The separation energy B_Λ is defined by

$$B_\Lambda = - \left[E\left({}^{A}_{\Lambda}Z\right) - E({}^{A-1}Z) \right] \quad . \tag{51}$$

C. The three-body bound state - two particles of equal mass: Phenomenology

Alpha-particle models of light nuclei will continue to be a testing ground for both effective interactions and for microscopic thoeries of composite particle interactions. For example, Bao and collaborators have presented calculations of 3α systems, $\alpha+\alpha+\Lambda$ systems, and $\alpha+\alpha+n$ systems, which are then compared with properties of the corresponding nuclei and hypernuclei.[27,34,35] These calculations are said to be made with the harmonic oscillator expansion, but no details have ever been given. Thus, this phenomenology is difficult to evaluate. A molecular Born-Oppenheimer approach to the $n+\alpha+\alpha$ system is being developed here in Lisboa.[36] The formalism developed in Section 4.A might also be useful in these problems.

Our interest in the three-body bound state with two equal mass particles is already evident from our choice of model problems to compare calculational techniques. We have been intrigued by the new knowledge of spectroscopy in light hypernuclei[37] and the development of more "realistic" potentials between composite particles.[25,31] For example, the $\Lambda\alpha$ potentials have traditionally been obtained by folding an attractive Gaussian ΛN interaction into the nucleon density distriubtion of the alpha particle.[28] The strength is then adjusted to fit the separation energy (~3.1 MeV) of the Λ from the α core of a rigid-core two-body model of ${}^{5}_{\Lambda}He$. Recent $\Lambda\alpha$ potentials differ from these by a central repulsion and stronger attraction at larger r. Two of these newer potentials fold into the α density an effective ΛN interaction obtained from hard-core ΛN potentials by nuclear matter methods.[38,39] A third potential of this type is not obtained by folding, but from the solution of multiple scattering equations with hard-core ΛN and NN potentials in which the incident particle (Λ) and the target nucleus (${}^4He$) are explicitly treated as a correlated (N+1)-body system.[31] It is our hope that a comparison of the predictions of the low-lying O^+ and 2^+ states of the $\alpha+\alpha+\Lambda$ system with the experimental properties[37] of ${}^{9}_{\Lambda}Be$ will aid in the selection of a phenomenological $\alpha\Lambda$ potential. After that is settled, one could learn about the $\Lambda\Lambda$ interaction via a study of the $\alpha+\Lambda+\Lambda$ model of ${}^{6}_{\Lambda\Lambda}He$.

Our progress in this program has been reported in Ref. 40. In this lecture, we content ourselves with an attempt to address a question posed in Ref. 39: "Can ${}^{9}_{\Lambda}Be$ be described as an $\alpha+\alpha+\Lambda$ system with realistic two-body potentials, or does one need a three-body force?" For this phenomenological problem, we solved the equations of Section 4.A with the Chein-Brown $\alpha\alpha$ force, which includes the Coulomb force and gives the best fit to the $\alpha\alpha$ scattering and bound-state data. Matrix elements are displayed in the Appendix of Ref. 17. The central $\Lambda\alpha$ potentials obtained by folding, such as Tang-Herndon A and B (TH-A, TH-B), appear less than realistic because they are based on monotonic, attractive ΛN potentials. One expects the ΛN interaction to have a strong short-range repulsion, as do most models of the NN interaction. In addition, it has been suggested[41] that the pionic decay rate in ${}^{5}_{\Lambda}He$ is very sensitive to the Λ-density distribution in ${}^{5}_{\Lambda}He$. The effect of hard cores in the ΛN interaction is to produce a central repulsion in the $\Lambda\alpha$ potential, which suppresses the Λ-density distribution at the center. This in turn enhances the pionic decay rate toward agreement with experiment.[42] For these reasons, it is interesting to examine the predictions of the $\Lambda\alpha$ potential with the strongest central repulsion, the "Isle" potential of (49). Our results are in Table VI.

Table VI. Properties of ${}^{9}_{\Lambda}Be$ with different $\Lambda\alpha$ potentials.

$V_{\Lambda\alpha}$	E(MeV)	$R_{\alpha\alpha}$(fm)	$R_{\Lambda\text{-}\alpha\alpha}$(fm)
TH-A	-5.33	3.46	2.45
TH-B	-3.56	3.51	2.63
Isle	-7.33	3.57	2.47

The distance between the α particles of ${}^{9}_{\Lambda}Be$ is rather larger than twice the individual rms matter radius of 1.45 fm, indicating no breakdown of the assumptions of the model.

The experimental separation energy B_Λ of ${}^{9}_{\Lambda}Be$ is 6.71±0.04 MeV. To a first approximation, this is just the negative of eigenvalues E of Tables V and VI, since B_Λ is defined by (51), and the core nucleus ${}^{8}Be$ is unbound by 0.092 MeV. One can tentatively conclude that a three-body model with a realistic $\Lambda\alpha$ interaction and an $\alpha\alpha$ interaction that fits phase shifts can give a reasonable description of the ground state of ${}^{9}_{\Lambda}Be$. This conclusion differs from that of Ref. 39, which emphasized a need for three-body forces to solve the ${}^{9}_{\Lambda}Be$ binding energy problem.

5. Conclusions

In this lecture, we have described the harmonic oscillator expansion variational method as used in two-body and three-body bound states. (The development of this method for the four-body bound state and applications to the 4α model of ${}^{16}O$ can be found in Refs. 3, 4, and 43.) We have analyzed convergence properties of the expansion and given an indication of the labor involved in setting up the equations. We have shown that the method is reliable for energies and wave functions for common two-body problems with the exception of the Coulomb potential, and even then it is not too bad. The Coulomb interaction pervades all of physics, and the few-body problem is no exception. The alert reader may have noticed that the model systems chosen by other few-body groups and re-examined here have not included the Coulomb interaction. That is why we have grouped them under the heading "models" rather than "phenomenology." The harmonic oscillator expansion in the three-body bound state is competitive with other methods for boson problems with either neutral or charged particles. Indeed, in our opinion, the convergence theorems of a systematic expansion such as the HO expansion imply a confidence in the results which can, in turn, be used to discuss other methods.

We break with tradition in this final paragraph by revealing the failures we have encountered with this tool. We were not able to solve the He trimer problem, probably because the short-range atom-atom potential is even more singular than the nucleon-nucleon potential. So far, we have been unable to obtain the first excited state[37] of $^{9}_{\Lambda}$Be with the available potentials. It is not clear yet whether this is a helpful remark on the quality of the potentials or whether we simply cannot get large enough matrices into our computer. Recently we have gained access to a supercomputer. We will be interested to learn if these failures are inherent in the method or are an artifact of our present facilities.

Acknowledgments

This Brazilian-American collaborative effort was supported by NSF Grant PHY86-06368 and the Conselho Nacional de Pesquisas (Brazil). We would like to thank our teachers, collaborators, and friends V.C. Aguilera-Navarro, D. A. Agrello, P. S. C. Alencar, Z. M. O. Shokranian, and J. P. Vary (that's right, he is not Portuguese speaking) for many helpful conversations.

References

1. L. M. Delves, in Advances in Nuclear Physics, ed. M. Baranger and E. Vogt (Plenum Press, New York, 1972) p. 1.
2. M. Moshinsky, The Harmonic Oscillator in Modern Physics: From Atoms to Quarks (Gordon and Breach, New York, 1969).
3. V. C. Aguilera-Navarro, M. Moshinsky, and W. W. Yeh, Rev. Mex. di Fisica **12** (1968) 241.
4. D. A. Agrello,V. C. Aguilera-Navarro, and J. N. Maki, Rev. Brasileira de Fisica **11** (1981) 163.
5. C. Quigg and J. L. Rosner, Physics Reports **56** (1979) 168.
6. D. Gromes and I. O. Stamatescu, Nucl. Phys. **B112** (1976) 233.
7. J. Dias de Deus, A. B. Henriques, and J. M. R. Pulido, Z. Physik **C7** (1981) 157; see also, J. Dias de Deus and A. B. Henriques, Portgal. Phys. **16** (1985) 105.
8. D. Eyre and J. P. Vary, Phys. Rev. D (October 1986).
9. K. J. Miller and M. G. Olsson, Phys. Rev. **D25** (1982) 2383.
10. J. R. Spence and J. P. Vary, to be published; J. P. Vary, private communication.
11. O. Portilho and Z. M. O. Shokranian, Rev. Brasileira de Fisica **14** (1984) 15.
12. Z. M. O. Shokranian and O. Portilho, J. Phys. G: Nucl. Phys. **12** (1986) 583.
13. O. Portilho and Z. M. O. Shokranian, submitted to Phys. Rev. D.
14. M. R. Strayer and P. U. Sauer, Nucl. Phys. **A231** (1974) 1.
15. V. C. Aguilera-Navarro and O. Portilho, Ann. Phys. **107** (1977) 126.
16. T. A. Brody and M. Moshinsky, "Tables of Transformation Brackets" (Gordon and Breach, New York, 1976).
17. O. Portilho and S. A. Coon, Z. Physik. **A290** (1979) 93.
18. O. Portilho, D. A. Agrello, and S. A. Coon, Phys. Rev. **C27** (1983) 2923.
19. P. Nunberg, D. Prosperi, and E. Pace, Nucl. Phys. **A285** (1977) 58.
20. S. Nakaichi-Maeda, Y. Akaishi, and H. Tanaka, Prog. Theor. Phys. **64** (1980) 1315.
21. S. Ali and A. R. Bodmer, Nucl. Phys. **80** (1966) 99.
22. C. Ciofi degli Atti and S. Simula, Lett. Nuovo Cimento **41** (1984) 101.
23. C. Ciofi degli Atti and S. Simula, Phys. Rev. **C32** (1985) 1090.
24. V. C. Aguilera-Navarro and O. Portilho,Lett. Nuovo Cimento **15** (1976) 169.
25. W. S. Chien and R. E. Brown, Phys. Rev. **C10** (1974) 1767.
26. J. F. Germond and C. Wilkin, Nucl. Phys. **A249** (1975) 457; Z. A. Khan and I. Ahmad, Pramana **8** (1977) 149.
27. C.-G. Bao, Nucl.Phys. **A373** (1982) 1.
28. Y. C. Tang and R. C. Herndon, Nuovo Cimento **46B** (1966) 117.
29. K. Ikeda, H. Bandō, and T. Motoba, Prog. Theor. Phys. (Supp.) **81** (1985) 147.
30. R. H. Dalitz and G. Rajasekaran, Nucl. Phys. **50** (1964) 450.
31. Y. Kurihara, Y. Akaishi, and H. Tanaka, Prog. Theor. Phys. **71** (1984) 561.

32. Y. Kurihara, private communication, Y. Akaishi, private communication.
33. A. R. Bodmer and S. Ali, Nucl. Phys. 56 (1964) 657.
34. E. W. Schmid, M. Orlowski, and Bao Cheng-guang, Z. Phys. **A308** (1982) 237.
35. M. C. L. Orlowski, Boo Cheng-guang, and Liu-yuen, Z. Phys. **A305** (1982) 249.
36. A. C. Fonseca, J. Revai, and A. Matveenko, Nucl. Phys. **A326** (1979) 182; M. T. Peña and A. C. Fonseca, private communication.
37. M. May et al., Phys. Rev. Lett. **51** (1983) 2085.
38. H. Bandō, Nucl. Phys. **A450** (1986) 217c.
39. A. R. Bodmer and Q. N. Usmani, Nucl. Phys. **A450** (1986) 257c.
40. O. Portilho, P. S. C. Alencar, and S. A. Coon, Nucl. Phys. **A450** (1986) 237c; contribution to INS International Symposium of Hypernuclear Physics, Tokyo, August 1986; and to be published.
41. R. H. Dalitz and L. Liu, Phys. Rev. **116** (1959) 1312.
42. Y. Kurihara, Y. Akaishi, and H. Tanaka, Phys. Rev. **C31** (1985) 971.
43. D. A. Agrello, V. C. Aguilera-Navarro, and J. N. Maki, Lett. Nuovo Cimento 28 (1980) 310; D. A. Agrello and O. Portilho, Phys. Rev. **C23** (1981) 1898.

RESONATING GROUP CALCULATIONS IN LIGHT NUCLEAR SYSTEMS

Hartmut M. Hofmann
Institute for Theoretical Physics
University of Erlangen-Nürnberg, Erlangen, Germany

1. INTRODUCTION

Already in the early times of nuclear physics Wheeler /WH 37/ invented the Resonating Group Method (RGM). In close analogy to the molecular binding he studied nuclear few body system. His idea becomes most transparent by considering the H_2^+-ion (neglecting for the momentum spin-degrees of freedom and the identity of particles): The binding of this three body system can then be understood as the strong binding of the electron to one of the protons forming a group of 2 particles and the weak binding of the remaining proton by polarizing the neutral atom. Since no proton is distinguished, we could also start our considerations with the second proton forming a strongly bound group consisting of the electron and the second proton and the weakly bound first proton. In practice the electron will be considered as jumping resonantly from one configuration to the other so that a variational ansatz for the total wave function will consist of a linear combination of the two configuration. As interactions serve the basic two body potentials.

This simple example elucidates already the essential ideas of the RGM in nuclear physics. A solution for the total wave function is sought as linear combination of strongly bound substructures (groups of particles) times relative motion wave function and two-body interactions are employed. This idea of strongly bound substructures lead to the α-particle model of nuclei /BA 80/.

At first glance the RGM appears to be most suited to describe the relative motion of groups, i. e. collective motion or processes which are dominated by compound nucleus formation. Due to the Pauli principle however, this is not so. This important point has been particular emphasized by Wildermuth /WI 79/. Below we will give an example which demonstrates that the RGM is capable of describing in one system with increasing energy compound nuclear processes and direct processes as well as the transition region. In a multichannel formulation elastic and inelastic scattering and reaction are described equally well. This flexibility of the RGM results from the following essential characteristics:

1) It is a microscopic formulation which takes explicitely substructures (clusters) into account.

2) The Pauli principle is fully accounted for by employing totally antisymmetric wave functions.

3) The center of mass motion is treated correctly.

4) Nucleon-nucleon potentials are used which reproduce the essential features of the two nucleon problem.

5) The nuclear bound-states, scattering and reaction problems are treated within the same framework utilizing one potential only.

6) The formulation is based on variational principles, hence, the results can be improved by expanding the model-space considered in the calculation.

Taken these points together we see that RGM is perfectly suited for studying examples where arbitrary composite nuclei interact with each other. Most formulations, however, are restricted to two-body dynamics only, see /HA 85/ for the contrary. It is unavoidable that the flexibility of the model raises the question of its technical feasibility at all. At the moment there exist essentially three different methods which allow numerical studies without further approximations and are applied to cases with more than one cluster decomposition: Two of them the complex-generator coordinate technique (CGCT) /TA 81/ and the Bargman transformation /SE 76, HO 76/, employ single particle coordinates, whereas the third one, the refined resonating group model (RRGM) /HA 73/ works with Jacobi coordinates. Since by the use of single particle coordinates far developed shell model techniques are available for the calculation of many-particle matrix elements (ME), these methods are well suited for applications to heavier nuclei. On the other hand in light nuclear system, where complicated internal wave functions are necessary to describe details of the reaction process the RRGM bears many advantages. For a connection between RRGM and Bargman transformations see /HA 77, ZA 81, SU 83/. I will in the sequel restrict my considerations mainly to the RRGM because no concise manuscript exists till now for this method and in my opinion all essential ideas and advantages of the RGM can be most pedagogically presented in this framework, using simple examples. Furthermore this method in its basic approach is feasible only for few body system and is therefore more closely related to the topic of this school. On the other hand detailed manuscripts exist for the CGCT (see e. g. /TA 81/) and the Bargman transform /ZA 81, SU 83, FU 84/. Furthermore, the latter being mostly used in connection with the non-relativistic quark model additional degrees of freedom like color or antiparticles have to be introduced, such that the simple structure of the RGM is no more apparent.

In the next section we apply RGM to the bound and scattering potentials problem, to show its differences and similarities and the connection to variational principles. The formulation of the RRGM is described in section 3. The classification and actual calculation of matrix elements is given in section 4. The essential point of all practical calculations the evaluation of the high dimensional spatial integrals is exploited in this section. In section 5 we discuss the notion of Pauli forbidden and partially Pauli forbidden states, the interpretation of RGM wave functions and the extraction of potentials. Some applications employing more realistic forces form the final section.

2. VARIATIONAL PRINCIPLES FOR THE POTENTIAL PROBLEM

2.1 A glimpse on the bound state problem

First we start with a brief review of the bound state problem in a potential model in order to set the notations and illustrate the method by a simple example, such that the differences to the potential scattering problem become evident.

The Hamiltonian for a spinless particle has the simple form

$$H = p^2/2M + V(R) = -\hbar^2\,\Delta/2M + V(R) \tag{2.1}$$

where the potential V(R) should be short ranged and nonsingular. For a central potential V the wave function ψ can be expanded into partial waves

$$\psi(\underline{R}) = \sum_{LM} \psi_{LM}(\underline{R}) = \sum_{LM} u_L(R)/R \; Y_{LM}(\hat{\underline{R}}) \tag{2.2}$$

with real u_L. Here, as everywhere vectors are underlined and unit vectors carry additionally a hat ^. This expansion leads to the well known radial Schrödinger equation

$$H_L \, u_L(R) = E \, u_L(R) \tag{2.3}$$

with

$$H_L = \frac{-\hbar^2}{2M}\left(\frac{d^2}{dr^2} - L(L+1)/R^2\right) + V(R) = h_L + V(R) \tag{2.4}$$

Schrödinger's equation eq. (2.3) can be obtained from the variational principle

$$\delta\left[\int_0^\infty dR \; u_L(R)\,(H_L - E)\,u_L(R)\right] = 0 \tag{2.5}$$

Varying u_L, we have

$$\lim_{r\to\infty} \int_0^r dR\,[\delta u_L(R)(H_L-E)u_L(R) + u_L(R)(H_L-E)\,\delta u_L(R)] = \\ = \lim_{r\to\infty} 2\int_0^r dR \;\; \delta u_L(R)(H_L-E)\,u_L(R) = 0 \tag{2.6}$$

where H_L being selfadjoint on the square integrable functions u_L has been taken into account during shifting the action of the operators from δu_L to u_L. Hence, because δu_L is arbitrary, we obtain the required equation (2.3). In the many particle case, as we will see, the spatial integrals eq. (2.5) are only feasible for Gaussian functions. Therefore we will illustrate the convergence of the variational principle eq. (2.5) by choosing a potential V(R) of Gaussian form and allow for the trial functions u_L linear combinations of Gaussians times R^L. To give some connection to later examples let us consider a proton and a neutron interacting via a central potential

$$V(R) = V_0 \, \exp(-\,b\,R^2) \tag{2.7}$$

where the parameters V_0 = -66.327 MeV and b = 0.410125 fm^{-2} are chosen to reproduce the deuteron binding energy E_D = -2.225 MeV. In table 1

Number of Gaussians	width parameters from variation	kinetic energy	potential energy	binding energy
1	0.2055	12.783	-12.792	-0.009
2	0.5187 0.7382	11.105	-13.173	-2.068
3	0.7307 0.1831 0.03662	10.736	-12.948	-2.212
4	0.9099 0.3179 0.09036 0.02336	10.687	-12.911	-2.224
5	1.064 0.4625 0.1725 0.06090 0.01898	10.682	-12.907	-2.225

Table 1: Convergence study of the variational principle eq. (2.6) using various numbers of Gaussians for a pure central potential with parameters given below eq. (2.7).

we give the results of the non-linear variations for varying number of Gaussians. The convergence for the boundstate energy is rapid, already for two Gaussians the variations lie within 10 % of the numerical exact value. To achieve high precision, however, many terms are necessary. The approximate wavefunctions u are displayed in fig. 1 together with the numerically calculated one. We see that the approximate solutions wiggle around the exact one and fall off too fast with increasing R. Only for four and more Gaussians agree the two wave functions in the region displayed within the width of the lines drawn. From this simple example we can draw the conclusion, that we need about four Gaussian width parameters in order to reproduce a boundstate wave function up to 20 fm.

2.2 Potential scattering as variational problem

In this section we review briefly potential scattering following roughly along the lines of /HA 73/. The scattering wave function $u_L(R)$ is now no more normalized to unity but its asymptotic form consists of

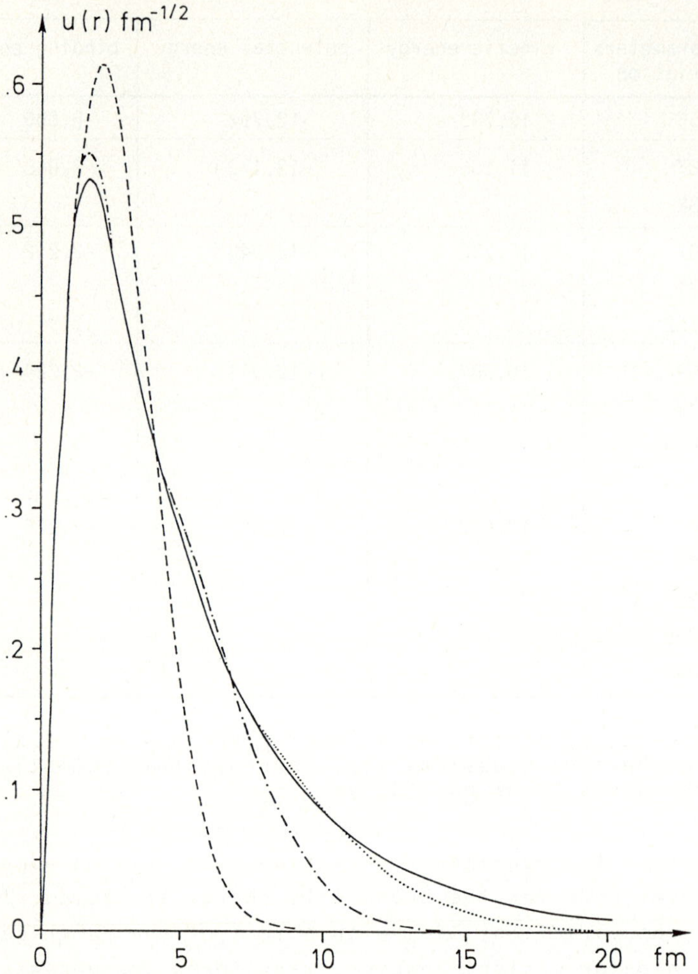

Fig. 1: Comparison of the exact wavefunction (full line) and the variational approximations using one Gaussian function (dashed line), two Gaussians (dashed-dotted line) and three Gaussians (dotted line). The approximations with four and more Gaussians are no more distinguishable from the full line.

a linear superposition of only two independent solutions of the free Hamiltonian h_L.

We will use here regular $F_L(R)$ and irregular solutions $G_L(R)$, because then all wave functions are real, such simplifying the numerical calculations. Even though the normalization of the total scattering wave function is irrelevant for determining physical quantities like cross sections etc., we normalize the total scattering wave function u_L in the following form

$$u_L(R) = \sqrt{M/\hbar^2 k}\ (F_L(R) + a\ \tilde{G}_L(R) + \sum_m b_{mL}\ X_{mL}(R)) \qquad (2.9)$$

Where the momentum k is related to the energy by $E = \hbar^2 k^2/2M$ and for uncharged particles the regular and irregular wave functions are just the Riccati-Bessel functions /AB 64/ which are given for L = 0 as

$$F_o(R) = \sin kR \quad \text{and} \quad G_o(R) = \cos kR. \qquad (2.10)$$

The X_{mL} are taken from a complete set of square integrable functions with $X_{mL}(R) \propto R^{L+1}$ for $R \to 0$.

In order $u_L(R)$ being regular, the irregular $\tilde{G}_L$ has to be regularized such that

$$\tilde{G}_L(R) = T_L(R)\ G_L(R) \qquad (2.11)$$

with

$$T_L(R) \propto R^{2L+1} \quad \text{for } R \to 0 \qquad (2.12)$$

and

$$T_L(R) \to 1 \quad \text{for } R \to \infty \qquad (2.12)$$

A convenient choice of the regularisation factor T_L is /HA 73/

$$T_L(R) = \sum_{i=2L+1}^{\infty} (\beta_o R)^i/i!\ \exp(-\beta_o R) = 1 - \sum_{i=0}^{2L} (\beta_o R)^i/i!\ \exp(-\beta_o R) \qquad (2.13)$$

where both forms make the asymptotic values of T_L apparent for $R = 0$ and $R \to \infty$. The parameter β_o is chosen in such a way that $T_L(R)$ approaches its asymptotic value of 1 just beyond the interaction region, an often used value is $\beta_o = 1.1\ fm^{-1}$ /HO 84/.

The third term in eq. (2.9) has to account for the difference between the exact solution of the scattering problem and the ansatz determined by the asymptotic form of the solution. Furthermore in the region where T_L differs from 1 this term has to compensate the difference between $\tilde{G}_L$ and G_L. With the standard choice of the parameter β_o this means that this term is different from zero only in the interaction region and somewhat beyond, hence it can be well approximated even by a finite number of square integrable terms.

Since the scattering problem has a solution for every energy, the only variational parameters left in eq. (2.9) are the (later) reactance matrix a and the expansion coefficients b_{mL}, after the set of square integrable functions X has been chosen. We will always choose the X_{mL} in the following form

$$X_{mL}(R) = R^{L+1}\exp(-\beta_m R^2) \tag{2.14}$$

The variational principle (2.5) will not lead to the Schrödinger equation, as we will show now.

Varying u_L as given in eq. (2.9) in eq. (2.5) we have

$$\lim_{r\to\infty} \int_0^\infty dR\ [\delta u_L(R)(H_L - E)u_L(R) + u_L(R)(H_L - E)\ \delta u_L(R)]. \tag{2.15}$$

In order to bring δu_L to the left of H_L we have to perform two partial integrations for the kinetic energy term, because on this scattering functions H_L is no more selfadjoint. Explicitely we find

$$\lim_{r\to\infty} \int_0^\infty dR\ [u_L(R)\ (\frac{-\hbar^2}{2M}\ \frac{d^2}{dR^2})\ \delta u_L(R)] = \tag{2.16}$$

$$= (\frac{-\hbar^2}{2M})\lim_{r\to\infty} \left\{u_L(R)\ \frac{d}{dR}\delta u_L(R)\Big|_0^r - \delta u_L(R)\ \frac{d}{dR}\ u_L(R)\Big|_0^r + \int dR\delta u_L(R)\frac{d^2}{dR^2}\ u_L(R)\right\}$$

Inserting (2.16) into eq. (2.15) we get

$$\delta\int_0^\infty dR u_L(R)(H_L - E)u_L(R) = 2\int_0^\infty dR\delta u_L(H_L - E)u_L - \tag{2.17}$$

$$- \frac{\hbar^2}{2M} \lim_{r\to\infty} (u_L\ \frac{d}{dR}\ \delta u_L - u_L\ \frac{d}{dR}\ u_L)\Big|_0^r$$

The variational principle would give the desired Schrödinger equation, if the second term in eq. (2.17) would vanish. Since u_L is regular and all terms in eq. (2.9) are regular too, $u_L(0) = 0$ and $\delta u(0) = 0$. Thus the lower bound does not contribute. Noticing that the X_{mL} do not contribute to the upper bound, we find

$$-\hbar^2/2M \lim_{R\to\infty} (u_L(R)\ \frac{d}{dR}\ \delta u_L(R) - \delta u_L(R)\ \frac{d}{dR}\ u_L(R)) =$$

$$= \frac{-\hbar^2}{2M}\ \frac{M}{\hbar k}\ \delta a_L \lim_{R\to\infty}\ F_L\ \frac{d}{dR}\ G_L - G_L\ \frac{d}{dR}\ F_L\ = \tag{2.18}$$

$$= -\delta a_L/2k \lim_{R\to\infty}\ \sin(kR - L\pi/2)\ \frac{d}{dR}\ \cos(kR - L\pi/2) - \cos(kR - L\pi/2)\cdot$$

$$\cdot\ \frac{d}{dR}\ \sin(kR - L\pi/2) = \delta a_L/2$$

where the asymptotic form of the Bessel functions has been used. In case of charged particles, where F_L and G_L are Coulomb wavefunctions the Wronskian leads to the identical result.

In order to regain the Schrödinger equation we can therefore start from a modified variational principle

$$\delta [\int dR \; u_L(R)(H_L - E) \; u_L(R) - \frac{1}{2} a_L] = 0 \qquad (2.19)$$

which is known as Kohn's variational principle. A much more rigorous and general derivation can be found in /GE 83/. Our normalization factor in eq. (2.9) resulted in the simple factor $\frac{1}{2}$ in eqs. (2.18) and (2.19).

Since F_L and G_L are solutions of the free Hamiltonian h_L to the correct energy it is easy to show, /HA 73/ that all integrals appearing in eq. (2.19) are of short range.

If we consider charged particles then h_L consists in addition to the kinetic energy also of the Coulomb term and the functions F_L and G_L are instead of Bessel functions Coulomb functions. All considerations, however, remain unchanged.

The solution of eq. (2.19) is not given here, because in the next section we discuss the solution of the variational equation in the general case, which is only slightly more complicated.

3. VARIATIONAL APPROACH TO MANY BODY SCATTERING

3.1 Variational equations and their solution

In this section we consider N nucleons which interact via nuclear and Coulomb forces. Two body forces are assumed throughout the text, but three-body forces are no principle problem, they can be treated along analogous lines as the two-body forces. Breakup channels into three or more fragments, however, pose a serious problem, because of the ansatz for the scattering wave function eq. (2.9), and will not be discussed in the following. In our presentation we follow along the lines of /HA 73, HO 84/.

With the assumption of two-body forces only the Hamiltonian of a N-particle system is given by

$$H(1,\dots,N) = \sum_i T_i + \frac{1}{2} \sum_{i \neq j} V_{ij} \qquad (3.1)$$

The c.m. kinetic energy can be separated off, because of momentum conservation, yielding

$$\sum_{i=1}^{N} T_i = T_{CM} + \frac{1}{2mN} \sum_{i<j}^{N} (\underline{p}_i - \underline{p}_j)^2 \qquad (3.2)$$

where m is the mass of the nucleons, assumed to be the same for neutrons and protons and p_i is the momentum of nucleon i. Since we restrict our considerations to two-fragment channels only, the translationally invariant Hamiltonian H' can be decomposed into the internal Hamiltonian for both fragments and the relative motion part

$$H'(1,\dots,N) = H_1(1,\dots,N_1) + H_2(N_1+1,\dots,N) + $$

$$+ T_{rel} + \sum_{\substack{i \in \{1,\dots,N_1\} \\ j \in \{N_1+1,\dots,N\}}} V_{ij} \qquad (3.3)$$

By adding and subtracting the point Coulomb interaction between the fragments $Z_1Z_2e^2/R$ the potential term becomes shortranged, where R is the relative coordinate between the two fragments:

$$H'(1,\dots,N) = H_1(1,\dots,N_1) + H_2(N_1+1,\dots,N) + $$

$$+ \Big(\sum_{\substack{i \in \{1,\dots,N_1\} \\ j \in \{N_1+1,\dots,N\}}} V_{ij} - Z_1Z_2e^2/R \Big) \qquad (3.4)$$

$$+ T_{rel} + Z_1Z_2e^2/R$$

Thus the translationally invariant part of the Hamiltonian is split into the internal Hamiltonians of the two fragments, the shortranged interaction of the fragments and the relative motion part of the two fragments with charges Z_1 and Z_2. This decomposition now allows an ansatz for the total wave function ψ in the following form

$$\psi_1 = A \sum_{k=1}^{n_k} \psi^k_{kan} \, \psi^{1k}_{rel} \quad , \qquad (3.5)$$

where A denotes the antisymmetriser, n_k the numbers of channels, ψ^k_{kan} the channel wave function in channel k which will be described below and ψ_{rel}^{1k} the relative motion wave function analogous to eq. (2.9)

$$\psi_{rel}^{1k}(R) = \delta_{1k}F_k(R) + a_{1k}G_k(R) + \sum_m b_{1km}\ X_{km}(R) \tag{3.6}$$

Here F_k and G_k are now the regular and regularised irregular Coulomb wave functions including the normalization factor of eq. (2.9). The index 1 on the total wave function ψ_1 is a reminder of the boundary condition chosen, namely regular waves only in channel 1. The sum k over channels runs over physical channels, open or closed, but it may also contain "distortion channels" which consist only of the square integrable part of eq. (3.6). Such distortion channels are especially needed if the number of physical channels is low, one or two, to allow for enough freedom for the variation.

The quantities a_{1k} and b_{1km} are the variational parameters, which are determined from Kohn's principle, analogously to eq. (2.19). As we saw in the previous section, the interference of regular and irregular free scattering waves results in a contribution of the endpoints of integration when switching over the operator (H' - E) from $\delta\psi$ to ψ itself, hence Kohn's principle now reads /HA 73/.

$$\delta\ (< \psi_1\ |H' - E|\ \psi_1 > - \frac{1}{2}\ a_{11}) = 0\ . \tag{3.7}$$

The solution of the variational problem is described in the following. In order to simplify notation, we combine the individual terms of the relative motion wavefunction with the channel function such that

$$\psi_1 = A\left\{\sum_k (f_k\ \delta_{k1} + a_{1k}\ g_k\ + \sum_m b_{1km}\ X_m')\right\} \tag{3.8}$$

The last term consists of square integrable functions only, hence, the Hamiltonian H' eq. (3.4) can be diagonalised in this function space. Let us assume, that this diagonalisation has been performed, then we can switch over to new square integrable functions Γ_ν with

$$< \Gamma_\nu\ |\ \ A\,\Gamma_\mu > = \delta_{\nu\mu} \tag{3.9}$$

and

$$< \Gamma_\nu\ |\ H'\ |\ A\ \Gamma_\mu > \ = \ e_\nu\ \delta_{\nu\mu}$$

Note: Since 1 und H' commute with the antisymmetriser A, it is enough to apply A on one wavefunction only, see also section 4.1. The total wave function can now be represented as

$$\psi_1 = A\left\{\sum_k (f_k \delta_{1k} + a_{1k} g_k) + \sum_m d_{1m} \Gamma_m\right\} \tag{3.10}$$

where now the variational parameters are a_{1k} and d_{1m}. Performing the variation, eq. (3.7) yields the following equations:

$$< g_k|\hat{H}|A f_1 > + \sum_{k'} < g_k|\hat{H}|A g_{k'}> a_{1k'} + \sum_m < g_k|\hat{H}|A\Gamma_m> d_{1m} = 0 \tag{3.11a}$$

$$< \Gamma_m|\hat{H}|Af_1> + \sum_{k'} <\Gamma_m|\hat{H}|Ag_{k'}> a_{1k'} + \sum_{m'} <\Gamma_m|\hat{H}|A\Gamma_{m'}>d_{1m'} = 0 \tag{3.11b}$$

where $\hat{H}$ ist an abbreviation of $H' - E$. Equation (3.11b) can be solved for $d_{1m'}$, taking eq. (3.9) into account

$$d_{1m} = (E - e_m)^{-1}(<\Gamma_m|\hat{H}|Af_1> + \sum_{k'} <\Gamma_m|\hat{H}|Ag_{k'} >a_{1k'}) \tag{3.12}$$

Defining the operator $\tilde{H}$ as

$$\tilde{H} = \hat{H} - \sum_m \frac{\hat{H}| A\Gamma_m><\Gamma_m|\hat{H}}{e_m - E} \tag{3.13}$$

and inserting eq. (3.12) into (3.11a) yields

$$\sum_{k'} < g_k|\tilde{H}|Ag_{k'}> a_{1k'} = - <g_k |\tilde{H}| A f_1 > \tag{3.14}$$

In obvious matrix notation eq. (3.14) reads

$$< G|\tilde{H}|G > a^T = - < G|\tilde{H}|F > \tag{3.15}$$

where a^T denotes the transposed matrix a. Equation (3.15) can be easily solved

$$a = - < G|\tilde{H}|F >^T < G|\tilde{H}|G >^{-1} \tag{3.16}$$

Now, if the matrix elements are known, the parameters of ψ_1 are determined in eq. (3.16) and consecutively in (3.12), hence, the total wave function is known.

Obviously, the reactance matrix a_{1k} in eq. (3.16) is not symmetric in the general case, therefore also the S-matrix computed from a_{1k} via

$$S = (1 + i a)(1 - i a)^{-1} \quad (3.17)$$

is not symmetric, thus violating time-reversal invariance. In general even unitarity is not guaranteed. To enforce unitarity we have to have a symmetric reactance matrix a. This goal is achieved by the so-called Kato correction /KA 51/ which can be understood most easily following /JO 71/.

Instead of the ansatz (3.10) we choose another boundary condition

$$\psi_l' = A\left\{\sum_k (a_{lk}' f_k + \delta_{lk} g_k) + \sum_m d_{lm}' \Gamma_m\right\} . \quad (3.18)$$

Following along the lines of eqs. (3.11) - (3.16) yields

$$a' = - < F|\tilde{H}|G >^T < F|\tilde{H}|F >^{-1} \quad (3.19)$$

again with an apparently unsymmetric a'. Since the special boundary condition choosen does not affect observables, we should have

$$a = a'^{-1} \quad (3.20)$$

Taking into account the properties of the channel wave function, discussed below, and the relative motion wave function it is easy to derive /JO 71/

$$< F|\tilde{H}|G > = < G|\tilde{H}|F >^T + \frac{1}{2} \mathbb{1} \quad (3.21)$$

by performing the partial integrations analogous to eq. (2.16) in order to switch the Hamiltonian $\tilde{H}$ onto F. Inserting eq. (3.21) into (3.20) yields

$$- < G|\tilde{H}|F >^T < G|\tilde{H}|G >^{-1} = - < F|\tilde{H}|F >(< G|\tilde{H}|F > + \frac{1}{2}\mathbb{1})^{-1} \quad (3.22a)$$

or

$$< F|\tilde{H}|F > = < G|\tilde{H}|F >^T < G|\tilde{H}|G >^{-1} (< G|\tilde{H}|F > + \frac{1}{2}\mathbb{1}) \quad (3.22b)$$

or

$$\begin{aligned} - < G|\tilde{H}|F >^T < G|\tilde{H}|G >^{-1} &= a = \\ &= - 2 (< F|\tilde{H}|F > - < G|\tilde{H}|F >^T < G|\tilde{H}|G >^{-1} < G|\tilde{H}|F >) \end{aligned} \quad (3.22c)$$

where the last expression for a is obviously symmetric. This expression has been derived in /HA 73/ as second order correction and as stationary condition in /GE 83/.

Employing eq. (3.17) we can now calculate cross sections etc., provided we are able to determine all matrix elements entering into eq. (3.22c). Before we discuss, how these matrix elements are calculated in praxi, we define the channel wave functions and discuss the existence of integrals.

3.2 Ansatz for the wave functions

The total wave function ψ_1 eq. (3.5) contains the channel functions Ψ_{kan} and the relative motion wave function $\Psi_{rel}(R)$. In the following we restrict our consideration to just one term of eq. (3.5) and study the action of the antisymmetriser A below. The channel function ψ_{kan} has the structure

$$\psi_{kan} = \left[Y_1(\hat{\underline{R}}) \; R^{-1} \times [\phi_1^{J_1} \times \phi_2^{J_2}]^{S_c} \right]^J \tag{3.23}$$

Here $\phi_i^{J_i}$ denotes the translationally invariant wave function of fragment i (i = 1, 2) with spin J_i. The individual spins J_1 and J_2 are coupled to the channel spin S_c. The orbital angular momentum 1 and S_c are coupled to total angular momentum J. The coupling is indicated by the square brackets. The vector $\underline{R}$ denotes the relative vector between fragment 1 and 2 in case of a scattering wave function and the corresponding Jacobi coordinate in case of a bound state wave function. In the latter case the coupling to total J can be omitted.

The individual fragment wave function consist of the spatial function X and the spin (isospin) function Ξ, e. g.

$$\phi_1^{J_1} = \sum_{1_I, S, m} \left[C_m^{1_I L S} \quad X_m^{1_I}(\underline{\rho}) \; \Xi^{S,T} \right]^{J_1} \tag{3.24}$$

where 1_I denotes the set of internal orbital angular momenta of fragments consisting of more than one cluster. We use the expression "cluster" only for groups of particles without internal orbital angular momenta, i. e. the largest cluster can contain only 4 particles, two protons and two neutrons with opposite spin projections. The spin (-isospin) functions Ξ are coupled to total spin S and may be coupled to good isospin. For a fragment consisting of n_c clusters the coefficients $C_m^{1_I L S}$ include the Clebsch-Gordan coefficients for the coupling of the orbital angular momenta $1_1, 1_2, \ldots, 1_{n_c-1}$ to total orbital angular momentum L, the coupling of the spin (and isospin) to S (and T) and additional coefficients, which allow a superposition of different radial dependencies. The spatial function X_m^{lI} is constructed from cluster internal functions $X_{m,int}$ and cluster relative functions $X_{m,rel}$ in the form

$$X_m^{lI} = \left(\prod_{h=1}^{n_c} X_{m,h,int}\right) \left(\prod_{k=1}^{n_c-1} X_{m,k,rel}^{l_k}\right) \tag{3.25}$$

where the internal functions consist of a single Gaussian function

$$X_{m,h,int} = \exp\left(-\beta_{mh} \sum_{i<j}^{n_h} (\underline{r}_i - \underline{r}_j)^2 / n_h\right) \tag{3.26}$$

with n_h denoting the number of nucleons (≤ 4) inside cluster h, which has a width β_{mh}. In case of the cluster containing just one nucleon $X_{m,h,int} \equiv 1$. The cluster relative functions $X_{m,k,rel}^{l_k}$ contain in addition to the Gaussian-function solid spherical harmonics $\mathcal{Y}_{l_k}$ /ED 64/

$$X_{m,k,rel}^{l_k} = \exp\left(-\gamma_{mk}\, \rho_k^2\right) \mathcal{Y}_{l_k}(\underline{\rho}_k) \tag{3.27}$$

where ρ_k denotes the Jacobi coordinate between cluster k + 1 and the center-of-mass of the cluster 1 to k, see fig. 2.

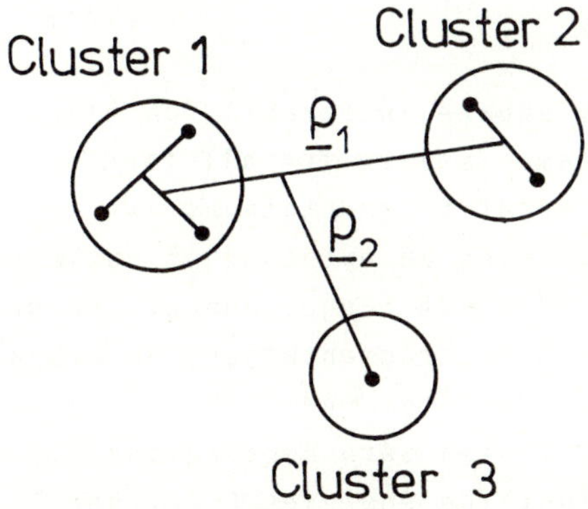

Fig. 2:
Schematic illustration of the intercluster coordinates ρ used in eq. (3.27).

The wavefunctions $\phi_i^{J_i}$ are determined from the Ritz variational principle analogous to eq. (2.5) once the fragmentation and number of radial functions has been chosen. Therefore, in the function space chosen $\phi_i^{J_i}$ is most deeply bound with e. g. for fragment 1

$$< \phi_1^{J_1} | H'(1,\ldots,N_1) | A_1 \phi_1^{J_1} > = E_1 < \phi_1^{J_1} | A \phi_1^{J_1} > \tag{3.28}$$

where A_1 is the antisymmetriser for the first N_1 nucleons. Denoting analogously the antisymmetriser for the remaining group of $N - N_1$ nucleons by A_2, we can write the total antisymmetriser as

$$A = A_3 A_1 A_2 = \sum_{P_3} (-1)^{P_3} P_3 A_1 A_2 \tag{3.29}$$

where the sum P_3 runs over all permutations which exchange particles between the two fragments, including the identity.

In order to show that the solution of the variational problem is more than just a formal one, we must prove that the functional of eq. (3.7) exists. If we take the kinetic energy ε_k in channel k to be

$$\varepsilon_k = E - E_{1,k} - E_{2,k} \tag{3.30}$$

then the Hamiltonian eq. (3.4) can be decomposed as

$$\begin{aligned} H'(1,\ldots,N) - E = {} & (H_1(1,\ldots,N_1) - E_{1,k}) + (H_2(N_1+1,\ldots,N) - E_{2,k}) + \\ & + \Big(\sum_{\substack{i\in\{1,\ldots,N_1\} \\ j\in\{N_1+1,\ldots,N\}}} V_{ij} - Z_1 Z_2 e^2/R \Big) + \\ & + T_{rel} + Z_1 Z_2 e^2/R - \varepsilon_k \end{aligned} \tag{3.31}$$

All terms in eq. (3.7) which contain only square integrable functions cannot lead to divergent integrals. The same is true for all terms in which the functions of channels with different fragmentations are connected. Furthermore, all integrals involving an exchange of nucleons between the fragments, i. e. $P_3 \neq 1$, are of short range. Hence, the only possibly dangerous integrals involve identical fragmentations in bra and ket of the matrix elements and P_3 being the identity. Under these circumstances the first line of eq. (3.31) contributes zero because the internal function $\phi_i^{J_i}$ are solutions to just the energies $E_{i,k}$, see eq. (3.28). The second line results in a short-ranged integral by construction. If and only if, the Coulomb function F_k and G_k in eq. (3.6) are chosen as eigenfunctions of the last line of (3.31), the related integrals are finite. This implies that the threshold energies are completely fixed by the binding energies of the fragments and therefore one has no possibility to adjust those energies.

Since we have now convinced ourselves that all integrals are shortranged, we can expand within the relevant region the regular and regularised irregular Coulomb function in terms of Gaussian functions, e. g. the square integrable function X of eq. (3.6). By this expansion we are left with integrals of the Hamiltonian or unity between translationally invariant wavefunctions consisting besides the spin-isospin part only of linear combinations of Gaussians and solid spherical harmonics.

In the next section we will describe how to calculate such a matrix-element.

4. EVALUATION OF THE MATRIX ELEMENTS

4.1 Reduction to reduced matrix elements

For the calculation of the matrix elements, we have to apply the anti-symmetriser A onto the wavefunctions in bra and ket as

$$< A\,\psi^{J'M'} \,|\, \sum_{i<j} w_{ij} \,|\, A\,\psi^{J\,M} > \qquad (4.1)$$

where $\Sigma\, w_{ij}$ is either the Hamilton operator or just the unit operator. Since the operator A commutes with the symmetric operator $\Sigma\, w_{ij}$ and we are calculating expectation values only, we can reduce the expression (4.1) to

$$ME = < \psi^{J'M'} \,|A \sum_{i<j} w_{ij} \,|\, \psi^{JM} > \qquad (4.2)$$

by using the proportionality between A^2 and A. For the further calculation we decompose A into the sum over all permutations acting on spatial, spin and isospin coordinates

$$A = \sum_{P} (-1)^P\, P \qquad (4.3)$$

For each permutation eq. (4.2) has to be calculated. Here ψ^{JM} denotes a single term of the total wavefunction ψ eqs. (3.5) and (3.6) allowing for the various fragmentations of the channels, the possible different components of the internal functions, eqs. (3.23) and (3.24), and the various square integrable functions X in eq. (3.6). Because of the rotational invariance of the Hamiltonian the operator $\Sigma\, w_{ij}$ can be decomposed into a product of orbital and spin space spherical tensors /ED 64/

$$\sum_{i<j} w_{ij} = \sum_{i<j} \sum_{kq} (-1)^q \, w_{ij}^{o}(k,q) \, w_{ij}^{S}(k,-q) \tag{4.4}$$

where k denotes the rank of the interaction, e. g. $k = 2$ for the tensor potential.

The wavefunction ψ^{JM} consists of orbital and spin-isospin part according to

$$\psi^{JM} = \sum_{\alpha m \sigma} C_{\alpha}^{LSJ} \, (Lm\, S\sigma | JM) \, \psi_{\alpha}^{Lm}(\text{space}) \;\; \psi_{\alpha}^{S\sigma}\, (\text{spin}) \tag{4.5}$$

where the Clebsch-Gordan coefficient $(Lm\, S\sigma|JM)$ explicitely gives the coupling of total orbital angular momentum L and total spin S to J. The index α is a short notation of all other quantum numbers, e. g. the fragmentation. Using Racah algebra we find for fixed i, j and k for eq. (4.2)

$$M_{ijk} = \sum_{P} < \psi^{J'M'} | \sum_{q} (-1)^P (-1)^q \, P \, w_{ij}^{o}(kq) \, w_{ij}^{S}(kq) | \psi^{JM} >$$

$$= \delta_{JJ'} \delta_{MM'} \sum_{\alpha\alpha' P} (C_{\alpha}^{L'S'J})^* \, C_{\alpha}^{LSJ} (-1)^P (-1)^{L+2S+S'-J} \begin{Bmatrix} S & L & J \\ L' & S' & k \end{Bmatrix}$$

$$< L'\alpha' \, || \, Pw_{ij}^{o}(k) \, || \, L\alpha > \, < S'\alpha' \, || \, Pw_{ij}^{S}(k) \, || \, S\alpha > \tag{4.6}$$

where $\begin{Bmatrix} S & L & J \\ L' & S' & k \end{Bmatrix}$ is a 6j-coefficient /ED 64/.

4.2 Calculation and classification of the spin-isospin matrix elements

According to eq. (4.6) we have to calculate reduced spin matrix elements which is done by using Wigner-Eckart theorem

$$< S'\alpha' \, || \, Pw_{ij}^{S}(k) \, || \, S\alpha > = (2S'+1)^{1/2} \, < S'S'\alpha' | \, Pw_{ij}^{S}(k, S'-S) | SS\alpha > / \; (S\, S\, k\; S' - S | S'S') \tag{4.7}$$

We use here maximal projections of the spin functions so the ME is different from zero, if the triangular conditions are fulfilled at all.

The coupled spinfunctions $|SS\alpha>$ can be decomposed into linear combinations of products of elementary single particle spin function by using again Clebsch-Gordan coefficients. Analogously consists w_{ij}^{S} of products of the isospin operators 1 resp. $\underline{\tau}_i \cdot \underline{\tau}_j$ with the spin operators 1 (for norm and central potential), $\underline{\sigma}_i \cdot \underline{\sigma}_j$ (for central potential), $(\underline{\sigma}_i + \underline{\sigma}_j)_q$ spherical component q (for spin-orbit potential) and $\sigma_{iq}\,\sigma_{jq'}$ (for the tensor potential). These operators acting on product wave functions again give product wavefunctions. The permutation P can be easily applied onto this product and the ME can be easily evaluated. Since usually many of these matrix elements vanish, it is more economic to start from product functions in bra and ket and determine all permutations P which give a ME different from zero, for details see /HA 73, HO 84/. To find the reduced matrix elements (4.7) itself, only the summation over the known Clebsch-Gordan coefficients have to be performed.

As will be shown below, the calculation of the spatial matrixelements is much more complicated than that of the spin matrix elements. Therefore the symmetry of the spatial wavefunctions is utilized. Here the cluster decomposition plays the essential role. From the ansatz for the internal coordinate space wave functions eq. (3.25) to (3.27) it is evident that the function in coordinate space is symmetric under exchange of nucleons inside of clusters, but not if particles are exchanged beyond cluster boundaries. Hence, different permutations may yield identical orbital matrix elements.

In order to exploit this symmetry one considers instead of the symmetric group S_n itself subgroups $S_{n_1} x S_{n_2} x$ ... with $\Sigma\, n_n = N$ in bra and ket, where n_i is the number of nucleons in cluster i /SE 75/. The group S_n is decomposed into double cosets of these subgroups, and each double coset can be characterised uniquely by one permutation. Restricting, for the moment, our considerations to the norm, where $w_{ij} = 1$, we find for permutations belonging to one double coset always the same orbital ME. Therefore, the known spin MEs belonging to one double coset can be summed up, including the sign of the permutation. This allows to reduce the sum over all permutations in eq. (4.6) appreciably to a sum over double cosets only

$$M_{ijk} = \delta_{JJ'}\delta_{MM'} \sum_{\alpha\alpha'} (C_{\alpha}^{L'S'J})^* \, C_{\alpha}^{LSJ} (-1)^{L+2S+S'-J} \begin{Bmatrix} S & L & J \\ L' & S' & k \end{Bmatrix}$$

$$\sum_{dc} < L'\alpha' \,||\, P_{dc}\, w_{ij}^{o}(k) \,||\, L\alpha > C_{dc}^{SS'\alpha\alpha' ij} \qquad (4.8)$$

where $C_{dc}^{SS'\alpha\alpha' ij}$ contains the sum over spin MEs and P_{dc} is any permutation representing the double coset dc. If the ME contains an interaction $w_{ij} \neq 1$ one has also to mark the interacting particles i and j and to extend the double coset decomposition.

In the following we will not elaborate on a general treatise on double cosets, this can be found in /SE 75, LU 81/, but rather give an illustrative example /HO 84/. Let us consider the ^{7}Be nucleus, which can be described very well in a fragmentation ^{4}He and ^{3}He, but small components of a framentation ^{6}Li and p allow to reproduce experimental data better /ME 86/, see also final section.

The decomposition into double cosets can be illustrated by symbols of matrix form, which are called dc-symbols /SE 75/. In the ^{7}Be case we have for example a decomposition of the S_7 into $S_4 \times S_3$, for the main component of the wave function, and $S_4 \times S_2 \times S_1$ for the small component. Note that the ^{6}Li, containing 6 nucleons, has to contain at least 2 clusters, the main component being ^{4}He - ^{2}H (for further details see /ME 86/).

	S_4	S_3
S_4	3	1
S_2		2
S_1	1	

In the different sites of the scheme we have entered the number of particles which are exchanged by the permutation from the cluster given on the left side into the cluster stated above . The sum of one row (column) is always the number of nucleons in that cluster, as can be seen in the above example. The second row illustrates that 2 particles from the ^{2}H cluster are exchanged into the ^{3}He-cluster.

A dc-symbol can be associated with each permutation and permutations having the same dc-symbol belong to the same double coset /SE 75/. On the other hand, a permutation representing the double coset can be constructed uniquely from the dc-symbol. To achieve this we write the digits 1 to N rowwise into the dc-symbol, as many digits as indicated per site, and then read this scheme columnwise. If one writes the digits found by this procedure below the digits 1 to N in natural order one finds a permutation representing the double coset. In our example this will read

	S_4	S_3
S_4	123	4
S_2		56
S_1	7	

$$= \begin{pmatrix} 1 & 2 & 3 & 4 & 5 & 6 & 7 \\ 1 & 2 & 3 & 7 & 4 & 5 & 6 \end{pmatrix} = P_{dc}$$

One has to discriminate the orbital MEs, in addition to which particles are interacting. For this we mark the permuted digit of the interacting particles with a point. For two body-interactions we find for interacting particles 4 and 6 the digits in our above example.

$$7\ (= P_{dc}(4)) \quad \text{and} \quad 5\ (= P_{dc}(6))$$

	S_4	S_3
S_4	123	4
S_2		$5^{\bullet}$ 6
S_1	$7^{\bullet}$	

The group configurations indicated are often omitted for convenience. Thus we arrived at a new classification scheme for matrix elements of any two-body interaction in terms of 2-point dc-symbols. The example tells us, that also an interaction between particle 4 and 7 would yield the same coordinate space ME. Therefore one adds up all spin ME belonging to the same 2-point dc-symbol. Analogously the sum over dc in eq. (4.8) runs over all 2-point dc symbols in case of an interaction. Extending this classification scheme to one-body or more-body interactions is straight forward. It remains only to calculate the ME in coordinate space, which will be described in the following subsection.

4.3 Evaluation of the orbital matrix elements

The reduced MEs of eq. (4.8) are converted into usual integrals analogous to eq. (4.7). Since the coupling scheme of the various orbital angular momenta just introduces trivial linear combinations, we disregard this complication in the sequel and start with the bare orbital functions, which are for the right hand side of the ME (marked by the index r) of the structure

$$|L_r\ \alpha\rangle = \prod_{i=1}^{N-1} \exp\left(-\beta_i\ \underline{s}_i^r \cdot \underline{s}_i^r\right) \prod_{j=1}^{n_{cr}} \mathcal{Y}_{l_j m_j}\left(\underline{s}^r_{N-n_{cr}+j}\right) \qquad (4.9)$$

where the numbering of the Jacobian coordinates $\underline{s}_i$ starts with the internal coordinates, see fig. 3, and the single particle coordinates of eq. (3.26) have been eliminated in favour of the Jacobians. Note, that because of translational invariance the c. m. coordinate, proportional to $\underline{s}_N$, is absent in eq. (4.9); the number of cluster is denoted by n_{cr}. The function on the left hand side $|L_l\alpha'\rangle$ can be expressed

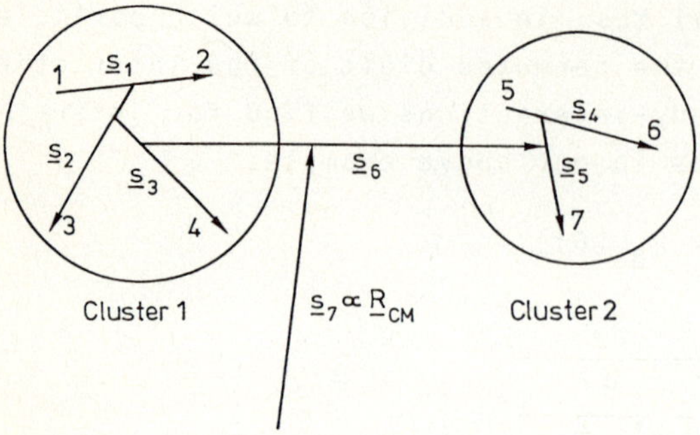

Fig. 3: Schematic diagramm illustrating the numbering of the Jacobi coordinates of eq. (4.9) for a cluster decomposition into 4 and 3 particles. Note, that the arrows shown are only proportional to the Jacobi vectors.

analogously by Jacobian coordinates of the left hand side (index 1), which will be different from the right hand side in the general case. Starting from eq. (4.8) we want to show how to evaluate a ME of the type

$$J^{ij}_{L_1\alpha',\, L_r\alpha}\,(P_{dc}) = <L_1\alpha'|\; P_{dc}\; w^{o}_{ij}(k)\,|L_r\alpha> \qquad (4.10)$$

The operators w_{ij} contain coordinates in the form of eq. (4.9) or in addition to that also differential operators may occur. In order to keep the presentation as transparent as possible, we restrict our considerations in the following to the norm and refer for the other operators to the literature /HA 70, HO 84/.

Choosing the Jacobian coordinates of the left hand side as independent variables and suppressing their index 1 in the following we can express the ME eq. (4.10) by

$$J^{ij}_{L_1\alpha' L_r\alpha}\,(P_{dc}) = \int d\underline{s}_1 \ldots d\underline{s}_{N-1}\;\exp\,(-\sum_{\mu\mu'}^{N-1}\rho_{\mu\mu'}(P_{dc})\underline{s}_\mu\cdot\underline{s}_{\mu'})\prod_{n=1}^{Z}\mathcal{Y}_{L_nM_n}(\underline{Q}_n) =$$

$$= \Gamma_{L_1M_1\ldots L_ZM_Z} \qquad (4.11)$$

Since in the norm no particles interact, we have suppressed the indices i and j on the symbol Γ. The coefficients $\rho_{\mu\mu'}$ result from expressing the Jacobians of the r. h. s. by those of the left hand side, applying the permutation P_{dc} first. The Q_n vectors are the intercluster coordinates after applying P_{dc} to the r. h. s. and can be expressed as linear combinations of the Jacobians $\underline{s}$. In case of an interaction w^{o}_{ij}, its radial dependence in Gaussian form has to be included into ρ. See for an example of other radial dependencies in the Appendix A. The combined number Z of orbital angular momenta is

$$Z = n_{cl} - 1 + n_{cr} - 1 + n_w \qquad (4.12)$$

where $n_w = 1$ for the tensor interaction and $n_w = 0$ otherwise. Except for the solid spherical harmonics, the ME in eq. (4.11) is just a multidimensional Gaussian integral, which is straightforward to evaluate by bringing $\rho_{\mu\mu'}$ to diagonal form. In order to utilize this procedure we introduce the generating function for the spherical harmonics /RO 57/

$$(\underline{b}\cdot\underline{r})^L = b^L \sum_{m=-L}^{L} C_{Lm}\, b^{-m}\, \mathcal{Y}_{Lm}(\underline{r}) \tag{4.13}$$

employing the vector $\underline{b} = (1 - b^2,\ i(1 + b^2),\ -2b)$, with the property $\underline{b}\cdot\underline{b} = 0$. The coefficient C_{Lm} are given by

$$C_{Lm} = (-2)^L\, L!(4\pi/((2L+1)(L-m)!(L+m)!)^{1/2} \tag{4.14}$$

Instead of eq. (4.11) we now consider the generating integral

$$I(a_1 b_1 \ldots a_z b_z) = \int d_{\underline{s}_1} \ldots d_{\underline{s}_{N-1}}\, \exp(-\sum_{\mu\mu} \rho_{\mu\mu'} \underline{s}_\mu \cdot \underline{s}_{\mu'} + \sum_{n=1}^{Z} a_n \underline{b}_n \cdot \underline{Q}_n)$$

(4.15)

Expanding the expression $\exp(\Sigma\, a_n \underline{b}_n \cdot \underline{Q}_n)$ into a power series in a_n and b_n and taking eq. (4.13) into account we relate the generating integral to the desired ones

$$I(a_1 b_1 \ldots a_z b_z) = \sum_{l_1 m_1 \ldots l_z m_z} (\prod_{n=1} C_{l_n m_n}/l_n!)\, a^{l_n} b^{l_n - m_n}\, \Gamma_{l_1 m_1 \ldots l_z m_z}$$

(4.16)

On the other hand the generating integral eq. (4.15) can be evaluated explicitely and afterwards expanded into a power series. To accomplish this we first transform the expression $\rho_{\mu\mu'}\, \underline{s}_\mu \cdot \underline{s}_{\mu'}$ onto diagonal form by

$$\underline{s}_\mu = \sum_{\lambda=\mu}^{N-1} T_{\mu\lambda}\, \underline{t}_\lambda \qquad \text{with } T_{\lambda\lambda} = 1 \tag{4.17}$$

which yields

$$\sum_{\mu\mu'} \rho_{\mu\mu'}\, \underline{s}_\mu \cdot \underline{s}_{\mu'} = \sum_\lambda \beta_\lambda\, t_\lambda^2 \tag{4.18}$$

and

$$\underline{Q}_n = \sum_\mu P_{n\mu}\, \underline{t}_\mu \tag{4.19}$$

Inserting eq. (4.18) and (4.19) reduces eq. (4.15) to the form

$$I(a_1b_1\ldots a_zb_z) = \int d\underline{t}_1\ldots d\underline{t}_{N-1}\ \exp(-\sum_\lambda(\beta_\lambda t_\lambda^2 - \sum_n P_{n\lambda}a_n\,\underline{b}_n\cdot\underline{t}_\lambda)) \quad (4.20)$$

where the condition on $T_{\lambda\lambda}$ has been utilized. Employing the method of completing squares the integral amounts to

$$I(a_1b_1\ldots a_zb_z) = (\prod_{r=1}^{N-1}(\pi/\beta_n)^{3/2}\exp(\tfrac{1}{4}\sum_{n,n'}\sigma_{nn'}a_na_{n'}\underline{b}_n\cdot\underline{b}_{n'}) \quad (4.21)$$

where we defined

$$\sigma_{nn'} = 2\sum_\lambda P_{n\lambda}\,P_{n'\lambda}\ /\ \beta_\lambda \quad (4.22)$$

Computing the scalar product $\underline{b}_n\cdot\underline{b}_{n'}$ and restricting the summation over n and n' allows to reduce the exponent in eq. (4.21) to

$$\sum_{n>n'}\sigma_{nn'}\,a_na_{n'}\,(b_nb_{n'} - \tfrac{1}{2}(b_n^2 + b_{n'}^2)) \quad (4.23)$$

Expanding the exponential in eq. (4.21) into a power series and collecting terms in the form of eq. (4.16) yields the final result /HA 70, HO 84/

$$\Gamma_{l_1m_1\ldots l_zm_z} = (\prod_{r=1}^{N-1}(\pi/\beta_r)^{3/2}(\prod_{j=1}^{Z}l_j!/C_{l_jm_j})\sum_{g_{nn'}h_{nn'}k_{nn'}}(-1/2)^{h_{nn'}+k_{nn'}}$$

$$\prod_{n>n'}^{Z}\sigma_{nn'}^{g_{nn'}+h_{nn'}+k_{nn'}}/(g_{nn'}!\ h_{nn'}!\ k_{nn'}!) \quad (4.24)$$

The sums $g_{nn'}$, $h_{nn'}$ and $k_{nn'}$ run over all possible combinations of nonnegative integers, which fulfill the following relations:

$$2\sum_{n>n'}(g_{nn'} + h_{nn'} + k_{nn'}) = \sum_{j=1}^{z} l_j \quad (4.25a)$$

$$\sum_{n\neq n'}(g_{nn'} + h_{nn'} + k_{nn'} + g_{n'n} + h_{n'n} + k_{n'n}) = l_{n'} \quad (4.25b)$$

$$\sum_{n\neq n'}(h_{n'n} - h_{nn'} - k_{n'n} + k_{nn'}) = m_{n'} \quad (4.25c)$$

In these relations $g_{nn'} = h_{nn'} = k_{nn'} = 0$ if $n < n'$. The eqs. (4.25) allow in the case of more than two clusters in bra or ket many solutions, which have to be found by trial and error. A very efficient computer program, based on a scheme given in /ST 70/, accomplishes this task.

We are now in the position to calculate the norm matrix element from eq. (4.24), with the help of eqs. (4.14), (4.18), (4.19), (4.22) and (4.25). In addition to that the ME of the central potential is also given by eq. (4.25), proviso the radial dependence of the force has been included into $\rho_{\mu\mu'}$ of eq. (4.11), which then modifies accordingly the elements β and $\sigma_{nn'}$. Analogous considerations apply for the tensor force except that the number of orbital angular momenta has to be increased by one. With the relation /ED 60/

$$4\pi/(2l+1) \sum_{m=-l}^{l} (-1)^m \mathcal{Y}_{lm}(\underline{r})\, \mathcal{Y}_{l-m}(\underline{r}) = r^{2l} \tag{4.26}$$

one can now utilize more complicated orbital wave functions which may then consist of Gaussian times solid spherical harmonics times polynomials in r^2 at the expense of two additional angular momenta for each monomial in eq. (4.9) and following. The calculation of other operators follows along the lines of eq. (4.9) to (4.25). The results can always be expressed by normintegrals times factors which can be easily calculated. The explicit expressions can be found in the literature for the kinetic energy /HH 70, HO 84/, the spin-orbit potential /HO 84/ and for electromagnetic transition operators /ME 86/.

The method described above can only be efficiently applied, if the radial dependence is given in terms of Gaussians. Besides the many effective nucleon-nucleon potentials given in Gaussian form, there are two realistic NN-potentials /EI 71, KE 86/ with full operator structure but as radial dependence only sums of Gaussians. These potentials reproduce the NN data up to the pion threshold. Other operators can be expressed in terms of Gaussians /BU 85/, e. g. the 1/r-dependence of the Coulomb potential may be written as

$$1/r = 2\sqrt{\beta/\pi} \int_0^\infty dk\, \exp(-k^2\beta \mathbf{r}^2) \tag{4.27}$$

which is again of Gaussian form and can be treated by the above method. Since in the framework of the RRGM the Coulomb force can be treated exactly contrary to many other approaches, we derive the ME in the Appendix.

With the methods described till now, we can evaluate all matrix elements and thus attack any problem. Before applying these methods to actual cases we study in the following section general properties of the RGM.

5. RGM WAVE FUNCTION AND EQUIVALENT LOCAL POTENTIALS

5.1 Interpretation of the RGM wave function

In section 3 the variational equations were solved for the reactance matrix a, which is the only observable quantity for scattering systems. If we are, however, interested in properties of bound states or in electromagnetic transitions then we need the variational wavefunction itself, i. e. we need the coefficients d_{lm} in eq. (3.10) which were eliminated by eq. (3.12). Let us for simplicity concentrate on the bound state problem, since transitions are only slightly more complicated and give no new inside.

The many body bound state problem can be solved with the variational procedure described in section 3, by reducing the ansatz for the wave function eq. (3.8) to square integrable functions only

$$\psi = A \sum_m b_m X_m \tag{5.1}$$

In order to have a transparent notation we number the functions consecutively and specify no more the channel k from which they originate. Since we restricted our considerations to the bound state problem, we can start from the analogue of eq. (2.5) in the many body case. Performing the variation we find

$$\sum_j (H_{ij} - EN_{ij})\, b_j = 0 \tag{5.2}$$

with

$$H_{ij} = \langle X_i | H | AX_j \rangle \tag{5.3a}$$

and

$$N_{ij} = \langle X_i | A | X_j \rangle \tag{5.3b}$$

Noting that the matrix N_{ij} is positive semi-definite standard methods to solve the general eigenvalue problem /WI 65/ eq. (5.2) can be applied.

The symmetric matrix N can be diagonalised by an orthogonal matrix B yielding a positive diagonal matrix. For the moment we exclude the possibility of a zero eigenvalue, but return to this point later on . Thus we can write the matrix N as

$$N = B^T D D B \tag{5.4}$$

With this expression we find from eq. (5.2)

$$H - EN = B^TD(D^{-1}BHB^TD^{-1} - E)DB = B^TD(P - E)DB \tag{5.5}$$

The symmetric matrix P has a complete set of eigenvectors z_i which may be taken orthogonal. We therefore have

$$P\underline{z}_i = e_i \underline{z}_i \tag{5.6}$$

giving

$$H(B^TD^{-1}\underline{z}_i) = e_iB^TD\underline{z}_i = e_iB^TDDBB^TD^{-1}\underline{z}_i = e_iN(B^TD^{-1}\underline{z}_i) \tag{5.7}$$

Hence $\underline{x}_i = B^TD^{-1}\underline{z}_i$ is an eigenvector of the generalised eigenvalue problem eq. (5.2), which fulfills the relations (3.9). We mention in passing that all solutions e_i of eq. (5.2) which are below the first threshold energy E_{th} are an upper bound for the energy of bound states. The energies $e_i > E_{th}$ have no intuitive physical meaning and cannot be interpreted as resonances /HA 73/.

Equation (5.2) is the matrix version of the standard RGM equation using integrals kernels /TA 81/

$$\int(\mathcal{H}(\underline{R}',\underline{R}) - E\mathcal{N}(\underline{R}',\underline{R}))\, F(\underline{R})\, d\underline{R} = 0 \tag{5.8}$$

This is obviously not of the usual form of the Schrödinger equation, therefore we cannot interpret F(R) as probability amplitude, but rather $\mathcal{N}^{1/2}F(R)$. This suggest by analogy that we can interprete $N^{1/2}\underline{x}_i$ straightforwardly as probability amplitudes too. Because our basis functions X_m are neither normalised nor orthognal $N^{1/2}$ is not just DB, but some care has to be taken in defining $N^{1/2}$ /BU 86/.

The RRGM differs in another point again from standard RGM: In case of equal width parameters for both fragments in the internal wave function eq. (3.26) and (3.27) we may have a zero norm eigenvalue.

In the standard RGM these redundant states /TA 81/ are a stringent test of the correct calculation of the norm and hamiltonian kernels. In the RRGM, however, we have to avoid such states. Since we have to divide by the square root of the norm eigenvalues during the transformation of H the standard routines for solving the general eigenvalue problem would fail. The redundant states are also called Pauli forbidden states. In case of different width parameters in both fragments, a zero norm eigenvalue is no more possible, but norm eigenvalues might be small. The corresponding eigenvectors are then often called almost Pauli forbidden state.

The existence of these Pauli forbidden (PV) states obviously prevents the direct physical interpretation of the solutions of eq. (5.2), because inside the range in which the PV are different from zero the solution is arbitrary. With the factor $N^{1/2}$, however, these arbitrary components are projected onto zero. The number of PV can be easily determined by considering the corresponding oscillator shell model states /HO 77/, examples are given in /TA 81/. On the other side these PV form the basis for the orthogonality condition model /SA 69/.

The RGM wave functions contain a further source of arbitraryness. In case of coupled channels we may consider different fragmentations, e.g. in the A = 6 case $^4He + d$ and $^3He + {}^3H$. Since the resulting channel wave functions eq. (3.23) may be non-orthogonal, we cannot answer the question of the admixture probability in the total wave function uniquely. This can be done only if solely orthogonal channel are taken into account, which can be done by applying some orthogonalisation procedure onto the non-orthogonal channels /SC 85/. These channels, however, do no more consist of the physical particles only. Compared to the full microscopic wave function eq. (3.5) these orthogonal channels give no deeper physical insight. Considering, however, only the relative motion part, the orthogonalised channel approach allows to draw well defined conclusions, for an example see /SP 86/.

5.2 Extraction of equivalent local potentials

As we have shown in the previous sections, the RGM provides an approximate solution of the relative motion wave function of two complex fragments with proper antisymmetrisation taken into account. Because of the exchange terms, P_3 of eq. (3.29) being different from the identity, the resulting kernels in eq. (5.8) are non-local.

Since local potentials are much easier to handle, one searches for such a local potential, containing the antisymmetrisation proper, which can be applied in much cruder reaction models like DWBA. Because of the problem associated with the non-orthogonality of channels discussed in the previous section, I restrict in the following my considerations to the single channel case only, see, however, /YA 85/ for the coupled channel approach.

Two different methods are used to define such a local potential. The easiest to understand is based on the potential Schrödinger equation eq. (2.3). Since we are now left with a function of the relative coordinate only, which we known from eq. (5.2) or (3.8) together with eq. (3.12) and (3.16) resp. (3.22c) we can now solve eq. (2.3) for the potential V finding

$$V_L^{loc}(R) = E - \frac{\hbar^2}{2M} L(L+1)/R^2 + \hbar^2/2M \; u_L''(R)/u_L(R) \tag{5.9}$$

Despite its simplicity eq. (5.9) might be ill-defined in the neighbourhood of zeros of u_L because of the unavoidable inaccuracies in forming the second derivative. The problems resulting from this division by zero, however, are often in such regions, where the potential is negligible anyhow. It is obvious from eq. (5.6) that V_L is energy dependent in general, since u_L depends on energy too. The second approach is based on the semiclassical WKB method. In the following we list the essential points, for details see /HO 80/. The starting point is a decomposition of the antisymmetriser A_3, eq. (3.29), into the identity and nontrivial exchanges, which leads to direct and exchange terms /TA 81/. Rewriting eq. (5.8) in these terms yields

$$\left(-\frac{\hbar^2}{2M}\nabla^2 + V_D(R) - E\right)F(R) = -\int G(R,R')F(R')dR' \tag{5.10}$$

where G contains contributions from the kinetic energy, the interaction and the norm. Defining the Wigner transform of G by

$$G_W(R^2,p^2,(\underline{R}\cdot\underline{p})^2) = \int d\underline{s}\, \exp\left(\frac{i}{\hbar}\,\underline{s}\cdot\underline{p}\right)\; G(\underline{R}-\underline{s}/2,\underline{R}+\underline{s}/2) \tag{5.11}$$

we have used the symmetry and rotational invariance of G to show explicitely the functional dependence of G_W. The effective local potential V_{eff} follows from G_W via the transcendental equation /HO 80/.

$$V_{eff}(R) = G_W(R^2, 2M(E - V_{eff}(R)), 2MR^2(E - V_{eff}(R)\; \frac{-\hbar^2(L+1/2)^2))}{2MR^2} \tag{5.12}$$

Examples are numerous, for an application to the non-relativistic quark model see /FU 86/.

6. ILLUSTRATIVE EXAMPLES

In this section we try to demonstrate the flexibility of the RRGM described previously by way of example. In order to keep the presentation concise we will consider only the seven-nucleon system. We will not discuss well-known effects, like antisymmetrisation effects, or single channel results but refer mostly to recent work.

6.1 Scattering results over a wide energy range

In the ^{7}Be system the ^{4}He - ^{3}He is the lowest threshold and then follows the ^{6}Li - p threshold. In order to explain all low-lying resonances, in addition the ^{5}Li - d fragmentation has to be added /HO 83, HO84a/. Recently these calculation were extended /HE 87/ to include also further fragmentations like ^{6}Be + n to allow for a larger range of energies in cross sections and polarizations. In fig. 4 we see a typical compound nucleus behaviour at the low energy, whereas the high energy corresponds almost to diffractive scattering due to the many maxima and minima. The data of fig. 5 cover the intermediate energy range. All the gross structures are well reproduced thus demonstrating that the RRGM can reproduce complex data over a wide energy range employing one potential /ME 86/ only. More details, like polarisations and reaction cross section will be published elsewhere /HE 87/.

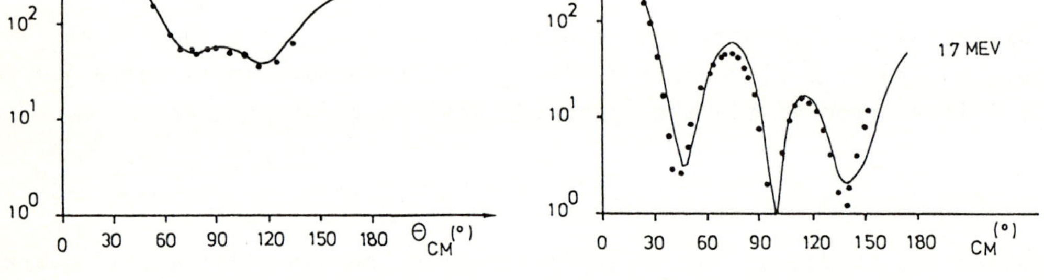

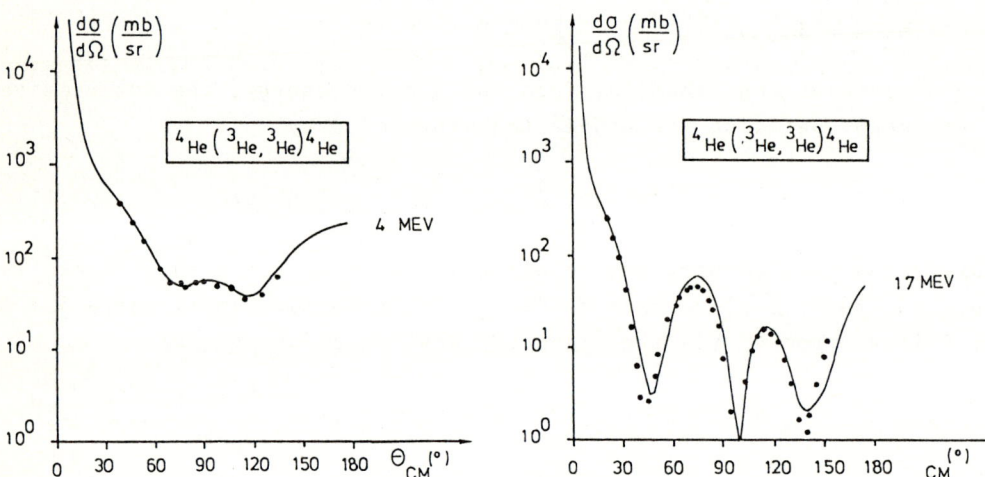

Fig. 4: Comparison of calculated ^{3}He - ^{4}He elastic cross sections for one small and one large energy with data from /LU 78/

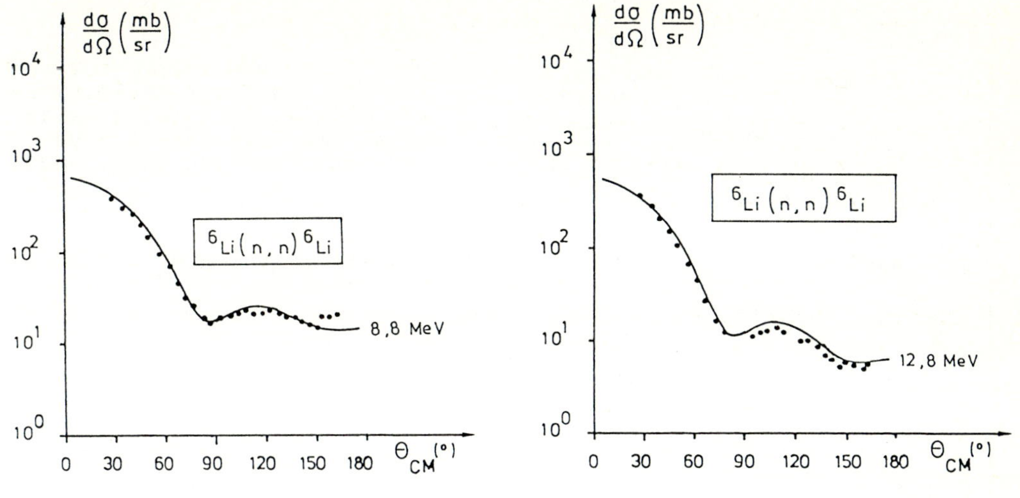

Fig. 5: Comparison of elastic neutron scattering from ^6Li for two intermediate energies with data from /HO 79/.

6.2 Expansion of the scattering wave function

In section 4 we described how to calculate ME between Gaussian functions only. From section 3, however, we know that we need ME with Coulomb functions too. As discussed in section 3 these integrals are all of short range due to the choice of kinetic energies. In fig. 6 the relative motion part of the ^4He - ^3He wave function, eq. (3.6) is displayed, together with the expansion in terms of 15 /ST 77/ and 20 /ME 86/ Gaussian functions, where the smallest width parameters are 0.0015 fm^{-2} and 0.0001 fm^{-2} respectively.

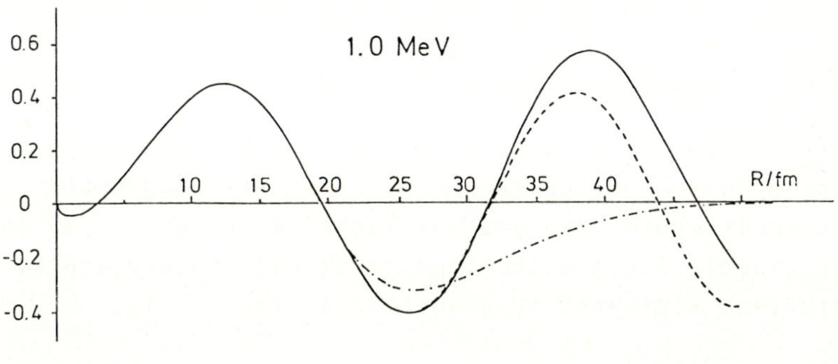

Fig. 6: Comparison of the relative motion wave function eq. (3.6) for exact Coulomb functions (full line) with expansions of the Coulomb functions into 15 Gaussians (dashed dotted line) and into 20 Gaussians (dashed line).

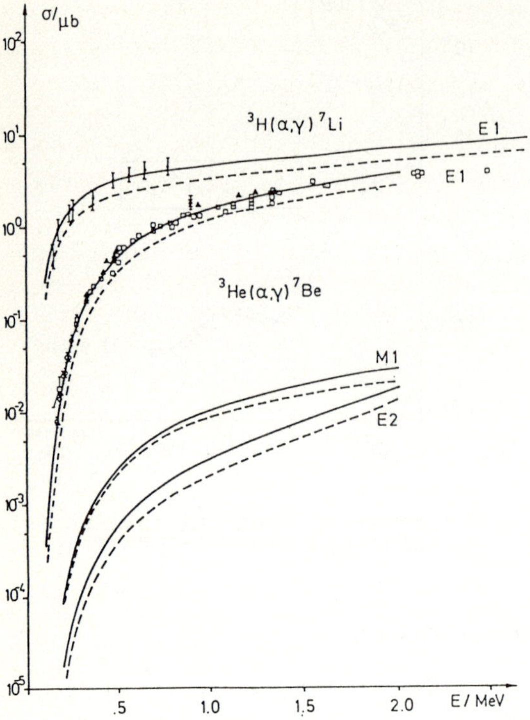

Fig. 7:
The E1-capture cross sections for the reactions $^3He(\alpha,\gamma)^7Be$ and $^3H(\alpha,\gamma)^7Li$ together with data. The full curves give the cross section into ground and first excited state, the dashed curves only into the groundstate. For the ^3He capture also the M1 and E2 contributions are displayed.

By way of example we see that the Gaussian expansion reproduces the correct wavefunction up to 20 and 30 fm resp. This range is even enough for electromagnetic transition operators as is shown in fig. 7, where the radiative capture cross sections $^4He(^3He,\alpha)^7Be$ and $^4He(^3H,\gamma)^7Li$ are compared to experimental data. The agreement is almost perfect, thus ruling out the ^4He - ^3He capture reaction as source of the missing solar neutrinos. For more details see /ME 86/.

6.3 Admixtures of different fragmentations

For nuclear reactions it is well known, that different fragmentations are often the origin of resonances /HA 72/ or the occurences of resonances in channels where they are not anticipated /HO 83, HO 84a/. We give here an example for electromagnetic transition operators, where again admixture of different fragmentations play a role.

With the existing polarised lithium ion-source sub-coulomb scattering of polarised lithium ions became feasible /EG 80/. These data allowed to determine the quadrupole moment Q and the BE2-value but also the polarisability τ simultaneously /WE 85/. The polarisability τ can be understood as virtual excitations of the ^7Li-nucleus via E1-transitions, as indicated in fig. 8 a . Whereas the BE2-value and Q are easily reproduced, poses the polarisability τ a serious problem /KA 86/.

In fig. 8 b we display the integrand of the polarizability, $\tau = \int \rho \, dE$, as function of energy. As can be seen the integrand peaks just above the ^4He - ^3H threshold, thus falsificating the idea of exciting the giant resonances. Besides the ^4He - ^3He channel the ^6Li - n channel contributes another 50 percent /ME 86/, but still the calculated result is only half the experimental datum /ME 86/.

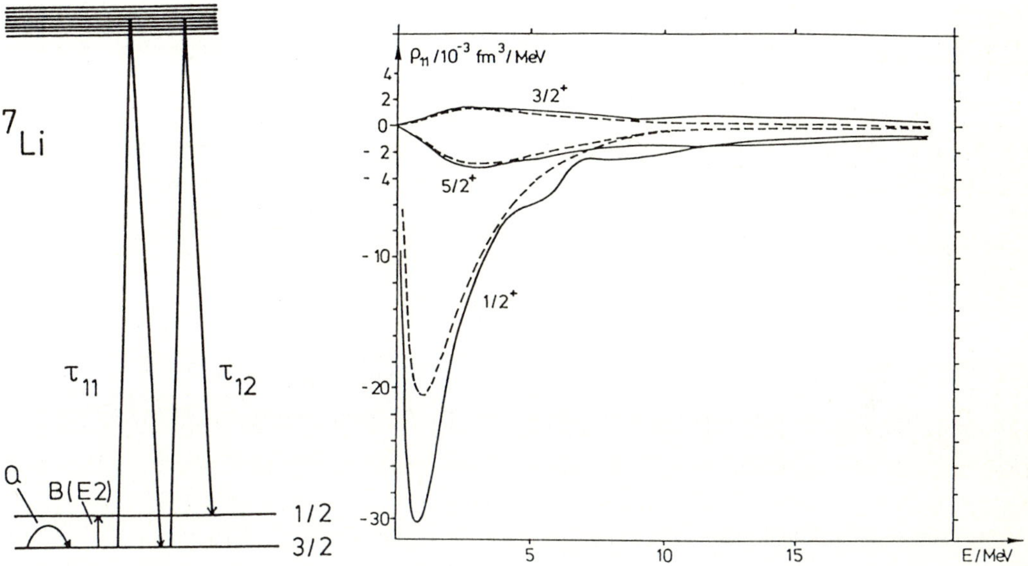

Fig. 8a: Schematic illustration of the groundstate reorientation effect (Q), the real E2-excitation (B(E2)) and the polarizability (τ_{11} and τ_{12}) via virtual E1-excitation.

b: The single partial wave contributions to the integrand ρ of the polarizability. For the dashed curves only the $^4He + {}^3H$ waves are taken into account, whereas the ^6Li + n channels are also considered in the full curve.

CONCLUSION

In the previous section it was shown that the microscopic RRGM method allows to describe gross structure of nuclear reactions over a wide energy range without adjustable parameters. The electromagnetic properties and the radiative capture cross sections demonstrate, that not only the asymptotic part of the wave function, but also the wave function in the interaction region are well under control. Therefore this method should be an excellent tool to study also other properties of light nuclei, e. g. the d-state admixture in ^4He. A typical example how detailed nuclear structure information can be extracted is the tiny quadrupol moment of ^6Li /ME 84 , ME 86 , HO 86/. The application of the RGM to the nonrelativistic quark model allows to study the nucleon-nucleon force from a more fundamental point. Application in this direction are still numerous and very promising.

Acknowledgement
I would like to express my sincere gratitude to all colleagues who have contributed over the years. In particular, I wish thank my coworkers J. Burger, T. Mertelmeier and W. Zahn.

This work was supported in part by the Deutsche Forschungsgemeinschaft and the Bundesministerium für Forschung und Technologie, Bonn.

APPENDIX A: EVALUATION OF THE COULOMB MATRIX ELEMENTS

In the following we present a derivation of the Coulomb matrix elements /BU 85, HO 84/ which relies on ideas of H. Stöwe /ST 81/. The Coulomb potential is given by

$$V_{CB} = \sum_{i<j} e_i e_j / |\underline{r}_i - \underline{r}_j| = \sum_{i<j} e_i e_j \, Q^{ij}_{z+1} \tag{A.1}$$

where e_i is the charge of particle i and Q^{ij}_{z+1} is just the difference vector $\underline{r}_i - \underline{r}_j$. Writing 1/Q as

$$1/Q = 2\sqrt{\beta/\pi} \int_0^\infty dk \, \exp(-k^2 \beta Q^2) \tag{A.2}$$

the matrix elements of the Coulomb potential are again of the form eq. (4.11) except for the additional variable Q_{z+1} and the integration over k. β is a free parameter, which is chosen appropriately later on. The integral we have to calculate now is

$$J^{ij}_{CB\,L'_\alpha L_\alpha}(P_{dc}) = 2\sqrt{\beta/\pi} \int_0^\infty dk \int d\underline{s}_1 \ldots d\underline{s}_{N-1}$$

$$\exp(-\sum_{\mu\mu'} \rho_{\mu\mu'} \underline{s}_\mu \cdot \underline{s}_{\mu'} - k^2 \beta Q^{i'j'2}_{z+1}) \prod_{n=1}^{z} \mathcal{Y}_{L_n M_n}(\underline{Q}_n) \tag{A.3}$$

The permutation P_{dc} has been applied onto the wave function to the right and onto Q^{ij}, the primes on $Q^{i'j'}$ are just a reminder of that fact. Defining the generating integral analogous to eq. (4.15) and following along the lines of eqs. (4.17) to (4.20) we find

$$I_{CB}(a_1b_1\ldots a_zb_z) = 2\sqrt{\beta/\pi}\int dk \int d_{s_1}\ldots d_{s_{N-1}}$$

$$\exp\Big(-\sum_\mu \beta_\mu t_\mu^2 - k^2\beta \sum_{\tau\tau'} P_{z+1,\tau}\, P_{z+1,\tau'}\, \underline{t}_{\tau'}\cdot \underline{t}_\tau + \sum_{n=1}^{z} a_n \underline{b}_n\cdot \underline{Q}_n\Big) \tag{A.4}$$

with the relation $\underline{Q}_{z+1} = \sum_\tau P_{z+1,\tau}\, \underline{t}_\tau$ being used. In order to make the evaluation of eq. (A.4) more transparent we introduce a more compact matrix notation by defining

$$\underset{\sim}{t} := (\underline{t}_1,\ldots,\underline{t}_{N-1})$$

$$\underset{\sim}{c} := \Big(\sum_{n=1}^{z} a_n\underline{b}_n P_{n1},\ldots,\ \sum_{n=1}^{z} a_n\underline{b}_n P_{n,N-1}\Big) \tag{A.5}$$

and

$$B(k)_{\tau\tau'} = \beta_\tau(\delta_{\tau\tau'} + k^2\beta\ P_{z+1,\tau}P_{z+1,\tau'}/\beta_\tau\ ) =: \beta_\tau\ C_{\tau\tau'}$$

Inserting eqs. (A.5) into eq. (A.4) yields

$$I_{CB}(a_1b_1\ldots,a_zb_z) = 2\sqrt{\beta/\pi}\int dk \int d\underline{t}_1\ldots d\underline{t}_{N-1}\ \exp(-\underset{\sim}{t}^T B\underset{\sim}{t} + \underset{\sim}{c}^T\cdot\underset{\sim}{t})$$

$$= 2\sqrt{\beta/\pi}\int_0^\infty dk\ \frac{\pi^{3(N-1)/2}}{(\det(B(k)))^{3/2}}\exp(\underset{\sim}{c}^T B(k)^{-1}\underset{\sim}{c}/4) \tag{A.6}$$

The solution is found by the methods of completing squares. We could also use this general solution in the derivation of eq. (4.21), but there the determinant and the inverse B^{-1} are trivial to calculate; therefore we renounced to introduce new quantities. For the Coulomb potential, however, this task is left to us.

Let us first consider the determinant of B. Noting that the second term of B in eq. (A.5) is a dyadic product, the determinant is given by

$$\det B = \Big(\prod_{\alpha=1}^{N-1}\beta_\alpha\Big)\Big(1 + k^2\beta\sum_{\alpha=1}^{N-1} P^2_{z+1,\alpha}/\beta_\alpha\Big) \tag{A.7}$$

Now we can obviously choose the free parameter β such that eq. (A.7) simplifies to

$$\det B = \left(\prod_{\alpha=1}^{N=1} \beta_\alpha\right)(1 + k^2) \tag{A.8}$$

For the evaluation of the inverse of B we again make use of the dyadic property by considering matrix C of eq. (A.5). This matrix can be written formally as

$$C = 1 + k^2 * \text{const} * \text{projector}, \tag{A.9}$$

hence, the inverse of C is of the structure

$$C^{-1} = 1 + f(k) * \text{projector} \tag{A.10}$$

where the function f(k) is determined to

$$f(k) = - 2k^2/(1 + k^2)\, \sigma_{z+1,z+1} \tag{A.11}$$

by demanding $C \cdot C^{-1} = 1$. Note that we have extended the range of indices of σ, eq. (4.23) in a natural way.
The inverse of B is now given by

$$(B^{-1})_{\alpha\alpha'} = \delta_{\alpha\alpha'}/\beta_\alpha - \frac{2k^2}{(1+k^2)\sigma_{z+1,z+1}} \cdot \frac{P_{z+1,\alpha}\, P_{z+1,\alpha'}}{\beta_\alpha \quad \beta_{\alpha'}} \tag{A.12}$$

Inserting the determinant, eq. (A.8) and the inverse of B, eq. (A.12), into eq. (A.6) we can write the generating integral as

$$I_{CB}(a_1 b_1, \ldots, a_z b_z) = 2\sqrt{2/\pi\, \sigma_{z+1,z+1}} \int_0^\infty dk \prod_{\lambda=1}^{N-1} (\pi/\beta_\lambda)^{3/2} (1 + k^2)^{-3/2}$$

$$\exp\left(\sum_{n>n'}^{z} a_n a_{n'} (b_n b_{n'} - (b_n^2 + b_{n'}^2)/2)\, \sigma_{nn}^{CB}\right) \qquad /A.13)$$

where we fixed the free parameter β as in eq. (A.8) and introduced the abbreviation

$$\sigma_{nn'}^{CB} = \sigma_{nn'} - k^2(1 + k^2)^{-1}\, \sigma_{n,z+1}\sigma_{n',z+1}/\sigma_{z+1,z+1} \tag{A.14}$$

to illustrate the correspondence between the norm integral eq. (4.21) and the Coulomb integral. This correspondence indicates how to proceed. Expanding the exponential in eq. (A.13) in powers of a_n and b_n one finds the analogous expression to eq. (4.24), except that it now contains powers of σ^{CB}. The k-integration is accomplished by expanding the powers of σ^{CB} into a binomial series.

Performing these steps the Coulomb ME are given by

$$\Gamma^{CB}_{1_1 m_1 \cdot\cdot 1_z m_z} = 2\sqrt{2/\pi\,\sigma_{z+1,z+1}}\ \prod_{r=1}^{N-1} (\pi/\beta_r)^{3/2} \prod_{j=1}^{z} (1_j!/C_{1_j m_j})$$

$$\sum_{g_{nn'},h_{nn'},k_{nn'}} (-1/2)^{h_{nn'}+k_{nn'}} \prod_{n>n'} 1/(g_{nn'}!h_{nn'}!k_{nn'}!) \sum_{\varepsilon'_{nn'}, \varepsilon_{nn'}} \sigma_{nn'}^{\varepsilon_{nn'}} \binom{\varepsilon_{nn'} + \varepsilon'_{nn'}}{\varepsilon_{nn'}}$$

$$\left(\frac{\sigma_{n,z+1}\ \sigma_{n',z'+1}}{\sigma_{z+1,z+1}}\right)^{\varepsilon'_{nn'}} \int_0^\infty dk(1+k^2)^{-3/2} \prod_{n>n'} \left(\frac{-k^2}{1+k^2}\right)^{\varepsilon'_{nn'}} \qquad (A.15)$$

Here the $g_{nn'}$, $h_{nn'}$ and $k_{nn'}$ are determined by eqs. (4.25) and the exponents ε and ε' fulfill the relations

$$\varepsilon_{nn'} + \varepsilon'_{nn'} = g_{nn'} + h_{nn'} + k_{nn'} \qquad n > n' \qquad (A.16)$$

The integrals over k are given in /GR 58/

$$\int_0^\infty dk(1+k^2)^{-3/2}\ (k^2/(1+k^2))^e = 1/(1+2e) \qquad (A.17)$$

With these equations we have an exact analytic treatment of the Coulomb potential. Combining the procedure of treating monomials in r^2 with this $1/r$ expressions, we can calculate ME of all positive powers of r. For negative powers in r additional expressions are given in /BU 85/.

APPENDIX B: THREE NUCLEON GROUND STATE WAVE FUNCTION

In order to illustrate the methods developed in the previous sections, we discuss here a simple model problem, which, however, gives already a glimpse of the ideas and techniques employed in more complicated systems. Considering the ^3H groundstate we avoid the Coulomb force and by applying the effective potential eq. (2.7) we circumvent angular momentum coupling problems.

Since this model force has no spin dependence the total spin S and total orbital angular momentum L are already good quantum numbers, hence, we need only consider states with $S = \frac{1}{2}$ and $L = 0$. The simplest wave function to comply with these requirements are the two neutrons coupled to spin zero with the additional proton leading to $S = 1/2$ and no orbital angular momentum between the nucleons. Thus we may put all three nucleons into one cluster, or the two neutrons into one cluster and the remaining proton into another one or put each nucleon into a separate cluster.

Since there is no orbital angular momentum between the two neutrons the last two fragmentations lead to identical results.

In the following we consider the fragmentation into two neutrons and the remaining proton. Let us now first consider the spin ME. The total spin function is of the form

$$|1/2 \quad 1/2 > = (n(+)n(-)-n(-)n(+)) \; p(+)/\sqrt{2} \qquad (B.1)$$

where we have explicitely given the spin projections. The spin ME eq. (4.7) for the norm and the potential is given by

$$ME = \; < 1/2 \quad 1/2 \; || \, P \, || \; 1/2 \quad 1/2 \; > \qquad (B.2)$$

since the proton is distinguishable from the neutrons, only permutations of the two neutrons may contribute, namely P = identity and P = (12), the exchange of the two neutrons, hence, the only dc-symbol contributing corresponds to the identity.

The orbital ME is also easily calculated, by noting that in eq. (4.9) no spherical harmonics occur and from all permutations only the identity is left. Hence, eq. (4.11) is just a multidimensional Gaussian integral and in eq. (4.24) only the first two products differ from one. We mention in passing that for the fragmentation considered, i. e. only two orbital angular momenta, eq. (4.25) yields $k_{12} = 1$ for a relative angular momentum 1 of the proton to the first cluster.

To determine now the ground state energy, we perform the variation eq. (2.5) or (2.19). In the potential example considered in section 2.1 we found agreement between the variational approximation and the exact numerical result when using five Gaussian width parameters.

Therefore to get an idea in this three nucleon case, we again use five Gaussian width parameters for the internal function eq. (3.26), fixed to those given in table 1, and another five for the relative motion function X in eq. (3.6). If we allow all possible 25 combinations the variation yields - 21.775 MeV binding energy, if we fix the internal nn-wave function to the one yielding - 2.225 MeV binding, we find only - 13.423 MeV in the ^3H case. We mention in passing that putting all three nucleons into one cluster only and again allowing for five free Gaussian parameters yields - 21.582 MeV binding energy, just a little less than the much more complicated wave function.

From this simple example we can conclude, that in a scattering calculation the results would be bad if the internal function would be fixed to the free one. In that case additional distortion channels, where the two neutrons are in orthogonal two body states, would improve the results appreciably.

Adding a component with l = 2 on both relative nucleon-nucleon coordinates in a three cluster fragmentation yields another 0.4 MeV binding energy, thus demonstrating that the pure S-wave structure provides already most of the binding for such an effective force.

Contrary to that is a pure S-wave fragmentation not bounded at all for a realistic force, like /EI 71/, where the tensor potential contributes appreciably to the binding energy. In this case a two cluster configuration including a d-wave is only a poor approximation to the exact result. A three cluster configuration, however, is within 10 percent of the exact binding energy /KE 86/.

REFERENCES

/AB 64/ Abramowitz, M. and I. A. Stegun (editors), Handbood of Mathematical Fucntions (National Bureau of Standards, 1972)

/BA 80/ Bauhoff, W., H. Schultheis and R. Schultheis, Phys. Rev. C22 (1980) 861

/BU 85/ Burger J., Ph. D. thesis, Erlangen 1985

/BU 86/ Burger J. and H. M. Hofmann, to be published

/ED 60/ Edmonds, A. R., Angular Momentum in Quantum Mechanics, Princeton University Press, 1960

/EG 80/ Egelhof, P., W. Dreves, K.-H. Möbius, E. S. Steffens, G. Tungate, P. Zupranski, D. Fick, R. Böttger and F. Roesel, Phys. Rev. Lett. 44 (1980) 1380

/EI 71/ Eikemeier, H. and H. H. Hackenbroich, Nucl. Phys. A169 (1971) 407

/FU 84/ Fujiwara, Y. and Y. C. Tang, University of Minnesota report UM-RGM2 (1984)

/FU 86/ Fujiwara, Y. and K. T. Hecht, Phys. Lett. 171B (1986) 17

/GE 83/ Gerjuoy, E., A. R. P. Rau and L. Spruch, Rev. Mod. Phys. 55 (1983) 725

/GR 58/ Gröbner, W. and N. Hofreiter, Bestimmte Integrale, Springer Verlag, Wien 1958

/HA 70/ Hackenbroich, H. H., Z. Phys. 231 (1970) 216

/HA 72/ Hackenbroich, H. H. and T. H. Seligman, Phys. Lett. 41B (1972) 102

/HA 73/ Hackenbroich, H. H., in The Nuclear Many-Body Problem, eds. F. Calogero and C. Cioffi Degli Atti, Editrice Compositori, Bologna 1973

/HA 77/ Hackenbroich, H. H., T. H. Seligman and W. Zahn, Helv. Phys. Acta 50 (1977) 723

/HA 85/ Hahn, K., E. W. Schmid and P. Doleschall, Phys. Rev. C31 (1985) 325

/HE 87/ Herman, M. and H. M. Hofmann, to be published
/HO 76/ Horiuchi, H., Progr. Theor. Phys. 55 (1976) 1448
/HO 77/ Horiuchi, H., Progr. Theor. Phys. (Suppl.) 62 (1977) 90
/HO 79/ Hogue, H. H., P. L. von Behren, D. W. Glasgow, S. G. Glendinning, P. W. Lisowski, C. E. Nelson, F. O. Purser, W. Tornow, C. R. Gould and L. W. Seagondollar, Nucl. Sci. and Eng., 69 (1979) 22
/HO 80/ Horiuchi, H., Progr. Theor. Phys. 64 (1980) 184
/HO 83/ Hofmann, H. M., T. Mertelmeier and W. Zahn, Nucl. Phys. A410 (1983) 208
/HO 84/ Hofmann, H. M. and T. Mertelmeier, Interner Bericht, Erlangen 1984
/HO 84a/ Hofmann, H. M., Nucl. Phys. A416 (1984) 363c
/HO 86/ Hofmann, H. M., T. Mertelmeier and D. Sachsenweger, Proc. 11th Int. Conf. Few Body Systems, Tokyo 1986, p. 258
/JO 71/ John, G., BMBW-FB K71-20, ZAED, Leopoldshafen 1971
/KA 51/ Kato, T., Prog. Theor. Phys. (Japan) 6 (1951) 394
/KA 86/ Kajino, T. and K.-I. Kubo, Proc. 11th Int. Conf. Few Body Systems, Tokyo 1986, p. 256
/KE 86/ Kellermann, H. and H. M. Hofmann, Few Body Systems, to be publ.
/LU 78/ Lui, Y.-W., O. Karban, A. K. Basak, C. O. Blyth, J. M. Nelson, S. Roman, Nucl. Phys. A297 (1978) 189
/LU 81/ Ludwig, A., Diploma thesis, Erlangen 1981, unpublished
/ME 84/ Merchant, A. C. and N. Rowley, Phys. Lett. 150B (1984) 35
/ME 86/ Mertelmeier, T. and H. M. Hofmann, Nucl. Phys. in press
/RO 57/ Rose, M. E., Elementary Theory of Angular Momentum, Wiley, New York 1957
/SA 69/ Saito, S., Prog. Theor. Phys. 41 (1969) 705
/SC 85/ Schmid, E. W. and G. Spitz, Z. Phys. 321 (1985) 581
/SE 75/ Seligman, T. H., Couble Coset Decomposition of Finite Groups, Burg Verlag, Basel 1975
/SE 76/ Seligman, T. H. and W. Zahn, J. Phys. G2 (1976) 79
/SP 86/ Spitz, G. and E. W. Schmid, Few Body Systems 1 (1986) 37
/ST 70/ Stöwe, H., Diploma thesis, Cologne 1970 unpublished
/ST 77/ Stöwe, H. and W. Zahn, Nucl. Phys. A289 (1977) 317
/ST 81/ Stöwe, H., priv. communication
/SU 83/ Suzuki, Y., Nucl. Phys. A405 (1983) 40
/TA 81/ Tang, Y. C., in Topics in Nuclear Physics, eds. T. T. S. Kuo and S. S. M. Wong, Lect. Notes in Phys. 145, Springer Heidelberg 1981
/WE 85/ Weller, A., P. Egelhof, R. Caplar, O. Karban, D. Krämer, K.-H. Möbius, Z. Moroz, K. Rusek, E. Steffens, G. Tungate, K. Blatt, I. König and D. Fick, Phys. Rev. Lett. 55 (1985) 480
/WH 37/ Wheeler, J. A., Phys. Rev. 52 (1937) 1083, 1107
/WI 65/ Wilkinson, J. H., The Algebraic Eigenvalue Problem, Clarendon Press, Oxford 1965
/WI 79/ Wildermuth, K. and E. J. Kanellopoulos, Rep. on Progr. in Physics 42 (1979) 1719
/YA 85/ Yabana, K., Prog. Theor. Phys. 73 (1985) 516
/ZA 81/ Zahn, W., Habilitationsschrift Erlangen 1981, Burg Monographs in Science 11, Burg Basel 1981

THE HYPERSPHERICAL EXPANSION METHOD

M. Fabre de la Ripelle

Division de Physique Théorique*, Institut de Physique Nucléaire,
91406, Orsay Cedex, France

Abstract

This lecture is divided in four main sections. In the first one we study the general properties of harmonic polynomials, we derive various hyperspherical harmonic basis and we explain how to construct antisymmetric harmonic polynomials.

In the second part we introduce the Potential Harmonics for systems of bosons and for fermions, and we derive the coupled equations enabling one to describe the two-body correlations. In the third section it is shown that the infinite system of coupled differential equations of the Potential Harmonic expansion method can be reduce to a single integro-differential equation in two variables.

In the last section we present the Adiabatic Approximation in which the radial and orbital motions are decoupled and which provides a method for solving scattering states.

Introduction

In the many-body problem it happens very often that the equations of motion are known but the solution cannot be found analytically. We are then obliged to rely on models without in many case being able to check the validity of our approximations.

When the model gives agreement with experiments we are tempted to believe that it describes the reality, but when we are not able to understand the relation between the model and the exact (unknown !) solution of our equations this position can be misleading. In most of these models parameters are introduced to fit experiments in such a way that an already questionable model is artificially ajusted to the data.

This kind of procedure of constructing models is a usefull attempt when the laws governing the behaviour of the system is either completely or partially unknown as for example in high energy physics, but when the equations describing the system of particles are known it might lead to hide important physical properties of the investigated states.

For instance the exchange meson effect, in which the mesons exchanged between nucleons are "seen" by electron scattering, has been extracted from the charge and magnetic form factors of few-nucleons systems, because we are able to solve the few-body bound states with a very good accuracy. Otherwise

IPNO/TH 86-81

*Laboratoire associé au C.N.R.S.

it would be easy to find a wave function, very similar to the realistic one, which fits exactly the experimental form factors, but this agreement would be meaningless. Unfortunately, it is what happens when we have to deal with more than few bodies where models are substituted for realistic solutions of the many-body problem. We are, for instance, led to question the meaning of agreements reached with such models like the widely used independent particle model (IPM) which does not produce any binding when used as a trial function in connection with the Schrödinger equation describing the motion of nucleons interacting through realistic potentials.

It is obviously very important to find a realistic solution of the many-body problem including the full effect of the interaction because only the agreement, or difference, between results obtained from a good solution describing a system of particles in interaction and the experimental data has a meaning.

Indeed, it is from the difference between the true solution and the experimental data that new effects can be found.

As most of interactions act between only two particles it is unavoilable to introduce two-body correlations in the solution . This task has been approach at the beginning by introducing Jastrow-type correlation functions which improve the IPM for small interparticle separations.

The energy is obtained by a variational calculation in which the parameters in the Jastrow function are ajusted to give, according to the Rayleigh-Ritz variational principle, the smallest eigen energy for describing the ground state. As the Jastrow function $f(r_{ij})$ is assumed to be the same for any pair (i,j) of identical particles, the wave function Ψ becomes the product of $f(r_{ij})$ for all pairs (i,j) and a suitably symmetrized function of the individual particle states $\phi_k(x_i)$ where i, j and k = 1, 2, ..., A. Antisymmetry is required for identical fermions. The average variational energy

$$E = \langle \Psi | H | \Psi \rangle$$

where H is the hamiltonian of the system can be accurately calculated for few-body systems but the help of a Monte-Carlo procedure is needed in the calculation of many-dimensional integrals for large systems.

Another approach to the problem has been proposed in the sixties by Faddeev for three bodies and extended by Yakubovsky to four-body systems. It requires the solution of integro-differential equations which are difficult to calculate with accuracy.

The solution of this problem, first obtained with separable potentials, has been later on reached with local potentials. But the extension of this last approach to a large number of particles is very complicated because various channels have to be taken into account in the description of the many-body system. Already the treatment of the Faddeev equation for three-body problem with realistic nuclear potentials which require the solution of a large number ($\simeq 40$) of coupled integro-differential equations lead to difficult numerical calculations.

Another type of approach is based on an expansion of the wave function in terms of a specific basis. The harmonic oscillator basis has been much used for this purpose. The wave function is expanded in a serie of basis elements, each one associated with an unknown coefficient, and the Schrödinger equation is projected on this basis and is transformed in an infinite set of coupled equations linear in the coefficients. The set of equations is truncated in order to be treated numerically and to provide the unknown coefficients.

Here we face another difficulty : the one of the huge degeneracy of the basis which require that a large number of basis elements be included in the expansion of the wave function and then a large set of equations be solved to obtain a good accuracy.

The number of equations to be solved for a fixed precision is closely related to the number of terms needed to obtain a good accuracy in a Fourier expansion of the potential. Practically for increasing number of particles this method becomes untractable because too many terms are needed to describe properly the two-body correlations, at least if a selection of the most significant basis elements is not performed in order to reduce the number of terms in the expansion.

Another procedure consists after quantizing some degrees of freedom to treat the remainder. For example we describe the motion of a particle in a spherical well by introducing the spherical harmonics, eigenfunction of the angular part of the kinetic energy operator, enabling one to reduce the Schrödinger equation to a set of (uncoupled) radial differential equations, one for each orbital ℓ, which can easily be solved.

The same idea can be extended to more than one particle in a well, or two-bodies with a mutual interaction.

The scheme is the one used for solving the motion of one particle in a non central well where an expansion of the wave function in a serie of spherical harmonics is performed followed by the projection of the Schrödinger equation on the same basis. It transforms this equation in an infinite set of second order coupled differential equations in the radial coordinate r which is truncated for a numerical solution.

This mathematical techniqueextended to more than two particles is called "Hyperspherical Harmonic Expansion Method" (H.H.E.M.). It is based on the mathematical property that any continuous function $\Psi(\vec{x})$, where $\vec{x}$ is the coordinate of a point in a D dimensional space, can be completely expanded in a serie of harmonic polynomials. For atomic systems where the center of mass is not eliminated, D = 3Z is three times the number Z of electrons. For systems of A identical particles submitted to mutual forces the center of mass $\vec{X}$ is eliminated and D = 3(A-1).

This method, where the full basis is introduced, leads to an exact asymptotic solution but cannot be used without care. Indeed, the harmonic polynomial basis, contains an infinite number of elements, and the expansion of $\Psi(\vec{x})$ must be truncated in order to generate a finite set of differential equations obtained by projection of the Schrödinger equation on the truncated basis, but a large degeneracy of the basis for a fixed degree of the polynomials prevents even to take all the low degree polynomials into account in the expansion of the wave function.

One is therefore obliged to select the polynomials providing the most significant contribution in the expansion. These polynomials are generated from the expansion of the product of the potential and the lowest degree harmonic polynomial in the wave function : They are called potential harmonics (P.H.).

In this lecture, the first part is devoted to a report of the general properties of harmonic polynomials, and recipes for the construction of properly symmetrized low degree harmonic polynomials are given.

In the second part the method for generating P.H. is carefully studied and the polynomials are given in closed form.

In the third part it is shown that the solution of the infinite set of coupled equations obtained from the P.H. expansion of the wave function can also be obtained from the one of a single (or a set of coupled) integro-differential equation(s) in two variables : the hyperradius and another one describing the two-body correlations. Incidentally it is proved that for three-body in S state this equation is the Faddeev equation written by Noyes for S state.

In the fourth section we introduce Adiabatic methods for solving the coupled differential equations. This procedure enables one to decouple the equations and to define the various channels occuring in the description of the many-body wave function. Each channel is associated with a fixed partition in terms of the clusters which can be constructed from the particles.

By using the so-called "Adiabatic basis", generated in the Adiabatic method, a formalism for describing scattering can be formulated.

The H.H.E.M. which, in principle, can be used for solving any many-body problem, has been utilised mainly in the calculation of few-body bound states in either atomic, nuclear and quark physics.

Photonuclear desintegration calculations which require the knowledge of continuum states have also been carried out with inclusion of final state interaction. In the last section various examples of applications of H.H.E.M. are given.

I. Harmonic Polynomials

Harmonic polynomials (H.P.) are homogeneous polynomials of the linear coordinates $\vec{x}_1, \vec{x}_2, \ldots, \vec{x}_n$, solution of the Laplace equation

$$\left(\sum_{i=1}^{n} \frac{d^2}{dx_i^2}\right) H_{[L]}(x_1,x_2,\ldots,x_n) = \nabla^2 H_{[L]}(\vec{x}) = 0 \tag{1.1}$$

where $\vec{x}$ stands for the set $(x_1, x_2, \ldots, x_n)$ and where the Laplace operator ∇^2 is the square of $\vec{\nabla}$, the gradient operator $(\frac{\partial}{\partial x_1}, \frac{\partial}{\partial x_2}, \ldots, \frac{\partial}{\partial x_n})$. Each polynomial is characterized by a set $[L]$ of 3n-1 numbers including the degree L.

One defines the length $r = (\sum_{i=1}^{n} x_i^2)^{\frac{1}{2}}$ called hyperradius.

The homogenous polynomial $H_{[L]}(\vec{x})$ can be written in polar coordinate as :

$$H_{[L]}(\vec{x}) = r^L Y_{[L]}(\Omega) \tag{1.2}$$

where Ω is a set of 3n - 1 coordinate over the unit hypersphere r = 1. The $Y_{[L]}(\Omega)$ which is the value of $H_{[L]}(\vec{x})$ on the surface of the unit hypersphere (r=1) is called a "Hyperspherical Harmonic" (H.H.) for n>3.

When n = 3, it is a spherical harmonic, usually denoted by $Y_{\ell,m}(\omega)$, where ω is the set of the two angular coordinates θ, φ at the surface of the unit sphere r = 1.

From the linear structure of the Laplace operator one deduces that the product $H_{[L_\alpha]}(\vec{x}_\alpha) H_{[L_\beta]}(\vec{x}_\beta)$ of two H.P. of two different (disconnected) sets of variables $\vec{x}_\alpha$ and $\vec{x}_\beta$ is also a H.P.. The H.P. basis can be chosen in such a way to fulfil the orthonormal condition over the surface of the unit hypersphere

$$\int Y^*_{[L]}(\Omega) Y_{[L']}(\Omega) \, d\Omega = \delta_{[L],[L']} \tag{1.3}$$

where $d\Omega$ is the surface element and the δ function is one when the two sets of quantum numbers [L] and [L'] are identical and zero otherwise.

Any homogeneous polynomial of degree L $\mathcal{H}_L(\vec{x})$ can be written as a sum of H.P.

$$\mathcal{H}_L(\vec{x}) = \sum_{n=0}^{[L/2]} r^{2n} H_{L-2n}(\vec{x}) \tag{1.4}$$

where $H_m(\vec{x})$ is a H.P. of degree m and L/2 is L/2 for L even and (L-1)/2 for L odd. The decomposition is unique.

As a consequence any function $F(\vec{x})$ which can be expanded in a power serie of the linear coordinates $\vec{x}$, can also be expanded in a serie of H.P.

$$F(\vec{x}) = \sum_{[L]=0}^{\infty} H_{[L]}(\vec{x}) \, f_{[L]}(r^2) = \sum_{[L]=0}^{\infty} Y_{[L]}(\Omega) \, r^L f_{[L]}(r^2) \qquad (1.5)$$

where the sum is taken over all the quantum numbers [L] for L running from 0 to infinity.

- <u>Parity</u> -

The parity operator applied to a (homogeneous) H.P. of degree L gives

$$H_{[L]}(-\vec{x}) = (-1)^L H_{[L]}(\vec{x}) = r^L (-1)^L Y_{[L]}(\Omega) \qquad (1.6)$$

and proves the parity $(-1)^L$ of the H.H. $Y_{[L]}(\Omega)$ associated with a H.P. of degree L.

- Corollary -

If $F(\vec{x})$ has a definite parity only harmonics of the same parity occur in the H.H. expansion.

- <u>Laplace Operator in Polar Coordinates</u> -

The position of a point $\vec{x}$ is given in polar coordinates by r, the hyperradius, and Ω the set of angular coordinates at the surface of the unit hypersphere. With these coordinates the Laplace operator becomes :

$$\nabla^2 = \frac{1}{r^{D-1}} \frac{\partial}{\partial r} r^{D-1} \frac{\partial}{\partial r} + \frac{L^2(\Omega)}{r^2} \qquad (1.7)$$

in a D dimensional space.

The second order differential operator $L^2(\Omega)$ is called grand orbital (or grand angular) operator. Its analytical expression depends upon the choice of angular coordinates Ω. From the Laplace equation

$$\nabla^2 H_{[L]}(\vec{x}) = \nabla^2 r^L Y_{[L]}(\Omega) = 0$$

one deduces the eigen equation for the H.H. :

$$\left(L^2(\Omega) + L(L + D - 2)\right) Y_{[L]}(\Omega) = 0 \qquad (1.8)$$

The L is called the "grand orbital" quantum number . For D = 3 one finds the usual equation for spherical harmonics :

$$\left(\ell^2(\omega) + \ell(\ell+1)\right) Y_{\ell,m}(\omega) = 0$$

It is often convenient to eliminate the first derivative $\frac{D-1}{r} \frac{\partial}{\partial r}$ occuring in the Laplace operator in polar coordinates. It can be achieved by using the formula :

$$\nabla^2\, \mathcal{H}(\Omega)\, r^{-(D-1)/2} u(r) \tag{1.9}$$
$$= r^{-(D-1)/2} \left[\frac{\partial^2}{\partial r^2} + \frac{\mathcal{L}^2(\Omega)}{r^2} \right] \mathcal{H}(\Omega)\, u(r)$$

where the $\mathcal{L}^2(\Omega)$ applied to a H.H. gives

$$\left[\mathcal{L}^2(\Omega) + \mathcal{L}(\mathcal{L}+1) \right] Y_{[L]}(\Omega) = 0 \tag{1.10}$$

where $\mathcal{L} = L + (D-3)/2$, ($\mathcal{L} = L$ for $D = 3$).

- Analytical Expression for Hyperspherical Harmonics -

In principle one needs to solve either the Laplace equation $\nabla^2 H_{[L]}(\vec{x}) = 0$ or the equivalent angular equation

$$[\, L^2(\Omega) + L(L+D-2)\,] Y_{[L]}(\Omega) = 0$$

for obtaining analytical expressions for H.H.. This method has been used in most of the original papers dealing with the H.H. basis. It is the standard method.

The analytical expression of $L^2(\Omega)$ is given in Appendix for a standard choice of angular coordinates /1/. The tree method of Vilenkin et al./ 2 / enables one to construct H.H. for various other choices of hyperspherical coordinates.

A more elegant and simple derivation is obtained by writing that the H.H. constitute a complete orthogonal set of functions $\{ Y_{[L]}(\Omega) \}$ at the surface of the unit sphere $r = 1$, and fulfil the orthonormal equations

$$\int_{(r=1)} H^*_{[L]}(\vec{x}) H_{[L]}(\vec{x})\, d\Omega = \int Y^*_{[L]}(\Omega) Y_{[L']}(\Omega)\, d\Omega = \delta_{[L],[L']} \tag{1.11}$$

This equation can be used for constructing antisymmetrical H.H., a case where the standard method fails to apply easily. In order to calculate the H.H. basis one can use a recursion method : when the complete set of H.H. is known in a D dimensional space, one uses the orthonormal equation to derive the complete set in the (D + 1) dimensional space.

Let us begin by the two-dimensional space. The polar coordinates (ρ,φ) are related to the linear coordinates (x,y) of a point in the plane by

$$x = \rho \cos\varphi\ ,\quad y = \rho \sin\varphi\ ,\quad \rho^2 = x^2 + y^2$$

The "surface" element is

$$d\Omega = d\varphi \qquad 0 < \varphi < 2\pi .$$

There is only one quantum number : the degree $L \equiv m$ of the H.P.
We write the orthonormal equation

$$\int^{\pi} Y^*_m(\varphi)\, Y_{m'}(\varphi)\, d\varphi = \delta_{m,m'}$$

and find

$$Y_m(\varphi) = \frac{1}{\sqrt{\pi}} \begin{cases} \sin m\varphi \\ \cos m\varphi \end{cases} \quad \text{or} \quad Y_m(\varphi) = \frac{1}{\sqrt{2\pi}}\, e^{im\varphi}$$

where m is an integer which can be negative when the exponential representation is used.

The associated H.P. is

$$H_m(\vec{x}) = \rho^m Y_m(\varphi) = \frac{1}{\sqrt{2\pi}}(x \pm iy)^{|m|} \qquad (1.12)$$

- <u>Three dimensional space</u> - (D = 3) -

We use the traditional polar coordinate system

$$x = \rho\cos\varphi \quad y = \rho\sin\varphi \quad \rho = r\sin\Theta \qquad z = r\cos\Theta$$

leading to the surface element

$$d\Omega = d\omega = \sin\Theta d\Theta d\varphi.$$

This time we have two quantum numbers related to the two degrees of freedom φ and Θ. One of them is the degree $L = \ell$ of the H.P. This polynomial is the product of a H.P. of degree m in the D-1 = 2 dimensional space and a polynomial of degree ℓ-m of the new coordinate z :

$$H_{\ell,m}(\vec{x}) = \frac{1}{\sqrt{2\pi}}(x \pm iy)^{|m|} P_{\ell-|m|}(z)$$

It is of degree ℓ in the coordinates $\vec{x} \equiv (x, y, z)$. By introducing $H_{\ell,m}(\vec{x})$ in the orthonormal equation for r = 1 we obtain :

$$\frac{1}{2\pi}\int_0^{2\pi} e^{i(m-m')\varphi} d\varphi \int_0^{\pi} (\sin\Theta)^{|m|+|m'|} P_n^{|m|}\cos\Theta\, P_{n'}^{|m'|}\cos\Theta\sin\Theta d\Theta$$

$$= \delta_{\ell,\ell'}\,\delta_{m,m'}$$

where $n = \ell - |m|$.

By using the variables $u = \cos\Theta$ this equation becomes

$$\int_{-1}^{1}(1-u^2)^{|m|} P_{\ell-|m|}^{|m|}(u)\, P_{\ell'-|m|}^{|m|}(u)\, du = \delta_{\ell,\ell'}$$

The polynomials $P_n^{|m|}(u)$ associated with the weight function $(1 - u^2)^{|m|}$ are the Gegenbauer polynomials $C_n^{|m|+\frac{1}{2}}(u)$.

By using the normalisation of the Gegenbauer polynomials we obtain the spherical harmonics (D = 3)

$$Y_{\ell,m}(\omega) = \frac{1}{\sqrt{2\pi}} e^{im\varphi} P_\ell^m(\cos\Theta) \qquad (1.13)$$

where

$$P_\ell^m(\cos\Theta) = (-1)^m h_{\ell-|m|}^{-\frac{1}{2}} (\sin\Theta)^{|m|} C_{\ell-|m|}^{|m|+\frac{1}{2}}(\cos\Theta) \qquad (1.14)$$

is called an associated Legendre function of the first kind.
The phase $(-1)^m$ is chosen in such away that

$$Y_\ell^{m*}(\omega) = (-1)^m Y_\ell^{-m}(\omega) \qquad (1.15)$$

and h_n is the normalisation constante of the Gegenbauer polynomial $C_n^{|m|+\frac{1}{2}}$:

$$h_n = \frac{\sqrt{\pi}}{n+|m|+\frac{1}{2}} \; \frac{|m|!}{\Gamma(|m|+\frac{1}{2})} \binom{2|m|+n}{n} \qquad (1.16)$$

where $\binom{a}{b}$ is a binomial coefficient.

To go further to more dimensional spaces (D>3) we have to specify the hyperspherical coordinates. Assume that the H.P. of the n variables x_1, x_2, x_n, $H[L_n](x_1, x_2, \ldots, x_n)$ are known and that they are requested for one more variable x_{n+1}. One sets

$$\rho = \left[\sum_{i=1}^{n} x_i^2\right]^{\frac{1}{2}} \qquad \text{and } r^2 = \rho^2 + x_{n+1}^2$$

i.e. $\qquad \rho = r\sin\phi_{n+1} \quad$ and $\quad x_{n+1} = r\cos\phi_{n+1}$.

The volume element in the n dimensional space is

$$dV_n = dx_1.dx_2.\ldots.dx_n = \rho^{n-1}d\rho d\Omega_n$$

then

$$dV_{n+1} = dV_n dx_{n+1} = r^n dr\, d\Omega_n (\sin\phi_{n+1})^{n-1} d\phi_{n+1}.$$

We write the H.H. as the product

$$H[L_{n+1}](\vec{x}) = H[L_n](x_1, x_2, \ldots, x_n)\, P_m(x_{n+1})$$

and we write $H[L_n]$ in hyperspherical coordinates

$$H[L_n](x_1,\ldots,x_n) = \rho^{L_n} Y_{[L_n]}(\Omega_n) = r^{L_n}(\sin\phi_{n+1})^{L_n} Y_{[L_n]}(\Omega_n)$$

that we introduce in the orthonormal equation leading to the condition :

$$\int_0^{\pi/2} (\sin\phi_{n+1})^{2L_n} P_m(\cos\phi_{n+1}) P_{m'}(\cos\phi_{n+1}) (\sin\phi_{n+1})^{n-1} d\phi_{n+1} = \delta_{m,m'}$$

The polynomials associated with the weight function $(1-u^2)^{L_n+(n/2)-1}$ are Gegenbauer polynomials $C_m^{L_n+(n-1)/2}$.

The H.P. in the n+1 dimensional space are

$$Y_{[L_{n+1}]}(\Omega_{n+1}) = h^{-\frac{1}{2}}_{(L_{n+1}-L_n)} (\sin\phi_{n+1})^{L_n} C^{L_n+(n-1)/2}_{(L_{n+1}-L_n)}(\cos\phi_{n+1})\, Y_{[L_n]}(\Omega_n) \tag{1.17}$$

where h_n is the normalisation constante of the Gegenbauer polynomial.

Starting from the spherical harmonics (i.e. from D = 3) it is easy to generate a complete H.H. basis by recurrence in using this formula.

But this kind of H.H. is not suitable for physical applications because the knowledge of the behaviour of each particle under rotation is generally required.

This behaviour is not defined in the previously derived H.H. To achieve this task we define 3 dimensional vectors $\vec{\xi}_1, \vec{\xi}_2, \ldots, \vec{\xi}_N$, which are related to the particle coordinates $\vec{x}_i$.

It can be for instance $\vec{\xi}_i = \sqrt{2}\, \vec{x}_i$ or, when the center of mass must be eliminated, it can be the Jacobi coordinates :

$$\vec{\xi}_N = \vec{x}_2 - \vec{x}_1 \tag{1.18}$$

$$\vec{\xi}_{N-1} = \sqrt{3}(\vec{x}_3 - \vec{X}_3)$$

$$\vec{\xi}_{N-i+1} = \sqrt{2\frac{i+1}{i}}\,(\vec{x}_{i+1} - \vec{X}_{i+1}) = \sqrt{\frac{2i}{i+1}}\,(\vec{x}_{i+1} - \vec{X}_i)$$

$$\ldots \quad \vec{\xi}_1 = \sqrt{\frac{2A}{A-1}}\,(\vec{x}_A - \vec{X}) \qquad N = A-1$$

where $\vec{X}_j = \frac{1}{j}\sum_{i=1}^{j}\vec{x}_i$ is the center of mass of the subsystem $(\vec{x}_1, \vec{x}_2, \ldots, \vec{x}_j)$ and $\vec{X}$ is the center of mass. The normalisation of the $\vec{\xi}$ coordinates has been adjusted in such a way that the hyperradius becomes

$$r^2 = \sum_{i=1}^{N=A-1} \xi_i^2 = 2\sum_{i=1}^{A}(\vec{x}_i-\vec{X})^2 = \frac{2}{A}\sum_{i,j>1} r_{ij}^2 \tag{1.19}$$

where $\vec{r}_{ij} = \vec{x}_i-\vec{x}_j$, and that the Laplacian be written

$$\frac{1}{2}\sum_{i=1}^{A}\nabla^2_{x_i} = \frac{1}{2A}\nabla^2_X + \sum_{i=1}^{N=A-1}\nabla^2_{\xi_i}$$

where $\vec{\nabla}_X$ refers to the center of mass momentum operator. The kinetic energy operator for equal mass particles written with the translationaly invariant $\vec{\xi}$ coordinates becomes

$$T = -\frac{\hbar^2}{2m}\sum_i^A \nabla^2_{x_i} = -\frac{\hbar^2}{2mA}\nabla^2_X - \frac{\hbar^2}{m}\nabla^2_0 \tag{1.20}$$

$$\nabla^2_0 = \sum_{i=1}^{N}\nabla^2_{\xi_i}$$

To construct the H.H. associated with the $\vec{\xi}$ coordinates we make use of the property that a product of H.P. of disconnected variables is also a H.P..

Here the disconnected variables are the spherical coordinates ω_i of each vector $\vec{\xi}_i$ in such a way that the product

$$H_{\ell_1 m_1,\ldots,\ell_N m_N}(\vec{\xi}_1,\ldots,\vec{\xi}_N) = \prod_{j=1}^{N} \xi_j^{\ell_j}\, Y_{\ell_j,m_j}(\omega_j) \tag{1.21}$$

where $\xi_j = |\vec{\xi}_j|$, is a H.P..

Indeed each $\xi_j^{\ell_j} Y_{\ell_j,m_j}(\omega_j)$ is itself a H.P.. This polynomial $H_{[L]}(\vec{\xi})$ depends on the 2N quantum numbers ℓ_i, m_i $(i = 1, \ldots, N)$ and has the degree $L = \sum_{i=1}^{N}\ell_i$. The others N-1 quantum numbers depends on the definition of the hyperspherical coordinates related to the ξ_i.

A standard choice of angular coordinates already used fifty years ago by Zernike and Brinkman is the following /3/ :

$$\xi_N = r\cos\phi_N \tag{1.22}$$

$$\xi_{N-1} = r\sin\phi_N\cos\phi_{N-1}$$

$$\ldots$$

$$\xi_j = r\sin\phi_N \ldots \sin\phi_{j+1}\cos\phi_j .$$

$$\ldots$$

$$\xi_2 = r\sin\phi_N \ldots \sin\phi_3\cos\phi_2$$

$$\xi_1 = r\sin\phi_N \ldots \sin\phi_3\sin\phi_2$$

The relation $r^2 = \sum_1^N \xi_j^2$ is easy to check. By using a recurrence method associated with the orthonormal equation, in a way very similar to the one explained above, one finds easily the H.H. derived by differential equation procedure by Zernike and Brinkman /3/

$$Y_{[L]}(\Omega) = Y_{\ell_1,m_1}(\omega_1) \prod_{j=2}^{N} Y_{\ell_j,m_j}(\omega_j) \; {}^{(j)}P_{L_j}^{\ell_j,L_{j-1}}(\phi_j) \tag{1.23}$$

where

$$^{(j)}P_{L_j}^{\ell_j,L_{j-1}}(\phi_j) = \left[\frac{2\nu_j\Gamma(\nu_j-n_j)n_j!}{\Gamma(\nu_j-n_j-\ell_j-\tfrac{1}{2})\Gamma(n_j+\ell_j+3/2)}\right]^{1/2} \tag{1.24}$$

$$\times\,(\cos\phi_j)^{\ell_j}(\sin\phi_j)^{L_{j-1}}\,P_{n_j}^{\nu_{j-1},\ell_j+\frac{1}{2}}(\cos 2\phi_j)$$

$\nu_j = \nu_{j-1} + 2n_j + \ell_j + 3/2 = L_j + (3j)/2 - 1 = L_{j-1} + 2n_j + \ell_j + (3j)/2 - 1$ and where $P_n^{a,b}$ is a Jacobi polynomial.

The quantum numbers $[L]$ are the individual quantum numbers ℓ_j, m_j and the partial grand orbital quantum numbers L_j, $j = 2, \ldots, N$ associated with the $\phi_2, \ldots, \phi_N$ degrees of freedom. The grand orbital is $L = L_N$ and we have the recurrence relation

$$L_i = L_{i-1} + 2n_i + \ell_i \qquad \text{and then}$$

$$L = \ell_1 + \sum_{j=2}^{N}(2n_j + \ell_j)$$

For isolated systems the total angular momentum ℓ is a good quantum number, therefore for bosons the spherical harmonics Y_{ℓ_j,m_j} must be coupled to produce a definite ℓ, m. It can be obtained by using Clebsh-Gordan and associated coefficients for building the coupling in $Y_{[L]}(\Omega)$ which then will be denoted $Y_L^{\ell,m}(\Omega)$.

- Addition Theorems -

Before going further to more details concerning the various kinds of H.P. we intend to give general relations between H.H.. Let $\vec{x}_1$ and $\vec{x}_2$ be two vectors in the D dimensional space . We define the scalar product $\vec{x}_1 \cdot \vec{x}_2 = r_1 r_2 \cos\alpha$ where $r_1 = |\vec{x}_1|$ and $r_2 = |\vec{x}_2|$.

Assuming that the H.H. basis in the D dimensional space is orthonormalised we have the addition theorem

$$\sum_{[L]} Y^*_{[L]}(\Omega_1)\,Y_{[L]}(\Omega_2) = \frac{2L+1}{4\pi^{D/2}}\Gamma(D/2 - 1)\,C_L^{D/2-1}(\cos\alpha) \tag{1.25}$$

where the sum is taken, for L fixed, over all quantum numbers $[L]$, and C_n^{α} is a Gegenbauer polynomial. For D = 3 we recover the addition theorem for spherical harmonics :

$$\sum_m Y^*_{\ell,m}(\omega_1)\,Y_{\ell,m}(\omega_2) = \frac{2\ell+1}{4\pi}\,C_\ell^{\frac{1}{2}}(\cos\alpha)$$

When the two vectors are colinear (i.e. $\Omega_1 \equiv \Omega_2$) the addition theorem becomes

$$\sum_L Y^*_L(\Omega)\,Y_L(\Omega) = \frac{2\;\;+1}{4\pi^{D/2}}\Gamma(D/2 - 1)\binom{L + D - 3}{D - 3} \tag{1.26}$$

By integrating over the unit hypersphere and by taking the normalisation of the H.H. into account we obtain the number N(L) of independent (orthogonal) H.H. of grand orbital L :

$$N(L) = \frac{2L+D-2}{D-2}\binom{L+D-3}{D-3} \tag{1.27}$$

we used

$$\int d\Omega = Y_{[0]}(D)^{-2} = \frac{2\pi^{D/2}}{\Gamma(D/2)} \tag{1.28}$$

where $Y_{[0]}(D)$ is the H.P. (and the normalised H.H.) of degree zero in the D dimensional space.

- Introduction of Symmetry -

The H.H. basis do not exhibit in general any definite symmetry when particles are exchanged. It is the case for the Zernike-Brinkman (Z.B.) basis. The symmetry must be introduced on purpose and adapted to each particular system of particles in starting from a H.H. basis defined for a fixed partition of particles coordinates.

For examples for 3 identical particles let us chose the Jacobi coordinates in the partition (ij,k) :

$$\vec{\xi}_1 = \sqrt{3}(\vec{x}_k - \vec{X}) \quad \text{and} \quad \vec{\xi}_2 = \vec{x}_i - \vec{x}_j = \vec{r}_{ij} \tag{1.29}$$

with $\quad r^2 = \xi_1^2 + \xi_2^2 .$

The Z.B. coordinates are

$$\omega_{ij} , \omega_k \quad \text{and} \quad \cos\phi_2 = r_{ij}/r.$$

The corresponding Z.B. basis where the spherical harmonic of ω_{ij} and ω_k are coupled to a total angular momentum (ℓ,m) is

$$Y^{\ell;m}_{L,\ell_1,\ell_2}(\Omega_{ij,k}) = N_{L,\ell_1,\ell_2} \sum_{m_1 m_2} \langle \ell_1,\ell_2;m_1,m_2|\ell,m\rangle \tag{1.30}$$

$$Y^{m_1}_{\ell_1}(\omega_k)\, Y^{m_2}_{\ell_2}(\omega_{ij})(r^2-r^2_{ij})^{\ell_1/2}\, r^{\ell_2}_{ij} \,/\, r^{\ell_1+\ell_2}$$

$$P_n^{\ell_1+\frac{1}{2},\ell_2+\frac{1}{2}}(2r^2_{ij}/r^2-1)$$

$$L = 2n + \ell_1 + \ell_2$$

$$N_{L,\ell_1,\ell_2} = \left[\frac{2(L+2)(L+1-n)!n!}{\Gamma(n+\ell_1+\frac{3}{2})\Gamma(n+\ell_2+\frac{3}{2})}\right]^{\frac{1}{2}}$$

where $\langle \ell_1, \ell_2; m_1, m_2|\ ,m\rangle$ is a Clebsh-Gordan coefficient.

When the two coordinates $\vec{x}_i$ and $\vec{x}_j$ are exchanged the H.H. becomes

$$Y^{\ell,m}_{L,\ell_1,\ell_2}(\Omega_{ji,k}) = (-1)^{\ell_2}\, Y^{\ell,m}_{L,\ell_1,\ell_2}(\Omega_{ij,k}) \tag{1.31}$$

but for any other exchange of coordinates a new partition is obtained. There are three available partitions : (12, 3), (23, 1), (31, 2) obtained by cyclic permutations. Any kind of symmetry can be constructed with the H.H. in the various partitions. For each one we have a complete set of H.H.. The H.H. of one partition are not orthogonal to those of another partition and the overlapp between two H.H. of different partitions has been studied by Raynal and Revai / 4 /. A change of partition is equivalent to a rotation in the 6 dimensional space in which the degree L of the H.P. and the orbital and azimuthal quantum numbers ℓ and m are preserved The overlapping integral

$$\int Y^{\ell m*}_{L,\ell_1,\ell_2}(\Omega_{ij,k})\, Y^{\ell' m'}_{L,\ell'_1,\ell'_2}(\Omega_{jk,i})\, d\Omega \tag{1.32}$$

$$= \langle \ell_1 \ell_2 | \ell'_1\ \ell'_2 \rangle_{L,\ell}\, \delta_{\ell\ell'}\, \delta_{mm'}$$

is known as "Raynal-Revai coefficient".

- <u>The Simonov Basis / 5 /</u> -

For a system of 3 bosons with total angular momentum zero (ℓ= 0), Simonov constructed a basis in which the symmetry is related to a quantum number. Let us call it the S basis (S for Simonov and for S state). As it is a basis which has been much utilised we intend to show in details the connection between the S basis and the Z.B. basis for S states (i.e., ℓ = m = 0).

We start from the Jacobi coordinates :

$$\sqrt{3}(\vec{x}_3-\vec{X}) = \vec{\xi}_1 \qquad \vec{x}_1-\vec{x}_2 = \vec{\xi}_2 \tag{1.33}$$

$$\sqrt{3}(\vec{x}_1-\vec{X}) = -\tfrac{1}{2}(\vec{\xi}_1-\sqrt{3}\vec{\xi}_2) \qquad \vec{x}_2-\vec{x}_3 = -\tfrac{1}{2}(\vec{\xi}_2+\sqrt{3}\vec{\xi}_1)$$

$$\sqrt{3}(\vec{x}_2-\vec{X}) = -\tfrac{1}{2}(\vec{\xi}_1+\sqrt{3}\vec{\xi}_2) \qquad \vec{x}_3-\vec{x}_1 = -\tfrac{1}{2}(\vec{\xi}_2-\sqrt{3}\vec{\xi}_1)$$

and we define the two complex conjugate vectors :

$$\vec{z} = \vec{\xi}_1 + i\vec{\xi}_2 \quad \text{and} \quad \vec{z}^* = \vec{\xi}_1 - i\vec{\xi}_2 \tag{1.34}$$

By applying the permutation operators to $\vec{z}$ we obtain

$$P_{12}\vec{z} = \vec{z}^* \qquad P_{13}\vec{z} = \vec{z}^*\, e^{-2i\pi/3} \qquad P_{23}\vec{z} = \vec{z}^*\, e^{2i\pi/3} \tag{1.35}$$

and the complex conjugate equations.

Homogeneous polynomials invariant by rotation can be constructed from the scalar products z^2, $(z^*)^2$ and $\vec{z}.\vec{z}^*$ where

$$z^2 = \xi_1^2 - \xi_2^2 + 2i\vec{\xi}_1\vec{\xi}_2 \qquad \vec{z}.\vec{z}^* = \xi_1^2 + \xi_2^2 = r^2 \tag{1.36}$$

Besides the usual angular coordinates ϕ, φ defined by

$$\xi_1 = r\sin\phi\ ,\ \xi_2 = r\cos\phi \quad \text{and}\ \vec{\xi}_1.\vec{\xi}_2 = \xi_1\xi_2\cos\varphi$$

$$0 < \phi < \pi/2\ ,\quad 0 < \varphi < \pi$$

we introduce the new variables A and λ related to ϕ and φ by

$$z^2 = r^2[-\cos2\phi + i\sin2\phi\cos\varphi] = -A\, e^{-i\lambda} \tag{1.37}$$

where

$$A = [\, z^2(z^*)^2]^{\frac{1}{2}} = [\,\cos^2 2\phi + \sin^2 2\phi\cos^2\varphi]^{\frac{1}{2}}$$

and $\mathrm{tg}\,\lambda = \mathrm{tg}2\phi\cos\varphi$.

Let $u = \sin\Theta\cos\varphi$, $v = \sin\Theta\sin\varphi$ and $w = \cos\Theta$ be the coordinates of apoint on a unit sphere. We have obviously $0<(u^2+w^2)^{\frac{1}{2}}<1$ from which we deduce with $\Theta = 2\phi$ the range of variation $0<A<1$.

By applying the permutation operators to z^2 we obtain

$$P_{12}z^2 = (z^*)^2,\ P_{13}z^2 = (z^*)^2\, e^{2i\pi/3}\ ,\ P_{23}z^2 = (z^*)^2\, e^{-i2\pi/3} \tag{1.38}$$

Therefore A being invariant by permutation of any two coordinates :

$$P_{12}\lambda = -\lambda \qquad P_{13}\lambda = -\lambda - 2\pi/3 \quad P_{23}\lambda = -\lambda + 2\pi/3 \tag{1.39}$$

Let us now explain how to deduce the S basis from the orthonormal equation.

The surface element is

$$d\Omega = \pi^2\, A dA\, d\lambda \quad \text{for } 0<A<1\ ,\quad 0<\lambda<2\pi \tag{1.40}$$

The H.H. $e^{i\nu\lambda}$ (ν integer) is associated to $d\lambda$ and the H.P. $(r^2 A e^{\pm i\lambda})^{|\nu|}$ is of degree $2|\nu|$. The polynomial in A is associated with the weight function

$$(\cos\psi)^{2|\nu|+1}\, d(\cos\psi) \quad \text{where } A = \cos\psi \quad \text{and} \quad 0 < \psi < \pi/2.$$

With $\cos 2\psi = z$ we obtain the weight function $(1+z)^{|\nu|}$ associated with the Jacobi polynomials $P_n^{0,|\nu|}(z)$. Finally the S harmonics become

$$Y_{2K}^{(\nu)}(A,\lambda) = (-1)^n N_K^{(\nu)}\, e^{i\nu\lambda} (\cos\psi)^{|\nu|} P_n^{0,|\nu|}(\cos 2\psi) \tag{1.41}$$

$$= N_K^{(\nu)}\, e^{i\nu\lambda} A^{|\nu|} P_n^{|\nu|,0}(1-2A^2)$$

where the grand orbital is

$$L = 2K = 2(2n + |\nu|)$$

and the normalization constant is

$$N_K^{(\nu)} = \sqrt{(K+1)/\pi^3}$$

According to the permutation property of λ the combinations

$$\frac{1}{\sqrt{2}}\left(Y_{2K}^{(\nu)}(A,\lambda) \pm Y_{2K}^{(\nu)*}(A,\lambda)\right) \tag{1.42}$$

are respectively symmetric (+) and antisymmetric (-) when $\nu = 3n$ (n integer) for any exchange of two coordinates $\vec{x}_i$. The other values of ν are related to mixed symmetry states. Instead of the coordinates (A,λ) let us use the angular coordinates (ϕ,φ). The surface element becomes :

$$d\Omega = \pi^2 A dA d\lambda = \pi^2 A \begin{vmatrix} \frac{\partial A}{\partial \phi} & \frac{\partial A}{\partial \varphi} \\ \frac{\partial \lambda}{\partial \phi} & \frac{\partial \lambda}{\partial \varphi} \end{vmatrix} d\phi d\varphi \tag{1.43}$$

$$= 8\pi^2 (\sin\phi\cos\phi)^2\, d\phi\, \sin\varphi d\varphi$$

The Legendre polynomials $P_\ell(\cos\varphi)$ are associated with $\sin\varphi d\varphi$ and generate the homogeneous polynomials

$$(\xi_1 \cdot \xi_2)^\ell P_\ell(\cos\varphi) = r^{2\ell}(\sin\phi\cos\phi)^\ell P_\ell(\cos\varphi)$$

The orthonormal equation contains the weight function $(\sin\phi\cos\phi)^{2(\ell+1)}\, d\phi = \frac{1}{4}\left(\frac{1-z}{2}\right)^{\ell+\frac{1}{2}} dz$ for $z = \cos 2\phi$ associated with the Gegenbauer polynomials $C_n^{\ell+1}(\cos 2\phi)$. One deduces the corresponding H.H. :

$$Y_{2K,\ell}(\phi,\varphi) = N_{K,\ell}(\sin 2\phi)^\ell C_{K-\ell}^{\ell+1}(\cos 2\phi) P_\ell(\cos\varphi) \tag{1.44}$$

where $N_{K,\ell}$ is the norm.

By using the Z.B. basis with $\ell = m = 0$, the addition theorem, and the relation between Jacobi and Gegenbauer polynomials one proves easily that

$$Y_{2K,\ell}(\phi,\varphi) = Y_{2K,\ell,\ell}^{00}(\Omega_{12,3}) \tag{1.45}$$

The S basis and the Z.B. basis can be transformed into each other by

$$Y_{2K}^{(\nu)}(A,\lambda) = \sum_{\ell=0}^{K} \langle \ell|\nu\rangle_K\, Y_{2K,\ell}(\phi,\varphi) \tag{1.46}$$

where $\langle \ell|\nu\rangle_K$ is a coefficient.

- <u>H.H. in 2N Dimensional Space</u> -

When we have to deal with particles in two dimensional space like for particles moving in a plane we define H.H. in 2N dimensional spaces by using the volume element

$$dV = \prod_{i=1}^{N} d^2\xi_i = \prod_{i=1}^{N} \xi_i d\xi_i d\varphi_i \tag{1.47}$$

$$= r^{2N-1} dr\, d\varphi_1 \prod_{i=2}^{N} (\sin\phi_i)^{2i-3} \cos\phi_i d\phi_i d\varphi_i$$

where (ξ_i, φ_i) are the polar coordinate of $\vec{\xi}_i$ and where ϕ_i is related to the ξ_j by (1.22). The harmonics $e^{im_i\varphi_i}/\sqrt{2\pi}$ are associated with $d\varphi_i$ and by using the orthonormal equation one finds a H.H. basis similar to the Z.B. basis

$$Y_{[L]}(\Omega) = \frac{e^{im_1\varphi_1}}{\sqrt{2\pi}} \prod_{i=2}^{N} \frac{e^{im_i\varphi_i}}{\sqrt{2\pi}} \, {}^{(i)}P_{n_i}^{m_i, L_{i-1}}(\phi_i) \tag{1.48}$$

where

$${}^{(i)}P_{n_i}^{m_i, L_{i-1}}(\phi_i) = N_{n_i}^{m_i, L_{i-1}} (\sin\phi_i)^{L_{i-1}} (\cos\phi_i)^{m_i} P_{n_i}^{L_{i-1}+i-2, m_i}(\cos 2\phi_i) \tag{1.49}$$

$$L_i = L_{i-1} + 2n_i + m_i\,, \quad n_1 = 0,\ L_1 \equiv m_1\,,$$

$$L = m_1 + \sum_{i=2}^{N} (2n_i + m_i) \tag{1.50}$$

$$(N_{n_i}^{m_i, L_{i-1}})^{-2} = 2^{L_{i-1}+m_i+i-2}\, h_{n_i}^{L_{i-1}+i-2, m_i}$$

where $h_n^{\alpha,\beta}$ is the normalization constant of the Jacobi polynomial $P_n^{\alpha,\beta}$.

- H.H. in 4 Dimensions -

Kilpatrick and Larsen introduced symmetrical angular coordinates in 4 dimensional space and defined the two vectors $\vec{\xi}$ and $\vec{\eta}$ in two dimensional space by / 6 /

$$\begin{aligned} \xi_x &= r[\cos\theta\sin\phi\cos\psi - \sin\theta\cos\phi\sin\psi] \\ \xi_y &= r[\cos\theta\sin\phi\sin\psi + \sin\theta\cos\phi\cos\psi] \\ \eta_x &= r[\cos\theta\cos\phi\cos\psi + \sin\theta\sin\phi\sin\psi] \\ \eta_y &= r[\cos\theta\cos\phi\sin\psi - \sin\theta\sin\phi\cos\psi] \end{aligned} \tag{1.51}$$

where $\vec{\eta}$ is obtained from $\vec{\xi}$ by the rotation $\phi \to \phi + \pi$.

The corresponding volume element is

$$dV = r^3\, dr\, d\Omega$$

$$d\Omega = \cos 2\theta\, d\theta\, d\phi\, d\psi = \frac{1}{2} d\phi\, d\psi\, dx \tag{1.52}$$

for $0 < \phi < \pi$ $\quad 0 < \psi < 2\pi \quad -\pi/4 < \theta < \pi/4$ and $-1 < x = \sin 2\theta < 1$

The eigenfunctions for $d\phi d\psi$ are $e^{i\nu\phi}\, e^{i\lambda\psi}$ where ν and λ are integers.
We define the new coordinates

$\phi_\pm = \phi \pm \psi$ and the linear combinations

$\eta_x \pm \xi_y = r[\cos\theta \pm \sin\theta]\cos\phi_\mp$

$\eta_y \pm \xi_x = \pm r[\cos\theta \mp \sin\theta]\sin\phi_\pm$

where $\cos\theta \pm \sin\theta > 0$ for $-\pi/4 < \theta < \pi/4$.

We use $\cos\theta \pm \sin\theta = [(\cos\theta \pm \sin\theta)^2]^{\frac{1}{2}} = (1 \pm x)^{\frac{1}{2}}$.
to prove that

$$r(1+x)^{\frac{1}{2}} e^{i\phi_-} \quad \text{and} \quad r(1-x)^{\frac{1}{2}} e^{i\phi_+}$$

are polynomials of degree one in $\vec{\xi}$ and $\vec{\eta}$, therefore the products

$$[r(1-x)^{\frac{1}{2}}e^{i\phi_+}]^{\frac{|\lambda+\nu|}{2}} \cdot [r(1+x)^{\frac{1}{2}}e^{i\phi_-}]^{\frac{|\lambda-\nu|}{2}} = h_{\lambda,\nu}(\vec{\xi},\vec{\eta})$$

are H.P. in $\vec{\xi}$ and $\vec{\eta}$.
The H.H. basis can be written as the product

$$Y_L^{\nu\lambda}(\Omega) = h_{\nu,\lambda}(\vec{\xi},\vec{\eta})\, P_n(\vec{x})_{/r=1}$$

where $Y_L^{\nu\lambda}(\Omega)$ fulfil the orthonormal equation

$$\int (Y_L^{\nu\lambda}(\Omega))^* Y_{L'}^{\nu'\lambda'}(\Omega)\, d\Omega = \delta_{LL'}\,\delta_{\nu'\nu}\delta_{\lambda'\lambda}$$

leading to

$$\frac{1}{2}\int_{-1}^{1} P_n(x)P_{n'}(x)\,dx \int_{(r=1)} |h_{\nu,\lambda}|^2\, d\phi d\psi = \delta_{nn'}$$

and to

$$\int_{-1}^{1} P_n(x)P_{n'}(x)(1-x)^{\frac{|\lambda+\nu|}{2}}(1-x)^{\frac{|\lambda-\nu|}{2}}\,dx = \frac{1}{\pi^2}\,\delta_{nn'}$$

The $P_n(x)$ are the Jacobi polynomials associated with the weight function $(1-x)^{|\lambda+\nu|/2}(1+x)^{|\lambda-\nu|/2}$ and the normalised solution is

$$P_n(x) = [\pi\sqrt{h_n^{\alpha,\beta}}]^{-1}\, P_n^{\alpha,\beta}(x)$$

where $\alpha = \frac{|\lambda+\nu|}{2}$, $\beta = \frac{|\lambda-\nu|}{2}$ and $h_n^{\alpha,\beta}$ is the normalisation constant of the Jacobi polynomial $P_n^{\alpha,\beta}(x)$.

The grand orbital is $L = 2n + \max(|\lambda|, |\nu|)$. The Kilpatrick-Larsen (K.L.) basis

$$Y_L^{\nu,\eta}(\Omega) = N_L^{\nu,\eta}\, e^{i\frac{\lambda+\nu}{2}\phi_+}\, e^{i\frac{\lambda-\nu}{2}\phi_-} \tag{1.53}$$

$$\cdot\,(1-x)^{|\lambda+\nu|/4}(1+x)^{|\lambda-\nu|/4}\, P_n^{\frac{|\lambda+\nu|}{2},\frac{|\lambda-\nu|}{2}}(x) \quad , \quad x = \sin 2\theta$$

where $N_L^{\nu\lambda}$ is a normalisation constant can be written as a linear combination of the previously defined $Y_{L,m_1,m_2}(\Omega)$ basis in the 2N dimensional space for N = 2.

We can also apply the Simonov procedure to derive a S basis including the symmetry in the 4 dimensional space . We introduce the complex vectors (see (1.34))

$$\vec{z} = \vec{\xi} + i\vec{\eta} \text{ and } \vec{z}^* = \vec{\xi} - i\vec{\eta} \tag{1.54}$$

for which (1.33), (1.35), (1.36) and (1.38) hold again. Here only the angular coordinates θ, ϕ occur in $z^2 = r^2\cos 2\theta\, e^{-2i\phi}$ and according to (1.38)

$$P_{12}\phi = -\phi \quad P_{13}\phi = -\phi-\pi/3 \quad P_{23}\phi = -\phi+\pi/3 \tag{1.55}$$

must be substituted for (1.39).

The coordinate ψ does not occur in the surface element

$$d\Omega = \cos 2\theta\, d\theta\, d\phi = \frac{1}{2}\, dx\, d\phi \tag{1.56}$$

$$-1 < x = \sin 2\theta < 1 \text{ for } -\frac{\pi}{4} < \theta < \frac{\pi}{4}\,.$$

The polynomials $(r^2\cos 2\theta\ e^{\pm 2i\phi})^{|\nu|}$, where ν is an integer, are associated with $d\phi$. The Gegenbauer polynomials $C_n^{|\nu|+\frac{1}{2}}(x)$ are associated with the weight function $(\cos 2\theta)^{2|\nu|}\,dx = (1-x^2)^{|\nu|}dx$ leading to the H.H.

$$Y_{2K}^{(\nu)}(\Omega) = N_K^{(\nu)}(1 - x^2)^{|\nu|/2}\, C_{K-|\nu|}^{|\nu|+\frac{1}{2}}(x)\, e^{i2\nu\phi} \tag{1.57}$$

for the grand orbital $L = 2K$ where $N_K^{(\nu)}$ is a normalisation constant. The combinations $Y_{2K}^{(\nu)*} \pm Y_{2K}^{(\nu)}$ are respectively symmetric (+) and antisymmetric (-), for any exchange of two coordinates $\vec{x}_i$ when $\nu = 3m$, m integer.

- Antisymmetric Harmonic Polynomials -

When we have to deal with a system of bosons, all particles can be in the 1S state and the H.H. for L = 0 is included in the expansion of the wave function for S states, but when we have to deal with a system of A identical fermions, only two particles with opposite spin can be in the 1S state according to the Pauli exclusion principle, therefore $Y_{[0]}(D)$ is not allowed in the expansion of S states for A>2.

In order to generate antisymmetric H.P. we define the individual polynomials

$$p_{n_j \ell_j m_j}(\vec{x}_i) = x_i^{2n_j+\ell_j}\, Y_{\ell_j,m_j}(\omega_i)$$

Then we construct a Slater determinant where the element of the ith row and j^{th} column is $s_j^i\, p_{n_j\ell_j m_j}(\vec{x}_i)$ where s_j^i is the spin state of the particle (i) with $j = 1/2$ for spin up and $j = -1/2$ for spin down.

If we construct the determinant in such a way that for a given spin (up or down) and for fixed $\ell_j = \ell_\alpha$ and $m_j = m_\alpha$ all the quantum numbers n_j are used from $n_j = 0$ up to a maximal value $n_j = N(\ell_\alpha, m_\alpha)$ defined independently for each set ℓ_α, m_α, like in the following determinant where the first row only is written :

$$H_{[L]}(\vec{x}) = |\ldots s_{\frac{1}{2}}^1\, p_{0\ell_\alpha m_\alpha}(\vec{x}_1)\; s_{\frac{1}{2}}^1\, p_{1\ell_\alpha m_\alpha}(\vec{x}_1)\; \ldots\; s_{\frac{1}{2}}^1\, p_{N(\ell_\alpha,m_\alpha)\ell_\alpha m_\alpha}(\vec{x}_1)\; \ldots| \tag{1.58}$$

we obtain a H.P..

The determinant could have been constructed in j-j coupling with the polynomials $p_{n,\ell,j,m}(\vec{x}_i)$ where the spin is combined with the orbital to generate an angular momentum j. The rule for n holds for any fixed ℓ, j, m.

The elements $p_{N(\ell,m)\ell m}(\vec{x}_i)$ constitute the set of highest "occupied states" $n_j \ell_j m_j$ for all the ℓ, m used in the determinant. It is easy to show that $H_{[L]}(\vec{x})_N$ constructed according to the previous rule is a H.P. of degree $L = \sum_{i<j<A} (2n_j+\ell_j)$ where n_j, ℓ_j refer to the quantum numbers in the jth column. Indeed by applying the Laplace operator to the jth column of $H_{[L]}(x)$ we obtain for each i = 1, 2, ... , A

$$\nabla^2 x_i^{2n_j+\ell_j}\, Y_{\ell_j,m_j}(\omega_i) = 2n_j(2n_j + 2\ell_j + 1)\, x_i^{2n_j+\ell_j-2}\, Y_{\ell_j,m_j}(\omega_i) \tag{1.59}$$

and we generate for $n_j \neq 0$ a column proportional to the previous one cancelling the determinant.

The exchange of two particles permutes two rows in $H_{[L]}(x)$ which is therefore antisymmetric.

Let us denote $H_{[L_m]}(\vec{x})$ an homogeneous polynomial of lowest degree L_m antisymmetric in any exchange of two particles.

It is a H.P. because the (symmetrical) Laplace operator acting on an homogeneous polynomial of degree L_m generates another homogeneous polynomial of degree L_m-2 preserving the symmetry and since homogeneous antisymmetric polynomials of degree lower than L_m do not exist

$$\nabla^2 H[L_m](\vec{x}) = 0 \quad \text{and } H[L_m](\vec{x}) \text{ is harmonic.} \tag{1.60}$$

For the same reason $H[L_m](\vec{x})$ is translationaly invariant since

$$\sum_{i=1}^{A} \vec{\nabla}_i \, H[L_m](\vec{x}) = \vec{\nabla}_X H[L_m](\vec{x}) = 0 \tag{1.61}$$

where $\vec{\nabla}_X$ is the momentum operator of the center of mass $\vec{X}$. Translationaly invariant H.P. are denoted by $H_{[L]}(\vec{\xi})$ where ξ refers to a translationaly invariant set of coordinates, for instance the Jacobi one. If we substitute the individual harmonic oscillator (H.O.) eigenfunctions

$$\varphi_{n\ell m}(\vec{x}_i) = Y_{\ell,m}(\omega_i)\left[\frac{2n!}{\Gamma(n+\ell+3/2)b^3}\right]^{1/2}\left(\frac{x_i}{b}\right)^{\ell} L_n^{\ell+1/2}\left(\frac{x_i^2}{b^2}\right) e^{-x_i^2/2b^2} \tag{1.62}$$

where, L_n^{α} is a Laguerre polynomial, for the $p_{n\ell m}(\vec{x}_i)$ in $H_{[L]}(\vec{x})$ we obtain

$$D_{A[L]}^{H.O}(\vec{x}) = H_{[L]}(\vec{x})\, e^{-r^2/4b^2}\, e^{-AX^2/2b^2} \tag{1.63}$$

where $r^2 = 2\sum_{i=1} (\vec{x}_i - \vec{X})^2$ is the hyperradius and b is the H.O. parameter. Notice that $H[_L](\vec{x})$ is normalised here such that

$$\int |D_{[L]}^{H.O.}(\vec{x})|^2 d^3x_1 \ldots\ldots d^2x_A = A! \tag{1.64}$$

The relation (1.63) is a consequence of the cancellation of a determinant with two proportional columns. This relation between H.O. states and H.P. proceeds from the hypercentral nature of the H.O. potential

$$\sum_{i=1}^{A} x_i^2 = AX^2 + \sum_{i=1}^{A} (x_i - X)^2 = A X^2 + r^2/2 \tag{1.65}$$

Let $H_{[L]}^{J,M}(\vec{x})$ be a H.P. of total angular momentum J. Any $H_{[L]}^{J,M}(\vec{x})$ can be expanded in terms of the spherical harmonics of the center of mass

$$H_{[L]}^{J,M}(\vec{x}) = \sum_{\ell=0}^{L-L_m} X^{\ell} \sum_{m,\mu} \langle \lambda, \ell; \mu, m | J, M \rangle Y_{\ell,m}(\omega_X) H_{[L-\ell]}^{\lambda,\mu}(\vec{\xi}) \tag{1.66}$$

where $\langle \lambda, \ell; \mu m | J, M \rangle$ is a Clebsh-Gordan coefficient. If $H_{L_m}^{J,M}(\vec{x})$ is an homogeneous polynomial of minimal degree L_m for the total angular momentum J it is a H.P. because the Laplace operators ∇^2 preserves J and M then

$$\nabla^2 H_{[L_m]}^{J,M}(\vec{x}) = 0 \tag{1.67}$$

II. The Potential Harmonic Basis

The purpose of the H.H.E.M. is to solve the Schrödinger equation written here for equal mass particles

$$(T + V(\vec{x}) - E)\Psi(\vec{x}) = 0 \qquad T = -\frac{h^2}{2m}\nabla^2 \tag{2.1}$$

where $V(\vec{x})$ can be any many-body potential. For particles with different masses the Schrödinger equation can be written like (2.1) by introducing a suitable set of renormalised coordinates generating a symmetric Laplace operator ∇^2. In the H.H.E.M. the wave function is expanded a serie of H.H.

$$\Psi(\vec{x}) = \sum_{[L]=0}^{\infty} Y_{[L]}(\Omega)\, u_{[L]}(r)/r^{(D-1)/2} \tag{2.2}$$

and (2.1), projected on the same basis, is converted into an infinite set of second order coupled differential equations defining $u_{[L]}(r)$:

$$\{\frac{h^2}{m}[-\frac{d^2}{dr^2}+\frac{\mathcal{L}(\mathcal{L}+1)}{r^2}]-E\}u_{[L]}(r)+\sum_{[L']=0}^{\infty} V_{[L]}^{[L']}(r)\,u_{[L']}(r)=0 \qquad (2.3)$$

where $\mathcal{L} = L + (D - 3)/2$ and where

$$V_{[L]}^{[L']}(r) = \langle Y_{[L]}|V(\vec{x})|Y_{[L']}\rangle = \int Y^*_{[L]}(\Omega)\,V(\vec{x})Y_{[L']}(\Omega)\,d\Omega \qquad (2.4)$$

is the potential matrix.

The sum is taken over all quantum numbers $[L']$ for L' running from zero to infinity. The expansion (2.2) must be truncated to a maximal value L_{max} in order to generate a finite number of coupled equations (2.3) subsequently treated numerically.

But according to (1.27) with D = 3(A-1) the number N(L) of independent H.P. of degree L increases dramatically with A the number of particles. When the H.H. expansion of the wave function in truncated to L_{max} the number of coupled equations to be solved becomes rapidely too large for increasing A to be computed and the problem becomes untractable. A selection of the most significant H.H. occuring in the expansion basis becomes unavoidable as soon as more than few-body systems are considered.

Here we study the case where the interaction

$$V(\vec{x}) = \sum_{i<j<A} V(\vec{r}_{ij}) \;,\; \vec{r}_{ij} = \vec{x}_i - \vec{x}_j \qquad (2.5)$$

is a sum of pair wise local potentials. We assume for the sake of simplicity that we have to deal with a system of identical particles where $V(\vec{r}_{ij})$ is the same for any pair (i,j). The problem can be generalised easily to different kinds of particles. Let

$$A(A-1)/2\,V_0(r) = \int V(\vec{x})\,|Y_0(D)|^2\,d\Omega \qquad (2.6)$$

be that part of $V(\vec{x})$ invariant by rotation in the D dimensional space and let us write the wave function

$$\Psi(\vec{x}) = \sum_{i,j>i} \psi_{ij}(\vec{x}) \qquad (2.7)$$

as a sum, where $\psi_{ij}(\vec{x})$ is defined independly for each pair (i,j).

When $\psi_{ij}(\vec{x})$ is a solution of equations

$$(T + \frac{A(A-1)}{2}V_0(r) - E)\,\psi_{ij}(\vec{x}) = -\,[V(\vec{r}_{ij}) - V_0(r)]\,\Psi(\vec{x}) \qquad (2.8)$$

one finds by summing over all pairs (i,j) that $\Psi(\vec{x})$ given by (2.7) is a solution of the Schrödinger equation (2.1).

The residual interaction $V(\vec{r}_{ij}) - V_0(r)$ disappears for a H.O. two body potential r_{ij}^2 according to (1.19).

In the first order approximation / 7 / we assume that the residual interaction is negligible and we find that the solution

$$\psi_{ij}(\vec{x}) = H_{[L]}(\vec{x})\phi(r),\; \phi(r) = u_{L,n}(r)/r^{\mathcal{L}+1} \qquad (2.9)$$

is the product of a suitably symmetrized H.P. which define the state and a function of r where $u_{L,n}(r)$ is a solution of the radial equation

$$\{\frac{h^2}{m}\Big[-\frac{d^2}{dr^2}+\frac{\mathcal{L}(\mathcal{L}+1)}{r^2}\Big]+\frac{A(A-1)}{2}V_0(r)-E_{L,n}\}\,u_{L,n}(r)=0 \qquad (2.10)$$

where n, the number of nodes of the radial wave, is finite for bound states.

In this approximation the particles move in a common central well $A(A-1)/2\ V_0(r)$ in the D dimensional space.

The solution (2.9) does not contain any correlation since it is not sensitive to the distance r_{ij} between two particles. This property is well known for the H.O. interaction where $\Psi(\vec{x})$ can be written either like (2.9) where $\phi(r)$ is a gaussian function of r or as a sum of product of individual H.O. states (1.62). The correlations are generated by the residual interactions.

Assuming that $V(\vec{r}_{ij})-V_0(r)$ is small we substitute the first order solution (2.9) for Ψ in (2.8) to find

$$(T + \frac{A(A-1)}{2} V_0(r) - E)\psi_{ij}(\vec{x}) = -[V(\vec{r}_{ij}) - V_0(r)]\,A(A-1)/2\ H_{[L]}(\vec{x})\phi(r) \tag{2.11}$$

In this approximation the partial wave ψ_{ij} becomes the product

$$\psi_{ij}(\vec{x}) = H_{[L]}(\vec{x})\, F(\vec{r}_{ij}, r) \tag{2.12}$$

of a H.P., which define the state, and a function $F(\vec{r}_{ij}, r)$, describing the correlations, solution of

$$(T + \frac{A(A-1)}{2} V_0(r) - E)\, H_{[L]}(\vec{x})\, F(\vec{r}_{ij}, r) \tag{2.13}$$

$$= -[V(\vec{r}_{ij}) - V_0(r)]\, H_{[L]}(\vec{x}) \sum_{k,\ell>k} F(\vec{r}_{k\ell}, r)$$

obtained by substituting (2.12) for $\psi_{ij}(\vec{x})$ and the corresponding expression for the other pairs in (2.8).

For solving (2.13) we expand $H_{[L]}(\vec{x})\, F(\vec{r}_{ij}, r)$ in a serie of H.H. and we project this equation on the same basis for generating a system of coupled equations (2.3) computed numerically.

- <u>Bosons in Ground State</u> -

Let us begin by the simplest problem : the one of solving the A bosons ground state with a local central two-body interaction. The ground state in the first order approximation is described by a H.P. of minimal degree L_m / 7 /. The proof is the following : Let us assume that $H_{[L]}(\vec{x})\, u(r)/r^{(\ +1)}$ is the ground state wave function and $H_{[L_m]}(\vec{x})\, u(r)/r^{(\ _m+1)}$ is a trial function. The difference between the energy calculated with $H_{[L]}(\vec{x})$ and the one calculated with $H_{[L_m]}(\vec{x})$ is according to (2.10)

$$E_L - E_{L_m} = \frac{h^2}{m}[\mathcal{L}(\mathcal{L}+1) - \mathcal{L}_m(\mathcal{L}_m+1)] \int^\infty (u(r)/r)^2 dr$$

therefore $E_L > E_{L_m}$: The ground state is reached when the degree L of the H.P. is minimal ($L = L_m$) according to the Rayleigh-Ritz principle. For a system of bosons where all particles can be in the 1 S state, $L_m = 0$ and $H_{[0]}(\vec{x}) = Y_0(D)$.

In order to expand $F(\vec{r}_{ij}, r)$ we need a H.H. basis complete for the expansion of any function of $\vec{r}_{ij}$ only.

Let us choose a system of Jacobi coordinates similar to (1.18) but with $\vec{\xi}_N = \vec{r}_{ij}$. The required H.P. basis does not depend on $\vec{\xi}_j$ for $j \neq N$ and can be written

$$H_{[L]}(\vec{r}_{ij}, r) = Y_{[0]}(D-3)\, P_\lambda(\vec{r}_{ij}, r)$$

where λ is a set of 3 quantum numbers for the three degrees of freedom $\vec{\xi}_N$. By using the polar coordinates (r_{ij}, ω_{ij}) of $\vec{r}_{ij}$ we write

$$P_\lambda(\vec{r}_{ij}, r) = r_{ij}^{\ell}\, Y_{\ell,m}(\omega_{ij})\, P_n(r_{ij}, r)$$

where the polynomial P_n is obtained from the orthonormal equation

$$\int_0^{\pi/2} P_n(\cos\phi_N)\,P_{n'}(\cos\phi_N)(\sin\phi_N)^{D-4}\cos^{2(\ell+1)}\phi_N\,d\phi_N = \delta_{nn'}$$

with $r_{ij} = r\cos\phi_N$ and $0<\phi_N<\pi/2$.
We use $z = \cos 2\phi_N = 2r_{ij}^2/r^{2} - 1$ do find the "Potential Harmonic" (P.H.) basis :

$$\mathcal{P}^{\ell,m}_{2K+\ell}(\Omega_{ij}) = Y_{[0]}(D-3)\,Y_{\ell,m}(\Omega_{ij})\;{}^{(N)}P_K^{\ell,0}(\phi_N) \tag{2.14}$$

where ${}^{(N)}P_K^{\ell,0}(\phi_N)$ is given by (1.24) with

$$\cos\phi_N = r_{ij}/r,\ \ell_N = \ell,\ L_{N-1} = 0,\ n_N = K$$

The grand orbital is $L = 2K+\ell$. The P.H. are normalised to

$$\int(\mathcal{P}^{\ell,m}_{2K+\ell}(\Omega_{ij}))^*\,\mathcal{P}^{\ell,m}_{2K'+\ell'}(\Omega_{ij})\,d\Omega = \delta_{KK'}\delta_{\ell\ell'}\delta_{mm'} \tag{2.15}$$

The P.H. are obtained from the Z.B. basis by setting in (1.23) $n_j = \ell_j = m_j = L_j = 0$ for $j<N$ and $\ell_N = \ell$, $m_N = m$, $n_N = K$, $\omega_N = \omega_{ij}$ and $\cos\phi_N = r_{ij}/r$. For a fixed angular momentum (ℓ,m) there is only one P.H. for each grand orbital $2K+\ell$. We reduced the degeneracy of the H.H. basis to only one H.P. of degree $2K+\ell$ for each fixed (ℓ,m) but we have to keep in mind that the P.H. constitutes a complete basis for the expansion of any function of $\vec{r}_{ij}$ and r only.

This basis can also be used in the expansion of any function $F(\vec{z},r)$ where $\vec{z}$ is a three dimensional vector, for instance a linear combination of Jacobi coordinates $\vec{\xi}_i$.

Let us define the "kinematic rotation vector"

$$\vec{z}(\varphi) = \vec{\xi}_N\cos\varphi_N + \vec{\xi}_{N-1}\sin\varphi_N\cos\varphi_{N-1} + \dots + \vec{\xi}_{N-i+1}\sin\varphi_N\sin\varphi_{N-1} \tag{2.16}$$

$$\dots\sin\varphi_{N-i+2}\cos\varphi_{N-i+1} + \dots + \vec{\xi}_1\sin\varphi_N\dots\sin\varphi_2$$

where φ is the set of parameters $\varphi_2, \varphi_3, \dots\varphi_N$.

The associated P.H.

$$\mathcal{P}^{\ell,m}_{2K+\ell}(\Omega_z) = Y_{[0]}(D-3)\,Y_{\ell,m}(\omega_z)\;{}^{(N)}P_K^{\ell,0}(\phi_z) \tag{2.17}$$

where $|\vec{z}(\varphi)| = z(\varphi) = r\cos\phi_z$, is a function of the angular parameters φ. If we choose the Jacobi coordinates (1.18) there is a set of parameters φ^{ij} such that for any pair (i,j)

$$\vec{z}(\varphi^{ij}) = \vec{r}_{ij} \tag{2.18}$$

The P.H. for two linear combinations $\vec{z}(\varphi)$ and $\vec{z}(\varphi') = \vec{z}'$ are not orthogonal and we have

$$\mathcal{P}^{\ell,m}_{2K+\ell}(\Omega_{z'}) = \langle z|z'\rangle_{K,\ell}\,\mathcal{P}^{\ell,m}_{2K+\ell}(\Omega_z) + p_{K\ell}\,\mathcal{P}^{\ell,m}_{2K+\ell}(\Omega_{zz'}) \tag{2.19}$$

where

$$\langle z|z'\rangle_{K,\ell} = \int\mathcal{P}^{\ell,m*}_{2K+\ell}(\Omega_z)\,\mathcal{P}^{\ell,m}_{2K+\ell}(\Omega_{z'})\,d\Omega \tag{2.20}$$

is an overlapp coefficient.

The residual part $\mathcal{P}^{\ell,m}_{2K+\ell}(\Omega_{zz'})$ is a H.H. orthogonal to $\mathcal{P}^{\ell,m}_{2K+\ell}(\Omega_z)$.

To prove it we multiply (2.19) by $r^{2K+\ell}$ to generate the H.P.$\,r^{2K+\ell}\,\mathcal{P}^{\ell,m}_{2K+\ell}(\Omega_z)$ and the same for z' and we apply the Laplace operator to find

$$\nabla^2 r^{2K+\ell}\,\mathcal{P}^{\ell,m}_{2K+\ell}(\Omega_{zz'}) = 0$$

If the residual part $\mathcal{P}^{\ell,m}_{2K+\ell}(\Omega_{zz'})$ is normalised the coefficient $p_{K\ell}$ is

$$p_{K\ell} = [\,1 - (\langle z|z'\rangle_{K\ell})^2]^{\frac{1}{2}} \tag{2.21}$$

Let $\vec{z}$ refers to the case $\varphi_N = 0$ (i.e. $\vec{z} = \vec{\xi}_N$) the overlapp coefficient of $\mathcal{P}^{\ell,m}_{2K+\ell}(\Omega_z)$ and $\mathcal{P}^{\ell,m}_{2K+\ell}(\Omega_{z'})$ is

$$\langle z|z'\rangle_{K\ell} = (\cos\varphi_N)^{\ell}\; P_K^{(D-5)/2,\ell+\frac{1}{2}}(\cos 2\varphi_N)/\,P_K^{(D-5)/2,\ell+\frac{1}{2}}(1) \tag{2.22}$$

where φ_N refers to $\vec{z}' = \vec{\xi}_N \cos\varphi_N + \dots$. A particularly important case refers to $\vec{\xi}_N = \vec{r}_{ij}$ and $\vec{z}(\varphi^{p,q}) = \vec{r}_{pq}$ for which

$$\mathcal{P}^{\ell,m}_{2K+\ell}(\Omega_{pq}) = \langle ij|pq\rangle_{K\ell}\,\mathcal{P}^{\ell,m}_{2K+\ell}(\Omega_{ij}) + p_{K\ell}\,\mathcal{P}^{\ell,m}_{2K+\ell}(\Omega_{ij,k\ell}) \tag{2.23}$$

where $\langle ij|pq\rangle_{K\ell}$ is given by (2.22) with $\varphi_N = \varphi_N^{pq}$.

Assuming that the approximations leading to (2.13)are good, the wave function describing the ground state of a system of A identical bosons with a local central two-body interaction is

$$\Psi(\vec{x}) = \sum_{i,j>i} F(r_{ij},r) \tag{2.24}$$

where $F(r_{ij},r)$ is a solution of

$$(T + \frac{A(A-1)}{2}V_0(r) - E)F(r_{ij},r) = -\,[V(r_{ij}) - V_0(r)]\Psi(\vec{x}) \tag{2.25}$$

For solving (2.25) we expand $F(r_{ij}, r)$ in a serie of P.H.

$$F(r_{ij},r) = \sum_K \mathcal{P}^0_{2K}(\Omega_{ij})\,u_K(r)/r^{(D-1)/2} \tag{2.26}$$

and we project (2.25) on the same basis. The P.H. basis is complete for the expansion of $_{2K}^{0}(\Omega_{ij})V(r_{ij})$ since it is a function of r_{ij} and r only, and $\Psi(\vec{x})$ given by (2.24) and (2.26) is projected on the $(\vec{r}_{ij})$ space for S states by using (2.23)

$$\Psi(\vec{x}) = \sum_K \{ \sum_{k,\ell>k} \langle ij|k\ell\rangle_{K0}\,\mathcal{P}^0_{2K}(\Omega_{ij}) \tag{2.27}$$

$$+ C_{K0}\,\mathcal{P}^{\perp}_{2K}(\Omega_{ij})\}\,u_K(r)/r^{(D-1)/2}$$

where $\mathcal{P}^{\perp}_{2K}(\Omega_{ij})$ is a H.H. orthogonal to any function of r_{ij} and r only and C_{K0} is a constant. The projection generates the system of coupled equations

$$\{\frac{h^2}{m}(-\frac{d^2}{dr^2} + \frac{\mathcal{L}_K(\mathcal{L}_K+1)}{r^2}) + \frac{A(A-1)}{2}V_0(r) - E\}u_K(r) \tag{2.28}$$

$$= -\sum_{K'} f_{K'}^2\, V_K^{K'}(r)\,u_{K'}(r) \quad , \quad \mathcal{L}_K = 2K + (D-3)/2$$

where

$$V_K^{K'}(r) = \int \mathcal{P}^0_{2K}(\Omega_{ij})[V(r_{ij}) - V_0(r)]\,\mathcal{P}^0_{2K'}(\Omega_{ij})\,d\Omega \tag{2.29}$$

is the potential matrix and where

$$f_K^2 = \sum_{k,\ell>k} \langle ij|k\ell\rangle_{K0} \tag{2.30}$$

$$= 1 + [2(A-2)P_K^{\alpha,\frac{1}{2}}(-\tfrac{1}{2}) + (A-2)(A-3)/2P_K^{\alpha,\frac{1}{2}}(-1)]/P_K^{\alpha,\frac{1}{2}}(1)$$

with $\alpha = (D-5)/2$, in agreement with (2.22) where $\varphi_N = \pi/3$ for connected pairs like (i,j) and (j,k), $(k \neq i)$, and $\varphi_N = \pi/2$ for disconnected pairs like (i,j) and (k, ℓ) with k and ℓ ≠ of i and j / 1 /.

The solution obtained

$$\Psi(\vec{x}) = \sum_K \Big[\sum_{i,j>i} \mathcal{P}^0_{2K}(\Omega_{ij}) \Big] u_K(r)/r^{(D-1)/2} \tag{2.31}$$

where $u_K(r)$ is a solutions of (2.28) is an approximation, but how close it is from the exact solution ? To answer we first notice that in (2.25) we introduced the hypercentral potential $V_0(r)$ to isolate $V(r_{ij}) - V_0(r)$,the part of the potential generating the correlations, but that whatever $V_0(r)$ is the sum of (2.25) over the pairs (i,j) reproduces the Schrödinger equation

$$(T + \sum_{i,j>i} V(r_{ij}) - E)\Psi(\vec{x}) = 0 \tag{2.32}$$

In our discussion we cancel $V_0(r)$ and we consider only the equation

$$(T - E)\psi_{ij}(\vec{x}) = -V(r_{ij})\Psi(\vec{x}) \tag{2.33}$$

where $\psi_{ij} = F(r_{ij},r)$ and $\Psi(\vec{x})$ is (2.24). By equating to zero the projection of (2.33) on the r_{ij} space we did not take, according to (2.27), the term

$$V(r_{ij})\, r^{-(D-1)/2} \sum_K C_{K0}\, \mathcal{P}^{\perp}_{2K}(\Omega_{ij}) u_K(r) \tag{2.34}$$

orthogonal to the potential basis $\mathcal{P}^0_{2K}(\Omega_{ij})$ into account in the right hand side of (2.33). The inaccuracy of (2.31) can originates from this term only.

- <u>Three Bosons in S State</u> -

If the Z.B. basis (1.45) is used for S states with the partition (ij,k), the P.H. basis is

$$\mathcal{P}^0_{2K}(\Omega_{ij}) = Y^{0\,0}_{2K,0,0}(\Omega_{ij,k})$$

The symmetrical combinations of P.H. can be expanded in a serie of Z.B. basis for S states

$$\sum_{k,\ell>k} \mathcal{P}^0_{2K}(\Omega_{k\ell}) = \sum_{\substack{\ell=0 \\ \ell\ \text{even}}}^{K} a^{\ell}_K\, Y^{00}_{2K,\ell,\ell}(\Omega_{ij,k}) = a^0_K \mathcal{P}^0_{2K}(\Omega_{ij}) + C_{K0}\, \mathcal{P}^{\perp}_{2K}(\Omega_{ij}) \tag{2.35}$$

where $a^0_K = f^2_K$ according to (2.27) and (2.30). When we have to deal with a S state projected local potential, (like for the nuclear Reid soft core potential / 8 /), the product $V(r_{ij})\mathcal{P}^{\perp}_{2K}(\Omega_{ij})$ is cancelled since $\mathcal{P}^{\perp}_{2K}(\Omega_{ij})$ contains only $\ell \neq 0$ states and the term (2.34) disappears. In this case the solution of (2.33) projected on the $\vec{r}_{ij}$ space for S states is the exact solution of the Schrödinger equation (2.32). The corresponding coupled equations are

$$\Big\{\frac{h^2}{m}\Big[-\frac{d^2}{dr^2} + \frac{\mathcal{L}_K(\mathcal{L}_K+1)}{r^2}\Big] - E\Big\}\, u_K(r) \tag{2.36}$$

$$+ \sum_{K'} \Big[1 + \frac{2}{K'+1}\,\frac{\sin 2(K'+1)\pi/3}{\sin 2\pi/3}\Big]\, V^{K'}_K(r)\, u_{K'}(r) = 0$$

where

$$V^{K'}_K(r) = \frac{1}{\pi}\int_0^{\pi}[\cos|K-K'|\theta - \cos(K+K'+2)\theta]\, V(r\cos\tfrac{\theta}{2})\, d\theta\ , \quad \mathcal{L}_K = 2K+3/2$$

We notice that $\sin 2(K+1)\pi/3 / \sin 2\pi/3 = 1, -1, 0$ respectively for K = 3n, 3n + 1, 3n + 2, n integer. With the same potential, but for more than three bosons in S state, some contributions can araise from the part of $\mathcal{P}^{\perp}_{2K}(\Omega_{ij})$ which describe the motion of 3 particles through the Jacobi coordinates $\vec{r}_{ij}$ and $\sqrt{3}(\vec{x}_k - \vec{X})$ where the orbital motion of both $\vec{r}_{ij}$ and $\vec{x}_k - \vec{X}$ is in S state.

- Symmetrical Potential Matrix -

The matrix $f^2_{K'}V^{K'}_K(r)$ in (2.28) is not symmetric. Numerical calculations are easier to perform with a symmetrical matrix which can be obtained by the following procedure. In (2.31) each P.H. is normalised but the symmetrical combination is not. In order to expand in terms of a normalised basis we substitute the new elements

$$B^{(S)}_K(\Omega) = C_K \sum_{i,j>i} \mathcal{P}^0_{2K}(\Omega_{ij}) \tag{2.37}$$

for the symmetric combination of P.H. in (2.31) where C_K is determined by

$$\int B^{(S)*}_K(\Omega)\, B^{(S)}_{K'}(\Omega)\, d\Omega = \delta_{KK'} \tag{2.38}$$

For this choice the normalisation constant is

$$C_K = \sqrt{2/(A(A-1))}\; / f_K \tag{2.39}$$

where f_K is given by (2.30).

The expansion of $F(r_{ij},r)$ becomes

$$F(r_{ij},r) = r^{-(D-1)/2} \sqrt{2/A(A-1)} \sum_K \frac{1}{f_K} \mathcal{P}^0_{2K}(\Omega_{ij}) u_K(r) \tag{2.40}$$

leading to the wave function

$$\Psi(\vec{x}) = \sum_{i,j>i} F(r_{ij},r) = \sum_K B^{(S)}_K(\Omega) u_K(r)/r^{(D-1)/2} \tag{2.41}$$

The norm $N_K = \int_0^\infty (u_K(r))^2 dr$ is the one of $B^{(s)}_K(\Omega)$ in the H.H. expansion (2.41) which is normalised when $\sum_{K=0}^{\infty} N_K = 1$.

By introducing (2.40) and (2.41) in (2.25) we obtain by projection on the P.H. basis

$$\left\{\frac{h^2}{m}\left[-\frac{d^2}{dr^2} + \frac{\mathcal{L}_K(\mathcal{L}_K+1)}{r^2}\right] + \frac{A(A-1)}{2} V_0(r) - E\right\} u_K(r) \tag{2.42}$$

$$= -\sum_{K'} U^{K'}_K(r)\, u_{K'}(r)$$

where the potential matrix

$$U^{K'}_K(r) = U^K_{K'}(r) = f_K f_{K'} \int_0^{\pi/2} {}^{(N)}P^{00}_{2K'}(\phi)\,[V(r\cos\phi) - V_0(r)] \tag{2.43}$$

$$\times\; {}^{(N)}P^{00}_{2K}(\phi)(\sin\phi)^{D-4} \cos^2\phi\, d\phi = f_K f_{K'} [h_K h_{K'}]^{-\frac{1}{2}} \int_{-1}^{1} P^{\alpha,\frac{1}{2}}_K(z)\, P^{\alpha,\frac{1}{2}}_{K'}(z)$$

$$\times\, [V(r\sqrt{\tfrac{1+z}{2}}) - V_0(r)](1-z)^\alpha (1+z)^{\frac{1}{2}} dz \;, \qquad \alpha = (D-5)/2$$

where h_K is the normalisation of the Jacobi polynomial $P^{\alpha,\frac{1}{2}}_K$, is a symmetrical matrix.

- Fermions in Ground State -

The P.H. expansion method can be applied to fermions in ground state, but this time the degree L_m of the H.P. describing the state is not zero.

The H.P. is constructed for identical fermions according to the rule generating H.P. with the structure (1.58) where the quantum numbers ℓ_α,

m_α, $N(\ell_\alpha, m_\alpha)$ are chosen to produce a determinant of degree minimum L_m.

When we operate in j-j coupling we have to use instead the quantum numbers ℓ_α, j_α, m_α, $N(\ell_\alpha, j_\alpha, m_\alpha)$.

The ground state is described by a single determinant for closed shell nuclei and by a sum of determinants for open shells. Any way a sum of H.P. is itself a H.P. therefore the notation $H[L_m](\vec{x})$ for the H.P. describing our state is used. It can be either a single or a sum of determinants (1.58) antisymmetric for any exchange of two fermions.

The P.H. are obtained as a solution of the orthonormal equation /1,9/

$$\int_{(r=1)} |H[L_m](\vec{x})|^2 P_{[\lambda]}(\vec{r}_{ij},r)P_{[\lambda']}(\vec{r}_{ij},r)d\Omega = \delta_{[\lambda],[\lambda']} \tag{2.44}$$

where $[\lambda]$ is a set of three quantum numbers. Let $\vec{r}_{ij} = \vec{\xi}_N$ be the last Jacobi coordinate of the set $(\vec{\xi}_1, \vec{\xi}_2, \dots, \vec{\xi}_N)$ similar to (1.18), we choose the hyperspherical coordinates

$$(\omega_{ij}, \phi_N ; \Omega_{N-1}) \tag{2.45}$$

where ω_{ij} are the spherical coordinates θ_{ij}, φ_{ij} of $\vec{r}_{ij}$ and Ω_{N-1} is the set of angular coordinates for $(\vec{\xi}_1, \dots, \vec{\xi}_{N-1})$ and ϕ_N is defined by

$$r_{ij} = r \cos\phi_N \tag{2.46}$$

The surface element becomes

$$d\Omega = 2^{-D/2}(1-z)^{(D-5)/2}(1+z)^{\frac{1}{2}} dz\, d\omega_{ij} d\Omega_{N-1} \tag{2.47}$$

where $z = \cos 2\phi_N$.

The polynomial $P_{[\lambda]}(\vec{r}_{ij}, r)$ on the unit hypersphere $r = 1$ is a function of ω_{ij} and z

$$P_{[\lambda]}(\vec{r}_{ij},r)/_{r=1} = P_{[\lambda]}(z,\omega_{ij}) \tag{2.48}$$

called "Potential Polynomial" (P.P.). In order to solve (2.44) we first integrate over $d\Omega_{N-1}$ to obtain the weight function

$$W[L_m](z,\omega_{ij}) = 2^{-D/2}(1-z)^{(D-5)/2}(1+z)^{1/2}\int_{(r=1)} |H\, L_m\, (\vec{x})|^2 d\Omega_{N-1} \tag{2.49}$$

The integral is performed by using the H.O., representation (1.63) of $H[L_m](\vec{x})$:

$$D^{H.O.}_{[L_m]}(\vec{x}) = H_{[L_m]}(\vec{x})\, e^{-r^2/4b^2}\, e^{-AX^2/2b^2}$$

where b is the H.O. parameter.

Let $\langle D^{H.O.}_{[L_m]}(\vec{x})|e^{i\vec{k}.\vec{r}_{ij}}|D^{H.O.}_{[L_m]}(\vec{x})\rangle$ be the Fourier transform for the relative coordinate $\vec{r}_{ij}$. We have shown / 10/ that if $([L_m]|n,\ell)$ are the coefficient in the expansion

$$\langle D^{H.O.}_{[L_m]}(\vec{x})|e^{i\vec{k}.\vec{r}_{ij}}|D^{H.O.}_{[L_m]}(\vec{x})\rangle = \sqrt{4\pi} \sum_{\substack{n \\ \ell\ \mathrm{even}}} (-1)^{\ell/2}([L_m]|n,\ell)\, Y_{\ell,0}(\omega_k)\, y^{2n+\ell}\, e^{-y^2} \tag{2.50}$$

where $y = kb/\sqrt{2}$ and (k,ω_k) are the polar coordinates of $\vec{k}$, the weight function is

$$W_{[L_m]}(z,\omega_{ij}) = (1-z)^{\mathcal{L}_m-2\ell_m-1}(1+z)^{\frac{1}{2}}\rho_{[L_m]}(z;\omega_{ij}) \tag{2.51}$$

where $\rho[L_m](z,\omega_{ij})$ is the polynomial of degree $2\ell_m$ in z :

$$\rho_{[L_m]}(z,\omega_{ij}) = 1/\pi\ \Gamma(\mathcal{L}_m+3/2)/2^{(\mathcal{L}_m+1/2)} \tag{2.52}$$

$$\sum_{\substack{n \\ \ell\ \text{even}}} \frac{(-2)^n\ n!}{\Gamma(\mathcal{L}_m-\ell/2)} ([L_m]|n,\ell)\ Y_{\ell,0}(\omega_{ij})(1+z)^{\ell/2}(1-z)^{2\ell_m-n-\ell/2}$$

$$P_n^{\mathcal{L}_m-n-\ell/2-1,\ell+1/2}(z)\ , \quad \mathcal{L}_m = L_m + (D-3)/2$$

The weight function is normalised to

$$\int W_{[L_m]}(z,\omega_{ij})\, dz\, d\omega_{ij} = 1 \tag{2.53}$$

when $D^{H.O}_{[L_m]}(\vec{x})$ is normalised to

$$\int |D^{H.O.}_{L_m}(\vec{x})|^2 d^3 \underset{x}{A} = 1 \tag{2.54}$$

leading to $([L_m]|0,0) = 1$.
The P.P. are solution of

$$\int_{-1}^{1} dz \int_0^{\pi} d\theta_{ij} \int_0^{2\pi} d\varphi_{ij}\, W_{[L_m]}(z,\omega_{ij})\ P_{[\lambda]}(z,\omega_{ij}) P_{[\lambda']}(z,\omega_{ij}) = \delta_{[\lambda],[\lambda']} \tag{2.55}$$

According to (2.44) they are defined in such a way that the product $Y[L_m](\Omega)P_\lambda(z,\omega_{ij})$ is a H.H. The simplest case occurs when $H[L_m](\vec{x})$ is spherically symmetric (a S state) generating a weight function $W[L_m](z)$ independent of ω_{ij}. It happens in particular for closed shell or closed subshell systems (e.g. nuclei). The P.P. is then the product

$$P_K^{\ell,m}(z,\omega_{ij}) = Y_{\ell,m}(\omega_{ij})(1+z)^{\ell/2} P_K^{[L_m],\ell}(z) \tag{2.56}$$

where the polynomials $P_K^{[L_m],\ell}(z)$ are associated with the weight function

$$W^{\ell}_{[L_m]}(z) = (1+z)^{\ell} W_{[L_m]}(z) \tag{2.57}$$

$$W_{[L_m]}(z) = (1-z)^{\mathcal{L}_m-2\ell_m-1}(1+z)^{1/2} \rho_{[L_m]}(z)$$

The H.H.

$$\mathcal{P}^{\ell,m}_{[L_m+2K+\ell]}(\Omega_{ij}) = Y_{[L_m]}(\Omega)\, Y_{\ell,m}(\omega_{ij})(1+z)^{\ell/2} P_K^{[L_m],\ell}(z) \tag{2.58}$$

$z = 2r_{ij}^2/r^2 - 1$, is a P.H. of grand orbital $L_m + 2K + \ell$, when $L_m = 0$ one finds again the P.H. (2.14).

The basis (2.58) is complete for an expansion of any function $H[L_m](\vec{x})F(\vec{r}_{ij},r)$.

According to (2.49) the weight function is positive evrywhere, therefore the zeros of $\rho[L_m](z,\omega_{ij})$ are out of the range of variation of ω_{ij} and of $-1 < z < 1$.

When the weight function is (2.57) the P.P. are given by the Chritoffel's formula

$$P_K^{[L_m],\ell}(z) = \frac{N_K^{\ell}}{\rho_{[L_m]}(z)}\ P^{\alpha\beta}\begin{pmatrix} K, & K+1, & \ldots, & K+n \\ z, & z_1, & \ldots, & z_n \end{pmatrix} \tag{2.59}$$

with the standard notation

$$P^{\alpha\beta}\begin{pmatrix} K,K+1, & \dots, K+n \\ z,z_1,\dots,z_n \end{pmatrix} = \begin{vmatrix} P_K^{\alpha,\beta}(z) & P_{K+1}^{\alpha,\beta}(z)\dots & P_{K+n}^{\alpha,\beta}(z) \\ P_K^{\alpha,\beta}(z_1) & & P_{K+n}^{\alpha,\beta}(z_1) \\ \dots & \dots & \dots \\ P_K^{\alpha,\beta}(z_n) & \dots & P_{K+n}^{\alpha,\beta}(z_n) \end{vmatrix}$$

for the determinant of Jacobi polynomials $P_m^{\alpha,\beta}$ where $\alpha = \mathcal{L}_m - 2\ell_m - 1$, $\beta = \ell + \frac{1}{2}$, $n = 2\ell_m$ and where z_1, z_2, ..., z_n are the zeros of $\rho_{[L_m]}(z)$. The normalisation constant is fixed by

$$\int_{-1}^{1} (P_K^{[L_m],\ell}(z))^2 \, W^{\ell}_{[L_m]}(z)dz = 1 \qquad (2.60)$$

Let us come back to (2.13) with a local two-body central potential $V(r_{ij})$, which for $L = L_m$ is the equation for fermions ground state.

We expand the partial wave (2.12) for the pair (i,j) in a serie of P.H.

$$\psi_{ij}(\vec{x}) = H_{[L_m]}(\vec{x})F(r_{ij},r) = \sum_K \mathcal{P}^0_{[L_m+2K]}(\Omega_{ij})\, u_K(r)/r^{(D-1)/2} \qquad (2.61)$$

and we project (2.13) on the same basis to generate a system of coupled equations similar to (2.28) with

$$\mathcal{L}_K = L_m + 2K + (D-3)/2 \qquad (2.61)$$

and where $\mathcal{P}^0_{[L_m+2K]}(\Omega_{ij})$ has been substituted for $\mathcal{P}^0_{2K}(\Omega_{ij})$ in (2.29) and $P_K^{[L_m],0}(z)$ for $P_K^{\alpha,\frac{1}{2}}(z)$ in (2.30). From (2.49) we deduce

$$V_K^{K'}(r) = \int_{-1}^{1}[\, V(r\sqrt{\tfrac{1+z}{2}}\,) - V_0(r)]\, P_K^{[L_m],0}(z)\, P_{K'}^{[L_m],0}(z)\, W_{[L_m]}(z)\, dz \qquad (2.63)$$

where $P_K^{[L_m],0}(z)$ are normalised according to (2.60). It is suitable to normalise the weight function to

$$\int_{-1}^{1} W_{[L_m]}(z)dz = 1 \quad \text{then } P_0^{[L_m],0} = 1 \qquad (2.64)$$

The potential matrix

$$U_K^{K'}(r) = f_K f_{K'} V_K^{K'}(r) \qquad (2.65)$$

must be used in (2.42) with (2.63). It is identical to (2.43) when $L_m = 0$.

- Relation between Potential Polynomials -

The computation of the integrals (2.63) occuring in the potential matrix (2.65) is time consuming and any simplification in the calculation of the P.P. which can reduce this time is welcomed. The P.P. (2.59) are associated with the weight function (2.57) and according to general properties of orthogonal polynomials any three consecutive polynomials are connected by a linear relation. They fulfil the recurrence formula

$$P_{K+1}(z) = (A_K z + B_K)P_K(z) - C_K P_{K-1}(z) \qquad (2.66)$$

where the superscript $[L_m]$, ℓ has been omitted for simplicity.

The coefficients A_K, B_K, C_K, can be obtained by computing P_{K-1}, P_K and P_{K+1} for three values, for example $z = 0, \pm 1$ and by solving the 3 linear equations obtained.

By this method the determinants $P^{\alpha,\beta}$ are computed for only three values of z for each K, with the help of the recurrence formula for Jacobi polynomials. The normalisation N_K needed to compute P_K with (2.59) is given in

Appendix. For other z values the recurrence formula (2.66) enables one to compute $P_K(z)$ by starting from $P_0 = 1$ when the weight function is normalised to one (see eq. (2.64)).

- Calculation of Excited States -

Our first task is to define independent states in connection with the H.H.E.M.. In our method the states are defined by suitably symmetrized H.P.. They are (anti) symmetric for (fermions) bosons for any exchange of two identical particles. In our scheme two independ states described by two H.P. $H_{[L]}(\vec{x})$ and $H_{[L']}(\vec{x})$ fulfil the conditions :

$$\langle H_{[L]}(\vec{x})|\,\mathcal{P}^{\ell,m}_{2K+\ell}(\Omega_{ij})\,|H_{[L']}(\vec{x})\rangle = 0 \qquad (2.67)$$

for any K, ℓ , m where the bracket means an integration over the unit hypersphere r = 1. Let us assume indeed that we include the $H_{[L']}(\vec{x})$ polynomial in the wave function describing the $H_{[L]}(x)$ state which becomes

$$\Psi(\vec{x}) = H_{[L]}(\vec{x}) \sum_{i,j>i} F(\vec{r}_{ij},r) + H_{[L']}(\vec{x})\phi(r) \qquad (2.68)$$

We introduce (2.68) in (2.13) and project on the P.H. basis to generate (2.28) in which the coupling between the partial waves $u_K(r)$ in (2.61) and $\phi(r)$ is

$$\langle\mathcal{P}^{\ell,m}_{[L+2K+\ell]}(\Omega_{ij})\,|V(r_{ij})|H_{[L']}(\vec{x})\rangle \qquad (2.69)$$

But $\mathcal{P}^{\ell,m}_{L+2K+\ell}(\Omega_{ij})\ V(r_{ij})$ can be completely expanded with the P.H. basis for $H_{[L]}(\vec{x})$, which can inturn be expanded in a serie of P.H. for bosons since $(1+z)^{\ell/2}P_K^{[L],\ell}(z)$ in (2.56) can be expressed as a serie of the ${}^{(N)}P^{\ell,0}_{K'}(\phi_N)$ occuring in (2.14). Assuming (2.67) the matrix element (2.69) is cancelled, the two polynomials $H_{[L]}(\vec{x})$ and $H_{[L']}(\vec{x})$ are decoupled in (2.28) and describe independent states.

The independent states are classified by starting from $H_{[L_m]}(\vec{x})$ the lowest states. Each $H_{[L]}(\vec{x})$ must fulfil (2.67) for all H.P. with L'<L.

In practice it is easier to verify that

$$\langle\mathcal{P}^{\ell,m}_{[L+2K+\ell]}(\Omega_{ij})|H_{[L]}(\vec{x})\rangle = 0 \quad \text{for } L'>L \qquad (2.70)$$

for the single case L' = L+2K+ℓ since two H.P. of different degree are orthogonal.

For each state $H_{[L]}(\vec{x})$ we apply the method used for the ground state to generate the coupled equations (2.42) where the potential matrix is given by (2.65), (2.63) and (2.59) where [L] has been substituted for $[L_m]$. The weight function associated with the P.P. is calculated as for the ground state by using the correspondance between H.P. and H.O. Slater determinants for fermions.

III. A bridge between H.H.E.M. and Integro-differential approaches

The calculation of the many-body problem by the H.H.E.M. require the computation of an infinite system of coupled differential equations. This system must be truncated in order to be treated numerically.

For strongly repulsive core potentials and large systems the number of equations to be treated to obtain a good accuracy becomes rapidely very large and practically untractable. For example with a purely central two-body local nuclear interaction, and for a P.H. expansion which reduce to the minimum the number of significant coupled equations, about 15 equations must be treated for few-body systems in the symmetrical state and at least 40 equations for ^{40}Ca to obtain a converged solution.

About twice more have to be treated when the tensor force is included, and still more when mixed symmetry states are taken into account.

The difficulty is not actually the one of solving a large system of coupled equations, which can be performed with the efficient and powerfull Adiabatic Approximation method, but the one of the time needed to compute the potential matrix, the size of which grows as the square of the number of coupled equations. The integrals (2.63) require indeed much time to be computed for all the r values needed to solve the coupled equations. For all these reasons it would be a drastic simplification to be able to sum up the infinite system of coupled differential equations (2.28) and to transform it into a single equation, avoiding the calculation of the potential matrix and the study of the convergence of the solution in terms of a growing number of coupled equations (2.28).

It can be done at the expense of introducing besides r another variable describing the two-body correlations included in the P.H. expansion of the wave function. This transformation is performed by starting from (2.13). For the sake of simplicity let us begin by bosons in ground state.

We have seen that our coupled equations (2.28) are obtained by projection of (2.25) on the $\vec{r}_{ij}$ space. In this projection the terms $\mathcal{P}^{\perp}_{2K}(\Omega_{ij})$ in (2.27) disappear.

We use the sytem of coordinates (2.45), i.e.

$$(\omega_{ij},\phi_N;\Omega_{N-1}) \text{ with } z = \cos 2\phi_N = 2r_{ij}^2/r^2 - 1 , \qquad (3.1)$$

for which the kinetic energy operators /1/ Appendix/

$$T = -\frac{h^2}{m}\nabla_0^2 = -\frac{h^2}{m}\left[\frac{\partial^2}{\partial r^2} + \frac{D-1}{r}\frac{\partial}{\partial r} + \frac{L^2(\Omega)}{r^2}\right] \qquad (3.2)$$

$$L^2(\Omega) = 4\left\{\frac{1}{W_{[0]}(z)}\frac{\partial}{\partial z}(1-z^2)W_{[0]}(z)\frac{\partial}{\partial z} + \frac{1}{2}\frac{\ell^2(\omega_{ij})}{1+z} + \frac{1}{2}\frac{L^2(\Omega_{N-1})}{1-z}\right\} \qquad (3.3)$$

where

$$W_{[0]}(z) = (1-z)^{\alpha}(1+z)^{\beta} , \quad \alpha = (D-5)/2 , \ \beta = 1/2 \qquad (3.4)$$

is the weight function associated with the P.H. for bosons in S state. The grand orbital operator $L^2(\Omega_{N-1})$ refers to the coordinates $(\vec{\xi}_1, \dots, \vec{\xi}_{N-1})$. The partial wave $F(r_{ij}, r)$ for S states (2.26) is written in terms of z and r :

$$F(r_{ij},r) = Y_{[0]}(D-3)P(z,r)/r^{(D-1)/2} \qquad (3.5)$$

it is independent of Ω_{N-1} (and of ω_{ij} for S states). We premultiply (2.25) by $Y_0(D-3)$ and we integrate over $d\Omega_{N-1}$ to generate the kinetic energy operator and the hypercentral part :

$$r^{-(D-1)/2}\left\{\frac{h^2}{m}\left[-\frac{\partial^2}{\partial r^2} - \frac{4}{r^2}\frac{1}{W_{[0]}(z)}\frac{\partial}{\partial z}(1-z^2)W_{[0]}(z)\frac{\partial}{\partial z} + \frac{\mathcal{L}_0(\mathcal{L}_0+1)}{r^2}\right] + \frac{A(A-1)}{2}V_0(r) - E\right\} P(z,r) \qquad (3.6)$$

The partial wave $u_K(r)$ in (2.26) are given according to (3.5) and (2.14) by

$$u_K(r) = r^{(D-1)/2}\int \mathcal{P}^0_{2K}(\Omega_{ij})\, F(r_{ij},r)\, d\Omega \qquad (3.7)$$

$$= \int_{-1}^{1} P(z',r)\; P_K^{\alpha,\beta}(z')(h_K^{\alpha,\beta})^{-\frac{1}{2}}\; W_{[0]}(z')dz$$

where $h_K^{\alpha\,\beta}$ is the norm of the Jacobi polynomial $P_K^{\alpha,\beta}(z)$.
The projection on the $\vec{r}_{ij}$ space of the partial waves with the pairs $(k,\ell)\neq(i,j)$ in

$$\Psi(\vec{x}) = F(r_{ij},r) + \sum_{\substack{k,\ell>k \\ (k,\ell)\neq(i,j)}} F(r_{k\ell},r) \tag{3.8}$$

calculated with (2.27) and (2.30) gives

$$\sum_K (f_K^2-1)\mathcal{P}^0_{2K}(\Omega_{ij})\,u_K(r) \tag{3.9}$$

$$= Y_{[0]}(D-3)\int_{-1}^{1} P(z',r)\,[\sum_K (f_K^2-1)\,P_K^{\alpha,\beta}(z)\,P_K^{\alpha,\beta}(z')/h_K^{\alpha,\beta}]\,W_{[0]}(z')dz'$$

where the integral in (3.7) and the sum over K in (2.27) have been permuted. The sum

$$f_{[0]}(z,z') = W_{[0]}(z')\sum_K (f_K^2-1)\,P_K^{\alpha,\beta}(z)\,P_K^{\alpha,\beta}(z')\,/\,h_K^{\alpha,\beta} \tag{3.10}$$

is called "the Projection Function" for S states $[0]$. With (3.6), (3.9) and (3.10) equations (2.28) which are the projection on the $\vec{r}_{ij}$ space of the P.H. expansion of bosons ground state become equivalent to the integro-differential equation in two variables :

$$\{\frac{h^2}{m}[-\frac{\partial^2}{\partial r^2} + \frac{\mathcal{L}_0(\mathcal{L}_0+1)}{r^2} - \frac{4}{r^2}\frac{1}{W_{[0]}(z)}\frac{\partial}{\partial z}(1-z^2)W_{[0]}(z)\frac{\partial}{\partial z}] \tag{3.11}$$

$$+\frac{A(A-1)}{2}V_0(r) - E\}\,P(z,r) = -\,[V(r\sqrt{\frac{1+z}{2}}) - V_0(r)]$$

$$\times\{P(z,r) + \int_{-1}^{1} f_{[0]}(z,z')\,P(z',r)dz'\}$$

The solution of the single integro-differential equation (3.11) is identical to theone of the infinite system of coupled differential equations (2.28). Indeed reversing the procedure used to obtain (3.11) by expanding the solution $P(z,r)$ of (3.11) in a serie of normalised Jacobi polynomials $P_K^{\alpha,\beta}(z).(h_K^{\alpha,\beta})^{-\frac{1}{2}}$ and by projecting this equation on the same basis, which is complete for an expansion of any function of z, we generate again (2.28). We established by this procedure a bridge between the H.H.E.M. and the integro-differential approachs to the many-body problem.

- Integro-differential equation for 3 bosons in S state -

We have seen that for three identical bosons in S state the limit of the expansion of the wave function (2.31) in a serie of P.H. for $K\to\infty$ provides the exact solution of the Schrödinger equation when the interaction is a S state projected local central potential.

It seems worthwile to investigate the structure of the equivalent integro-differential equation (3.11) for three bosons in S state.

We operate like previously by starting from (2.33) where $V_0(r)$ is cancelled. For A = 3 in a D = 3(A-1) = 6 dimensional space the weight function for the ground state

$$W_{[0]}(z) = (1-z^2)^{\frac{1}{2}} = \sin 2\phi \quad (\text{with } z = \cos 2\phi),\ 0\leqslant\phi\leqslant\pi/2,$$

is associated with the Gegenbauer polynomials

$$P_K^{\frac{1}{2},\frac{1}{2}}(z)(h_K^{\frac{1}{2},\frac{1}{2}})^{-\frac{1}{2}} = \sqrt{\frac{2}{\pi}}\;C_K^1(z) = \sqrt{\frac{2}{\pi}}\;\frac{\sin 2(K+1)\phi}{\sin 2\phi}\,. \tag{3.12}$$

Here we have to deal with two connected pairs (j,k) and (k,i) only for which the kinematic rotation angular parameters in (2.18) are $\varphi^{jk} = \varphi^{ki} = \pi/3$.

Then according to (2.28), (2.30) and (2.35)

$$f_K^2 - 1 = \frac{2}{K+1} \sin 2(K+1)\pi/3 \Big/ \sin 2\pi/3$$

The projection function (3.10) is calculated from the trigonometric representation (3.12)

$$f_{[0]}(z,z') = \frac{4}{\pi}(\sin 2\pi/3 \sin 2\phi)^{-1} \sum_{K=0}^{\infty} \sin 2(K+1)\pi/3 . \sin 2(K+1)\phi . \sin 2(K+1)\phi'/(K+1) \tag{3.13}$$

The calculation is performed by reducing the product of sin in a sum and by using

$$\sigma = \sum_{n=1}^{\infty} \sin 2nw/n = i \operatorname{Im} \ln(1-e^{2iw})$$

we have for a suitable choice of phase :

$$\sigma = (n + \tfrac{1}{2})\pi - w \quad \text{for} \quad n\pi < w < (n+1)\pi$$

where n = -1, 0, 1.
We obtain the projection function

$$f_{[0]}(z,z') = (\sin 2\pi/3 \sin 2\phi)^{-1} = 2/\sqrt{3(1-z^2)} \tag{3.14}$$

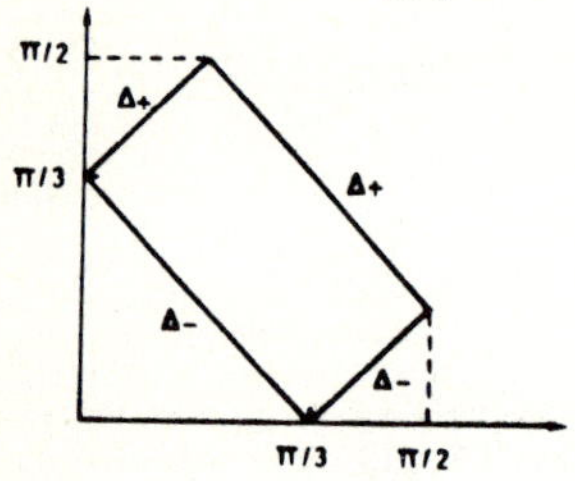

for ϕ' inside the rectangle of fig.1 and zero otherwise. The integro-differential equation equivalent to (2.36) is

$$\left\{\frac{h^2}{m}\left(-\frac{\partial^2}{\partial r^2} + \frac{15}{4r^2} - \frac{4}{r^2} \cdot \frac{1}{\sqrt{1-z^2}} \frac{\partial}{\partial z}(1-z^2)^{3/2}\frac{\partial}{\partial z}\right) - E\right\} P(z,r) \tag{3.15}$$

$$= -V\left(r\sqrt{\frac{1+z}{2}}\right)\left[P(z,r) + \frac{2}{\sqrt{3(1-z^2)}} \int_{z_-}^{z_+} P(z',r)\,dz'\right]$$

where

$$z_\pm = (-z \pm \sqrt{3(1-z^2)})/2$$

Let us introduce the change of variable from z to $\theta = 2\phi$ and of function

$$P(z,r) = \sqrt{r}\{U(r,\theta)/\sin\theta\} \tag{3.16}$$

The new function $U(r,\theta)$ is a solution of

$$\left\{\frac{h^2}{m}\left(\frac{\partial^2}{\partial r^2} + \frac{1}{r}\frac{\partial}{\partial r} + \frac{4}{r^2}\frac{\partial^2}{\partial\theta^2}\right) + E\right\} U(r,\theta)$$

$$= V(r\cos\tfrac{\theta}{2})\left\{U(r,\theta) + \frac{2}{\sqrt{3}}\int_{\Delta_-}^{\Delta_+} U(r,\theta')\,d\theta'\right\} \tag{3.17}$$

where the integral is taken inside the rectangle of the figure constructed by substituting θ, θ', $2\pi/3$ and π respectively for ϕ, ϕ', $\pi/3$ and $\pi/2$ in figure 1.

It is the Faddeev equation written by Noyes /11/ for S states. It provides an exact solution of the Schrödinger equation for S state projected central local potential as it is well known. The same property holds obviously for the solution of (3.15) in the z variable and for the one of the equivalent system of coupled equations (2.36) as it has been shown already in section 2.

- Integro-differential equation for bosons in S states -

The utilization of the integro-differential equation (3.11) require the knowledge of the projection function (3.10). In general the serie in (3.10) cannot be summed up analytically but this calculation can be achieved for bosons in S state where $\beta = \frac{1}{2}$.

According to (2.30) we have to consider two cases :

i) the projection of the 2(A-2) connected pairs with $P_K^{\alpha,\frac{1}{2}}(-\frac{1}{2})$

ii) the projection of the (A-2)(A-3)/2 disconnected pairs with $P_K^{\alpha,\frac{1}{2}}(-1)$.

We have therefore to sum up the series

$$S^{\alpha}(z,z';\delta) = \sum_{K=0}^{\infty} P_K^{\alpha,\frac{1}{2}}(\cos 2\delta)\, P_K^{\alpha,\frac{1}{2}}(z)\, P_K^{\alpha,\frac{1}{2}}(z') \,/\, (h_K^{\alpha,\frac{1}{2}}\, P_K^{\alpha,\frac{1}{2}}(1)) \tag{3.18}$$

for $\delta = \pi/3$ and $\pi/2$.

We use the relation between Jacobi and Gegenbauer polynomials where the Pochammer symbols $(a)_n$ means $(a)_n = \Gamma(a+n)/\Gamma(a)$:

$$P_K^{\lambda-\frac{1}{2},\frac{1}{2}}(2x^2-1) = (\tfrac{1}{2})_{K+1}/(\lambda)_{K+1}\, C_{2K+1}^{\lambda}(x)/x \tag{3.19}$$

with $2x^2-1 = \cos 2\phi = z$, $2x'^2-1 = z'$, $x = \cos\phi = r_{ij}/r$, $\lambda = \alpha+\frac{1}{2} = D/2-2$, to find

$$S^{\alpha} = \frac{1}{2\cos\delta}\;\frac{1}{\cos\phi}\;\frac{1}{2^{\lambda}\sqrt{\pi}}\;\frac{\Gamma(\lambda+\frac{1}{2})}{\Gamma(\lambda)} \tag{3.20}$$

$$\times \sum_{n\ \mathrm{odd}} \left\{ \frac{n!}{\Gamma(2\lambda+n)}\;\frac{\sqrt{\pi}\,\Gamma(2\lambda)\Gamma(\lambda)}{\Gamma(\lambda+\frac{1}{2})}\; C_n^{\lambda}(\cos\phi)\, C_n^{\lambda}(\cos\delta) \right\} C_n^{\lambda}(x')/(x'\, h_n^{\lambda})$$

where h_n^{λ} is the norm of the Gegenbauer polynomial C_n^{λ}. We notice that S^{α} belongs in (3.11) to an integrand which is an even function of x' in such a way that the contribution of any even Gegenbauer polynomial $C_n(x')$ (for n even) in (3.20) disappears when the weight function

$$W_{[0]}(z')\,dz' = 2^{\lambda+1}(1-x'^2)^{\lambda-\frac{1}{2}}\, x'^2 dx$$

is taken for $-1<x'<1$.

Then S^{α} can be calculated for all n values without change in the result.

The addition theorem provides the relation /12/

$$\frac{n!}{\Gamma(2\lambda+n)}\;\frac{\sqrt{\pi}\,\Gamma(2\lambda)\Gamma(\lambda)}{\Gamma(\lambda+\frac{1}{2})}\; C_n^{\lambda}(\cos\phi)\, C_n^{\lambda}(\cos\delta) \tag{3.21}$$

$$= \int_0^{\pi} C_n^{\lambda}(\cos\Theta)(\sin\varphi)^{2\lambda-1}\, d\varphi$$

where

$$\cos\Theta = \cos\phi\cos\delta + \sin\phi\sin\delta\cos\varphi \tag{3.22}$$

leading to

$$S^{\alpha}(z,z';\delta)W_{[0]}(z')dz' = \frac{1}{\sqrt{\pi}}\;\frac{\Gamma(\lambda+\frac{1}{2})}{\Gamma(\lambda)}\;\frac{(1-x'^2)^{\lambda-\frac{1}{2}}}{\cos\delta\cos\phi}\; x'\, dx'$$

$$\times \int_0^{\pi} \left[\sum_n C_n^{\lambda}(\cos\Theta)\, C_n^{\lambda}(x')/h_n^{\lambda}\right] (\sin\varphi)^{2\lambda-1}\, d\varphi. \tag{3.23}$$

The product of the sum over n and the weight function $(1-x'^2)^{\lambda-\frac{1}{2}}$ associated with

$C_n^\lambda(x')$ is a singular δ function $\delta(x'-\cos\Theta)$ in such a way that the integral with the projection function for connected pairs $f_c(z,z') = S^\alpha(z,z';\delta)W_{[0]}(z')$ and with $z' = 2x'^2-1$ becomes after permutation of the two integrals :

$$\int_{-1}^{1} f_c(z,z')\, P(z',r)dz' = \frac{1}{\sqrt{\pi}} \frac{\Gamma(\lambda+\frac{1}{2})}{\Gamma(\lambda)} \frac{1}{\cos\delta\cos\phi}$$

$$\times \int_0^\pi P(2\cos^2\Theta -1,r)\cos\Theta\,(\sin\varphi)^{2\lambda-1}\, d\varphi \tag{3.24}$$

Let us now substitute the variable $u = \cos\Theta$ for φ, where according to (3.22) $a = \cos(\phi+\delta) < u < \cos(\phi-\delta) = b$ for $0 < \varphi < \pi$. We have the transformation

$$\cos\varphi = (u-\cos\phi\cos\delta)/(\sin\phi\,\sin\delta)$$

$$d(\cos\varphi) = -\sin\varphi d\varphi = du/(\sin\phi\sin\delta) \tag{3.25}$$

leading to /13 / :

$$\int_{-1}^{1} f_c(z,z')P(z',r)dz' = \frac{1}{\sqrt{\pi}} \frac{\Gamma(\lambda+\frac{1}{2})}{\Gamma(\lambda)} \frac{1}{\cos\phi\cos\delta}\left(\frac{1}{\sin\delta\sin\phi}\right)^{2\lambda-1} \tag{3.26}$$

$$\times \int_a^b [(u-a)(b-u)]^{\lambda-1} P(2u^2-1,r)\,u du$$

where $\cos\phi = \sqrt{\frac{1+z}{2}}$, $\sin\phi = \sqrt{\frac{1-z}{2}}$, $\delta = \pi/3$ and $\lambda = D/2-2$. One checks with $P = 1$ the normalization :

$$\int_{-1}^{1} f_c(z,z')dz' = 1 \tag{3.27}$$

We notice that for $A = 3$, $D = 6$, $\lambda = 1$ and with the change of variable from u to $z' = 2u^2-1$, one recovers with 2 pairs and $\cos 2\phi = z$, the integral part of (3.15).

The projection function $f_d(z,z')$ for the disconnected pairs is obtained for $\delta = \pi/2$. For this value $\cos\delta = 0$ in the denominator of (3.26), we have therefore to calculate the limit of this expression for $\delta\to\pi/2$. For $\delta\to\pi/2$ we have $a\to -\sin\phi$ and $b\to\sin\phi$ with an odd integrand, the limit is of the type 0/0.

By starting from

$$\lim_{\delta\to\pi/2} (u-\cos(\phi+\delta))(\cos(\phi-\delta)-u) = (\sin^2\phi-u^2) + 2\,u\cos\phi\cos\delta$$

we obtain the limit of (3.26) for $\delta\to\pi/2$.

$$\int_{-1}^{1} f_d(z,z')P(z',r)dz' = \frac{2}{\sqrt{\pi}} \frac{\Gamma(\lambda+\frac{1}{2})}{\Gamma(\lambda-1)}\left(\frac{1}{\sin\phi}\right)^{2\lambda-1} \tag{3.28}$$

$$\times \int_{-\sin\phi}^{\sin\phi} (\sin^2\phi - u^2)^{\lambda-2}\, P(2u^2-1,r)\,u^2 du$$

where $\sin^2\phi = (1-z)/2$.

The integral (3.28) can be converted in the variables z and $z' = 2u^2-1$:

$$\int_{-1}^{1} f_d(z,z')P(z',r)dz' = \frac{2}{\sqrt{\pi}} \frac{\Gamma(\lambda+\frac{1}{2})}{\Gamma(\lambda-1)} (1-z)^{\frac{1}{2}-\lambda} \tag{3.29}$$

$$\times \int_{-1}^{-z} [-(z+z')]^{\lambda-2}\sqrt{1+z'}\, P(z',r)dz'$$

Both projection functions for the connected and the disconnected pairs can be written as an integral with fixed limits ± 1 by a suitable change of variables

$$\int_{-1}^{1} f_c(z,z')\,P(z',r)\,dz' = 1/\sqrt{\pi}\,.\Gamma(\lambda+\tfrac{1}{2})/\Gamma(\lambda) \qquad (3.30)$$

$$\int_{-1}^{1} (1-v^2)^{\lambda-1}\, P[(\sqrt{1+z}\cos\delta + v\sqrt{1-z}\sin\delta)^2 - 1, r]$$

$$\times [1 + v\sqrt{\frac{1-z}{1+z}}\,\mathrm{tg}\delta\; dv]$$

$$\int_{-1}^{1} f_d(z,z')\,P(z',r)\,dz' = 2/\sqrt{\pi}\,.\Gamma(\lambda+\tfrac{1}{2})/\Gamma(\lambda-1) \qquad (3.31)$$

$$\times \int_{-1}^{1} (1-v^2)^{\lambda-2} P[(1-z)v^2-1\;, r]\, v^2\, dv$$

We notice that for $\lambda \to \infty$ (i.e. $A\to\infty$) the function in the integrand operates like a $\delta(v)$ function leaving only a function

$$\lim_{\lambda\to\infty} \int_{-1}^{1} f_c P(z',r)\,dz' = P[(z-3)/4, r] \qquad (3.32)$$

$$\lim_{\lambda\to\infty} \int_{-1}^{1} f_d P(z',r)\,dz' = P(-1,r)$$

The projection function

$$f_{[0]}(z,z') = 2(A-2)\, f_c(z,z') + (A-2)(A-3)/2\, f_d(z,z') \qquad (3.33)$$

contains δ functions when $\lambda\to\infty$ converting the integro-differential equation (3.11) in a differential equation in z and r.

- Integro-differential equation for $L \neq 0$ states -

When we have to deal with either fermions in ground state or with excited states, the grand orbital L is not zero and the partial wave is written

$$\psi_{ij} = H_{[L]}(\vec{x})P(z,r)/r^{(\mathcal{L}+1)}\;,\; \mathcal{L} = L+(D-3)/2 \qquad (3.34)$$

where $z = 2r_{ij}^2/r^2 - 1$.

In order to obtain an equation simular to (3.11) valid for any state, when the states are defined by (2.67), we project (2.8) on the $\vec{r}_{ij}$ space. We assume that the H.P. $H_{[L]}(\vec{x}) = H_{[L]}(\vec{\xi})$ is translationaly invariant.

The kinetic energy is calculated from the Laplace operator $\nabla^2 = \sum_{i<A} \nabla^2_{\xi_i}$ by using (3.2) and (3.3) with $\vec{\xi}_N = \vec{r}_{ij}$.

Three terms contribute to the kinetic energy (2.8) :

$$\nabla^2 H_{[L]}(\vec{\xi})P(z,r)r^{-(\mathcal{L}+1)} = P(z,r)r^{-(\mathcal{L}+1)}\,\nabla^2 H_{[L]}(\vec{\xi})$$

$$+ H_{[L]}(\vec{\xi})\nabla^2 P(z,r)r^{-(\mathcal{L}+1)} + 2\vec{\nabla}P(z,r)r^{-(\mathcal{L}+1)}.\vec{\nabla}\, H_{[L]}(\vec{\xi})$$

The first term is cancelled because $H_{[L]}(\vec{\xi})$ is an H.P.. The second term is obtained from (3.2) and (3.3) with $\ell^2(\omega_{ij})P = L^2(\Omega_{N-1})P = 0$.

For the calculation of the third term we use the following formulae :

$$\vec{\nabla}F(z,r).\vec{\nabla} = \sum_{i=1}^{N}\left[\frac{\partial F}{\partial z}(\vec{\nabla}_i z.\vec{\nabla}_i) + \frac{\partial F}{\partial r}(\vec{\nabla}_i\, r.\vec{\nabla}_i)\right] \qquad (3.35)$$

$$\sum_{i=1}^{N} \vec{\nabla}_i r.\vec{\nabla}_i = \frac{1}{r} \sum_{i=1}^{N} \vec{\xi}_i.\vec{\nabla}_i$$

$$\sum_{i=1}^{N} \vec{\nabla}_i z.\vec{\nabla}_i = \frac{4\vec{\xi}_N^2}{r^2}[\vec{\nabla}_N - \frac{1}{r^2}\sum_{i=1}^{N}\vec{\xi}_i.\vec{\nabla}_i] \tag{3.36}$$

$$\vec{\xi}_N.\vec{\nabla}_N = \xi_N \frac{\partial}{\partial \xi_N} = (1 - z^2)\frac{\partial}{\partial z} + \frac{1+z}{2} r \frac{\partial}{\partial r}$$

and

$$\sum_{i=1}^{N} \vec{\xi}_i.\vec{\nabla}_i H_{[L]}(\vec{\xi}) = r\frac{\partial}{\partial r} H_{[L]}(\vec{\xi}) = L H_{[L]}(\vec{\xi}) \tag{3.37}$$

since $H_{[L]}(\vec{\xi})$ is a homogeneous polynomial of degree L, and we find

$$\nabla^2 H_{[L]}(\vec{\xi})P(z,r)r^{-(\mathcal{L}+1)} = r^{-(\mathcal{L}+1)}\Big\{H_{[L]}(\vec{\xi})\left(\frac{\partial^2}{\partial r^2} - \frac{\mathcal{L}(\mathcal{L}+1)}{r^2}\right) \tag{3.38}$$

$$+ \frac{4}{r^2}[2(1-z^2)\frac{\partial}{\partial z} H_{[L]}(\vec{\xi}) + H_{[L]}(\vec{\xi})((1-z^2)\frac{\partial^2}{\partial z^2} + (3-\frac{D}{2}(1+z))\frac{\partial}{\partial z}))]\Big\} P(z,r)$$

The kinetic energy term is obtained by premultiplying (3.38) by $r^{-L}H^*_{[L]}(\vec{\xi})$ and by integrating over all variables but z. We use (2.49) where $[L]$ has been substituted for $[L_m]$ and

$$\int_{r=1} H_{[L]}(\vec{\xi})\frac{\partial}{\partial z} H_{[L]}(\vec{\xi})d\Omega_{N-1} = \frac{1}{2}\frac{\partial}{\partial z}\int_{(r=1)} |H_L(\vec{\xi})|^2 d\Omega_{N-1}$$

to finally obtain

$$2^{-D/2}(1-z)^{(D-5)/2}(1+z)^{1/2}\int r^{-L} H^*_{[L]}(\vec{\xi})\nabla^2 H_{[L]}(\vec{\xi})P(z,r)\, r^{-(\ +1)} d\omega_{ij} d\Omega_{N-1}$$

$$= r^{-(\mathcal{L}+1)} W_{[L]}(z)\Big\{\frac{\partial^2}{\partial r^2} - \frac{\mathcal{L}(\mathcal{L}+1)}{r^2} + \frac{4}{r^2}\frac{1}{W_{[L]}(z)}\frac{\partial}{\partial z}(1-z^2)W_{[L]}(z)\Big\}P(z,r) \tag{3.39}$$

where $W_{[L]}(z)$ is given by (2.49) integrated over ω_{ij}. The calculation of $W_{[L]}(z)$ is achieved like previously through the H.O. representation of H.P. and by using (2.50), (2.51) and (2.52) with $[L]$ instead of $[L_m]$. The P.P. are given by (2.59).

The projection of $\Psi(\vec{x})$ is performed through a P.H. expansion of the partial waves

$$\psi_{k\ell}(\vec{x}) = \sum_K \mathcal{P}^0_{[L+2K]}(\Omega_{k\ell})u_K(r)/r^{(D-1)/2} \tag{3.40}$$

where

$$\mathcal{P}^0_{[L+2K]}(\Omega_{k\ell}) = r^{-L}H_{[L]}(\vec{\xi})P_K^{[L]}(2r_{k\ell}^2/r^2-1) \tag{3.41}$$

where the P.P. of degree K is associated with the weight function $W_{[L]}(z)$.

The projection on the $\vec{r}_{ij}$ space of the P.H. (3.41) is obtained by a decomposition in which we isolate the term of highest degree in $\xi_N = r_{ij}$ in the H.P.

$$r^{2K+L}\mathcal{P}^0_{[L+2K]}(\Omega_{k\ell}) = H_{[L]}(\vec{\xi})r^{2K}P_K^{[L]}(2z^2(\varphi^{k\ell})/r^2-1) \tag{3.42}$$

where $\vec{z}(\varphi^{k\ell}) = \vec{r}_{k\ell}$ in terms of the kinematic rotation vector

$$\vec{z}(\varphi) = \vec{\xi}_N \cos\varphi_N + \vec{\xi}_{N-1} \sin\varphi_N \cos\varphi_{N-1} + \dots$$

where we set $\vec{z}(0) = \vec{\xi}_N = \vec{r}_{ij}$. The coefficient of the term of highest degree ξ_N^{2K} in

$$r^{2K} P_K^{[L]}(2z^2(\varphi)/r^2 - 1) = \sum_{n=0}^{K} a_n^K (z^2(\varphi))^{K-n} r^{2n} \tag{3.43}$$

where $r^2 = \sum_{i=1}^{N} \xi_i^2$ is

$$\sum_{n=0}^{K} a_n^K (\cos^2\varphi_N)^{K-n} = P_K^{[L]}(\cos 2\varphi_N) \tag{3.44}$$

It is $P_K^{[L]}(1)$ for $\varphi=0$ i.e. for $\vec{z}(0) = \vec{r}_{ij}$. We can therefore decompose (3.42) into the following two terms :

$$H_{[L]}(\vec{\xi}) r^{2K} P_K^{[L]}(2r_{k\ell}^2/r^2 - 1) = H_{[L]}(\vec{\xi}) r^{2K} P_K^{[L]}(\cos 2\varphi_N^{k\ell})/P_K^{[L]}(1) P_K^{[L]}(z) + H^{\perp}_{[L+2K]}(\vec{\xi}) \tag{3.45}$$

where $H^{\perp}$ is an H.P. orthogonal to the P.H. of the pair (i,j). The first term at right gives the projection of the P.H. for the pair (k,ℓ) on the pair (i,j)

$$\langle \mathcal{P}^0_{[L+2K]}(\Omega_{ij}) | \mathcal{P}^0_{[L+2K]}(\Omega_{k\ell}) \rangle = P_K^{[L]}(\cos 2\varphi_N^{k\ell})/P_K^{[L]}(1) \tag{3.46}$$

where $\varphi_N^{k\ell} = \pi/3$ for connected pairs and $\varphi_N^{k\ell} = \pi/2$ for disconnected pairs in a system of identical particles.

We construct the projection function by applying the same procedure as for eqs.(3.7)-(3.10) :

$$f_{[L]}(z,z') = W_{[L]}(z') \sum_K [(f_K^{[L]})^2 - 1] \; P_K^{[L]}(z) P_K^{[L]}(z')/h_K^{[L]} \tag{3.47}$$

where $h_K^{[L]}$ is the norm of the P.P.

$$h_K^{[L]} = \int_{-1}^{1} (P_K^{[L]}(z))^2 W_{[L]}(z) dz \tag{3.48}$$

and where

$$(f_K^{[L]})^2 - 1 = (A-2)[2P_K^{[L]}(-\tfrac{1}{2}) + (A-3)/2 \; P_K^{[L]}(-1)] / P_K^{[L]}(1) \tag{3.49}$$

By operating as for the integro-differential equation for bosons we obtain now the general equation valid for any "state" [L] :

$$\begin{aligned} &\Big\{ \frac{h^2}{m} \Big[-\frac{\partial^2}{\partial r^2} + \frac{\mathcal{L}(\mathcal{L}+1)}{r^2} - \frac{4}{r^2} \frac{1}{W_{[L]}(z)} \frac{\partial}{\partial z} (1-z^2) W_{[L]}(z) \frac{\partial}{\partial z} \Big] \\ &+ \frac{A(A-1)}{2} V_0(r) - E \Big\} P(z,r) \\ &= -\Big[V\Big(r\sqrt{\tfrac{1+z}{2}}\Big) - V_0(r)\Big] \Big(P(z,r) + \int_{-1}^{1} f_{[L]}(z,z') P(z',r) dz'\Big) \end{aligned} \tag{3.50}$$

The point which might seem the most difficult in this procedure is the calculation of the serie occuring in the projection function.

Actually only a limited number of terms are needed to obtain a very good accuracy because the coefficients $f_K^2 - 1$ decrease faster than $2(A-2)/4^K$.

As an example the first values of $f_k^2 - 1$ for bosons in ground state for A = 4, 16, 40, 80 and 140 are exhibited in table 1 for K>2.

For K = 0 and K = 1 the values are $f_0^2 = A(A-1)/2$ and $f_1 = 0$

A\K	2	3	4	5	6	7
4	.0158	.0488	.106	-.2007	.1815	-.1056
16	1.3665	-.122	.0088	-.004	.0016	.0004
40	4.3459	.566	.0892	-.008	.0004	
80	9.3395	1.798	.365	.067	.0114	.0017
140	16.8368	3.666	.818	.175	.0362	.0072

The coefficients decrease nearly as fast for systems of fermions as for bosons as it is shown in table 2 for ^{16}O and ^{40}Ca in ground state

A\K	2	3	4	5	6
16	1.3859	-.0423	.0058	-.003	.0006
-40	4.360	.6947	.1307	.0207	.0029

From table 1 and 2 it appears that only a few polynomials are needed to obtain the projection function with a good accuracy, indeed if P(z,r) were also expanded in terms of the same basis the coefficient for each K would be 1 and by neglecting the coefficients for K>7 we bring a negligible error in the solution.

As only low degree polynomials are needed we can start from the calculation of the moments to generate these polynomials by a standard procedure /16/ . It is easier to calculate the moments for $\cos^2\phi_N = (1+z)/2$. The moments are calculated from (2.51) and (2.52) integrated over ω_{ij} /15/

$$c_n = \int_{-1}^{1} \left(\frac{1+z}{2}\right)^n W_{[L]}(z)dz \tag{3.51}$$

$$= \frac{2}{\sqrt{\pi}} \frac{\Gamma(\quad +3/2)}{\Gamma(\quad +n+3/2)} n!\Gamma(n+\frac{3}{2}) \sum_{N<n} (-1)^N \frac{([L_m]|N,0)}{(n-N)!}$$

and the polynomials are the determinants

$$P_K^{[L]}(z) = C_K^{[L]} \begin{vmatrix} c_0 & c_1 & \cdots c_n \\ c_1 & c_2 & \cdots c_{n+1} \\ \cdots & \cdots & \cdots \\ c_{n-1} & c_n & \cdots c_{2n-1} \\ 1 & \frac{1+z}{2} & \cdots \left(\frac{1+z}{2}\right)^n \end{vmatrix} \tag{3.52}$$

where $C_K^{[L]}$ is a normalisation constant.

IV. The Adiabatic Approximation Method

Let $\vec{x}$ be the set of Jacobi coordinates $(\vec{\xi}_1, \vec{\xi}_2, \ldots, \vec{\xi}_N)$ for the positions of A = N+1 particles in the center of mass frame.

The Schrödinger equation describing the motion of A identical particles in hyperspherical polar coordinates (r,Ω) in the D = 3(A-1) dimensional space is

$$(T+V(\vec{x})-E)\Psi(\vec{x}) = \left\{-\frac{h^2}{m}\left(\frac{\partial^2}{\partial r^2} + \frac{\mathcal{L}^2(\Omega)}{r^2}\right) + U(r,\Omega) - E\right\} u(r,\Omega) = 0 \tag{4.1}$$

for the wave function

$$\Psi(\vec{x}) = u(r,\Omega)/r^{(D-1)/2} \tag{4.2}$$

In the Adiabatic Approximation we assume that the radial and orbital motions in the D dimensional space are decoupled in such a way that the wave function can be written as the product

$$u(r,\Omega) = B_\lambda(\Omega,r)u_\lambda(r) \quad (4.3)$$

of a function $B_\lambda(\Omega,r)$ of the angular coordinates Ω, where r is a parameter, and a function of r only. /17,18/.

The set $\{B_\lambda(\Omega,r)\}$ constitutes the Adiabatic basis eigenfunction of

$$\left\{-\frac{\hbar^2}{m}\mathcal{L}^2(\Omega)/r^2 + U(r,\Omega) - U_\lambda(r)\right\}B_\lambda(\Omega,r) = 0 \quad (4.4)$$

where λ is a set of quantum numbers characterising each basis element and where $U_\lambda(r)$ is the associated eigenpotential.

The orthonormal equation

$$\int B_\lambda^*(\Omega,r)B_{\lambda'}(\Omega,r)d\Omega = \langle B_\lambda | B_{\lambda'}\rangle = \delta_{\lambda\lambda'} \quad (4.5)$$

is fulfilled for any r.

According to (4.3) and (4.1) we should have

$$\left(-\frac{\hbar^2}{m}\left(\frac{\partial^2}{\partial r^2} + \frac{\mathcal{L}^2(\Omega)}{r^2}\right) + U(r,\Omega) - E\right)B_\lambda(\Omega,r)u_\lambda(r) = 0 \quad (4.6)$$

If we neglect the variation of $B_\lambda(\Omega,r)$ with r, then the radial wave is a solution of

$$\left(-\frac{\hbar^2}{m}\frac{d^2}{dr^2} + U_\lambda(r) - E_E\right)u_\lambda^E(r) = 0 \quad (4.7)$$

This Approximation is good as long as the derivatives of $B_\lambda(\Omega,r)$ with respect to r is very small. The corresponding solution $u_\lambda^E(r)$ is called the Extreme Adiabatic Approximation. In order to obtain a better solution by taking the change of B (Ω,r) with r into account we premultiply (4.6) by $B_\lambda^*(\Omega,r)$ and we integrate over the unit hypersphere to obtain the radial equation

$$\left\{-\frac{\hbar^2}{m}\left[\frac{d^2}{dr^2} + \langle B_\lambda(\Omega,r)|\frac{d^2}{dr^2}B_\lambda(\Omega,r)\rangle\right] + U_\lambda(r) - E^U\right\}u_\lambda^U(r) = 0 \quad (4.8)$$

The solution (4.3) with $u_\lambda(r)$ is called the Uncoupled Adiabatic Approximation. It fulfils the Rayleigh-Ritz variational principle since it has been obtained from $B_\lambda(\Omega,r)u_\lambda(r)$ used as a trial function in (4.1).

We used the orthonormal equation (4.5)

$$\langle B_\lambda | B_\lambda\rangle = 1 \quad \text{to obtain} \quad \langle B_\lambda|\frac{d}{dr}B_\lambda\rangle = 0 \quad (4.9)$$

and

$$\langle B_\lambda|\frac{d^2}{dr^2}B_\lambda\rangle = -\langle\frac{dB_\lambda}{dr}|\frac{dB_\lambda}{dr}\rangle < 0 .$$

From (4.9) we deduce that $U_\lambda(r) - \frac{\hbar^2}{m}\langle B_\lambda|\frac{d^2}{dr^2}B_\lambda\rangle$, the effective potential for the uncoupled approximation, is always larger than $U_\lambda(r)$ and that for ground states the Extreme Adiabatic binding energy is always below the Uncoupled Adiabatic binding energy which is an upperbound of the exact binding energy. More generally one proves that we have the inequalities

$$E^E < E_{exact} < E^U. \quad (4.10)$$

The Adiabatic Basis can be calculated from a H.H. expansion

$$B_\lambda(\Omega,r) = \sum_{[L]=0}^{\infty} b^\lambda_{[L]}(r)\, Y_{[L]}(\Omega) \qquad (4.11)$$

where according to (4.4) the coefficients $b^\lambda_{[L]}(r)$ are a solution of the linear equations

$$\left(\frac{\hbar^2}{m}\mathcal{L}(\mathcal{L}+1)/r^2 - U_\lambda(r)\right) b^\lambda_{[L]}(r) + \sum_{[L']=0}^{\infty} U^{[L']}_{[L]}(r)\, b^\lambda_{[L']}(r) = 0 \qquad (4.12)$$

where $= L + 3A/2 - 3$ and where the potential matrix is defined by

$$U^{[L']}_{[L]}(r) = \langle Y_{[L]}(\Omega) | U(r,\Omega) | Y_{[L']}(\Omega) \rangle \qquad (4.13)$$

The large degeneracy of the complete H.H. basis prevents to use the expansion (4.11) with the full basis and in most of applications one is obliged to restrict (4.11) to the P.H. expansion

$$B^{[L]}_\lambda(\Omega,r) = H_{[L]}(\vec{x}) \sum_{K=0}^{\infty} b^{[L],\lambda}_K(r) \sum_{i<j\leqslant A} P^{[L]}_K(2r^2_{ij}/r^2-1) \qquad (4.14)$$

where the eigen vectors $\{b^{[L],\lambda}_K(r)\}$ are solution of the linear equations

$$\left[\frac{\hbar^2}{m}\mathcal{L}_K(\mathcal{L}_K+1)/r^2 - U_\lambda(r)\right] b^{[L],\lambda}_K(r) + \sum_{K'} f^2_{K'}\, V^{K'}_K(r)\, b^{[L],\lambda}_{K'}(r) = 0 \qquad (4.15)$$

where the symbols have the same meaning as in (2.62) and (2.63) with $V_0 = 0$ and $L_m = L$.

With the same P.H. expansion approximation one can use the integro-differential approach to obtain $B_\lambda(\Omega,r)$ by writing

$$B^{[L]}_\lambda(\Omega,r) = H_{[L]}(\vec{x}) \sum_{i<j\leqslant A} P^{[L]}_\lambda(2r^2_{ij}/r^2-1,r) \qquad (4.16)$$

where $P^{[L]}_\lambda$ is a solution of the integro-differential equation for the single variable z obtained from (3.50) by setting $\partial^2/\partial r^2 = 0$ and $V_0(r) = 0$ and by substituting $P^{[L]}_\lambda(z,r)$ for $P(z,r)$. The $\{B^{[L]}_\lambda(\Omega,r)\}$ constitutes an orthonormal basis which fulfil the orthonormal equation

$$\int_{-1}^{1} dz\, W_{[L]}(z) P^{[L]}_\lambda(z,r) \int_{-1}^{1} f_{[L]}(z,z')\, P^{[L]}_{\lambda'}(z',r)\,dz' = \delta_{\lambda,\lambda'} \qquad (4.17)$$

In order to give an idea about the quality of the Adiabatic Approximations the binding energy of the ground state of the nuclei ^4He, ^{12}C, ^{16}O, ^{28}S, ^{40}Ca has been calculated with the central part of the Gogny-Pires-de Tourreil N-N supersoft core potential / 19/.

In table 3, the increase of binding energy in MeV $\Delta E = E_0 - E$ proceeding from the contribution of the two-body correlations is shown together with the energy difference $\delta E = E^E - E^U$ between the Extreme and Uncoupled adiabatic solution (i.e. from (4.7) and (4.8))

Nuclei	^4He	^{12}C	^{16}O	^{28}Si	^{32}S	^{40}Ca
ΔE(MeV)	4.1	20	33	61	72	107
δE(MeV)	.4	.16	.1	.07	.07	.05

The energy E_0 is obtained from the first order approximation (see (2.10)) where $V_0(r)$ only is considered. The contribution ΔE of the correlations to the binding energy is roughly proportional to A, the number of nucleons, conversely the error δE in the binding energy decreases as 1/A. The larger is the number of particles the better is the Adiabatic Approximation. The interpolation formula $E = E^U - 0.3\delta E$ provides a very accurate estimate of the binding energy.

The contribution of $B_0(\Omega,r)u_0(r)$, the Adiabatic partial wave, to the total wave function of the ^3H ground state has been found to be .9995 ± .0001 for various N-N nuclear potentials /18/. Therefore, both the binding energy and the wave function are obtained with a good accuracy by the Adiabatic Approximation method.

The main property of the Adiabatic basis is to decouple the orbital and radial motions. When the hyperradius r increases indefinitely the various possible channels are decoupled and each B_λ describes a particular cluster decomposition, each one in a definite state. The eigenpotential $U_\lambda(r)$ becomes constant for $r \to \infty$ and $U_\lambda(r)$ is the total binding energy of the cluster decomposition described by $B_\lambda(\Omega, r \to \infty)$ where all clusters are at rest.

From $U_\lambda(r)$ either in (4.7) or (4.8) the bound as well as the scattering states can be calculated /20,21/.

In order to illustrate this important property we have chosen to solve the Positronium-Ion in which two electrons interact with one positron. We used the Faddeev decomposition :

$$\Psi(\vec{x}) = \psi_e(r_{12},r) + \psi_p(r_{23},r) + \psi_p(r_{31},r) \tag{4.18}$$

where r_{12} is the distance between the two electrons (1) and (2) and where (3) refers to the positron. The partial waves are a solution of

$$\begin{aligned}(T-E)\psi_e(r_{12},r) &= -1/r_{12}\,\Psi\\ (T-E)\psi_p(r_{23},r) &= 1/r_{23}\,\Psi\end{aligned} \tag{4.19}$$

in atomic units, and a similar equation for $\psi_p(r_{31},r)$. For solving (4.19) we use the Adiabatic method with a P.H. expansion /22/

$$\begin{aligned}\psi(r_{ij},r) &= B_n^{(ij)}(\Omega,r)\,u(r)/r^{5/2}\\ B_n^{(ij)}(\Omega,r) &= \sum_K b_{K,n}^{i,j}\,\mathcal{P}_{2K}^{0}(\Omega_{ij})\end{aligned} \tag{4.20}$$

where $b^{23} = b^{31}$.
The eigen vectors $b^{ij}(r)$ are solution of

$$[\mathcal{L}_K(\mathcal{L}_K+1)/r^2 - U_n(r)]\,b_{K,n}^{12}(r) \tag{4.21}$$

$$= (-1)^{K+1}\,\frac{16}{\pi r}\sum_{K'=0}^{\infty}(-1)^{K'}\,S_K^{K'}\left[\,b_{K',n}^{12} + \frac{O(K')}{K'+1}\times\left(b_{K',n}^{23}(r) + b_{K',n}^{31}(r)\right)\right]$$

$$[\mathcal{L}_K(\mathcal{L}_K+1)/r^2 - U_n(r)]\,b_{K,n}^{23}(r)$$

$$= (-1)^{K}\,\frac{16}{\pi r}\sum_{K'=0}^{\infty}(-1)^{K'}\,S_K^{K'}\left[b_{K',n}^{23} + \frac{O(K')}{K'+1}\left(b_{K',n}^{31}(r) + b_{K',n}^{12}(r)\right)\right]$$

where $O(K) = 1, -1, 0$ for $K = 3m, 3m+1, 3m+2$ (m integer), $\mathcal{L}_K = 2K + 3/2$ and where

$$S_K^{K'} = \sum_{n=|K-K'|}^{K+K'} (n+1)\Big/[(2n+1)(2n+3)] \tag{4.22}$$

for $n = |K-K'|, |K-K'|+2, \ldots, K+K'$.

The lowest eigenpotentials $U_n(r)$ are shown on fig.2 and fig.3. Each one has an horizontal asymptote corresponding to one of the S excited states of the positronium of eigenenergy $-[2(n+1)]^{-2}$ where n is the number of nodes of the S wave.

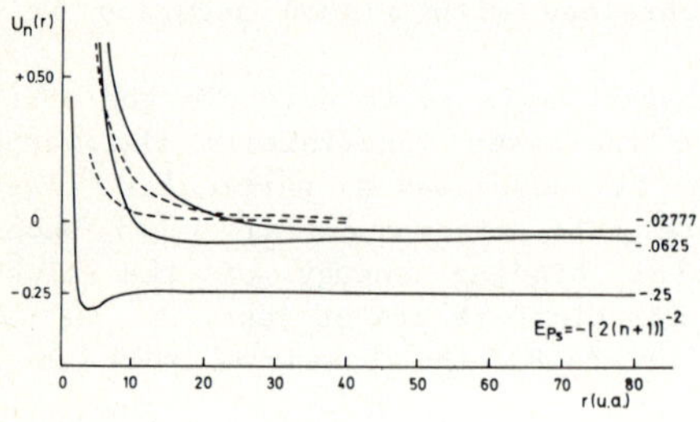

Fig.2

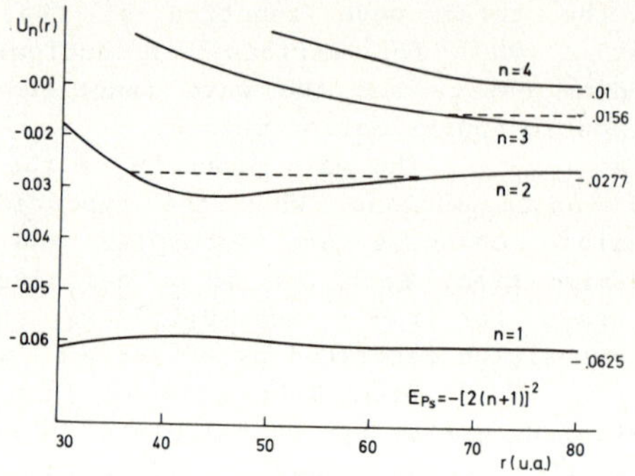

Fig.3

From the eigenpotential $U_n(r)$ can be calculated the S phase shifts in the scattering of one electron on the positrium in the (n+1)S state.

- Applications -

The H.H.E.M. has been applied to various domains in Physics. A detailed review of applications to atomic physics is given in ref./21/. In nuclear physics applications to three nucleons ground state with a general review of references is given in ref./23/ including a calculation of the two-body photodesintegration of ^3He. We have nevertheless to notice that all these calculations have been done with the Schrödinger equation but not with the Faddeev-like decomposition (2.7) of the wave function. This last procedure which is more simple and more powerfull enabled us to solve the ground state of ^4He, ^{16}He and ^{40}Ca with central potentials /19/ and can be applied in fact to any "state" and any many-body system.

References

/1/ M. Fabre de la Ripelle, Ann. Phys. (N.Y.) **147**(1983) 281

/2/ N. Ya. Vilenkin, G.L. Kuznetsov and Ya. R. Smorodinsky, Sov. J. Nucl. Phys. **2**(196)645

/3/ F. Zernike and H.C. Brinkman, Proc. Kon. Ned. Acad. Wensch **33**(1935)3

/4/ J. Raynal and J. Revai, Nuovo Cim. **A68**(1970)612.

/5/ Yu. A. Simonov, Sov. J. Nucl. Phys **3**(1966)461

/6/ J.E Kilpatrick and S.Y. Larsen to be published in Few-body systems.

/7/ M. Fabre de la Ripelle and J. Navarro, Ann. Phys. (N.Y.) **123**(1979)185

/8/ R. Reid, Ann. Phys. (N.Y.) **50**(1968)411

/9/ M. Fabre de la Ripelle, C. R. Acad. Sci. Paris **269** Serie II (1983)1027 ; Phys. Lett. **B135**(1984)5

/10/ M. Fabre de la Ripelle, H. Fiedeldey and G. Wiechers, Ann. Phys. (N.Y.) **138**(1982)275

/11/ H.P. Noyes in : Three-body problem in nuclear and particle physics eds. J.S.E McKee and P.M. Rolph (North-Holland Amsterdam 1970)

/12/ A. Erdelyi et al. in "Higher transcendental functions" Vol.II, p.178 eq.(34) (Ed. California Int. of Tech.)

/13/ M. Fabre de la Ripelle, H. Fiedeldey and S. Sofianos, Proc. XI Int. Conf. Few-body problems Sendai 1986.

/14/ M. Fabre de la Ripelle, C. R. Acad. Sci. Paris **299** Serie II (1984)839 ; M. Fabre de la Ripelle and H. Fiedeldey, Phys. Lett. **B171**(1986)325

/15/ M. Fabre de la Ripelle, C. R. Acad. Sci. Paris **292** Serie II (1981)275

/16/ A. Erdelyi et al. in Higher transcendental functions Vol.2 p.157.

/17/ M. Fabre de la Ripelle, C. R. Acad. Sc. Paris **274**(1972)104

/18/ J.L. Ballot, M. Fabre de la Ripelle and J.S. Levinger, Phys. Rev. **C26**(1982)2301

/19/ J.L. Ballot, M. Fabre de la Ripelle and J. Navarro , Phys. Lett. **B143**(1984)19

/20/ C.D. Lin, Phys. Rev. **A129**(1975)493

/21/ Review in U. Fano Rep. Prog. Phys. Vol. **46**(1983)97

/22/ M. Fabre de la Ripelle and S. Y. Larsen, Proceeding of XI Int. Conf. on Atomic Physics and Few-body systems (Sendai 1986).

/23/ J.L. Ballot and M. Fabre de la Ripelle, Ann. Phys. (N.Y.) **127**(1980)62

THE ATMS METHOD IN FEW-BODY PHYSICS

Yoshinori Akaishi
Department of Physics, Hokkaido University, Sapporo 060, Japan

ATMS is the abbreviation of the "Amalgamation of Two-body correlations into Multiple Scattering process", and is a method for treating few-body systems with realistic interactions. The basic idea, the formulation and the accuracy of the method are presented. How to use the ATMS method is explained through the calculations of three- and four-nucleon systems with central NN interactions, the Coulomb three-body system (dtμ) and momentum distributions in ^{4}He.

1. Introduction

1.1. ^{4}He Nucleus

What happens when two nucleons come very close to each other in the nucleus? Are they excited to the Δ-isobar state, or melted into a six-quark state? The alpha particle is the most suitable nucleus to investigate the short-range behavior of the nuclear wave function, as discussed below.

Let's see the nucleon density distribution of ^{4}He, which is given by employing a harmonic oscillator model as

$$\rho(r) = 4\left(\frac{4\beta}{3\pi}\right)^{3/2} \exp\left(-\frac{4}{3}\beta r^2\right), \tag{1}$$

$$\hbar\omega = \frac{\hbar^2}{M}\beta = 21.6 \text{ MeV}, \tag{2}$$

where the center-of-mass motion is removed. As seen in Fig. 1, the central density of ^{4}He is very high compared with the normal density $\rho_0 = 0.18$ fm^{-3}, that is, the central density of heavy nuclei. The density ρ exceeds 1.5 ρ_0 (2 ρ_0) in about 17 % (4 %) of the total volume. This feature is essentially retained for the realistic density

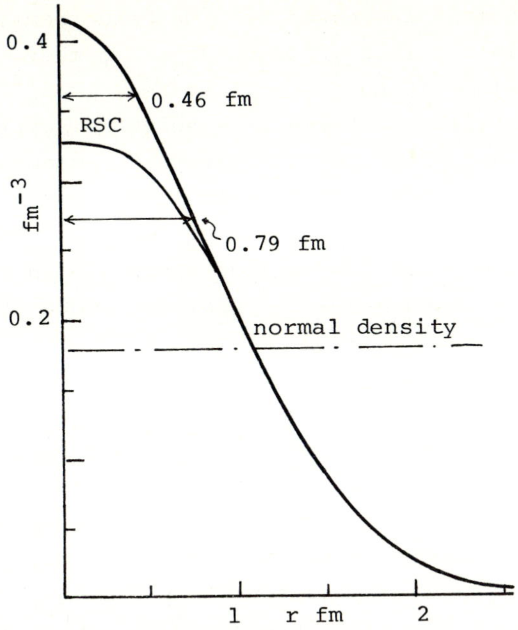

Fig. 1.
The density distributions of ^{4}He from the harmonic oscillator model and from the Reid soft core potential.

distribution obtained with the Reid soft-core (RSC) NN potential [1]. If an "exotic" component such as a six-quark state could exist in a high-density region, the ^{4}He nucleus would be the best one to be studied.

A more suitable parameter for the short-range component than the density is the probability P of finding two nucleons within a short distance d;

$$P = \langle \Phi_0 | \sum_{(ij)} \exp(-(r_{ij}/d)^2) \, P_{ij}^S | \Phi_0 \rangle, \tag{3}$$

where the projection operator P_{ij}^S is introduced by assuming that only S-state nucleon-pairs could melt into "exotic" states. The results with harmonic oscillator wave functions are given in Fig. 2. The

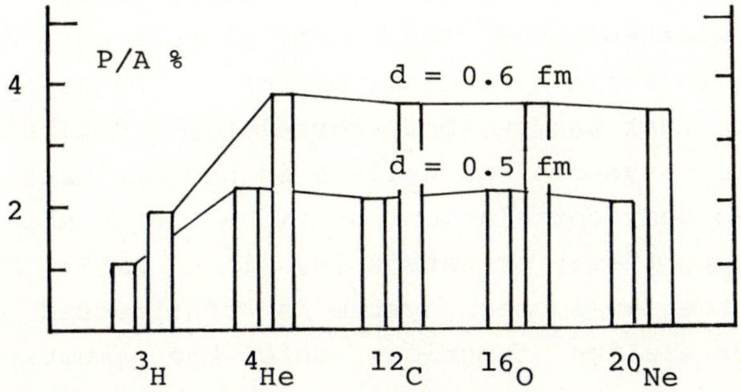

Fig. 2.
The probability of finding two nucleons within a short distance d.

probability per nucleon P/A abruptly increases at ^{4}He and remains almost constant to heavier nuclei. This is because the number of relative-S nucleon-pairs is not A(A-1)/2 but 3(^{3}H), 6(^{4}He), 27(^{12}C), 42(^{16}O) and 54.75(^{20}Ne). The alpha particle is a solvable system with the realistic interaction, and might contain "exotic" components as much as heavy nuclei do. Therefore, ^{4}He could give decisive information about what happens in the central region of nuclei. This property distinguishes ^{4}He from the other few-nucleon systems, ^{2}H and ^{3}He. Thus, the alpha particle takes a special position among the whole of nuclei.

1.2. Background of ATMS

In addition to the above-mentioned properties, the four-nucleon system is of fundamental interest to nuclear physics: The ^{4}He nucleus is the lightest system to exhibit the basic nuclear structure property of excited states and is the most important cluster unit in nuclei. In order to investigate structures of realistic four-body systems the present author and others proposed the ATMS method twelve years ago [2], which makes it possible to treat the four-nucleon system as well as the three-nucleon system on the basis of realistic nuclear interactions.

A pioneering work for ^{4}He had been done by Schmid, Tang and Herndon [3] for the central NN potential with a hard core. They used a Jastrow-type wave function and performed multiple numerical integrations by the Monte-Carlo method. Some progresses have been made along the variational approach: The ATMS wave function is the natural extension of the Jastrow one to include more dynamics of the realistic interaction such as the effect of the tensor force. Tanaka and Nagata [4] introduced in ATMS calculations the quasi-random number (QRN) method for numerical integrations which is much more effective than the Monte-Carlo one. Thus, Sakai et al.[5] succeeded by ATMS to treat the realistic four-nucleon system.

The ATMS method originates from the nuclear matter theory of Brueckner et al.[6,7] which established the essential role of the independent-pair correlation in infinite nuclear matter. Nagata and the present author [8] found that the two-body correlation dominates even in three- and four-nucleon systems as well as in nuclear matter. Thus, the idea of using two-body correlation functions is combined with the multiple scattering theory of Watson [9] into ATMS: The realistic wave function of the few-nucleon system is constructed by amalgamating two-body correlation functions into the multiple scattering process.

ATMS is reformulated by Tanaka [10] on the basis of the generalized wave-matrix theory [11]. This general formulation suggests that ATMS may be applicable not only to the few-nucleon system but also to various few-body systems. In fact, hypernuclei and few-atom molecules are successfully treated by Shinmura [12], Kurihara [13], Nakaichi-Maeda [14] and others. ATMS possesses wider applicability [15] than considered earlier.

2. The ATMS Method

ATMS constructs the wave function of the few-body system in the following form;

$$\Psi = F \Phi_0 , \tag{4}$$

where Φ_0 is a reference (model) wave function and F is a correlation function which is obtained by solving the multiple scattering equation with a few kind of two-body correlation functions. Here we give the foundation of the method.

2.1. Derivation of Multiple Scattering Equation

Consider the Schrödinger equation of an N-body system

$$H |\Psi_0\rangle = E_0 |\Psi_0\rangle \tag{5}$$

with the Hamiltonian

$$H = T + V = (\sum_i t_i - T^{cm}) + \sum_{(ij)} v_{ij} , \tag{6}$$

where (ij) denotes a distinguishable particle-pair. In order to get a bound-state solution of Eq.(5), we start from a reference system easier to treat:

$$H^R |\Phi_0\rangle = E_0^R |\Phi_0\rangle , \tag{7}$$

$$H^R = T + V^R = (\sum_i t_i - T^{cm}) + \sum_{(ij)} v^R_{ij} , \tag{8}$$

where v^R_{ij} is an appropriately chosen two-body interaction (reference interaction). The reference state $|\Phi_0\rangle$ is normalized as

$$\langle \Phi_0 | \Phi_0 \rangle = 1. \tag{9}$$

The exact solution $|\Psi_0\rangle$ of the Schrödinger equation can be connected with the reference state $|\Phi_0\rangle$ as

$$|\Psi_0\rangle = \hat{F}\,|\Phi_0\rangle\ , \tag{10}$$

$$\hat{F} = 1 + \frac{Q}{e}\tilde{V}\hat{F} \tag{11}$$

with $Q = 1 - |\Phi_0\rangle\langle\Phi_0|$, (12)

$$e = \mathcal{E} - H^R\ , \tag{13}$$

$$\tilde{V} = V - V^R - E_0 + \mathcal{E}\quad . \tag{14}$$

From Eqs.(10), (11) and (12) the state $|\Psi_0\rangle$ is normalized to be

$$\langle \Phi_0 | \Psi_0 \rangle = 1. \tag{15}$$

The energy E_0 is obtained from

$$E_0 = E_0^R + \langle \Phi_0 | V - V^R | \Psi_0 \rangle, \tag{16}$$

as is easily seen from $V - V^R = H - H^R$.

The proof of Eq.(10) is given by operating e on it and by using Eqs.(15) and (16):

$$\begin{aligned}(\mathcal{E} - H^R)|\Psi_0\rangle &= (\mathcal{E} - E_0^R)|\Phi_0\rangle + (V - V^R - E_0 + \mathcal{E})|\Psi_0\rangle \\ &\quad - |\Phi_0\rangle\langle\Phi_0|(V - V^R - E_0 + \mathcal{E})|\Psi_0\rangle \\ &= (V - V^R - E_0 + \mathcal{E})|\Psi_0\rangle,\end{aligned}$$

Therefore, $(E_0 - H)|\Psi_0\rangle = 0$.

Equations (16) and (10) produce a perturbation series;

$$E_0 = E_0^R + \langle\Phi_0|V^{res}|\Phi_0\rangle + \langle\Phi_0|V^{res}\frac{Q}{e}V^{res}|\Phi_0\rangle + \cdots, \tag{17}$$

$$V^{res} = V - V^R\ .$$

This series is the Brillouin-Wigner one for $\mathcal{E} = E_0$ and the Rayleigh-Schrödinger one for $\mathcal{E} = E_0^R$. The reference interaction V^R must be, of course, chosen so that the series converges. In ATMS the energy $\mathcal{E}$ is flexibly treated as a variational parameter.

Here, we define the reaction matrix g_{ij}:

$$g_{ij} = \tilde{v}_{ij} + \tilde{v}_{ij} \frac{Q}{e} g_{ij} , \tag{18}$$

$$\tilde{v}_{ij} = v_{ij} - v^R{}_{ij} - (E_0 - \mathcal{E})/n_P , \tag{19}$$

where n_P is number of particle pairs $N(N - 1)/2$ and $\tilde{v}_{ij}$ is chosen so that

$$\tilde{V} = \sum_{(ij)} \tilde{v}_{ij} . \tag{20}$$

Next, we introduce an operator $\hat{F}_{ij}$ defined by

$$g_{ij} \hat{F}_{ij} = \tilde{v}_{ij} \hat{F} . \tag{21}$$

The left-hand side of this equation is converted with Eq.(18) to

$$\tilde{v}_{ij} (1 + \frac{Q}{e} g_{ij}) \hat{F}_{ij}$$

and the right-hand side is done with the aid of Eqs.(11),(20) and (21) to

$$\tilde{v}_{ij} (1 + \sum_{(k\ell)} \frac{Q}{e} g_{k\ell} \hat{F}_{k\ell}) .$$

Thus, we get

$$\hat{F}_{ij} = 1 + \sum_{(k\ell)}' \frac{Q}{e} g_{k\ell} \hat{F}_{k\ell} , \tag{22}$$

where the prime means the pair (ij) is omitted from the summation. Equation (11) is rewritten by the use of Eqs.(20) and (21) as

$$\hat{F} = 1 + \sum_{(ij)} \frac{Q}{e} g_{ij} \hat{F}_{ij} . \tag{23}$$

Equations (22) and (23) mean that the operator $\hat{F}$ describes the multiple scattering process in the N-body system. This can be seen from Fig. 3. Wavy lines denote two-body scatterings by the reaction matrices g. Let $\hat{F}$ be the level after complicated multiple two-body scatterings and $\hat{F}_{ij}$ a level just before the (ij) scattering starts.

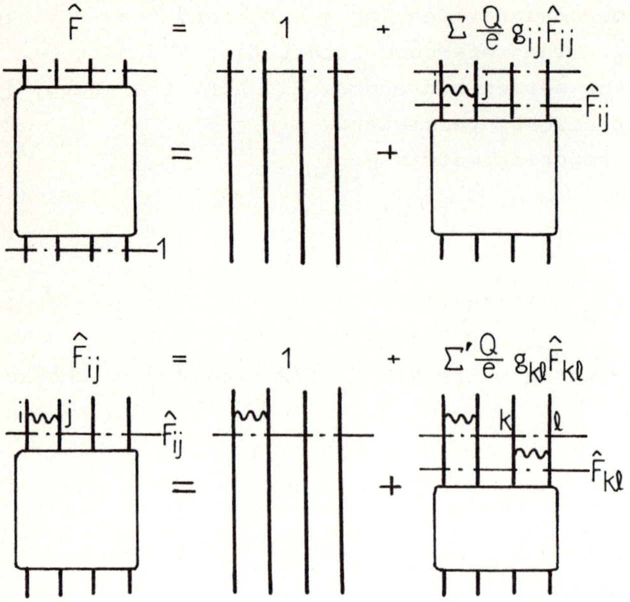

Fig. 3.
The multiple scattering process. Two-body scattering by g (wavy line) propagates upwards through Q/e.

The level $\hat{F}$ consists of a level with no scattering and such levels as that after the (ij) scattering. The level $\hat{F}_{ij}$ is also decomposed in a similar way. Thus, Fig. 3 corresponds to Eqs.(22) and (23).

In summary, the exact bound-state $|\Psi_0\rangle$ is produced without any approximation from the reference state $|\Phi_0\rangle$ as Eq.(10) by the multiple scattering operator of Eqs.(22) and (23).

2.2. ATMS Wave Function

Now, let's proceed to obtain a solution of the multiple scattering equation. In order to solve the reaction-matrix equation (18) we introduce the wave matrix Ω_{ij} by

$$g_{ij} = \tilde{v}_{ij} \ \Omega_{ij} \tag{24}$$

and define the state $|\Psi_{ij}\rangle$ by

$$|\Psi_{ij}\rangle = \Omega_{ij}\,|\Phi\rangle. \tag{25}$$

Then, the reaction-matrix equation on $|\Phi\,\rangle$ turns to the following equation;

$$|\Psi_{ij}\rangle = |\Phi\,\rangle + \frac{Q}{e}\,\tilde{v}_{ij}\,|\,\Psi_{ij}\rangle, \tag{26}$$

which corresponds to the Bethe-Goldstone equation in the nuclear matter theory. Since Eq.(26) is rewritten as

$$(1 + \frac{Q}{e} g_{ij})|\Phi\rangle = |\Psi_{ij}\rangle, \tag{27}$$

the scattering operator $(1 + \frac{Q}{e} g_{ij})$ may be replaced by a suitable correlation function u_{ij} as

$$(1 + \frac{Q}{e} g_{ij}) \longrightarrow u_{ij} = \Psi_{ij}/\Phi. \tag{28}$$

The essential idea of ATMS is to represent effectively the scattering operator with only a few kind of correlation functions, though mathematically infinite number of such functions are required. In the standard ATMS two kinds of correlation functions are used, that is,

$$u^{on}{}_{ij} \text{ for the on-shell case,} \quad u_{ij} \text{ for the off-shell case,} \tag{29}$$

where the on-shell and the off-shell are distinguished, according as the starting state $|\Phi\rangle$ of the (ij) scattering is the reference state $|\Phi_0\rangle$ or not.

The multiple scattering equations (22) and (23) are combined into

$$\hat{F} = \hat{F}_{ij} + \frac{Q}{e} g_{ij} \hat{F}_{ij} . \tag{30}$$

By summing it over all pairs, we get

$$n_p \hat{F} = \sum_{(ij)} \hat{F}_{ij} + \sum_{(ij)} \frac{Q}{e} g_{ij} \hat{F}_{ij} = \sum_{(ij)} \hat{F}_{ij} + (\hat{F} - 1)$$

and, therefore,

$$\hat{F} = 1 + \frac{1}{n_p - 1} \sum_{(ij)} (\hat{F}_{ij} - 1) . \tag{31}$$

Equation (30) is rewritten as

$$\hat{F} = (1 + \frac{Q}{e} g_{ij}) (\hat{F}_{ij} - 1) + (1 + \frac{Q}{e} g_{ij}) . \tag{32}$$

Since $\hat{F}$ operates on the reference state $|\Phi_0\rangle$, the replacement of Eq. (28) is done as follows

$$F = u_{ij} (F_{ij} - 1) + u^{on}{}_{ij} , \tag{33}$$

where the off-shell and the on-shell of Eq.(29) are taken into account and carets on F's are dropped because they are no longer operators but functions. By substituting into Eq.(31) the relation

$$(F_{ij} - 1) = (u_{ij})^{-1} (F - u^{on}_{ij}), \tag{34}$$

we get the ATMS representation of the multiple scattering operator;

$$F = D^{-1} [\prod_{(k\ell)} u_{k\ell}][\sum_{(ij)} u_{ij}^{-1} u^{on}_{ij} - (n_P - 1)], \tag{35}$$

where

$$D = [\prod_{(k\ell)} u_{k\ell}][\sum u_{ij}^{-1} - (n_P - 1)]. \tag{36}$$

It should be stressed that the replacement of various off-shell (ij) scatterings with a single u_{ij} is only one approximation made so far.

Thus, the ATMS wave function is obtained in the form of Eq.(4) with F of Eq.(35) which is obtained by amalgamating correlation functions into the multiple scattering process. The total correlation function F vanishes whenever any two particles come closer than the hard core radius of the interaction if exists. Thus, F is a natural extension of the Jastrow correlation which is a special case of Eq.(35) with $u^{on} = u$ and $D = 1$.

2.3. Two-Body Correlation Function

In ATMS the functions u_{ij} and u^{on}_{ij} are treated as the two-body functions by taking account of the donimance of the independent-pair correlation found in the reaction-matrix theory.

When the harmonic oscillator model is used as the reference state of Eq.(7), the on-shell correlation function is obtained to be of two-body one without any approximation. The reference state of the $(0s)^N$ ground configuration is

$$\Phi_0 = C \exp[-\frac{1}{2} \beta \sum_{i=1}^{N} (\vec{r}_i - \vec{R}^{cm})^2], \qquad C = (\frac{\beta^{N-1}}{\pi^{N-1} N})^{3/4}, \tag{37}$$

and the reference two-body potential of Eq.(8) is

$$v^R_{ij} = \frac{1}{2N} M \omega^2 (\vec{r}_i - \vec{r}_j)^2 - \frac{1}{n_P} U_0, \qquad U_0 = (N-1)\frac{3}{2}\hbar\omega - E^R_0. \tag{38}$$

In the reaction-matrix equation (26), two-body relative functions are factorized out as

$$\Phi_0 = \phi_0(\vec{r})\ \tilde{\Phi}_0(\vec{\xi}_2,\cdot\cdot,\vec{\xi}_{N-1}), \qquad \Psi_{12} = \psi_0(\vec{r})\ \tilde{\Phi}_0, \tag{39}$$

where the Jacobi coordinate

$$\vec{\xi}_n = \vec{r}_{n+1} - \frac{1}{n}\sum_j \vec{r}_j, \qquad n = 1, 2, \cdot\cdot, N-1 \tag{40}$$

is used. Hereafter, we often use $\vec{r} = \vec{\xi}_1$. Then, the many-body equation (26) is reduced to the two-body equation of

$$[\ \hat{h} - \frac{\hbar^2}{M}\gamma]\ \psi_0(\vec{r}) = \phi_0(\vec{r})\ \langle\phi_0|\ \hat{h} - \frac{\hbar^2}{M}\gamma\ |\psi_0\rangle\ , \tag{41}$$

$$\text{with } \hat{h} = -\frac{\hbar^2}{M}\vec{\nabla}_r^2 + v(\vec{r}) + v^{ho}(\vec{r}),$$

$$v^{ho}(\vec{r}) = \frac{1}{2}(\frac{1}{2} - \frac{1}{N})\ M\ \omega^2\ \vec{r}^2, \tag{42}$$

$$\frac{\hbar^2}{M}\gamma = (\frac{1}{2} - \frac{1}{N})\ 3\hbar\omega + \frac{1}{n_P}E_0^R\ , \tag{43}$$

where ε is chosen so as to eliminate the unknown E_0. Equation (41) might seem to be an eigenvalue problem, but it has solution for any γ. In fact, the solution of following equation;

$$[\ \hat{h} - \frac{\hbar^2}{M}\gamma]\ \psi_0(\vec{r}) = \phi_0(\vec{r}) \tag{44}$$

is the solution of Eq.(41). The reference energy E_0^R is regarded as a variational parameter in the ATMS calculation and its optimum value would be found around E_0. Thus, the two-body on-shell correlation function is obtained by

$$u^{on}(\vec{r}) = \psi_0(\vec{r})/\ \phi_0(\vec{r}) \tag{45}$$

with the solution of Eq.(44). In the case of the off-shell correlation, we need approximations. Fortunately, because the off-shell is less important than the on-shell, we can treat it well by assuming a two-body function with variational parameters.

2.4. Procedure for ATMS Calculation

The ATMS calculation is carried out through the following four steps.

First step: We introduce a reference (model) wave function which describes dominant symmetries of the system, where the isospin-spin part is antisymmetric and the spatial part is symmetric. The reference function should be free from a spurious center-of-mass motion which causes serious troubles in the few-body calculation.

Second step: We prepare two-body correlation functions by solving the reaction-matrix equation such as Eq.(44) or the ATMS-Euler equation in Section 3.1. The dispersion e and the exclusion Q are the many-body effects included in the equation.

An essential point of ATMS is to represent the operator with two kinds of correlation functions as Eq.(29). ATMS succeeds in taking account of the dynamical property of the system through these two-body correlation functions which well reflect the realistic interaction.

Third step: We construct the total correlation function based on the multiple scattering equations (22) and (23) as Eq.(35).

Final step: We refine the ATMS wave function under the Ritz variational principle. We search optimum values of variational parameters such as β in $|\Phi_0\rangle$ and E_0^R by minimizing the expectation value of the original Hamiltonian

$$E_U = \langle \Psi H \Psi \rangle / \langle \Psi | \Psi \rangle. \tag{46}$$

An advantage from the final step is that the ATMS energy becomes a strict upper-bound to the exact ground-state energy of H.

In the ATMS method, both the many-body dynamical treatment and the variational treatment are unified: The former substantially reduces the number of variational parameters. The latter determines variationally ambiguous parts of the many-body treatment. ATMS with few parameters at a "deeper level" stands comparison with the conventional variation method with a lot of parameters introduced "directly" into the trial function itself.

3. Method for Calculation

3.1. ATMS-Euler Equation

The two-body correlation function is obtained as the solution of the reaction-matrix equation in Section 2.3. When the ATMS-type wave function is set up, there exists another way to determine the two-body correlation function, that is, Euler-Lagrange's equation.

To put it simply, we consider a system of three equal-mass particles interacting via a central potential and assume a Jastrow-type (the simplest case of ATMS) wave function;

$$\Psi = \prod_{(ij)} f(r_{ij}), \tag{47}$$

where $f = u\,\phi$; $\Phi = \prod_{(ij)} \phi(r_{ij})$. As the coordinate we use interparticle distances r_{ij} and three Euler angles. In the three-body case, since the interparticle distances are expressed as

$$\begin{aligned} r_{12} &= \sqrt{\vec{r}^2}, \\ r_{23} &= \sqrt{\vec{\xi}_2^2 + \vec{r}^2/4 - \vec{r}\vec{\xi}_2}, \\ r_{31} &= \sqrt{\vec{\xi}_2^2 + \vec{r}^2/4 + \vec{r}\vec{\xi}_2}, \end{aligned} \tag{48}$$

the followings hold for any function f of only r_{ij}:

$$\begin{aligned} \vec{\nabla}_r f &= \left[\frac{\vec{r}}{r_{12}} \frac{\partial}{\partial r_{12}} + \frac{1}{2} \frac{\vec{r}_{23}}{r_{23}} \frac{\partial}{\partial r_{23}} + \frac{1}{2} \frac{\vec{r}_{31}}{r_{31}} \frac{\partial}{\partial r_{31}} \right] f, \\ \vec{\nabla}_{\xi_2} f &= \left[- \frac{\vec{r}_{23}}{r_{23}} \frac{\partial}{\partial r_{23}} + \frac{\vec{r}_{31}}{r_{31}} \frac{\partial}{\partial r_{31}} \right] f, \end{aligned} \tag{49}$$

where $\vec{r}_{23} = \vec{r}/2 - \vec{\xi}_2$ and $\vec{r}_{31} = \vec{r}/2 + \vec{\xi}_2$. The kinetic energy operator

$$T = -\frac{\hbar^2}{M} \vec{\nabla}_r^2 - \frac{3\hbar^2}{4M} \vec{\nabla}_{\xi_2}^2, \tag{50}$$

is expressed by the derivatives with respect to r_{ij}'s. The four-body case is similarly treated.

Thus, the Hamiltonian is written for the wave function of only r_{ij}'s as

$$\begin{aligned} H = &- (\hbar^2/M) \Big[\sum_{(ij)} \Big\{ \frac{\partial^2}{\partial r_{ij}^2} + \frac{2}{r_{ij}} \frac{\partial}{\partial r_{ij}} \Big\} \\ &+ \sum_{(ijk)} \Big\{ \cos\theta_j \frac{\partial}{\partial r_{ij}} \frac{\partial}{\partial r_{jk}} + \cos\theta_k \frac{\partial}{\partial r_{jk}} \frac{\partial}{\partial r_{ki}} + \cos\theta_i \frac{\partial}{\partial r_{ki}} \frac{\partial}{\partial r_{ij}} \Big\} \Big] \\ &+ \sum_{(ij)} v(r_{ij}), \end{aligned} \tag{51}$$

$$\cos\theta_j = (r_{ij}^2 + r_{jk}^2 - r_{ki}^2)/(2\, r_{ij} r_{jk}),$$

where the summation on (ij) or (ijk) is done over all distinguishable pairs or trios.

When the wave function is set up, the best two-body function f is determined by the variational principle

$$\delta \left[\langle \Psi \; H \; \Psi \rangle - \lambda \langle \Psi \,|\, \Psi \rangle \right] = 0, \tag{52}$$

where the variation δ is done with respect to f, and λ is Lagrange's multiplier to conserve the norm. The Euler equation is

$$[-\frac{\hbar^2}{M}\left\{\frac{d^2}{dr^2} + \left(\frac{2}{r} + \frac{1}{S(r)}\frac{dS}{dr}\right)\frac{d}{dr}\right\} + U(r) + v(r)]\, f(r)$$

$$= \lambda\, f(r), \tag{53}$$

Here, we show explicit exressions of S and U only for the case of three-body system:

$$S(r) = \int d\vec{\xi}_2 \quad [\, f(r_{23})\, f(r_{31})\,]^2_{\xi_1=r} \tag{54}$$

$$\begin{aligned} S(r)\, U(r) = \int d\vec{\xi}_2 \quad & f(r_{23})\, f(r_{31}) \\ & \times [-\frac{\hbar^2}{M}\left\{ \left\{ f''(r_{23}) + \frac{2}{r_{23}} f'(r_{23}) \right\} f(r_{31}) \right. \\ & \quad + \left\{ f''(r_{31}) + \frac{2}{r_{31}} f'(r_{31}) \right\} f(r_{23}) \\ & \quad \left. + \frac{r_{23}^2 + r_{31}^2 - r^2}{2\, r_{23}\, r_{31}} f'(r_{23})\, f'(r_{31}) \right\} \\ & + \left\{ v(r_{23}) + v(r_{31}) \right\} f(r_{23})\, f(r_{31})\,]_{\xi_1=r} \end{aligned} \tag{55}$$

Equation (53) is derived with the aid of such relations as

$$[\frac{\partial r_{23}}{\partial r}] = \frac{1}{2}\,\frac{\vec{r}\,\vec{r}_{23}}{r\, r_{23}} = \frac{1}{2}\,\frac{r^2 + r_{23}^2 - r_{31}^2}{2r\, r_{23}}, \tag{56}$$

where $\vec{\xi}_2$ and the angles of $\vec{r}$ are fixed at [15].

It has a physical meaning to exclude the third term from Eq.(53) by an "off-shell" transformation;

$$g(r) = \sqrt{S(r)}\, f(r). \tag{57}$$

Then, Eq.(53) becomes

$$[-\frac{\hbar^2}{M}\left\{\frac{d^2}{dr^2} + \frac{2}{r}\frac{d}{dr}\right\} + U^{av}(r) + v(r)]\, g(r) = \lambda\, g(r) \tag{58}$$

with $$U^{av}(r) = U(r) + \frac{\hbar^2}{M}\left\{\frac{1}{rS}\frac{dS}{dr} + \frac{1}{2S}\frac{d^2S}{dr^2} - \frac{1}{4S^2}\left(\frac{dS}{dr}\right)^2\right\}. \tag{59}$$

The function g(r) is just the two-body wave function of the system;

$$|g(r)|^2 = \int d\vec{\xi}_2 \quad |\Psi(\vec{r}, \vec{\xi}_2)|^2 \tag{60}$$

and conserves the norm of the total wave function;

$$\int d\vec{r}\, |g(r)|^2 = \langle \Psi | \Psi \rangle. \tag{61}$$

The present method is easily extended to general cases. In fact, results in Sections 4 and 5 are obtained by the ATMS-Euler equation.

3.2. Numerical Calculation

A following transformation of the variable is recommended;

$$x = 1 - \exp(- r/a) ; \tag{62}$$

then x varies from 0 to 1 as r ranges from 0 to ∞. This makes it easy to impose a boundary condition on g(r) at $r = \infty$ and to integrate quantities up to infinity. Equation (58) is expressed in the normal form on x;

$$- \frac{d^2}{dx^2} W(x) + \left\{ P(x) - \lambda\, Q(x) \right\} W(x) = 0 , \tag{63}$$

$$P(x) = \left\{ \frac{M}{\hbar^2} a^2 \,(U^{av} + v) - \frac{1}{4} \right\} / (1 - x)^2 ,$$

$$Q(x) = \frac{M}{\hbar^2} a^2 / (1 - x)^2,$$

where $W(x) = \sqrt{1 - x}\;\, r\, g(r)$. (64)

Here, we employ the Fox-Goodwin method [16]: The range of x is split up by a number of points $x_0, x_1, \cdot\cdot, x_M, x_{M+1}$ at equal intervals h, and Eq.(63) is written as

$$- \frac{1}{h^2} \left\{ \hat{\delta}^2 + \frac{1}{240} \hat{\delta}^6 + \cdot\cdot \right\} W(x)$$

$$+ (1 + \frac{1}{12} \hat{\delta}^2) \left\{ P(x) - \lambda\, Q(x) \right\} W(x) = 0. \tag{65}$$

with the central difference operator $\hat{\delta}$ defined by

$$\hat{\delta}\, y(x) = y(x + \frac{1}{2} h) - y(x - \frac{1}{2} h).$$

By neglecting $\hat{\delta}^6$ and higher terms we get a matrix eigenvalue problem;

$$\sum_{m=1}^{M} [A_{n,m} - \lambda\, B_{n,m}]\, W_m = 0 , \tag{66}$$

$$\begin{cases} A_{n,n} = \frac{5}{6} h^2 P_n + 2 , & A_{n-1,n} = A_{n+1,n} = \frac{1}{12} h^2 P_n - 1, \\ B_{n,n} = \frac{5}{6} h^2 Q_n , & B_{n-1,n} = B_{n+1,n} = \frac{1}{12} h^2 Q_n, \\ \text{otherwise} = 0, \end{cases} \tag{67}$$

where $P_m = P(x_m)$, $Q_m = Q(x_m)$ and $W_m = W(x_m)$. This eigenvalue problem can be solved by iterating the following with a guess Λ ;

$$\sum_{m=1}^{M} [A_{n,m} - \Lambda B_{n,m}] W_m^{(k)} = \sum_{m=1}^{M} B_{n,m} W_m^{(k-1)}, \quad k = 1, 2, \cdots \tag{68}$$

Then, the eigenvalue λ nearest to Λ is obtained by

$$\lambda = \Lambda + (W_m^{(k-1)}/W_m^{(k)}) \tag{69}$$

together with the corresponding eigenvector W.

3.3. Multi-Dimensional Integration

In realistic four-body calculations we have to carry out 6- or 9-dimensional numerical-integrations. The quasi-random number (QRN) method [17] applies to such multi-dimensional integrations and, in many cases, works more effectively than the Monte-Carlo (MC) method.

Suppose an integral

$$I = \int d\vec{\xi}_1 \cdots \int d\vec{\xi}_{N-1} \; Y(\vec{\xi}_1, \vec{\xi}_2, \cdots, \vec{\xi}_{N-1}). \tag{70}$$

We connect the Jacobi coordinates with the variables x_i of [0, 1], for example, as

$$(\vec{\xi}_1, 2\vec{\xi}_2, \cdots, (N-1)\vec{\xi}_{N-1}) = (X_1, X_2, \cdots, X_{3(N-1)}), \tag{71}$$

$$X_i = c \tan \pi (x_i - \frac{1}{2}), \tag{72}$$

and write the integral as

$$I = \int_0^1 dx_1 \cdots \int_0^1 dx_d \; y(x_1, x_2, \cdots, x_d), \tag{73}$$

where $d = 3(N-1)$ and

$$y = [(N-1)!]^{-3} \prod_{i=1}^{d} [c\pi(1 + (X_i/c)^2)] \tag{74}$$

Here, we extend the integrand y to a periodic function z with the period 2 as

$$z(x_1, x_2, \cdots, x_d) = y(|x_1|, |x_2|, \cdots, |x_d|),$$
$$z(\vec{x} + 2\vec{n}) = z(\vec{x}), \tag{75}$$

where $\vec{x} = (x_1, \cdots, x_d)$ with $x_i = [-1, 1]$ and $\vec{n} = (n_1, n_2, \cdots, n_d)$ with any integer n_i. The function z can be expanded into a Fourier series of the period 2;

$$z(\vec{x}) = \sum_{\vec{m}} a(\vec{m}) \exp(i\pi\, \vec{m}\, \vec{x}). \tag{76}$$

Then, $I = 2^{-d} \int_{-1}^{1} \cdots \int_{-1}^{1} dx\, z(\vec{x}) = a(\vec{0}),$ (77)

where the higher Fourier amplitudes integrate to zero.

QRN is a method to pick out the amplitude $a(\vec{0})$. Starting from a set of irrational numbers $\vec{\alpha} = (\alpha_1, \alpha_2, \cdots, \alpha_d)$, we detemine sampling points, of which distribution is not truly random but quasi-random, and make a summation as

$$S_1(N) = \sum_{n=-N}^{N} z(n\,\vec{\alpha})$$
$$= (2N + 1)\, a(\vec{0}) + D_1(N), \tag{78}$$

where $D_1(N) = \sum_{\vec{m}}{}' a(\vec{m}) \sin(\pi(N + \frac{1}{2})\, \vec{m}\, \vec{\alpha})/\sin(\frac{1}{2}\pi\, \vec{m}\, \vec{\alpha})$.

We make a further summation

$$S_2(N) = \sum_{n=0}^{N} S_1(n),$$
$$= (N + 1)^2\, a(\vec{0}) + D_2(N), \tag{79}$$

$$D_2(N) = \sum_{\vec{m}}{}' a(\vec{m})\, [\sin(\tfrac{1}{2}\pi(N + 1)\, \vec{m}\, \vec{\alpha})/\sin(\tfrac{1}{2}\pi\, \vec{m}\, \vec{\alpha})]^2.$$

Haselgrove [17] proved that D_1 and D_2 remain to be $O(N^0)$ as N tends to infinity, if the derivative $\frac{\partial^{2d}}{\partial x_1^2 \cdots \partial x_d^2} z$ is of bounded variation. Thus, the double summation formula of QRN is obtained;

$$I = S_2(N)/(N + 1)^2 + O(N^{-2}). \tag{80}$$

The QRN error of $O(N^{-2})$ decreases very rapidly for large N compared with the Monte-Carlo error which is $O(N^{-1/2})$ and is statistical (that is, the probability of getting I within one standard deviation is 68 %). In this respect QRN gains an advantage over MC: The accuracy of QRN is investigated and compared with that of MC by employing a simple function [4]. The double summation of QRN comprises the flattening and the sampling of the integrand and thus assures a good convergence [15].

It should be mentioned that the usefulness of QRN depends on the choice of irrational numbers and on such transformations as Eqs.(71) and (72). In the present calculations the sets of irrational numbers are chosen from

$$\alpha_i = \text{the decimal of } \sqrt{N_i}\,, \qquad N_i = 181,\ 271,\ 23,\ 487,\ 787,\ 691,\ 919,\ 631,\ 709,\ 13,\ 137,\ 883, \tag{81}$$

The set seems to be safe against abrupt changes in the QRN convergence behavior which appear when $\vec{m}\,\vec{\alpha}$ in $D_1(N)$ or $D_2(N)$ extremely closes to any even integer.

The value of parameter c in Eq.(72) should be chosen so that the distribution of sampling points well resembles the integrand. Figure 4 is the distribution of sampling points on the interparticle distance obtained from Eqs.(71) and (72). We can choose a suitable c by comparing the two-body part of the integrand with this distribution. When the matching is good between them, the integrand becomes effectively flat and a rapid convergence of QRN realizes.

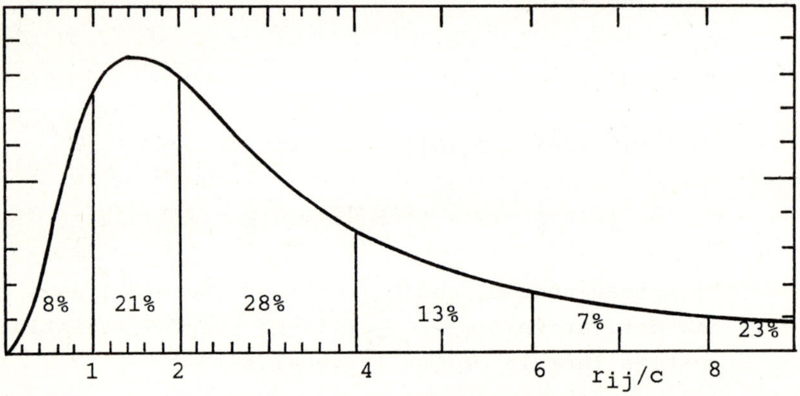

Fig. 4. The distribution of sampling points in interparticle distances. The numbers denote ratios of points within each interval.

4. Accuracy of ATMS

The accuracy of ATMS can be discussed unambiguously by the use of variational upper and lower bounds. Temple's lower bound [18] is

$$E_L = \langle H \rangle - (\langle H^2 \rangle - \langle H \rangle^2)/(E_1 - \langle H \rangle), \tag{82}$$

where $\langle\ \rangle$ denotes the expectation value with respect to the normalized variational wave function and E_1 is the first-excited-state energy.

4.1. Three- and Four-Nucleon Systems

For few-nucleon systems we make a comparison of various methods by employing the Malfliet-Tjon (M-T) NN potential [19];

$$v(r) = 1458.05 \exp(-3.11r)/r - 578.09 \exp(-1.55r)/r, \tag{83}$$

where v is in MeV and r in fm. Table 1 compares the results.

Table 1. Energies (MeV) of three- and four-nucleon systems obtained from the M-T potential with $(\hbar^2/M)=41.47$ MeV fm^2.

	Three-nucleon	Four-nucleon
VMC	-8.22±0.02	-31.19±0.05
Hu-Fi	-8.238	
CVHO	-8.244	
FCS	-8.25	
CCM(4)		-31.36
GFMC	-8.26±0.01	-31.3 ±0.2
ATMS	-8.26(E_U) -8.9(E_L)	-31.36(E_U) -32.8(E_L)

The variational Monte-Carlo method (VMC) by Lomnitz-Adler, Carlson, Pandharipande and Smith [20,21] uses symmetrized Jastrow correlation functions and the Monte-Carlo method for numerical integrations. The correlated variational harmonic-oscillator method (CVHO) by Ciofi degli Atti and Simula [22] is an extension of Strayer and Sauer's expansion method on the harmonic-oscillator basis [23]. Hu and Fink (Hu-Fi) [24] uses a polynomial basis with exponential weight functions. The Faddeev calculation in the coordinate space (FCS) is performed by Friar, Gibson and Payne [25]. The Green function Monte-Carlo method (GFMC) by Kalos, Zabolitzky and Schmidt [26,27] integrates the Shrödinger equation of the simple few-body system by means of Green's function and is accepted as the exact energy though its error bar is not small enough. The coupled cluster method (CCM) by Kümmel, Lührmann and Zabolitzky [28] provides the results consistent with GFMC. ATMS is the upper- and lower-bound energies obtained by the Euler-equation treatment. It is known that the ratio $(E_0 - E_L)/(E_U - E_0)$ becomes very large for the potential with a singular repulsion [3]. This fact enables us to estimate the exact energy from two ATMS bounds with a much smaller error than the GFMC's.

It is interesting to observe that both CVHO and ATMS describe the wave function in the form of Eq.(4). In CVHO, the correlation function F is of simple Jastrow-type with a few variational parameters but Φ_0 is expanded on a large enough harmonic-oscillator basis. In ATMS, though Φ_0 is taken to be simple, the correlation F is dynamically determined. One could find a more practical method in between.

For the convenience of Reader's calculational check we give the detailed results obtained with the central three-range Gaussian potential (C3G) [29];

$$v(r) = 2000 \exp(-(r/0.447)^2) - 318 \exp(-(r/0.942)^2) - 5 \exp(-(r/2.5)^2), \tag{84}$$

where v is in MeV and r in fm. Compared to M-T, C3G is more realistic and its finite core does not cause any trouble in the calculation.

The ATMS calculation is performed as follows: The transformation of Eq.(62) is done with a = 3.0 fm and that of Eq.(72) with c = 0.6 to 0.7 fm. The Euler equation is converted into a 90 × 90-matrix eigenvalue problem of Eq.(66) and a self-consistent g(r) is obtained by 15 times iterations. Multi-dimensional integrations are carried out by 70,000 QRN estimates. The first-excited-state energy E_1 of three- and four-nucleon systems are -0.281MeV and E_U of the three-nucleon case, respectively.

The energy results are summarized in Table 2. It is interesting to see how the two-body wave function of Eq. (57) behaves in the nucleus. The wave function is in general

$$|g(r)|^2 = \int d\vec{\xi}_2 \cdots \int d\vec{\xi}_{N-1} \left| \Psi(\vec{r}, \vec{\xi}_2, \cdots, \vec{\xi}_{N-1}) \right|^2. \tag{85}$$

with the normalized wave function Ψ . Figure 5 shows the two-body wave functions for C3G in the three- and four-nucleon systems together

Table 2. Energies (MeV) of three- and four-nucleon systems for C3G.

	Three-nucleon	Four-nucleon
Total energy E_U	-8.07	-32.03
Kinetic energy	36.88	86.18
Potential energy	-44.95	-118.21
Lower bound E_L	-10.2	-36.3
Rms radius (fm)	1.62	1.36

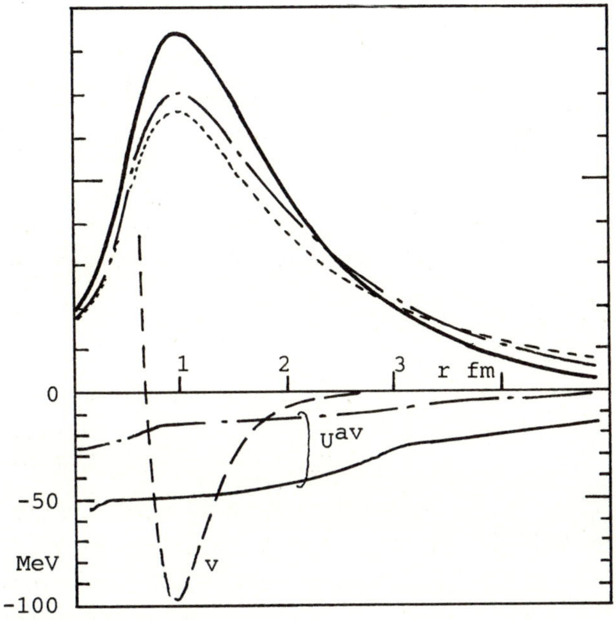

Fig. 5.
The two-body functions in the three-nucleon system (dot-dashed line) and in the four-nucleon system (solid line). The dotted line is the free one.

with the free NN wave function. We can see a confining effect in the nucleus, which becomes stronger in the four-nucleon system. This confining effect comes from U^{av} and λ of Eq.(58). The average potentials U^{av} , shown in Fig. 5 together with the original C3G, are of rather long range. The value of $|\lambda - U^{av} - v|$ is responsible for the curvature of the wave function, which is larger in the four-nucleon system.

4.2. Accuracy of Wave Function

The variational method has often been criticized that the wave function is rather poor even when the very accurate energy (E_U) is obtained. We can show that the variational wave function is very reliable when the gap G

$$G \equiv E_U - E_L \tag{86}$$

is small enough [30].

Suppose that the varational wave function Ψ is expanded on the exact eigenfunctions Ψ_n with the eigenvalue E_n of the Hamiltonian H;

$$\Psi = \sum_{n=0}^{\infty} a_n \Psi_n , \qquad \sum_{n=0}^{\infty} |a_n|^2 = 1. \tag{87}$$

We define the impurity in the wave function from mixing of eigenstates higher than a given energy E by

$$B(E) = \sum_{n=1}^{\infty} |a_n|^2 \quad \Theta(E_n - E) \tag{88}$$

where Θ is 0 for $E_n < E$ and 1 for $E_n \geqq E$. We see immediately that the following relations hold;

$$B(E_k)\cdot(E_k - E_U)^2 \leqq \sum_{n=k}^{\infty} |a_n|^2 (E_n - E_U)^2 \leqq C_0 ,$$
$$C_0 = \sum_{n=0}^{\infty} |a_n|^2 (E_n - E_U)^2. \tag{89}$$

With the aid of Eqs.(82), (87) and

$$\langle H \rangle = E_0 + \sum_{n=1}^{\infty} (E_n - E_0) \; |a_n|^2 , \tag{90}$$

$$\langle H^2 \rangle = E_0{}^2 + \sum_{n=1}^{\infty} (E_n{}^2 - E_0{}^2) \; |a_n|^2 , \tag{91}$$

we rewrite C_0;

$$C_0 = \langle H^2 \rangle - \langle H \rangle^2 = G\cdot(E_1 - E_U). \tag{92}$$

From Eqs.(89) and (92) we get the variational upper bound B_U to the impurity

$$B(E) \leqq B_U(E),$$

$$B_U(E) \equiv G\cdot(E_1 - E_U)/(E - E_U)^2. \tag{93}$$

The total impurity is given by taking $E = E_1$.

The accuracy of ATMS wave function is investigated for the system of three point-α particles interacting via the potential [31];

$$v(r) = 500 \exp(-(0.7r)^2) - 130 \exp(-(0.475r)^2), \tag{94}$$

where v is in MeV and r in fm. The upper-bound energy E_U of -5.18 MeV and the lower-bound energy E_L of -5.31 MeV are obtained with $\hbar^2/M$ = 41.467/4 MeV fm^2, and E_1 is -1.37 MeV. Therefore, the total impurity in the ATMS wave function is 3.4 %. Equation (93), however, tells us the impurity decreases drastically as E increases. For example, the impurity from mixing of eigenstates higher than 17 MeV becomes less than 0.1 %. Figure 6 shows the error of the charge form factor. The inaccuracy at high momentum transfers comes mostly from the contamination of high-energy eigenstates which is restricted to be small enough. The error is bounded in the shaded area of Fig. 6.

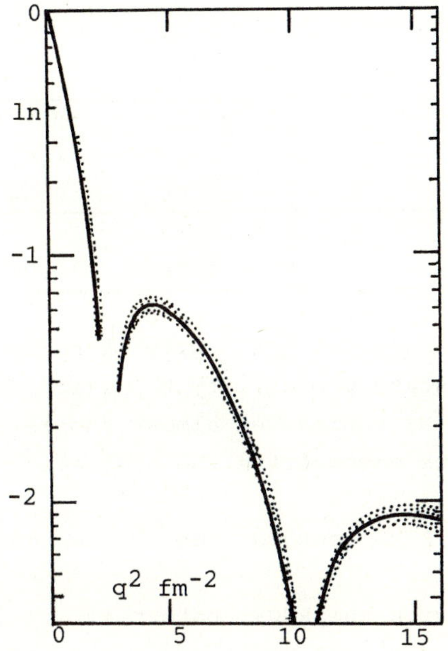

Fig. 6.
The error of the charge form factor by ATMS, which is confined into the shaded area.

We should pay attention to the Temple lower bound if we want to see the accuracy of the wave function: Because Eq.(82) involves $\langle \vec{\nabla}^4 \rangle$ the lower bound is extremely sensitive to the curvature in the wave function and gives a severe check to how well the short-range correlation is described.

4.3. Realistic Case

The accuracy of ATMS is investigated by Morita with the realistic nuclear interaction. The NN potential employed are the Reid soft-core V_8 model ($RSCV_8$) used in the VMC calculation [20] and the Reid soft core $^1S_0$, $^3S_1$-$^3D_1$ potential (RSC5) used in the Faddeev 5-channel calculations [32,33,34]. The results are shown in Table 3. In the case of $RSCV_8$, ATMS gives the upper-bound energy lower by about 0.27 MeV than the result of VMC.

Table 3. Energies (MeV) of ^3H with RSC potentials.

NN potential	$RSCV_8$		RSC5	
Method	ATMS	VMC	ATMS	Faddeev
Total energy	-7.13	-6.86±0.08	-7.02	-7.023
Kinetic energy	49.76	47.81	49.17	
Potential energy	-56.89	-54.68	-56.19	
Central	-14.15		-13.85	
Tensor	-43.75		-43.45	
LS	1.01	0.95	1.11	
P(S) (%)	89.75		89.63	88.91
P(S') (%)	1.47		1.71	1.67
P(D) (%)	8.79		8.66	9.34
Mass rms radius (fm)	1.81		1.83	
Charge rms rad. (fm)	1.68		1.70	1.70

In the case of RSC5, several Faddeev calculations give consistent results for ^3H. Except for the D-state probability to which the total energy is rather insensitive, ATMS reproduces almost completely the Faddeev result. These comparisons demonstrate the high accuracy of ATMS.

It should be stressed that the ATMS method can be extended

Table 4. Energies (MeV) of ^4He with the $RSCV_8$ potential.

Method	ATMS	VMC	Exp.
Total energy	-21.8	-22.9±0.5	-28.3
Kinetic energy	102.9	106.9	
Potential energy	-124.7	-129.8	
Central	-37.6		
Tensor	-89.1		
LS	2.0	4.05	
P(S) (%)	88.98		
P(S') (%)	0.17		
P(D) (%)	10.86		
Rms radius (fm)	1.54		1.47

straightforwardly to the four-nucleon system. Table 4 shows an preliminary ATMS result for the $RSCV_8$ potential, where the number of two-body correlation functions introduced is small compared to the three-nucleon case. The final calculation will be soon accomplished.

In conclusion, we can say that ATMS is a powerful method to solve few-body systems with high accuracy. It should be mentioned that the form of Eq.(4) is a successful description of the physical content of the realistic few-body system.

5. Coulomb Few-Body System

Three-body systems interacting via the Coulomb interaction are the important subjects of the few-body physics: Recently, the muonic molecule (dtμ) has aroused considerable attention in relation to the realization of a useful muon-catalyzed fusion [35]. The negative positronium ($e^-e^+e^-$) is of special interest as the lightest ternary lepton system. Owing to the long-range nature of the Coulomb interaction some difficulty arises in the momentum space Faddeev calculation. The Coulomb three-body bound problem provides a severe testing ground for the ATMS method, because both attractive and repulsive pairs must be simultaneously treated with high accuracy. It is interesting to see how well ATMS works.

5.1. Three Kinds of ATMS Wave Functions

Here, we introduce three kinds of wave functions derived from different approximations of the multiple scattering equations (22) and (23).

The simplest one is obtained by converting, with Eq.(28), the operator equation

$$\hat{F}_{ij} = 1 + \sum_{(k\ell)}{}' \frac{Q}{e} g_{k\ell}\ \hat{F}_{k\ell} \tag{95}$$

into

$$F_{ij} = 1 + \sum_{(k\ell)}{}'(u_{k\ell} - 1)\ F_{k\ell}\,. \tag{96}$$

By solving this coupled algebraic equation with respect to F_{ij}'s we get the "0th-kind ATMS",i.e. the Jastrow, wave function

$$\Psi = \prod_{(ij)} f_{ij}(r_{ij}), \tag{97}$$

where $f_{ij} = u_{ij}\ \phi_{ij}$ with the initial state of $\Phi = \prod_{(ij)} \phi_{ij}(r_{ij})$. We put D of Eq.(35) to be unity: The reason will be discussed later.

The second one is the usual ATMS. Rewriting Eq.(95) as

$$[\hat{F}_{ij} - 1] = \sum_{(k\ell)}{}' \frac{Q}{e} g_{k\ell} + \sum_{(k\ell)}{}' \frac{Q}{e} g_{k\ell} \, [\hat{F}_{k\ell} - 1], \tag{98}$$

we replace the operator $\frac{Q}{e} g_{k\ell}$ with an on-shell $(u_{k\ell} - 1)$ in the first term and with an off-shell $(\bar{u}_{k\ell} - 1)$ in the second term on the right-hand side. The resultant equations are solved with respect to $[\cdot\cdot]$, and the D = 1 prescription is applied, once all pairs are correlated. Thus, we get

$$\begin{aligned} \Psi = {} & \bar{f}_{31}(r_{31})\, \bar{f}_{23}(r_{23})\, f_{12}(r_{12}) + \bar{f}_{12}(r_{12})\, \bar{f}_{31}(r_{31})\, f_{23}(r_{23}) \\ & + \bar{f}_{23}(r_{23})\, \bar{f}_{12}(r_{12})\, f_{31}(r_{31}) - 2\, \bar{f}_{31}(r_{31})\, \bar{f}_{23}(r_{23})\, \bar{f}_{12}(r_{12}), \end{aligned} \tag{99}$$

which is called the "1st-kind ATMS" wave function.

The third one is derived by rewriting Eq.(98) once more:

$$\begin{aligned} [\hat{F}_{ij} - 1 - \sum_{(k\ell)}{}' \frac{Q}{e} g_{k\ell}] = {} & \sum_{(k\ell)}{}' \frac{Q}{e} g_{k\ell} \sum_{(mn)}{}' \frac{Q}{e} g_{mn} \\ & + \sum_{(k\ell)}{}' \frac{Q}{e} g_{k\ell} \, [\hat{F}_{k\ell} - 1 - \sum_{(mn)}{}' \frac{Q}{e} g_{mn}]. \end{aligned} \tag{100}$$

On the right-hand side the first term is replaced with the on-shell and the off-shell functions as $\sum'(\bar{u}_{k\ell} - 1) \sum'(u_{mn} - 1)$, and the second term with a far-off-shell one as $\sum'(\tilde{u}_{k\ell} - 1)$ $[\cdot\cdot]$ because scatterings have occurred at least twice there. By solving the algebraic equation with respect to $[\cdot\cdot]$ we get

$$\begin{aligned} \Psi = {} & \tilde{f}_{31}(r_{31})\, \bar{f}_{23}(r_{23})\, f_{12}(r_{12}) + \tilde{f}_{23}(r_{23})\, \bar{f}_{31}(r_{31})\, f_{12}(r_{12}) \\ & + \tilde{f}_{12}(r_{12})\, \bar{f}_{31}(r_{31})\, f_{23}(r_{23}) + \tilde{f}_{31}(r_{31})\, \bar{f}_{12}(r_{12})\, f_{23}(r_{23}) \\ & + \tilde{f}_{23}(r_{23})\, \bar{f}_{12}(r_{12})\, f_{31}(r_{31}) + \tilde{f}_{12}(r_{12})\, \bar{f}_{23}(r_{23})\, f_{31}(r_{31}) \\ & - \tilde{f}_{31}(r_{31})\, \tilde{f}_{23}(r_{23})\, [f_{12}(r_{12}) + 2\, \bar{f}_{12}(r_{12})] \\ & - \tilde{f}_{12}(r_{12})\, \tilde{f}_{31}(r_{31})\, [f_{23}(r_{23}) + 2\, \bar{f}_{23}(r_{23})] \\ & - \tilde{f}_{23}(r_{23})\, \tilde{f}_{12}(r_{12})\, [f_{31}(r_{31}) + 2\, \bar{f}_{31}(r_{31})] \\ & + 4\, \tilde{f}_{31}(r_{31})\, \tilde{f}_{23}(r_{23})\, \tilde{f}_{12}(r_{12}), \end{aligned} \tag{101}$$

which we call the "2nd ATMS" wave function. This 2nd ATMS reduces to 1st ATMS if $\tilde{f}_{ij} = \bar{f}_{ij}$, and to 0th ATMS if $\tilde{f}_{ij} = \bar{f}_{ij} = f_{ij}$.

The prescription

$$D \text{ of Eq.(35)} = 1 \qquad (102)$$

is a procedure of obtaining Day's solution [36] of the multiple scattering equation. The two-body correlation function in the denominator D is of far-off-shell. Owing to the large starting energy of the far-off-shell scattering, the correlation function can not deviate so much from unity as long as the potential is not so strong and, therefore, D can not do. It would be interesting to note that the Jastrow wave function can be theoretically derived under some conditions.

5.2. (dtμ) Molecule

The ground-state energy of the (dtμ) molecule is calculated by using three kinds of ATMS wave functions. The calculation is done as follows: r_{ij} is transformed to x_{ij} by Eq.(62) with a = 900 fm, and x_{ij} is devided into 100 mesh points. Integrations are carried out by the use of Simpson's formula. Two-body functions are solved self-consistently by 15 to 24 times iterations.

The results are shown in Table 5. Though the total energies are similar, the physically important energies are those measured from the (tμ)+d threshold -2711.27 eV and are different among three ATMS's.

Table 5. Ground-state energy (eV) of the (dtμ) molecule.

	Total energy	Energy from (tμ)+d
0th ATMS	-2932.11	-220.84
1st ATMS	-3023.91	-312.64
2nd ATMS	-3030.45	-319.18

The 1st ATMS is superior to the 0th ATMS. The 2nd ATMS attains a still higher accuracy by the better treatment of the multiple scattering process. The 2nd ATMS result can satisfactorily be compared to the best value -319.15 eV of Vinitskiĭ et al.[37] obtained from an very elaborate adiabatic calculation. Thus, the ATMS method can treat with enough accuracy the Coulomb three-body system by improving the wave function systematically.

Let's see the wave function. We define the relative wave function between particles 1 and 2 as

$$\psi_{12}(\vec{r}) = \sqrt{\int d\vec{\xi}_2 \,|\, \Psi(\vec{r},\vec{\xi}_2)|^2}. \tag{103}$$

Figure 7 shows the results obtained by 2nd ATMS. The d-t relative wave function has a large zero-point motion centered around 500 fm which brings about a chance of the nuclear fusion at $r \sim 0$. The rms distances $\langle r^2 \rangle^{-1/2}$ are calculated to be 736 fm for d-t, 621 fm for d-μ, 595 fm for t-μ, which are compared with 468 fm of the free (dμ) and 460 fm of the free (tμ).

The fusion cross section of (dtμ) is proportional to the probability that d and t come closer to the range of nuclear interaction. The d-t relative wave function by 2nd ATMS is well reproduced by Eq.(58) with $U^m = U^{av} + v^C$ of

$$U^m(r) = v^C(r)\,(1 + r/a_0)\,\exp(-2r/a_0) - D\,\exp(-r/a) \tag{104}$$

$$v^C(r) = e^2/r \quad \text{for } r \geq r_0, \qquad e^2(3r_0^2 - r^2)/(2r_0^3) \quad \text{for } r < r_0,$$

where a_0 = 231 fm, a = 480 fm, r_0 = 2.5 fm and D = 0.002 MeV.

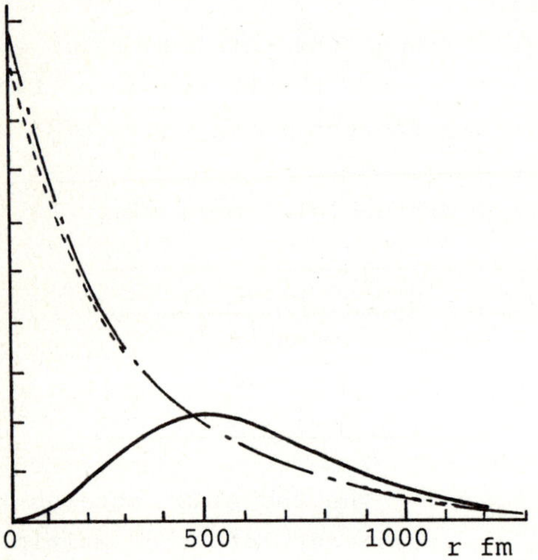

Fig. 7. The relative wave functions in (dtμ). The solid, the dot-dashed and the dotted lines are those of d-t, t-μ and d-μ, respectively.

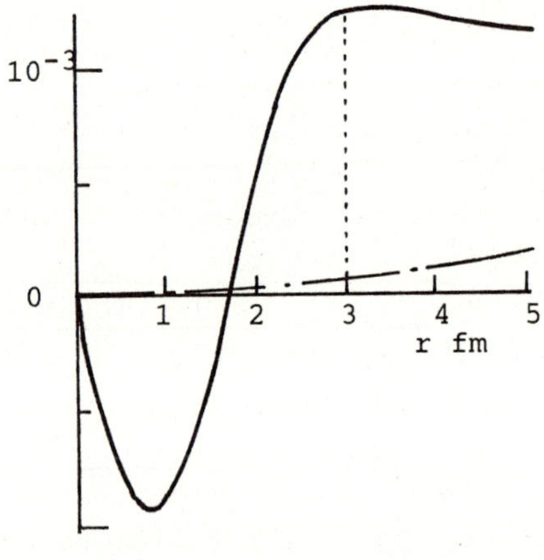

Fig. 8. The d-t relative wave function at short-distances. The solid and the dot-dashed lines are the cases with and without the nuclear interaction, respectively.

Now, we can investigate the effect of the nuclear interaction between d and t, which has a near-threshold resonance at +66 keV. The nuclear potential is given by Kamimura as

$$V^n(r) = -V_0 \,/[1 + \exp((r - r_0)/b)] \tag{105}$$

with b = 0.3 fm, r_0 = 2.5 fm and V_0 = 58.7 MeV. Figure 8 shows the effect of the nuclear interaction. The near-threshold nuclear resonance strongly couples with the ($dt\mu$) molecular state. The solid line is the case including the nuclear potential: The wave function is greatly enhanced in the nuclear interaction range, and the fusion rate becomes more than 200 times as large as that of the only Coulomb-interaction case denoted by the dot-dashed line.

5.3. Intermolecular Potential

The resonant formation of the ($dt\mu$) molecule occurs through the shallowest bound level of ($dt\mu$) [35], of which energy is about -0.64 eV. Then, between the total energy and the threshold energy, 4 digits cancels out. Here, we present another approach of using an intermolecular potential between ($t\mu$) and d.

Let's divide the Hamiltonian of the ($dt\mu$) system into the internal part ($t\mu$) and the external part;

$$H = H^{in} + T_\rho + V^{ex} \, , \tag{106}$$

$$H^{in} = T_r + V_{t\mu} \, , \qquad V^{ex} = V_{dt} + V_{d\mu} \, . \tag{107}$$

where $\vec{r} = \vec{r}_\mu - \vec{r}_t$ and $\vec{\rho} = \vec{r}_d - (\vec{r}_\mu + \vec{r}_t)/2$. The Schrödinger equation of the system is

$$H|\Psi\rangle = E|\Psi\rangle \, , \tag{108}$$

where $|\Psi\rangle$ is of the ground or excited state and is normalized as $\langle\Psi|\Psi\rangle = 1$. The ground state of ($t\mu$) satisfies

$$H^{in}|\phi_0) = \varepsilon_0 |\phi_0) \tag{109}$$

with a normalization $(\phi_0|\phi_0) = 1$. In this section a round ket $|\;)$ denotes a state vector in the internal space.

We define an external operator $\hat{F}^{ex}$ [38];

$$\hat{F}^{ex} = 1 + \frac{1 - |\phi_0)(\phi_0|}{E - H^{in} - T_\rho} V^{ex} \hat{F}^{ex}. \tag{110}$$

When E is below the $(t\mu)$+d threshold energy ε_0, the denominator $E - H^{in} - T_\rho$ never takes zero. An important relation follows from the definition Eq.(110);

$$(H - E) \hat{F}^{ex} = H^{in} + T_\rho + |\phi_0)(\phi_0| V^{ex} \hat{F}^{ex} - E . \tag{111}$$

This is proved as follows:

$$\begin{aligned}(H - E) \hat{F}^{ex} &= (H^{in} + T_\rho + V^{ex} - E) \hat{F}^{ex} \\ &= (H^{in} + T_\rho + V^{ex} - E) + |\phi_0)(\phi_0| V^{ex} \hat{F}^{ex} \\ &\quad - V^{ex} \hat{F}^{ex} + V^{ex} \frac{1 - |\phi_0)(\phi_0|}{E - H^{in} - T_\rho} V^{ex} \hat{F}^{ex} . \\ &= \text{right-hand side of Eq.(111)},\end{aligned}$$

where we used Eq.(110) twice.

Now, let's consider the following equation;

$$[\, H^{in} + T_\rho + |\phi_0)(\phi_0| V^{ex} \hat{F}^{ex} - E \,]|\Phi\rangle = 0 . \tag{112}$$

By expanding $|\Phi\rangle$ on the complete set $\{|\phi_n)\}$;

$$\langle\vec{r}, \vec{\rho}|\Phi\rangle = \sum_{n=0}^{\infty} (\vec{r}|\phi_n) \langle\vec{\rho}|\chi_n\rangle, \tag{113}$$

we get a set of equations for $|\chi_n\rangle$:

$$[\, T_\rho + \delta_{n0} (\phi_0| V^{ex} \hat{F}^{ex} |\phi_0) - (E - \varepsilon_n) \,] |\chi_n\rangle = 0 .$$

Since we are treating the state below the threshold, $(E - \varepsilon_n)$ is negative and therefore, only $|\chi_0\rangle$ can satisfy the boundary condition. From this fact and Eq.(111) it follows that the solution of the Shrödinger equation (108) can be expressed as

$$|\Psi\rangle = \hat{F}^{ex} |\phi_0 \;\; \chi_0\rangle , \tag{114}$$

where the function $|\chi_0\rangle$ satisfies

$$[\ T_\rho + U - \tilde{E}\ ]\,|\,\chi_0\rangle = 0\ , \tag{115}$$

where $U = (\phi_0|\ V^{ex}\ \hat{F}^{ex}\ |\,\phi_0)$, (116)

$$\tilde{E} = E - \varepsilon_0 . \tag{117}$$

Equation (116) is the definition of our intermolecular potential. The function $\chi_0(\vec{\rho})$ is a spectroscopic amplitude as seen from

$$\chi_0(\vec{\rho}) = \langle\ \vec{\rho},\ \phi_0|\ \Psi\ \rangle \tag{118}$$

which follows from Eqs.(114) and (110). Thus, Eq.(115) with the intermolecular potential Eq.(116) determines, under the boundary condition, the spectroscopic amplitude and the energy $\tilde{E}$.

Since the spectroscopic amplitude does not conserve the norm of the total wave function, we make a following "off-shell" transformation;

$$|\tilde{\chi}_0\rangle = \sqrt{J}\ |\,\chi_0\rangle \tag{119}$$

with

$$J = (\ \phi_0|\ \hat{F}^{ex\dagger}\ \hat{F}^{ex}\ |\,\phi_0) . \tag{120}$$

This transformed function is just the relative wave function as seen from

$$|\tilde{\chi}_0(\vec{\rho})|^2 = \int d\vec{r}\ \ |\,\Psi(\vec{r},\vec{\rho})|^2\ , \tag{121}$$

and conserves the norm as

$$\langle\Psi\ |\ \Psi\rangle = \langle\ \tilde{\chi}_0|\ \tilde{\chi}_0\rangle . \tag{122}$$

The external scattering operator $\hat{F}^{ex}$, as shown in Eq.(114), describes exactly the polarization and the deformation of the $(t\mu)$ atom caused by the third particle d. The operator $\hat{F}^{ex}$ can be decomposed into the multiple scattering process. By introducing the reaction matrix;

$$g^{ex}_{ij} = v^{ex}_{ij} + v^{ex}_{ij}\,\frac{Q}{e}\,g^{ex}_{ij}\ , \tag{123}$$

$$Q = 1 - |\phi_0)(\phi_0|\ , \qquad e = E - H^{in} - T_\rho\ ,$$

we can derive analogously to Section 2.1 the followng equations;

$$\hat{F}^{ex} = 1 + \sum_{(ij)} \frac{Q}{e} g^{ex}_{ij} \hat{F}^{ex}_{ij} ,$$
$$\hat{F}^{ex}_{ij} = 1 + \sum_{(k\ell)}' \frac{Q}{e} g^{ex}_{k\ell} \hat{F}^{ex}_{k\ell} . \tag{124}$$

This external multiple scattering equations can be solved in a good approximation by the ATMS technique, that is, by replacing the operator $(Q/e)g^{ex}_{ij}$ with a few kinds of two-body correlation functions [38].

First, let's see the case of no polarization and no deformation, i.e., $\hat{F}^{ex} = 1$. Then, U of Eq.(116) is nothing but a folding potential;

$$U^F(\vec{\rho}) = \int d\vec{r} \;\; \phi_0^*(\vec{r}) \; (v_{dt} + v_{d\mu}) \; \phi_0(\vec{r}) .$$
$$= \frac{e^2}{\rho} \left\{ \exp(-y_1)\cdot(1 + \frac{1}{2} y_1) - \exp(-y_2)\cdot(1 + \frac{1}{2} y_2) \right\},$$
$$y_1 = \frac{2}{a_0} \frac{M_t + M_\mu}{M_t} \rho , \quad y_2 = \frac{2}{a_0} \frac{M_t + M_\mu}{M_\mu} \rho , \tag{125}$$

where a_0 is the Bohr radius. This folding potential is positive everywhere as easily understood by Gauss's theorem, and therefore has no bound level.

In Fig. 9 the calculated intermolecular potential U is shown by

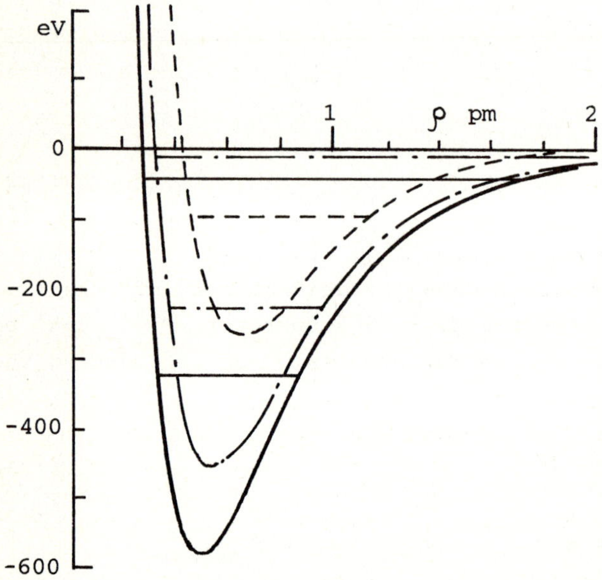

Fig. 9.
The intermolecular potential between ($t\mu$) and d. The solid curve is that for the ground state obtained by 2nd ATMS. The dot-dashed and the dashed curves are approximate ones for L=1 and L=2. The horizontal bars denote the positions of the levels.

the solid line. This potential is not for the relative wave function but for the spectroscopic amplitude to the $(t\mu)$ ground state. We can see the attraction of about -600 eV, which is the polarization effect of $(t\mu)$ described by $\hat{F}^{ex} - 1$. The L=0 potential can be simulated by a Morse-potential form;

$$U(\rho) = D\,[\exp(-2a(\rho - \rho_0)) - 2\exp(-a(\rho - \rho_0))] \tag{126}$$

with D = 580.3 eV, ρ_0 = 0.509 pm and a = 3.0136 pm^{-1}.

Although the present framework should separately be applied to each excited state of the $(dt\mu)$ system, here we simply assume that the intermolecular potential U for the ground state can be commonly used for the other states. The dot-dashed and dashed lines in Fig. 9 are the cases of L=1 and L=2. Then, we have, in addition to the ground state, four more excited levels as shown by horizontal lines. The present framework is a hopeful approach to the excited-state problem.

Finally, it would be interesting to note a "doubly resonant" feature that the $(dt\mu)$ molecular states of size $\sim$pm resonantly couples with D_2 molecular states of $\sim$Å and with the d-t nuclear state of $\sim$fm.

5.4. $(e^-e^+e^-)$ Molecule

The negative positronium $(e^-e^+e^-)$ is the lightest ternary lepton system. ATMS calculates the total energy to be -7.126 eV, which is lower than the energy -6.803 eV of (e^+e^-) by the amount of -0.323 eV. Thus, the $(e^-e^+e^-)$ system has a particle-stable bound state.

A variety of possible few-body molecules may be classified into two types: One is the H^- type like $(e^-p^+e^-)$ where the charge center coincides with the mass center. Another is the H^+ type like $(e^-p^+e^+)$ where the charge center deviates from the mass center. As an example, let's see the following two branches;

$$\begin{array}{lllll}
 & \nearrow \begin{array}{c}(e^-\mu^+e^-)\\ -0.71\ \text{eV}\end{array} & \longrightarrow & \begin{array}{c}(e^-p^+e^-)\\ -0.74\ \text{eV}\end{array} & ;\ H^-\ \text{branch},\\
\begin{array}{c}(e^-e^+e^-)\\ -0.323\ \text{eV}\end{array} & & & & \\
 & \searrow \begin{array}{c}(e^+\mu^-e^-),(e^-\mu^+e^+)\\ (\text{unbound?})\end{array} & \longrightarrow & \begin{array}{c}(e^-p^+e^+)\\ \text{unbound}\end{array} & ;\ H^+\ \text{branch},
\end{array} \tag{127}$$

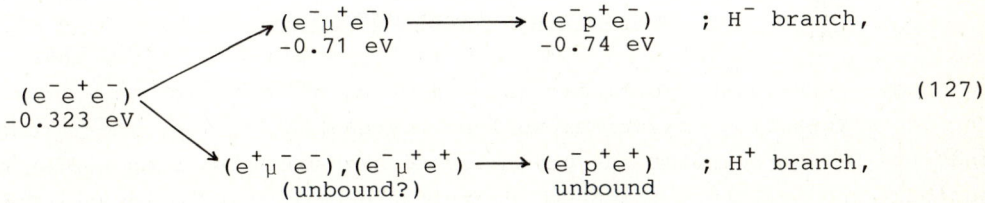

where the figures are the calculated energies measured from the thresholds. The H^+ branch seems hard to be bound. We can discuss the possible existence of such molecules in terms of the intermolecular potential. Figure 10 illustrates the cases of the negative positronium and of the negative hydrogen ion. Compared with the (e^+e^-)-e^- potential, the (pe^-)-e^- potential is strongly attractive around the origin due to the presence of the heavy attractive center p^+, but is of shorter range due to the tightly binding between p and e^-.

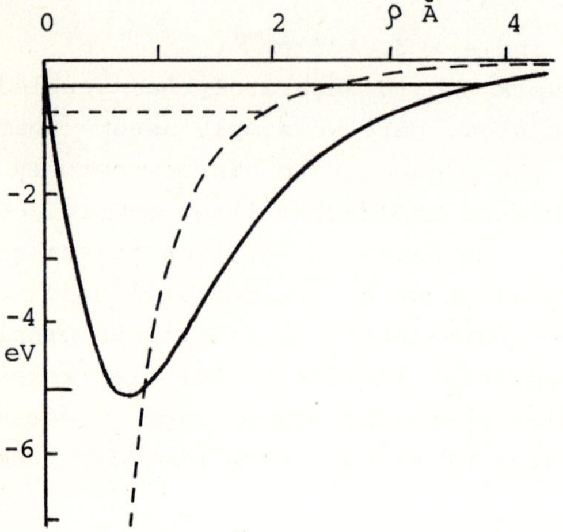

Fig. 10.
The intermolecular potentials between (e^+e^-) and e^- (solid line) and between (pe^-) and e^- (dashed line).

As these effects are opposite, the binding energy do not change so much along the H^- branch. On the other hand, if we enter the H^+ branch, the large-mass proton, for example, makes the (pe^-)-e^+ potential repulsive at the origin and, simultaneously, of short-range. Therefore, the bound state suddenly disappears along the H^+ branch.

It would be interesting to discuss the possible existence of various muonic molecules or to proceed to Coulomb four-body systems. The $(e^-e^+e^-e^+)$ molecule may now be within the range of ATMS.

6. Momentum Distributions in ^{4}He

Now we have the realistic wave function of ^{4}He which is a source of rich information on various nuclear correlations: The short-range and the D-wave correlations give rise to high-momentum components of nucleon- and cluster- momentum distributions in the nucleus. Such high-momentum nucleons inside the nucleus may play an essential role in intermediate-energy nuclear phenomena. Our purpose is to show the real amount of the high-momentum components in ^{4}He.

6.1. Momentum Distribution of Single Nucleon

We define the momentum distribution of a single nucleon by

$$W^{SN}(\vec{p}) = (2\pi)^{-3}\int d\vec{r}\int d\vec{r}\,' \exp(i\vec{p}\,(\vec{r}-\vec{r}\,'))\,\rho(\vec{r},\vec{r}\,'),$$

$$\rho(\vec{r},\vec{r}\,') = \int d\vec{\xi}_3\int d\vec{\xi}_3{}' \int d\vec{\xi}_1\int d\vec{\xi}_2\ \Psi^*(\vec{\xi}_1,\vec{\xi}_2,\vec{\xi}_3)\,\Psi(\vec{\xi}_1,\vec{\xi}_2,\vec{\xi}_3{}') \times (4/3)^3\,\delta(\vec{\xi}_3-(4/3)\vec{r})\,\delta(\vec{\xi}_3{}'-(4/3)\vec{r}\,'), \tag{128}$$

where Ψ is the ^4He wave function. This definition is free from the center-of-mass motion of ^4He, and the normalization is taken to be

$$\int d\vec{p}\ W^{SN}(\vec{p}) = 1. \tag{129}$$

The momentum distribution is the Fourier transformation of the "density correlation", whereas the form factor given by

$$F(\vec{p}) = \int d\vec{r}\ \exp(i\vec{p}\,\vec{r})\,\rho(\vec{r},\vec{r}) \tag{130}$$

is that of the "density fluctuation". The former is more sensitive to NN correlations than the latter [39].

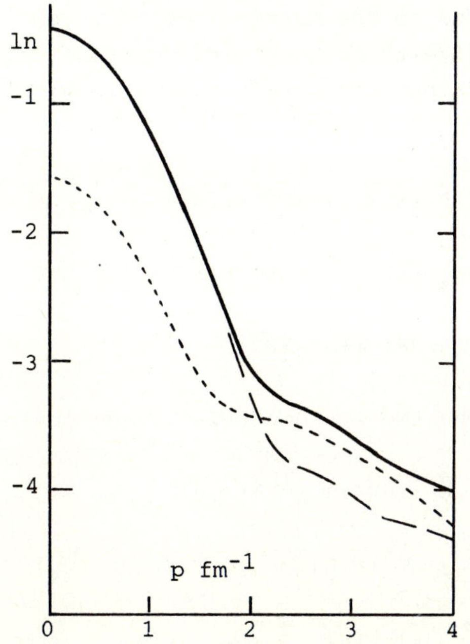

Fig. 11.
The single-nucleon momentum distribution in ^4He. The dashed and the dotted lines are the S-wave and the D-wave contributions, respectively.

In order to see the low-momentum property we calculate W^{SN} by the harmonic oscillator model of Eq.(37): The distribution is

$$W_0^{SN}(\vec{p}) = \left(\frac{3}{4\pi\beta}\right)^{3/2} \exp\left(-\frac{3}{4\beta}\vec{p}^2\right). \tag{131}$$

and decreases very rapidly as the momentum p increases.

Figure 11 shows the realistic momentum distribution obtained with the $RSCV_8$ potential, which deviates from Eq.(131) and clearly shows the existence of the marked high-momentum component. The short-range correlation manifests itself around 2 fm^{-1} (0.4 GeV/c) as seen from the behavior of the S-state contribution. The D-wave correlation is of almost same magnitude in this momentum region. The kinetic energy of the nucleus is obtained from W^{SN} [15] by

$$K.E. = (N-1)[(N-1)/N] \int d\vec{p}\; W^{SN}(\vec{p})\; (\hbar^2/2M)\; \vec{p}^2. \tag{132}$$

The high-momentum component enhances the kinetic energy by about 70 %. Half of enhancement comes from the short-range correlation and half from the D-state correlation.

6.2. Momentum Distribution of Two-Nucleon Cluster

Fujita and Hüfner [40] pointed out that the two-nucleon cluster plays an important role in the backward proton-nucleus collision where the momentum transfer is, compared to the energy transfer, so large that a single nucleon could not "digest" such unbalanced transfers.

We define the momentum distribution of the two-nucleon-cluster center by

$$\begin{aligned} W^{TNC}(\vec{P}) &= (2\pi)^{-3} \int d\vec{R} \int d\vec{R}'\; \exp(i\vec{P}\,(\vec{R}-\vec{R}'))\; \rho^{TNC}(\vec{R},\vec{R}'), \\ \rho^{TNC}(\vec{R},\vec{R}') &= \int d\vec{u} \int d\vec{u}' \int d\vec{r}_{12} \int d\vec{r}_{34}\; \Psi^*(\vec{r}_{12},\vec{u},\vec{r}_{34})\; \Psi(\vec{r}_{12},\vec{u}',\vec{r}_{34}) \\ &\quad \times 8\, \delta(\vec{u}-2\vec{R})\; \delta(\vec{u}'-2\vec{R}') \end{aligned} \tag{133}$$

and the momentum distribution of the two-nucleon relative-motion by

$$\begin{aligned} W^{TNR}(\vec{k}) &= (2\pi)^{-3} \int d\vec{r} \int d\vec{r}'\; \exp(i\vec{k}\,(\vec{r}-\vec{r}'))\; \rho^{TNR}(\vec{r},\vec{r}'), \\ \rho^{TNR}(\vec{r},\vec{r}') &= \int d\vec{r}_{12} \int d\vec{r}_{12}' \int d\vec{u} \int d\vec{r}_{34}\; \Psi^*(\vec{r}_{12},\vec{u},\vec{r}_{34})\; \Psi(\vec{r}_{12}',\vec{u},\vec{r}_{34}) \\ &\quad \times \delta(\vec{r}_{12}-\vec{r})\; \delta(\vec{r}_{12}'-\vec{r}'), \end{aligned} \tag{134}$$

where $\vec{r}_{ij} = \vec{r}_i - \vec{r}_j$ and $\vec{u} = (\vec{r}_1 + \vec{r}_2)/2 - (\vec{r}_3 + \vec{r}_4)/2$.

Figure 12 shows the calculated momentum distribution of the two-nucleon-cluster center. We can see that there exists a pronounced high momentum component. It should be mentioned that this high-momentum component comes from many-body correlations not included in the independent-pair model of the nucleus [7]. Figure 13 is the momentum distribution of the two-nucleon relative-motion, where we can see again a high-momentum component. It is interesting to compare it with the momentum distribution in the deuteron. There are two differences: The deuteron one has a larger low-momentum component due to its long tail of the wave function. The deuteron momentum distribution around 2 fm^{-1} consists of almost the pure D-wave contribution, whereas the present W^{TNR} does of the short-range and the D-wave ones.

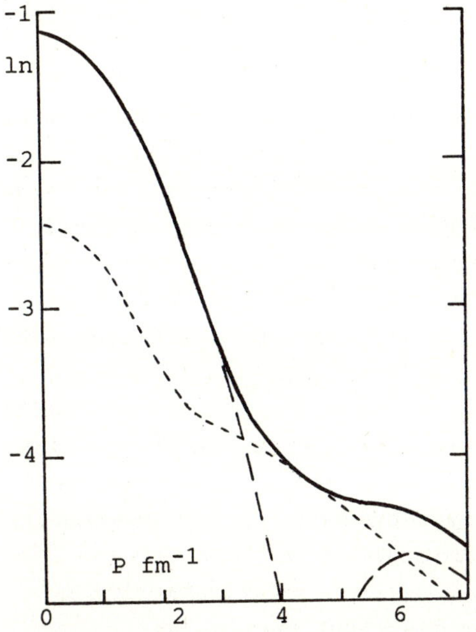

Fig. 12. The momentum distribution of the two-nucleon-cluster center.

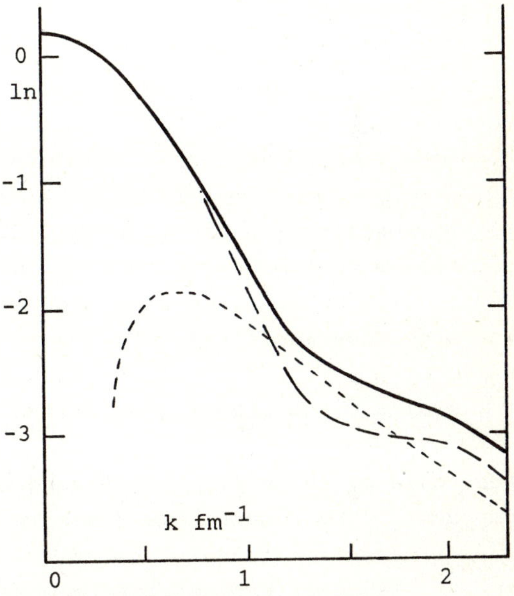

Fig. 13. The momentum distribution of the two-nucleon-cluster internal-motion.

An interesting example is the reaction $^3He(p, \pi^+)^4He_{gr}$ done with E_p = 800 MeV by Höistad et al.[41]. The experimental data in Fig. 14 shows a slope change of the cross section at a momentum transfer 650 MeV/c. By assuming the reaction mechanism depicted in Fig. 14, we can see that the slope change may correspond to the appearance of the high-momentum component of the two-nucleon-cluster center.

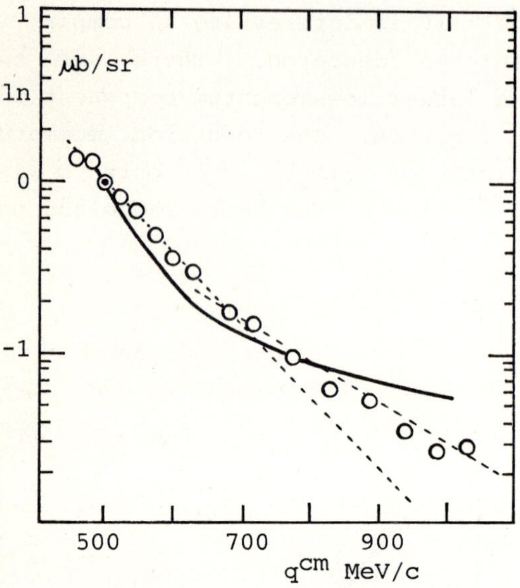

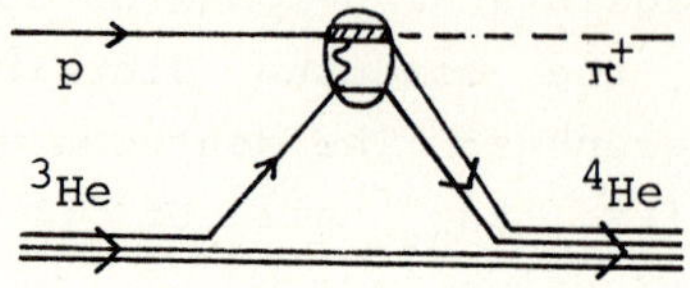

Fig. 14. The differential cross section of the reaction $^3He(p,\pi^+)^4He_{gr}$. The data is given by the circles and the dotted lines are drawn to see a slope change. The solid curve is the theoretical result normalized at q^{cm}= 500 MeV/c. The assumed process is depicted.

6.3. Parametrization of Momentum Distributions

The momentum distributions of the single nucleon and the NN cluster in ^4He obtained from ATMS calculations are parametrized into two-range Gaussian form;

$$W(q) = C\,[\exp(-Bq^2) + s\,\exp(-Bq^2/t)], \tag{135}$$

where q is p, P or k. The second term represents the high-momentum component. The results are given in Table 6.

Table 6. Parametrized momentum distributions in ^4He.

	B	β (fm^{-2})	t	s
W^{SN}	$3/(4\beta)$	0.42	12	0.00286
W^{TNC}	$1/(4\beta)$	0.42	12	0.00286
W^{TNR}	$2/\beta$	0.42	9	0.00634

In order to know what really happens when nucleons come close together, we have to remove the contribution of the D-wave correlation from the observed high-momentum component. This can be done by the coincidence experiment ^{4}He(e, e'p): From the spectral function [42] of ^{4}He we can get crucial information on the nuclear short-range correlation. This point is discussed in Ref.[43].

7. Concluding Remarks

ATMS makes it possible to treat few-nucleon systems on the basis of realistic nuclear interactions. The accuracy of the method is unambiguously demonstrated by the small gap between the upper and the lower bounds to the energy. An upper bound formula to the impurity in the wave function is derived and ATMS wave function is shown to be highly accurate especially for the high-momentum components.

The momentum distributions inside ^{4}He show the pronounced high-momentum components due to the nuclear correlations. The high-momentum nucleons play important roles in intermediate-energy phenomena, and the study of nuclear correlations becomes inebitable. ^{4}He, which has the very high central density and the strong short-range correlation, is a source of rich information on the realistic behavior of nucleons at short NN distances.

ATMS is also applicable to Coulomb few-body systems. Hypernuclei form a fruitful field of the few-body physics [43]. Thus, the ATMS method can treat a variety of few-body systems of current interests.

Acknowledgement

The author would like to thank Prof. H. Tanaka, Dr. T. Katayama, Dr. S. Nakaichi-Maeda, Dr. S. Shinmura, Dr. Y. Kurihara and Mr. H. Morita for valuable discussions.

References

1. Reid,Jr.,R.V.: Ann. of Phys. 50, 411 (1968).
2. Akaishi,Y., Sakai,M., Hiura,J., Tanaka,H.: Prog. Theor. Phys. Suppl. No. 56, 6 (1974).
3. Schmid,E.W., Tang,Y.C., Herndon,R.C.: Nucl. Phys. 42, 95 (1963); 65, 203 (1965).
4. Tanaka,H., Nagata,H.: Prog. Theor. Phys. Suppl. No.56, 121 (1974).
5. Sakai,M, Shimodaya,I., Akaishi,Y., Hiura,J., Tanaka,H.: Prog. Theor. Phys. Suppl. No.56, 32 (1974).

6. Brueckner,K.A., Levinson,C.A.: Phys. Rev. 97, 1344 (1955).
Bethe,H.A.: Phys. Rev. 103, 1353 (1956).
Bethe,H.A.: Ann. Rev. Nucl. Sci. Vol.21, 93 (1971).
7. Gomes,L.C., Walecka,J.D., Weisskopf,V.F.: Ann. of Phys. 3, 241 (1958).
8. Akaishi,Y., Nagata,S.: Prog. Theor. Phys. 48, 133 (1972).
9. Watson,K.M.: Phys. Rev. 89, 575 (1953).
10. Tanaka,H.: ATMS Method and Its Application. International Symposium in Nanning, 1985.
11. Tanaka,H.: Prog. Theor. Phys. 13, 497 (1955).
12. Shinmura,S., Akaishi,Y., Tanaka,H.: Prog. Theor. Phys. 71, 546 (1984).
13. Kurihara,Y., Akaishi,Y., Tanaka,H.: Phys. Rev. C31, 971 (1985).
14. Nakaichi-Maeda,S., Lim,T.K., Akaishi,Y., Tanaka,H.: J. Chem. Phys. 71, 4430 (1979).
15. Akaishi,Y., International Rev. Nucl. Phys. Vol.4. Singapore: World Scientific 1986.
16. Fox,L., Goodwin,E.T.: Proc. Camb. Phil. Soc. 45, 373 (1949).
Robertson,H.H.: Proc. Camb. Phil. Soc. 52, 538 (1956).
17. Haselgrove,C.B.: Math. Comp. 15, 323 (1961).
18. Temple,G.: Proc. Roy. Soc. A119, 276 (1928).
19. Malfliet,R.A., Tjon,J.A.: Nucl. Phys. A127, 161 (1969).
20. Lomnitz-Adler,J., Pandharipande,V.R., Smith,R.A.: Nucl. Phys. A361, 399 (1981).
21. Carlson,J., Pandharipande,V.R.: Nucl. Phys. A371, 301 (1981).
22. Ciofi degli Atti,C., Simula,S.: Nuovo Cim. 41, 101 (1984).
23. Strayer,M.R., Sauer,P.U.: Nucl. Phys. A231, 1 (1974).
24. Hu,C.Y., Fink,S.: Phys. Rev. C30, 715 (1984).
25. Friar,J.L., Gibson,B.F., Payne,G.L.: Phys. Rev. C24, 2279 (1981).
26. Kalos,M.H.: Phys. Rev. 128, 1791 (1962).
27. Zabolitzky,J.G., Kalos,M.H.: Nucl. Phys. A356, 114 (1981).
Zabolitzky,J.G., Schmidt,K.E., Kalos,M.H.: Phys. Rev. C25, 1111 (1982).
28. Kümmel,H., Lührmann,K.H., Zabolitzky,J.G.: Phys. Rep. 36C, 1 (1978).
29. Tamagaki,R.: Prog. Theor. Phys. 39, 91 (1968).
30. Akaishi,Y., Nakaichi-Maeda,S., Schmid,E.W.: Prog. Theor. Phys. 66, 211 (1981).
31. Ali,S., Bodmer,A.R.: Nucl. Phys. 80, 99 (1966).
32. Hajduk,Ch., Sauer,P.U.: Nucl. Phys. A369, 321 (1981).
33. Ishikawa,S., Sasakawa,T., Sawada,T., Ueda,T.: Phys. Rev. Lett. 53, 1877 (1984).
34. Chen,C.R., Payne,G.L., Friar,J.L., Gibson,B.F.: Phys. Rev. C31, 2266 (1985).
35. Bracci,L., Fiorentini,G.: Phys. Rep. 86, 169 (1982).
36. Day,B.D.: Phys. Rev. 187, 1269 (1969).
37. Vinitskiĭ,S.I., Melezhik,V.S., Ponomarev,L.I., Puzynin,I.V., Puzynina,T.P., Somov,L.N., Truskova,N.F.: Soviet Phys. JETP 52, 353 (1980).
38. Kurihara,Y., Akaishi,Y., Tanaka,H.: Prog. Theor. Phys. 67, 1483 (1982).
39. Zabolitzky,J.G., Ey,W.: Phys. Lett. 76B, 527 (1978).
40. Fujita,T., Hüfner,J.: Nucl. Phys. A314, 317 (1979).
41. Höistad,B., Gazzaly,M., Aas,B., Igo,G., Rahbar,A., Whitten,C., Adams,G.S., Whitney,R.: Phys. Rev. C29, 553 (1984).
42. Ciofi degli Atti,C.: Prog. Part. Nucl. Phys. 3, 163 (1980).
Ciofi degli Atti,C., Pace,E., Salmè,G.: Phys. Lett. 127B, 303 (1983).
43. Akaishi,Y.: Few-Body Calculations at Sapporo. European Workshop in Rome, 1986.

VARIATIONAL AND GREEN'S FUNCTION MONTE CARLO CALCULATIONS OF FEW-BODY SYSTEMS

K.E. Schmidt
Courant Institute of Mathematical Sciences and
Department of Chemistry
New York University
New York, NY 10012

1. INTRODUCTION

Monte Carlo methods are a powerful way to calculate the many dimensional sums and integrals that appear in many-body theory. Typically the Monte Carlo method is useful when the number of dimensions is greater than three or four. Even three-body calculations can usefully employ Monte Carlo methods.

In this paper I will describe the variational and Green's function Monte Carlo (GFMC) methods. The variational method is a rather straightforward application of standard Monte Carlo techniques, and several references describe various calculations in detail.[1-4] The GFMC method[1,5-9] is somewhat more complicated; it will be described in detail and explicit algorithms given. Both methods are primarily useful for calculating ground-state properties. However, they can be extended to some extent to calculate excited states and even some scattering properties.[10,11]

In what follows, a spin-isospin independent (i.e. purely central interaction) will be written formally as

$$H = -\nabla^2 + V(R), \tag{1.1}$$

and a spin-isospin dependent Hamiltionian as

$$H_{ss'} = -\nabla^2 \delta_{ss'} + V_{ss'}(R), \tag{1.2}$$

where R indicates the 3N coordinates of the N particles, and s and s' indicate the total spin-isospin state of the system.

Typically the potential terms are a sum of two and sometimes three body potentials,

$$V(R) = \sum_{i<j} v_2(r_{ij}) + \sum_{i<j<k} v_3(r_{ij}, r_{ik}, r_{jk}), \tag{1.3}$$

where $r_{ij} = |\mathbf{r}_i - \mathbf{r}_j|$. The effect of the potential being different in different channels is conveniently written in terms of spin-isospin operators and angular momentum operators. A typical form might be a v_6 form with 6 operators. The two-body potential would be written as

$$\sum_p v_p(r_{ij})O^p_{ij},$$

where $v_p(r_{ij})$ are functions of r_{ij} and O^p_{ij} are operators. For the v_6 potential the operators are 1, $\vec{\sigma}_i\cdot\vec{\sigma}_j$, $\vec{\tau}_i\cdot\vec{\tau}_j$, $\vec{\sigma}_i\cdot\vec{\sigma}_j\vec{\tau}_i\cdot\vec{\tau}_j$, t_{ij}, and $t_{ij}\vec{\tau}_i\cdot\vec{\tau}_j$, where $\vec{\sigma}$ and $\vec{\tau}$ are the Pauli spin and isospin matrices, and t is the tensor operator. More complicated interactions may contain spin orbit operators and perhaps operators corresponding to delta states.

The Monte Carlo method is reviewed in appendix A. Readers who are not familiar with the basic concepts of the Monte Carlo method are urged to read this appendix first since many of the results demonstrated there are assumed in the body of this article.

2. THE VARIATIONAL METHOD

The variational method consists of calculating the expectation value of the Hamiltonian using a suitably chosen trial wave function. The trial function is varied to minimize the expectation value of H since

$$E_0 \leq \frac{\langle\Psi_T|H|\Psi_T\rangle}{\langle\Psi_T|\Psi_T\rangle}. \qquad 2.1$$

For a purely central potential, the spin-isospin part of the problem drops out and only the spatial part of the wave function needs to be calculated. In this case the expectation value can be written as

$$E_0 \leq \frac{\int dR\Psi_T^*(R)H\Psi_T(R)}{\int dR\Psi_T^2(R)} = \frac{\int dR\left[\frac{\Psi_T^*(R)H\Psi_T(R)}{\Psi_T^2(R)}\right]\Psi_T^2(R)}{\int dR\Psi_T^2(R)}. \qquad 2.2$$

The Monte Carlo method consists of sampling M positions R_i for the N particles from

$$P(R) = \frac{\Psi_T^2(R)}{\int dR\Psi_T^2(R)}, \qquad 2.3$$

and calculating

$$\langle H\rangle \approx \frac{1}{M}\sum_{i=1}^{M}\left[\frac{\Psi_T^*(R)H\Psi_T(R)}{\Psi_T^2(R)}\right] \qquad 2.4$$

as the expectation value of the Hamiltonian. Typically, the Metropolis et. al. method[12] is used to do the sampling as described in appendix A. The selection of P(R) from Eq. 2.3 has two useful effects. First it eliminates the need to calculate the ration of two integrals. The calculation of the ratio of two integrals with the Monte Carlo method is necessarily biased. That is the average of the ratio for a finite number of samples does not equal the ratio of the average, so the central limit theorem does not apply. Second, it reduces the variance which determines the statistical error. This can be easily seen by considering the the case where the exact ground state wave function is used as a trial function. In this case, $\frac{\Psi_T^*(R)H\Psi_T(R)}{\Psi_T^2(R)} = E_0$ is a constant independent of the position R_i. Therefore $\langle H\rangle$ is calculated with zero variance. When a good trial function is used the variance is low.

Most of the effort in a variational calculation goes into devising a good trial function which contains the physics necessary to describe the ground state and is reasonably efficient to compute. A typical trial function for the deuteron, triton or alpha using a central potential is of the form

$$\prod_{i<j} f_{ij} |\Phi_0> \qquad 2.5$$

or

$$\prod_{i<j} f_{ij} \prod_{k} f_k |\Phi_0> \qquad 2.6$$

Here f_{ij} is a function of the interparticle distance, and $|\Phi_0>$ is a model state which contains the spin-isospin part of the wave function. For purely central potentials, this function drops out of the expectation values. In Eq. 2.6, a one-body correlation f_k is also included. It is actually a function of $|\mathbf{r}_i - \mathbf{r}_{CM}|$ where $\mathbf{r}_{CM}$ is the center of mass of the system $\mathbf{r}_{CM} = \frac{1}{N}\sum_{i=1}^{N} \mathbf{r}_i$, so f_k can be viewed as an N-body correlation, albeit a very simple one. Two physical effects that it is necessary to include in a variational wave function are: first, when two particles are close together, their relative wave function should satisfy the corresponding two-body equation; second when a particle is far from the others, the wave function should decay exponentially like $\exp(-kr)/r$ where $k = (E_b 2m^*/\hbar^2)^{\frac{1}{2}}$, where $m^* = m(N-1)/N$ and E_B is the separation energy for the system. The f_{ij} is usually chosen to be a solution to the two-body equation for small r_{ij}. If no one-body term is included in the wave function, f_{ij} is chosen so that the wave function has the correct property at large r_{ij}. That is $f_{ij} \approx (\exp(-kr)/r)^{\frac{1}{N-1}}$. An alternative is to choose f_{ij} to heal smoothly to one at large separations, and include the one-body correlation f_k, which is chosen to decay like $\exp(-kr)/r$.

Variational calculations with realistic potentials can also be done.[2-4] The main differences are that the spin-isospin part of the wave function does not trivially drop out, and both the potential and a good trial function contain spin-isospin dependence. A typical trial wave function contains correlations induced by the potential. When the potential contains spin-isospin operators, these correlations can be represented as correlation operators. A trial wave function then could be

$$\Psi_T(R,s) = S\prod_{i<j}\left[\sum_p f_{ij}^p O_{ij}^p\right] |\Phi_0>. \qquad 2.7$$

A one-body term can be added exactly as in the purely central case if desired. In Eq. 2.7, S is a symmetrizing operator necessary because the O_{ij}^p do not commute, and s indicates a spin-isospin component of the wave function implicit in the $|\Phi_0>$ and the operators O_{ij}^p. $|\Phi_0>$ can be represented as a column vector of spin-isospin states. For the triton $|\Phi_0>$ would have 24 components (8 spin x 3 isospin) and for the alpha particle 96 components (16 spin x 6 isospin). Each of the operators O_{ij}^p are matrices in this representation. However, they are very sparse matrices. For v_6 operators, the only nonzero elements of the matrices are those that 1) leave spins and isospins the same, 2) flip spin i, 3) flip spin j, 4) flip both i and j, and 5) an isospin exchange associated with each of these if the isospin of particle i is different from that of particle j. It is therefore very easy to efficiently program the multiplication of these operators times a spin-isospin vector. The expectation value that needs to be calculated is

$$<H>=\frac{\int dR\sum_{s,s'}\Psi_T^*(R,s')H_{s's}\Psi_T(R,s)}{\int dR\sum_{s}\Psi_T^2(R,s)} \qquad 2.8$$

At first glance, it might seem that both the spin-isospin sums and the spatial integral should be done using the Monte Carlo method. However, this does not lead to a low variance solution even in the limit that Ψ_T is the exact ground-state wave function Ψ_0. The exact ground-state wave function satisfies

$$\sum_{s'}H_{ss'}\Psi_0(R,s')=E_0\Psi_0(R,s). \qquad 2.9$$

To make a low variance solution, at least the sum corresponding to the sum in Eq. 2.9 needs to be done explicitly. The expectation value can be written as

$$<H>=\frac{\sum_{s}\int dR\left(\frac{\sum_{s'}\Psi_T^*(R,s')H_{ss'}\Psi_T(R,s)}{\Psi_T^2(R,s)}\right)\Psi_T^2(R,s)}{\sum_{s}\int dR\Psi_T^2(R,s)}. \qquad 2.10$$

If R and s are sampled from $\Psi_T^2(R,s)/\sum_{s}\int dR\Psi_T^2(R,s)$ and the term in brackets evaluated at these points, a low variance algorithm results, since the term in brackets is E_0 if Ψ_T is the exact ground-state wave function. Typically in few-body problems, the calculation of the Ψ_T requires the calculation of many spin-isospin sums by itself, and therefore very little is saved by doing the one spin-isospin sum by Monte Carlo methods. Therefore, the expectation value is written as

$$<H>=\frac{\int dR\left(\frac{\sum_{ss'}\Psi_T^*(R,s')H_{ss'}\Psi_T(R,s)}{\Psi_T^2(R,s)}\right)\sum_{s}\Psi_T^2(R,s)}{\sum_{s}\int dR\Psi_T^2(R,s)}. \qquad 2.11$$

and R is sampled from the distribution $\sum_{s}\Psi_T^2(R,s)/\int dR\sum_{s}\Psi_T(R,s)$. One small correction has to be made to Eq. 2.11. For wave functions like Eq. 2.7, the symmetrizing operators can lead to a large increase in the computer time required to calculate the value of the wave function. These symmetrizing operators are implemented by simply summing over all possible orderings of the operators. However, they contribute relatively little to the structure of the wave function. It seems likely, and turns out to be true, that doing the symmetrizing sums with the Monte Carlo method can reduce the computer time required by a lot, while only increasing the variance a little. Finally, the expectation value can be written as

$$<H>=\sum_{\theta_1\theta_2}\int dRf(R,\theta_1,\theta_2)P(R,\theta_1,\theta_2) \qquad 2.12$$

where

$$f(R,\theta_1,\theta_2)=\frac{\mathrm{Re}[<\Phi_0|\prod_{i<j\theta_1}(\sum_p f_{ij}^p O_{ij}^p)H\prod_{i<j\theta_2}(\sum_p f_{ij}^p O_{ij}^p)|\Phi_0>]}{\mathrm{Re}[<\Phi_0|\prod_{i<j\theta_1}(\sum_p f_{ij}^p O_{ij}^p)\prod_{i<j\theta_2}(\sum_p f_{ij}^p O_{ij}^p)|\Phi_0>]} \qquad 2.13$$

and

$$P(R,\theta_1,\theta_2)=\frac{\mathrm{Re}[<\Phi_0|\prod_{i<j\theta_1}(\sum_p f_{ij}^p O_{ij}^p)\prod_{i<j\theta_2}(\sum_p f_{ij}^p O_{ij}^p)|\Phi_0>]}{\int dR\sum_{\theta_1\theta_2}\mathrm{Re}[<\Phi_0|\prod_{i<j\theta_1}(\sum_p f_{ij}^p O_{ij}^p)\prod_{i<j\theta_2}(\sum_p f_{ij}^p O_{ij}^p)|\Phi_0>]} \quad 2.14$$

where θ_1 and θ_2 specify the order of the operators in the operator products, and Re specifies the real part as usual. The integral in Eq. 2.12 is then done by sampling R, θ_1, θ_2 using the Metropolis et. al. Monte Carlo method. The moves made in such a walk would consist of some combination of moving particles and taking different operator orders.

The wave functions are chosen in the same spirit as those in the purely central case. The operator dependence of the correlations is obtained by solving approximate two-body coupled channel equations. Similar boundary conditions are also employed.

An interesting calculation has been done by Wiringa et. al.[4] In this calculation, the trial wave function was that obtained from the solution of the Fadeev equations. This wave function can then be substituted directly into Eq. 2.11 to calculate the expectation value of the Hamiltonian. Wiringa et. al. used this Fadeev wave function as a starting point for studying the effect of three-body potentials. The inclusion of three-body potentials is trivial using the Monte Carlo method since the calculations are done directly in the 3N dimensional configuration space. Similarly, the inclusion of three or more body correlations in the wave function is easy.

Results of variational calculations with the Reid v_8 potential by Lomnitz-Adler et. al. and with the Maffliet-Tjon V central potential by Pandharipande and Carlson are shown in Table I below.

Table I. The Binding energy in MeV of the triton and alpha particle calculated variationally with the Reid v_8 (v_8) potential from reference 2. and with the and Maffliet-Tjon V (MTV) potential from reference 3.

Potential	System	Binding Energy
v_8	^{3}He	6.86±0.08
v_8	^{4}He	22.9±0.5
MTV	^{3}He	8.22±0.02
MTV	^{4}He	31.19±0.05

3. FORMAL ANALYSIS OF THE GFMC METHOD

The Green's function Monte Carlo method[1,5–9,13] is based on integrating the Schroedinger equation in imaginary time. The time dependent Schroedinger equation

$$H\Phi(R,t) = -\frac{\partial\Phi(R,t)}{i\partial t}, \tag{3.1}$$

becomes

$$H\psi(R,\tau) = -\frac{\partial\psi(R,\tau)}{\partial\tau} \tag{3.2}$$

in imaginary time. Eq. 3.2 is recognized as a diffusion equation in 3N dimensions where N is the number of particles. The $-\nabla^2$ in the Hamiltonian causes the local density to diffuse, while the potential causes the density to grow or decay. These processes can be modeled on a computer, and by simulating the diffusion, the Schroedinger equation in imaginary time can be solved. To see how this is done, I will write Eq. 3.2 as an integral equation and show how the integrals can be done using the Monte Carlo method. The results of this integration can be used to calculate the properties of the quantum system.

A formal solution of Eq. 3.2 is

$$|\psi(\tau+\Delta\tau)\rangle = \exp(-H\Delta\tau)|\psi(\tau)\rangle. \tag{3.3}$$

Expanding $\psi(0)$ the initial wave function in terms of the eigenstates of H gives

$$|\psi(0)\rangle = \sum_n a_n|\psi_n\rangle, \tag{3.4}$$

where

$$a_n = \langle\psi_n|\psi(0)\rangle. \tag{3.5}$$

Substituting Eq 3.4 into 3.3,

$$|\psi(\tau)\rangle = a_0\exp(-E_0\tau)\left[|\psi_0\rangle + \sum_{n\neq 0}\exp(-(E_n-E_0)\tau)\frac{a_n}{a_0}|\psi_n\rangle\right], \tag{3.6}$$

where E_n is the eigenvalue of H corresponding to $|\psi_n\rangle$, and E_0 is the lowest eigenvalue whose eigenvector has nonzero overlap with $|\psi(0)\rangle$. For example, if the starting function $|\psi(0)\rangle$ is antisymmetric under particle interchange and since the Hamiltonian is symmetric, only a_n where $|\psi_n\rangle$ is a purely antisymmetric state will be nonzero. Only these states will contribute even though the Hamiltonian may have lower energy states with other symmetries. This is not to say that there are no problems in calculating states that change sign.[5] I will have more to say about this later. Unless otherwise specified, the following analysis is for spin and isospin independent interactions and spatially symmetric states. That is the ground state of the deuteron, triton and alpha particle with purely central forces.

For large τ the second term in brackets in Eq. 3.6 is exponentially small compared to the first so the solution is proportional to $|\psi_0\rangle$ the desired ground-state wave function. The coefficient in front of $|\psi_0\rangle$ can be made nearly constant by subtracting a constant energy E_T from the Hamiltonian so that $E_0-E_T=0$. Since E_0 is not known, E_T is a trial value for the ground-state energy.

Many functions of H other than the exponential will produce the ground-state wave function when operated repeatedly on an initial trial wave function. It is sufficient that the value of the operator be larger for the ground-state eigenvalue than for any other eigenvalue. Another convenient operator is

$$\frac{E_T+E_C}{H+E_C}, \tag{3.7}$$

which results in the iterative equation

$$|\psi^{(n)}\rangle = \frac{E_T + E_C}{H + E_C} |\psi^{(n-1)}\rangle \qquad \mathbf{3.8}$$

where E_C is a constant added to H to make the eigenvalue spectrum positive. As before, E_T is a trial energy that should be adjusted to approximately the ground-state energy. Both Eqs. 3.3 and 3.8 can be written as integral equations which can be iterated to obtain the ground-state energy and wave function. Both methods have been used successfully for a variety of problems. The two methods are simply related since

$$\frac{1}{H + E_C} = \int_0^\infty \exp(-(H + E_C)\tau)d\tau. \qquad 3.9$$

It can be shown from Eq. 3.9 that the $(H + E_C)^{-1}$ propagator corresponds to diffusion with a Poisson distribution of diffusion times with average time $(E_C + E_0)^{-1}$. Physical insight gained from the diffusion analogy can be used with either the time integrated or time dependent method. Often the τ integral in Eq. 3.9 is done using the Monte Carlo method, which makes the methods even more similar in actual practice.

4. CALCULATION OF EXPECTATION VALUES

With either the time integrated or time dependent technique, the Monte Carlo method is used to calculate the resulting integrals. As explained in Appendix A, the Monte Carlo method corresponds to converting functions to a sum of delta functions. In the GFMC method, the ground-state wave function is therefore represented as a sum of delta functions. I will now show typical methods used to extract useful information from this representation of the wave function.

One of the primary quantities of interest is the energy. Two related energy estimates are the mixed and growth energies. For the time integrated method, the mixed energy is

$$E_{MIXED} = \frac{\langle\psi_T|H|\psi^{(n)}\rangle}{\langle\psi_T|\psi^{(n)}\rangle}, \qquad 4.1$$

where $|\psi_T\rangle$ is a trial ground state wave function. Typically, $|\psi_T\rangle$ is found by a variational calculation. By rewriting Eq 3.8 as

$$H|\psi^{(n)}\rangle = (E_T + E_C)|\psi^{(n-1)}\rangle - E_C|\psi^{(n)}\rangle, \qquad 4.2$$

and substituting into Eq. 4.1, the growth energy is

$$E_{GROWTH} = \frac{\langle\psi_T|\psi^{(n-1)}\rangle}{\langle\psi_T|\psi^{(n)}\rangle}(E_T + E_C) - E_C. \qquad 4.3$$

The growth energy gets its name because it is calculated from essentially the ratio of the normalization of $|\psi^{(n-1)}\rangle$ and $|\psi^{(n)}\rangle$ which in the Monte Carlo simulation corresponds to the ratio of the number of delta functions representing the wave function at steps n-1 and n. In the time dependent method the corresponding expressions are

$$E_{MIXED} = \frac{\langle\psi_T|H|\psi(\tau)\rangle}{\langle\psi_T|\psi(\tau)\rangle} \qquad 4.4$$

and

$$E_{GROWTH}=\frac{-1}{\Delta\tau}\ln\frac{<\psi_T|\psi(\tau+\Delta\tau)>}{<\psi_T|\psi(\tau)>}+E_T. \qquad 4.5$$

Since the mixed and growth estimates or so closely related, the values calculated from them are not statistically independent. Both estimates are upper bounds to the exact ground-state energy at every step of the calculation if the starting wave function for the iteration ($|\psi^{(0)}>$ or $|\psi(0)>$) is the trial function used to calculate the energy. I will assume that $|\psi^{(0)}>=|\psi(0)>=|\psi_T>$ throughout the rest of this article. The propagator is a real function of H and so it is hermitian and commutes with H. Therefore,

$$E_{MIXED}=\frac{<\psi_T|H|\psi^{(n)}>}{<\psi_T|\psi^{(n)}>}=\frac{<\psi_T|H\left(\frac{E_T+E_C}{H+E_C}\right)^n|\psi_T>}{<\psi_T|\left(\frac{E_T+E_C}{H+E_C}\right)^n|\psi_T>}=\frac{<\psi^{(\frac{n}{2})}|H|\psi^{(\frac{n}{2})}>}{<\psi^{(\frac{n}{2})}|\psi^{(\frac{n}{2})}>}\geq E_0. \qquad 4.6$$

Similarly, Eq. 4.3-4.5 are upper bounds. This property is useful in cases where calculations cannot be run long enough to be certain of convergence.[5,14]

Properties other than the energy are somewhat more difficult to extract. The obvious integral to do for the expectation value of the operator O is

$$<O>=\frac{<\psi^{(n)}|O|\psi^{(n)}>}{<\psi^{(n)}|\psi^{(n)}>}. \qquad 4.7$$

In the case where O is the Hamiltonian, Eq. 4.6 shows that the mixed energy calculates this quantity. In fact the expression

$$<O>_{MIXED}=\frac{<\psi_T|O|\psi^{(n)}>}{<\psi_T|\psi^{(n)}>}. \qquad 4.8$$

will give the expectation value of O with $|\psi^{(\frac{n}{2})}>$ for any operator that commutes with H. Normally the eigenvalue of such operators such as the magnitude of the angular momentum is built directly into the trial wave function and the expectation value is uninteresting. For the majority of operators which do not commute with the Hamiltonian, the expectation value must be calculated directly from Eq. 4.7 or approximated. Direct calculations with Eq. 4.7 are possible but are difficult. Such calculations are done by first iterating $|\psi_T>$ n times with the propagator $\frac{E_T+E_C}{H+E_C}$ at which time O (for the numerator) or 1 (for the denominator) is operated. The propagator is then operated n more times, and the overlap of the result with the trial function calculated. Although this method is simple, a straightforward application usually gives large variance. In the diffusion analogy this can be understood as follows. The absorbtion and creation of walkers causes fluctuations in their number. For a large number of steps, a few walkers from the step where O is operated will have the bulk of the progeny which have survived the additonal n steps. That means that for a large calculation as n increases, the number of independent O values decreases and the variance of the calculation increases. This effect can be reduced by using good importance sampling as discussed later.

Typically expectation values are calculated by an extrapolation procedure which has proved accurate. The trial function is written as

$$|\psi_T>=\sqrt{1-\alpha^2}|\psi_0>+\alpha|\delta>, \qquad 4.9$$

where $|\delta>$ is orthogonal to $|\psi_0>$. Eq. 4.8 becomes with $n\to\infty$, so $|\psi^{(n)}>=|\psi_0>$

$$O_{MIXED}=\frac{<\psi_0|O|\psi_0>}{<\psi_0|\psi_0>}+\alpha\frac{<\delta|O|\psi_0>}{<\psi_0|\psi_0>}+\mathrm{Order}(\alpha^2). \qquad 4.10$$

The variational expression is

$$O_{VAR}=\frac{<\psi_T|O|\psi_T>}{<\psi_T|\psi_T>}=\frac{<\psi_0|O|\psi_0>}{<\psi_0|\psi_0>}+2\alpha\frac{<\delta|O|\psi_0>}{<\psi_0|\psi_0>}+\mathrm{Order}(\alpha^2). \qquad 4.11$$

Combining Eqs. 4.10 and 4.11,

$$<O>=2O_{MIXED}-O_{VAR}+\mathrm{Order}(\alpha^2). \qquad 4.12$$

For a good trial function the α^2 term is small and an estimate of the ground-state expectation value is given by the first two terms above. Eq. 4.12 can be used in a self consistent manner by calculating an extrapolated expectation value of O from Eq. 4.12 ignoring the Order(α^2) terms, and then modifying the trial function to give approximately this value. Another calculation is then done and typically O_{MIXED} and O_{VAR} will be very close so that presumably the extrapolation is accurate. Alternatively, the accuracy can be checked by calculating results with several different trial functions. In reference 8 the results of the extrapolation of the point proton density in the triton and alpha particle are shown. Typical changes in the extrapolation are a few percent even with trial functions whose point proton densities vary by a factor of two. The difference is quite small even for large changes in the trial function.

5. THE GREEN'S FUNCTION EQUATIONS

To iterate the GFMC equations Eqs. 3.3. or 3.8, an expression for the propagator or Green's function must be known. Typically, the unknown Green's function is expanded in terms of a set of known Green's functions of a simpler Hamiltonian H_U,

$$\exp[-(H-E_T)\tau]=\exp[-(H_U-E_T)\tau]$$

$$+\int_0^\tau d\tau'\exp[-(H-E_T)(\tau-\tau')](H_U-H)\exp[-(H_U-E_T)\tau'] \qquad 5.1$$

or

$$\frac{1}{H+E_C}=\frac{1}{H_U+E_C}+\frac{1}{H+E_C}(H_U-H)\frac{1}{H_U+E_C} \qquad 5.2$$

Eq. 5.1 can be easily verified by doing the integral, Eq. 5.2 by cancelling terms on the right hand side. Most of the rest of this article will be used to develop the GFMC method using the time integrated method. However, a number of calculations have used an approximation to Eq. 5.1 known as the short time approximation.[15,18] This approximation will be explained next and importance sampling introduced before moving on to the exact sampling of Eq. 5.2. I should mention that a corresponding large E_C approximation exists for Eq. 5.2.

6. THE SHORT TIME APPROXIMATION

To make further progress, I write Eq. 5.1 in configuration space with Green's function $\mathcal{G}$ and $\mathcal{G}_U$ defined to be the configuration space matrix elements

$$\mathcal{G}(R,R';\tau) = \langle R|\exp(-(H-E_T)\tau)|R'\rangle, \quad 6.1$$

and

$$\mathcal{G}_U(R,R';\tau) = \langle R|\exp(-(H_U-E_T)\tau)|R'\rangle. \quad 6.2$$

These satisfy the equations,

$$(H-E_T)\mathcal{G}(R,R';\tau) = \frac{-\partial \mathcal{G}(R,R';\tau)}{\partial \tau}, \quad 6.3$$

and

$$\mathcal{G}(R,R';\tau=0) = \delta(R'-R) \quad 6.4$$

and similar equations for $\mathcal{G}_U$ and H_U. The Hamiltonian is spin independent

$$H = -\nabla^2 + V(R). \quad 6.5$$

Typically, H_U is taken to be a Hamiltonian with a constant potential,

$$H_U = -\nabla^2 + U. \quad 6.6$$

If H_U is allowed to operate over all space, $\mathcal{G}_U$ is easily found to be

$$\mathcal{G}_U(R,R';\Delta\tau) = (4\pi\Delta\tau)^{-\frac{3N}{2}} \exp(-\frac{(R-R')^2}{4\Delta\tau} - (U-E_T)\Delta\tau). \quad 6.7$$

Eq. 5.1 becomes in configuration space,

$$\mathcal{G}(R,R';\Delta\tau) = \mathcal{G}_U(R,R';\Delta\tau)$$

$$+ \int_0^{\Delta\tau} d\tau' \int dR'' \mathcal{G}(R,R'',\Delta\tau-\tau')(U-V(R''))\mathcal{G}_U(R'',R',\tau'). \quad 6.8$$

Notice this equation is valid for any value of U. For a given value of R′, any value of U may be chosen. In particular, U may be chosen to be V(R′). Notice this corresponds to expanding $\mathcal{G}$ using an infinite set of $\mathcal{G}_U$'s, one for each value of V(R′). In this case, the method is very closely related to a path integral. And though a perturbation expansion is used for $\mathcal{G}$, since different $\mathcal{G}_U$'s are used at different R′ values, the method is nonperturbative.

If $\Delta\tau$ is now chosen to be very small, Eq. 6.8 shows that the displacement $|R-R'|$ is proportional to $\sqrt{\Delta\tau}$. Taylor series expansion of V(R″) around U=V(R′) in Eq. 6.7 shows that the second term is of order $\sqrt{\Delta\tau}$. The short time approximation corresponds to dropping this term and taking

$$\mathcal{G}_{ST}(R,R';\Delta\tau) = \mathcal{G}_{U=V(R')}(R,R';\Delta\tau). \quad 6.8$$

The unknown Green's function $\mathcal{G}$ is now known in the limit $\Delta\tau \to 0$ and is given by the simple expression in Eq. 6.7.

Let's see how a short time GFMC calculation could be implemented. First, we need to sample a set of walkers from the starting wave function. Since it is known, a simple application of the Metropolis et. al. method as in the variational procedure will sample a set of walkers from the trial wave function. As explained in Appendix A, sampling corresponds to replacement of a function by a sum of delta functions. The algorithm can be written as:

1. Sample $\psi(R,\tau=0) = \psi_T(R)$. This corresponds to $\psi_T(R,\tau=0) \to \sum_{i=1}^{M(\tau=0)} \delta(R-R_i)$ where R_i

are the sampled values and are considered to be the initial positions of the diffusing $M(\tau=0)$ walkers.

The wave function at time $\tau+\Delta\tau$ is given by Eqs. 3.3 and 6.1 to be

$$\psi(R,\tau+\Delta\tau)=\int dR'\mathcal{G}(R,R';\Delta\tau)\psi(R',\tau). \qquad 6.9$$

2. With $\psi(R,\tau)$ given by $\sum_{i=1}^{M(\tau)}\delta(R-R_i)$, substituting into Eq. 6.9 and integrating

$$\psi(R',\tau+\Delta\tau)=\sum_{i=1}^{M(\tau)}\mathcal{G}(R',R_i;\tau)$$

$$=\sum_{i=1}^{M(\tau)}(4\pi\Delta\tau)^{-\frac{3N}{2}}\exp(-\frac{(R'-R_i)^2}{4\Delta\tau}-(V(R_i)-E_T)\Delta\tau), \qquad 6.10$$

where I have introduced the short time approximation.

3. Sample a new R value for each R_i value from the normalized gaussian $(4\pi\Delta\tau)^{-\frac{3N}{2}}\exp(-\frac{(R-R_i)^2}{4\Delta\tau})$, and take m_i copies of this R where $m_i=\text{int}(\exp(-(V(R_i)-E_T)\Delta\tau)+\xi_i)$ and ξ_i indicates an independent random number uniformly sampled on the unit interval as in appendix A and int indicates the integer part. This makes $\psi(R',\tau+\Delta\tau)=\sum_{i=1}^{M(\tau+\Delta\tau)}\delta(R'-R_i)$, where $M(\tau+\Delta\tau)=\sum_{i=1}^{M(\tau)}m_i$.

This is just one of many possible ways to sample $\psi(\tau+\Delta\tau)$. Other ways will change the variance but not the average. The sampling of the normalized gaussian can be carried out by sampling the change in each of the 3N coordinates from an independent gaussian using, for example, the Box-Muller scheme. The selection of m_i shown is a simple way to convert the weight of a walker into an integer. The average expected number of walkers is $\exp(-(V(R_i)-E_T)\Delta\tau)$ as desired. The completion of step three leaves $\psi(R,\tau+\Delta\tau)$ represented as a sum of delta functions which can be iterated to $\tau+2\Delta\tau$ by repeating step two with this new set of walkers. Since $\psi(R,\tau)$ is then represented at each step by a sum of delta functions, the integrals in the growth and mixed energies or the mixed estimator of any other operator can be easily done.

The method as just described generally suffers from high variance. High variance in the energy can be traced to fluctuations in m_i. These fluctuations occur because walker moves to new positions are given by the gaussian which moves walkers equally well in all directions. A move in an undesired direction is signaled after the move by a small weight in the next iteration. On average fewer than one copy of the new walker position is kept in the next iteration. Similarly, a desired move results in multiple copies. A lower variance method can equalize the weights by sampling from a different function that tends to move walkers into desired regions more often and undesired regions less often. This is of course just the importance sampling technique discussed in appendix A.

The proper way to apply importance sampling to an integral equation is to use as an importance function, the solution to the adjoint equation run backwards in time. Without going into the details, the result is that the importance function is the ground-state wave function. Since the ground-state wave function is not known, it is approximated by the trial function. This approximation of the ground-state wave function by the trial function will increase the variance over that given with perfect importance sampling, but hopefully not too much. In any case the introduction of importance sampling will not change the result. If the trial wave function is a good approximation to the ground-state wave function, dramatic decreases in the variance over that with no importance function result.

The importance sampled equation is

$$\psi_T(R)\psi(R,\tau+\Delta\tau)=\int dR'\left[\frac{\psi_T(R)}{\psi_T(R')}\mathcal{G}(R,R',\Delta\tau)\right]\psi_T(R')\psi(R',\tau), \qquad 6.11$$

which is the analogue of Eq. 3.3. Note that this equation is Eq. 6.9 multiplied on both sides by $\psi_T(R)$. Before applying the short time approximation to Eq. 6.11, let's look at the weight it would give. If $\psi_T(R')\psi(R',\tau)$ is represented as a sum of delta functions, the integral in Eq. 6.11 can be easily done. Following the method used above without importance sampling, I should first sample a new walker position from the normalized probability density

$$\frac{\dfrac{\psi_T(R)}{\psi_T(R')}\mathcal{G}(R,R',\Delta\tau)}{\int dR\dfrac{\psi_T(R)}{\psi_T(R')}\mathcal{G}(R,R',\Delta\tau)} \qquad 6.12$$

and take

$$w=\int dR\frac{\psi_T(R)}{\psi_T(R')}\mathcal{G}(R,R',\Delta\tau) \qquad 6.13$$

copies of this walker position. Using the eigenfunction expansion of $\mathcal{G}$

$$\mathcal{G}(R,R_i;\Delta\tau)=\sum_m \psi_m(R)\psi_m^*(R_i)\exp(-(E_m-E_T)\Delta\tau), \qquad 6.14$$

and doing the integral in Eq. 6.13 with $\psi_T(R)=\psi_0(R)$

$$w=\exp(-(E_0-E_T)\Delta\tau). \qquad 6.15$$

With the correct importance function, w is independent of the position R_i, and the growth estimate is known with zero variance. If $E_T=E_0$, then $w=1$; every input walker produces exactly one output walker. In practical calculations the weight w fluctuates, but the inclusion of $\psi_T(R)$ as an importance function drastically reduces the variance.

In general, sampling from the importance sampled Green's function can be difficult. Within the short time approximation, $\mathcal{G}$ is replace by $\mathcal{G}_{U=V(R')}$. Since this replacement is valid only for small $\Delta\tau$, $\psi_T(R)$ can be expanded around $\psi_T(R')$ and only terms of the same order in $\Delta\tau$ as the potential terms kept. This is easily accomplished by writing,

$$\frac{\psi_T(R)}{\psi_T(R')}\mathcal{G}_{U=V(R')}(R,R',\Delta\tau)$$

$$=(4\pi\Delta\tau)^{-\frac{3N}{2}}\exp(-\frac{(R-R')^2}{4\Delta\tau}-(V(R')-E_T)\Delta\tau+\ln(\psi_T(R))-\ln(\psi_T(R')), \qquad 6.16$$

and expanding

$$\ln(\psi_T(R))-\ln(\psi_T(R')=$$

$$=\sum_{i=1}^{3N}\frac{\partial\ln(\psi_T(R'))}{\partial x_i}(x_i-x'_i)+\frac{1}{2}\sum_{i,j=1}^{3N}\frac{\partial^2\ln(\psi_T(R'))}{\partial x_i\partial x_j}(x_i-x'_i)(x_j-x'_j)+\cdots \qquad 6.17$$

where x_i i=1 to 3N are the 3N coordinates of the N particles. Substituting Eq. 6.17 into Eq. 6.16 gives a short time approximation to the importance sampled $\mathcal{G}$. A simple way to sample this distribution is to define,

$$P(R,R')=\prod_i (4\pi\tau_i^*)^{-\frac{1}{2}}\exp\left(-\frac{(x_i-x'_i-\tau\frac{d\ln(\psi_T(R'))}{dx'_i})^2}{4\tau_i^*}\right) \qquad 6.18$$

with

$$\tau_i^*=\tau(1+2\tau\frac{d^2\ln(\psi_T(R'))}{dx'^2_i}). \qquad 6.19$$

I write

$$\frac{\psi_T(R)}{\psi_T(R')}\mathcal{G}_{U=V(R')}(R,R',\Delta\tau)=\left[\frac{\psi_T(R)}{\psi_T(R')P(R,R')}\mathcal{G}_{U=V(R')}(R,R',\Delta\tau)\right]P(R,R'). \qquad 6.20$$

The new walker position is sampled from P(R,R') and the term in brackets becomes the weight of that walker. Notice that any errors in P(R,R') are corrected by this weight.

For completeness, I give the algorithm for an importance sampled short time approximation calculation.

1. Sample an initial set of walkers from $\frac{\psi_T^2(R)}{\int dR\psi_T^2(R)}\rightarrow\sum_{i=1}^{M(0)}\delta(R-R_i)$. Notice that this is exactly what the usual variational Monte Carlo calculation does.
2. With $\psi_T(R)\psi(R,\tau)$ given by $\sum_{i=1}^{M(\tau)}\delta(R-R_i)$, substitute into Eq. 6.11, and use Eq. 6.20 for the short time approximation to get a sum of terms,

$$\psi_T(R)\psi(R,\tau+\Delta\tau)=\sum_i\left[\frac{\psi_T(R)}{\psi_T(R_i)P(R,R_i)}\mathcal{G}_{U=V(R_i)}(R,R_i,\Delta\tau)\right]P(R,R_i). \qquad 6.21$$

3. Sample a new R value for each R_i value from the normalized gaussians given by $P(R,R_i)$ and take m_i copies of this R'_i where

$$m_i=\text{int}\left(\frac{\psi_T(R'_i)}{\psi_T(R_i)P(R'_i,R_i)}\mathcal{G}_{U=V(R_i)}(R'_i,R_i,\Delta\tau)+\xi_i\right) \qquad 6.22$$

which makes $\psi_T(R)\psi(R,\tau+\Delta\tau)=\sum_{i=1}^{M(\tau+\Delta\tau)}\delta(R-R'_i)$ where $M(\tau+\Delta\tau)=\sum_{i=1}^{M(\tau)}m_i$.

7. EXACT SAMPLING OF THE GREEN'S FUNCTION

Both the time integrated and time dependent methods can be made exact by including all terms in Eq. 5.1 or 5.2 and therefore sampling the exact Green's function. Since Eqs. 5.1 and 5.2 define infinite series, the Green's function must be sampled from this infinite series. The basic technique for sampling from an infinite series is explained in appendix A. I will now concentrate on sampling from the time integrated Green's function Eq. 5.2. Sampling from Eq. 5.1 is covered in reference 19 and is very similar.

The right hand side of Eq. 5.2 has a term which contains H_U-H. Since we wish to interpret all terms as probabilities in the Monte Carlo method, this term must be made positive definite. In what follows, the H_U will be the Hamiltonian of a constant potential.

If $V(R)$ is not bounded above, $H_U - H$ cannot be made positive definite. To circumvent this problem, as well as to make the method more efficient in the general case, the domain of H_U is restricted. Equivalently, outside this Domain D the potential in H_U is taken to be positive infinity. Defining,

$$G_U(R,R') = \langle R | \frac{1}{H_U + E_C} | R' \rangle \quad R,R' \in D, \tag{7.1}$$

or

$$(H_U + E_C) G_U(R,R') = \delta(R - R') \quad R,R' \in D, \tag{7.2}$$

with

$$H_U = -\nabla^2 + U, \tag{7.3}$$

U a constant, and similar equations for G. Note that $G_U = 0$ for R or R′ outside the domain D. Eq. 5.1 becomes,

$$G(R,R') = G_U(R,R') + \int_D dR''\, G(R,R'')(U - V(R''))G_U(R'',R')$$

$$+ \int_S dR''\, G(R,R'')[-\hat{n}\cdot\vec{\nabla} G_U(R'',R')] \tag{7.4}$$

The last term on the right hand side is integrated over the surface of the domain of G_U and the second term is integrated over the domain. The term $[-\hat{n}\cdot\vec{\nabla} G_U(R'',R')]$ is minus the normal derivative at the surface.

Importance sampling can be used in exactly the same way as for the short time approximation and for the same reason. From the eigenfunction expansion for $G(R,R')$ it is easy to show that

$$\int dR (E_0 + E_C) \frac{\psi_0(R)}{\psi_0(R')} G(R,R') = 1. \tag{7.5}$$

So again, the approximately importance sampled $G(R,R')$ is $\frac{\psi_T(R)}{\psi_T(R')} G(R,R')$. The importance sampled equations become,

$$\psi_T(R)\psi^{(n)}(R) = (E_T + E_C) \int dR' [\frac{\psi_T(R)}{\psi_T(R')} G(R,R')] \psi_T(R') \psi^{(n-1)}(R'), \tag{7.6}$$

and,

$$(E_T + E_C)\frac{\psi_T(R)}{\psi_T(R')} G(R,R') = (E_T + E_C)\frac{\psi_T(R)}{\psi_T(R')} G_U(R,R')$$

$$+ \int_D dR'' (E_T + E_C) \frac{\psi_T(R)}{\psi_T(R'')} G(R,R'')(U - V(R'')) \frac{\psi_T(R'')}{\psi_T(R')} G_U(R'',R')$$

$$+ \int_S dR'' (E_T + E_C) \frac{\psi_T(R)}{\psi_T(R'')} G(R,R'') \frac{\psi_T(R'')}{\psi_T(R')} [-\hat{n}\cdot\vec{\nabla} G_U(R'',R')] \tag{7.7}$$

As in the short time approximation, U can be chosen to have a value dependent on R′. Similarly, the domain can be chosen differently for each value of R′. The only restriction on the domains is that it must be possible to propagate from any valid part of configuration space to any other. The U value is selected to be an upper bound to $V(R'')$ inside the domain and $E_C > E_0$ so that all terms are positive.

Sampling is now accomplished in a similar manner to that given for the importance sampled short time approximation. The exception is that G(R,R′) is given by an infinite series of terms. Using the result of Eq. 7.5, for exact importance sampling, the propagator Eq. 7.7 is a normalized probability density and can be sampled via the methods for an infinite series described in appendix A.

To see how this is done in a specific simple case, I will describe the algorithm for solving for the ground state of one particle in a one dimensional harmonic oscillator potential. In this case,

$$H = -\frac{d^2}{dx^2} + x^2 \qquad 7.8$$

and

$$H_U = -\frac{d^2}{dx^2} + U \quad x' - L < x < x' + L. \qquad 7.9$$

where the domain for $G_U(x,x')$ has been taken to be symmetric about x' and of width 2L. $G_U(x,x')$ is easily calculated to be

$$G_U(x,x') = \frac{\sinh[\sqrt{U+E_C}\,|x-x'|-L]}{2\sqrt{U+E_C}\cosh[\sqrt{U+E_C}L]} \qquad 7.10$$

with normal derivative at $x = x' \pm L$,

$$-\hat{n}\cdot\frac{d}{dx}G_U(x' \pm L, x') = \frac{1}{2\cosh[\sqrt{U+E_C}L]} \qquad 7.11$$

The trial function is taken to be a gaussian

$$\psi_T(x) = \exp(-\frac{\alpha x^2}{2}).$$

The choice $\alpha = 1$ corresponds to the exact ground state. The effects of various choices for a realistic importance function can be seen by selecting other α values . A simple way (but not the necessarily the optimum) to select U and L is to take $U = V(x') + \epsilon$, where ϵ is a positive constant. L is then chosen to make $V(x) < U$ in the domain. For the harmonic oscillator this becomes,

$$U = x'^2 + \epsilon \qquad 7.12$$

$$L = \sqrt{U} - |x'|. \qquad 7.13$$

In this simple one dimensional case, it is possible to sample the equations optimally. However, this is not normally possible for many particles in many dimensions. Particularly, the inclusion of the importance sampling is usually done only approximately. As before, this does not mean that the results are approximate. It just means that the separation of the terms in the integrands into a sampled part and a weight will not give the lowest possible variance. The correct weight is always used to give the correct answer. Typically, the effect of sampling the many-body wave function is included by using linear terms in a Taylor series expansion around the current position. This will be explained in more detail later. I will do this one-dimensional problem in the same spirit so that the generalization to more particles in three dimensions is easier.

A normalized probability density is given by $\dfrac{G_u(x,x')}{\int_D dx G_U(x,x')}$ for fixed x'. The normalization is

$$\int_D dx G_U(x,x') = \frac{1}{U+E_C}\left[1 - \frac{1}{\cosh(\sqrt{U+E_C}L)}\right] \qquad 7.14$$

The distance $\Delta x = |x - x'|$ can be sampled from $\frac{G_U(x,x')}{\int_D dx G_U(x,x')}$ by a straightforward transformation method for sampling from the sinh (see appendix A),

$$y = \frac{1}{\cosh(\sqrt{U+E_C}L} + \xi\left(1 - \frac{1}{\cosh(\sqrt{U+E_C}L)}\right),$$

$$\Delta x = L + \left[-\ln(\cosh(\sqrt{U+E_C}L)) - \ln\left(y + \left\{y^2 - \frac{1}{\cosh(\sqrt{U+E_C}L)}\right\}^{\frac{1}{2}}\right)\right](U+E_C)^{-\frac{1}{2}}. \quad 7.15$$

An algorithm to iterate Eq. 7.6 is

1. Sample an initial set of walkers from $\frac{\psi_T^2(x)}{\int dx \psi_T^2(x)} \rightarrow \sum_{i=1}^{M(0)} \delta(x - x_i)$ and push onto the old stack.

   Here I assume a stack structure for storing the walker positions. That is, walkers positions are pushed onto or popped off a stack. The positions on the old stack must be propagated by $(E_T+E_C)\frac{\psi_T(x)}{\psi_T(x_i)}G(x,x_i)$ before they can be placed on the new stack.

2. Pop a walker x_i off the old stack. If the old stack is empty, the propagation of all walkers to the new stack is complete.
3. Substitute the delta function from step 2 into Eq. 7.6. This walker's contribution is $(E_T+E_C)\frac{\psi_T(x)}{\psi_T(x_i)}G(x,x_i)$. New x values need to be sampled from this function. Calculate U and L from Eqs. 7.12 and 7.13. These values now define the terms in Eq. 7.7.

The importance sampled Green's function, $(E_T+E_C)\frac{\psi_T(x)}{\psi_T(x_i)}G(x,x_i)$, is a sum of three terms. If I assume that the sum is a normalized probability density, I can sample from the sum by first selecting a single term according to its normalization, and then sampling a new walker position from that term. To calculate the normalizations exactly requires knowledge of the true Green's function which is unknown. However, if $\psi_T \approx \psi_0$ and $E_T \approx E_0$, then using Eq. 7.5, $\int dx(E_T+E_C)\frac{\psi_T(x)}{\psi_T(x_i)}G(x,x_i) \approx 1$. The normalization of the surface term is then

$$\int dx \int_S dx''(E_T+E_C)\frac{\psi_T(x)}{\psi_T(x'')}G(x,x'')\frac{\psi_T(x'')}{\psi_T(x_i)}\left[-\hat{n}\cdot\frac{d}{dx''}G_U(x'',x_i)\right]$$

$$\approx \int_S dx''\frac{\psi_T(x'')}{\psi_T(x_i)}\left[-\hat{n}\cdot\frac{d}{dx''}G_U(x'',x_i)\right]$$

$$\approx \int_S dx''\left[-\hat{n}\cdot\frac{d}{dx''}G_U(x'',x_i)\right] \qquad 7.16$$

where I have used Eq. 7.5 and also assumed in the last line that the average value of

$\frac{\psi_T(x'')}{\psi_T(x_i)}$ over S is approximately one. The right hand side of Eq. 7.16 is an approximation to the probability of sampling from the surface term. For the one dimensional case, the surface is just the two endpoints $x=x_i\pm L$. The next step in the algorithm is then

4. With probability

$$P_S=\int_S dx''[-\hat{n}\cdot\frac{d}{dx''}G_U(x'',x_i)]=\frac{1}{\cosh(\sqrt{U+E_C}L)}$$

take a surface step (below) else go to step 5. That is

4a. If $\xi_1<P_S$ then $\Delta x=L$. Use the relative value of the trial function to select whether the move is $+\Delta x$ or $-\Delta x$. That is if $\xi_2<\frac{\psi_T(x_i+\Delta x)}{\psi_T(x_i+\Delta x)+\psi_T(x_i-\Delta x)}$ then $x'_i=x_i+\Delta x$ else $x'_i=x_i-\Delta x$.

4b. Calculate the weight of x'_i. $w=\frac{\psi_T(x_i+\Delta x)+\psi_T(x_i-\Delta x)}{2\psi_T(x_i)}$, and push $n=\mathrm{int}(w+\xi_3)$ copies of the walker position x'_i onto the *old* stack and go to step 2.

The reason for pushing x'_i onto the old stack is because only the first part of the surface term has been sampled. These walkers have to be further propagated by the full importance sampled Green's function before they can be placed on the new stack. However, this is just the definition of the walkers that have been placed on the old stack already, so the x'_i can be placed on the old stack as well. This is an example of a recursive sampling procedure for an infinite series mentioned in appendix A. Notice that exactly as in the short time procedure, the ratio of the integrand to what was sampled is included as a weight for the walker.

If the surface term is not selected, one of the other two must be. These terms both involve sampling from $\frac{\psi_T(x)}{\psi_T(x_i)}G_U(x,x_i)$. Integrating over the R variable in Eq. 7.5 gives for the probabilities.

$$\int_D dx(E_T+E_C)\frac{\psi_T(x)}{\psi_T(x_i)}G_U(x,x_i),$$

and

$$\int_D dx''\int dx(E_T+E_C)\frac{\psi_T(x)}{\psi_T(x'')}G(x,x'')(U-V(x''))\frac{\psi_T(x'')}{\psi_T(x_i)}G_U(x'',x_i)$$

$$\approx\int_D dx''(U-V(x''))\frac{\psi_T(x'')}{\psi_T(x_i)}G_U(x'',x_i), \qquad 7.17$$

and their ratio is $\approx\frac{E_T+E_C}{U-V(x'')}$. From this, step 5 becomes:

5. Sample a magnitude $\Delta x=|x''-x_i|$ from

$$\frac{G_U(x'',x_i)}{\int dx''G_U(x'',x_i)}\equiv\frac{G_U(x'',x_i)}{\mathrm{Gnorm}}.$$

And as in the surface step, use the relative value of $\psi_T(x+\Delta x)$ and $\psi_T(x-\Delta x)$ to select whether x'_i is taken to be $x_i+\Delta x$ or $x_i-\Delta x$. Use the relative values of E_T+E_C and $U-V(x'_i)$ to determine whether the first or second term in Eq. 7.7 is sampled. Calculate the weight and push the walker position onto the correct stack. This becomes:

5a. Sample Δx as indicated in Eq. 7.15

5b. If $\xi_4 < \frac{\psi_T(x_i+\Delta x)}{\psi_T(x_i+\Delta x)+\psi_T(x_i-\Delta x)}$ take $x'_i = x_i + \Delta x$ else take $x'_i = x_i - \Delta x$.

5c. Calculate the weight

$$w = \frac{\psi_T(x_i+\Delta x)+\psi_T(x_i-\Delta x)}{2\psi_T(x_i)}(E_T+E_C+U-V(x'_i))\mathrm{Gnorm}/(1-P_S),$$

and set $n = \mathrm{int}(\xi_5 + w)$.

5d. If $\xi_6 < \frac{E_T+E_C}{E_T+E_C+U-V(x'_i)}$ then push n copies of x'_i onto the *new* stack, else push n copies of x'_i onto the *old* stack, and go to step 2.

The reader can verify that these steps correctly sample the exact Green's function by noting that the probability of each kind of step multiplied by its weight is the integrand corresponding to that step in Eq. 7.16.

Appendix B is a listing of a FORTRAN code that implements the algorithm given above for the one dimensional harmonic oscillator. The reader should be warned that this is not a production quality code. No provision has been made to make sure that the stacks do not overflow or underflow for example. Also the variance is very high if the α parameter in the trial function is taken less than one. This is a simple example of what bad importance sampling can do. If guiding the random walk towards high probability regions lowers the variance, then guiding the walk away from these regions naturally raises the variance. Another problem with this code is that it assumes when the error bars are calculated that the energy value at each step is uncorrelated with the others. This is not usually true, and the printed error bars will tend to be too small. A production code would block the averages over enough steps to ensure that the results were uncorrelated. However, this little program can be used to demonstrate the method. It can show, for example, that if E_C is chosen to be a large positive constant, that the Green's function equation is dominated by the first term much as in the short time approximation. If the ϵ parameter is picked very small, then the domain tends to be small and the Green's function iteration is dominated by surface terms. If the ϵ is chosen large, the Green's function iteration is dominated by the $U-V(R'')$ term. In a production code, the goal is to optimize these choices so that the convergence takes the smallest amount of computer time.

The problems mentioned above can all be easily solved. The resulting code for N particles in three dimensions is a few thousand lines of FORTRAN. The logic and sampling routines contain fewer than 500 lines of FORTRAN. This is very small compared to most *ab initio* codes. I will now indicate the kinds of changes that must be made to produce an efficient three dimensional many particle code.

8. N PARTICLES IN THREE DIMENSIONS

The direct generalization of the Green's function of Eq. 7.10 to many dimensions can be done in several ways. A hyperspherical domain in 3N dimensions was used by Kalos in a number of his early investigations.[6,7] This Green's function is rather inefficient for many-particle systems, although it should work fine for few-body systems. For example, if a pair of particles is within a distance d of their hard core diameter, the radius of the domain can be at most d. The average distance moved by each of the N particles will be proportional to $d/(2N)^{\frac{1}{2}}$. For large N this can seriously degrade efficiency. An improved Green's function developed by Kalos, Levesque and Verlet[13] is given by letting each particle move in a sphere centered at its current position. The average distance is then proportional to d/2 and independent of N. The disadvantage of this new domain is that Eq. 7.1 cannot be easily solved analytically. This is because $(H+E_C)^{-1}$ cannot be factored into a product of one-body operators. However, Eq. 3.9 can be used to express G_U in terms of the time dependent $\mathcal{G}_U$. A time integration is then needed to calculate G_U, which is naturally done using the Monte Carlo method. This entails sampling a time. The $\mathcal{G}_U$ can be factored,

$$\mathcal{G}_U(R,R';\tau)=\langle R|\exp(-(H_U+E_C)\tau)|R'\rangle$$

$$=\exp(-(U+E_C)\tau)\prod_{i=1}^{N}\langle \vec{r}_i|\exp(\nabla_i^2\tau)|\vec{r}\,'_i\rangle \qquad 8.1$$

where

$$\langle \vec{r}_i|\exp(\nabla_i^2\tau)|\vec{r}\,'_i\rangle \equiv a_i^{-3}g\left(\frac{|\vec{r}_i-\vec{r}\,'_i|}{a_i},\frac{\tau}{a_i^2}\right) \qquad 8.2$$

and

$$g(\rho,t\to 0)=\delta(\rho) \qquad 8.3$$

$$-\nabla_\rho^2 g(\rho,t)=-\frac{\partial g(\rho,t)}{\partial t} \quad 0<\rho<1.$$

$$g(\rho=1,t)=0 \qquad 8.4$$

The sphere radius for particle i is a_i, and $\vec{r}_i$ and $\vec{r}\,'_i$ are the three dimensional positions of particle i. G_U is

$$G_U(R,R')=\int_0^\infty d\tau e^{-(U+E_C)\tau}\prod_{i=1}^{N}a_i^{-3}g\left(\frac{|\vec{r}_i-\vec{r}\,'_i|}{a_i},\frac{\tau}{a_i^2}\right). \qquad 8.5$$

It should be noted that the second argument in g is really $\hbar\tau/2m_ia_i$, where m_i is the mass of particle i, but I have assumed $\hbar=1$ and $\hbar^2/2m=1$ with all the masses the same. This change of the scaling factor is the only change necessary to calculate with unequal mass particles. The g is easily calculated from the eigenfunction expansion of Eq. 8.4 to be

$$g(\rho,t)=\sum_n\frac{n\sin(n\pi\rho)\exp(-\pi^2n^2t)}{2\rho}$$

$$=(4\pi t)^{-\frac{3}{2}}\exp\left(-\frac{\rho^2}{4t}\right)$$

$$+(4\pi t)^{-\frac{3}{2}}\rho\sum_n\left[(\rho+2n)\exp\left(-\frac{(\rho+2n)^2}{4t}\right)\right.$$

$$\left.-(\rho-2n)\exp\left(-\frac{(\rho-2n)^2}{4t}\right)\right] \qquad 8.6$$

The second line is the result of using the Poisson summation formula on the eigenfunction expansion. The result of the Posisson formula can be recognized as the contributions for a free source at the center of the sphere and a set of image sources outside the sphere to maintain the boundary conditions. The first expression converges rapidly for large t, the second for small t. An expression that can be checked to be accurate to machine precision on 64 bit computers is

$$g(\rho,t)=\sum_{n=1}^{5}\frac{n\sin(n\pi\rho)\exp(-\pi^2n^2t)}{2\rho} \qquad t>.11$$

$$=(4\pi t)^{-\frac{3}{2}}\exp(-\frac{\rho^2}{4t})$$

$$+(4\pi t)^{-\frac{3}{2}}\rho\sum_{n=1}^{2}\Big[(\rho+2n)\exp(-\frac{(\rho+2n)^2}{4t})$$

$$-(\rho-2n)\exp(-\frac{(\rho-2n)^2}{4t})\Big] \qquad t<.11 \qquad 8.7$$

The integral over $d^3\rho$ of $g(\rho,t)$ is necessary to sample from Eq. 8.5. This is, to the same precision as Eq. 8.7,

$$h(t)=\int_0^1 d^3\rho g(\rho,t)=\begin{cases}2\sum_{n=1}^{4}(-1)^{n-1}\exp(-\pi^2n^2t) & t>.16\\ 1-2(\pi t)^{-\frac{1}{2}}\sum_{n=1}^{2}\exp(-\frac{(2n-1)^2}{4t}) & t<.16\end{cases} \qquad 8.8$$

The surface term $-\hat{n}\cdot\nabla G_U(R,R')$ is calculated using Eq. 8.7 as

$-\hat{n}\cdot\nabla G_U(R,R')=$

$$\int_0^\infty d\tau e^{-(U+E_C)\tau}\sum_{i=1}^{N}-\hat{n}_i\cdot\nabla_i a_i^{-3}g(\frac{|\mathbf{r}_i-\mathbf{r}'_i|}{a_i},\frac{\tau}{a_i^2})\Big|_{|\mathbf{r}_i-\mathbf{r}'_i|=a_i}\prod_{j\neq i}a_j^{-3}g(\frac{|\mathbf{r}_j-\mathbf{r}'_j|}{a_i},\frac{\tau}{a_j^2}). \qquad 8.9$$

Note that surface pieces with two or more particles each on the surface of their sphere are measure zero. Since g is spherically symmetric, the normal derivative of g is a constant. Its normalization is

$$-\frac{dh(t)}{dt}=\int d^3\rho[-\frac{\partial g(\rho,t)}{\partial t}]=\int d^3\rho[-\nabla_\rho^2 g(\rho,t)]=\int_S d^2\rho[-\hat{n}\cdot\nabla_\rho g(\rho,t)], \qquad 8.10$$

which is just a consequence of current conservation in the related diffusion problem. The flux through the boundary is just the negative of the density change in the domain. With these results, the normalizations of G_U and $-\hat{n}\cdot\nabla G_U$ are known and each of these distributions can be sampled once the time τ is chosen. The time integral is of course done by the Monte Carlo method. The time sampling and the choice of which term to select in Eq. 7.7 can be combined. In the diffusion analogy, the time can be calculated for each particle to be absorbed by the change in the potential U-V, or to be absorbed by the trial energy E_T+E_C, or to be absorbed at the surface of its sphere. Clearly, whatever event happens first governs the physical process that actually takes place. To see how this comes about assume that the times τ are sampled from the probability density

$$P(\tau)=-\frac{d}{d\tau}[\prod_i h(\frac{\tau}{a_i^2})e^{-\tilde{U}\tau}], \qquad \mathbf{8.11}$$

which has a probability distribution function given by the term in brackets and is clearly normalized. $\tilde{U}$ is for now an arbitrary positive constant. Eq. 8.11 can be sampled by taking the minimum of the times sampled from each $-\frac{d}{d\tau}h(\frac{\tau}{a_i^2})$ and $\tilde{U}e^{-\tilde{U}\tau}$, as can be easily verified. (This is an example of a combinatorial sampling method discussed in appendix A). The former can be sampled by direct application of the transformation method

$$\tau=a_i^2h^{-1}(\xi_1), \qquad 8.12$$

and the latter from

$$\tau=-\frac{\ln(\xi_2)}{\tilde{U}} \qquad 8.13$$

Let's look at the proability that the τ sampled from $-\frac{d}{d\tau}h(\frac{\tau}{a_i^2})$ is the smallest. It is the probability of sampling τ between τ and $\tau+d\tau$ times the probability of all the other τ's being larger, which is

$$-\frac{d}{d\tau}h(\frac{\tau}{a_i^2})d\tau\prod_{j\neq i}h(\frac{\tau}{a_j^2})e^{-\tilde{U}\tau}. \qquad 8.14$$

Eq 8.14 is the normalization of the $-\hat{n}\cdot\nabla''_i\mathcal{G}_U(R'',R';\tau)$ term if $\tilde{U}=U+E_C$. Similarly, if the τ sampled from $\tilde{U}e^{-\tilde{U}\tau}$ is smallest, the probability is

$$\tilde{U}e^{-\tilde{U}\tau}d\tau\prod_i h(\frac{\tau}{a_i^2}), \qquad 8.15$$

which is the normalization of $\tilde{U}\mathcal{G}_U(R'',R',\tau)$. To sample from G_U, I can choose τ to be the smallest of the τ's from Eq. 8.12 and 8.13. If the τ from $\tilde{U}e^{-\tilde{U}\tau}$ is smallest, sample an R'' from $\frac{\mathcal{G}_U(R'',R',\tau)}{\int dR''\mathcal{G}_U(R'',R',\tau)}$. This latter sampling is simply done by sampling a ρ for each particle from $g(\rho,\frac{\tau}{a_i^2})/h(\frac{\tau}{a_i^2})$ and letting $|\mathbf{r}_i-\mathbf{r}'_i|=a_i\rho$. Sampling from the surface term is very similar. If the smallest τ was sampled from $-\frac{d}{d\tau}h(\frac{\tau}{a_i^2})$, then to sample the surface derivative of G_U, the position of particle i must be sampled at the surface of its sphere, and the other particles sampled as before. Since the normal derivative of $g(\rho,t)$ is independent of position on the surface, this can be easily accomplished by sampling ρ values as above and simply promoting particle i to the surface of its sphere.

Normally, the $\tilde{U}$ term is not chosen to be $U+E_C$, but something like

$$\tilde{U}=[E_T+E_C+U-V(R')] \qquad 8.17$$

with appropriate factors included in the weight terms to correct for this difference. This change tends to increase the number of volume steps selected which corresponds to picking the first term in Eq. 6.7.

The inclusion of some effect of the importance function terms like $\frac{\psi_T(R')}{\psi_T(R)}$ is done by using them to select the direction of $\mathbf{r}'_i-\mathbf{r}_i$. This is accomplished by expanding

$$\frac{\psi_T(R')}{\psi_T(R)} \approx \prod_i (1+|\nabla_i \ln\psi(R)||\mathbf{r}'_i-\mathbf{r}_i|\cos\theta_i)$$

$$\approx \prod_i (1+\min(|\nabla_i \ln\psi(R)||\mathbf{r}'_i-\mathbf{r}_i|,1)\cos\theta_i), \quad 8.18$$

where θ_i is the angle between the gradient and $\mathbf{r}'_i-\mathbf{r}_i$. The min function constrains the expression to be positive definite so that is can be used as a probability density. The right hand side of Eq 8.18 can be recognized as a sum of a uniform and linear distribution in $\cos(\theta)$ since the solid angle element is $d\cos\theta d\phi$. The ϕ is sampled uniformly. After sampling from Eq. 8.18, as usual I must include an additional correction factor in the weight for the walker which is simply

$$W_\psi = \frac{\psi_T(R')}{\psi_T(R)} \frac{1}{\prod_i (1+\min(|\nabla_i \ln\psi(R)||\mathbf{r}'_i-\mathbf{r}_i|,1)\cos\theta_i}. \quad 8.19$$

The choice of $\tilde{U}$, E_C, and the a_i's is done to maximize approximately the average time τ per step[20,21]. A similar method to that given for the harmonic oscillator below will work adequately for few-body systems. Typically, there is a large range of values that give similar results. The complete algorithm for a 3N dimensional code is:

1. Sample a set of walkers from $\psi_T^2(R) \rightarrow \sum_i \delta(R-R_i)$ and push onto the old stack.
2. Pop a walker off the old stack. If it is empty all walkers have been propagated and the iteration of the equation is complete. Calculate its U , $V(R_i)$, a_i's and the trial function value and gradient. Usually these have already been calculated and have been stored with the walker position.
3. Sample the smallest τ from the N terms $-\frac{d}{d\tau}h(\frac{\tau}{a_i^2})$ and from $\tilde{U}e^{-\tilde{U}\tau}$ with $\tilde{U}=[E_T+E_C+U-V(R_i)]$
4. Sample the distances $|\mathbf{r}_m-\mathbf{r}'_m|=\rho a_m$ from $g(\rho,\frac{\tau}{a_m^2})d^3\rho$ with τ the smallest from step 3.
5. If surface time of particle k was smallest make its $|\mathbf{r}'_k-\mathbf{r}_k|=a_k$
6. Sample a set of angles θ_i from right hand side of Eq. 8.18, and calculate W_ψ. This now defines the new position R'_i of the walker.
7. If a surface time was smallest, push $n=\mathrm{int}(w+\xi_1)$, $w=W_\psi\exp((E_T-V(R_i))\tau)$, copies of R'_i onto the old stack and go to step 2. If no surface time was smallest go to step 8.
8. The volume time sampled from $\tilde{U}e^{-\tilde{U}\tau}$ is smallest. With probability $\frac{(E_T+E_C)}{\tilde{U}}$ push $\mathrm{int}(w+\xi_2)$, $w=W_\psi\exp((E_T-V(R_i))\tau)$, copies of R'_i onto the new stack, else push $\mathrm{int}(w+\xi_3)$, $w=\frac{U-V(R'_i)}{U-V(R_i)}W_\psi\exp((E_T-V(R_i))\tau)$, copies of R'_i onto the old stack and go to step 2.

Routines to sample ρ from $g(\rho,\tau)d^3\rho$ and τ from $-\frac{dh(\tau)}{d\tau}$ are given in appendix C.

9. RESULTS FOR CENTRAL POTENTIALS

Results for the binding energy using the GFMC method on the triton and alpha particle for the Mafflíet-Tjon V potential are given in Table II below. All energies are in MeV. Readers interested in other quantities such as the point proton densities and rms radii should refer to references 8 and 9.

Table II. The Binding energy in MeV of the triton and alpha particle calculated with Green's function Monte Carlo for the Mafflíet-Tjon V (MTV) potential ^{3}He result from reference 9, and ^{4}He result from reference 8.

System	Binding Energy
^{3}He	8.26±0.01
^{4}He	31.3±0.2

10. FERMION PROBLEMS

I will try to briefly describe the fermion problem. Interested readers should read reference 5 which the reviewed fermion problem and some solutions. I have restricted the GFMC calculations described to those with a spatially symmetric ground-state. Very few interesting problems have this property, and no nucleus is in this class if realistic interactions are included. To calculate the ground-state with a realistic interaction, the various parts of the wave function corresponding to different spin-isospin states have different orbital angular momentum and therefore must be complex. The problem of applying the Monte Carlo method to these complex wave functions in an efficient way is unsolved.

For spin-isospin independent forces, the nuclei up to the alpha particle have spatially symmetric ground-states. For five or more nucleons, the wave function must have some spatial antisymmetry. The spatial wave function must change sign but can be taken to be real. The sign changes make the Monte Carlo calculation difficult. The basic problem is that since the wave function can be negative, it cannot be viewed as a probability density. If the negative sign is kept as a weight for those walkers corresponding to a negative value of the wave function, the negative and positive populations independently converge exponentially to the symmetric state of the Hamiltonian with the lowest energy. This occurs even those this state is not allowed physically. The difference between the positive and negative populations is still the correct antisymmetric state, but it is soon buried in the fluctuations of the unwanted symmetric state. Eq. 4.6 shows that at every step of the calculation, the calculated energy is an upper bound, so some information can some times be extracted before the variance is overwhelming. A calculation of this type is known as a transient estimation method.

Another method is to keep the walkers from crossing the nodes of the trial wave function. This leads to an upper bound to the ground-state energy, and is often a quite good approximation. Since all quantities can be made positive definite, no variance problems arise. Excellent results in liquid ^{3}He and atomic structure calculations are possible within the fixed node approximation.

The transient estimation method has been tried using direct spin summations and the GFMC as described here for the triton interacting via a Reid V_6 potential. The result was a binding energy of 8 ± 1 MeV using about one half hour of Cray 1 time. To lower the variance to a useful level by using more computer time would be too expensive. An improved G_U Green's function which includes a U that is spin isospin dependent may allow a transient estimation calculation to be done on the alpha particle and triton. Another possible method to tackle the realistic potential problem is to develop a fixed phase approximation analogous to the fixed node method. Work is in progress in these directions.

11. CONCLUSION

I have tried to give a fairly complete description of the details of variational and GFMC methods. I hope that this article will allow readers to use these methods when the need arises. I wish to thank Profs. P.A. Whitlock, J.W. Moskowitz, and Dr. J. Carlson for helpful conversations on this material. I wish especially to thank Prof. M.H. Kalos for his help and guidance which has lead to some understanding on my part of the Monte Carlo method.

This work was supported by the Division of Nuclear Physics of the Office of High Energy and Nuclear Physics, United States Department of Energy under contract DE-AC02-79ER10353.

APPENDIX A

In this appendix, I would like to give a concise review of the basic principles used in the Monte Carlo method. The interested reader who wishes a more complete exposition should refer to references 22 and 23.

A probability density is defined to be a function $p(x)\geq0$ on an interval $a\leq x\leq b$, normalized to one,

$$\int_a^b p(x)dx=1. \tag{A.1}$$

A value x_i is said to be sampled from the probability density $p(x)$ if the probability of selecting a value x_i between x and $x+dx$ is $p(x)dx$ in the limit $dx\to0$. Two related functions are the cumulative probabilty distribution function,

$$p_{cdf}(x)=\int_a^x p(x)dx, \tag{A.2}$$

and the characteristic function which is simply the fourier transform of p(x). The probability of sampling a value x_i less than x is just $p_{cdf}(x)$. The normalization condition Eq. A.1 is equivalent to requiring that the probability of sampling an x somewhere in the interval [a,b] is one.

The central limit theorem is the corner stone of Monte Carlo calculations. In this section I take the interval for p(x) to be from $-\infty$ to ∞ without loss of generalization since p(x) can be taken to be zero outside the interval [a,b]. The main result that we need can be stated simply. Take an average of a large number N of values x_i individually sampled from a probability density p(x). Repeat this step many times to get an ensemble of these average values. The values X_j obtained will have a gaussian probability density,

$$P(X)=(2\pi\sigma_{av}^2)^{-\frac{1}{2}}\exp(-\frac{(X-\overline{X})^2}{2\sigma_{av}^2}) \qquad \text{A.3}$$

with average value $\overline{X}$ equal to $\overline{x}$ the average of x,

$$\overline{x}=\int_{-\infty}^{\infty} x\,p(x)dx, \qquad \text{A.4}$$

and standard deviation σ_{av} equal to $\frac{\sigma}{\sqrt{N}}$ where σ is the standard deviation of x,

$$\sigma^2=\int_{-\infty}^{\infty}(x-\overline{x})^2p(x)dx. \qquad \text{A.5}$$

A nonrigorous demonstration of this result can be given as follows. The probability density for the average of N x_i values to be X is given by

$$P(X)=\int_{-\infty}^{\infty}\prod_{i=1}^{N}dx_ip(x_i)\delta(X-\frac{1}{N}\sum_{m=1}^{N}x_m). \qquad \text{A.6}$$

Fourier transforming Eq. A.6 gives

$$\tilde{P}(K)=[\tilde{p}(\frac{K}{N})]^N, \qquad \text{A.7}$$

where $\tilde{p}(k)$ and $\tilde{P}(k)$ are the fourier transforms of p(x) and P(X) respectively. In the limit of large N, the argument of $\tilde{p}(\frac{K}{N})$ becomes small for fixed K. We examine the power series expansion of $\tilde{p}$

$$\tilde{p}(k)=1+ik\overline{x}-\frac{1}{2}k^2(\sigma^2+\overline{x}^2)+O(k^3) \qquad \text{A.8}$$

Where $\overline{x}$ and σ are defined by Eq. A.4 and A.5. Without loss of generality, I am free to choose the origin of for x such that $\overline{x}=0$ i.e. make the transformation to a new coordinate $x'=x-\overline{x}$ and then drop the primes. This simplifies the algebra somewhat. With this change substituting Eq. A.8 into Eq. A.7 gives

$$\tilde{P}(K)=[1-\frac{K^2}{2N^2}\sigma^2+O(\frac{K^3}{N^3})]^N. \qquad \text{A.9}$$

Using the definition,

$$\exp(x)=\lim_{N\to\infty}[1+\frac{x}{N}]^N, \qquad \text{A.10}$$

gives for large N,

$$\tilde{P}(K)=\exp(-\frac{(K\sigma)^2}{2N}). \qquad \textbf{A.11}$$

Fourier transforming this expression back to X space gives Eq A.3.

The importance of the central limit theorem is that it tells us that if we have a probability density that we can sample, the average of this density can be estimated by averaging many samples from the distribution. As long as σ is finite, the error in this procedure is gaussian distributed with a width that shrinks like $\frac{1}{\sqrt{N}}$ and can be estimated in the same way.

Numerical integrations can be written in the form,

$$I=\int_a^b f(x)p(x)dx \qquad \text{A.12}$$

where p(x) has the properties of a probability density. The integral may be broken up into p(x) and f(x) in many ways. Accurate numerical work using either standard numerical methods or the Monte Carlo method requires that the f(x) term be as smooth as possible with most of the fluctuations placed in p(x). A standard numerical integration formula can be written as

$$I\approx\sum_{i=1}^{N} f(x_i)w(x_i), \qquad \text{A.13}$$

where x_i and $w(x_i)$ are the points and corresponding weights derived for example from a gauss point formula for the weighting function p(x). It is useful to regard Eq. A.13 as a consequence of making the replacement

$$p(x)\approx\sum_{i=1}^{N} w(x_i)\delta(x-x_i). \qquad \text{A.14}$$

Substituting Eq. A.14 into A.12 reproduces Eq. A.13. In this way standard numerical methods can be viewed as replacing the function p(x) with a weigthed sum of delta functions. Similarly, a Monte Carlo calculation of the integral can be taken as the replacement

$$p(x)\approx\frac{1}{N}\sum_{i=1}^{N} \delta(x-x_i) \qquad \text{A.15}$$

where the x_i have been sampled from the probability density p(x). This replacement then leads to the approximation of the integral

$$I\approx\frac{1}{N}\sum_{i=1}^{N} f(x_i) \qquad \text{A.16}$$

The view that sampling a continuous distribution is equivalent to replacing it with a sum of delta functions is particularly useful in the solution of the Schroedinger equation via the diffusion analogy.

The central limit theorem can now be used to define the error of replacing the integration by the sum in Eq. A.16. Take a fixed value for the number of sampled points N and calculate the approximate value of the integral from Eq. A.15. If N is large, the central limit theorem applies. Each sample from p(x) and evaluation of f(x) can be viewed as implicitly sampling possible values of the integral I. The average of this distribution is just the true value of the integral Eq. A.12 and the standard deviation is given by

$$\sigma^2 = \int_a^b (f(x) - I)^2 p(x) dx. \qquad \text{A.17}$$

Typically, both I and σ are calculated using the Monte Carlo method. Since an ensemble of calculated I values would be gaussian distributed, and a normalized gaussian is completely determined by its average and standard deviation only these quantities are reported, usually in the form $I \pm \frac{\sigma}{\sqrt{N}}$. The $\sqrt{N}$ factor coming from the central limit theorem.

The result that the error in a Monte Carlo calculation is $\frac{\sigma}{\sqrt{N}}$ contains both the great virtue as well as the main disadvantage of the Monte Carlo method. The great virtue is that the result is independent of the dimensionality of the integration, as can be easily verified by repeating the above steps in many dimensions. This is in contrast to standard numerical methods where the effort required to maintain a constant error grows exponentially with the dimensionality of the integral. The problem is that the error only falls like $\frac{1}{\sqrt{N}}$ so that it takes a large increase in computational effort to substantially lower the error.

A random number generator is necessary to sample a distribution. Most computer languages contain a built in pseudorandom number generator that samples uniformly on the unit interval, i.e. the probability density $p(x) = 1$ for $0 < x < 1$. Pseudorandom means that a deterministic algorithm is used so that it is possible to predict the next random number given some set of previous random numbers, but produce the same results as true random numbers for most applications. Unfortunately, the quality of these random number generators varies greatly, and there is no standard random number generator.

The Quantum Monte Carlo calculations described here tend to be rather insensitive to the quality of the random number generator. The calculations here have all been performed with the multiplicative congruential random number generator given by

$$i_{n+1} = i_n 11^{13} \bmod 2^{48},$$

$$\xi_n = i_n 2^{-48}, \qquad \text{A.18}$$

where the i_m are 48 bit integers and ξ_n is the n th generated pseudorandom number. Good results are obtained if the initial seed i_0 is taken to be 1 or the last value produced by a previous run. Many other pseudorandom number generators exist.

To carry out Monte Carlo calculations, we must be able to sample from the probability density appropriate to our problem. Only certain classes of distributions can be easily sampled. I will try to indicate the sampling techniques that are needed to sample the distributions that arise in Quantum Monte Carlo calculations. A useful compilation of sampling methods can be found in reference 24.

The easiest distribution to sample is of course the uniform distribution on the unit interval since the output of the random number generator is sampled from this distribution. Througout this article ξ_1, ξ_2, etc. will indicate independent random numbers sampled from this distribution. All the sampling techniques given here will be methods of converting these random numbers into values sampled from other distributions.

One way to sample other distributions is by transformation. Given a monotonic function T(y) define a set of sampled values x_i from the equation

$$x_i = T(\xi_i). \qquad \text{A.19}$$

To see from what probability density the x_i values are sampled, calculate the probability that x_i is between x and x+dx. This is clearly the probability that ξ_i is between $T^{-1}(x)$ and $T^{-1}(x+dx)$. Since the ξ values are by definition uniformly distributed on the unit interval, the probability that one of them is between two values is just the difference in the two values. So the probability density of sampled x values p(x) is

$$p(x)dx=[T^{-1}(x+dx)-T^{-1}(x)]$$

$$=\frac{dT^{-1}(x)}{dx}dx. \qquad A.20$$

This can be considered the defining equation for T(x). This expression is somewhat simpler if we integrate both sides over x. The left hand side gives the cumulative probability distribution function defined earlier and the result is

$$p_{cdf}(x)=T^{-1}(x). \qquad A.21$$

The x values can now be seen to be calculated from the expression

$$p_{cdf}(x)=\xi \qquad A.22$$

The right hand side can be recognized as the cumulative probability distribution function for the uniform distribution. The result can be generalized so that if we can sample y values from a cumulative probability distribution $p_{1\,cdf}(y)$, x values sampled from a cumulative probability distribution $p_{2\,cdf}(x)$ can be generated from

$$p_{2\,cdf}(x)=p_{1\,cdf}(y). \qquad A.23$$

A useful probability density to sample is the exponential density on the interval $0\leq x\leq\infty$

$$p(x)=\alpha e^{-\alpha x}. \qquad A.24$$

The cumulative probability distribution function for this density is

$$p_{cdf}(x)=1-e^{-\alpha x}. \qquad A.25$$

Equating this with the probability density for the random numbers on the unit interval and solving for x yields

$$x=-\ln(1-\xi)/\alpha, \qquad A.26$$

or noting that $1-\xi$ is distributed identically with ξ,

$$x=-\ln(\xi)/\alpha. \qquad A.27$$

Similarly, a gaussian can be sampled by using the inverse of the error function. There are, in fact, many ways of sampling a gaussian distribution, not the least of which is to sum many ξ values and use the central limit theorem to see that this gives an approximately gaussian distribution. Probably the most popular method is that of Box and Muller. This method is a transformation method based on the standard method of evaluating a gaussian integral by calculating in two dimensions. To sample from the two-dimensional gaussian

$$\sqrt{2\pi\sigma}\exp(-\frac{x^2+y^2}{2\sigma^2})=\sqrt{2\pi\sigma}\exp(-\frac{\rho^2}{2\sigma^2}), \qquad A.28$$

where as usual $x=\rho\cos(\phi)$ and $y=\rho\sin(\phi)$, first sample ϕ uniformly on the interval from 0 to 2π, and then sample ρ using the cumulative probability distribution

$$p_{cdf}(\rho)=(2\sigma)^{-2}\int_{-\infty}^{\rho} e^{-\frac{\rho'^2}{2\sigma^2}}\rho' d\rho'. \qquad A.29$$

The integral in Eq. A.29 is easily performed and by following the same technique used above for sampling from an exponential ρ is easily sampled. The result for sampling x and y is

$$x=-\sqrt{2}\sigma\ln(\xi_2)\cos(2\pi\xi_1),$$

$$y=-\sqrt{2}\sigma\ln(\xi_2)\sin(2\pi\xi_1), \qquad \text{A.30}$$

which gives two independent samples x and y from a gaussian of width σ.

The transformation method can be applied to many distributions. It can also be used in conjunction with the rejection method.

In the rejection method, a random variate is sampled from the probability density $p_1(x)$ and the resulting value is accepted with probability f(x). If the value is rejected, the process is repeated until a value is accepted. The resulting probability density $p_2(x)$ is normalized (since in the end a value is always accepted) and proportional to the product of f(x) and $p_1(x)$. Since f(x) is a probability, it must be between zero and one. Therefore it can be chosen to be

$$f(x)=\frac{p_2(x)}{Bp_1(x)} \qquad \text{A.31}$$

where B is an upper bound on $p_2(x)/p_1(x)$. The acceptance probability will be maximized if B is the least upper bound.

A simple example of the rejection method is the sampling of the probability density 2x on the interval $0<x<1$. Take $p_1(x)=1$ the uniform distribition. With this choice, f(x) can be taken to be x. The algorithm then becomes generate two random numbers ξ_1 and ξ_2. Take ξ_1 as the sampled value if $\xi_2<\xi_1$ else generate two more random numbers and repeat the whole process.

For another example of the rejection method, suppose we wish to sample x and y values uniformly on a circle i.e. with the constraint $x^2+y^2=1$. We could transform to cylindrical coordinates and sample ϕ uniformly on the interval $0<\phi<2\pi$, and let $x=\cos(\phi)$, $y=\sin(\phi)$. On the other hand, a rejection technique can be used without calculating trigonometric functions. First calculate trial x and y values from

$$x_0=2\xi_1-1,$$

$$y_0=2\xi_2-1. \qquad \text{A.32}$$

Set $a^2=x_0^2+y_0^2$, and if $a^2<1$ accept $x=x_0/a$ and $y=y_0/a$ otherwise repeat the process.

Random variates can be combined in various ways to produce new distriubtions. Consider taking the maximum of the values sampled from two cumulative probability distributions. The probability that the resulting value is less than x, i.e. the cumulative probability distribution of the resulting x values, is just the product of the probability that the first value is less than x times the probability that the second is less than x. The resulting cumulative probability distribution is just the product of the two sampled cumulative probability distributions. The maximum of two values sampled uniformly on the unit interval is therefore sampled from the cumulative probability distribution function x^2. Taking the derivative the probability density is 2x.

The Metropolis, Rosenbluth, Rosenbluth, Teller and Teller Monte Carlo method $M(RT)^2$ is a general method to sample probability densities.[12] Its main disadvantage is that the values sampled using this method are autocorrelated. That is the value of the normalized autocorrelation function ζ_i

$$\zeta_i=\langle x_j x_{i+j}\rangle-\langle x_j\rangle^2, \qquad \text{A.33}$$

is nonzero where the < > indicates an ensemble average. In this case since sampled values are not independent, the central limit theorem does not directly apply. I will have more to say about this later.

The original application for the $M(RT)^2$ method was in statistical mechanics problems, and the motivation for the $M(RT)^2$ method is particularly easy to undestand there. In these systems, an ensemble average of a microscopic quantity is desired. For example, the average of the potential energy is

$$<V>=\frac{\int V(\vec{R})\exp[-\beta H(\vec{P},\vec{R})]d\vec{R}d\vec{P}}{\int \exp[-\beta H(\vec{P},\vec{R})]d\vec{R}d\vec{P}}, \qquad \text{A.34}$$

where the Hamiltonian is

$$H(\vec{P},\vec{R})=\frac{\vec{P}^2}{2M}+V(\vec{R}). \qquad \text{A.35}$$

Here, $\vec{R}$ and $\vec{P}$ are the coordinates and momenta of the N particles, and β is the usual inverse of the boltzman constant times the temperature. Usually, the $\vec{P}$ space integral in Eq. A.34 ia performed analytically leaving only the $\vec{R}$ space integral. If the probability density $P(\vec{R})$ is defined as

$$P(\vec{R})=\frac{\exp[-\beta V(\vec{R})]d\vec{R}}{\int \exp[-\beta V(\vec{R})]d\vec{R}}, \qquad \text{A.36}$$

then the ensemble average of an operator $F(\vec{R})$ is given by

$$<F>=\int F(\vec{R})P(\vec{R})d\vec{R}, \qquad \text{A.37}$$

which is of the form of our general Monte Carlo integral Eq A.12. Initially, it seems that it would be very difficult to come up with a general method to sample $P(\vec{R})$. However, most physical systems with a large number of degrees of freedom are ergodic. That is, the time average of a quantity is the same as its ensemble average. The time average can be easily calculated numerically by integrating Newton's equations for the Hamiltonian of Eq. A.35 using the molecular dynamics method. The result is of course that of the microcanonical ensemble rather than the desired canonical ensemble of Eq. A.36, but the molecular dynamics method can be modified to give results for the canonical ensemble. Clearly, many kinds of dynamics have the same time averages. The $M(RT)^2$ method is designed to give a time average that is the same as the canonical ensemble average.

Given a current position, the $M(RT)^2$ method consists of making a trial transition to a new position with some known probability. This new position is then either accepted and kept as the next postition, or rejected in which case the old position is kept as the next position. The mathematical statement of this dynamics is as follows. Define P_I^n as the probability of being in the state I (i.e. the position R_i) at step n. Define $T_{I\to J}$ as the probability of making a trial move from the state I to state J. Define $A_{I\to J}$ as the probability of accepting a trial move from I to J. The $M(RT)^2$ can then be written as

$$P_I^{n+1}=\sum_J [A_{J\to I}T_{J\to I}P_J^n+(1-A_{I\to J})T_{I\to J}P_I^n]. \qquad \text{A.38}$$

This states that the probability of being in the state I at step n+1 is given by the sum of two terms. The first is the probability of starting in any state J and making a trial move to I and then accepting that move. The other possibility is to start in the state I and make a trial move to any state J and reject this move thereby staying in state I. For convenience I choose to deal with a discrete system in what follows. To convert the result to a continuous system simply replace the P and T probabilities by probability densities and change the sums to integrals in the usual way. The probability for a trial transition is chosen at the outset. A typical choice is to move a particle uniformly in a small box around its current position. After many moves, i.e. for large n, P_I^n must be equal to the desired probability. The $M(RT)^2$ method is a way of choosing the $A_{I\to J}$ to accomplish this. Take the limit $n\to\infty$ in Eq. A.38, and use the fact that the probability of making a trial transition somewhere is one. Eq. A.38 becomes

$$0=\sum_J [T_{J\to I}A_{J\to I}P_J^\infty - T_{I\to J}A_{I\to J}P_I^\infty], \qquad \text{A.39}$$

which is known as the balance equation. Clearly, we have many $A_{I\to J}$ values to choose, but only this one equation to satisfy. A convenient, but in principle more restrictive than necessary, way of satifying Eq. A.39 is to satisfy it term by term. This is called the detailed balance condition and gives

$$T_{J\to I}A_{J\to I}P_J^\infty = T_{I\to J}A_{I\to J}P_I^\infty, \qquad \text{A.40}$$

or rearranging

$$\frac{A_{I\to J}}{A_{J\to I}} = \frac{T_{J\to I}P_J^\infty}{T_{I\to J}P_I^\infty}. \qquad \text{A.41}$$

The acceptance probabilities satisfy Eq. A.41, and are maximized by the $M(RT)^2$ choice of

$$A_{I\to J} = \min[1, \frac{T_{J\to I}P_J^\infty}{T_{I\to J}P_I^\infty}]. \qquad \text{A.42}$$

It can be shown that under very general conditions, the $M(RT)^2$ dynamics will converge to the desired probability distribution P_I^∞. The requirements on the transition probability $T_{I\to J}$ are small. They are that every possible state of the system must be reachable by some combination of moves from every other state, and if $T_{I\to J}$ is nonzero then $T_{J\to I}$ must be nonzero. For many applications, the $T_{I\to J}$ is chosen to be equal to $T_{J\to I}$ and may then be cancelled in Eq. A.42. Examples when this was not true can be found in many references for example references 25 and 26.

In the Green's function Monte Carlo method, a series expansion of the Green's function is used. This series is summed exactly using the Monte Carlo method. In this section, I will attempt to show how exact series summation can be done using the Monte Carlo method.

Consider sampling the following finite series,

$$S(x)=a_1p_1(x)+a_2p_2(x)+a_3p_3(x), \qquad \text{A.43}$$

where each of the $p_i(x)$ is normalized. For $S(x)$ to be normalized, $a_1+a_2+a_3=1$. Further assume that all the a_i are positive. The sampling of this series is straightforward. With probability a_1 sample from $p_1(x)$, etc. A possible algorithm is as follows. If $\xi_1<a_1$ sample from $p_1(x)$, else if $a_1<\xi_1<a_1+a_2$, sample from $p_2(x)$, else sample from $p_3(x)$.

Sampling from an infinite series can be done in a similar way. Rewrite the infinite series

$$S(x)=\sum_{i=1}^{\infty} a_ip_i(x) \qquad \text{A.44}$$

as

$$S(x)=a_1p_1(x)+(1-a_1)\sum_{i=2}^{\infty}\frac{a_i}{1-a_1}p_i(x). \qquad \mathbf{A.45}$$

This can now be interpreted as follows. With probability a_1 (if $\xi_1<a_1$) sample from $p_1(x)$ else (if $\xi_1>a_1$) sample from the infinite series

$$\sum_{i=2}^{\infty}\frac{a_i}{1-a_1}p_i(x). \qquad \text{A.46}$$

This defines a recursive method of sampling from the first infinite series. Let's look at how to sample a series where the coefficients a_i are those of the geometric series $a_i=2(\frac{1}{3})^i$. The recursive algorithm can be implemented as follows.

1. Examine the first term of the series. The coefficients of the series are 2/3, 2/9, 2/27 etc. which sum to 1 as required. Generate a new random number on the unit interval, ξ, and if $\xi<2/3$ sample the first term of the series. and stop. This samples the first term with probability 2/3.
2. If $\xi>2/3$ sample one of the next terms in the series. However, the probability of $\xi>2/3$ is 1/3 so the following coefficients must be divided by this probability. The new coefficients of $p_2(x)$, $p_3(x)$, $p_4(x)$ etc. are respectively 2/3, 2/9, 2/27 etc. Repeat step 1 with this series and a new random number.

Eventually, the random number generated in step 1 will satisfy the inequality in step 1 and one of the $p_i(x)$ will be sampled.

In the Green's function Monte Carlo method, the series that must be sampled is the expansion of the many body Green's function. This series will not normally be in the form of Eq. A.44 since the sum of the coefficients of the terms will not be known and will not be equal to one. This can be taken care of by using what are known as weights or branching. A typical kind of integral will be of the form

$$I=\frac{\int dxf(x)g(x)}{\int dxg(x)}, \qquad \text{A.47}$$

where g(x) is an infinite sum, but is not normalized,

$$g(x)=\sum_{i=1} g_i(x). \qquad \text{A.48}$$

In general, the integrals of the $g_i(x)$ may be very difficult to calculate, so that normalizing g(x) is impractical. The integral of Eq. A.47 can still be done as follows. Write g(x) as

$$g(x)=a_1p_1(x)\frac{g_1(x)}{a_1p_1(x)}+(1-a_1)\sum_{i=2}\frac{g_i(x)}{1-a_1}, \qquad \text{A.49}$$

where $p_1(x)$ is an approximation to $\frac{g_1(x)}{\int dxg_1(x)}$, and a_1 is an approximation to the relative norm of $g_1(x)$. As above, either the first term is sampled with probability a_1 or the remaining terms are sampled. The remaining terms are sampled recursively as above if needed. If the first term is selected, x is sampled from $p_1(x)$. If g(x) from Eq. A.49 is substituted into Eq. A.47, it can be easily seen that we need to include the weight term $\frac{g_1(x)}{a_1p_1(x)}$ with f(x) in the numerator, and also with 1 in the denominator.

The disadvantage to this technique is that the normalization integral in Eq. A.47 is done with Monte Carlo, and the answer is a ratio of two Monte Carlo calculated integrals.

This is known as a biased estimator of the integral, which simply means that the expected value of the ratio of the integrals depends on the number of Monte Carlo samples taken, and will be correct only in the limit of a large number of samples. This is a consequence of the average of a quotient not being equal to the quotient of the averages. That is, the central limit theorem only applies to each of the integrals in the numerator and denominator alone, and not their quotient. In practice, this means that a large number of samples must be taken, and the results tested to make sure that varying the number of samples does not effect the result. The use of importance sampling so that a_i and $p_i(x)$ are chosen to be good approximations to the relative norm and the normalized $g_i(x)$ respectively, can lower the variance of the denominator integral and therefore reduce the bias.

APPENDIX B

A simple code to calculate the ground state of the one-dimensional harmonic oscillator is given below.

```
      program gfhold
      parameter (zero=0.e0,one=1.e0,two=2.e0,three=3.e0,four=4.e0)
      parameter (five=5.e0,six=6.e0,seven=7.e0,eight=8.e0,anine=9.e0)
      parameter (ten=10.e0,tenth=.1e0,half=.5e0,third=1.e0/3.e0)
c
c program to compute the ground-state energy of a one dimensional
c hamiltonian. A harmonic oscillator potential is used in this
c version
c
      parameter (nconmx=8000)
      common /param/ etrial,e0,eps,alpha
      common /config/ x1(nconmx),x2(nconmx),n1,n2
      dimension iseed(4)
c
c common block variables
c  /param/ parameters to be read in
c     etrial = trial energy for harmonic oscillator, the
c              exact enrgy is 1
c     e0     = a constant to be added to the hamiltonian
c     eps    = a constant to determine the domain size by
c              making U-V = eps at the initial walker position
c              the domain size is chosen to make U-V > 0
c     alpha  = trial function = exp(-.5*(alpha*x)**2) alpha = 1
c              gives the exact answer. alpha must be chosen
c              less than 1 in this code.
c /config/ walker stacks
c     x1     = array to hold walker postions
c     n1     = number of entries on x1
c     x2,n2    same for second stack
```

```
c
c read in data
c    nconf = number of walkers to startc
c    nit   = number of iterations to take
c    neq   = number of equilibrating iterations
c   iseed  = random number seed
c
      write (6,'(1x,"number of configurations ?")')
      read (5,*) nconf
      write (6,'(1x,"number of iterations ?")')
      read (5,*) nit
      write (6,'(1x,"number of equilibration its ?")')
      read (5,*) neq
      write (6,'(1x,"trial energy ?")')
      read (5,*) etrial
      write (6,'(1x,"constant added to hamiltonian ?")')
      read (5,*) e0
      write (6,'(1x,"epsilon to determine domain ?")')
      read (5,*) eps
      write (6,'(1x,"trial function parameter ?")')
      read (5,*) alpha
      write (6,'(1x,"random number seed - input 4 numbers n"
     +," 0 <= n <= 4095 ")')
      read (5,*) iseed
      call setrn(iseed)
c
c get initial configurations from psi-trial**2
c
      call init(nconf)
c
c calculate initial energy
c
      call energy(x1,n1,e)
      nit2=nit/2
c
c zero out estimators
c en,eng are the mixed and growth estimators
c en2,eng2 will accumulate squared averages for variance
c
      en=zero
      en2=zero
      den=zero
      eng=zero
      eng2=zero
c
c loop to do gfmc iterations
c
      do 10 i=1,nit2
c
c walk configurations on list 1 and put them on list 2
```

```
c
      call iter(x1,x2,n1,n2)
c
c calculate growth and mixed energies
c
      eg1=float(n1)*(etrial+e0)/float(n2)-e0
      call energy(x2,n2,e1)
c
c walk from 2 back to 1
c
      call iter(x2,x1,n2,n1)
c
c calculate energies
c
      eg2=float(n2)*(etrial+e0)/float(n1)-e0
      call energy(x1,n1,e2)
c
c if equilibration steps are done then collect averages
c
      if (2*i.gt.neq) then
        en=en+e1+e2
        en2=en2+e1**2+e2**2
        eng=eng+eg1+eg2
        eng2=eng2+eg1**2+eg2**2
        den=den+two
        endif
   10 continue
c
c write out final energies
c
      write (6,'(1x,"mixed estimate = ",f10.5," +- ",f10.5)')
     +   en/den,sqrt(abs(en2/den-(en/den)**2)/den)
      write (6,'(1x,"growth estimate =",f10.5," +- ",f10.5)')
     +   eng/den,sqrt(abs(eng2/den-(eng/den)**2)/den)
      end
      subroutine init(nconf)
      parameter (zero=0.e0,one=1.e0,two=2.e0,three=3.e0,four=4.e0)
      parameter (five=5.e0,six=6.e0,seven=7.e0,eight=8.e0,nine=9.e0)
      parameter (ten=10.e0,tenth=.1e0,half=.5e0,third=1.e0/3.e0)
c
c routine to sample x1 values from gaussian trial function
c using Box Muller method
c
      parameter (nconmx=8000)
      common /param/ etrial,e0,eps,alpha
      common /config/ x1(nconmx),x2(nconmx),n1,n2
      pi=four*atan(one)
      n1=nconf
      do 10 i=1,nconf
      temp1=cos(two*pi*rannyu(0))
```

```
      temp2=sqrt(-log(rannyu(0)))
   10 x1(i)=temp1*temp2/alpha
      return
      end
      subroutine energy(x,n,e)
      parameter (zero=0.e0,one=1.e0,two=2.e0,three=3.e0,four=4.e0)
      parameter (five=5.e0,six=6.e0,seven=7.e0,eight=8.e0,nine=9.e0)
      parameter (ten=10.e0,tenth=.1e0,half=.5e0,third=1.e0/3.e0)
c
c routine to calculate the value of the mixed energy for n walkers
c in array x
c
      common /param/ etrial,e0,eps,alpha
      dimension x(1)
      xn=zero
      xd=zero
      do 10 i=1,n
      xn=xn+x(i)**2
   10 xd=xd+one
      xn=xn*(one-alpha**4)+xd*alpha**2
      e=xn/xd
      write (*,'(1x,"energy now = ",f10.5," n = ",i10)') e,n
      return
      end
      subroutine iter(xold,xnew,nold,nnew)
      parameter (zero=0.e0,one=1.e0,two=2.e0,three=3.e0,four=4.e0)
      parameter (five=5.e0,six=6.e0,seven=7.e0,eight=8.e0,nine=9.e0)
      parameter (ten=10.e0,tenth=.1e0,half=.5e0,third=1.e0/3.e0)
      parameter (nconmx=8000)
c
c routine to walk walkers from xold to xnew
c
      common /param/ etrial,e0,eps,alpha
      dimension xold(nconmx),xnew(nconmx)
c
c statement function for value of trial wave function
c
      psit(x)=exp(-half*(alpha*x)**2)
c
c initialize stack pointers
c
      nnew=0
      io=nold
c
c main loop which is broken out of when old stack is empty
c
      do 10 i=1,99999
c
c calculate u
c
```

```
      u=xold(io)**2+eps
c
c calculate el the distance from the current position to the
c edge of the domain such that u-v > 0 in the domain
c
      el=sqrt(u)-abs(xold(io))
c
c add e0 to u since it is added to all energies
c
      u=u+e0
c
c calcuate the inverse of the trial wave function at the
c old position for importance sampling
c
      psioi=one/psit(xold(io))
      rtu=sqrt(u)
c
c calculate 1/cosh(sqrt(u)*el) for the derivative of the green's
c function at surface
c
      expul=exp(rtu*el)
      expuli=one/expul
      cshi=two/(expul+expuli)
c
c calculate probability of moving to either of the surface points
c
      probs=cshi
c
c take surface step with probability probs
c
      if (rannyu(0).lt.probs) then
c
c calculate the value of the wave function at the two surface points
c
        xs1=xold(io)-el
        xs2=xold(io)+el
        psis1=psit(xs1)
        psis2=psit(xs2)
c
c now taking surface step - step left or right according to
c the relative value of the trial function
c
        if ((psis1+psis2)*rannyu(0).lt.psis2) then
          xs=xs2
        else
          xs=xs1
          endif
        wt=half*(psis1+psis2)*psioi
        npart=wt+rannyu(0)
        io=io-1
```

```
        if (npart.gt.0) then
           do 15 j=1,npart
           io=io+1
   15      xold(io)=xs
           endif
      else
c
c not taking surface step so must be volume step
c calculate stuff to sample green's function
c dx is the distance to move
c
         y=cshi+rannyu(0)*(one-cshi)
         dx=el+(log(cshi)-log(y+sqrt(y**2-cshi**2)))/rtu
c
c distance now chosen so calculate value of the trial wave
c function to see which direction to move
c
         xp=xold(io)+dx
         xm=xold(io)-dx
         psip=psit(xp)
         psim=psit(xm)
         prob=psip/(psip+psim)
c
c decide using relative value of wave function which direction
c
         if (rannyu(0).lt.prob) then
            xn=xp
         else
            xn=xm
            endif
c
c new position is now decided - evaluate potential
c
         vnew=xn**2
c
c check to make sure u was a bound for this v
c
         if (u.lt.vnew+e0) write (*,'(1x," u wrong ")')
c
c caluclate factors that are not sampled to include in the weight
c
         etot=etrial+u-vnew
         gnorm=(one-cshi)/u
         wt=half*etot*gnorm*(psip+psim)*psioi/(one-probs)
c
c convert weight into an integer
c
         npart=wt+rannyu(0)
c
c decrement old pointer since configuration has now been moved
```

```
c
      io=io-1
c
c if npart is less than zero then get next configuration
c otherwise decide whether xn goes on new stack or old
c
      if (npart.gt.0) then
        if (etot*rannyu(0).lt.etrial+e0) then
c
c walk has ended put xn on new stack npart times
c
          do 20 j=1,npart
          nnew=nnew+1
 20       xnew(nnew)=xn
        else
c
c walk must continue put xn on old stack npart times
c
          do 30 j=1,npart
          io=io+1
 30       xold(io)=xn
          endif
        endif
      endif
c
c if old stack is empty return
c
      if (io.le.0) go to 40
 10   continue
 40   continue
      return
      end
      function rannyu(x)
c
c function to get random numbers sampled uniformly on 0 to 1
c using multiplicative congruential method.
c seed and multiplier are 48 bit integers stored as 4 twelve
c bit integers. multiplier is 11**13. seed is set in setrn
c
      common /rnyucm/ m1,m2,m3,m4,l1,l2,l3,l4
      data itwo12 /4096/
      i1=l1*m4+l2*m3+l3*m2+l4*m1
      i2=l2*m4+l3*m3+l4*m2
      i3=l3*m4+l4*m3
      i4=l4*m4
      l4=mod(i4,itwo12)
      i3=i3+i4/itwo12
      l3=mod(i3,itwo12)
      i2=i2+i3/itwo12
      l2=mod(i2,itwo12)
```

```
      l1=mod(i1+i2/itwo12,itwo12)
      rannyu=2.**(-12)*(float(l1)+
     +       2.**(-12)*(float(l2)+
     +       2.**(-12)*(float(l3)+
     +       2.**(-12)*(float(l4)))))
      return
      end
      subroutine setrn(iseed)
c
c routine to set the random number seed and make sure it is an
c odd integer
c
      common /rnyucm/ m(4),l(4)
      dimension iseed(4)
      do 10 i=1,4
   10 l(i)=iseed(i)
      l(4)=2*(l(4)/2)+1
      return
      end
      subroutine savern(iseed)
c
c routine to return the current seed
c
      common /rnyucm/ m(4),l(4)
      dimension iseed(4)
      do 10 i=1,4
   10 iseed(i)=l(i)
      return
      end
      block data random
      common /rnyucm/ m(4),l(4)
      data m / 502, 1521, 4071, 2107/
      data l /   0,    0,    0,    1/
      end
```

APPENDIX C

Routines to sample the time variable tau and the Green's function of a particle propagating in a spherical domain. Rannyu is the random number generator as in appendix B, and gauray(gausig,x,n) returns n gaussian distributed random variates with $\sigma=$ gausig in x(1) through x(n). Tau is the time sampled from tscig a2 is a set of sphere radii squared and s is the gaussian width. Sampled coordinate changes are returned in tar.

```
      subroutine samg(s,tau,a2,tar)
c
```

```
c Routine to sample from the green's function in a
c domain given by the outer product of non overlapping
c spheres. The time tau has been sampled already. This
c routine samples from r**2*g(r,t)dr. Two series are used
c one for short times and one for long.
c    t < .11
c  g(r,t) = (1/(4*pi*t)**1.5)*exp(-r**2/(4t))
c        + (1/(4*pi*t)**1.5*r) sum (on n)( (r+2n)exp(-(r+2n)**2/4t)
c                                  +(r-2n)exp(-(r-2n)**2/4t) )
c    for n up to 2
c
c    t > .11
c  g(r,t) = (1/2r) sum (on n) (n*sin(n*pi*r)*exp(-pi**2*n**2*t)
c    for n up to 5
c
c   a = 1 for above formulas. r's are scaled by a later.
c
      parameter (zero=0.e0,one=1.e0,two=2.e0,three=3.e0,four=4.e0)
      parameter (five=5.e0,six=6.e0,seven=7.e0,eight=8.e0,nine=9.e0)
      parameter (ten=10.e0,tenth=.1e0,half=.5e0,third=1.e0/3.e0)
      parameter (twelve=12.e0,sixtn=16.e0,twfive=25.e0)
      parameter (tswtch=.11e0)
      common /const/ npart,pi,hb,etrial,e0,hbi,el2
      dimension tau(1),a2(1),tar(3,1)
      gausig=s
      do 10 i=1,npart
      t=tau(i)
      at=a2(i)
      if(t.gt.tswtch) go to 20
      tinv=one/t
   30 call gauray(gausig,tar(1,i),3)
      rsq=zero
      do 40 ic=1,3
   40 rsq=rsq+tar(ic,i)**2
      if(rsq.gt.at) go to 30
      r=sqrt (rsq/at)
      rinv=one/r
c
c expressions here are analytically the same but have better
c roundoff characteristics when used as below
c
      if (r.gt.t) then
         gratio=one+((one+two*rinv)*exp(-(one+r)*tinv)
     +            +((one-two*rinv)*exp(-(one-r)*tinv)
     +            +((one+four*rinv)*exp(-(four+two*r)*tinv)
     +            +(one-four*rinv)*exp(-(four-two*r)*tinv))))
      else
         rtinv=r*tinv
         gratio=one+(exp(-tinv)*(two*cosh(rtinv)
     +           -four*sinh(rtinv)*rinv)
```

```
     +           +exp(-four*tinv)*(two*cosh(two*rtinv)
     +           -eight*sinh(two*rtinv)*rinv))
      endif
      if (rannyu(1).gt.gratio) go to 30
      go to 10
   20 continue
      call gauray(gausig,tar(1,i),3)
c
c generate the third largest of four random numbers.  it has pdf
c 12x**2*(1-x)   this x is then mapped into an r drawn from
c pi*r*sin(pi*r).
c
      r4=rannyu(0)
      r3=zero
      do 50 j=1,3
      rtest=rannyu(0)
      r3=min(r4,max(r3,rtest))
   50 r4=max(r4,rtest)
c
c map onto pi*r*sin(pi*r) by equating integral (12x**2*(1-x) to
c integral (pi*r*sin(pi*r)) and solving for r using Newton
c Raphson iteration.
c
      rand3=r3**3
      rand4=r3*rand3
      fl=four*rand3-three*rand4
      r=r3
      do 60 k=1,4
      s=sin(pi*r)
      c=cos(pi*r)
   60 r=r+(pi*(fl+r*c)-s)/(pi**2*r*s)
      pisq=pi*pi
      ex=exp(-pisq*t)
      ex3=ex**3
      ex8=(ex*ex3)**2
      ex15=ex8*ex*(ex3)**2
      ex24=ex8*ex15*ex
      ex3=four*ex3
c
c find maximum value of ratio of true g to sample
c
      sup=one+ex3+nine*ex8+sixtn*ex15+twfive*ex24
c
c find ratio
c
      gc=one+ex3*c+ex8*(twelve*c**2-three)
     +  +sixtn*ex15*c*(two*c**2-one)
     +  +five*ex24*(sixtn*c**4-twelve*c**2+one)
c
c reject with probabilty proportional to ratio
```

```
c
      if (sup*rannyu(3).gt.gc)   go to 20
c
c use direction defined by old tar 3-vector to get new one length r
c
      rsq=0.
      do 70 l=1,3
   70 rsq = rsq+ tar(l,i)**2
      r = r*sqrt(at/rsq)
      do 80 l=1,3
   80 tar(l,i) = tar(l,i)*r
   10 continue
      return
      end
      function tscig(csi)
c
c Routine to sample the time for a particle to be absorbed by
c the surface of it's spherical domain.
c Sampling is done by solving the equation csi=1-h(tau) for a time
c tau given a random number csi. H(tau) is the probability of
c still being inside the domain at time tau. It is the
c integral of the green's function over the domain. Therefore
c -d(h(tau))/d(tau) is the probability of being absorbed by the
c surface at time tau.
c
c Two series are used for h(tau) they are:
c At t > .16
c    h(t)=2*sum(on n) ( (-1)**(n+1)*exp(-pi**2*n**2*t) )
c    n up to 4
c At t < .16
c    h(t)=1- (2/sqrt(pi*t))*sum(on n) ( exp(-(2n-1)**2/(4t)) )
c    n up to 2
c
      parameter (zero=0.e0,one=1.e0,two=2.e0,three=3.e0,four=4.e0)
      parameter (five=5.e0,six=6.e0,seven=7.e0,eight=8.e0,nine=9.e0)
      parameter (ten=10.e0,tenth=.1e0,half=.5e0,third=1.e0/3.e0)
      parameter (sixtn=16.e0)
      parameter (tswtch=.16d0)
      save pi2,rtpi,csimat,jump
      data jump /0/
c
c following stuff done only first time through
c
      if (jump.eq.0) then
          jump=1
          pi=four*atan(one)
          pi2=pi**2
          rtpi=one/sqrt(pi)
c
c match short and long time solutions at tswtch=.16
```

```
c get the value of the random number that corresponds
c to this time
c
         t=tswtch
         ex=exp(-pi2*t)
         ex4=ex**4
         ex9=ex*ex4**2
         ex16=ex4**4
         csimat=two*(ex-ex4+ex9-ex16)
         endif
c
c decide whether to use long or short time solutions to the
c h(tau) series
c
      if (csi.lt.csimat) then
c
c now using long time series
c
         t=-log(csi*half)/pi2
         do 10 i=1,3
         ex=exp(-pi2*t)
         ex4=ex**4
         ex9=ex*ex4**2
         ex16=ex4**4
         h=two*(ex-ex4+ex9-ex16)
         hp=-two*pi2*(ex-four*ex4+nine*ex9-sixtn*ex16)
         t=t+(csi-h)/hp
   10    continue
      else
c
c now using short time series
c
         t=one/(four*log(four*pi/(one-csi)))
         do 20 i=1,5
         ex=exp(-one/(four*t))
         ex9=ex**9
         thalf=one/sqrt(t)
         h=one-two*(ex+ex9)*thalf*rtpi
         hp=rtpi*thalf**3*((ex+ex9)-half*thalf**2*(ex+nine*ex9))
         t=t+(csi-h)/hp
   20    continue
      endif
c
c return time
c
      tscig=t
      return
      end
```

REFERENCES

1. D.M. Ceperley and M.H. Kalos, In *Monte Carlo Methods in Statistics Physics,* ed. by K. Binder, Topics in Current Physics, 7 (Springer, Berlin, Heidelberg, New York, 1979) Chapter 4.
2. J. Lomnitz-Adler, V.R. Pandharipande, and R.A. Smith, Nucl. Phys. A361 399 (1981).
3. J. Carlson and V.R. Pandharipande, Nucl. Phys. A371 301 (1981)
4. R.B. Wiringa, J.L. Friar, B.F. Gibson, G.L. Payne, and C.R.Chen, Phys. Lett. 143B 273 (1984).
5. K.E. Schmidt and M.H. Kalos, In *Applications of the Monte Carlo Method in Statistical Physics,* ed. by K. Binder, Topics in Current Physics Vol. 36 (Springer, Berlin, Heidelberg, New York, 1984).
6. M.H. Kalos, Phys. Rev. 128 1891 (1962).
7. M.H. Kalos, Nucl. Phys. A126 609 (1969).
8. J.G. Zabolitzky and M.H. Kalos, Nucl. Phys. A356 114 (1981).
9. J.G. Zabolitzky, K.E. Schmidt, M.H. Kalos, Phys. Rev. C25 1111 (1982).
10. J. Carlson, V.R. Pandharipande, and R.B. Wiringa, Nucl. Phys. A424 427 (1984).
11. J. Carlson, K.E. Schmidt, and M.H. Kalos, *Condensed Matter Theories* Vol. 1, Ed. F.B. Malik (Plenum, New York, 1986).
12. N. Metropolis, A.W. Rosenbluth, H. Rosenbluth, A. Teller, E. Teller, J. Chem. Phys. 21 1087 (1953).
13. M.H. Kalos, D. Levesque, L. Verlet, Phys. Rev. A9 2178 (1974).
14. M.A. Lee, K.E. Schmidt, M.H. Kalos, G.V. Chester, Phys. Rev. Lett. 47 807 (1981).
15. J.W. Moskowitz, K.E. Schmidt, M.A. Lee, M.H. Kalos, J. Chem. Phys. 77 349 (1982).
16. D.M. Ceperley and B.J. Alder, Phys. Rev. Lett. 45 566 (1980).
17. P.J. Reynolds, D.M. Ceperley, B.J. Alder, W.A. Lester, J. Chem Phys. 77 5593 (1982).
18. J.W. Negele, J. Stat. Phys. 43 991 (1986).
19. D.W. Skinner, J.W. Moskowitz, M.A. Lee, P.A. Whitlock, and K.E. Schmidt, J. Chem. Phys. 83 4668 (1985).
20. D.M. Ceperley, J. Comp. Phys. 51 404 (1983).
21. J.W. Moskowitz and K.E. Schmidt, J. Chem. Phys. 85 2868 (1986).
22. J.M. Hammersley and D.C. Handscomb, *Monte Carlo Methods* (Methuen, London, 1964).
23. M.H. Kalos and P.A. Whitlock, *Monte Carlo Methods Volume I: Basics,* (John Wiley, New York, 1986).
24. C.J. Everett and E.D. Cashwell, *A Third Monte Carlo Sampler,* Los Alamos National Laboratory Report LA-9721-MS (1983).
25. D.M. Ceperley, M.H. Kalos, G.V. Chester, Phys. Rev. B16 3081 (1977),
26. M. Rao and B.J. Berne, J. Chem. Phys. 77 129 (1979).
27. K.E. Schmidt, Phys. Rev. Lett. 51 2175 (1983).

PART III

RELATIVISTIC FEW-BODY EQUATIONS
AND
BAG MODELS

BETHE-SALPETER EQUATION AND THE NUCLEON-NUCLEON INTERACTION

J.A. Tjon

Institute for Theoretical Physics
Princetonplein 5, P.O. Box 80.006
3508 TA Utrecht, The Netherlands

1. Introduction

Numerous investigations have been made in the past to determine the basic nature of the nuclear force and from this the properties of nuclei. In order to carry out this program dynamical equations have to be solved. For the few nucleon system much progress has been made in recent years. In particular exact and realistic quantum mechanical calculations can now be done in the case of the tri-nucleon system [1]. To describe the dynamical behaviour of such systems it is customary to employ the non-relativistic Schrödinger equation. At intermediate energies and large momentum transfers however the kinetic energies of the moving nucleons become comparable in size to the rest masses. As a consequence a Lorentz covariant description is needed.

In recent years various relativistic approaches have been explored. Hamiltonian extensions in terms of the generators of the Poincaré group have been examined in detail [2,3]. They are based on the pioneering work of Dirac [4], in which it is implicitly assumed that the particles are on the mass shell. A different approach which can be adopted is that of a relativistic quantum field theory based on Feynman diagrams. In these lectures a review is given of work which has been done within such a framework to build an effective relativistic theory of the nucleon-nucleon interaction at intermediate energy.

The quantum field theory can be analyzed in terms of Feynman graphs. Since we will be interested in systems of strongly interacting particles we have to go in some way beyond perturbation theory. In particular, the study of the properties of composite systems makes it necessary. Requirements on the theory like unitarity i.e. conservation of probability plays thereby an important role. The field theoretical Bethe-Salpeter equation for two particles is a natural candidate for discussing the strong interaction problem. It has the nice property that it exhibits the two-particle unitarity aspect in a clear way. In these lectures we will concentrate on this and the related quasi potential approaches.

In section 2 we review some of the properties of the Bethe-Salpeter equations in the simple case of scalar particles. Soluble models in a $\phi 3$ and $\phi 4$ theory are discussed in section 3. In most actual applications of the equations we have to rely on its numerical solutions. In view of the many singularities present in the equations a form of regularisation has to be found. In section 4 we describe a way using the Wick rotation in the relative energy plane to reduce the equations essentially to one of the Fredholm type. Section 5 deals with some often used approximations to the Bethe-Salpeter equation where the relative energy variable is eliminated so that the resulting equations have the same structure as the nonrelativistic potential scattering equations.

For nucleon-nucleon scattering we have to include the relativistic spin structure in our analysis. This is done in section 6 and applied in section 7 to the relativistic one boson exchange model for nucleon-nucleon interaction. In section 8 the relativistic deuteron model and its em form factors are treated, while the consistency problem of dynamics and em interaction is discussed in section 9. It is in particular shown, that the pair term contribution is cancelled by a dynamical contribution. In sections 10 and 11 we study the nucleon-nucleon interaction at intermediate energies. The importance of the isobar degrees of freedom are emphasized and various models for the Δ are discussed. Also it is demonstrated that the Bethe-Salpeter equations can be extended through the inclusion of the renormalisation of the internal nucleon lines by the lowest order πN bubble graph to also satisfy three particle unitarity above the one-pion production threshold, but below the two-pion one. Some attention is paid to the description in these models of the resonantlike structures found experimentally in spin polarized pp elastic scattering. Finally in the last section we review some of the studies carried out in the treatment of relativistic effects in the three nucleon system.

2. Scalar model

Our starting point is the Bethe-Salpeter (BS) equation for elastic scattering of two particles. Historically it has first been introduced by Nambu [5] in the ladder approximation. Its general form has been derived by Salpeter and Bethe [6]. To show some of the essential features we consider the simple case of the scattering of two scalar particles a and b with masses m_a and m_b. The corresponding scattering Green's function, which is defined in coordinate space by a time ordered product

$$G(x_a',x_b';x_a,x_b) = \langle 0|T[\phi_a(x_a')\phi_b(x_b')\phi_a^+(x_a)\phi_b^+(x_b)]|0\rangle \tag{2.1}$$

where ϕ_a and ϕ_b are the field operators of the particles a and b respectively, can be expanded in the standard way in a covariant Feynman perturbation series. By rearranging this series in terms of two-particle irreducible graphs we readily obtain the BS equation. A connected graph is called two-particle irreducible if it cannot be made disconnected by cutting only one a and one b particle line internally. The BS equation has formally the form

$$G = \bar{G}_o + \bar{G}_o I G \tag{2.2}$$

Here I is the set of all two-particle irreducible graphs while $\bar{G}_o$ is the free two-particle Green's function. In the momentum representation it is given by

$$\bar{G}_o = - \Delta_F^{(a)}(k_a^2)\Delta_F^{(b)}(k_b^2) \tag{2.3}$$

where $\Delta_F^{(i)}$ is the single particle propagator

$$\Delta_F^{(i)}(k) = [k^2-m_i^2+i\varepsilon]^{-1} \tag{2.4}$$

Instead of G it is often more convenient to introduce the T-matrix, defined as

$$iT = \bar{G}_o^{-1}[G-\bar{G}_o]\bar{G}_o^{-1} \tag{2.5}$$

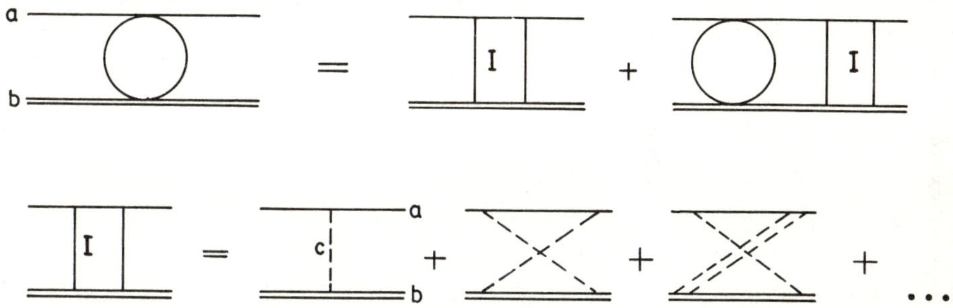

Figure 1. Diagrammatic representation of the BS equation for a ϕ^3 theory. I is the set of all two-particle irreducible diagrams.

In Fig. 1 is shown the diagrammatic representation of the corresponding BS equation for T in the case of a ϕ^3 theory i.e. the interaction Lagrangian is such that particles a and b can emit or absorb a scalar meson c with mass μ. The resulting equation for the T-matrix becomes

$$T = V + V\, G_o T \tag{2.6}$$

with $iV = I$ and $G_o = i\bar{G}_o$. Although it looks formally the same as the well-known Lippmann-Schwinger (LS) equation for nonrelativistic two-particle scattering [7], there are distinct differences. Due to translational symmetry in time and space we have conservation of the total four momentum P. Moreover, it is a fully relativistic covariant equation. Defining $k_a = \mu_a P + p$ and $k_b = \mu_b P - p$ with the mass ratio $\mu_i = \frac{m_i}{m_a+m_b}$, we see that Eq. (2.6) reduces to a four dimensional integral equation in the relative momentum p. For the physical scattering process the particles should be on the mass shell i.e. $k_{io} = E_i(k_i)$ with $E_i(k)=\sqrt{\vec{k}^2 + m_i^2}$ and $P_o = E_a(k_a) + E_b(k_b)$. Therefore the particles in the intermediate state are not only off the energy shell, but also off the mass shell. It is precisely the presence of the relative energy p_o (or the related relative time variable) which makes this equation richer than the corresponding nonrelativistic one. From Eq. (2.6) we can in principle determine the fully off shell T-matrix. Besides the total invariant mass square $P^2 = s$ it also depends on the initial and final off shell relative momenta p and p' respectively.

Since the BS equation is fully off shell it can obviously be used to investigate the boundstate problem. Assuming that the interaction does support a boundstate with mass M_B the BS amplitude must have a pole at $P^2 = M_B^2$. Near the boundstate pole the BS amplitude factorizes as

$$iG(p',p;P) = \sum_m \frac{\psi^{(m)}(p',P)\bar{\psi}^{(m)}(p,P)}{M_B^2-P^2} + \text{regular terms} \tag{2.7}$$

where ψ is the boundstate wavefunction and $\bar{\psi}$ is its conjugate, depending on the polarization m and the four momenta p and P, subject to the condition $P^2 = M_B^2$. Moreover, they satisfy the corresponding homogeneous integral equations

$$\begin{aligned} \psi^{(m)} &= \bar{G}_o I \psi^{(m)} \\ \bar{\psi}^{(m)} &= \bar{\psi}^{(m)} I \bar{G}_o \end{aligned} \tag{2.8}$$

with $P^2 = M_B^2$ and where we have assumed that time reversal invariance holds. Much attention has been paid to the derivation of the boundstate wavefunction normalization condition [8]. In particular the relation with current conservation has extensively been studied. One simple way to derive this condition is to follow essentially the nonrelativistic derivation. From Eq. (2.2) we infer that a formal solution is given by

$$G = [\bar{G}_o^{-1} - I]^{-1} \tag{2.9}$$

Differentiating this with respect to s yields

$$\frac{\partial G}{\partial s} = - G[\frac{\partial \bar{G}_o^{-1}}{\partial s} - \frac{\partial I}{\partial s}]G \tag{2.10}$$

Next substituting Eq. (2.7) into Eq. (2.10) we find that Eq. (2.10) has a double pole at $P2 = M_B^2$. Its residue is given by

$$i \sum_{m} \psi^{(m)}\bar{\psi}^{(m)} = - \left(\sum_{m} \psi^{(m)}\bar{\psi}^{(m)}\right) \left[\frac{\partial \bar{G}_o^{-1}}{\partial s} - \frac{\partial I}{\partial s}\right] \left(\sum_{m'} \psi^{(m')}\bar{\psi}^{(m')}\right) \tag{2.11}$$

Hence

$$i\bar{\psi}^{(m)}\left[\frac{\partial \bar{G}_o^{-1}}{\partial s} - \frac{\partial I}{\partial s}\right]\psi^{(m')} = \delta_{mm'} \tag{2.12}$$

The operator I in Eq. (2.2) is in general dependent on the total four momentum. However, in the ladder approximation i.e. keeping only the one meson exchange diagram for the driving force I becomes independent of s. For that particular case instead of Eq. (2.12) the normalization condition reads

$$i\bar{\psi}^{(m)} \frac{\partial \bar{G}_o^{-1}}{\partial s} \psi^{(m')} = \delta_{mm'} \tag{2.13}$$

We now turn to discuss the scattering case. The BS equation takes a relatively simpler form if we are in the overall two-particle center of mass system $P = (\vec{0},\sqrt{s})$. In the scattering region, $\sqrt{s} > m_a + m_b$, on shell scattering is given by the kinematics $\hat{p} = |\vec{p}| = |\vec{p}\,'|$, $\sqrt{s} = E_a(\hat{p}) + E_b(\hat{p})$ and $\hat{p}_o = p_o = p_o' = \mu_a E_a(\hat{p}) - \mu_b E_b(\hat{p})$. For energies below the production threshold of one c meson one can readily show that two-particle unitarity is satisfied by making use of the cutting rules [9]. They are as follows. To determine the contributions to the imaginary part of the T-matrix consider a given diagram of T. Cut it internally so that it becomes disconnected. Every cut line corresponds to a particle which is put on the positive mass shell. We get a nonvanishing contribution to Im(T) only if the intermediate state formed by the cut lines is energetically allowed. Carry out the cutting in all possible ways. Note that this procedure immediately yields the result that all two-particle irreducible graphs do not contain any imaginary parts in the elastic region $m_a + m_b < \sqrt{s} < m_a + m_b + \mu$. Unitarity holds if the graphs generated by reconnecting the cut lines in all possibly allowed ways are already contained in the set of graphs of the T-matrix. In the elastic region the only allowed cutting is the one corresponding to the two-particle intermediate state of a particle a and b. Reconnecting the cut lines in this case yields the same graph back. Hence the T-matrix is indeed unitary in the elastic region.

Since there is rotational symmetry the BS equations can be partial wave decomposed. In the two-particle center of mass system the free Green's function has the simple form

$$\bar{G}_o(q,q_o;\sqrt{s}) = - \left[(\mu_a\sqrt{s}+q_o)^2 - E_a(q)^2 + i\varepsilon\right]^{-1} \left[(\mu_b\sqrt{s}-q_o)^2 - E_b(q)^2 + i\varepsilon\right]^{-1} \tag{2.14}$$

Taking the Legendre polynomial expansion of the T-matrix

$$T(\vec{p}',p_o';\vec{p},p_o;P) = 4\pi \sum_{\ell=o}^{\infty} (2\ell+1)P_\ell(\cos\theta)T_\ell(|\vec{p}'|,p_o';|\vec{p}|,p_o) \tag{2.15}$$

where θ is the center of mass scattering angle between $\vec{p}$ and $\vec{p}'$, the BS equation reduces to the two-dimensional integral equation

$$T_\ell(p',p_o';p,p_o) = V_\ell(p',p_o';p,p_o) +$$

$$+ \frac{i}{\pi}\int_o^\infty q^2\, dq \int_{-\infty}^{\infty} dq_o\, V_\ell(p',p_o';q,q_o)\bar{G}_o(q,q_o;\sqrt{s})T_\ell(q,q_o;p,p_o) \tag{2.16}$$

In the ladder approximation we have

$$V_\ell(p',p_o';p,p_o) = \frac{4\pi\lambda}{2pp'}\, Q_\ell\left(\frac{p^2+p'^2-(p_o-p_o')^2+\mu^2-i\varepsilon}{2pp'}\right) \tag{2.17}$$

The relation between the partial wave amplitude T_ℓ and the scattering phase shift δ_ℓ and inelasticity parameter η_ℓ is up to a kinematic factor the same as in the nonrelativistic case

$$T_\ell(\hat{p},\hat{p}_o;\hat{p},\hat{p}_o) = \frac{i}{p}\,[E_a(\hat{p})+E_b(\hat{p})][1-\eta_\ell e^{2i\delta_\ell}] \tag{2.18}$$

Sometimes complex phase shifts are used where the inelasticity is expressed as the imaginary part of the phase shift. We have the relation

$$\eta_\ell = \exp(-2\mathrm{Im}\delta_\ell) \tag{2.19}$$

The validity of two-particle unitarity in the elastic region implies that $\eta_\ell = 1$ or $\mathrm{Im}(\delta_\ell) = 0$.

3. Soluble models

Before discussing some specific models let us first consider the nonrelativistic limit of the BS equation. This is obtained by letting $c \to \infty$. Since the relative energy variable p_o carries implicitly a factor of $1/c$, we may set in the driving force and the T-matrix these variables to zero. Taking for simplicity the equal mass case, the BS equation becomes in the overall center of mass system

$$T_\ell(p',0;p,0) = V_\ell(p',0;p,0) + \int_o^\infty q^2\, dq\, V_\ell(p',0;q,0)\tilde{G}_o(q)T_\ell(q,0;p,0) \tag{3.1}$$

with

$$\tilde{G}_o(q) = -\frac{1}{\pi^2}\int_{-\infty}^{\infty} [(\sqrt{s}/2+q_o)^2-E(q)^2+i\varepsilon]^{-1}\,[(\sqrt{s}/2-q_o)^2-E(q)^2+i\varepsilon]^{-1}\,dq_o \tag{3.2}$$

The integration in $\tilde{G}_o$ can explicitly be evaluated to give

$$\tilde{G}_o(q) = \frac{-1}{2\pi E(q)\sqrt{s}}\left[\frac{1}{\sqrt{s}-2E(q)+i\varepsilon} + \frac{1}{\sqrt{s}+2E(q)}\right] \tag{3.3}$$

Using now the nonrelativistic relations $\sqrt{s} = 2mc^2 + \hat{p}^2/m$ and $E(q) = mc^2 + q^2/2m$ we get for c large that the second term on the rhs of Eq. (3.3) can be neglected. As a result we get

$$\tilde{G}_o(q) = \frac{1}{4\pi m}\,\frac{1}{q^2-\hat{p}^2-i\varepsilon} \tag{3.4}$$

Hence Eq. (3.1) reduces to the nonrelativistic LS equation. In the ladder approximation the potential is given by the Yukawa potential

$$V_\ell(p,0;q,0) = \frac{4\pi\lambda}{2pq}\,Q_\ell\left(\frac{p^2+q^2+\mu^2}{2pq}\right) \tag{3.5}$$

We have clearly followed the convention here that the potential in Eq. (3.5) is attractive if $\lambda > 0$.

The BS equations cannot in general be solved in closed form. One situation is known where explicit solutions exist for total four momentum $P \neq 0$. This is the case when the c meson has zero mass and furthermore the ladder approximation is made. Going to the center of system the homogeneous BS equation for the Green's function in the case of equal masses $m_a = m_b = m$ reads

$$-i\eta\psi(p,P) = \left[\left(\frac{\sqrt{s}}{2} + p_o\right)^2-E(p)^2+i\varepsilon\right]^{-1}\left[\left(\frac{\sqrt{s}}{2} - p_o\right)^2-E(p)^2+i\varepsilon\right]^{-1}\int \frac{d^4q}{(p-q)^2}\psi(q,P) \tag{3.6}$$

One can easily verify that for the special case $\sqrt{s} = 0$, Eq. (3.6) admits a solution

$$\psi(p,0) = [p^2-m^2]^{-3} \tag{3.7}$$

with eigenvalue $\eta = (2m^2\pi^2)^{-1}$. Since the wavefunction does not have a node in $|\vec{p}|$, it corresponds to the groundstate. The above model has been studied extensively by Wick [10] and shown to be reducible to an eigenvalue problem of a Sturm-Liouville type. A complete solution has been given in an accompanying paper by Cutkosky [11]. Similarly as in the nonrelativistic Coulomb problem an O(4) symmetry is present. Apart from the hydrogenic levels additional boundstate solutions are found, which do not have a nonrelativistic analog. Some of these solutions exhibit the property that the normalization condition can become zero or negative [8]. One other interesting point to note is that it can explicitly be verified that in the Wick-Cutkosky model the one body limit is not recovered i.e. in the limit that $m_b \to \infty$, keeping m_a fixed, the motion of the a particle is not the one corresponding to a particle in an external potential field. This has recently also been discussed by Gross [12]. From now on we shall confine ourselves to the equal mass case.

Another model which can be solved in closed form is the Zachariasen model [13]. Keeping only the set of sausage-like diagrams shown in Fig. 2 for a ϕ^4 theory and assuming that the vertex function is of a separable form i.e.

$$V(p',p_o';p,p_o) = \sum_{n=1}^{N} \lambda_n g_n(p',p_o') g_n(p,p_o) \tag{3.8}$$

we find immediately, that the solution of Eq. (2.16) is given by

$$T_\ell(p',p_o';p,p_o) = \sum_{m,n} \lambda_m g_m(p',p_o') \tau_{mn}(\sqrt{s}) g_n(p,p_o) \tag{3.9}$$

where the two-particle propagator τ satisfies an algebraic equation

$$\tau_{mn}(\sqrt{s}) = \delta_{mn} + \sum_{k=1}^{N} \lambda_k K_{mk}\, \tau_{kn}(\sqrt{s}) \tag{3.10}$$

with

$$K_{mk} = -\frac{i}{\pi^2} \int_0^\infty q^2\, dq \int_{-\infty}^{\infty} dq_o \frac{g_m(q,q_o) g_k(q,q_o)}{\{(\sqrt{s}/2+q_o)^2 - E(q)^2 + i\varepsilon\}\{(\sqrt{s}/2-q_o)^2 - E(q)^2 + i\varepsilon\}} \tag{3.11}$$

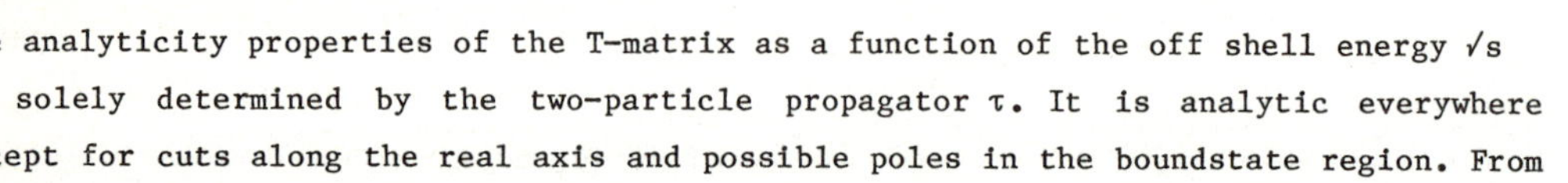

Figure 2. Diagrammatic representation of the BS equation for the separable Zachariasen model.

The analyticity properties of the T-matrix as a function of the off shell energy $\sqrt{s}$ is solely determined by the two-particle propagator τ. It is analytic everywhere except for cuts along the real axis and possible poles in the boundstate region. From Eq. (3.10) we may explicitly determine in the case of N=1 the discontinuity of the square root branch cut starting at $\sqrt{s}$ = 2m. It is given by

$$\text{Im}\, \tau^{-1} = -\frac{2}{\pi} \int_0^\infty q^2 dq \int_{-\infty}^{\infty} dq_o\; g(q,q_o)^2$$

$$\delta^{(+)}[(\sqrt{s}/2+q_o)^2 - E(q)^2]\; \delta^{(+)}[(\sqrt{s}/2-q_o)^2 - E(q)^2] \tag{3.12}$$

The two δ-functions corresponds precisely to putting both particles in the intermediate state on the mass shell. From this one can immediately verify that two-particle unitarity is satisfied for all energies provided no other singularities are present in the form factors. It has the form

$$T - T^{+} = 2iTT^{+} \tag{3.13}$$

which also holds for the more general form of the interaction. Taking simple generalizations of nonrelativistic form factors such as of the Yamaguchi type as an example

$$g_{Yam}(p,p_o) = [p^2-p_o^2+\beta^2-i\varepsilon]^{-1} \tag{3.14}$$

we can get additional branchpoints. In the case of Eq. (3.14) we have one at $\sqrt{s} = 2m + \beta$, which corresponds to the production of a "β" particle.

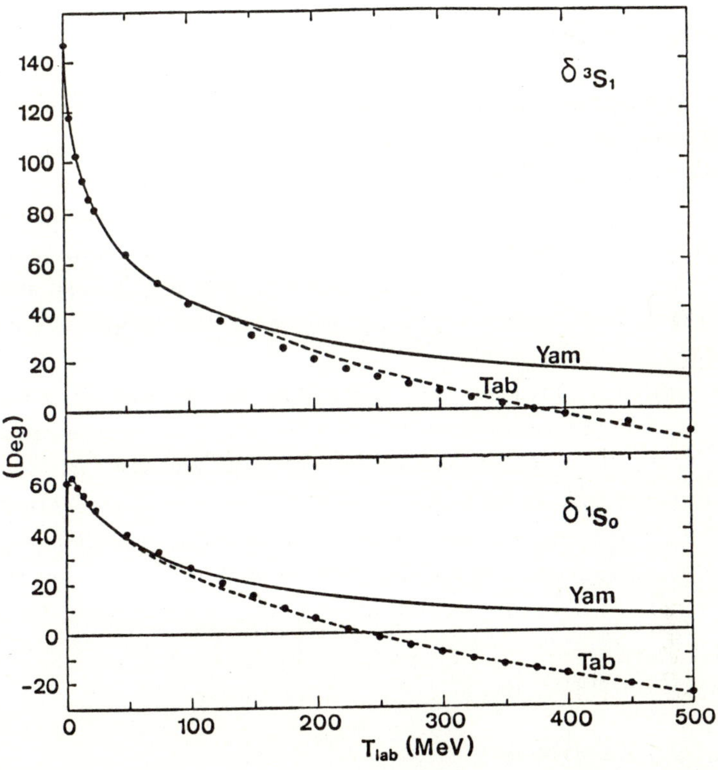

Figure 3. The ^1S_o and $^3S_1$ phase shifts calculated in the separable Zachariasen model using Yamaguchi and Tabakin like form factors. Experimental data are from Ref. 94.

As is known the nucleon-nucleon s-wave phase shifts changes sign at energies of about a few hundred MeV. Interactions like the one-term Yamaguchi form factor have in general the property that the zero in such phase cannot be reproduced. However by some suitable choice of the form factor as has been shown by Tabakin [14] this can be achieved. In Fig. 3 are shown some fits to the NN s-wave phase shifts using the above model. As can be seen we indeed get a very good fit to the experimental data for the case of the Tabakin like interaction. It has the form

$$g_{Tab}(p,p_o) = \frac{p^2-p_o^2+\alpha^2}{p^2-p_o^2+\gamma^2-i\varepsilon}\,\frac{p^2-p_o^2-p_c^2}{(p^2-p_o^2+\beta^2-i\varepsilon)^2} \tag{3.15}$$

However such interactions induce some pathologies in the three particle system such as the effective collapse of the three-particle groundstate [15-17]. This is due to the presence of a continuum discrete state at a positive energy.

4. Wick rotated equations

Except for the few cases as discussed in the previous section we have to rely on numerical studies of the BS equation. In practical applications it has only been solved in the ladder approximation. Direct matrix inversion of the discretized integral equation does not work because of the presence of singularities in the kernel. Rather accurate solutions have been obtained in the boundstate and elastic region using variational methods. In particular Schwarz and Zemach [18] studied in coordinate space the elastic scattering solutions using the Schwinger variational method [19].

Another approach which is also applicable in the inelastic region below two-meson production is to reduce the BS equation first to a nonsingular equation and then to solve it by standard methods [20-23]. To see how we can regularize the equations we assume that we only have to account for the various singularities in the kernel. Let us first confine ourselves to the boundstate case with $0 < \sqrt{s} < 2m$. From Eq. (2.14) we see that the two-particle free Green's function has four poles in the complex q_o plane. They are located at

$$\begin{aligned} q_o &= \sqrt{s}/2 \pm E(q) \mp i\varepsilon \\ q_o &= -\sqrt{s}/2 \pm E(q) \mp i\varepsilon \end{aligned} \tag{4.1}$$

Varying the three momentum q these poles never cross each other. We may now apply a Wick rotation on p_o' and q_o by replacing these variables by $e^{i\alpha}p_o'$ respectively $e^{i\alpha}q_o$ and letting α go from 0 to $\pi/2$ [10,24]. This rotation is allowed since none of the singularities (4.1) and the logarithmic singularities in the interaction (2.17) are encountered in the process of rotation. The resulting homogeneous equation becomes

$$T_\ell(p',ip'_4) = -\frac{1}{\pi^2}\int_0^\infty q^2dq \int_{-\infty}^\infty dq_4\ V_\ell(p',ip'_4;q,iq_4)\bar{G}_o(q,iq_4;\sqrt{s})T_\ell(q,iq_4) \tag{4.2}$$

with p'_4 and q_4 real. From Eqs. (2.14) and (2.17) we see that the integral equation (4.2) has a real valued kernel along the imaginary q_o axis and that it is free of any singularities. Therefore it can be solved by standard discretization methods.

In the scattering region below one meson production threshold the two poles located at

$$\begin{aligned} q_o &= \sqrt{s}/2 - E(q) + i\varepsilon \\ q_o &= -\sqrt{s}/2 + E(q) - i\varepsilon \end{aligned} \tag{4.3}$$

can cross each other and therefore a pinching of the path of integration can occur. This happens precisely at the point that both particles in the intermediate states are on the mass shell. The Wick rotation can again be carried out provided we deform the integration path (see Fig. 4) to pick up the additional contribution from the poles once they have crossed. As a result we get

$$\begin{aligned} T_\ell(p',ip'_4;\hat{p},0) &= V_\ell(p',ip'_4;\hat{p},0) + \\ &-\frac{1}{\pi^2}\int_0^\infty q^2dq \int_{-\infty}^\infty dq_4\ V_\ell(p',ip'_4;q,iq_4)\bar{G}_o(q,iq_4;\sqrt{s})T_\ell(q,iq_4;\hat{p},0) \\ &+\frac{1}{2\pi\sqrt{s}}\int_0^{\hat{p}} q^2dq\ \{V_\ell(p',ip'_4;q,\omega(q)) + V_\ell(p',ip'_4;q,-\omega(q))\}\ \frac{T_\ell(q,\omega(q);\hat{p},0)}{\omega(q)E(q)} \end{aligned} \tag{4.4}$$

with $\omega(q) = \sqrt{s}/2 - E(q) + i\varepsilon$ and where the initial state is taken on shell. Eq. (4.4) has to be supplemented by another equation for the T-matrix with p_o real. One readily verifies that it is given by

$$\begin{aligned} T_\ell(p',\omega(p');\hat{p},0) &= V_\ell(p',\omega(p');\hat{p},0) + \\ &-\frac{1}{\pi^2}\int_0^\infty q^2dq \int_{-\infty}^\infty dq_4\ V_\ell(p',\omega(p');q,iq_4)\bar{G}_o(q,iq_4;\sqrt{s})T_\ell(q,iq_4;\hat{p},0) \\ &+\frac{1}{2\pi\sqrt{s}}\int_0^{\hat{p}} q^2dq\{V_\ell(p',\omega(p');q,\omega(q))+V_\ell(p',\omega(p');q,-\omega(q))\}\ \frac{T_\ell(q,\omega(q);\hat{p},0}{\omega(q)E(q)} \end{aligned} \tag{4.5}$$

The only singularity which is still present in the elastic region is the pole of the Green's function in the one-dimensional integral of Eqs. (4.4) and (4.5) at the on shell value $\hat{q} = [s/4-m^2]^{\frac{1}{2}}$. Using a Kowalski-Noyes subtraction method [25,26] this can be regularized. The resulting equations can be solved by standard methods like direct

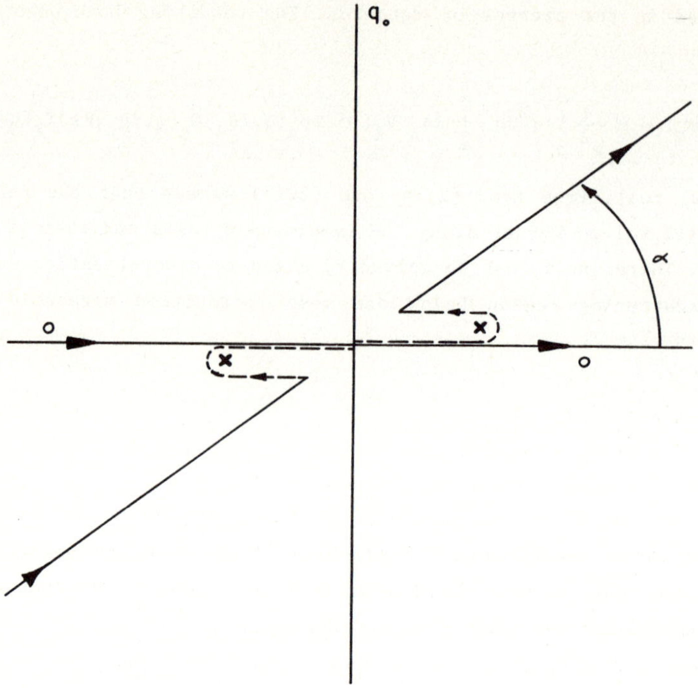

Figure 4. Wick rotation in the q_0-plane over an angle α with the integration path deformed to pick up the contributions of the poles (4.3) after they have pinched.

matrix inversion. Above one meson production threshold, but below the two-meson one, the only additional complication is that the interaction I has logarithmic branch-points which can lie on the integration path of the one-dimensional integral in Eq. (4.5). These singularities are of an integrable type and no special precautions need to be taken, at least for the scalar model.

5. Quasi potential equations

Although the BS approach is in principle a good candidate for describing relativistic quantum dynamics, it is unfortunately not trivial to perform actual computations. In particular, including higher order graphs beyond the ladder approximation may be a major effort. Therefore it is interesting to try to construct relativistic theories which are closer to nonrelativistic potential theories. Such formulations have also an important advantage that it makes a direct comparison possible and as a result relativistic corrections can be more easily discussed.

The relativistic quasi potential (QP) equations can in general be formulated by essentially giving a prescription how to chose the free Green's function and then subsequently to calculate the potential using information from quantum field theory. For a given free Green's function G_{QP} in which the relative energy variable is eliminated by taking it in general to be a function of the relative three momentum, the T-matrix is defined to satisfy a LS-type of equation

$$T(s) = W + W\, G_{QP}\, T(s) \tag{5.1}$$

The procedure is carried out so that the resulting equations are relativistically covariant. Comparing Eq. (5.1) with the BS equation we see that the new interaction W can be expressed in terms of the BS kernel as

$$W = V + V[G_o - G_{QP}]W \tag{5.2}$$

Various choices have been examined in the literature. One is where the two-particle unitarity is manifestly imposed by writing down a dispersion relation for the QP propagator subject to the condition that its discontinuity is the same as the BS free two-particle Green's function [27,28]

$$G_{BBS}(q) = 2\pi \int_{4m^2}^{\infty} \frac{ds'}{s'-s-i\varepsilon}\, \delta^{(+)}[(\frac{P'}{2} + q)^2 - m^2]\delta^{(+)}[(\frac{P'}{2} - q)^2 - m^2] \tag{5.3}$$

For equal mass particles it treats both particles on equal footing and as a result the symmetry for identical particles can easily be satisfied. Using the scalar model of section 2 it is found that the QP approximation yields in general more attraction than the BS equation [21,29]. In Fig. 5 is shown the position of the first boundstate as a function of the coupling constant λ for the BS case and the Blankenbecler-Sugar (BBS) QP prescription, Eq. (5.3). For a given λ the BBS result is clearly more bound.

Another often used choice is that due to Gross [30]. It essentially puts one of the particles on the mass shell i.e.

$$G_{Gross} = -\, 2\pi[(\frac{P}{2} + q)^2 - m^2 + i\varepsilon]^{-1}\, \delta^{(+)}[(\frac{P}{2} - q)^2 - m^2] \tag{5.4}$$

The major advantage of this choice is that the description of the electromagnetic (em) properties of the system can be done in a fully Lorentz covariant way. In that case it is rather natural to put the spectator particle on the mass shell. For the case of describing the dynamics in systems with unequal mass particles the heavier particle is taken to be onshell. This prescription leads automatically to having a correct one-body limit. In the case of equal mass particles the symmetry of identical particles is violated. This in principle can be restored by symmetrizing the intermediate state in some way. In so doing nonphysical singularities can occur in the driving force.

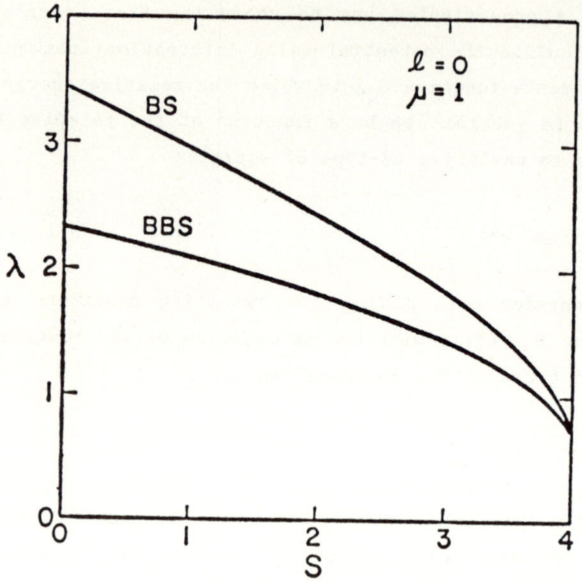

Figure 5. The position of first bound state in the equal mass scalar BS and BBS model.

In an analysis by Gammel and Menzel [31] of the groundstate of e^+e^- using the BS equation in the one photon exchange approximation it is found that the energy of the groundstate differs by a few percent from the Balmer formula, suggesting that this difference may be of the order of the fine structure constant α. Including the correction from the two-photon cross box diagram might remove this difference. For unequal masses with one of them becoming infinitely heavy, one can readily show that on the one loop level a cancellation takes place between the two-photon box and cross box diagrams, so that the quasi potential result holds [12]. It should be interesting to see whether this also happens in the equal mass case.

6. Partial wave analysis for particles with spin

Up to now we have neglected the complication that the particles can have a spin structure. In the case of a point nucleon, it can relativistically be described as a Dirac particle. As a consequence if we want to describe nucleon-nucleon scattering relativistically many coupled channels occur in the dynamical equations. Since the Dirac space of a spin one half particle consists of two positive and two negative energy spinors, the maximal number of channels is given by sixteen. Parity conserving forces reduce it to eight channels.

The partial wave decomposition as discussed in section 2 can be generalized to the case of particles with spin. Here we only consider the case of nucleon-nucleon scattering [32]. The analysis is performed by using the helicity states as basis states. For this purpose the work of Goldberger et al. [33] has to be extended to also include negative energy spinor states. The positive and negative energy states of a spin one half particle can be characterized by the spin quantum number as being $+\frac{1}{2}$ and $-\frac{1}{2}$ respectively [34]. Using the notion of this "energy" spin the symmetry properties of the two-nucleon states can be classified by the spatial parity P and the so-called exchange parity P_H, which is the symmetry under interchange of all coordinates. It represents the generalized Pauli principle for off shell states. The operator P_H is given in terms of the parity operator π_o in the relative energy variable by

$$P_H = (-1)^L\,(1+\vec{\sigma}.\vec{\sigma}')(1+\vec{\rho}.\vec{\rho}')\pi_o/4 \tag{6.1}$$

with $\vec{\sigma}$ and $\vec{\rho}$ being the Pauli spin and energy spin operators of the two nucleons and L the orbital angular momentum.

We now briefly sketch how the partial wave reduction is done. Following Jacob and Wick [35] we introduce in the ordinary spin space the helicity states for the two nucleons. As a result a complete set of two particle states which can be used to reduce the BS equation is given by

$$|\lambda_1\rho_1;\lambda_2\rho_2\rangle = U_{\lambda_1}^{\rho_1}(1)U_{\lambda_2}^{\rho_2}(2) \tag{6.2}$$

where $U_{\lambda_n}^{\rho_n}(n)$ are the single particle spinor states with energy spin ρ_n and helicity λ_n of particle number n, satisfying the free Dirac equation. Assuming parity conservation we need only eight of the helicity states $\langle\lambda_1'\lambda_2'|T|\lambda_1\lambda_2\rangle$, which are taken to be

$$\begin{aligned}
\phi_1 &= \langle{++}|T|{++}\rangle \quad , \quad & \phi_5 &= \langle{++}|T|{+-}\rangle \\
\phi_2 &= \langle{++}|T|{--}\rangle \quad , \quad & \phi_6 &= \langle{++}|T|{-+}\rangle \\
\phi_3 &= \langle{+-}|T|{+-}\rangle \quad , \quad & \phi_7 &= \langle{+-}|T|{++}\rangle \\
\phi_4 &= \langle{+-}|T|{-+}\rangle \quad , \quad & \phi_8 &= \langle{-+}|T|{++}\rangle
\end{aligned} \tag{6.3}$$

Going to the two-particle center of mass system we write a partial wave expansion for the helicity amplitudes

$$\phi_n = \sum_J (2J+1)\phi_n^J d_{\lambda\lambda'}^J(\theta) \tag{6.4}$$

with $\lambda = \lambda_1 - \lambda_2$, $\lambda' = \lambda_1' - \lambda_2'$ and d^J are the known rotation matrices [36]. To exhibit more clearly the symmetry of the generalized Pauli principle we form the states

$$\phi^{+}_{\lambda_1\lambda_2} = U^{(+)}_{\lambda_1}(1)U^{(+)}_{\lambda_2}(2)$$

$$\phi^{-}_{\lambda_1\lambda_2} = U^{(-)}_{\lambda_1}(1)U^{(-)}_{\lambda_2}(2)$$

$$\phi^{e}_{\lambda_1\lambda_2} = \frac{1}{\sqrt{2}}\left[U^{(+)}_{\lambda_1}(1)U^{(-)}_{\lambda_2}(2)+U^{(-)}_{\lambda_1}(1)U^{(+)}_{\lambda_2}(2)\right] \tag{6.5}$$

$$\phi^{o}_{\lambda_1\lambda_2} = \frac{1}{\sqrt{2}}\left[U^{(+)}_{\lambda_1}(1)U^{(-)}_{\lambda_2}(2)-U^{(-)}_{\lambda_1}(1)U^{(+)}_{\lambda_2}(2)\right]$$

Under interchange $\rho_1 \leftrightarrow \rho_2$ the +, - and e states are even and o is odd. Another step which has to be taken is to construct helicity amplitudes between states with a definite parity i.e. for a given total angular momentum J we take as states

$$|J;r\lambda_1\lambda_2\rangle = \frac{1}{\sqrt{2}}\left[|J;\lambda_1\lambda_2\rangle + r|J;-\lambda_1-\lambda_2\rangle\right] \tag{6.6}$$

with $r = \pm 1$.

Using the above representation one finds that the various partial wave states which are coupled to each other in the BS equation separate into three categories, which can be labeled according to which physical states they contain. We have (i) singlet states with L=J, (ii) triplet states with L=J, and (iii) coupled triplet channel states with $L = J \pm 1$. In table 1 the angular momentum states are listed. Also is given the relative energy parity of the various states. As an example if we want to calculate the scattering in the $^3S_1^+$ channel, we see from the table, using the spectroscopic notation $^{2S+1}L_J^{\rho}$ that the following states are present as intermediate states in the BS equation

$$n=1 : {}^3S_1^+ \; ; \; n=2 : {}^3D_1^+ \; ; \; n=3 : {}^3S_1^- \; , \; n=4 : {}^3D_1^-$$

$$n=5 : {}^1P_1^e \; , \; n=6 : {}^3P_1^o \; , \; n=7 : {}^1P_1^o \; , \; n=8 : {}^3P_1^e \tag{6.7}$$

Table 1

The eight partial wave states (S,L,ρ) which are coupled together in the BS equations for the three cases. Also is given the relative energy parity of π_o of the states.

singlet	π_o	triplet	π_o	coupled triplet	π_o
(s,L=J,+)	+	(t,L=J,+)	+	(t,L=J-1,+)	+
(s,L=J,-)	+	(t,L=J,-)	+	(t,L=J+1,+)	+
(t,L=J-1,e)	+	(t,L=J-1,0)	+	(t,L=J-1,-)	+
(t,L=J+1,e)	+	(t,L=J+1,0)	+	(t,L=J+1,-)	+
(t,L=J,+)	-	(s,L=J,+)	-	(s,L=J,e)	+
(t,L=J,-)	-	(s,L=J,-)	-	(t,L=J,0)	+
(t,L=J-1,0)	-	(t,L=J-1,e)	-	(s,L=J,0)	-
(t,L=J+1,0)	-	(t,L=J+1,e)	-	(t,L=J,e)	-

The first six states are even in the relative energy while the last two are odd. Furthermore, notice that the physical $^3S_1^+$ state is coupled to odd space parity (L=1) states in the negative energy states sector. Parity conservation is not violated in view of the opposite intrinsic parity of the positive and negative energy spinor states.

7. The relativistic OBE model

Nucleon-nucleon scattering at energies up to 300 MeV has been described successfully in a nonrelativistic Schrödinger theory using the one boson exchange (OBE) potential to characterize the nuclear force. In view of this, to explore a relativistic description in terms of the BS equation or a QP approach, it is natural to parameterize the driving force by a sum of relativistic meson exchange diagrams. In analogy with the nonrelativistic situation we take the exchange of $\pi,\rho,\omega,\varepsilon,\eta$ and δ mesons. The BS equation for the OBE model reads [37]

$$\phi(p,P) = V_{OBE}(p,\hat{p}) - \frac{i}{4\pi^3}\int d_4q\, V_{OBE}(p,q)S(q)\phi(q,P) \tag{7.1}$$

with

$$V_{OBE}(p,q) = \sum_B \frac{V_B^{(1)}V_B^{(2)}}{(\vec{p}-\vec{q})^2-(p_o-q_o)^2+\mu_B^2-i\varepsilon} \tag{7.2}$$

and the two-particle propagator

$$S(q) = [\frac{\not{P}}{2} + \not{q}-m]^{-1}\,[\frac{\not{P}}{2} - \not{q}-m]^{-1} \tag{7.3}$$

Performing the partial wave analysis as described in the previous section we get

$$\phi(p,p_o,\alpha) = V_{OBE}(p,p_o,\alpha;\hat{p},0,1) - \frac{i}{2\pi^2}\int_o^\infty q^2dq \int_{-\infty}^{\infty} dq_o$$

$$\sum_{\beta,\gamma} V_{OBE}(p,p_o,\alpha;q,q_o,\beta)S(q,q_o,\beta,\gamma)\phi(q,q_o,\gamma) \tag{7.4}$$

The two-nucleon propagator is independent of spin indices and is given in the energy spin space by

$$S_{++} = [(E-E(q)+i\varepsilon)^2-q_o^2]^{-1}$$

$$S_{--} = [(E+E(q)-i\varepsilon)^2-q_o^2]^{-1}$$

$$S_{eo} = S_{oe} = \frac{1}{\sqrt{s}}\left[\frac{E-E(q)}{(E-E(q)+i\varepsilon)^2-q_o^2} + \frac{E+E(q)}{(E+E(q)-i\varepsilon)^2-q_o^2}\right] \tag{7.5}$$

$$S_{ee} = S_{oo} = -\frac{q_o}{\sqrt{s}}\left[\frac{1}{(E-E(q)+i\varepsilon)^2-q_o^2} - \frac{1}{(E+E(q)-i\varepsilon)^2-q_o^2}\right]$$

with $2E = \sqrt{s}$. The vertex operators $V_B^{(n)}$ are determined from the meson-nucleon interaction Lagrangians, given by

$$L_{PS} = i\, g_{PS}\, \bar{\psi}\gamma_5\psi\phi$$

$$L_S = g_S\, \bar{\psi}\psi\phi \tag{7.6}$$

$$L_V = g_V\, \bar{\psi}\gamma_\mu\psi\phi^\mu - i\,\frac{g_T}{4m_N}\,\bar{\psi}\sigma_{\mu\nu}\psi(\partial^\mu\phi^\nu-\partial^\nu\phi^\mu)$$

for the pseudoscalar, scalar and vector mesons respectively. Due to the singular behaviour of the kernels at high momenta we have to introduce a form factor at each vertex. For the meson-nucleon vertex function we use the monopole form

$$F(t) = \frac{-\Lambda^2}{t-\Lambda^2} \tag{7.7}$$

where the cutoff mass Λ is assumed to be the same for all the mesons. The calculation of the partial wave matrix elements of the driving force is straightforward but extremely tedious. The algebra was carried out by the symbolic manipulation programs "SCHOONSCHIP" [38] and "REDUCE" [39]. The coupling constants were taken approximately the same as those employed in Ref. [40]. To solve the resulting equations a Wick rotation was performed. Due to the presence of the spin channels the dimension of the matrices involved becomes prohibitively large so that the methods as described for the scalar case are not very practical. For that reason we have studied the applicability of Padé approximants on the multiple scattering series. Iterating the BS equation we determine the Born series solution of the integral equation and then use the Padé sequence to find the exact solution. In table 2 is shown as an example the rate of convergence of the diagonal Padé approximants at E_{lab} = 100 MeV for the ^1S_o phase shift.

The relativistic OBE model has been used to calculate the various partial wave channels. In Fig. 6 we show the result for the ^1S_o channel together with a calculation using the Blankenbecler-Sugar (BBS) QP prescription with

$$\mathrm{Disc}(S_{++}) = -\,2\pi^2\,\delta(q_o)\delta(E-E(q)) \tag{7.8}$$

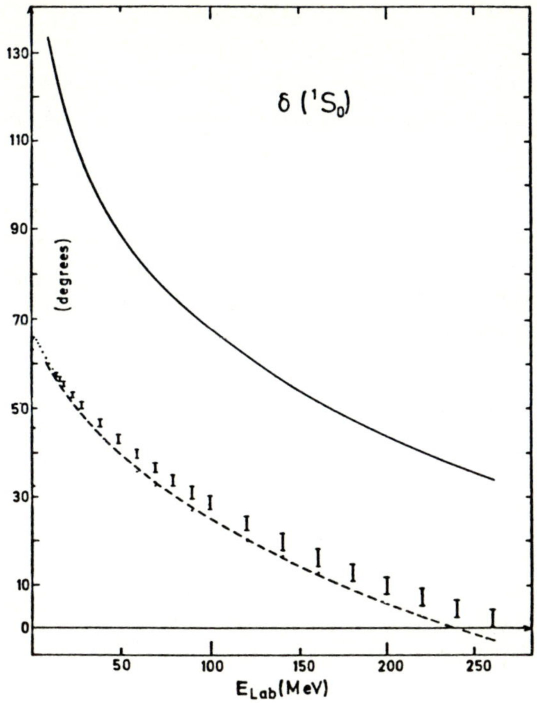

Figure 6. The $^1S_0$ phase shift for the BS (solid curve) and BBS (dashed curve) OBE model using the pseudoscalar πN interaction and the same coupling parameters for both models. Experimental points are from Ref. [93].

and keeping only the positive energy spinor states. We see that the calculated phase shifts are very different from the BS results. This is due to the use of the pseudo scalar interaction for the pion, which gives a strong coupling between positive and negative energy states.

In attempting to fit simultaneously the low partial waves we did not succeed in finding an acceptable description [37]. This can be traced back to the too strong coupling to the negative energy states. Some effective pair suppression is needed in the intermediate states. Such a suppression is also suggested in a chiral symmetric theory. It can be realized in our OBE model by introducing for the pion a pseudo vector interaction. The corresponding Lagrangian is given by

$$L_{PV} = g_{PV}\,\bar{\psi}\gamma_\mu\gamma_5\psi\partial^\mu\phi \qquad (7.9)$$

In so doing we succeed in obtaining a reasonable fit up to 250 MeV lab energy for the various partial waves [41]. As an example, in Fig. 7 is shown the calculated results

Table 2

The convergence rate of the diagonal Padé sequence for tan δ of the ^1S_o phase shift. The individual terms of the Born series are given in the last two columns.

N	[N/N]	2N	2N+1
1	$-4.553\ 10^{-3}$	9.14	$-4.63\ 10^{1}$
2	$2.676\ 10^{-1}$	$2.43\ 10^{2}$	$-1.28\ 10^{3}$
3	$2.151\ 10^{-1}$	$6.71\ 10^{3}$	$-3.52\ 10^{4}$
4	$4.341\ 10^{-1}$	$1.85\ 10^{5}$	$-9.71\ 10^{5}$
5	$4.732\ 10^{-1}$	$5.10\ 10^{6}$	$-2.68\ 10^{7}$
6	$4.803\ 10^{-1}$	$1.41\ 10^{8}$	$-7.38\ 10^{8}$
7	$4.806\ 10^{-1}$	$3.88\ 10^{9}$	$-2.04\ 10^{10}$
8	$4.750\ 10^{-1}$	$1.07\ 10^{11}$	$-5.61\ 10^{11}$
9	$4.807\ 10^{-1}$	$2.95\ 10^{12}$	$-1.55\ 10^{13}$
10	$4.807\ 10^{-1}$	$8.12\ 10^{13}$	$-4.27\ 10^{14}$

for the ^1S_o and $^3S_1$-$^3D_1$ channels.
From this we see that the QP calculations yield an additional attraction, which is much smaller than in the pseudoscalar case, due to the effect of pair suppression in the intermediate states.

8. Relativistic treatment of the deuteron

The relativistic OBE model can be used to study the properties of the deuteron. Assuming that it is a boundstate in the two-nucleon system, it can be found by solving the homogeneous BS equation. Numerically this is done by looking for a pole in the T-matrix at the position $P^2 = M_D^2$. Near the pole we have

$$\phi(p',p;P) = \sum_M \frac{\psi^{(M)}(p';P)\bar{\psi}^{(M)}(p;P)}{P^2-M_D^2} + \text{regular terms}$$

where M is the polarization of the deuteron. For the OBE model the normalization condition for the npD vertex function $\psi^{(M)}$ and its conjugate $\bar{\psi}^{(M)}$ takes the form

$$2P_\mu \delta_{MM'} = \frac{i}{4\pi^3} \int d^4p\ \bar{\psi}^{(M)}(p;P)\ [\frac{\partial}{\partial P_\mu} \mathcal{S}(p,P)]_{P^2=M_D^2}\ \psi^{(M')}(p;P) \qquad (8.2)$$

Since the deuteron is a spin one particle and it occurs as a boundstate in the $^3S_1$ channel, the angular dependence of ψ is given in the two-particle center of mass system by

$$\bar{U}^{\rho_1}_{\lambda_1}(1)\bar{U}^{\rho_2}_{\lambda_2}(2)\psi^{(M)}(p,P) = \sqrt{\frac{2J+1}{2}}\, D^{J}_{M\lambda}(\Omega_{\vec{p}})\phi(|\vec{p}|,p_o;n) \tag{8.3}$$

with J=1 and where n denotes the basis set (6.7). Eq. (8.2) holds in the center of mass system where the BS equation is solved. The normalization condition becomes

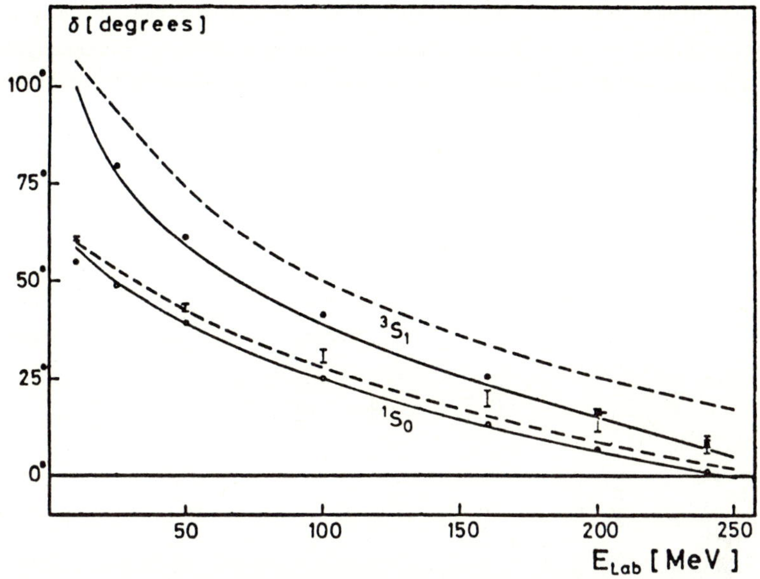

Figure 7. The ^1S_o and $^3S_1$ phaseshifts for the BS (solid curves) and BBS (dashed curves), OBE model using the pseudo vector πN interaction and the same coupling parameters for both models. Experimental points are from Ref. [93].

$$2M_D = \frac{i}{4\pi^2}\int_{-\infty}^{\infty} dp_o \int_{o}^{\infty} p^2 dp \sum_n \bar{\phi}(p,p_o;n)\left[\frac{\partial S(p,p_o;n)}{\partial E}\right]_{E=\frac{1}{2}M_D} \phi(p,p_o;n) \tag{8.4}$$

where $p = (\vec{0},2E)$. In view of Eq. (8.4) it is natural to define the "n-state" probability P_n for the deuteron as

$$P_n = \frac{i}{8\pi^2 M_D}\int_{-\infty}^{\infty} dp_o \int_{o}^{\infty} p^2 dp \quad \bar{\phi}(p,p_o;n)\left[\frac{\partial S(p,p_o;n)}{\partial E}\right]_{E=\frac{1}{2}M_D} \phi(p,p_o;n) \tag{8.5}$$

For the positive energy states it reduces to the proper nonrelativistic limit. Strictly speaking Eq. (8.5) does not represent the probability to find the deuteron in a definite state since it becomes negative for the negative energy state components. This is a reflection of the fact that P_n essentially measures the charge of the corresponding state. Our actual model supports a deuteron, which has a positive energy D-state probability of 4.7% while the negative energy state probabilities are a factor of 100 smaller.

With the npD vertex function determined from the BS equation we may study the em deuteron current. In the relativistic impulse approximation (Fig. 8) it can be written down schematically as

$$\langle P'M'|J_\mu^D|PM\rangle = \bar{\psi}'^{(M')} S_1' \Gamma_\mu^{(1)} S_1 S_2 \psi^{(M)} + \bar{\psi}'^{(M')} S_2' \Gamma_\mu^{(2)} S_1 S_2 \psi^{(M)} \qquad (8.6)$$

where S_n is the single particle propagator of nucleon n and the prime means that the arguments corresponds to those after absorption of the photon. Furthermore q^2 is the four momentum square of the photon and $P' = P+q$. Both terms in Eq. (8.6) yield exactly the same contribution. We assume that the em nucleon formfactors are given by the on shell form

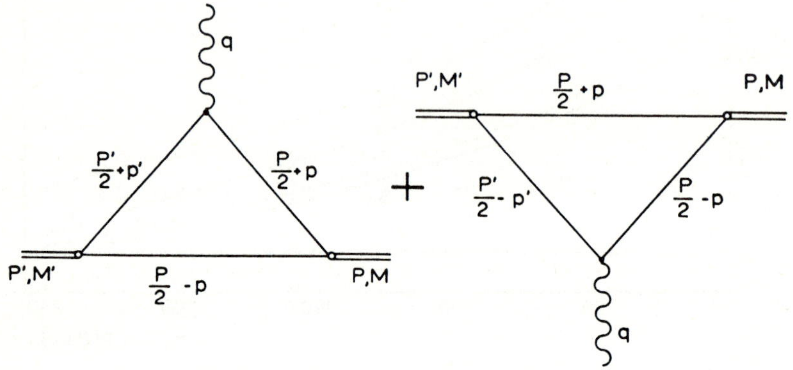

Figure 8. The em deuteron current matrix element in the relativistic impulse approximation.

$$\Gamma_\mu(q) = \gamma_\mu F_1(q^2) - \frac{1}{2m_N} \sigma_{\mu\nu} q^\nu F_2(q^2) \qquad (8.7)$$

Only the isoscalar part contributes because the deuteron is an isospin zero state. We first show that the current satisfies gauge invariance

$$q^\mu \langle P'M'|J_\mu^D|PM\rangle = 0 \qquad (8.8)$$

Using the Ward identity [42]

$$q^\mu \gamma_\mu^{(n)} = S_n^{-1}(p_n+q) - S_n^{-1}(p_n) \qquad (8.9)$$

we get for the current matrixelement J_μ^D

$$q^{\mu}\langle P'M'|J_{\mu}^{D}|PM\rangle = 2F_1\bar{\phi}'^{(M')}\,[S_1S_2-S_1'S_2]\phi^{(M)} \tag{8.10}$$

Since ϕ satisfies the homogeneous BS equation

$$\phi^{(M)} = VS_1S_2\phi^{(M)} \tag{8.11}$$

we may rewrite the second term in Eq. (8.10) as

$$\bar{\phi}'^{(M')}S_1'S_2\phi^{(M)} = \bar{\phi}'^{(M')}S_1'S_2VS_1S_2\phi^{(M)} \tag{8.12}$$

Since for the OBE model V has the property that it only depends on the difference between the relative momentum of initial and final state, we may explicitly verify that

$$\bar{\phi}'^{(M')}S_1'S_2V = \bar{\phi}'^{(M')} \tag{8.13}$$

Hence Eq. (8.10) is identically zero. Assuming more generally, that we are dealing with nonzero isospin composite objects, a similar analysis in this case for the isospin one current shows that an additional term in the current is needed when we want to eliminate the kernel by letting it act to the left. This corresponds to the current contribution from the meson-photon vertex. Such an analysis also shows that the various one particle em form factors are related to each other by gauge invariance.

A fully relativistic analysis of the em deuteron current can be carried out [43]. Covariance implies that it can be written as

$$\langle P'M'|J_{\mu}^{D}|PM\rangle = -\frac{e}{2M_D}\,e_{\rho}^{*}(\vec{P}',M')J_{\mu}^{\rho\sigma}e_{\sigma}(\vec{P},M) \tag{8.14}$$

where the spin one polarizat'on vectors are defined as

$$\sum_{M} e_{\mu}^{*}(\vec{P},M)e_{\nu}(\vec{P},M) = -\,g_{\mu\nu} + \frac{P_{\mu}P_{\nu}}{M_D^2} \tag{8.15}$$

with

$$\begin{aligned} &e_{\mu}^{*}(\vec{P},M)e^{\mu}(\vec{P},M') = -\,\delta_{MM'} \\ &P_{\mu}e^{\mu}(\vec{P},M) = 0 \end{aligned} \tag{8.16}$$

Using time reversal invariance and Lorentz covariance we get

$$J_\mu^{\rho\sigma} = (P'_\mu + P_\mu)\left[g^{\rho\sigma}F_1(q^2) - \frac{q^\rho q^\sigma}{2M_D^2}F_2(q^2)\right] + I_{\mu\nu}^{\rho\sigma}q^\nu G(q^2) \tag{8.17}$$

where $I_{\mu\nu}^{\rho\sigma}$ are the infinitesimal generators of the Lorentz group. Instead of the three invariant form factors in Eq. (8.17) one usually introduces the charge, quadrupole and magnetic form factors

$$\begin{aligned} F_c &= F_1 + \tfrac{2}{3}\eta\,(F_1+(1+\eta)F_2+G_1) \\ F_Q &= F_1 + (1+\eta)F_2 + G_1 \\ F_M &= G_1 \end{aligned} \tag{8.18}$$

with $\eta = -q^2/4M_D^2$. By calculating the em current matrixelements in the Breit system, defined as $\vec{P} + \vec{P}' = 0$, we may immediately relate the elements to the form factors [44]. Taking q along the z-axis we get

$$\begin{aligned} \langle M'|J_0^D|M\rangle &= e(1+\eta)^{\frac{1}{2}}\left[F_1\delta_{MM'} + 2\eta\{F_1+(1+\eta)F_2+G_1\}\delta_{mo}\delta_{M'o}\right] \\ \langle M'|J_1^D|M\rangle &= e\,\frac{q}{2M_D}\left(\frac{1+\eta}{2}\right)^{\frac{1}{2}} G_1[\delta_{M',M+1} - \delta_{M',M-1}] \\ \langle M'|J_2^D|M\rangle &= e\,\frac{q}{2M_D}\left(\frac{1+\eta}{2}\right)^{\frac{1}{2}} G_1[\delta_{M',M+1} + \delta_{M',M-1}] \\ \langle M'|J_3^D|M\rangle &= 0 \end{aligned} \tag{8.19}$$

To calculate the em deuteron current in the Breit system we need to rewrite it in terms of center of mass quantities. This is done by applying the boost transformations $\mathcal{L}$ and $\mathcal{L}'$ to the initial and final states respectively. As a result we may write

$$\begin{aligned} \langle P'M'|J_\mu^D|PM\rangle &= \frac{ie}{8\pi^3 M_D}\int d_4k \\ &\bar{\psi}^{(M')}(k';P_{cm})S_1^{(1)}\left(\frac{P_{cm}}{2}+k'\right)\tilde{\Gamma}_\mu^{(1)}(q)S(k,P_{cm})\psi^{(M)}(k;P_{cm}) \end{aligned} \tag{8.20}$$

where

$$\tilde{\Gamma}_\mu^{(1)}(q) = \Lambda^{-1}(\mathcal{L}')\Gamma_\mu^{(1)}(q)\Lambda(\mathcal{L}) \tag{8.21}$$

with Λ the boost operator in spin space and $k = \mathcal{L}^{-1}p$ and $k' = \mathcal{L}^{-1}p'$ with p and p' the relative momenta of the initial and final states in the Breit system. Since the wavefunction is only known in Wick rotated form we apply a Wick rotation on Eq.

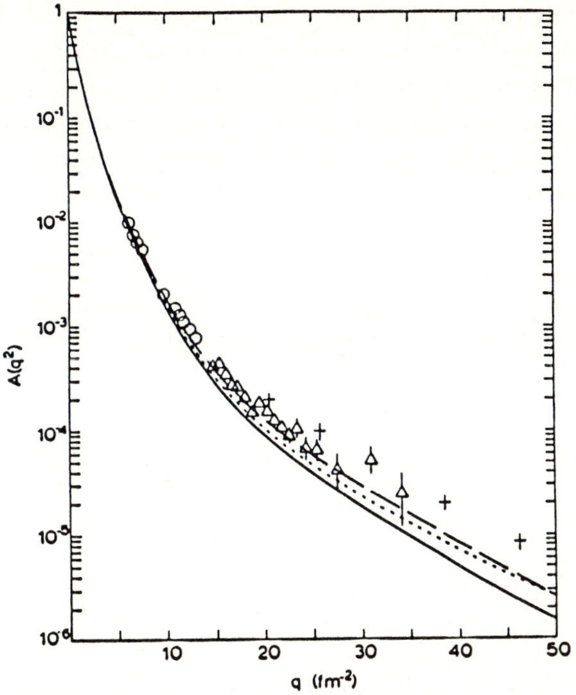

Figure 9. Calculated electric form factor $A(q^2)$ using the BS OBE model. For comparison is shown the static approximation (...) and the nonrelativistic RSC result (---). The data labeled 0, Δ and + are from Ref. [95], [96] and [97] respectively.

(8.21). As a result the relative energy variable argument in the wavefunction becomes in general complex due to the presence of the boost transformation. To simplify the calculation we have accounted for this by making a Taylor expansion around the imaginary k_o-axis for ψ and keeping only the terms up to second order. Since the second order contributions are small we may hope that the procedure converges fast. Apart from the above approximation all other effects have been exactly accounted for. In Fig. 9 is shown the calculated result for the electric form factor $A(q^2)$ defined as

$$A(q^2) = F_c^2 + \frac{8}{9}\eta^2 F_Q^2 + \frac{2}{3}\eta\ F_M^2 \qquad (8.22)$$

Also is shown the result for the nonrelativistic Reid soft core (RSC) potential.

In general it is not easy to make direct comparison with a nonrelativistic calculation because of the occurrence of the relative energy variable. An approximation which can be considered close to the nonrelativistic case is the so-called static approximation. It consists of neglecting all the boost effects in the arguments and the spin space to order q^2 and dropping the negative energy state contributions. As is seen in Fig. 9 the result is close to the RSC calculation. The various relativistic corrections relative to the static approximation can be determined. They are shown in Fig. 10. At small momentum transfer there are significant contributions from the boost on the one-nucleon propagator and the arguments of the deuteron vertex function, while at larger momentum transfer the one-nucleon propagator correction is the most dominant one. This leads to an interesting structure in the relativistic correction as a function of the momentum transfer. A similar structure is also found by Friar [46] and Gross [47] using different approaches.

9. Consistent treatment of dynamics and em interaction

A perturbational approach is usually adopted in the study of mesonic exchange current (MEC) and relativistic effects. Implicitly it is thereby assumed that the dynamics of the composite system and the em interaction can be treated independently. As a consequence one can make use of the boundstate wavefunctions obtained in a non-relativistic manner.

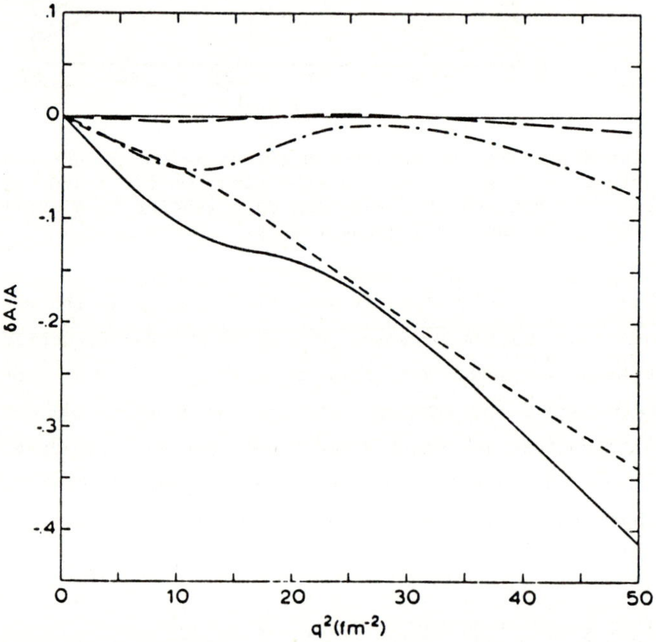

Figure 10. The various contributions from special relativity to $A(q^2)$ relative to the static approximation. Total result is given by the solid line. It consists of boost effects in the arguments of the deuteron vertex function (dash-dotted line) and of the single nucleon propagator S_1' (dashed line) and the negative spinor state contributions (long dashed line).

The relativistic OBE model may serve as an excellent testing ground for these ideas. The results obtained for the em form factors of the deuteron in the BS calculations indicate that they are not compatible with those obtained using the perturbational estimates [48]. To study this in detail one may rely on a relativistic potential analysis where the spectator nucleon in the photon-deuteron vertex is put on the mass shell. Using this Gross prescription we may recalculate the form factors, which are found to be close to the BS results. This QP model can now be analyzed perturbationally [49]. Since the deuteron vertex function satisfies the homogeneous QP equation and the dominant contribution to the em current comes from the one pion exchange contribution we may calculate the current matrix element by the once iterated one pion graph as shown in Fig. 11. This can be rewritten as a sum of three terms, two of which are the nonrelativistic impulse approximation (11c) and the so-called pair term (11d). In the perturbative approach it is precisely these two contributions which are usually considered [48]. The third contribution (11e) contains essentially the deformation of the deuteron due to relativistic effects in the intermediate states. Hence it entails the genuine relativistic dynamics of the composite system. Explicit calculation of these effects shows that the pair term and dynamical corrections are of the same order and that they tend to cancel each other at intermediate momentum transfers. This is shown in Fig. 12.

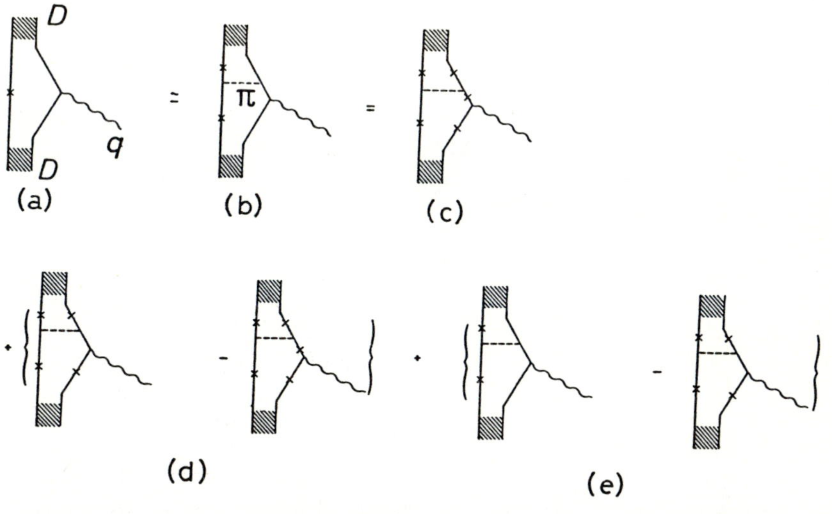

Figure 11. Perturbative analysis of the various corrections to the em charge form factors. (11c) is the nonrelativistic approximation, while (11d) and (11e) are the pair term and dynamical corrections.

In considering the dynamical correction in more detail, we find that the dominant contribution comes from the off shell behaviour of the hit nucleon. The effect of the presence of the negative energy spinor states is small. This is due to the use of pseudo vector coupling. Within a pseudo vector theory one can understand qualitatively why this cancellation takes place. For this purpose consider the one pion exchange graph as given in Fig. 13 and calculate its matrix element between positive energy spinors. Assuming that particle 2 is on mass shell and using the Dirac equation, it can be written as

$$0 = \frac{g^2}{4\pi}\left[\gamma_5^{(1)}\gamma_5^{(2)} - \frac{p_{10}-E(p)}{2m_N}\gamma_o^{(1)}\gamma_5^{(1)}\gamma_5^{(2)} + \frac{q_{10}-E(q)}{2m_N}\gamma_o^{(1)}\gamma_5^{(1)}\gamma_5^{(2)}\right][k_\pi^2-m_\pi^2]^{-1} \tag{9.1}$$

For on shell nucleons, $p_{10} = E(p) = q_{10} = E(q)$ only the first term survives, demonstrating the well known equivalence theorem of pseudo vector and pseudo scalar theory. When we use the Gross prescription we have $p_{10} = \sqrt{s} - E(p)$ and $q_{10} = \sqrt{s} - E(q)$. As a result we see that the second and third term tend to cancel. In the usual pair term analysis where this graph enters with the nucleon being off shell solely between the points where the photon is absorbed and the pion emitted only one of these terms contribute. In the process of neglecting certain contributions gauge invariance can be violated. From the above considerations we see that one should be careful in treating MEC and relativistic effects. Dynamics and em properties have to be determined in a consistent way within the same model, so that gauge invariance is not violated.

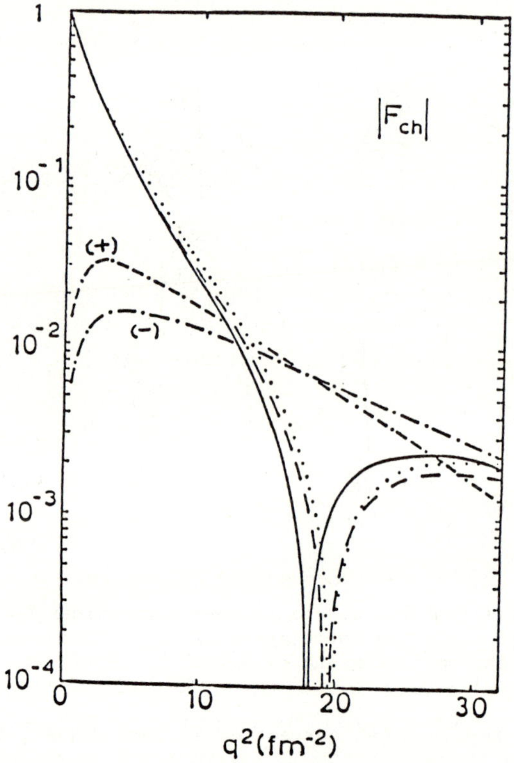

Figure 12. The pair term (-.-) and dynamical (--.--) corrections to the non-relativistic impulse approximation (....) of the charge form factor of the deuteron in the QP model. Total result is given by the solid curve.

In a recent experiment at Saclay [50] the magnetic form factor $B(q^2)$ of the deuteron has been measured. It is related to the invariant form factors of Eq. (8.18) by

$$B(q^2) = \frac{4}{3}\eta\,(1+\eta)F_M^2 \tag{9.2}$$

Taking the conventional pair term estimate together with the MEC contribution from the $\pi\rho\gamma$ graph [51] into account a remarkable agreement is found with the experiment.

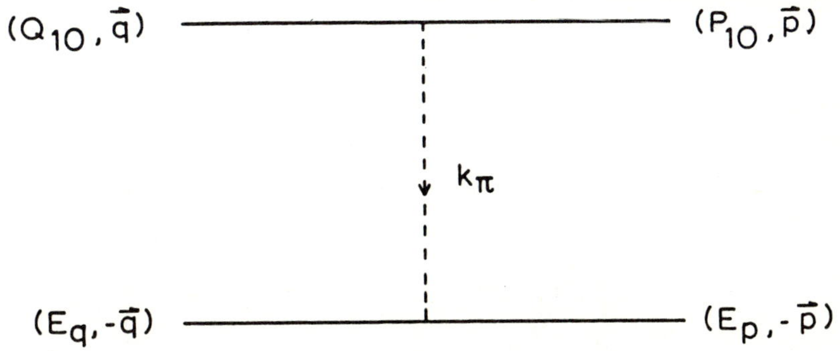

Figure 13. The one pion exchange diagram with particle 2 put on the mass shell.

However the proper treatment of the relativistic effects yields a result which devi ations significantly from the experiment as can be seen in Fig. 14 even after including the $\pi\rho\gamma$ contribution. Two relativistic calculations have been carried out [44,52]. The difference between the two are predominantly due to the chosen model for the nuclear force. A possible cause of the discrepancy might be the presence of $\Delta\Delta$ state in the deuteron and the relativistic terms in the $\pi\rho\gamma$ MEC contribution, which have been neglected.

10. Inclusion of isobar degrees of freedom

We return to the nucleon-nucleon scattering problem. Much interest has been paid in recent years to describing the nucleon-nucleon interaction at intermediate energies. In particular the experimental discoveries [53] with spin polarized protons of resonant-like structures at around 600 MeV lab kinetic energy stimulated much theoretical activity. Apart from suggestions that these structures may be interpreted as possible exotic dibaryons of the underlying quark structure, various detailed dynamical calculations have been carried out based on a meson theoretical framework.

With increasing energy pion production becomes possible. At intermediate energies up to 1 GeV the dominant inelastic process is the production of the P_{33} πN

resonance. Considering the energy at which this isobar channel opens it is in the neighbourhood of where the resonantlike structures are found. Therefore it is tempting to try to explain these structures as a threshold effect in coupled channels. Since the inelastic process is dominated by one-pion production many theoretical attempts have centered on using manifestly three-body unitary formulations which are of a Faddeev type [54-58]. Since the nucleon which is emitting the pion plays a special role in these three-body approaches, there is a drawback that the Pauli principle for fermions cannot be satisfied in a simple way. Also the two-nucleon interaction obtained from a meson theory has to be modified in order to satisfy three-particle unitarity. Another approach is to extend the nucleon-nucleon models based on two particle unitarity to also include isobar degrees of freedom [59-62]. It does not have the above shortcomings, but neglects the nonresonant pion production effects.

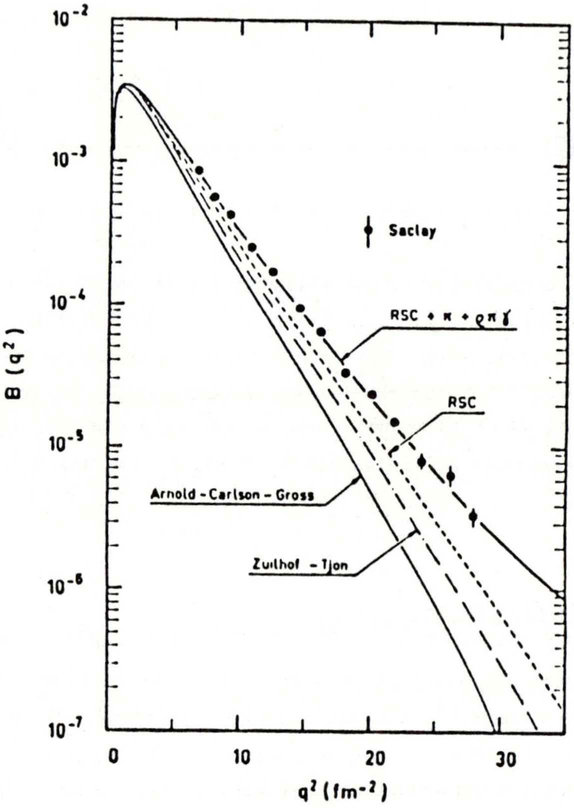

Figure 14. The magnetic form factor of the deuteron for the RSC interaction and the full result including the contributions from the πργ and pair term diagrams. Also is shown the QP relativistic results from Refs. [48,52], where the πργ contribution is not included. See for details Ref. [50].

At intermediate energies the kinetic energies of the mesons and nucleons become comparable in size to the rest masses of the particles. As a result relativistic formulations of the dynamical equations become necessary. Therefore the study of the relativistic OBE model including the isobar states as intermediate states is an interesting one. In Fig. 15 is shown the diagrammatic representation of the isobar model we studied [61]. The additional intermediate states included are the NΔ and ΔΔ states. For the transition interaction between NN and NΔ states we use π and ρ meson exchange diagrams. We have neglected the direct ΔΔ coupling of the mesons, because virtually nothing is known about it. As a result no direct interaction is present between the ΔΔ states.

Because the effect of the negative NN energy states can mostly be corrected for by modifying the meson coupling constants, in most calculations these states were neglected. Neglecting also the negative energy Δ states, the Δ propagator is taken to be

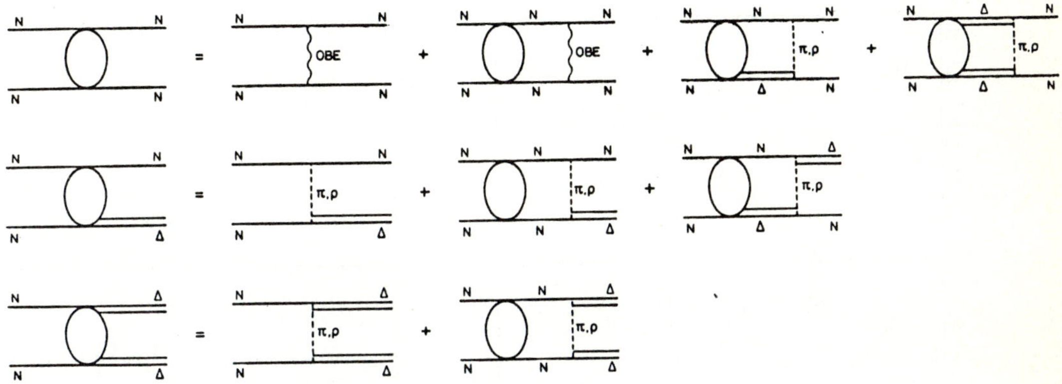

Figure 15. Diagrammatic representation of the relativistic OBE model with isobars.

$$P^{\mu\nu}(\vec{p},p_o) = [p_o-(\vec{p}^2+m_\Delta^2)^{\frac{1}{2}}]^{-1} \sum_\sigma \Delta^\mu(\vec{p},\sigma)\Delta^\nu(\vec{p},\sigma) \tag{10.1}$$

where Δ^μ are the positive energy Rarita-Schwinger spinors [63] for spin 3/2 particles and the spin indices $\sigma = \pm 1/2, \pm 3/2$. The Δ mass is complex and chosen to be

$$m_\Delta = m_o - i\Gamma(q)/2 \tag{10.2}$$

with m_o = 1236 MeV and the width is parameterized using the Bransden-Moorhouse form [64], which gives a good description of the P_{33} πN phase shift up to 1.3 GeV πN center of mass energy. The πN three momentum is related to the invariant πN mass square according to

$$q^2 = \frac{[s_{\pi N}-(m_\pi-m_N)^2][s_{\pi N}-(m_\pi+m_N)^2]}{s_{\pi N}} \tag{10.3}$$

Various options have been used to express $s_{\pi N}$ in terms of the total four momentum of the two-nucleon system. One choice is the fixed mass approximation [65] where it is assumed that the Δ receives the maximally allowed energy if it is embedded in the NΔ system i.e.

$$s_{\pi N} = (\sqrt{s}-m_N)^2 \tag{10.4}$$

Using the above model we may compute the resonating channels $^1D_2$ and $^3F_3$ [66]. For the case that the ρ exchange is neglected the calculated phase shifts δ and inelastic parameter η are shown in Fig. 16. For the uncoupled L=J channels ρ is related to the inelasticity η through

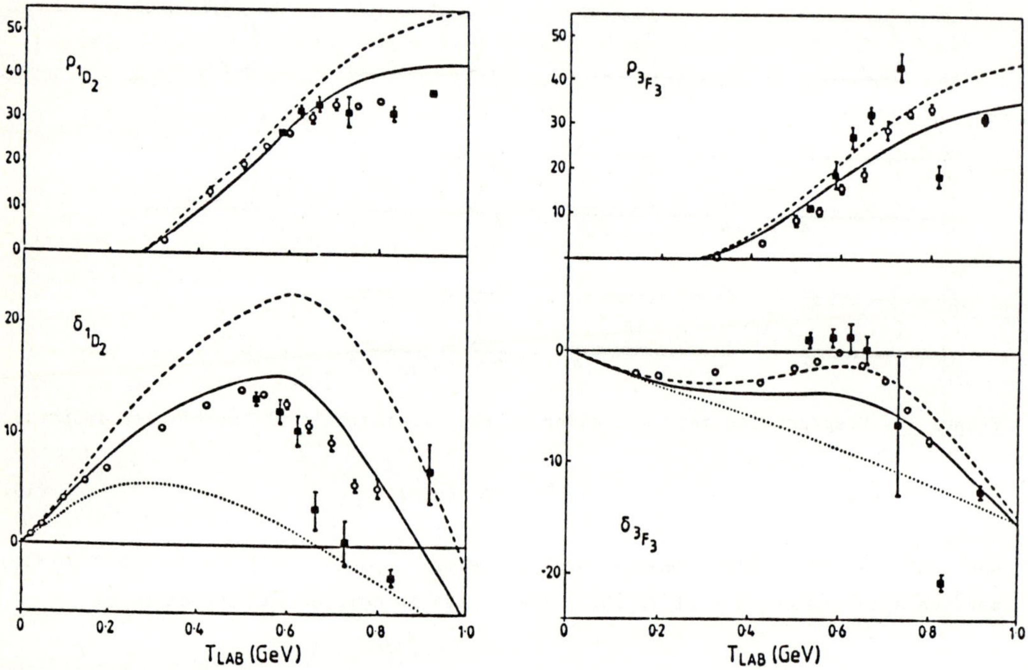

Figure 16. The calculated $^1D_2$ and $^3F_3$ phase parameters using the fixed mass prescription (10.4). The ΔΔ channels are neglected, while only the pion exchange is included in the transition interaction. The dotted, solid and dashed curves are with $f^2_{N\Delta\pi}/4\pi$ = 0, 0.23 and 0. 35 respectively. The data labeled 0 and ■ are from Ref. [94] and [98] respectively.

$$\eta = \cos 2\rho \tag{10.5}$$

A more complicated relation exists for the coupled channel case. From this figure we see that we indeed can in principle accommodate for the experimentally observed resonantlike structures. Much work has been devoted to the question whether these structures correspond to dynamical singularities in the S-matrix [67-72]. Although looping behaviour of the S-matrix in the Argand plot has been used as a criterium for a dynamical singularity, this may not hold in general. As an example one can show that a simple diagram like the NΔ box exhibits looping behaviour, whereas it does not have any dynamical singularities in the second Riemann sheet of the total energy variable. Due to the presence of various non-dynamical singularities from the coupled channels the situation can be quite complex. For a discussion in a soluble separable model, which elucidates on the physics of the problem see Ref. [71].

A more realistic choice than Eq. (10.4) is when we also include the recoil motion of the nucleon i.e.

$$s_{\pi N} = [\sqrt{s} - (\vec{q}^{\,2}+m_N^2)^{\frac{1}{2}}]^2 \qquad (10.6)$$

In Fig. 17 is shown the effect of the "smeared out" Δ width for the $^1D_2$ channel. We see that the inelasticity near the one-pion production threshold decreases considerably as compared to the fixed mass approximation. Similar results are found for the

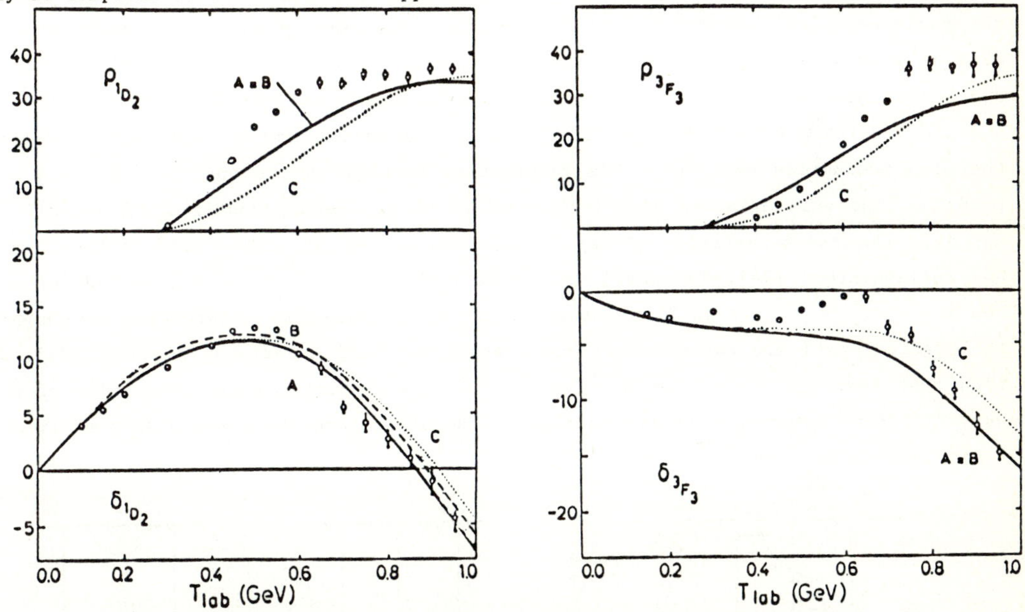

Figure 17. The calculated $^1D_2$ and $^3F_3$ phase parameters in the case that the ρ exchange is also included in the transition interaction. Curves A and B are for NN-NΔ and NN-NΔ-ΔΔ scattering respectively. Curve C is for NN-NΔ scattering using the smeared out Δ width prescription (5.1). The data are from Ref. [94].

high partial waves. In experimental phase shift analysis at lower energies it is customary to use as additional information that the peripheral high partial waves are well represented by the one pion exchange diagram. A similar constraint in the case of intermediate energies would be interesting. Unfortunately there is quite a model dependence present if we consider various dynamical models [73]. Much of the differences can be traced back to the above question how to treat the q dependence of the Δ width, although there is still a significant model dependence left over.

Considering the p-wave NN channels with the use of only pion exchange in the transition interaction we find that the state dependence in these channels is not well reproduced. To get a better overall description of the phase parameters the introduction of the ρ exchange in the transition interaction is needed. In Fig. 17 are shown the calculated $^1D_2$ and $^3F_3$ phase parameters when the ρ-exchange is included in the transition interaction. The coupling constant in the πNΔ vertex is taken $f^2_{\pi N\Delta}/4\pi = 0.35$. Comparing the results with those of Fig. 16, we see that the structure in the phase shift of the $^1D_2$ wave is less pronounced while there is less inelasticity present. Also the effect of the ΔΔ channel is shown in Fig. 17. In the I = 1 state it is in general small.

The above relativistic isobar model in which many of the essential features are built in on the basis of meson theory is expected to give a reasonable description of the nucleon-nucleon interaction at intermediate energies. Moreover it has also predicting power for the pion production amplitude, if we assume that the pion production is dominated by the production of the Δ isobar and its decay into a nucleon and pion. Using the NN → NΔ amplitude from this model one may in that case calculate the pion production amplitude. One interesting problem is the question of model dependence. The only existing dynamical model which has been investigated in detail is based on the three-particle unitary formulation of the NN interaction by Dubach and his collaborators [74]. The model is not as realistic in view of the neglect of the heavy mesons in the transition interaction. It is therefore interesting to compare the prediction of the two models. Such a study has been undertaken recently [75]. We find that there is a remarkable agreement in some kinematic regions in spite of the very different nature of the two models. On the other hand there are also some kine-

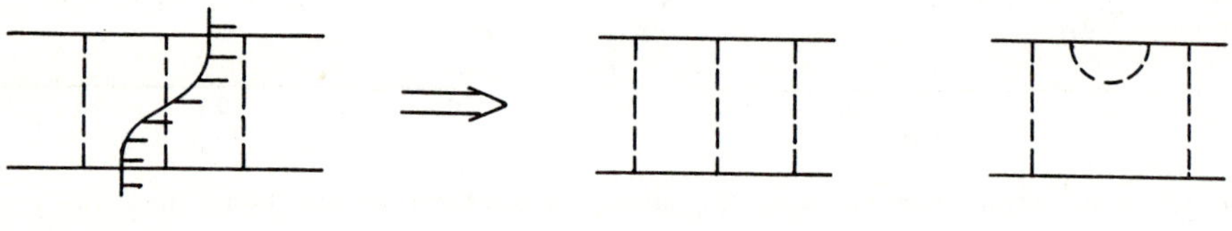

Figure 18. Example of a higher order diagram which has a discontinuity in the inelastic region due to the production of a pion. The cut diagram can be reconnected in two ways to yield the diagrams (b) and (c). To get a three-particle unitary theory we have to include also diagram (c) in the dynamical equation.

matic regions where the differences are as large as found with experiments. Consequently, the pion production processes may be good testing ground for models of the nucleon-nucleon interaction.

11. Unitary extension

In section 2 we have found that the ladder BS equation satisfies two-particle unitarity in the elastic region. Above the one-particle production threshold from the specific example of the diagram shown in Fig. 18, we see that the two-particle unitarity cannot be valid because it is possible to cut a higher order graph from the ladder series such that we have as intermediate state three particles on shell. Actual numerical calculation shows that the unitarity condition

$$\sigma_{elastic} \leqslant \sigma_{total} \tag{11.1}$$

i.e. $\eta_\ell \leqslant 1$ is even violated. For definiteness let us consider the scalar model with equal masses. In Fig. 19 is shown $\mathrm{Im}\delta_\ell$ for the s-wave case with m=1. For large enough coupling constant we can have $\eta_\ell > 1$. The critical coupling constant for having in this partial wave the first boundstate at threshold is given by $\lambda \simeq 0.8$ (see Fig. 5). Similar results for the inelasticity parameters hold in the relativistic OBE without isobars, as can be seen in Fig. 20. In this case η_ℓ becomes significantly greater than unity even for the $l > 0$ waves. Including the isobar degrees of freedom appears to mask strongly the violation of the unitarity condition. As a result we find that in the complete model with isobars that Eq. (11.1) in general holds.

A three-particle equation can be found by studying which class of graphs has to be added to the ladder series in order to restore unitarity. From the examples given in Fig. 18 we see with the aid of the cutting rules that there are two ways to reconnect the cut graph. One gives the ladder graph back while the other leads to a new graph where the bubble self energy appears in one of the internal lines. This suggests that three-particle unitarity should hold if we renormalize the internal lines by the set of bubble diagrams. It is simply achieved in the scalar model by replacing the single particle propagators in the free two-particle Green's function by

$$\bar{\Delta}_F(p) = [(p^2-m^2+i\varepsilon)\{1+(p^2-m^2)\Sigma(p^2)\}]^{-1} \tag{11.2}$$

where $\Sigma(p^2)$ represents the lowest order self energy contribution. It is given by

$$\Sigma(p^2) = \lambda \int_{4m^2}^{\infty} ds' \frac{\{[s'-4m^2]/s'\}^{1/2}}{(s'-m^2)^2(s'-p^2-i\varepsilon)} \tag{11.3}$$

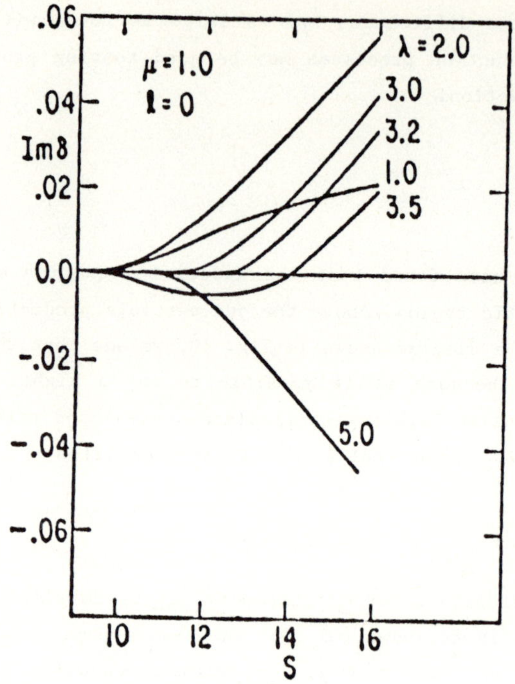

Figure 19. Imaginary part of the s-wave phase shift δ_ℓ for the scalar ladder model. for sufficiently large λ the unitarity condition Im $\delta_\ell > 0$ is violated.

The above conjecture about unitarity can be verified explicitly [76]. In a similar way a three-particle unitary theory can be constructed for the relativistic OBE model by renormalizing the nucleon propagators [77].

To describe pion production in a reliable way it is necessary to have a more detailed model for the Δ propagator than a phenomenological energy dependent width as we have used up to now. A possible dynamical model is to describe the Δ as a πN scattering process where the bare Δ is dressed by the πN interaction. A two-particle unitary amplitude can be obtained by summing the series of diagrams displayed in Fig. 21. For the πNΔ vertex function we assume that it is given by

$$F_{\pi N\Delta} = F_{OBE}(k^2)g_{sc}((p-2k)^2) \tag{11.4}$$

where F_{OBE} is the OBE vertex function and g_{sc} represents a scattering function, depending on the on shell relative πN momentum. It is assumed to have the form

$$g_{sc}(p^2) = [\frac{\Lambda_{sc}^2}{\Lambda_{sc}^2-p^2}]^{\frac{1}{4}} \tag{11.5}$$

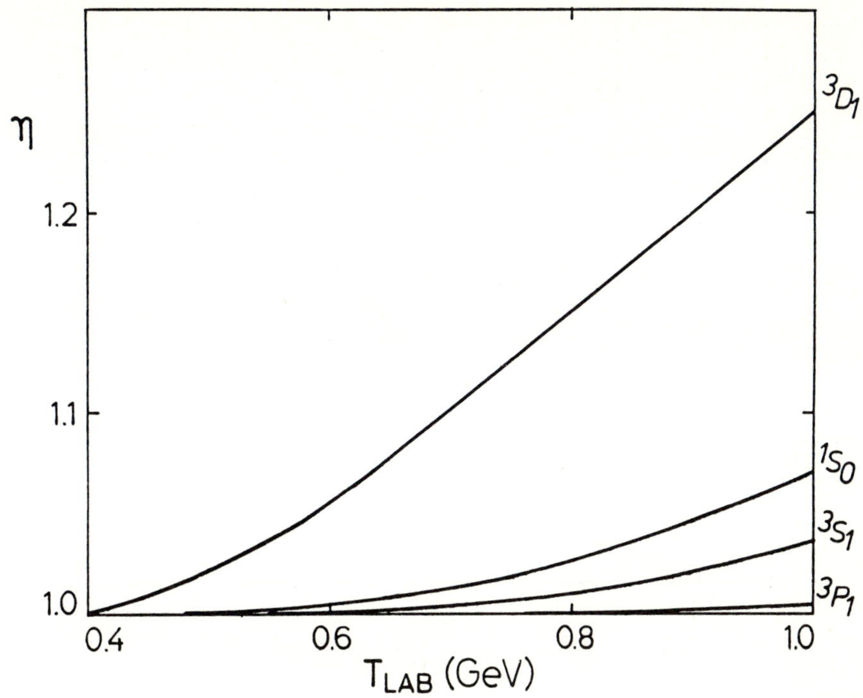

Figure 20. The inelastic parameter η_ℓ for the BS OBE model without isobars.

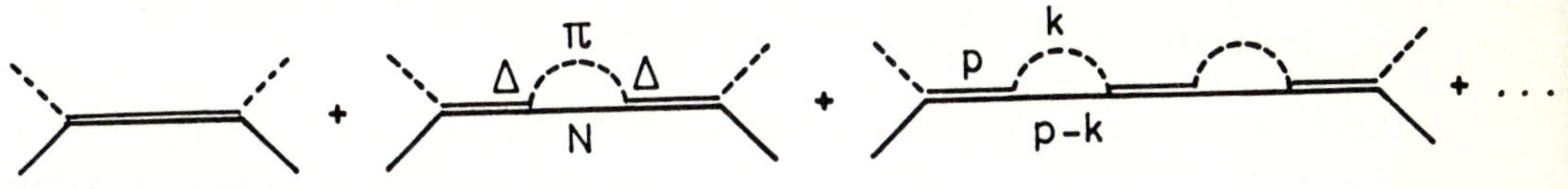

Figure 21. The scattering model for the P_{33} πN channel.

The P_{33} πN phase shift can be fitted accurately (see Fig. 22) by having a rather low cutoff mass of $\Lambda_{sc}^2 = 0.85\, m_N^2$ in the scattering function g_{sc}. This is needed to fit the effective range of the πN system and is consistent with a previous separable potential model study [73].

The unitary extension of the relativistic OBE model together with the two-particle model for the Δ has been examined recently. In Fig. 23 is exhibited the results for the $^1D_2$ and $^3F_3$ phase shifts. Similarly as in the smeared out Δ width calculation we find at lower energies that the inelasticity is considerably lower than experimentally found. The lack of inelasticity may be due to the neglect of the coupling to the pion deuteron channel [77]. Indeed if we correct for it by introducing an effective inelasticity parameter ρ_e, defined for a given wave as

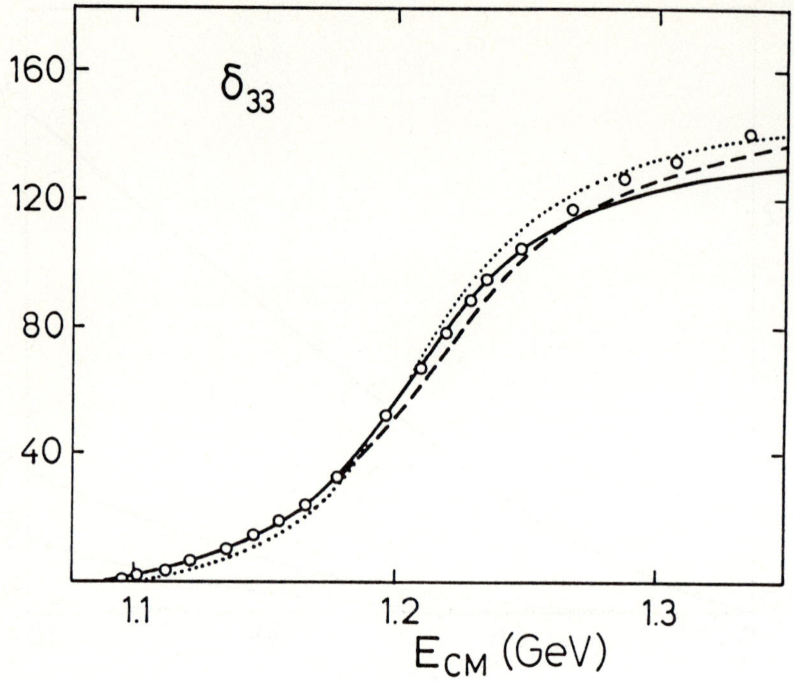

Figure 22. Calculated phase shifts of the P_{33} πN model shown in Fig. 21. For comparison are also shown the results from Refs. [54,64]. Experimental points are from Ref. [99].

$$\sigma_{inelastic} = (2J+1)\,\frac{\pi}{2q^2}\,(1-\cos^4\rho_e) + \sigma_{\pi D} \tag{11.6}$$

we find using the experimentally determined inelastic cross sections $\sigma_{\pi D}$ of NN to πD the results for the inelasticity in the $^1D_2$ channel improves considerably. It is less dramatic in the $^3F_3$ wave, where in addition the distinct energy dependent structure at around 600 MeV is not well reproduced. This is shown in Fig. 24.

Dynamical calculations have recently been done in the quasi potential approach where the πD is introduced as an additional channel reconfirming the above results [101]. From this we may conclude that to describe the NN dynamics at intermediate energies the degrees of freedom have to be included together for at least certain channels also the effect of the πD channel has to be considered. The gross features of NN scattering such as the presence of resonant-like structures are reproduced, although there seems to be a slight disbalance in the various waves. It should be emphasized however that no χ^2 fits on the coupling parameters have been carried out in these calculations.

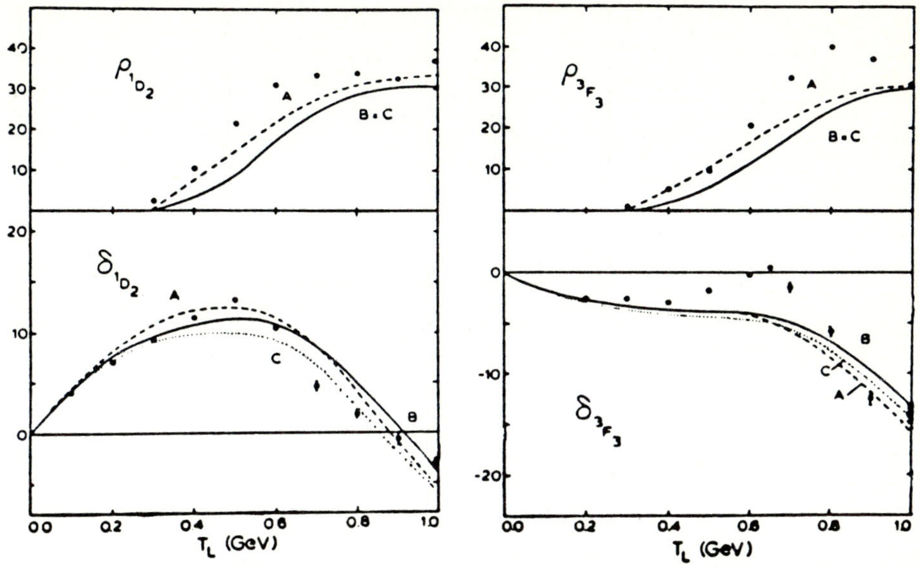

Figure 23. The calculated $^1D_2$ and $^3F_3$ phase parameters in the unitary model for two sets of coupling constants (curves B and C). See Ref. [77] for the choice of the coupling parameters. For comparison is shown the result with the fixed mass prescription (curve A). The data are from Ref. [94].

12. Three nucleon calculations

As compared to the two-body system virtually nothing has been done in the study of relativistic equations for three particles, except for including special relativity in a minimal way by essentially including the proper relativistic kinematics in the equations. In attempting to build a relativistic dynamical equation for three particles we may start from the quantum field theoretical formulation with Feynman graphs. The three-body scattering T-matrix satisfies a Bethe-Salpeter like equation neglecting the three-particle irreducible graphs, it can be written as

$$T = \sum_{k=1}^{3} V_k + \sum_{k=1}^{3} V_k \, G_k^{(o)} T \tag{12.1}$$

where V_k are all the irreducible graphs in which particle l and m are interacting, while particle k remains free (with $k \neq l \neq m$). Furthermore, G_k is the free Green's function of particles l and m which is given by

$$G_k^{(o)} = - i[k_\ell^2 - m^2 + i\varepsilon]^{-1} \; [k_m^2 - m^2 + i\varepsilon]^{-1} \tag{12.2}$$

Similarly as in the nonrelativistic situation we may recast Eq. (12.1) into a Faddeev like equation.

It has the standard form

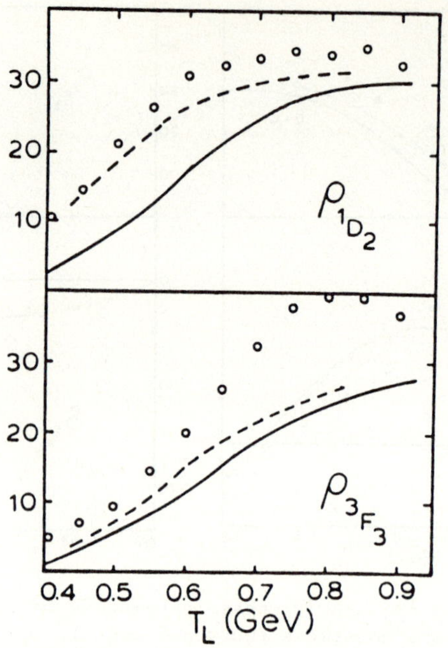

Figure 24. Calculated results for the effective inelastic parameter ρ_e for the $^1D_2$ and $^3F_3$ waves, using Eq. (11.6). The experimental data are taken from Ref. [94] and the πD amplitudes from Ref. [100].

$$T = \sum_{n=1}^{3} T^{(n)} \tag{12.3}$$

with $T^{(n)}$ satisfying

$$\begin{pmatrix} T^{(1)} \\ T^{(2)} \\ T^{(3)} \end{pmatrix} = \begin{pmatrix} T_1 \\ T_2 \\ T_3 \end{pmatrix} + \begin{pmatrix} 0 & T_1G_1^{(o)} & T_1G_1^{(o)} \\ T_2G_2^{(o)} & 0 & T_2G_2^{(o)} \\ T_3G_3^{(o)} & T_3G_3^{(o)} & 0 \end{pmatrix} \begin{pmatrix} T^{(1)} \\ T^{(2)} \\ T^{(3)} \end{pmatrix} \tag{12.4}$$

and T_n are the two-body T-matrices, which satisfies the two-body BS equations

$$T_n = V_n + V_n G_n^{(o)} T_n \tag{12.5}$$

As compared to the nonrelativistic case the Bethe-Salpeter-Faddeev (BSF) equations (12.1) contain as additional integration variables the relative energies. Looking at the kernel only one relative energy variable appears in the equation which is given by the one of the interacting pair.

To simplify the BSF equations we may attempt to generalize the quasi potential approximations [78,79]. For the case of the Blankenbecler-Sugar prescription the choice of the dispersion relation depends on which part of the kernel in the BSF equation is considered. The free Green's function is replaced by

$$G^{o}_{k,QP} = 2\pi \int_{4m^2}^{\infty} \frac{ds'}{s'-s}\, \delta^{(+)}[\{\tfrac{1}{2}(P'-k_k)^2+k^2_{\ell m}\}^2-m^2]\delta^{(+)}[\{\tfrac{1}{2}(P'-k_k)^2-k^2_{\ell m}\}^2-m^2] \qquad (12.6)$$

where $s = P^2$, $P' = (\frac{s'}{s})^{\frac{1}{2}} P$ and $k_{\ell m} = \frac{1}{2}(k_\ell - k_m)$ is the relative four momentum between particles ℓ and m. The prescription (12.4) has the virtue that the nonrelativistic property is maintained where the two-body T-matrix in the three-body Hilbert space can simply be obtained by a shift in the off-shell energy variable. However it has the basic problem that the separability condition is not satisfied [80]. This means that the S-matric does not factorize properly if one of the particles is taken infinitely far away from the other two particles in coordinate space. In practical cases the violation may however be small. The generalisation of the Gross prescription does not suffer from this problem, but strictly speaking leads to the presence of unphysical singularities in the driving force in a given approximation [79]. In practice these singularities may be far away and they can be removed ad hoc by making a principal value approximation to the potential.

The prescription (12.4) has been applied to the study of relativistic effects on the three nucleon ground state using the RSC interaction for the two-nucleon force [81]. The analysis has to be done consistently. Since relativistic dynamics affects also the two-body subsystem, the two body interaction has to be modified to describe two-nucleon scattering appropriately. In so doing the additional binding of the triton due to relativity was found to be small. It is of the order of 0.25 MeV and attractive. The smallness has been reconfirmed by various other studies [82-85], although the sign of the additional binding varies. In all these studies the relativistic spin structure is neglected.

Recently a calculation has been performed using the BSF equations for the cases that the two-nucleon interaction is described in terms of the Zachariasen model discussed in section 3 [86]. Because of the separable nature of the force the resulting equations become two-dimensional integral equations after partial wave decomposition. It can be shown that in the boundstate region the Wick rotation can be applied. The resulting equations which are singularity free can be solved using the standard ratio and Padé approximant methods [87].

As input to the calculations has been used s-wave two-nucleon interactions of the Yamaguchi and Tabakin like potentials in the $^1S_0$ and the $^3S_1$ channels. Similarly as in the nonrelativistic case a collapse of the groundstate takes place if we use in both channels Tabakin-like potentials. To have a comparison with the nonrelativistic situation we have made a quasi potential approximation to the BSF equations. By refitting the various coupling constants in the potentials almost identical phase shifts can be reproduced for the $^1S_0$ and $^3S_1$ waves as those of the BS solution. From a relativistic QP equation the phase equivalent nonrelativistic LS can immediately be obtained in the case of separable interactions by simply modifying the relativistic potential with a factor containing obvious kinematic factors. With this we are in a position to make comparisons between the various equations. The results are shown in Table 3 for the triton binding energy. From this we see that the old result obtained by various group is reconfirmed. The additional binding due to special relativity is small and attractive. The BSF results are found to be more attractive than the QP calculations, although again quite small.

Table 3

The calculated binding energies of triton in MeV for various choices of the two-nucleon form factors, using the LS, relativistic QP and BS equations.

$^3S_1$	$^1S_0$	NR	QP	BS
YAM	YAM	10.65	10.86	11.09
YAM	TAB	7.96	8.04	8.32
TAB	YAM	8.06	8.22	8.44

With regard to the elastic charge form factors an original calculation by Hammel et. al. [83] suggests that relativistic QP equations can give additional binding energy and at the same time more correlation is built in the relativistic wavefunction, so that there would be an increase in the secondary maximum of the em form factors. We have studied this in the case of the RSC interaction and did however not find any evidence for this [88]. A QP calculation with the prescription (12.4) for the relativistic OBE model, keeping only the s-wave parts of the two-nucleon T-matrix, also does not show significantly different results as can be seen in Fig. 25 [89]. The binding energy is found to be 7.5 MeV, considerably more than for the RSC in the same approximation. This is mostly due to the lower D-state probability of the OBE model.

In summary, relativistic tri-nucleon calculations are on a rather rudimentary level. Work is in progress to use within the Zachariasen model multi term separable interactions which gives a more realistic representation of the nuclear force and to which do not exhibit the pathologies of the Tabakin like interactions. Much work needs to be done on the proper inclusion of the relativistic spin structure. Results in backward pD scattering however seem to indicate that they may be small [90]. On the other hand the Dirac approach for elastic proton-nucleus scattering based on meson theory suggests strongly that it is important to include the Dirac structure in its full complexity [91, 92]. The simplest nucleus for which one can study this problem in detail is the deuteron. In that case it may be possible to carry out a full relavistic calculation. As a result we may hope to gain more insight about relativistic

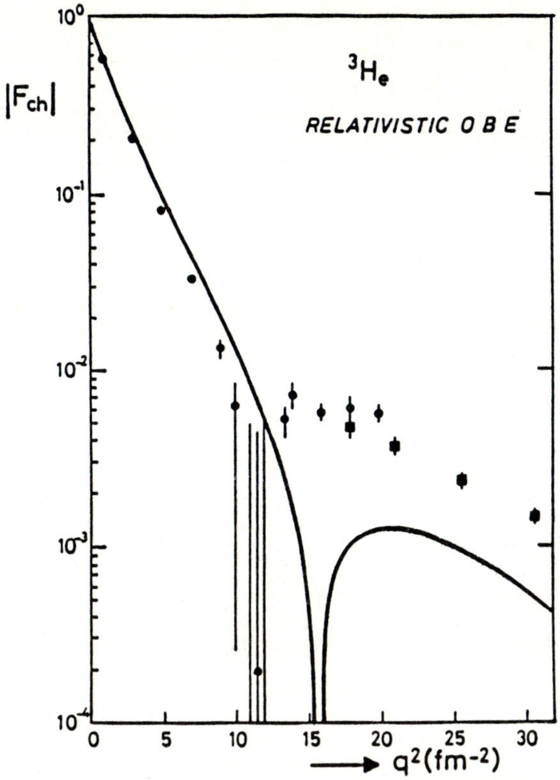

Figure 25. The calculated results of the ^3H_e charge form factor using the QP model with the relative OBE interaction without isobars. Experimental data are from Ref. [102].

effects in reaction theory and their possible role in more complex nuclei. Also it is not clear, in view of the arguments presented in the deuteron case, that the analysis of the em properties of the tri-nucleon system has been done consistently and in our opinion it should be reconsidered.

References

1. For a recent review see J.L. Friar, Proc. XII Int. Conf. on Few Body Problems, Tokyo-Sendai, August 1986.
2. For a review of relativistic Hamilton formulations see I. Todorov, SISSA report, Trieste (1980).
3. F. Coester and W. Polyzou, Phys. Rev. D26, 1348 (1982).
4. P.A.M. Dirac, Rev. Mod. Phys. 21, 392 (1949).
5. Y. Nambu, Prog. Theor. Phys. 5, 614 (1950).
6. E.E. Salpeter and H.A. Bethe, Phys. Rev. 84, 1232 (1951).
7. J.R. Taylor, Scattering Theory (Wiley, New York, 1972).
8. N. Nakanishi, Suppl. Prog. Theor. Phys., 43, 1 (1969).
9. R.E. Cutkosky, J. Math. Phys., 1, 429 (1960).
10. G.C. Wick, Phys. Rev. 96, 1124 (1954).
11. R.E. Cutkosky, Phys. Rev. 96, 1135 (1954).

12. F. Gross, Phys. Rev. C26, 2203 (1982).
13. F. Zachariasen, Phys. Rev. 121, 1851 (1961).
14. F. Tabakin, Phys. Rev. 174, 1208 (1968).
15. J.E. Beam, Phys. Lett. 30B, 76 (1969).
16. S. Sofianos, N.J. McGurk and H. Fiedeldey, Z. Phys. A286, 87 (1978).
17. G. Rupp, L. Streit and J.A. Tjon, Phys. Rev. C31, 2285 (1985).
18. C. Schwarz and C. Zemach, Phys. Rev. 141, 1454 (1966).
19. J. Schwinger, Harvard University (1947), unpublished; J.M. Blatt and J.D. Jackson, Phys. Rev. 76, 18 (1949).
20. J.G. Taylor, Nuov. Cim. Suppl., 1, 1002 (1963).
21. M.J. Levine, J. Wright, and J.A. Tjon, Phys. Rev. Lett. 16, 962 (1966); Phys. Rev. 154, 1433 (1967).
22. P.R. Graves-Morris, Phys. Rev. Lett. 16, 201 (1966).
23. A. Pagnamenta and J.G. Taylor, Phys. Rev. Lett. 17, 218 (1966).
24. W. Kemmer and A. Salam, Proc. Roy. Soc. (London) A230, 266 (1985).
25. K.L. Kowalski, Phys. Rev. Lett. 15, 798 (1965).
26. H.P. Noyes, Phys. Rev. Lett. 15, 538 (1965).
27. R. Blankenbecler and R. Sugar, Phys. Rev. 142, 1051 (1966).
28. A.A. Logunov and A.N. Tavkhelidze, Nuov. Cim. 29, 380 (1963).
29. M. Fortes and A.D. Jackson, Nucl. Phys. A175, 449 (1971).
30. F. Gross, Phys. Rev. 186, 1448 (1969).
31. J.L. Gammel and M.T. Menzel, Phys. Rev. A7, 858 (1973).
32. J.J. Kubis, Phys. Rev. D6, 547 (1972).
33. M.L. Goldberger et. al., Phys. Rev. 120, 2250 (1960).
34. J.L. Gammel, M.T. Menzel and W.R. Wortman, Phys. Rev. D3, 2175 (1971).
35. M. Jacob and G.C. Wick, Ann. Phys. (N.Y.) 7, 404 (1959).
36. A.R. Edmonds, Angular Momentum in Quantum Mechanics, Princeton University Press (1957).
37. J. Fleischer and J.A. Tjon, Nucl. Phys. B84, 397 (1986); Phys. Rev. D15, 2537 (1977).
38. M. Veltman, CERN report 1967; H. Strubbe, Comp. Phys. Comm. 8, 1 (1974).
39. A.C. Hearn, REDUCE user's manual, Rand Corporation Santa Monica (1984).
40. A. Gersten, R.H. Thomson and A.E.S. Green, Phys. Rev. D3, 2076 (1971).
41. J. Fleischer and J.A. Tjon, Phys. Rev. D21, 87 (1980).
42. J.D. Bjorken and S.D. Drell, Relativistic Field Theory (McGraw Hill, 1964).
43. M. Gourdin, Diffusion des electrons de haute energie, Masson & Cie. Paris (1966).
44. M.J. Zuilhof and J.A. Tjon, Phys. Rev. C22, 2369 (1980).
45. R.V. Reid, Ann. Phys., 50, 411 (1968).
46. J.L. Friar, Phys. Rev. C12, 695 (1975).
47. F. Gross, Phys. Rev. 142, 1025 (1966).
48. A.D. Jackson, A. Lande and D.O. Riska, Phys. Lett. 55B, 23 (1975).
49. J.A. Tjon and M.J. Zuilhof, Phys. Lett. 84B, 31 (1979).
50. S. Auffret et. al., Phys. Rev. Lett. 54, 649 (1985).
51. M. Gari and H. Hyuga, Nucl. Phys. A264, 409, (1976).
52. R.G. Arnold, C.E. Carlson and F. Gross, Phys. Rev. C21 1426 (1980).
53. G.A. Yokosawa, Phys. Rep. 64, 47 (1980).
54. W.M. Kloet and R.R. Silbar, Nucl. Phys. A338, 281 (1980); Phys. Rev. Lett. 45, 570 (1980)
55. M. Araki and T. Ueda, Nucl. Phys. A379, 449 (1982).
56. Y. Avishai and T. Mizutani, Phys. Rev. C27, 312 (1983).
57. A.W. Thomas and A.S. Rinat, Phys. Rev. C27, 312 (1979).
58. I.R. Afnan and R.J. McLead, Phys. Rev. C31, 1821 (1985).
59. A.M. Green and M.E. Sainio, J. Phys. G5, 503 (1979).
60. E.L. Lomon, Phys. Rev. D26, 576 (1972).
61. E. van Faassen and J.A. Tjon, Phys. Rev. C28, 2354 (1983);C30, 285 (1984).
62. T.-S.H. Lee, Phys. Rev. C29, 195 (1984).
63. W. Rarita and J. Schwinger, Phys. Rev. 59, 436 (1941).
64. B.H. Bransden and R.G. Moorhouse, The Pion-Nucleon System, Princeton University Press, Princeton (1973).
65. B.J. Verwest, Phys. Rev. C25, 482 (1982).
66. J.A. Tjon and E. van Faassen, Phys. Lett. 120B, 39 (1983).
67. B.J. Edwards and G.H. Thomas, Phys. Rev. D22, 2772 (1980).

68. B.J. Edwards, Phys. Rev. D23, 1978 (1981)
69. W.M. Kloet, J.A. Tjon and R.R. Silbar, Phys. Lett. 99B, 80 (1981).
70. I. Duck, Phys. Lett. 106B, 267 (1981).
71. W.M. Kloet and J.A. Tjon, Nucl. Phys. A392, 271 (1983).
72. T. Ueda, Phys. Lett. 119B, 281 (1982).
73. W.M. Kloet and J.A. Tjon, Phys. Rev. C28, 2354 (1983).
74. J. Dubach et. al., Phys. Lett. 106B, 29 (1981); preprint (1986).
75. J. Dubach et. al., Phys. Rev. C34, 955 (1986).
76. M. Levine, J. Wright and J.A. Tjon, Phys. Rev. 157, 1416 (1967).
77. E. van Faassen and J.A. Tjon, Phys. Rev. C33, 2105 (1986).
78. For some choices of quasi potential prescriptions A. Ahmadzadeh and J.A. Tjon, Phys. Rev. 147, 1111 (1966).
79. F. Gross, Phys. Rev. C26, 2226 (1982).
80. J.M. Namyslowski and H.J. Weber, Zeit. fur Phys. A295, 219 (1980).
81. A.D. Jackson and J.A. Tjon, Phys. Lett. 32B, 9 (1970).
82. V.S. Bhasin, H.J. Jacob and A.N. Mitra, Phys. Lett. 32B, 15 (1970).
83. E. Hammel, H. Baier and A.S. Rinat, Phys. Lett. 85B, 193 (1979).
84. L.A. Kondratyuk, J. Vogelzang and M.S. Fachenko, Phys. Lett. 98B, 405 (1981).
85. W.A. Glockle, T-S.H. Lee and F. Coester, Phys. Rev. C33, 709 (1986).
86. G. Rupp, L. Streit and J.A. Tjon, Proc. IX European Conf. on Few Body Problems in Physics, Tbilisi, 1984 (World Scientific, Singapore 1985); G. Rupp and J.A. Tjon, to be published.
87. J.A. Tjon in Padé Approximants and their Applications, ed. P.R. Graves-Morris, Academic Press, New York, p 241 (1973).
88. J.A. Tjon, Proc. VII International Conf. on Few Body Problems in Nuclear and Particle Physics, New Delhi 1975, p 567 (North-Holland, Amsterdam 1976).
89. J.A. Tjon, Proc. European Workshop, "Theoretical and Experimental Investigations of Hadronic Few-Body Physics", Rome, Oct. 7-11 (1986).
90. B.D. Keister and J.A. Tjon, Phys. Rev. C26, 578 (1982).
91. For a review see contributions to Proc. of LAMPF Workshop on Dirac Approaches to Nuclear Physics, ed. J.R. Shepard, Conf. Proc. LA-10438-C, Los Alamos (1985).
92. S.J. Wallace, Lecture Notes, Tokyo (1985); J.A. Tjon, New Vistas in Electronuclear Physics, 1985 NATO Advanced Study institute, Banff, Canada.
93. M.H. McGregor, R.A. Arndt and R.M. Wright, Phys. Rev. 182, 1714 (1969).
94. R.A. Arndt et al., Phys. Rev. D28, 97 (1983)
95. S. Galster et al., Nucl. Phys. B32, 221 (1971).
96. J.E. Elias et al., Phys. Rev. 177, 2075 (1969).
97. R.G. Arnold et al., Phys. Rev. Lett. 35, 776 (1975).
98. N. Woshizaki, Prog. Theor. Phys. 60, 1796 (1978).
99. R. Koch and E. Pietarinen, Nucl. Phys. A336, 331 (1980).
100. D.V. Bugg, Nucl. Phys. A437, 534 (1985).
101. E. van Faassen and J.A. Tjon, CEBAF Rept. Summer Studies (1986); to be published.
102. J.S. McCarthy, I. Sick and R.R. Whitney, Phys. Rev. C15, 1396 (1977).

BAG MODELS AND HADRON STRUCTURE *

by

P. González and V. Vento
IFIC (Universitat de València - CSIC)
Burjassot (València) Spain

Abstract

We review the fundamental ideas leading to the basic assumptions behind the bag model description of hadron structure and explore in some detail the so called MIT bag model. We discuss the relevance of chiral symmetry and incorporate it in a bag model scheme by adding a pion field. We show perturbative techniques of calculating gluonic and pionic effects. We discuss the consequences of the solitonic nature of the hedgehog solution of the pion field and introduce the skyrmion bag model. We end up by drawing some conclusions of our study and discussing recent developments in this area.

*Supported in part by CAICYT under *Plan Movilizador de la Física de Altas Energías*.

1. Introduction

The idea of quarks can be traced back to 1964. In that year Gell-Mann and, independently Zweig postulated the existence of elementary constituents of hadrons, called *quarks* by Gell-Mann. In their model, baryons are made up of three quarks, while mesons are quark-antiquark systems. The quarks relevant to our purposes here come in three flavors called u, d and s. The corresponding baryon and meson states are classified in SU(3) (flavor) multiplets according to their spectroscopic properties. A rather succesful phenomenology arises, the so-called *naive quark model*. However there is a fundamental difficulty met by the naive quark model related to the spin statistics theorem. A new quantum number by the name of *color* has to be introduced in order to avoid the breaking of this fundamental principle. Moreover at least three colors are necessary to allow for fully antisymmetrized baryon wavefunctions. The fact that there are just three colors can be experimentally confirmed, e.g., from the measurement of the ratio $R \equiv \sigma(e^+e^- \to \text{hadrons})/\sigma(e^+e^- \to \mu^+\mu^-)$ and $\Gamma\ (\pi^0 \to \gamma\gamma)$.

Up to this point color has been treated as another quantum number with no specific dynamical content. Now the next and decisive hypothesis is that color is the quantum number which governs the strong interaction dynamics, and that the field theory describing such interaction must be locally gauge invariant under SU(3) (color). This sets up the basis for what is nowadays considered the theory of the strong interactions, namely Quantum Chromodynamics (QCD) [MP78,YN83]. It is described by means of a Lagrangian Field Theory, in terms of quark ($q_i^f(x)$) and gluon ($A_\mu^a(x)$) fields, i.e.,

$$\mathcal{L}_{QCD}(x) = \bar{q}_i^f(x)\, i\, \not{D}_{ij}\, q_j^f(x) - \frac{1}{4} F_{\mu\nu}^a(x) F^{a\mu\nu}(x), \tag{1.1}$$

where

$$(D_\mu)_{ij} = \delta_{ij}\partial_\mu - ig(t^a)_{ij} A_\mu^a(x) \quad , \tag{1.2}$$

and

$$F_{\mu\nu}^a(x) = \partial_\mu A_\nu^a(x) - \partial_\nu A_\mu^a(x) + g f^{abc} A_\mu^b(x) A_\nu^c(x). \tag{1.3}$$

Here t^a are the generators of the SU(3) (color) Lie algebra satisfying

$$[t^a, t^b] = i f^{abc} t^c \quad , \quad (a,b,c = 1,\ldots 8), \tag{1.4}$$

where f^{abc} are the structure constants of the corresponding color group. The last term in Eq.(1.1) includes gluon self-interactions, which are a characteristic of a non-abelian theory. Equations (1.1) to (1.4) describe the QCD Lagrangian for massless quarks. In order to account for non-zero quark masses, we have to add a mass term,

$$\bar{q}_i^f(x)\, m_f\, q_i^f(x) \quad . \tag{1.5}$$

These masses are the so-called *current* masses, typically $m_u \simeq m_d \simeq 0$, $m_s \simeq 300$ MeV.

QCD has certain features which make it very attractive in order to understand the strong interactions. They are :

i) *Renormalizability* : all the ultraviolet infinities of the theory can be reabsorbed into the constants of the theory.

ii) *Universality* : Gauge invariance implies a unique coupling constant for all hadronic interactions.

iii) *Asymptotic freedom* : The effective coupling vanishes at short distances. This property not only justifies the Bjorken scaling of the deep inelastic lepton-hadron scattering data [CL79], but moreover corrections to scaling can be computed by perturbative QCD and a good agreement with experiment is found [PT84].

iv) *Confinement* : contrarily to what happens at short distances, the effective coupling constant, when perturbatively computed, increases as the distance between the quarks increases. This fact has been taken as a strong indication that QCD is a confining theory which does not allow free quarks and gluons to exist [HU82]. However, confinement has only been proved in the lattice approximation to QCD [KO79]. The reason for this lack of success is that confinement is a highly non perturbative phenomenon and QCD has escaped solution in the nonperturbative regime [SA77,NO78,MA86].

Besides the dynamical local symmetry, the QCD Lagrangian possesses other symmetries of global nature. In particular one has U(1) baryon number symmetry, and if one assumes n massless flavors QCD is invariant under chiral $SU(n)\otimes SU(n)$ symmetry.

2. The MIT Bag Model

The fact that QCD is in the low energy regime, highly nonperturbative and unsolved, has motivated the construction of models, whose aim is to substitute for the unknown solution of the theory [AF86]. One of the most popular of such models, and the starting point to a whole class of them is the so-called MIT bag model [CJ74a, CJ74b, DJ75, JO75]. The crucial idea behind the model is the implementation of confinement and asymptotic freedom. A hadron is described as a hypertube (the bag) which divides space-time in two very distinct regions The interior one, contains quarks and gluons whose dynamical behavior is described by perturbative QCD. Inside the tube one considers a constant energy density B, which allows this tube of perturbative vacuum to stabilize. The exterior region contains no color degrees of freedom and should describe the complicated nonperturbative QCD vacuum. The model is therefore formally describable in terms of a cavity field theory [LE79, MV83, HJ83].

We now develop in some detail these ideas. Initially we omit from our treatment gluons. Let $q^f_r(x)$ be quark field inside the bag, where f is the flavor index, and r is the color index (they will not appear unless required for clarity). The QCD Lagrangian (Eqs. (1.1) to (1.5)) implies that inside the bag

$$i\gamma^\mu \partial_\mu q^f(x) = m_f q^f(x) . \tag{2.1}$$

Outside the bag the quark fields must vanish as demanded by confinement. The covariant color current (inside) is

$$j^\mu_{rs}(x) = \bar{q}_r(x)\gamma^\mu q_s(x) . \tag{2.2}$$

In order that no color quantum numbers leave the bag, we have to impose on the bag surface,

$$n_\mu j^\mu{}_{rs}(x) = 0 , \quad i.e., \quad \bar{q}_r(x) n_\mu \gamma^\mu q_s(x) = 0 , \tag{2.3}$$

n_μ being a space-like unit fourvector normal to the surface. The bag is not static and thus its surface is continuously fluctuating. A possible way of realizing Eq. (2.3) is by demanding

$$i n_\mu \gamma^\mu q_r(x) = q_r(x) \quad \text{, on the surface,} \tag{2.4}$$

which is the *Linear Boundary Condition* (LBC) of the model.

The energy momentum tensor for Dirac particles inside the bag becomes

$$T_D{}^{\mu\nu}(x) = \sum_r \left[-\frac{i}{2} \bar{q}_r(x) \gamma^\mu \overleftrightarrow{\partial}^\nu q_r(x) \right] , \tag{2.5}$$

where as usual

$$A \overleftrightarrow{\partial}^{\nu} B \equiv A\, \partial^{\nu} B - (\partial^{\nu} A)\, B \,. \tag{2.6}$$

Evidently, inside the bag

$$\partial_{\mu} T^{\mu\nu}_{D}(x) = 0 . \tag{2.7}$$

If we want no energy-momentum flux to leave the bag, we have to impose $n_{\mu}T^{\mu\nu} = 0$ on the surface. By using Eq. (2.4) we obtain

$$n_{\mu} T^{\mu\nu}_{D}(x) = \frac{1}{2} \partial^{\nu}\Big(\sum_{r} \bar{q}_{r}(x)\, q_{r}(x) \Big) . \tag{2.8}$$

Since $\overline{q}_{r}(x)q_{r}(x) = 0$ on the surface, we shall have $n_{\mu}T^{\mu\nu}_{D} = n^{\nu}B$, where B has the dimensions of a pressure and is given by

$$B = \frac{1}{2} n_{\mu} \partial^{\mu}\Big(\sum_{r} \bar{q}_{r}(x) q_{r}(x) \Big) , \quad \text{on the surface,} \tag{2.9}$$

which is the so-called *nonlinear boundary condition* (NBLC) of the bag model. B plays the role of a pressure on the bag surface, which prevents it from collapsing. Therefore, the bag energy-momentum tensor is

$$T^{\mu\nu}_{B}(x) = \big[T^{\mu\nu}_{D} - B g^{\mu\nu} \big]\, \theta_{V}(x) , \tag{2.10}$$

where $\theta_{V}(x)$ is a step function describing the bag volume. Hence, the energy-momentum fourvector is

$$P^{\mu} \equiv \int d^{3}r\, T^{0\mu}_{B}(x) = \int_{V} d^{3}r \big[T^{0\mu}_{D}(x) - B g^{0\mu} \big] , \tag{2.11}$$

where the second integration is over the region occupied by the bag. Then, the total bag momentum and energy are

$$P^{i} = \int_{V} d^{3}r\, T^{0i}_{D}(x) , \tag{2.12}$$

$$E = P^{0} = \int_{V} d^{3}r \big[T^{00}_{D}(x) \big] + BV \tag{2.13}$$

V being the bag volume. We note that the *pressure* B does not contribute to the momentum, but it does to the energy, this contribution being proportional to the bag volume as expected from physical grounds. According to the MIT bag model philosophy

[DJ75,JO75], it is essential that B be a universal hadronic constant, as it is thought to arise from the actual structure of the physical (nonperturbative) vacuum.

There is a fundamental ingredient from QCD missing in the above description, namely glue. It is easy to extend the previous analysis to include the gluons. Using Eqs. (1.1) to (1.4) we derive the color current,

$$j^a_\mu(x) \equiv \partial^\nu F^a_{\mu\nu}(x) = g\left[\bar{q}^f(x)\, t^a \gamma_\mu\, q^f(x) + f^{abc} F^b_{\mu\nu}(x)\, A^{c\nu}(x)\right]. \qquad (2.14)$$

Gluons are colored objects and therefore should not leave the bag, thus

$$n_\mu F^{a\mu\nu}(x) = 0 \;, \quad \text{on the surface,} \qquad (2.15)$$

which is the LBC for gluons. The stress tensor including both the quark and the gluon contributions is

$$T^{\mu\nu}(x) = T^{\mu\nu}_D(x) + F^{a\mu\rho}(x)\, g^{\sigma\nu} F^a_{\rho\sigma}(x) - \frac{1}{4} g^{\mu\nu} F^{a\rho\sigma}(x)\, F^a_{\rho\sigma}(x). \qquad (2.16)$$

Then

$$n_\mu T^{\mu\nu}(x) = \frac{1}{2}\partial^\nu\left(\bar{q}^f(x)\, q^f(x)\right) - \frac{1}{4} n^\nu F^{a\rho\sigma}(x)\, F^a_{\rho\sigma}(x) \,. \qquad (2.17)$$

As in the previous case, we are looking for a tensor T_B satisfying $n_\mu T^{\mu\nu}(x) = 0$ on the surface. This is accomplished by taking

$$T^{\mu\nu}_B(x) = \left[T^{\mu\nu}(x) - B g^{\mu\nu}\right] \theta_V(x) \,, \qquad (2.18)$$

where $T^{\mu\nu}(x)$ is given in Eq. (2.17) and

$$B = \frac{1}{2} n_\mu \partial^\mu\left(\bar{q}^f(x)\, q^f(x)\right) - \frac{1}{4} F^{a\mu\nu}(x)\, F^a_{\mu\nu}(x) \,, \quad \text{on the surface,} \qquad (2.19)$$

is the new universal constant when gluons are included.

Next, we define the color current in the bag as

$$J^a_\mu(x) = j^a_\mu(x)\, \theta_V(x) \,, \qquad (2.20)$$

where $j^a_\mu(x)$ is given in Eq. (2.14). Because of Eqs. (2.4) and (2.15), we have

$$\partial_\mu J^{a\mu}(x) = n_\mu j^{a\mu}(x)\, \delta_S(x) = 0 \,, \qquad (2.21)$$

where $\delta_S(x)$ is a δ-function related to the bag surface. Therefore $J^a_\mu(x)$ is a conserved current. The constant color charge is then

$$Q^a \equiv \int d^3r \; J^{a0}(x) = \int d^3r \; \delta_S(x) \, n_\mu F^{a\mu 0} = 0 , \tag{2.22}$$

which tells us the total color charge inside the bag is zero, as demanded by QCD for physical hadron states.

In dealing with the MIT bag model, the first practical problem one encounters is to find the exact solutions to the bag equations. Indeed, no realistic solutions exist thus far. The conventional treatment is based on the so called static spherical cavity approximation. It means that one takes the bag as a static sphere of radius R. This certainly violates causality and translational invariance, but it is certainly the easiest possible calculation one can consider.

Before we proceed to a complete derivation of what is called *the mode expansion*, let us perform an exercise to present the simplicity and beauty of the bag model approach. Let us assume that we have n massless quarks inside the bag and that for the time being we disconnect the residual interaction (g=0). The lowest energy single particle wavefunctions are given by

$$q(x) = N \begin{pmatrix} i j_0(\omega r)\, \chi \\ - j_1(\omega r)\, \vec{\sigma}\cdot\hat{r}\, \chi \end{pmatrix} \exp(-i\omega t) . \tag{2.23}$$

Where N is a normalization constant, the j's are spherical Bessel functions and χ a Pauli spinor (complete details about the single particle wavefunctions will be given in Sect. 3). Within the static spherical cavity approximation (SSCA) the boundary condition for the lowest mode is given by

$$j_0(\omega R) = j_1(\omega R) , \tag{2.24}$$

leading to a first eigenfrequency with value ω_0=2.04../R. The mass of the corresponding hadronic groundstate will be

$$m_H \equiv P^0 = \int d^3r \left(T^{00}_D + B\right) = n \, \frac{2.04}{R} + \frac{4}{3}\pi R^3 B . \tag{2.25}$$

The nonlinear boundary condition, Eq.(2.9), implies

$$R = \left(\frac{2.04\, n}{4\pi B}\right)^{1/4} \tag{2.26}$$

and thus

$$m_H = \frac{4}{3}(4\pi B)^{1/4}(2.04n)^{3/4} = \frac{4}{3}n\omega_0 . \tag{2.27}$$

By taking n=3 and the nucleon mass as an input we obtain $R_N \simeq 1.7$ fm, certainly a rather extended object.

In this model the ratio of baryon to meson masses is simply $(3/2)^{3/4} \simeq 1.35$, which is rather accurate for m_ρ/m_N, but quite wrong for m_π/m_N. This is the first indication that pions play a special role in low energy phenomelogy.

It is worth to note that, according to this simple picture, the N and Δ states are degenerate. Their mass difference comes in the model from quark-quark interactions via gluon exchange inside the bag. Also it is important to stress, that the model leads to non trivial results even if one disconnects the quark gluon interaction. The dynamics generating the physics in such case is exclusively associated to the mechanism of confinement.

A similar procedure can be followed for excited states. The general formula for any hadron, whether in a groundstate or radially excited state is

$$m_H = \frac{4}{3}(4\pi B)^{1/4}\sum_j (\omega_j R)^{3/4} , \tag{2.28}$$

where ω_j (j=0,1,....) are the corresponding eigenmodes.

The first correction to the simple description presented above comes from taking into account quark masses. This amounts to just changing the eigenmode equation, which will read now

$$\omega R \cot(\omega R) = 1 - R\left[m + \sqrt{m^2+\omega^2}\,\right] . \tag{2.29}$$

Since $m_u R$, $m_d R \ll 1$, such correction is negligible for u and d quarks. In the case of the strange quark, taking m_s = 300 MeV, which leads to $m_\Lambda - m_N \simeq 170$ MeV, in good agreement with experiments, the mass correction has to be taken into account.

A less trivial correction comes from quark-quark perturbative interactions within the bag. It is quite difficult to deal correctly with such interactions because we have a field theory that is not defined in free space but in a cavity [LE79,MV83,HJ83,GH86]. Before discussing other observables we shall describe a formalism of performing perturbative calculations in the bag model.

3. Perturbative formalism in the Mode Expansion

3.1 Mode Expansion in the Static Spherical Cavity Approximation

We now describe a perturbative formalism that allows calculation of hadron observables. The approach starts by solving the equations of motion inside the cavity determined by the bag, subject to the appropriate boundary conditions. These solutions are called modes. One incorporates them into a second quantized scheme by defining the quark and gluon fields in terms of these modes and quantum creation and destruction operators. From these fields one obtains the Green's functions and thus a perturbative formalism to calculate hadron observables is easily derivable [MV83, HJ83, AF86].

We shall proceed to develop such a scheme in the static spherical cavity approximation. We first treat the quark problem. Recalling the equations of motion of Sect. 2, the mode problem reduces to solving the Dirac equation subject to the MIT boundary condition, i.e.,

$$i \gamma_\mu \partial^\mu q^f(x) = m_f q^f(x) \ , \tag{3.1}$$

and for the SSCA the surface boundary condition is given by,

$$-i \vec{\gamma} \cdot \hat{r}\, q^f(x) = q^f(x) \ , \qquad r = R . \tag{3.2}$$

The general stationary solutions to the Dirac equation can be conveniently labelled by a radial quantum number *n*, the total angular momentum *J*, and a quantum number $\lambda = \pm 1$ which distinguishes between the two orbital angular momentum states associated with a particular *J* [ME76]. For each set of quantum numbers $\alpha = (n, J, \lambda)$, there are two degenerate solutions:

$$u_\alpha(x) = u_\alpha(\vec{r}) \exp(-i\omega_\alpha t) = -N_\alpha \begin{pmatrix} i\lambda j_\ell(p_\alpha r) \\ \Omega_\alpha j_{\ell'}(p_\alpha r) \vec{\sigma} \cdot \hat{r} \end{pmatrix} \mathcal{Y}^M_{\ell J}(\hat{r}) \exp(-i\omega_\alpha t), \tag{3.3}$$

and

$$v_\alpha(x) = v_\alpha(\vec{r}) \exp(i\omega_\alpha t) = N_\alpha \begin{pmatrix} i\Omega_\alpha j_{\ell'}(p_\alpha r) \vec{\sigma} \cdot \hat{r} \\ \lambda j_\ell(p_\alpha r) \end{pmatrix} \mathcal{Y}^M_{\ell J}(\hat{r}) \exp(i\omega_\alpha t), \tag{3.4}$$

corresponding respectively to particle and anti-particle modes of energy ω_α and momentum $p_\alpha^2 = \omega_\alpha^2 - m^2$. In these expressions, $l = J + \lambda/2$ and $l' = J - \lambda/2$ are the upper and lower component (with respect to the particle solution) orbital angular momenta, N_α is a normalization constant, $\Omega_\alpha = p_\alpha/(\omega_\alpha + m)$, $j_l(p_\alpha r)$ is a spherical Bessel function, and $\mathcal{Y}_{lJ}^M(\hat{r})$ is the spinor spherical harmonic of total angular momentum J and projection M,

$$\mathcal{Y}_{lJ}^{M}(\hat{r}) = \sum_{m,\mu} (l\, m\, \tfrac{1}{2}\, \mu \,|JM)\, Y_{lm}(\hat{r})\, \chi_\mu \, . \qquad (3.5)$$

Inserting either solution into the boundary condition yields the eigenvalue equation

$$j_l'(p_\alpha R) = -\frac{\lambda}{\Omega_\alpha}\, j_l(p_\alpha R) \, , \qquad (3.6)$$

which fixes the mode energies and momenta.

Since the mode solutions form a complete set, the quark field can be expanded in modes

$$q(\vec{r},t) = \sum_\alpha \left[u_\alpha(\vec{r}) \exp(-i\omega_\alpha t)\, b_\alpha + v_\alpha(\vec{r}) \exp(i\omega_\alpha t)\, d_\alpha^* \right], \qquad (3.7)$$

and then quantized in the usual manner by elevating the classical coefficients, b_α and d_α, to the level of destruction operators (particle and anti-particle respectively) and requiring that these operators satisfy the standard anti-commutation relations. Taking the vacuum expectation value of the time-ordered product of q(x) and q(x') (with quantized fields) yields the fermion propagator

$$i S_F(x,x') = -\sum_\alpha \Big[u_\alpha(\vec{r})\, \bar{u}_\alpha(\vec{r}')\, \exp(-i\omega_\alpha (t-t'))\, \theta(t-t')$$
$$- v_\alpha(\vec{r})\, \bar{v}_\alpha(\vec{r}')\, \exp(i\omega_\alpha (t-t'))\, \theta(t'-t) \Big] \, . \qquad (3.8)$$

It is important to recognize that the validity of the mode expansion depends upon the use of a complete set of cavity eigenmodes; there is no possible justification for excluding some of the modes [MV83].

Let us now proceed with the gluon mode problem. To determine a set of gluon modes in a cavity, a gauge choice is required. A convenient choice is the Coulomb gauge

$$\vec{\nabla} \cdot \vec{A}^a = 0 \, , \qquad (3.9)$$

in which the space components of the gluon field are purely transverse and the time component of the gluon field reduces to a static, confined Coulomb field $\phi(\vec{r})$. Having chosen a gauge, the mode equations are derived by variation of the gluon kinetic term in the Lagrangian. One obtains for the transverse mode

$$(\nabla^2 + \omega_{n\ell}^2)\,\vec{A}^a(\vec{r}) = 0\,, \tag{3.10}$$

$$\vec{r}\cdot\vec{A}^a(\vec{r}) = 0\,,\quad r = R\,, \tag{3.11}$$

$$\hat{r}\times(\vec{\nabla}\times\vec{A}^a(\vec{r})) = 0\,,\quad r = R\,, \tag{3.12}$$

where $\omega_{n\ell}$ is the mode energy, and the first boundary condition is just the gauge condition on the surface.

For a particular choice of radial quantum number n, orbital angular momentum $\ell>0$, and projection m, (3.10) and (3.11) have two independent solutions, a transverse electric solution

$$\vec{A}^E_{n\ell m}(\vec{r}) = N^E_{n\ell}\, j_\ell(\omega^E_{n\ell} r)\,\vec{Y}_{\ell\ell m}(\hat{r})\,, \tag{3.13}$$

with parity $(-)^{\ell+1}$, and a transverse magnetic solution

$$\vec{A}^M_{n\ell m}(\vec{r}) = N^M_{n\ell}\,\vec{\nabla}\times[R\, j_\ell(\omega^M_{n\ell} r)\,\vec{Y}_{\ell\ell m}(\hat{r})]\,, \tag{3.14}$$

with parity $(-)^\ell$, where the normalizations are fixed by

$$\int d^3r\, |\vec{A}^{E(M)}_{n\ell m}|^2 = 1\,. \tag{3.15}$$

The vector spherical harmonic $\vec{Y}_{\ell\ell m}$ is defined in terms of an ordinary spherical harmonic $Y_{\ell m}$, and the unit spherical vector $\hat{e}_m$ by

$$\vec{Y}_{\ell\ell' m}(\hat{r}) = \sum_{m_1,m_2} (\ell' m_1\, 1 m_2 | \ell m)\, Y_{\ell' m_1}(\hat{r})\,\hat{e}_{m_2}\,. \tag{3.16}$$

The corresponding eigenvalue equations from (3.12) are

$$\frac{d}{dr}\left(r j_\ell(\omega^E_{n\ell} r)\right)\Big|_{r=R} = 0\,, \tag{3.17}$$

for the electric modes, and

$$j_\ell(\omega^M_{n\ell} R) = 0\,, \tag{3.18}$$

for the magnetic modes.

With a complete set of transverse modes specified, we can construct the corresponding field operators with the result

$$\vec{\mathcal{A}}^{a}_{K}(\vec{r},t) = \left[\sum_{n\ell m} A^{K}_{n\ell m}(\vec{r})\exp(-i\omega^{K}_{n\ell}t)\, a^{a}_{n\ell mK} + h.c.\right], \tag{3.19}$$

for K= E,M, where $a^{a}_{n\ell mK}$ is a transverse gluon destruction operator of color index a. Evaluating the vacuum expectation value of the time-ordered product $T(A^{a}_{K}(x)A^{a+}_{K}(x'))$ yields the time-dependent transverse propagator

$$-i\,D^{trans}_{ij}(x,x') = \sum_{n\ell mK} [2\omega^{K}_{n\ell}]^{-1} A^{K}_{n\ell m}(\vec{r})_i\, A^{K\,*}_{n\ell m}(\vec{r}')_j \cdot$$
$$\left[\exp(-i\omega^{K}_{n\ell}(t-t'))\,\theta(t-t') + \exp(i\omega^{K}_{n\ell}(t-t'))\,\theta(t'-t)\right] . \tag{3.20}$$

The energy factor here ensures that the propagator is properly normalized and has the correct dimensions.

The static Coulomb contribution is somewhat less straightforward to treat. In free space it assumes the simple form

$$\tilde{G}_c(\vec{r},\vec{r}') = \frac{1}{4\pi|\vec{r}-\vec{r}'|}, \tag{3.21}$$

satisfying the Green's function equation

$$\nabla^2 \tilde{G}_c(\vec{r},\vec{r}') = -\delta^3(\vec{r}-\vec{r}'), \tag{3.22}$$

and the Neumann boundary condition

$$\hat{r}\cdot\vec{\nabla}\,\tilde{G}_c(\vec{r},\vec{r}') = 0, \tag{3.23}$$

on an arbitrary surface at infinity surrounding the origin. For a confined system, the propagator is not defined on surfaces at infinity; hence, the boundary condition here, obviously cannot be satisfied. We must replace it by a boundary condition on the cavity surface, which will contain an inhomogeneity to remain consistent with the Green's function equation. For a spherical cavity the required boundary condition is

$$\hat{r}\cdot\vec{\nabla}\,\tilde{G}_c(\vec{r},\vec{r}') = -\frac{1}{4\pi R}, \quad r = R . \tag{3.24}$$

The confined propagator thus becomes

$$G_c(\vec{r},\vec{r}') = \frac{1}{4\pi|\vec{r}-\vec{r}'|} + \sum_{\substack{m \\ \ell>0}} \frac{\ell+1}{\ell(2\ell+1)} \frac{(rr')^\ell}{R^{2\ell+1}} Y_{\ell m}(\hat{r})\, Y^*_{\ell m}(\hat{r}') . \tag{3.25}$$

The identification of this as the confined Coulomb propagator presumes that a restriction to color singlet states has already been made at the outset [LE79].

We have thus completed the development of the confined propagators in terms of modes satisfying the appropriate boundary conditions. From the equations for the modes and the propagators it is straightforward to develop any calculaton at the level of tree diagrams. For loop diagrams a renormalization scheme is needed. Several have been proposed, but the loss of translational invariance hinders a complete solution [CK81, MV83, GH86]. Recently there are plausability arguments to impose a cutoff procedure, that cuts down the contribution of the high lying modes, which in principle should not be confined [GS86, HZ86, KH86].

We have introduced all the ingredients which are required to study the properties of hadrons (including glueballs) within the MIT bag model approach. In principle the procedure to be followed is to apply perturbation theory for the quark-gluon or gluon-gluon interactions. This will complete the study carried out in Sect. 2 where only the confining interaction was taken into account.

3.2 Calculation of baryon observables in a perturbative formalism

The formalism we now present is specially suited for the calculation of baryon observables. The first step consists in expressing the interacting fields in the form of integral equations in the zeroth-order quark fields, quantized as shown in Sect. 3.1. Time-ordering the second quantized operators appearing in the expressions for the observables and employing Wick's theorem to reduce the time-ordered products to normal-ordered products, permits the generation of all the second ordered diagrams in a systematic fashion. The procedure can be applied succesively to obtain any order in perturbation theory.

The baryon observables can all be expressed as expectation values of integrated one-body operators of the form

$$\Gamma = \int d^3r\, \bar{q}(x)\, \hat{\Gamma}\, q(x) , \tag{3.26}$$

where $\hat{\Gamma}$ is a field-independent operator, i.e., a collection of flavor and Dirac matrices and space-time functions appropriate to the observables in question, and q(x) is the time- and space-dependent interacting quark field inside the bag. Assuming the validity of perturbation theory within the bag, we can expand q(x) in powers of the effective QCD coupling constant, g, and insert the resulting perturbation series into the expression above. This gives, to order g^2,

$$\Gamma = \Gamma_0 + \delta\Gamma_2 + \mathcal{O}(g^4) , \tag{3.27}$$

with

$$\Gamma_0 \equiv \int d^3r \; \bar{q}_0(x) \, \hat{\Gamma} \, q_0(x) , \tag{3.28}$$

and

$$\delta\Gamma_2 = \int d^3r \left[\bar{q}_0(x) \, \hat{\Gamma} \, q_2(x) + \bar{q}_2(x) \, \hat{\Gamma} \, q_0(x) \right] , \tag{3.29}$$

where q_0 is the *bare* quark field, i.e., the quark field within the cavity but not interacting perturbatively with the gluons, and q_2 is the order g^2 correction to q_0.

As usual with perturbative expansions, the correctly normalized expectation value of Γ,

$$\langle \Gamma \rangle \sim \langle \int d^3r \, \bar{q} \, \hat{\Gamma} q \rangle / \langle \int d^3r \, q^\dagger q \rangle , \tag{3.30}$$

is obtained, keeping only the connected diagrams in the perturbation series; the disconnected diagrams are exactly cancelled in each order by contributions to the normalization integral in the denominator of the same order.

Beginning with the bare fields, the fully interacting quark fields within the cavity are generated by the multiple exchange of virtual confined gluons, which, in general, can interact among themselves as well. This can be represented by an integral equation that connects the interacting quark field with the confined fermion propagator, $S_F(x,x')$, and a source function involving the interacting gluon field, $A_\mu^a(x)$:

$$q(x) = q_0(x) + g \int d^4x' \, S_F(x,x') \gamma^\mu \lambda^a A_\mu^a(x') \, q(x') . \tag{3.31}$$

Here μ and a are Dirac and color indices respectively, and λ_a is the color SU(3) generator of index a. In principle, this expression should contain multiplicative renormalization factors, however such factors will disappear in the course of the normalization procedure outlined above and through the use of renormalized fields and masses, and, consequently, need not be considered explicitly. Still primitive divergences remain in the calculation which have to be renormalized to obtain their finite contributions. We shall not discuss the different renormalization schemes, nor pretend that this procedure is completely understood. Nevertheless one may simply argue that in the worst case one may be justified to eliminate the divergences simply by using appropriate cutoffs.

We shall only extend the formalism to conventional hadrons, which in lowest order, consist entirely of quarks. Thus, the gluon fields in (3.31) are always virtual and can be eliminated in favor of the confined gluon propagator, $D_{\mu\nu}(x,x')$, and the color source current, $J^a_\nu(x)$, through the integral equation:

$$A^a_\mu(x') = g\int d^4x \; D_{\mu\nu}(x',x)\, J^{a\nu}(x) \tag{3.32}$$

Equations (3.31) and (3.32) are the integral equations of motion in our formalism and are, in principle, valid independently of perturbation theory in the same sense as the LSZ formalism. Assuming a perturbative expansion to be meaningful, they may be employed to generate the corresponding series for q(x), the second term of which results when the interacting fields in Eqs. (3.31) and (3.32) are replaced by the *bare* ones.

The color current in (3.32) contains, in general, both quark and gluon components; however, the gluon components which arise from non-Abelian three- and four-gluon vertices in QCD, contribute to (3.31) with terms of at least fourth order in g and hence, may be omitted in a second-order calculation. The color current is then just

$$J^a_\mu(x) = \bar{q}(x)\,\gamma_\mu\,\lambda^a q(x) \,, \tag{3.33}$$

where now and hereafter we shall omit the subindex from the quark fields, since only bare ones appear. Inserting this relation into (3.32) and combining (3.31) and (3.32) with (3.29), we finally obtain the equations that define our formulation of the cavity perturbation theory:

$$\langle \delta\Gamma_2 \rangle = 2\langle \Gamma_{02} \rangle \quad , \tag{3.34}$$

and

$$\Gamma_{o2} = 4\pi\alpha_c \int d^3r\, \bar{q}(x)\, \hat{\Gamma} \int d^4x'\, S_F(x,x')\, \gamma^\mu \lambda^a q(x') \cdot \int d^4x''\, D_{\mu\nu}(x',x'')\, \bar{q}(x'')\, \gamma^\nu \lambda^a q(x'') \, , \tag{3.35}$$

with $\alpha_c = g^2/4\pi$.

The above expression for Γ_{o2} contains four quark field operators, each of them consisting of a particle piece and an anti-particle piece as shown in Eq. (3.7), i.e.,

$$q(x) = \sum_\alpha \left[u_\alpha(x)\, b_\alpha + v_\alpha(x)\, d_\alpha^+ \right] . \tag{3.36}$$

Here, b_α and d_α^+ are second quantized operators that destroy particles and create anti-particles respectively with quantum number α and $u_\alpha(x)$ and $v_\alpha(x)$ are the corresponding wave functions, Eqs. (3.3) and (3.4). Since we are explicitly treating the quantum aspects of the model, the operators in (3.35) must be time-ordered. The resulting expression can be reduced to normal-ordered form by using Wick's theorem yielding

$$\begin{aligned} T\{\Gamma_{o2}\} = -4\pi\alpha_c \int d^3r\, d^4x'\, d^4x'' [-iD_{\mu\nu}(x',x'')]\, \lambda^a \lambda^a \Big\{ \bar{q}^{(+)}(x)\, \hat{\Gamma} \\ [-iS_F(x,x')]\, \gamma^\mu : q^{(+)}(x')\, \bar{q}^{(+)}(x'') : \gamma^\nu q^{+}(x'') + \bar{q}^{(+)}(x'')\, \gamma^\nu [-iS_F(x'',x)] \\ \hat{\Gamma} [-iS_F(x,x')]\, \gamma^\mu q^{(+)}(x') + \bar{q}^{(+)}(x)\, \hat{\Gamma} [-iS_F(x,x')]\, \gamma^\mu [-iS_F(x',x'']\, \gamma^\nu \\ q^{(+)}(x'') - [-iS_F(x',x)]\, \hat{\Gamma} [-iS_F(x,x'')]\, \gamma^\mu\, \bar{q}^{(+)}(x'')\, \gamma^\nu q^{(+)}(x'') \\ - [-iS_F(x'',x)]\, \hat{\Gamma} [-iS_F(x,x')]\, \gamma^\mu [-iS_F(x',x'')]\, \gamma^\nu \Big\} . \end{aligned} \tag{3.37}$$

Here we have retained only the positive frequency parts of the field operators for simplicity and thus, this equation is only applicable to baryon states. Generalization to meson states is straightforward.

The fermion and gluon propagators appearing in (3.37) contain both forward and backward going contributions, so that each term in the expression represents a set of contributions to the amplitude. In particular, the first term, involving four field operators, contains the two-body contributions. The next two terms with two field operators, represent two-body vertex corrections and one-body quark self-energy corrections respectively. These one-body terms, together with the two body terms, are diagramatically illustrated in Fig. 3.1. Contributions from the fourth-order term in (3.37) vanish in the expectation value, since they require color singlet and color octet operators to be attached to the same $q\bar{q}$ loop, while the last term, representing vacuum polarization, is disconnected and therefore omitted from further consideration.

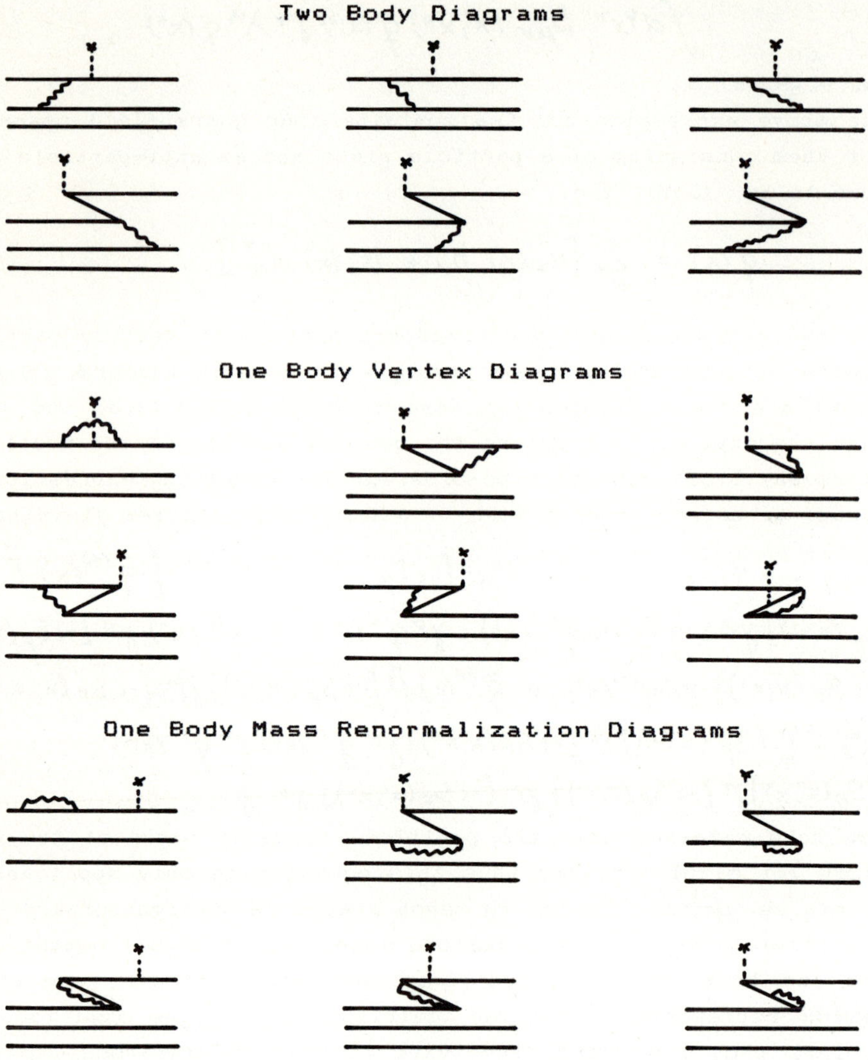

Fig. 3.1

Non-vanishing, time-ordered contributions to Γ_{02} with the flow of time directed left to right.

To evaluate the contributions to Γ_{02} arising from the diagrams discussed above, we must carry out the time and space integrations indicated in (3.35). The time integrations are easy once energy conservation is imposed, and yield the energy denominators associated with the various diagrams. To perform the space integrations, we employ standard partial wave decomposition techniques to effect a separation of the radial and angular integrations. Recasting the various vertex operators in terms of irreducible spherical tensor operators then reduces the angular integrations to familiar SU(2) matrix elements. The radial integrals require numerical treatment. Finally, one must evaluate the color-flavor matrix elements, appropriate to the external operator and hadron state considered.

The mechanics of this program are developed and discussed in several references: for the quark self-energy [CK81,BR82,GH86] and for other nucleon observables [CH81,KM79,MV83]

A final comment is necessary in order to justify the mode expansion just developed and the incompatibiliy of the higher modes to the NLBC of Sect.2. It is important to recognize that the mode expansion depends upon the use of a complete set of cavity eigenmodes. If any mode sum is cut off at some finite value of the orbital angular momentum without ensuring that partial wave contributions are negligible, the resulting propagator will not satisfy the equations that define the underlying Green function problem. Of course, quark cavity modes with $J>1/2$ do not satisfy the NLBC, an observation which has motivated some authors to exclude these modes in perturbative calculations of observables. We believe that two considerations justify a different point of view of this issue. First, the pressure balance equation is not an Euler-Lagrange equation derived from the dynamics but a subsidiary condition, which in principle, should be used to determine the cavity surface through a self-consistent procedure involving the full set of equations [JO75]. Such a self-consistent determination of the cavity surface is so enormously difficult in practice, that one has to resort to a very simple parametrization of the surface, a parametrization involving a single parameter, the bag radius. That this parametrization is too restricted to accomodate the local pressure-balance equation over the whole mode space does not justify cutting the mode space, but simply reflects the limitations of the SSCA. The second consideration has to do with the pressure balance equation itself. Locally, pressure balance is a classical concept. In a quantum mechanical system, one with quantum fluctuations, only the expectation value of the stress-tensor is well-defined, so that pressure balance should be viewed as a global, rather than local requirement. In that case, it is nothing more than the condition that the energy be minimized with respect to the bag parameters, a condition which will obviously not restrict intermediate states in any way.

Besides the mode expansion, there is another procedure which has also been applied to perturbative calculations known as the Multiple Reflection Expansion [HJ83,AF86]. Its aim is to separate out the propagator in the free one plus a series of terms which are implied by the boundary conditions. It is quite useful for formal manipulations, but calculations tend to be more complicated and less apt for numerical analysis than in the mode expansion.

3.3 Hadron properties in the MIT bag model

In Sect.2 we studied in a very simplified manner the spectrum of the confined hadrons by taking into account solely the mechanism of confinement. Under such circumstances the nucleon and the delta are degenerate in mass. Since this does not happen, quark interactions must be relevant to understand the hadronic masses. Our aim in what follows is not to present the latest up to date version of any model, but their conceptual foundations and main ingredients avoiding technicalities as much as possible.

As has been shown in the previous subsection it is quite difficult to deal correctly with the interactions between quarks and gluons , because we have a field theory which is not defined in *free* space but in a cavity. Therefore we shall restrict the results we will show to lowest order in perturbation theory [DJ75]. Corrections to these results by using any of the two techniques previously mentioned, should be looked upon in the given references, since they are too cumbersome, due to angular momentum algebra [AF86], to be shown here.

In analogy with the QED case, the first order gluon contribution can be expressed as a sum of color-electric and color-magnetic parts. Due to the bag boundary condition, the electric part practically vanishes. This statement assumes a naive interpretation of the quark self-energy [AF86]. The color-magnetic contribution is then computed [JA79] and it turns out to be

$$\Delta E_g^M = \frac{\rho \alpha_c}{R} \sum_{k>l} h(m_k R, m_l R)\, \vec{\sigma}(k)\cdot\vec{\sigma}(l)\,, \tag{3.38}$$

where k, l (= 1,2 for mesons; 1,2,3 for baryons) are ordinals for the quarks in the corresponding hadrons; ρ = 1,2 for baryons and mesons respectively; $\alpha_c = g^2/4\pi$ and h is a known function. In the limit m_u, $m_d \simeq 0$ one easily finds:

$$m_\Delta - m_N \simeq \frac{6\alpha_c}{R} h(0,0) \simeq 0.7 \frac{\alpha_c}{R} . \qquad (3.39)$$

A rough estimate with R = 1.0 fm gives $\alpha_c \simeq 2.0$

A problem which appears whenever an independent particle model is used, is removing the spurious center-of-mass (C.M.) effects. In the case of the bag model, one expects, out of dimensional reasons, a term proportional to 1/R in the energy of a quark system confined to a cavity of radius R [AF86]. Thus one is forced to introduce such a term in the mass formula of the MIT bag. For more specific information regarding this problem the reader is referred to [AF86].

There is another effect associated with the cavity, namely the so called Casimir effect, and it can be shown by semi-quantitative arguments [JO79] that it has a term proportional to 1/R. Consequently both effects (C.M. and Casimir) are collected in a term, $-Z_0/R$, where Z_0 is a parameter to be found by fitting the hadronic spectrum.

By putting together all the effects discussed up to now, we arrive at the MIT bag mass formula:

$$M(R) = \sum_j \frac{\Omega_j}{R} + \frac{4}{3}\pi R^3 B + \Delta E_g^M - \frac{Z_0}{R} , \qquad (3.40)$$

with $\Omega_j = \omega_j R$ (j=0,1...) and ΔE_g^M given by (3.38). There are four free parameters left, namely m_s, B, α_c, Z_0, since we will use the global pressure balance equation $(dM(R)/dR)_{R=R_0} = 0$.

DE GRAND and coworkers [DJ75] obtained the first fit to the hadronic spectrum with reasonable results, with the noteworthy exception of the pion (See Fig. 3.2). The values of the parameters they used are

$$B^{1/4} = 146 \text{ MeV}, \ Z_0 = 1.84, \ \alpha_c = 2.2, \ m_s = 279 \text{ MeV}.$$

The value of the radii of the hadrons which is determined by the pressure balance equation comes out to be of about 1.0 fm for baryons and 0.8 fm for mesons. Other fits have been obtained since then by taking into account in a better way some of the crude approximations of this first fit, but in general,they do not imply dramatic cuantitative changes.

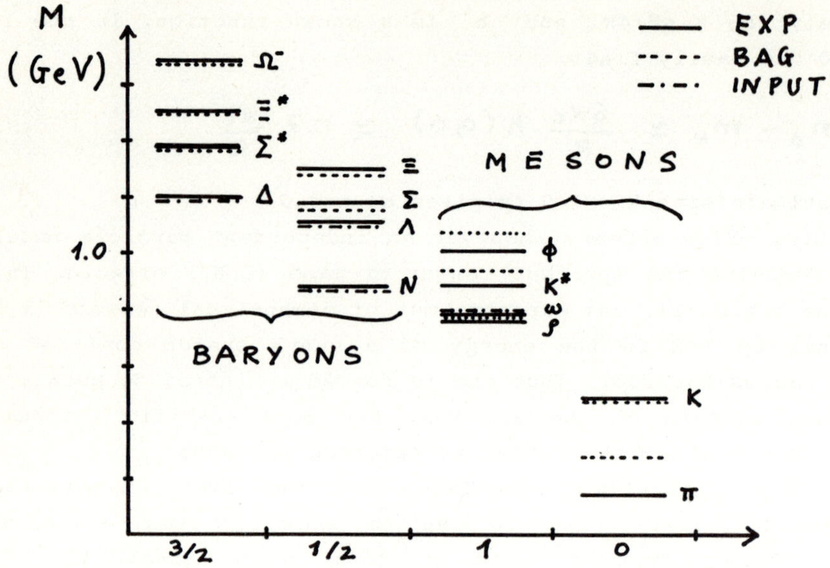

Fig. 3.2

Spectrum of the low lying hadrons obtained with the parameters mentioned in the text.

We may conclude this part by saying that two of the main inconviniences of the MIT bag model fit to hadron spectra is the large size of the perturbative phase and the large value of the coupling constant. These negative aspects have led to complement the model in two ways, one is by including additional degrees of freedom, the other by performing higher order calculations. The former has been more succesful than the latter.

In order to present some more formalism we now discuss other hadronic properties in the model. We select charge radii, magnetic moments and the axial coupling constant.

The mean square charge radius operator is defined in terms of the quark fields by

$$\langle r^2\rangle_{ch} = \sum_i Q_i \int_V d^3r\, q_i^+(\vec{r})\, r^2\, q_i(\vec{r}) \quad . \tag{3.41}$$

To lowest order, a straightforward calculation yields for the proton

$$\langle r_p^2\rangle_{ch} = R^2\left\{ \frac{\Omega_0^3}{2(\Omega_0-1)\sin^2\Omega_0} \int_0^1 du\, u^4\left[j_0^2(\Omega u) + j_1^2(\Omega u)\right]\right\} \simeq 0.57 R^2 fm^2 \tag{3.42}$$

For $R \simeq 1.0$ fm, we obtain $\langle r_p^2\rangle_{ch}^{1/2} \simeq 0.76$ fm, to be compared with the experimental value 0.82 fm; thus there is a reasonable agreement. This is not so for neutrons. A similar calculation leads to $\langle r_n^2\rangle_{ch} = 0$, while the experimental value is -0.116 fm^2. A more detailed analysis [MV83], which includes one-gluon exchange and ocean quark contributions, does not improve the situation.

In order to calculate the magnetic moments of the hadrons one starts from the well known expression

$$\vec{\mu} = \frac{1}{2} \int d^3r \, (\vec{r} \times \vec{j}_{em}) \,, \tag{3.43}$$

which in terms of the quark fields can be expressed as

$$\vec{\mu} = \frac{1}{2} e \int_V d^3r \, \vec{r} \times \sum_i Q_i \, q_i^+(\vec{r}) \, \vec{\alpha}_i \, q_i(\vec{r}) \,. \tag{3.44}$$

To lowest order it leads to

$$\vec{\mu} = \frac{4\Omega_0 - 3}{\Omega_0(\Omega_0 - 1)} \frac{eR}{12} \sum_i \vec{\sigma}(i) Q_i \equiv \mu_0 \sum_i \vec{\sigma}(i) Q_i \,. \tag{3.45}$$

For $\Omega_0 = 2.04$, we get $\mu_0 = 0.405 \, \mu_N m_N R$, where $\mu_N = e/2m_N$ is the nuclear magneton. For protons and neutrons we obtain (R = 1.0 fm): $\mu_p = \mu_0 = 1.9 \, \mu_N$; $\mu_n = -2\mu_p/3$. The corresponding experimental value for the proton moment is 2.78 μ_N, so the result is not particularly good. Nevertheless it is a remarkable fact that the ratios μ_B/μ_N for baryons computed with the model are in acceptable agreement with experiment [TH83]. This is a signature of the underlying SU(3) flavor-symmetry.

Finally let us comment briefly on the axial coupling constant in the present model. It can be computed in the MIT bag model by just taking the matrix element of the axial current $\vec{A}_\mu$ between bag states,

$$g_A = \langle p | \sum_i \int d^3r \, q_i^+(\vec{r}) \, \sigma_i^z \tau_i^3 \, q_i(\vec{r}) | p \rangle = \frac{5}{9} \frac{\Omega_0}{\Omega_0 - 1} \,. \tag{3.46}$$

One obtains to lowest order $g_A \simeq 1.09$. The experimental value is 1.24. This was considered as a big success of the model at the time since it improved the calculation of the so called non-relativistic quark model [AF86]. Nevertheless this result should not be taken too seriously since C.M. correction are not included in the previous calculation and more important the MIT bag model violates chiral symmetry, a major ingredient in the understanding of the properties of the axial current at low energies.

4. Chiral Symmetry and the Bag Model

4.1 Chiral symmetry in the MIT bag model

In the last section, we have discussed the value of g_A in the MIT bag model and have formulated some caveats in relation to the good agreement with experiment. One of the caveats refers to the fact that the MIT bag model violates chiral symmetry even in the case of massless quarks. This is certainly a major disagreement with QCD. Let us be more precise in our statement. The vector and axial vector current in the bag are obtained by conventional field theoretic procedures as

$$\vec{V}^{\mu} = \bar{q}(x)\,\gamma^{\mu}\,\frac{\vec{\tau}}{2}\,q(x)\;\theta_V\;, \tag{4.1}$$

and

$$\vec{A}^{\mu} = \bar{q}(x)\,\gamma^{\mu}\gamma_5\,\frac{\vec{\tau}}{2}\,q(x)\;\theta_V\,. \tag{4.2}$$

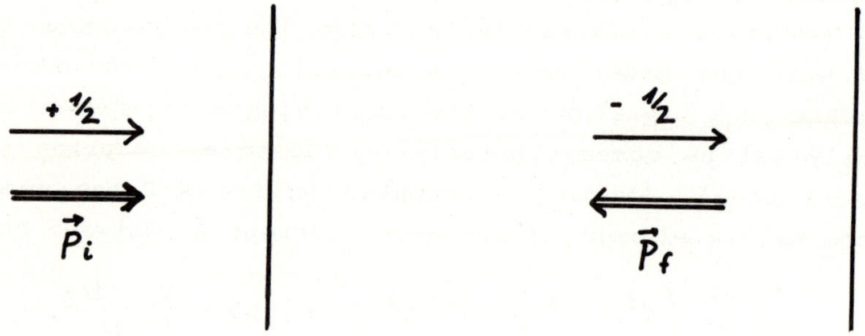

Fig. 4.1
Violation of helicity conservation by collision with a surface.

Because of the LBC (2.4), we obtain for massless quarks

$$\partial_{\mu}\,\vec{V}^{\mu}(x) = 0\;, \tag{4.3}$$

and

$$\partial_{\mu}\,\vec{A}^{\mu}(x) = -i\,\bar{q}(x)\,\gamma_5\,\frac{\vec{\tau}}{2}\,q(x)\;\delta_s\;\neq 0\,. \tag{4.4}$$

Hence the divergence of the axial current does not vanish on the bag surface and therefore chiral symmetry is violated.

One may look into this problem in a more physical way. Imagine a quark moving towards the surface with positive helicity, when it reaches the surface it bounces back loosing maybe some kinetic energy. Its momentum changes direction so its helicity changes sign, and therefore, helicity is not preserved in the process unless the surface is able to create some positive helicity (See Fig. 4.1). A mode with positive helicity has to be created at the surface to avoid this violation. In what follows, this mode will be the pion.

4.2 The Chiral Bag Model

The main idea behind chiral bag models is to incorporate besides asymptotic freedom and confinement, chiral symmetry as a fundamental ingredient. The *leitmotiv* of this proposal being, as initially emphasized by BROWN and RHO [BR79], that the success of the P.C.A.C. idea in hadron physics imposes a strong restriction on QCD as the theory of the strong interactions. QCD must possess spontaneous chiral flavor symmetry breaking in the medium-strong coupling regime and this has to be incorporated into any phenomenological model of hadrons. If we are guided by low energy phenomenology, the spontaneous breaking of chiral symmetry is intimately connected to the role of the pion as a Goldstone boson. Pionic degrees of freedom are therefore implicit into any mode to be incorporated to restore chiral symmetry. The two phase picture arises naturally in terms of the realization of chiral symmetry in the two possible modes: Wigner mode inside the confinement region and the Goldstone mode outside it.

There remains in the above picture an unaswered question: What is the pion ? Models have been developed in order to understand it [BB81, GP81, FM82, BW83, HZ86, KH86]. We shall in this lecture just ignore the question and proceed to accept a pseudoscalar Goldstone boson collective mode coupled to quarks in a manner that we shall explain shortly, and then we will pursue with the study of the structure of baryons.

There are several variants of chiral bag models, all of which result from the incorporation of chiral symmetry into the MIT bag model. They all have quite a similar basis: they result from the introduction of a pion field in a bag model of quarks and gluons and the development of a cavity field theory. We shall only describe in these lectures one of these models, namely the Chiral Bag Model. The other very popular model, the cloudy bag model, has been extensively reviewed in the literature and we refer to it [TH83,AF86]. However it is important to emphasized that, although the physics that comes out of the two models is quite different, the starting assumptions are closely related.

The Chiral Bag Model [BR79,CD79,BR80,VR80] has its origin in a two phase picture. The interior region is well represented by the MIT description. Inside the bag nearly free colored particles , quarks and gluons, exist. There is a constant energy density B, which represents in part the effect of the non-perturbative vacuum. This interior region is called the perturbative vacuum and in it the dynamics is determined by the QCD lagrangian density (See Sect.2).

The exterior region represents the complicated non-perturbative vacuum. Besides the effect associated with the constant B, one considers that there exists a pionic mode imposed by conservation of axial current. In this phase the global flavor SU(2)⊗SU(2) chiral symmetry is realized in the Goldstone mode via a phenomenological pion field and is spontaneously broken to SU(2).

Finally, a certain communication is required between the two phases and this is implemented by the sole requirement of chiral symmetry conservation.

A main feature of bag models is that the gluonic effects in the case of baryons can be treated perturbatively, therefore we omit them completely from this section, but make at this point the reader aware of their existence. Thus to zeroth order in the color coupling constant the above discussion is described by the following equations of motion:

i) for the quark field

$$i\gamma_\mu \partial^\mu q(x) = 0 \quad , \qquad \text{inside,} \tag{4.5}$$

and

$$i n_\mu \gamma^\mu q(x) = \exp(i\vec{\tau}\cdot\hat{\pi}\gamma_5\theta)q(x), \quad \text{on the surface.} \tag{4.6}$$

Here θ is the chiral angle which is related to the conventional sigma and pion fields by

$$\sigma = f_\pi \cos\theta \qquad , \qquad \vec{\pi} = f_\pi \sin\theta \; \hat{\pi} \; . \tag{4.7}$$

We are assumming a nonlinear realization of chiral symmetry, which implies that the *magic circle condition* is satisfied over all space time,

$$\sigma^2(x) + \pi^2(x) = f_\pi^2 \; , \tag{4.8}$$

as is evident from (4.7). Here f_π is the pion decay constant, whose value we take to be 95 MeV.

ii) for the pion field

$$\partial_\mu \cos^2\theta \; \partial^\mu (\tan\theta \; \hat{\pi}) = 0, \qquad \text{outside,} \tag{4.9}$$

and

$$f_\pi^2 \cos^2\theta \; n_\mu \partial^\mu (\tan\theta \, \hat{\pi}) = i\bar{q}\gamma_5 \vec{\tau} n_\mu \gamma^\mu q, \quad \text{on the surface.} \tag{4.10}$$

Here $\hat{\pi}$ represents a unit vector in the direction of the isospin.

We shall investigate two types of classical solutions to the above equations of motion: next the one characterized by $\partial_\mu \hat{\pi} = 0$, while later the one with $\hat{\pi} = \hat{r}$ or any rotation thereof.

4.3 The Perturbative Approach

Let us consider the first type of approach, i.e., $\partial_\mu \hat{\pi} = 0$. The equations of motion for the pion field become

$$\partial_\mu \partial^\mu \theta = 0 \; , \qquad \text{outside,} \tag{4.11}$$

and

$$n_\mu \partial^\mu \theta = i\bar{q}\gamma_5 \vec{\tau} n_\mu \gamma^\mu q, \qquad \text{on the surface.} \tag{4.12}$$

We take the conventional SSCA and solve (4.11) and (4.12) by Green's functions techniques, implementing à la Feynman, i.e., positive (negative) frequencies propagate forward (backward) in time, and obtain [G84]

$$\hat{\pi}\,\theta = \hat{\pi}\,\theta_0 + \frac{i}{f_\pi^2}\int dS'dt'\, G_F(\vec{r},\vec{r}',t-t')\,\bar{q}(x')\gamma_5\,\vec{\tau}\cdot\hat{r}'\vec{\gamma}\,q(x'). \tag{4.13}$$

Here θ_0 is a free-field solution satisfying $n_\mu\,\partial^\mu\theta_0 = 0$ on the boundary. G_F is the outside time-dependent Feynman Green's function, which can be constructed from the θ_0 modes in the standard way:

$$G_F(\vec{r},\vec{r}',t-t') = -\frac{i}{\pi}\int_0^\infty dK\,K\sum_{\ell,m}\varphi_\ell(Kr)\,\varphi_\ell^*(Kr')\,Y_{\ell m}(\hat{r})\,Y_{\ell m}(\hat{r}')\cdot$$
$$[\exp(-i\omega(t-t'))\,\theta(t-t') + \exp(i\omega(t-t'))\,\theta(t'-t)]\,, \tag{4.14}$$

where

$$\varphi_\ell(Kr) = [\,j_\ell'(KR)\,n_\ell(Kr) - n_\ell'(KR)\,j_\ell(Kr)]\,/\,[\,n_\ell'^2(KR) + j_\ell'^2(KR)]^{1/2}. \tag{4.15}$$

The θ_0 modes are given in terms of this notation by

$$\varphi_\ell(Kr)\,Y_{\ell m}(\hat{r})\,\exp(\pm i\omega t)\,, \tag{4.16}$$

where ω=k in the massless case. Finally in (4.13) $\hat{\pi}$ is a constant isospin vector which plays the role of an isospin polarization vector.

Let us proceed with the quark fields. In the SSCA an integral equation can be found in terms of the MIT quark modes:

$$q(x) = q_0(x) + \int dS'dt'\,S_F(\vec{r},\vec{r}',t-t')\,[\exp(i\vec{\tau}\cdot\hat{\pi}\gamma_5\theta) - 1]\,q(x') \tag{4.17}$$

where $q_0(\vec{r},t)$ are the MIT states and $S_F(r,r',t-t')$ is the Feynman Green's function constructed from them:

$$-i\,S_F(\vec{r},\vec{r}',t-t') = i\sum_\alpha[u_\alpha(\vec{r})\,\bar{u}_\alpha(\vec{r}')\exp(-i\omega_\alpha(t-t'))\,\theta(t-t') -$$
$$-\,v_\alpha(\vec{r})\,\bar{v}_\alpha(\vec{r}')\exp(i\omega_\alpha(t-t'))\,\theta(t-t')]\,, \tag{4.18}$$

where u_α and v_α are the MIT particle and anti-particle cavity modes described in Sect. 3.

The above integral equations solve the problem at the *classical* level, with the already mentioned peculiarity of using the Feynman propagator. By imposing canonical commutation or anticommutation rules we lay down the foundations of a time-dependent perturbation expansion.

In order to obtain baryonic observables we develop for the present problem similar techniques to those shown in Sect. 3. Let us consider the pion field to be small and drop all the terms beyond first order in (4.17); we thus obtain [GV84]

$$q(x) = q_0(x) + \int dS'dt'\, S_F(\vec{r},\vec{r}',t-t')\, i\,\theta\, \hat{\pi}\cdot\vec{\tau}\, q(x') . \tag{4.19}$$

We now develop an interaction scheme taking into account (4.14) and (4.15), i.e.,

$$\theta^{(0)} = \theta_0 \quad , \quad q^{(0)} = q_0 \; , \tag{4.20}$$

and then

$$\hat{\pi}\,\theta^{(1)} = - \int dt'dS'\, G_F(\vec{r},\vec{r}',t-t')\, \frac{i}{2f_\pi^2}\, \bar{q}_0 \gamma_5 \vec{\tau}\, q_0(x') \; , \tag{4.21}$$

and

$$q^{(1)} = \iint dt'dS'dt''dS''\, S_F(\vec{r},\vec{r}',t-t')\, G_F(\vec{r},\vec{r}',t-t') \frac{1}{2f_\pi^2}\, \bar{q}_0 \gamma_5 \vec{\tau}\, q_0(x'')\cdot\gamma_5 \vec{\tau}\, q_0(x') \; , \tag{4.22}$$

and so forth.

Once the fields have been calculated, we proceed to the calculation of observables following the same procedure previously described for the interior region. The pionic degrees of freedom lead to additional contributions. If one is considering the most general case of operator insertions, the contributions are of two types to lowest order: on the one hand, those due to the modification of the quark wavefunction arising from the exchange of pions, which we shall call quark type (See Fig. 4.2); on the other hand, those due to the modification of the purely mesonic contributions where the external insertions occur directly on the pion line; these we shall call of pion type and are of the form of meson exchange current contributions (See Fig. 4.3).

The first type of corrections can be described as in Sect. 3. The baryonic observables of quark type can be expressed as expectation values of operators of the form

$$\Gamma = \int d^3r\, \bar{q}(x)\, \hat{\Gamma}\, q(x)\,, \tag{4.23}$$

where $\hat{\Gamma}$ is a field-independent operator. To order $1/f_\pi^2$ one obtains

$$\Gamma = \Gamma_0 + \Gamma_2 \tag{4.24}$$

where

$$\Gamma_0 = \int d^3r\, \bar{q}_0(x)\, \hat{\Gamma}\, q_0(x)\,, \tag{4.25}$$

corresponds to the MIT results (See Subsect. 3.3), and

$$\Gamma_2 = \int d^3r \left[\bar{q}_0(x)\, \hat{\Gamma}\, q^{(1)}(x) + \bar{q}^{(1)}(x)\, \hat{\Gamma}\, q_0(x) \right], \tag{4.26}$$

represents the quark type corrections. Substituting Eq.(4.22) into (4.26) one obtains a sum of terms which are diagramatically shown in Fig. 4.2. The calculation hereafter is straightforward [MV83, GV84,GN84].

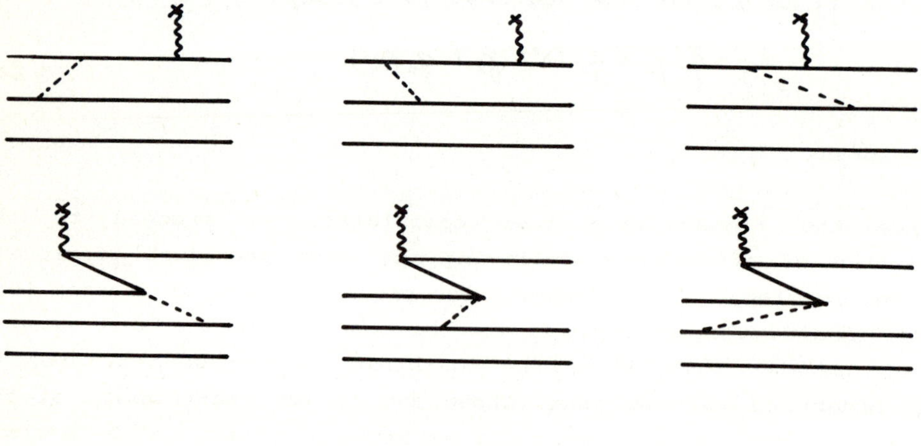

Fig. 4.2

Quark type contributions. The label A(B) has been chosen to characterize those diagrams that contain a fermionic particle (anti-particle) propagator.

The contributions of pionic type do not proceed via such a general scheme. As an example, we consider the calculation for the pionic contribution to the magnetic moment. The magnetic moment operator is given by

$$\mu^i = \frac{1}{2}\int d^3r \; \varepsilon_{ijk}\, x^j \,(\vec{\pi}\times\nabla^k\vec{\pi})_3 \, . \tag{4.27}$$

We now insert the values of the pion field to first order in θ using

$$\vec{\pi} = f_\pi \sin\theta \;\hat{\pi} \, , \tag{4.28}$$

and obtain

$$\mu^i = \frac{f_\pi^2}{2}\; \varepsilon_{ijk} \int d^3r \; x^j \,\theta^{(1)} (\hat{\pi}\times\nabla^k\hat{\pi})_3 \, . \tag{4.29}$$

The next step is to use (4.21) in this equation and one obtains a sum of terms shown diagramatically in Fig. 4.3.

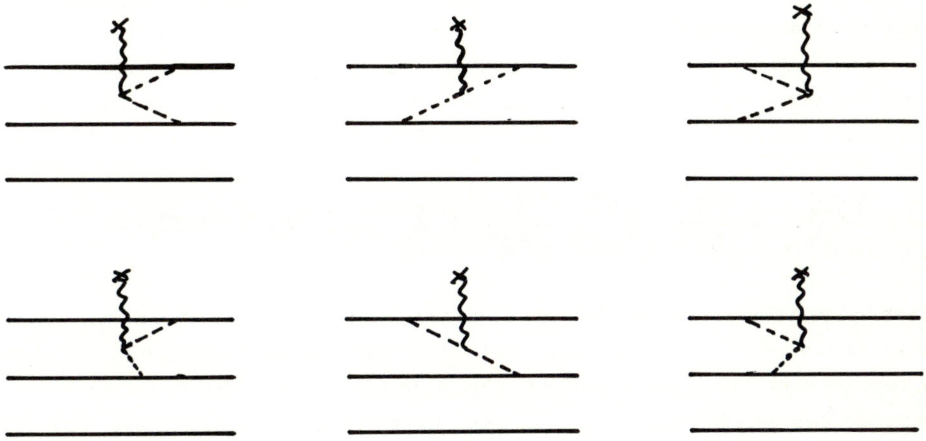

Fig. 4.3

Pionic type contributions.

To conclude, we stress that the scheme just developed considers the perturbative expansion about a topologically trivial solution of the pion field. One can easily realize that the adimensional parameter of this expansion is proportional to $1/f_\pi^2R^2$, which measures the intensity of the pion field at the surface of the bag.

4.4 Baryon properties in the Chiral Bag Model

The simplest application of the above scheme is the perturbative calculation to lowest order of baryon properties. Following the same lines of the discussion as in the MIT bag model (Subsect. 4.3), we start by describing the contribution of the pion cloud to the mass of the baryons. The calculation of the one pion exchange contribution to the energy is given by

$$\Delta E_\pi = \frac{1}{192\pi} \frac{1}{f_\pi^2 R^3} \frac{\Omega_0^2}{(\Omega_0-1)^2} \left\langle \sum_{i,j} \vec{\sigma}(i)\cdot\vec{\sigma}(j)\, \vec{\tau}(i)\cdot\vec{\tau}(j) \right\rangle . \tag{4.30}$$

The main result of the calculation is an SU(4) breaking term which contributes to the N-Δ mass splitting. Moreover since the gluonic corrections to this splitting may be included perturbatively, a fit to this experimental data as the performed earlier, implies a smaller value for the effective strong coupling constant α_c. As shown in Eq. (4.30) the contribution to the splitting is crucially dependent on the size of the system.

In order to stabilize the system one has to satisfy the global pressure balance equation. It changes due to a pionic contribution to the pressure exerted by the outside vacuum, and therefore, it is given by

$$B = \frac{n\Omega_0}{4\pi R^4} - \frac{1}{64\pi} \frac{\Omega_0^2}{(\Omega_0-1)^2} \frac{1}{f_\pi^2 R^6} \left\langle \sum_{i,j} \vec{\sigma}(i)\cdot\vec{\sigma}(j)\, \vec{\tau}(i)\cdot\vec{\tau}(j) \right\rangle . \tag{4.31}$$

The pionic pressure tends to shrink the bag. Thus, two of the major unwanted features of the MIT fit to the hadron properties, are corrected by the incorporation of the pionic degrees of freedom.

Another relative success of the first bag calculations was the explanation of the magnetic moments. As pointed out before, the magnitude of them was predicted too small. Additional contributions coming from pionic coupling do correct these results [TH83]. The contribution due to pion exchange currents and quark wavefunction correction are of the general structure given by

$$\vec{\mu}_\pi \sim \frac{1}{f_\pi^2 R}\, a(\Omega_0) \sum_{i,j} \left(\vec{\sigma}(i)\times\vec{\sigma}(j) \left[\vec{\tau}(i)\times\vec{\tau}(j) + b(\Omega_0)\, \vec{\tau}(i)\cdot\vec{\tau}(j) \right] \right) . \tag{4.32}$$

Since they add to the corresponding MIT results, they solve quite succesfully another unwanted feature of the primitive calculation.

Finally a main conflict arises with the weak axial coupling constant g_A. The MIT bag model calculation produced a far better result than the naive quark model value of 5/3 due to relativistic effects. When we add pions, as has been done in the present scheme, axial current conservation leads to a result which is approximately 5/3, even for $R \to \infty$. We shall discuss this feature in detail in the next Subsection, where results will be exact. Several mechanisms have been proposed to avoid this conflict. The most simple one has been to let the pion field penetrate inside the perturbative phase. This has given rise to the so called Cloudy Bag Model [TT81,TH83,AF86] and has been partially justified within the so called Skin Bag model [CM84,VE83,LV85]. Another proposal has been to abandon sphericity and incorporate deformation in the bag model scheme [VB81,VC83,MW84]. It leads to corrections to g_A in the right direction and moreover gives rise to a reanalysis of baryon observables which produce effects which measure such deformation [NV85]. Finally if one incorporates the full solitonic structure of the pion field and performs perturbations about that classical structure, one is also able to fit the axial coupling constant [BJ84].

5. The Little Bag

In the previous section we have studied perturbative solutions about a trivial classical pion field. These developments have been extremely succesful in reproducing the available data and have become quite popular. However, there is another approach which takes into account the full non trivial topological structure of the cavity and which aims at describing the physics mostly in terms of pionic degrees of freedom. This approach leads to the so called *Little Bag* [BR79].

5.1 The Hedgehog Solution

Let us now study the second type of solutions of Eqs. (4.5) to (4.10) which have non-trivial topology. Their structure is given by

$$\vec{\pi}(\vec{r}) = G(r)\,\hat{r} \,, \tag{5.1}$$

or any rotation of $\hat{r}$. This is the so-called hedgehog ansatz. A solution of this form preserves the spherical symmetry of the bag and has proven to be unique with this assumption. The hedgehog ansatz brings great simplification to the remaining boundary conditions [CT75, VR80, VR84]. In order to satisfy them, we have to construct quark states with new quantum numbers K, K_z defined from

$$\vec{K} = \vec{J} + \vec{I} \, , \tag{5.2}$$

in the usual way.

If the valence quarks are in the lowest possible angular momentum states $J^P = 1/2^+$, then the lowest energy occurs for

$$K^P = 0^+ \tag{5.3}$$

The quark wavefunction for this state can then be written as

$$q(x) = N \begin{pmatrix} i j_0(\omega r) \chi \\ -j_1(\omega r) \vec{\sigma}\cdot\hat{r} \, \chi \end{pmatrix} \exp(-i\omega t) \, , \tag{5.4}$$

with $(\vec{\sigma} + \vec{\tau})\chi = 0$, thus $\chi = \frac{1}{\sqrt{2}}(d\uparrow - u\downarrow)$, where the arrows refer to ordinary spin and u, d to isospin. There is an arbitrariness in this choice, namely that the isospin axes can be rotated relative to the spatial axes, but this choice does not change the results.

With the hedgehog ansatz for the pion field, the lowest energy baryon state that satisfies the boundary condition is made up of three quarks in the $K^P = 0^+$ configuration. Axial current conservation (4.10) implies then

$$f_\pi^2 \left.\frac{d\theta}{dr}\right|_{r=R} = \frac{3}{8\pi R^2} \, \frac{1+y^2}{1+y^2-\frac{2}{\Omega}y} \tag{5.5}$$

where $\Omega = \omega R$ and $y = j_1(\Omega)/j_0(\Omega)$. From the quark boundary condition (4.6) we obtain

$$\tan\theta(R) = \frac{1-y^2}{2y} \, . \tag{5.6}$$

The pionic field equation (4.9) can be reduced to

$$\frac{d^2\theta}{dr^2} + \frac{2}{r}\frac{d\theta}{dr} = \frac{\sin 2\theta}{r} \, . \tag{5.7}$$

There is no closed form solution for this differential equation, although its properties can be easily studied (VE80).

The coupling of quarks and pions is due to the boundary conditions (5.5) and (5.6). It is a striking feature of the hedgehog solution that this coupling occurs at the level of Ω as can be seen in Fig. 5.1. Therefore the single particle quark states (5.4) behave in a very peculiar way. As the pion field coupling becomes stronger, these plunge into the negative energy sea. The coupling of the pions is inversely related to the size of the core, and therefore, this peculiar behavior occurs for small bags. This was an unnoticed signature of the role of topology and the solitonic nature of the pion field.

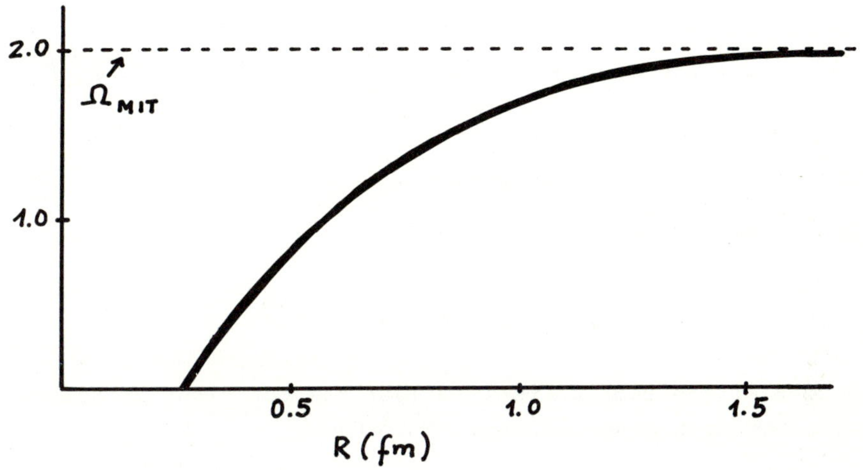

Fig. 5.1

The parameter Ω as a function of the bag radius R when calculated exactly.

5.2 The Goldberger-Treiman Relation

We investigate this result of the model in this case where the asymptotic behavior of the solution for the pion field is known. This discussion is useful to understand our comments regarding g_A in the previous section.

The axial current is given by

$$A_k^i = i\bar{q}\gamma_5\gamma_k \frac{\tau^i}{2} q\, \theta(R-r) + f_\pi \partial_k \pi^i\, \theta(r-R) . \tag{5.8}$$

The contribution from the quark sector is easily obtained and gives

$$(A^i_m)_q(K) = -\frac{3N^2}{8\pi}\int d^3r\, \exp(i\vec{K}\cdot\vec{r}) \,.$$
$$[(j_0^2(\omega r) + j_1^2(\omega r))\,\delta^i_m + 2 j_1^2(\omega r)\,\hat{r}^i\hat{r}^m] \,, \tag{5.9}$$

where ω=k in the massless case. Thus,

$$(A^i_m)_q(0) = -\frac{1}{2}\,\frac{1+y^2}{1+y^2-\frac{2}{\Omega}y}\,\delta^i_m \,, \tag{5.10}$$

which for y = 1 reproduces the MIT result.

From the pionic sector we have

$$(A^i_m)_\pi(K) = f_\pi^2\int d^3r\, \exp(i\vec{K}\cdot\vec{r})\left\{\left[\theta'^2 - \frac{\sin 2\theta}{r}\right]\hat{r}^i\hat{r}^m + \frac{\sin 2\theta}{2r}\,\delta^i_m\right\}. \tag{5.11}$$

After a careful integration we obtain

$$(A^i_m)_\pi(0) = -2\pi f_\pi^2 \left.\frac{d\theta}{dr}\right|_{r=R} R^3\,\delta^i_m$$
$$+ \lim_{K\to 0} 2\pi f_\pi^2 K\int_R^\infty dr\,\frac{d\theta}{dr}\,r^3 j_1(Kr)\,(\delta^i_m - \hat{K}^i\hat{K}^m) \,. \tag{5.12}$$

The first term of the pion current is just the contribution due to the quark fields but with opposite sign, therefore,

$$A^i_m(0) = (\delta^i_m - \hat{K}^i\hat{K}^m)\lim_{K\to 0} 2\pi f_\pi^2 K\int_R^\infty dr\,\frac{d\theta}{dr}\,r^3 j_1(Kr) \,. \tag{5.13}$$

This expression explicitly shows current conservation. Furthermore it tells us that

$$g_A^H = \lim_{K\to 0} 4\pi f_\pi^2 K\int_R^\infty dr\,\frac{d\theta}{dr}\,r^3 j_1(Kr) \,. \tag{5.14}$$

The integral can be evaluated exactly in the limit k→0 by using the pion field expansion

$$\theta = \sum_{n=0}^{\infty} \alpha_n \left(\frac{A}{2r^2}\right)^{2n+1} \,, \tag{5.15}$$

where A is some asymptotic normalization constant. One thus obtains

$$g_A^H = -4\pi f_\pi^2 A \,, \tag{5.16}$$

which is basically the Goldberger-Treiman relation. In order to see this more specifically, we compare the asymptotic value of the pion field

$$\hat{\pi}\theta \underset{r\to\infty}{\sim} \frac{A}{2}\frac{\hat{r}}{r^2} \,, \qquad (5.17)$$

with a classical Yukawa field, where $\vec{\sigma}.\hat{r}\vec{\tau}$ has been substituted by $\hat{r}$. By defining

$$g_A^H = -4\pi f_\pi^2 A \,, \qquad (5.18)$$

we obtain

$$\frac{g_A^H}{f_\pi} = \frac{g_{\pi HH}}{m_H} \qquad (5.19)$$

where $g_{\pi HH}$ is by definition the strong-coupling constant associated with this non-physical state.

From our point of view, (5.16) is precisely the Goldberger-Treiman relation, since it relates exactly the weak-coupling constants to the residue of the pion pole, i.e., the value of the pion field at the surface of infinity.

In order to get an estimate of the value of g_A as obtained by the model, we restrict the calculation to the so-called quasilinear approximation, i.e., $\theta = A/2r^2$. In this case,

$$g_A^H = \frac{3}{2} g_A^{H,q} \,, \qquad (5.20)$$

where $g_A^{H,q}$ is the value obtained purely from the quark sector. This result tends to overshoot the experimental value, if one does not include vacuum polarization effects [BJ84]. In order to relate the hedgehog with physical states, we go into the weak-coupling regime, where we expect the effects of the non-linearities to be small. Using the Godberger-Treiman relation for the nucleon we obtain

$$\frac{g_A^H m_H}{g_{\pi HH}} = \frac{g_A^N m_N}{g_{\pi NN}} = f_\pi \,, \qquad (5.21)$$

where by assuming $m_H = m_N$, which is cerrtainly true to the order we are working in (purely classical), one obtains

$$\frac{g_{\pi HH}}{g_{\pi NN}} = \frac{g_A^H}{g_A^N} = -\frac{9}{5} \,. \qquad (5.22)$$

This statement completes the calculation.

5.3 Calculation of the Hedgehog Mass

In this case one is able to obtain an exact expression for the energy function by making use of the differential equation for θ (5.7) and the boundary conditions (5.5) and (5.6). The result is

$$E = \frac{3\Omega}{R} + \frac{4\pi}{3} B R^3 + \frac{9}{32\pi f_\pi^2 R^3} \left(\frac{1+y^2}{1+y^2-\frac{2}{\Omega}y} \right)^2 - \left(\frac{1-y^2}{1+y^2} \right)^2 4\pi f_\pi^2 R. \tag{5.23}$$

Also one may calculate the pressure balance equation exactly,

$$B = \frac{3\Omega}{4\pi R^4} - \frac{9}{128\pi^2 f_\pi^2 R^6} \left(\frac{1+y^2}{1+y^2-\frac{2}{\Omega}y} \right)^2 + \frac{f_\pi^2}{R^2} \left(\frac{1-y^2}{1+y^2} \right)^2 . \tag{5.24}$$

What is the connection between the energy equation and the latter equation for this model? Let us write the energy function in terms of the value of the pion field at the surface,

$$E(R,\Omega,A) = \frac{4\pi}{3} B R^3 + \frac{3\Omega}{R} + 2 f_\pi^2 \left\{ R^3 \left[\left. \frac{d\theta}{dr} \right|_{r=R} \right]^2 - 2R \sin^2\theta(R) \right\} . \tag{5.25}$$

Here B and f_π are constants; R and A can be taken as variables, while Ω is related to these by the boundary conditions (5.5) and (5.6). In order to get stability with respect to the two variables, we impose on (5.25) to have a minimum, that is

$$\left(\frac{\partial E}{\partial R} \right)_A = 0 \qquad ; \qquad \left(\frac{\partial E}{\partial A} \right)_R = 0 , \tag{5.26}$$

and appropriate conditions on the higher derivatives. The first of these conditions gives rise to axial current conservation, while the second, if the former is satisfied, gives rise to the pressure balance equation. Thus, the boundary conditions together with the pressure balance equation state that the energy be stationary with respect to its variables.

A problem with the hedgehog is that its mass function has a positive slope for small radii, i.e., the system collapses [VE81]. This feature is intimately connected with the geometrical structure of this solution. The collapse feature can be avoided by either adding new terms to the lagrangian [SK61, VE82, UF83] or simply by boring a hole big enough where to put the quarks [BJ84].

In the last two sections we have introduced the main ideas associated with chiral bag models in the two phase scheme. The non-perturbative approach arose from the wish to implement the *Little Bag* idea. This point has recently developed very much, as the role of topology has been clarified. Thus in some way the *Little Bag* idea has been engulfed by the *Skyrmion* bag approach. We shall next discuss briefly this new development

5.4 The Skyrmion Bag

Let us start by discussing the spectrum of the hedgehog solution. As shown in Fig. 5.1, as R approaches the critical value R_c, Ω tends to go to zero; that is, a zero mode appears in the spectrum. It is known [JR76] that zero modes are associated with peculiar properties of the vacuum of a field theory. What has happened is that the coupling of the pion to the quarks has produced a quark spectrum which is non-symmetrical with respect to positive and negative energy states. This symmetry is recovered again at R_c, when the lowest positive energy state plunges into the negative energy sea. This feature is represented in Fig. 5.2. This lack of symmetry of the spectrum as well as the appearence of zero modes lead to a vacuum carrying baryon number [RG83, GJ83].

In general the baryon number of the vacuum in the cavity can be expressed by

$$B(\theta) = -\frac{1}{2} \lim_{t \to 0} \sum_n \varepsilon(E_n) \exp(-t/|E_n|), \tag{5.27}$$

where θ signals the dependence of the pion field and the sum is extended over all positive and negative energy states. This sum can be performed in our case [GJ83] and leads to

$$B(\theta) = \begin{cases} -\frac{1}{\pi}\left[\theta(R) - \frac{\sin 2\theta(R)}{2}\right], & 0 < \theta < \pi/2, \\ \\ 1 - \frac{1}{\pi}\left[\theta(R) - \frac{\sin 2\theta(R)}{2}\right], & \frac{\pi}{2} < \theta < \pi. \end{cases} \tag{5.28}$$

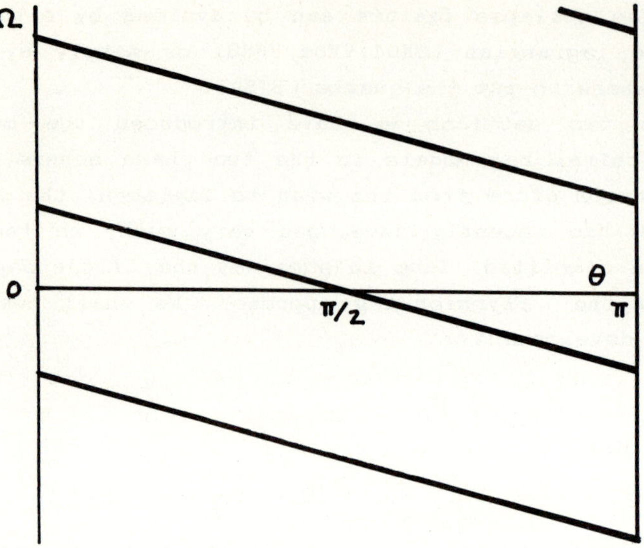

Fig. 5.2
The energy spectrum of the one dimensional bag is shown as a function of the chiral angle.

When Ω reaches the value zero, i.e. $\theta = \pi/2$, the baryon number of the vacuum becomes $-1/2$! Moreover, there is a jump in the baryon charge when one state plunges into the negative energy sea, as can be seen from (5.28). Since for a baryon the quarks inside carry baryon number 1/3, for a three quark state the baryon number for the system will be

$$B = 1 - \frac{1}{\pi}\left[\theta(R) - \frac{\sin 2\theta(R)}{2}\right], \quad 0<\theta<\pi. \tag{5.29}$$

The coupling of a topologically non-trivial pion field to a cavity leads to a situation where the interior phase diminishes its baryon number. Where is it gone?

It has been shown [SK61,WI83,RG83,GJ83] that a topologically non-trivial pion field carries a baryon number which is given by

$$B_\pi = \frac{1}{\pi}\left[\theta(R) - \frac{\sin 2\theta(R)}{2}\right], \quad 0<\theta(R)<\pi. \tag{5.30}$$

Certainly this is the amount needed to mantain the total baryon number equal to 1.

This discussion leads one to envisage the following scenario for a transition between the MIT bag model for large radii and Skyrme's model [SK61,AN83] at zero radius [GJ83]. As R decreases from large values $R \gg 1/f_\pi$ to small values $R \ll 1/f_\pi$, the pionic field grows from zero to infinity. The topological charge changes from zero to one. The baryon number for the inner phase starts at one and changes all the way to zero. As θ goes through π/2, the baryon number of the inner phase jumps in one unit as we reach a new negative state. The sum between the baryon charge outside and inside is always unity. For θ = π we have reached again a symmetrical vacuum without zero modes. All this can be understood by looking at Fig. 5.3, which is the exact solution for a 1+1 dimensional model [ZA84]. Realistic models have a similar behavior [ZM84]. This figure is easily interpreted if one remembers the expression for the baryon number of the vacuum

$$B = -\frac{1}{2} \langle c_0 c_0^+ \rangle - \frac{1}{2} \lim_{t \to 0} \sum_n \varepsilon(E_n) \exp(-t|E_n|), \tag{5.31}$$

where we now allow explicitly for the possibility of zero modes. If a single particle state is occupied then

$$B = B_{vacuum} + B_{particle} = B_- + 1. \tag{5.32}$$

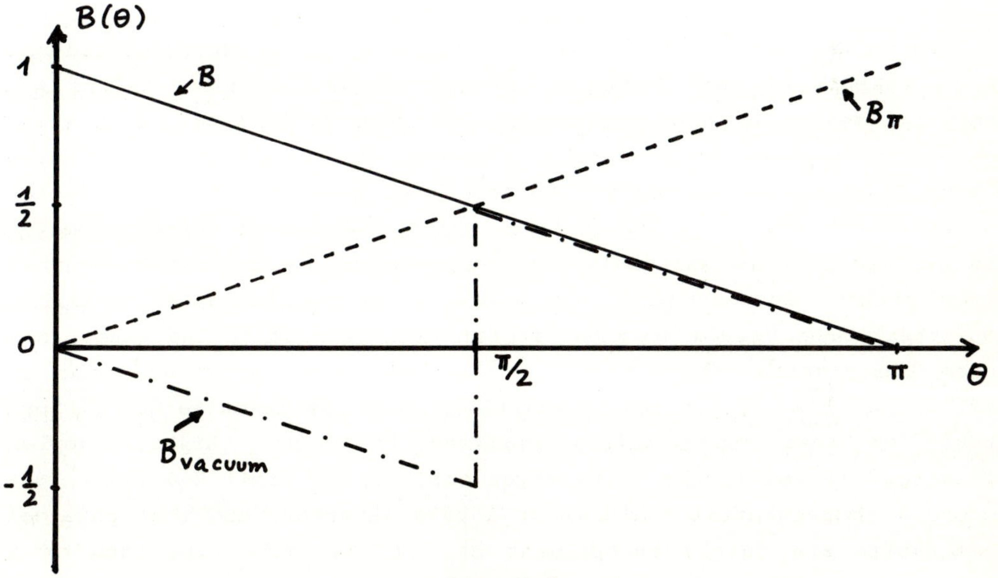

Fig. 5.3
Baryon number fractionation for the one dimensional bag as a function of the chiral angle

When the particle state jumps into the negative energy sea, since the Dirac sea is occupied, we have

$$B = B_{vacuum} = B_{+}, \tag{5.33}$$

therefore,

$$B_{+} - B_{-} = 1, \tag{5.34}$$

where $B_{\pm}$ is the baryon number of the vacuum and the ± signal its sign (See Fig. 5.3)

The procedure just described is the so-called charge fractionation. It is of crucial importance in our understanding of the bag model. The bag model was created to confine color. An unwanted result was that it also coinfined baryonic properties. Once the pion field is introduced, we are free from such oppression. The remaining problem is of technical origin; one has to include effects from the Dirac sea which although finite for the baryon number, are not that way for other observables [BJ84, VJ84, ZM84]. Much progress has occur recently in these calculations [JA86]. The outcome is that the physics should be independent of R in some kinematical domain. This has been proposed under the name of *Cheshire Cat Model* [NN85].

The chiral bag model discussed in these lectures can be considered as an example of this Cheshire Cat philosophy. As noted before the quark bag sector is weak-coupling in the color fine structure constant α_c, which, because of asymptotic freedom, would require an infinite number of meson fields [WI79]. On the other hand, the skyrmion sector is presumably weak coupling in $1/N_c$, which, because of chiral symmetry and confinement, is extremely complex in terms of microscopic quark-gluon fields. Consequently, the matching on the boundary is highly non-linear when viewed from one sector with respect to the other, and hence the physical quantities so calculated are highly nonlinear in both α_c and $1/N_c$. It is not at all clear that the Cheshire Cat picture should be even approximately realized in such a highly complex situation. It has though been proven that in the limit $R \to 0$ limit, one recovers the essential features of a pure skyrmion, and that physical observables are fairly independent of R [BJ84, JA86]. The idea of a Little Bag in this scheme, does not come as a unique picture but as a good approximation where probably the theory is simplest in the two sectors.

Acknowledgements

We have benefitted greatly from discussions with J. Bernabéu, J. Navarro, S. Noguera, V. Sanjosé and O.V. Maxwell. V. Vento acknowledges many illuminating conversations with R. F. Alvarez-Estrada, F. Fernández and J. L. Sánchez Gómez during the preparation of some lecture notes [AF86], which have influenced greatly this presentation. Some of the work described in these notes is the result of a long term collaboration of V. Vento with G.E. Brown, M. Rho and A.D. Jackson.

References

[AF86] R. F. Alvarez-Estrada, F. Fernández, J. L. Sánchez-Gómez and V. Vento, Lecture Notes in Physics vol.259, Springer Verlag, Heidelberg (1986).

[AN83] G. S. Atkins, C. P. Nappi and E. Witten, Nucl. Phys. **B228** (1983) 552.

[BB81] J. M. Blairon, R. Brout, F. Englert and J. Greensite, Nucl. Phys. **B180** (1981) 439.

[BJ84] G. E. Brown, A.D. Jackson, M. Rho and V. Vento, Phys. Lett. **140B** (1984) 285.

[BR79] G. E. Brown and M. Rho, Phys. Lett. **82B** (1979) 177.

[BR80] G. E. Brown, M. Rho and V. Vento, Phys Lett. **84B** (1980) 383.

[BR82] J. D. Breit, Nucl. Phys. **B202** (1982) 147.

[BW83] R. Brockmann, W. Weise and E. Werner,Phys. Lett. **122B** (1983) 201.

[CD79] C. G. Callan, R. F. Dashen and D. J. Gross, Phys. Rev. **D23** (1979) 2905.

[CH81] F. E. Close and R. R. Horgan, Nucl. Phys. **B185** (1981) 333.

[CJ74a] A. Chodos, R. L. Jaffe, K. Johnson, C. B. Thorn and V. F. Weisskopf, Phys. Rev. **D9** (1974) 3471.

[CJ74b] A. Chodos, R. L. Jaffe, K. Johnson and C. B. Thorn, Phys. Rev. **D10** (1974) 2599.

[CK81] S. A. Chin, A. K. Kerman and X. Y. Yang, MIT Preprint 1981.

[CM84] J. G. Cottingham, O. V. Maxwell and B. Loiseau, Orsay Preprint 1984.

[CT75] A. Chodos and C. B. Thorn, Phys. Rev. **D12** (1975) 2733.

[DJ75] T. de Grand, R. Jaffe, K. Johnson and J. Kiskis, Phys. Rev. **D21** (1975) 2060.

[FM82] J. R. Finger and J. E. Mandula, Nucl. Phys. **B199** (1982) 168.

[GH86] S. N. Goldhaber, T. H. Hansson and R. L. Jaffe, MIT Preprint, CTP # 1322 (1986).

[GJ83] J. Goldstone and R. L. Jaffe, Phys. Rev. Lett. **51** (1983) 1518.

[GN84] P. González, S. Noguera, J. Bernabéu and V. Vento, Nucl. Phys. **A423** (1984) 477.

[GP81] J. Greensite and J. Primack, Nucl. Phys. **B180** (1981) 170.

[GV84] P. González and V. Vento, Nucl. Phys. **A415** (1984) 413.

[HJ83] T. H. Hansson and R. L. Jaffe, Phys. Rev. **D28** (1983) 882; Annals of Physics **151** (1983) 204.

[HU82] K. Huang, Quarks, Leptons and Gauge Fields, World Scientific, Singapore (1982).

[HZ86] T. H. Hansson and I. Zahed, Stony Brook Preprint 1985 (Nucl. Phys. to be published).

[JA79] R. L. Jaffe, Proceedings of the *Ettore Majorana* School of Subnuclear Physics, Ed. A. Zichichi, Bologna (1979).

[JA86] A. D. Jackson, Stony Brook Symposium 1986 (to be published).

[JO75] K. Johnson, Acta Physica Polonica **B6** (1975) 865.

[JO79] K. Johnson, SLAC-PUB-2346 (1979) (Erratum January 1980).

[JR76] R. Jackiw and C. Rebbi, Phys. Rev. **D13** (1976) 3398; Phys. Rev. Lett. **36** 116.

[KH86] S. Klabucar, T.H. Hansson and I. Zahed, Stony Brook Preprint 1986.

[KM79] I. Yu Kobzarev, B. V. Marteniyanov and M. G. Shchepkin, Sov. J. Nucl. Phys. **30** (1979) 261.

[KO79] J. B. Kogut, Rev. Mod. Phys. **51** (1979) 659.

[LE79] T. D. Lee, Phys. Rev. **D19** (1979) 1802.

[LV85] H. K. Lee and V. Vento, Nuov. Cimento **90A** (1985) 135.

[MA86] P. Mansfield, Nucl. Phys. **B272** (1986) 439.

[ME76] A. Messiah, Quantum Mechanics, North Holland Pub. Com. (1976).

[MP78] W. Marciano and H. Pagels, Phys. Rep. **36C** (1978) 137.

[MV83] O. V. Maxwell and V. Vento, Nucl. Phys. **A407** (1983) 366.

[MW83] Z. Y. Ma and J. Wambach, Phys. Lett. **132B** (1983) 1.

[NN85] S. Nadkarni, H. B. Nielsen and I. Zahed, Nucl. Phys. **B253** (1985) 308; S. Nadkarni and H. B. Nielsen, Nucl. Phys. **B263** (1986) 1.

[NV85] J. Navarro and V. Vento, Nucl. Phys. **A440** (1985) 617.

[NO78] N. K. Nielsen and P. Olesen, Nucl. Phys. **B144** (1978) 376.

[PT84] P. Pascual and R. Tarrach, Lecture Notes in Physics 194, Springer Verlag, Heidelberg (1984).

[RG83] M. Rho, A. S. Goldhaber and G. E. Brown, Phys. Rev. Lett. **51** (1983) 747.

[SA77] G. K. Savvidy, Phys. Lett. **71B** (1977) 133.

[SK61] T. H. R. Skyrme, Proc. Roy. Soc. London **A260** (1961) 127.

[TH83] A. W. Thomas, Advances in Nuclear Physics **13** (1983) 1.

[TT81] A. W. Thomas, S. Theberge and G. A. Miller, Phys. Rev. **D24** (1981) 216.

[UF83] J. Urbano, M. Fiolhais and J. Pacheco, Proceedings of the 5th Topical School of Granada, World Scientific, Singapore 1983.

[VB81] V. Vento, G. Baym and A. D. Jackson, Phys. Lett. **102B** (1981) 97.

[VC83] R. D. Viollier, S. A. Chin and A. K. Kerman, Nucl. Phys. **A407** (1983) 269.

[VE80] V. Vento, Stony Brook Thesis, unpublished (1980)

[VE81] V. Vento, Lecture Notes in Physics **137** (1981) 203.

[VE82] V. Vento, Phys. Lett. **107B** (1981) 5.

[VE83] V. Vento, Phys. Lett. **121B** (1983) 370.

[VJ84] L. Vepstas, A. D. Jackson and A. S. Goldhaber, Phys. Lett. **140B** (1984) 280.

[VR80] V. Vento, M. Rho, E. M. Nyman, J. H. Jun and G. E. Brown, Nucl. Phys. **A345** (1980) 413.

[VR84] V. Vento and M. Rho, Nucl. Phys **A412** (1984) 413.

[WI79] E. Witten, Nucl. Phys. **B160** (1979) 57.

[WI83] E. Witten, Nucl. Phys. **B223** (1983) 423; Ibid 433.

[YN83] F. Yndurain, Quantum Chromodynamics, Springer Verlag, Heidelberg (1983).

[ZA84] I. Zahed, Nordita Preprint 1984

[ZM84] I. Zahed, U. G. Meissner and A. Wirzba, Phys. Lett. **145B** (1984) 117.

PART IV

ANTINUCLEON PHYSICS

ANTINUCLEON ANNIHILATIONS AT LOW ENERGIES AT LEAR

Ugo GASTALDI
EP Division, CERN, 1211 Geneva 23, Switzerland

ABSTRACT

This report is limited to a skeletal summary of the topics covered in the lecture. The arguments discussed were: physics interest of $N\bar{N}$ annihilations; S and P-wave $p\bar{p}$ annihilation at rest: new experimental techniques established with the ASTERIX detector at LEAR and physics prospects; $\bar{p}$ annihilation in nuclei: experimental findings of the Streamer chamber detector at LEAR and physics prospects; brief presentation of LEAR and summary of the physics programme planned starting from the end of 1987 when the antiproton collector (ACOL) will come into operation.

1. PHYSICS INTEREST OF $N\bar{N}$ ANNIHILATIONS

At low energies annihilation is the dominant reaction between particle and antiparticle and it produces the mediators of the forces acting between them in the elastic channel. $N\bar{N}$ annihilations produce the mediators of the strong force acting both at the nucleon level (mesons) and at the quark level (gluons).

Real $N\bar{N}$ annihilation is the sector in experimental studies of $N\bar{N}$ interactions with the richest phenomenology because of the numerous annihilation channels that are open. The various final states may contain pions, kaons and gammas with several multiplicities. In each exclusive final state with more than two particles are generally present intermediate resonant states. Virtual annihilations play an important role in non-annihilation reactions. Real and virtual annihilation reactions depend in principle on the angular momentum, the isospin, the energy and the spin of the initial $N\bar{N}$ state and on whether the target nucleon is free or bound into a nucleus. The dynamics of $N\bar{N}$ interactions will be understood when theory will have been able to make quantitative predictions on different exclusive final states and on their intermediate resonances for several initial states and when these predictions will have been checked experimentally for different initial states or compositions of initial states. $N\bar{N}$ interactions can be linked to NN interactions by G parity transformations if the $N\bar{N}$ annihilation effects are independently known. Experimental and theoretical work in the $N\bar{N}$ sector can therefore also shed light onto short range NN forces, that are at present poorly understood.

From a particle physics point of view $N\bar{N}$ annihilation is an advantageous instrument with unique features to study interactions between quarks and gluons. Indeed $N\bar{N}$ annihilation is a copious source of light mesons and mesonic resonances and it is a natural source of exotic mesonic resonances (like glueballs (gg,ggg), hybrids ($q\bar{q}g$) and multiquark ($qq\bar{q}\bar{q}$, $qqq\bar{q}\bar{q}\bar{q}$) structures), if they exist as it is expected in the framework of the Quantum Chromo Dynamic theory of strong interactions. These objects can be produced with the constituent quarks and antiquarks of the original $N\bar{N}$ pair that did not annihilate in the interaction and the gluons produced by the pairs of constituent quarks and antiquarks that did annihilate in the interaction. The determination of the existence of glueballs and hybrids is of fundamental importance as it would prove the basic point in QCD that gluons couple directly also to gluons. To this purpose the main physical advantages offered by $N\bar{N}$ annihilations at low energies are

a) high interaction rates

b) several possible initial states with different quantum numbers, that give access in formation experiments to many resonant final states with the same quantum numbers as the initial state and that can give access in production experiments to states with exotic quantum numbers.

c) extremely well defined energy of the initial state that permits in formation experiments to find and measure accurately narrow resonances (that cannot have exotic quantum numbers)

d) possibility of observing in production experiments also broad structures (that can have exotic quantum numbers) with differential measurements by comparing with the same detector identical decay channels (e.g. $\pi^+\pi^-$ and K^+K^-) of possible resonant states present in exclusive final states of the same type (e.g. final states with three pions and final states with a K^+K^- pair plus one pion) produced from different $N\bar{N}$ initial states, differing by angular momentum, isospin or spin[1-3]. Reviews covering topics mentioned in this section are indicated in ref. 3 and 4.

2. S AND P-WAVE $p\bar{p}$ ANNIHILATIONS AT REST

The second part of this lecture focused on the possibility of changing the angular momentum of the initial state of $p\bar{p}$ annihilations at rest without changing the total energy, the spin and isospin composition. This possibility[5-7] had been one of the leading motivations[8] for LEAR. It has been established as an operational experimental technique by the ASTERIX experiment, that has also produced clear evidence of dynamical effects appearing in exclusive final states when the angular momentum distribution of the initial states is changed[9,4,10].

The low momentum and small energy dispersion of the LEAR beam permits to stop efficiently antiprotons also into gas H_2 targets. Lowering the beam momentum the distribution of annihilation vertices concentrates around the target centre. Nearly all antiprotons annihilate at rest after having formed a $p\bar{p}$ atom.

The atomic cascade of protonium ($p\bar{p}$ atom) depends on the target density because of the combined effects of annihilation in S and P atomic levels and of collisions of protonium with the surrounding molecules[11)]. At normal temperature and pressure more than 10% of the $p\bar{p}$ atoms populate the 2P level[12)]. 2P levels decay by radiative transitions (with emission of a K_α X-ray) to the 1S ground state with less than 2% probability and annihilate with more than 98% probability. By using as central detector a cylindrical X-ray Drift Chamber[13)], XDC, (that has low mass, low energy threshold, large acceptance and granularity and can therefore also detect in coincidence two or more X-rays emitted by the same $p\bar{p}$ atom), the ASTERIX experiment is able to select P-wave annihilation events where an L X-ray populating the 2P level is measured in coincidence with the particles present in the final state. The XDC detects M, L and K $p\bar{p}$ X-rays to the n=3, n=2 and n=1 states. L X-rays (1.7 ÷ 3.2 KeV) are detected with typically 50% overall efficiency and the X-ray background in the L X-ray energy region is less than 10% of the L X-ray signal. The background is mainly constitued by X-rays produced by internal brehmsstrahlung[14)]. Consequently events with an X-ray in the L energy region undergo in more than 90% of the cases annihilation in P-wave in one of the 2P sublevels. The S-wave contamination of events with an X-ray in the L energy region is due mainly to events with an internal brehmsstrahlung X-ray simulating a L X-ray and the $p\bar{p}$ annihilation occuring in S-wave from any nS state. The fraction of annihilation in S-wave from all the nS levels is not yet determined exactly, however the comparison of bubble chamber data and ASTERIX preliminary data[10,15)] indicate that this fraction is in the region 20 to 50% of all annihilations, and that correspondingly the fraction of $p\bar{p}$ P-wave annihilation at rest in H_2 gas is in the region 50 to 80%. Taking into account also the possibility of 2P-1S radiative transitions, the contamination of the P-wave annihilations data sample selected with the L X-ray coincidences is then less than 7%.

Preliminary ASTERIX data in several exclusive final states (e.g. $p\bar{p} \to \pi^+\pi^-$, K^+K^-, $K^0_s K^0_s$, $\pi^+\pi^-\pi^0$) show experimentally that the branching ratios (BR) and ratios between branching ratios (RBR) for final states and intermediate resonant states change markedly when changing the fraction of initial states in S- and P-wave. We can also compare $p\bar{p}$ annihilations in liquid H_2 (Bubble Chamber data - dominant S-wave annihilation), $p\bar{p}$ annihilations in H_2 gas at NTP and $p\bar{p}$ annihilations with L X-ray in coincidence (dominant P-wave annihilation). This comparison in particular shows that resonant intermediate states hardly visible in bubble chamber data are present and dominant in P-wave annihilation.

A brief description has been given of the ASTERIX experimental set up[6,7,16] (fig. 1) and of the projection chamber of new geometry (Spiral Projection Chamber[17] SPC/XDC) that we used as central detector[18]. ASTERIX has been a step in the realization of an experimental programme that was proposed in 1978[5] to study the dependence of $p\bar{p}$ annihilations at rest on angular momentum. This programme required LEAR, plus a new type of large acceptance X-ray detector, plus a detector of annihilation products as complete and performant as possible. In ACOL time the research line of ASTERIX will be continued and extended by two experiments that plan to use an X-ray Drift Chamber[13] to measure $p\bar{p}$ X-rays and plan to improve substantially on the detection capability of the final state. One detector, OBELIX[3] (fig. 2) is based on the Open Axial Field spectrometer[19]. It will have good charged particles detection efficiency and momentum resolution and identification ($\Delta p/p \leqslant 2\%$, full $K^{\pm}$ identification from ~50 MeV/c up to 1 GeV/c), it will have high granularity and angular resolution for γ detection and will have trigger capabilities on angular momentum of the initial state and strangeness in the final state. The other detector, Crystal Barrel[20] (fig. 3), will have high efficiency and energy resolution for γ's and good charged particle momentum resolution. To study S-wave dominant $p\bar{p}$ annihilations the Crystal Barrel plans to use a liquid H_2 target and to remove the XDC. OBELIX will use only one detector configuration in order to factorize away detector and software efficiencies, acceptances and biases and will study at the same time S and P-wave annihilations to factorize detection and operation instabilities[1].

3. $\bar{p}$ ANNIHILATIONS IN NUCLEI

This topics is covered extensively in a recent review of Piragino[21]. The main results of the measurements made at LEAR by the Streamer Chamber Collaboration[22] are:

a) the principal mechanism of $\bar{p}$ absorption is the annihilation on quasi free nucleons of the nuclear surface.

b) non elastic reactions different from $\bar{p}$ annihilation are suppressed.

c) nuclei appear, in terms of the optical model, as black to antiprotons.

d) annihilations of antiprotons penetrating inside the nucleus have been observed and can be selected requiring high multiplicity of annihilation particles and nuclear fragments in the final state.

An extensive programme on $\bar{p}$ annihilations into nuclei is planned by the OBELIX collaboration. This programme includes:

a) survey of $\bar{p}$ annihilations in ^2H, ^3He, ^4He, N, Ne, Ar, Kr, Xe at rest and at a few momenta up to the highest LEAR energy.

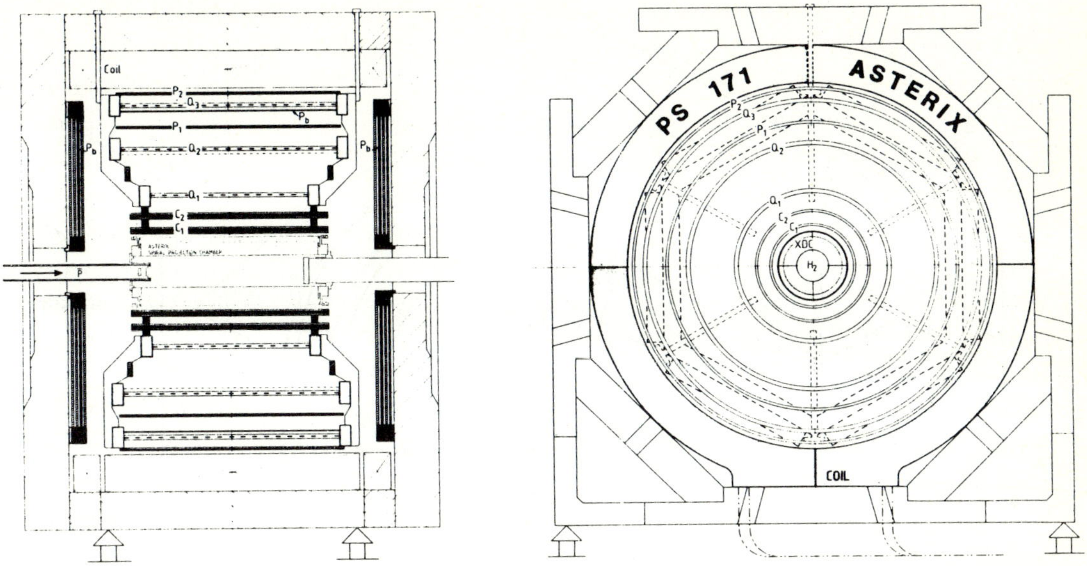

Fig. 1: ASTERIX detector. (XDC) X-ray Drift Chamber and SPC (C_1,C_2,Q_1,Q_2,Q_3). Cylindrical MWPCs with cathode read_out, (P_1,P_2) cylindrical MWPCs.

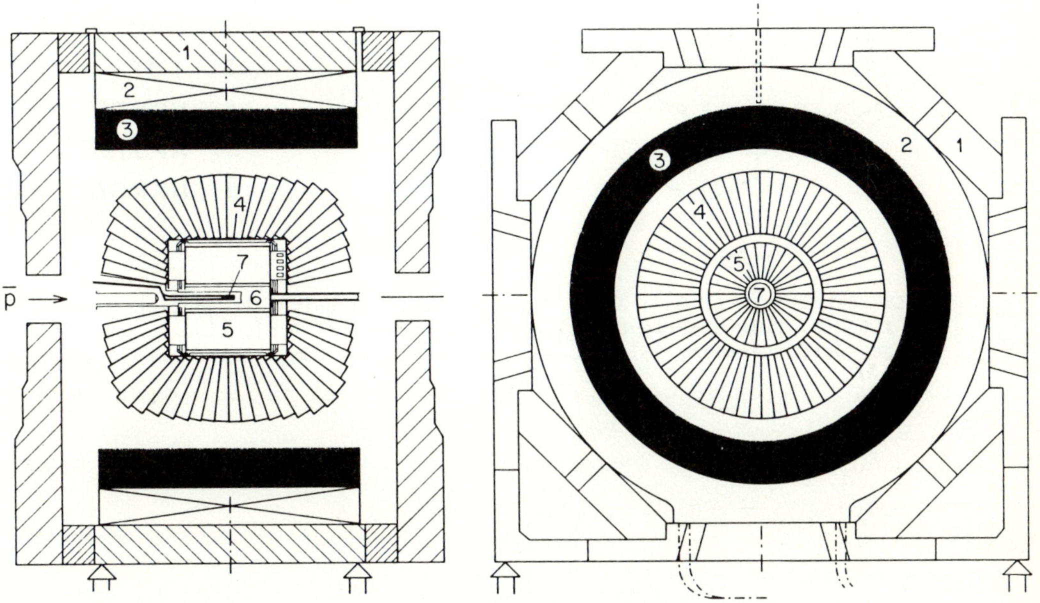

Fig. 3: Crystal Barrel detector: (1) Yoke, (2 & 3) Coils, (4) CsI Barrel, (5) Jet Drift Chamber, (6) PWC, (7) LH_2 target.

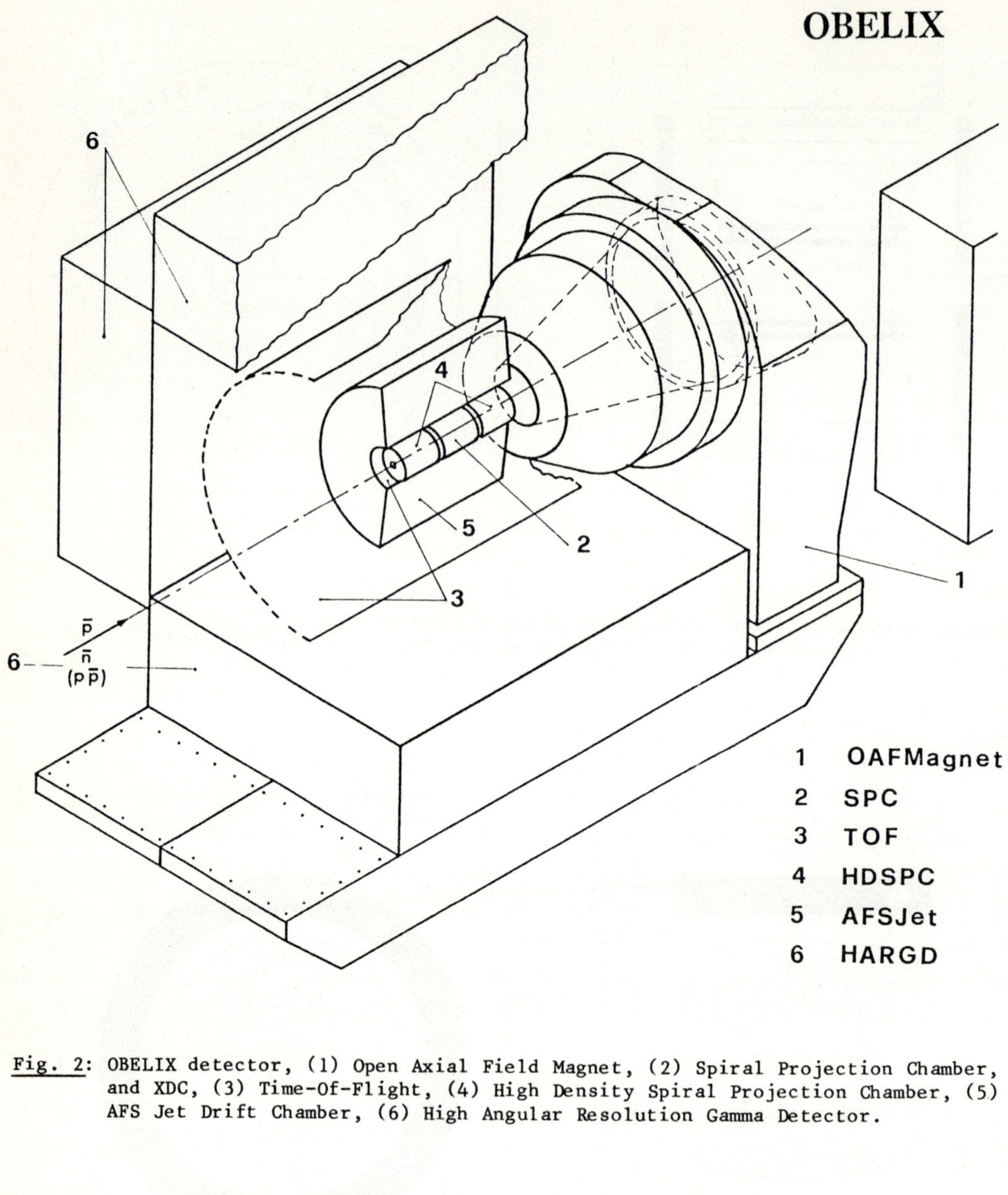

Fig. 2: OBELIX detector, (1) Open Axial Field Magnet, (2) Spiral Projection Chamber, and XDC, (3) Time-Of-Flight, (4) High Density Spiral Projection Chamber, (5) AFS Jet Drift Chamber, (6) High Angular Resolution Gamma Detector.

b) search for effects that could witness quark-gluon aspects of nuclear matter and the existence of highly exited or exotic states of nuclear matter. To this purposes dedicated triggers will be employed to search and measure i) pionless annihilation, ii) single pion annihilation and iii) strange particle production.

4. THE LEAR FACILITY

The reader is referred to the papers given at the last LEAR Workshop[23)] by R. Billinge, E. Jones, P. Lefèvre, D. Simon and D. Möhl who presented in turn the CERN Antiproton Complex, the new Antiproton Collector (ACOL) now under construction, the LEAR ring, the LEAR experimental areas and the LEAR machine developments necessary to accommodate new physics options. Fig. 4 shows schematically the LEAR facility with the experimental area in the 1986 configuration before the shutdown for ACOL.

LEAR delivered beams of antiprotons with no contamination at momenta between 105 and 1600 MeV/c with intensities going from up to 10^5 $\bar{p}$ sec^{-1} at 105 MeV/c to up to 10^6 $\bar{p}$ sec^{-1} at 600 MeV/c. The beam spill is continuous and lasts from 15min. to 5 hrs. Beam intensities are stable in the central momentum region and vary during one spill within typically a factor of 2 at low momenta. Up to three users can work simultaneously in parallel by the action of two beam splitters that divide the beam in the vertical plane. Table 1 summarizes 1986 LEAR operation for physics.

5. THE LEAR PHYSICS PROGRAMME IN THE ACOL TIME

A large number of intentions have been expressed since the third LEAR workshop and are documented in the proceedings[23)]. The experiments at LEAR in the ACOL time approved so far by the CERN Research Board are listed in table 2. An overview is given in ref. 24. The topics besides those already mentioned in this lecture ($N\bar{N}$ annihilation dynamics, spectroscopy of light mesons and exotics , $\bar{p}$ nuclei) include: precise measurements of the $\bar{p}$ inertial mass, measurement of the $\bar{p}$ gravitational mass, measurements of CP violation parameters, spin physics.

Table 1: Distribution of spills during 1986 with slow ejection (typically 1 hr long spills)

		SUD					CENTRE					NORD					
		1		2			1		2			1					2
USERS		Piragino	Smith	Davies	Bressani	V. Egidy	Dalpiaz	Tauscher	Bugg/Bradamante	Walcher		Simons	Gabrielse	Polikanov	Kilian	Uggerhoj	Asterix
		PS 179	PS 183	PS 174	PS 178	PS 186	PS 170	PS 182	PS 172	PS 173		PS 175	PS 196	PS 177	PS 185	PS 194	PS 171
Weeks	p (MeV/c)																
20,26,27,28	105	115		117										107		35	156
18,21,22,25	202	80				78				37		136	15				
19	250									18							
18	309									9							
29,30,31,32	352		98		95		113	53	36						44		
19	400									11							
17	420								28								
30	432		37						37								
17	530								14								
13,17,19	612						190		19								
33	800								21								
32	1100								37								
12,34	1300						14		15						14		
16	1360								13								
33	1369						16								16		
33	1407						15								15		
16	1425																
33	1426						16		17						16		
34	1435						18								18		
33	1444						14								14		
34	1453						17								17		
33	1472						15								15		
16	1475								7								
34	1491						12								12		
33	1506						10								10		
11,12	1512								44								
34	1568						12								12		
Total spills		195	135	117	95	78	462	53	288	75		136	15	107	203	35	156

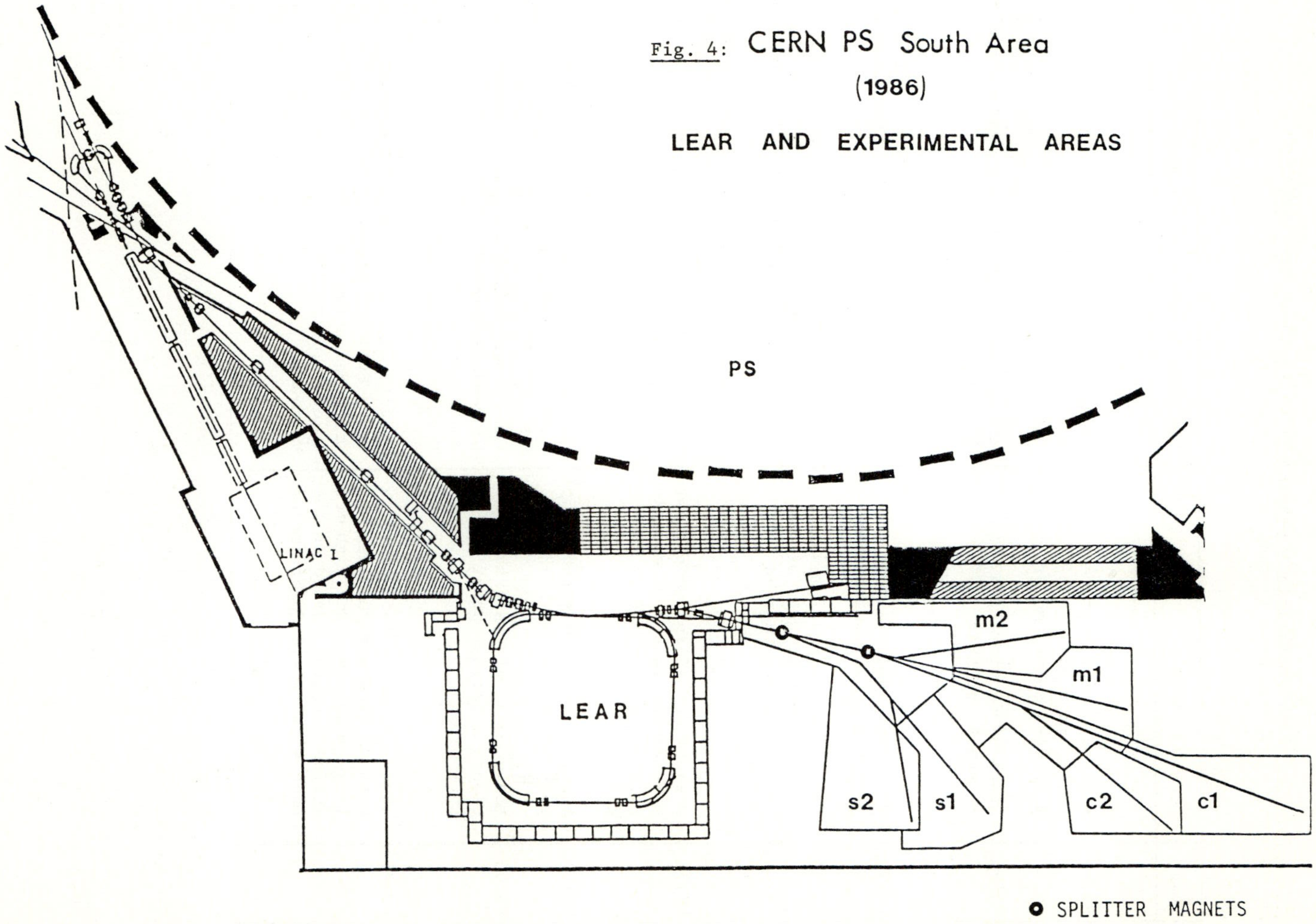

Fig. 4: CERN PS South Area
(1986)
LEAR AND EXPERIMENTAL AREAS

Table 2: Post-ACOL LEAR experiments (November 1986)

Exp	Title	Collaboration	Spokesman
PS189	High-precision measurement of the $p\bar{p}$ mass difference with a RF mass spectrometer	Orsay-CERN	C. Thibault
PS195	Tests of CP violation with K^0 and $\bar{K}^0$ at LEAR	Athens-Basle-CERN Fribourg-Liverpool-Saclay-SIN-Stockholm-Thessaloniki-Zurich	P. Pavlopoulos
PS196	Precision comparison of p & $\bar{p}$ masses in a Penning trap	Fermilab-Mainz-Washington	G. Gabrielse
PS197	The Crystal Barrel: Meson spectroscopy at LEAR with a 4π neutral and charged detector	Berkeley-Irvine-Karlsruhe-London-Mainz-Munich-Pennsylvania State-Strasbourg-Surrey-Vienna-Zurich	H. Koch
PS198	Measurement of spin-dependent observables in the $\bar{p}N$ elastic scattering from 300 to 700 MeV/c	Karlsruhe-Lyon-Saclay-SIN	R. Bertini
PS199	Study of the spin structure of $p\bar{p} \rightarrow n\bar{n}$ channel at LEAR	Cagliari-Geneva-Karlsruhe-Trieste-Turin	F. Bradamante
PS200	Measurement of the gravitational acceleration of the antiproton	Case Western-CERN-Genoa-Houston-Kent State-Los Alamos-NASA-Pisa-Texas A&M	M.V. Hynes
PS201	Study of antinucleon annihilation at LEAR with OBELIX, a large-acceptance and high-resolution detector, based on the Open Axial Field Spectrometer	Brescia-Cagliari-CERN-Dubna-Frascati-Legnaro-Orsay-Padua-Pavia-Trieste-Turin-Udine-Vancouver	U. Gastaldi

References

1. U. Gastaldi, CERN-OBELIX note 11, 1984.

2. OBELIX study group, R. Armenteros et al, in proc. 3rd LEAR Workshop on "Physics with Antiprotons at LEAR in the ACOL Era", Tignes, 1985, eds. U. Gastaldi, R. Klapisch, J.M. Richard and J. Tran Thanh Van (Editions Frontières, Gif-sur-Yvette, 1985), p. 369.

3. OBELIX collaboration, R. Armenteros et al, CERN/PSCC/86-4 (1986).

4. U. Gastaldi, in Atomic Physics 9, eds. R.S.Van Dick and E.N. Fortson (World Scientific, Singapore, 1984), p. 118.

5. U. Gastaldi, in Proc. 4th European Symposium on Antiproton Interactions, Barr (Strasbourg), 1978, ed. A. Friedman (CNRS, Paris, 1979), Vol 2, p. 607.

6. ASTERIX collaboration, R. Armenteros et al, CERN/PSCC/80-101 (1980).

7. ASTERIX collaboration, R. Armenteros et al, in Proc. 2nd LEAR Workshop on "Physics at LEAR with Low-Energy Cooled Antiprotons", Erice, 1982, eds. U. Gastaldi and R. Klapisch (Plenum, New York, 1985), p.109.

8. P. Dalpiaz, U. Gastaldi, K. Kilian and M. Schneegans, CERN Workshop on Intermediate Energy Physics, Geneva 1977.
U. Gastaldi, K. Kilian and G. Plass, A low-energy antiproton facility at CERN: physics possibilities and technical aspects, CERN/PSCC/79-17 (1977).

9. ASTERIX collaboration, S. Ahmad et al, S and P-wave $p\bar{p}$ annihilation at rest in a H_2 gas target, in preparation.

10. ASTERIX collaboration, S. Ahmad et al, in "Fundamental Interactions in low energy systems", eds. P. Dalpiaz, G. Fiorentini and G. Torelli (Plenum, New York, 1985), p. 279.

11. T.B. Day et al, Phys. Rev. 118 (1960)864.

12. ASTERIX collaboration, S. Ahmad et al, Phys. Lett. 157B(1985)33.

13. U. Gastaldi, Nucl. Instrum. Methods 157(1978)441.

14. ASTERIX collaboration, S. Ahamd et al, to be published in Proc. of the 7th European Symposium on $N\bar{N}$ interactions, Thessaloniki 1986.

15. ASTERIX collaboration, S. Ahamd et al, in same proceedings as ref. 2, p. 353.

16. ASTERIX collaboration, S. Ahamd et al, in same proceedings as ref. 7, p. 253.

17. U. Gastaldi, Nucl. Instrum. Methods 188(1981)459.

18. ASTERIX collaboration, U. Gastaldi et al, Construction and operation of the Spiral Projection Chamber of the ASTERIX experiment, to be submitted to Nucl. Instrum. Methods.

19. AFS collaboration, O. Botner et al, Nucl. Instrum. Methods 196(1982)315 and H. Gordon et al, Nucl. Instrum. Methods 196(1982)305.

20. Crystal Barrel collaboration, E. Aker et al, CERN/PSCC/85-56 (1985).

21. G. Piragino, in "Hadronic Physics at Intermediate Energy", Folgaria Winter School 1986, eds. T. Bressani and R.A. Ricci (North Holland, Amsterdam, 1986), p.293.

22. Streamer Chamber collaboration, F. Balestra et al, CERN/PSCC/80-78 (1980).

23. Same proceedings as in ref. 2, p. 13, 25, 33, 47 and 65.

24. R. Landua, CERN-EP/86-136 (1986).

PART V

FEW-BODY SYSTEMS WITH CHARGED PARTICLES
AND
CALCULATION OF ELECTROMAGNETIC OBSERVABLES

CHARGED-PARTICLE INTERACTIONS IN FEW-BODY SYSTEMS

L.P. Kok
Institute for Theoretical Physics
University of Groningen
P.O. Box 800, 9700 AV Groningen
The Netherlands

The quantum-mechanical scattering theory of charged paricles is reviewed. After some historical remarks we look back at classical scattering for long-range and singular potentials as an introduction to the difficulties that recently have been recognized for few-paricle scattering with charges. We review two-body scattering, effective-range theory, quantum-defect theory, three-body scattering, without and with charged particles. Emphasis is on the fundamental theoretical issues that were raised by the numerical results of practical implications of mathematically satisfactory theories during the past few years.

1. HISTORICAL PERSPECTIVE

1986 is a very special year. It happens to be three hundred years ago that Sir Isaac Newton completed his famous "Philosophiae Naturalis Principia Mathematica" (The Mathematical Principles of Natural Philosophy). The title page carries the date Julii 5. 1686, the preface that of May 8, 1686 (Motte 1729, Cajori 1946). The work was subsequently printed and published in MDCLXXXVII(Newton 1687). In the Ode Dedicated to Newton by Edmund Halley, prefixed to the principia of Newton, the second verse reads, translated form the Latin,

The inmost places of the heavens, now gained,
Break into view, nor longer hidden is
The force that turns the farthest orb. The sun
Exalted on his throne bids all things tend
Toward him by inclination and descent,
Nor suffers that the courses of the stars
Be straight, as through the boundless void they move,
But with himself as centre speeds them on
In motionless ellipses. Now we know
The sharply veering ways of comets, once

A source of dread, nor longer do we quail
Beneath appearances of bearded stars.

Book I, Propositions 1, 11-13, and 15, contains the mathematical foundations for his gravitational force ('inversely as the square of the distance') and for Johannes Kepler's three laws of planetary motion, publisehd in 1609 and 1619. In Proposition 1 Newton proves Kepler's second law, on the conservation of areal velocity: "The radius vector sweeps out equal areas in equal times". This is a general theorem for motion under a central force.

1986 is a very special year. It happens to be exactly 250 years ago that Charles Augustin de Coulomb was born, at 17 June 1736 at Angoulême, France.

1986 is a very special year. It happens to be two hundred years ago that Coulomb presented experimental evidence for the existence of the inverse-square law of forces for repulsive (in 1785) and attractive (1987) electrical charges to the Académie Royale des Sciences, Paris [Gillmor (1971): pp. 182 and 232].

1986 is a very special year. It happens to be one century ago that life started of one of the great founders of quantum mechanics, Erwin Schrödinger (*12 August 1887, Vienna). Moreover, one century ago was the time that Balmer (1885) had just published his empirically found power-law series of frequencies in the absorption and emission spectra of the hydrogen atom.

1986 is a very special year. It is sixty years ago that the Schrödinger equation (1926) was formulated. The Born approximation of the Coulomb scattering amplitude was calculated by Wentzel (1926). Schrödinger (1926) published the partial-wave and three-dimensional Coulomb scattering wave functions. Pauli (1926) found the discrete spectrum of the Coulomb Hamiltonian and thus the energy levels of the hydrogen atom.

1986 is a very special year. This spring, in April, Halley's comet returned to perihelion. It is the most famous of all comets and the one with the longest history of recorded observations. Its orbit consists of an ellipse with large eccentricity with the sun as one of its focal points, perturbed by gravity forces of objects other than the sun.

1986 is a very special year. The Eighth Autumn School, on "Models and Methods in Few-Body Physics" takes place in Lisbon.

The inverse-distance-square law of force is fundamental in nature. The force $\vec{F}$ (or potential V; $\vec{F} = -\vec{\nabla}V$) between electric charges can be attractive or repulsive. The Coulomb potential takes a special place in both classical and quantum mechanics. In both cases the two-body

Coulomb problem is exactly and analytically solvable. Much of the formalism described in the following (and much not described there) can be found in the book by van Haeringen (1985).

1.1. Specific Coulomb peculiarities

The two-body Coulomb problem is exactly and anlytically solvable, both in classical and quantum mechanics. There are several interesting differences between the Coulomb potential and an exponentially bounded potential. All of these can be traced back to its long range, by which is meant that it does not approach zero sufficiently fast for increasing distance. We mention the following:

1° The scattering orbits (classical) or states (quantum) do not approach free orbits or states, respectively;

2° The scattering wave function has a different asymptotic behavior ($kr - \gamma \ln(2kr)$ rather than kr, for notation see Secs. 2 and 3) and related to this, the phase shift has to be defined differently;

3° The scattering amplitude is singular in the forward direction, the differential cross section is not finitely integrable, and hence the total cross section is infinite;

4° The number of bound states is infinite;

5° The effective-range function fails to be meromorphic at zero energy;

6° The T matrix, considered as a complex function of the complex energy variable, possesses special singularities: It has
   (i) an essential singularity at zero energy, and
   (ii) branch-point singularities at the so-called half-shell points.

Some of these peculiarities, or "anomalies", are specific for the Coulomb potential, while some also occur in scattering by short-range potentials that are not exponentially bounded.

2. CHARGED-PARTICLE SCATTERING: TIME DEPENDENT

We shall briefly discuss time-dependent scattering for the classical and the quantum case, respectively.

2.1. Classical scattering

It is instructive to see the roots of time-dependent quantum scattering in classical mechanics. Typically, a classical scattering trajectory has three parts (1) the in part, along a(n almost) straight line, (2) the interaction-region part with possibly very complicated orbits, (3) the out part along some other approximately straight orbit.

In atomic and subatomic scattering only parts (1) and (3) are experimentally accessible. In and out parts of scattering orbits in general can be characterized by six real numbers $(\vec{d},\vec{v})$, e.g., $\vec{r}_{in}(t) = \vec{d}_{in} + \vec{v}_{in}t$. One expects that any real six numbers should represent a possible in asymptote, that uniquely defines a corresponding out asymptote, and vice versa. One must in general expect in addition to scattering orbits also bounded orbits. These have no in and out asymptotes.

The reasonable orbit properties sketched in the previous paragraphs do not hold for all potentials. Two important factors are (i) If the potential does not fall off sufficiently fast at infinity the scattering orbits have no straight-line asymptotes. The Coulomb potential is the notorious example for which the scattering orbits never approach free asymptotes, but instead

$$\vec{r}(t) \to \vec{d}_{out} + \vec{v}_{out}t + \vec{c}_{out}\log t \text{ for } t \to \infty. \tag{2.1}$$

(ii) If the potential is too attractive near $r = 0$ (e.g., $V = -1/r^3$) a particle coming in from infinity with a perfectly acceptable in asymptote may be caught in a spiralling orbit and never emerge; hence it has no out asymptote, vice versa. Conservation of energy in this case provides no argument against such an orbit. Newton (1982) Ch.5 and van Haeringen (1985) Ch.8 discuss spiralling orbits and corresponding potentials in more detail.

2.2. Quantum scattering

The description of quantum scattering is very analogous to the classic formalism outlined in the previous section. Instead of the classical orbit $\vec{r}(t)$ satisfying Newton's equation, we now have a state vector $\psi(t)$ satisfying the time-dependent Schrödinger equation

$$i \frac{d}{dt} \psi(t) = H \psi(t). \tag{2.2}$$

For conservative systems - which we shall always be considering - the Hamiltonian H = H or V is independent of t and the general solution is

$$\psi(t) = U(t)\psi(0) \equiv e^{-iHt}\psi(0), \tag{2.3}$$

where $U(t) := \exp(-iHt)$ is the so-called <u>evolution operator</u>, and $\psi(0) \equiv \psi$ is any vector in the appropriate Hilbert space $\mathcal{H}$. The solution $\psi(t)$ is called and <u>orbit</u> in analogy with the classical terminology. Let us suppose that the orbit $U(t)\psi$ decribes the evolution of some scattering process. This means that for $t \to -\infty$, and for $t \to +\infty$, $U(t)\psi$

represents a wave packet that is localized far away from the scattering center and, therefore, behaves like a free wave packet. The motion of a free particle is given by the "free evolution operator" $U_0(t) := \exp(-iH_0t)$, and therefore we expect that

$$U(t)\psi \to U_0(t)\psi_{in}, \quad t \to -\infty, \qquad U(t)\psi \to U_0(t)\psi_{out}, \quad t \to +\infty \tag{2.4}$$

for some vectors ψ_{in} and ψ_{out}. This is analogous to the classical case, and in analogy with the classical terminology the asymptotic free orbits in this equation are called the in and out asymptotes of the actual orbit $U(t)\psi$.

For a reasonable scattering theory one imposes important conditions, (C1) - (C3):

(C1) The asymptotic condition: "There is always a ψ satisfying (2.4), for every ψ_{in} and for every ψ_{out}".

As in the classical case we expect that not every orbit $U(t)\psi$ will have asymptotes, but rather that only certain scattering orbits will have asymptotes, and that the scattering states together with the bound states will span the space $\mathcal{H}$ of all states. This leads to two more conditions:

(C2) Orthogonality: "Any bound state is orthogonal to all states with in or out asymptotes", and

(C3) Asymptotic completeness: "The three classes: (i) all states with in asymptotes, (ii) all states with out asymptotes, and (iii) all states orthogonal to the bound states, are identical".

Thus the Hilbert space $\mathcal{H}$ is divided into two orthogonal parts: the subspace spanned by the bound states and the subspace of the scattering states, and $\mathcal{H}$ is the direct sum of these two subspaces. For every scattering state ψ the orbit $U(t)\psi$ describes a scattering process with in and out asymptotes according to (2.4), and every ψ_{in} (or ψ_{out}) in $\mathcal{H}$ labels the in (or out) asymptote of a unique actual orbit $U(t)\psi$. From (2.4) one obtains

$$\psi = \Omega_+ \psi_{in} = \Omega_- \psi_{out}, \tag{2.5}$$

where (s-lim signifies the strong limit (Taylor, 1972))

$$\Omega_\pm := \operatorname*{s-lim}_{t \to \pm\infty} U(t)^\dagger U_0(t) \tag{2.6}$$

are the Møller wave operators. [The choice of the subscripts ± for $t \to \pm\infty$ is convenient in the stationary scattering theory.] In the special case where H does not have bound states Ω_+ and Ω_- are unitary.

So far we have expressed the actual scattering state in terms of either of its two asymptotes. Our ultimate goal is to express the out asymptote in terms of the in asymptote without reference to the experimentally inaccessible actual orbit, and this we can now do. By multiplying on the left by $\Omega_-^\dagger$ we obtain from Eq. (2.5)

$$\psi_{out} = \Omega_-^\dagger \Omega_+ \psi_{in} = S\ \psi_{in}, \tag{2.6}$$

where

$$S := \Omega_-^\dagger\ \Omega_+ \tag{2.7}$$

is the scattering operator. The experimentally observable differential cross section can be expressed directly in terms of the matrix elements of S. An important property of the S operator is that it is unitary.

To give a physically reasonable scattering theory the potential must satisfy certain conditions. Unfortunately, the conditions under which some of the principal results have been proved are quite complicated; different proofs use different conditions. Further, a set of conditions that is both sufficient and necessary for all results is not known. For a local spherically symmetric potential a condition that is always necessary is that $\int_a^\infty |V(r)|\ dr < \infty$ for all positive a. Clearly this excludes the Coulomb potential. In fact, none of the principal results of "standard" scattering theory do hold for the Coulomb potential. Further, at the origin the potential should be $O(r^{\varepsilon-3/2})$, $\varepsilon > 0$. Potentials more singular than $r^{-3/2}$ are sometimes called "singular". (Repulsive singular potentials can be included in scattering theory, but with special treatment.) In discussions of the radial Schrödinger equation, the term "singular potential" is often used for a potential more singular than r^{-2} at $r = 0$, roughly speaking. However, in a general three-dimensional analysis of the scattering process the difficulties start at $r^{-3/2}$ [see Hunziker (1968) and references cited therein] . See further Taylor (1972) Ch.2 and Reed and Simon Vol. III (1979): Secs. XI.4, XI.5, XI.9.

2.3. Scattering à la Dollard

The existence of Møller operators and scattering states depends on whether the asymptotic dynamics is determined by the free Hamiltonian

H_0 alone ($H = H_0 + V$). For potentials $V = V_s + V_c$ with a Coulomb tail this is not the case: even at asymptotic times the 1/r tail is felt. Dollard (1964) has shown a way to take into account properly the asymptotic distortion, by introducing a modifield operator

$$\Omega_\pm^{mod} = \underset{t\to\mp\infty}{\text{s-lim}} \exp(iHt)\exp(-iH_{as}(t)), \tag{2.8}$$

with

$$H_{as}(t) = H_0 t - s\, H_0^{-\frac{1}{2}} \operatorname{sgn}(t)\ln(4H_0|t|), \tag{2.9}$$

leading to physical scattering states as before

$$\Omega_\pm^{mod}|\vec{k}\rangle = |\vec{k}\pm\rangle. \tag{2.10}$$

The pure Coulomb Møller operators $\Omega_\pm^C$ are defined as in (2.8) with H taken as

$$H_c = H_0 + V_c, \tag{2.11}$$

and give the Coulomb scattering states

$$\Omega_\pm^C\ |\vec{k}\rangle = |\vec{k}\pm\rangle_c, \tag{2.12}$$

which are explicitly known.

Chandler (1986) has recently reviewed attempts to include as much as tractable in the asymptotic dynamics. He writes (2.8) as

$$\Omega_\pm^D = \underset{t\to\pm\infty}{\text{s-lim}} \exp(iHt)U_D^\pm(t)\exp(-iH_0 t) \tag{2.8D}$$

where $U_D^\pm(t)$ obviously acts on a square integrable function $\psi(\vec{x})$ as

$$[U_D^\pm\psi](\vec{x}) \equiv (2\pi)^{-3/2}\int d^3k\langle\vec{k}\,|\psi\rangle\exp\{i\vec{k}.\vec{x}\pm i\phi^\pm\} \tag{2.9D}$$

with $\phi^\pm$ an appropriate phase derivable from (2.9).
He proceeds by listing the Mulherin-Zinnes (1970) wave operators, which, though differently defined are the same as those of Dollard. Chandler points out that proposals to include more asymptotic dynamics have been put forward by Hörmander (1976), leading to equations similar to (2.8D-9D) with D replaced by H, and by Isozaki and Kitada (1985). He stresses the potential importance of this work.

2.4. Scattering with screened Coulomb potentials

Screened Coulomb potentials do not have the difficulties associated with the long range of V_C. Therefore these difficulties should appear in the limit of unscreening. Let

$$V_R(r) = V_C(r)g(r/R) \tag{2.13}$$

with R the screening radius. The screening function g(x) should tend to zero for $x \to \infty$, and to 1 for $x \to 0$, for example $g(x) = \exp(-x)$. Replacing V_C by V_R Møller operators $\Omega_\pm^{(R)}$ can be defined via Eq. (2.6) for the full interaction

$$V^{(R)} = V_s + V_R. \tag{2.14}$$

Their existence is guaranteed by the short-range character of the interaction (2.14). The difficulty arises in the limit $R \to \infty$, because they do not have a well-defined limit, unless they are suitably renormalized (Dollard 1964; Gorshkov 1961; Taylor 1974). Indeed, the following limit does exist:

$$\underset{R\to\infty}{\text{s-lim}} \; \Omega_\pm^{(R)} \; Z_R^{\mp\frac{1}{2}} = \Omega_\pm^{\text{mod}} \tag{2.15}$$

with the r.h.s. as in Eq. (2.8). The renormalization factor Z_R can be calculated (Taylor 1974) for any screening function g(r/R): It gives a pure phase factor

$$(Z_R f)(\vec{k}) = \exp(2i\phi_R)f(\vec{k}), \tag{2.16}$$

which diverges in the limit $R \to \infty$, though of course its magnitude remains 1.

We refer to Alt's pedagogical review (1986) for the steps to the screened total and the screened Coulomb scattering operator $S^{(R)}$ and S^R, respectively. It shows that the following unscreening limits exist

$$S^{(R)} \; Z_R^{-1} \to S \qquad \text{for } R \to \infty \tag{2.17}$$

$$S^R \; Z_R^{-1} \to S_c \qquad \text{for } R \to \infty. \tag{2.18}$$

The important message is that time-dependent scattering theory, either appropriately modified as suggested by Dollard, or on the basis on the screening and renormalization approach, leads to the same results for the amplitude describing the scattering via potentials of the type $V_s + V_c$.

3. CHARGED-PARTICLE SCATTERING: STATIONARY

3.1. Transition form time-dependent to time-independent theory

The transition to the stationary formalism has been described in detail in the literature. Again the comprehensive review by Alt (1986) is a valuable guide. For the pure Coulomb potential $V = V_c = -2s/r \equiv 2k\gamma/r$ (γ is the Sommerfeld parameter, which is energy dependent) virtually all scattering quantities are explicitly known (see van Haeringen 1985, pp. 265-409 for a list of 145 pages of pure Coulomb formulas). We shall mention but a few.

The potential V_c has matrix elements

$$\langle\vec{p}|V_c|\vec{p}'\rangle = k\gamma\pi^{-2}|\vec{p} - \vec{p}'|^{-2}$$
$$\langle p|V_{c\ell}|p'\rangle = 2k\gamma(\pi pp')^{-1}Q_\ell(w), \qquad w = (p^2 + p'^2)(2pp'). \tag{3.1}$$

The Rutherford amplitude is

$$A_c(\theta) = -\tfrac{1}{2}\gamma k^{-1}\exp(2i\sigma_0)(\sin^2\tfrac{1}{2}\theta)^{-1-i\gamma}$$
$$= -\tfrac{1}{2}\gamma k^{-1}\exp(2i\sigma_0)(\tfrac{1}{2} - \tfrac{1}{2}\cos\theta)^{-1-i\gamma}. \tag{3.2}$$

The Rutherford differential cross section is

$$\frac{d\sigma}{d\Omega} = |A_c(\theta)|^2 = \frac{\gamma^2}{4k^2\sin^4\frac{1}{2}\theta} . \tag{3.3}$$

Exactly the same formula holds for the differential cross section for the Coulom potential in classical scattering theory, and also in first Bron approximation in quantum scattering theory.

The Coulomb phase shifts are given by

$$\sigma_\ell = (2i)^{-1}\ell n[\Gamma(\ell + 1 + i\gamma)/\Gamma(\ell + 1 - i\gamma)], \tag{3.4}$$

with σ_ℓ real when γ is real. Here $\Gamma(.)$ is the gamma function.

The S matrix is related to physical scattering states $|\vec{k}\pm\rangle$. This is clear from Eqs. (2.5-2.12), and application of the Møller operator onto plane waves $|\vec{k}\rangle$, which are eigenfunctions of H_0 with the eigenvalues $E = k^2 + i\varepsilon$, $\varepsilon \downarrow 0$. In fact

$$\langle\vec{k} -|\vec{k}' +\rangle = \langle\vec{k}|S|\vec{k}'\rangle \tag{3.5}$$

and p.w. projection gives for short-range potentials

$$\langle k\ell -|k'\ell +\rangle = \langle k\ell|S_\ell|k'\ell\rangle = k^{-2}\delta(k - k')\exp(2i\delta_\ell(k)), \quad (3.6)$$

where $\delta_\ell(k)$ is a phase shift. The relation between $|\vec{k}+\rangle$ and $|\vec{k}-\rangle$ is in general complicated, that between $|k\ell+\rangle$ and $|k\ell-\rangle$ is very simple:

$$|k\ell+\rangle = \exp(2i\delta_\ell(k))|k\ell-\rangle, \quad k > 0. \quad (3.7)$$

If the potential has a short-range one can extract a δ function from the S matrix:

$$\langle\vec{k}|S|\vec{k}'\rangle = \delta(\vec{k} - \vec{k}') - i\pi k^{-1}\delta(k - k')t(\vec{k},\vec{k}'). \quad (3.8)$$

Here t is continuous at $\vec{k} = \vec{k}'$. Because of the factor $\delta(k - k')$, $t(\vec{k},\vec{k}')$ is defined only on the (energy-)"shell" $\vec{k}'^2 = \vec{k}^2$. For this reason $t(\vec{k},\vec{k}')$ is called the transition (T) matrix on the energy shell, or just the on-shell T matrix.

It is closely related to the scattering amplitude A:

$$A(\vec{k},\vec{k}') = -2\pi^2 t(\vec{k},\vec{k}'), \quad k' = k. \quad (3.9)$$

It is customary to define an energy-dependent operator $T(k^2)$ whose matrix elements coincide with $t(\vec{k},\vec{k}')$ when $k = k'$. The matrix $\langle\vec{p}|T(k^2)|\vec{p}'\rangle$ is called the off-shell matrix. When either $p = k$ or $p' = k$ the T matrix is called the "half-off-shell" T matrix.
For short-range potentials one easily determines the relation between the on-shell p.w. projected T matrix and the phase shift:

$$\langle k|T_\ell(k^2)|k\rangle = -2/(\pi k)\exp(i\delta_\ell(k))\sin(\delta_\ell(k)), \quad k > 0. \quad (3.10).$$

In spite of the 1/r Coulomb tail, the Coulomb T operator T_C can perfectly be defined, however its (off-shell) matrix elements have no on-shell limit. Moreover, T_C and many quantities derivable from T_C have a so-called essential singularity at $k = 0$ (i.e., at zero energy). There are several ways around these difficulties in the stationary theory.
One possibility is to work with "distorted free Green's functions", or with "distorted free states". In this way one can guarantee well-defined on-shell limits for the resulting T matrix (Schwinger 1964; Okubo and Feldman 1960), proportional to the scattering amplitude. Similarly results have been proven for the p.w. T matrix (van Haeringen

and van Wageningen 1975; van Haeringen 1983; Dusek 1982; 1983), for scattering wave functions, and the off-shell Jost function (van Haeringen 1978; Talukdar et al. 1984).

Another possibility, when one insists on using the same basic definitions as for short-range potentials, is to use a renormalization procedure to perform the half- and on-shell limits which lead to the physically interesting quantities. For example, using Coulomb asymptotic states $|\vec{k}\,\infty\pm\rangle$ we have

$$A_C(\theta) = -2\pi^2 \langle \vec{k}'\,\infty - |T_C|\vec{k}\,\infty\rangle, \tag{3.11}$$

but also the renormalization function $\Omega(k,p,\gamma)$ can be used:

$$\Omega(k,p,\gamma) = \exp(\tfrac{1}{2}\pi\gamma)/\Gamma(1 - i\gamma) \lim_{\varepsilon\downarrow 0} [(p - k - i\varepsilon)/(p + k + i\varepsilon)]^{-i\gamma} \tag{3.12}$$

so that

$$\lim_{p\to k} \Omega(k,p,\gamma)\, T_{C\ell}|p\rangle = T_{C\ell}|k\ell\infty+\rangle_C = V_{C\ell}|k\ell+\rangle_C. \tag{3.13}$$

3.2. Solutions are known for several cases

For pure Coulomb scattering explicit solutions for many quantities are known. A systematic list of formulas is given in Ch.C van Haeringen (1985), pp. 265-409.

Some of the representations for the matrix elements of T_C that appear there are new. For Coulomb-like potentials $V_S + V_C$ solutions can be found only when V_S is separable.

There exists a vast literature on this subject. The most complete collection is contained presumably in Ch.CS of the above-mentioned book by van Haeringen (1985), pp. 411-447.

3.3. The screening approach in stationary scattering

For finite screening radius R (cf. Sec. 2.4) the standard formalism for short-range potentials can be used for the calculation of scattering amplitudes etc. The unscreening problem is manageable in practice. The phase $\phi_R(k)$ of the diverging factor (see Eq. (2.16) and Taylor 1974) can be calculated from

$$\phi_R = -\frac{1}{2k} \int_{1/2k}^{\infty} dr\, V_C(r) g(r/R). \tag{3.14}$$

In the limit of large R this agrees with the renormalization factor introduced in the time-dependent approach. Care has to be exercised when g is taken to be a sharp cut-off function (Goodmanson and Taylor 1980).

Of particular importance is the behavior of phase shifts when R increases (cf. Alt 1986). If $\delta_\ell(R)$ is the phase shift for the full screened potential, $\delta_{R,\ell}$ that for the screened Coulomb potential, then (Taylor 1974) for $R \to \infty$

$$\delta_{R,\ell} \sim \sigma_\ell + \phi_R \tag{3.15}$$

$$\delta_\ell(R) \sim \delta_\ell + \phi_R \tag{3.16}$$

so that the diverging phase ϕ_R cancels in the determination of the so-called Coulomb-modified short-range phase-shift,

$$\delta_{Rs,\ell} = \delta_\ell(R) - \delta_{R,\ell} \sim \delta_\ell - \sigma_\ell = \delta_{cs,\ell}\ . \tag{3.17}$$

3.4. Integral equations and screening

The experimentally measurable on-shell quantities cannot be computed directly from some integral equation. Instead, the appropriate off-shell quantities can be determined from well-behaved integral equations, after which the on-shell limit can be taken to compute the corresponding unknowns.

One has to distinguish two philosophies, that make use of the screening approach. On the one hand this method can be considered merely as a particularly convenient way to define the quantities of interest. Then taking advantage of the fact that all pure Coulomb entities are explicitly known, there remains the unscreened Coulomb-modified short-range operator T_{cs} (or related quantities) to be calculated. For two-body scattering this is practical, indeed. However, for three-body systems the situation is much more unfavorable.

As an alternative one may use the screening and renormalization approach as computational recipe. The basic idea is that for finite R short-range theory works, and the on-shell limit, say, $\langle\vec{k}|T^{(R)}(k^2 + i\varepsilon)|\vec{k}'\rangle$ exists. Multiplying the latter with the renormalization factor Z_R^{-1}, and repeating the computation for increasing values of R until an R-independent result is obtained "the unscreening limit is performed numerically".

In praxis one computes in a p.w. basis, and one repeats with increasing R values the calculations till the δ_{RS} of (3.17) is stable. One then sums the p.w. series, and adds explicitly the left-out known pure Coulomb amplitude. This strategy proves profitable especially in the three-body case.

4. CHARGED-COMPOSITE-PARTICLE SCATTERING: TIME DEPENDENT

If in a process more than two (charged) particles take part, its description becomes much more complicated for several reasons:
(i) If in an N-body system ($N \geqq 3$) at least some particles can form bound states, many dynamical processes can occur, even if only the experimentally accessible two-cluster incoming configurations are considered. In addition to elastic scattering, there are excitation, rearrangement, and breakup processes. These are discussed by other speakers at this meeting.
(ii) No pure Coulomb quantities like scattering wave function, Green's functions, etc., are known in analytical form for $N \geqq 3$. This is not surprising because all the dynamical complexities mentioned in (i) are present here too, and even in the two-body case many Coulomb quantities are not immediately simple.

4.1. Classical scattering

Classical N-body scattering for $N > 2$ is a well-developed branch of astronomy, where classical orbits of objects influencing mutual Newtonian r^{-1} interactions are computed.

4.2. Quantum scattering

First, one has to establish a notation. Usually one assumes that only pair interactions contribute to the total potential. Pair interactions are of the type discussed in Secs. 2 and 3; short-range plus possibly Coulomb. In N-body scattering different fragmentations can exist, labelled as channels "a". For each channel there is a channel-internal interaction V_a which consists of those pair potentials not connecting two fragments, and the corresponding channel Hamiltonian $H_a = H_0 + V_a$. The remaining part of the potential is the channel-external interaction.

4.3. Scattering à la Dollard

Cluster Møller operators can be defined, as was shown by Dollard (1971) in a generalization of his two-body results. For each channel an asymptotic Hamiltonian can be defined as in (2.9), that takes into account the asymptotic distortion of the motion of each of the charged clusters in the Coulomb field produced by the other. In the two-body case the phase in (2.9D) could be fixed by the requirement that all analytically known results are reproduced. Unfortunately, this is not possible for $N > 2$.

A variety of representations, more practical than that of Dollard has been proposed, see for example Alt (1986). It is possible to obtain two-fragment Møller operators acting on the relative motion of the clusters only, and not on their internal structure.

S operators, which now connect different channels can be introduced again.

4.4. Scattering with screened Coulomb potentials

We follow the structure of Sec. 2, and point out that the screening approach can be generalized to N-paricle scattering in several ways. Screened cluster Møller operators can again be connected to the full cluster Møller operators by introducing a renormalization factor $Z_{a,R}^{\pm \frac{1}{2}}$ in the unscreening limit, cf. Eq. (2.15). Similarly, screened S operators can be defined with well-defined limits using renormalization factors, e.g.,

$$Z_{b,R}^{-\frac{1}{2}} \, S_{ba}^{(R)} \, Z_{a,R}^{-\frac{1}{2}} \rightarrow S_{ba} \quad , \; R \rightarrow \infty, \tag{4.1}$$

which generalizes Eq. (2.17).

Concluding we remark that well-defined Møller and S operators exist, as can be seen in Dollard's as well as the unscreening approach. Both methods lead to the same answer.

The additional ingredient for the theory to be complete, the so-called asymptotic completeness (cf. Sec. 2.2) is much harder to prove (e.g., Enns, 1979; Merkuriev 1980). Numerical computation of three-body S operators is reported by Kröger and Slobodrian (1984).

5. CHARGED-COMPOSITE-PARTICLE SCATTERING: STATIONARY

5.1. Transition from time-dependent to time-independent theory

Again we refer to Alt's presentation (1986) for full details. It restricts N to 3, making room for more elaborate notation. Because in the following we shall treat one such case in detail we do not give the general formalism here. Suffices here to state that for charged-particle scattering when the complete-energy variable E approaches the real axis in the scattering region, specific Coulomb singularities occur. These, however, are understood and can be cured by renormalization factors

similar to those in the two-body case.
A variety of methods has been proposed both in coordinate-space formulations (in particular Merkuriev 1980, 1981), and using integral equations in momentum space below (Faddeev 1969; Veselova 1970, 1973, 1978) and above breakup threshold. For a more complete list of references see Alt (1986).

5.2. Analytical solutions are not known explicitly

As vast as the literature on the corresponding two-body case, as small is the list of known three-body Coulomb formulas.

5.3. The screening approach in stationary scattering

The screening approach has been used as a practical tool in three-body computations. The renormalization factors for $2 \rightarrow 2$ scattering are analogous to those of the two-particle scattering, cf. Secs. 3.3-4. Most of the actual computations have done for the p + d scattering system. In three-body scattering the screening approach seems to be competative in comparison with methods of direct calculation of scattering amplitdes. This is unlike the two-body case, where the latter method usually works faster. The reason for this is reportedly the non-Fredholm character of the integral equations for energies beyond the lowest breakup threshold.

6. TWO-BODY BOUND STATES

6.1. Poles of the T matrix

Two-body bound states may be found from the solution of the partial-wave (p.w.) Schrödinger equation in the r representation. For local potentials V(r) this requires solution of an eigenvalue differential equation. For nonlocal potentials it is often convenient to solve the problem in the momentum representation, as suggested first by Yamaguchi (1954). The resulting problem is the same as the one obtained form considering the general scattering problem through the Lippmann-Schwinger equations for T matrices or resolvent (Green) operators.
In these integral equations the boundary conditions to be imposed on scattering or bound-state solutions can easily be incorporated. In particular bound-state equations are usually the homogeneous versions of the scattering equations, i.e., obtainable by deleting the inhomogeneous term.
Simple separable potentials, defined by

$$V_s \; := -\lambda_\ell \; |g_\ell\rangle\langle g_\ell| , \tag{6.1}$$

$$\langle p|g_\ell\rangle \; := \; (2/\pi)^{\frac{1}{2}} \; p^\ell \; (p^2 + \beta^2)^{-\ell-1} , \qquad \beta > 0 , \tag{6.2}$$

and hence

$$\langle r|g_\ell\rangle = (\tfrac{1}{2}i)^\ell \; r^{\ell-1} \; \exp(-\beta r)/\ell! , \tag{6.3}$$

are very-easy-to-use models for short-range interactions. Bound states correspond to poles of the S or T matrix in the complex k plane on the positive imaginary axis, cf. Fig. 6.1.
Other poles of T are situated in the lower half k plane, and occur in mirror pairs (resonance poles and conjugates), or on the negative imaginary k axis near the origin (virtual states).

It is instructive to see the movement of the poles when the strength λ_ℓ of the interaction is varied (de Alfaro and Regge 1965; McVoy 1967; Burke 1977). Pioneering work here is that of Nussenzweig (1959),
For $\ell = 0$ there are virtual states just before the interaction is sufficiently strong to support bound states. For $\ell > 0$ two pair poles collide and scatter by $\pi/2$ at the origin to give a bound-state pole.
[Exercise: Calculate the trajectories of the poles for the interaction (6.1-2) in case $\ell = 0$, and $\ell = 1$].
The fact that the bound state appears when the resonance disappears is connected to interplay of repulsive centrifugal barrier and the short-range attraction.
For the Coulomb interaction ($\hbar = 1 = 2\mu$)

$$V_c(r) = V_{c\ell}(r) = -2s/r \equiv 2k\gamma/r . \tag{6.4}$$

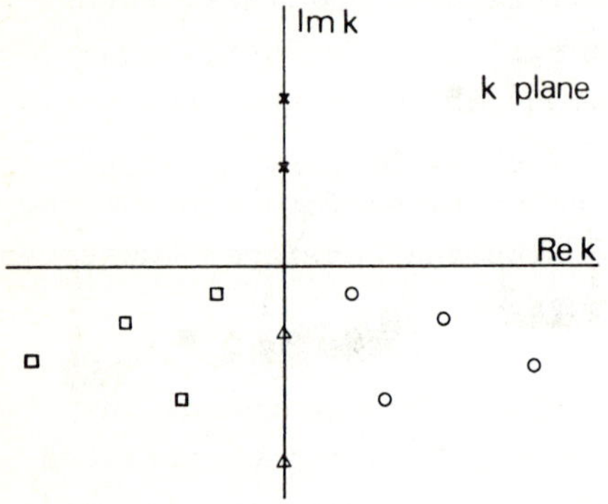

Fig. 6.1. Distribution of poles in the T matrix. Bound-state poles: x, Resonance poles: 0, Virtual-state poles Δ, Conjugate poles: □.

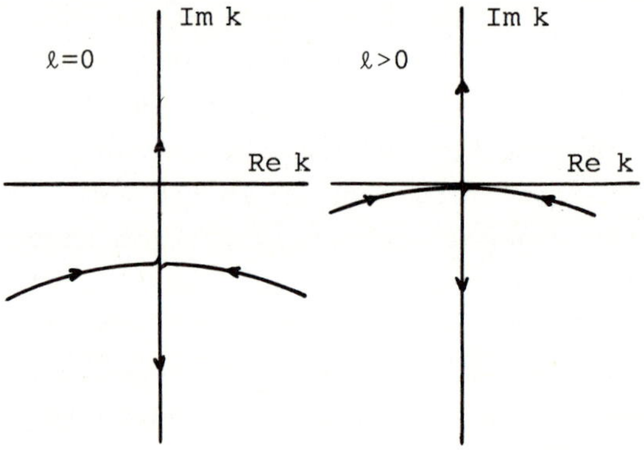

Fig. 6.2. Pole trajectories when λ increases.

The corresponding Coulomb T matrix T_C has an infinity of pure Coulomb poles. For attraction ($s > 0$) these correspond to all Coulomb bound states at $k = is/n$ ($n = \ell+1, \ell+2, \ldots$). The origin $k = 0$ is accumulation point of these poles, and moreover a branch point with branch cuts along the negative imaginary k axis. For Coulomb repulsion ($s < 0$) these poles lie in the lower half k plane: There are no bound states. When s varies the poles all move linear with s. It is also obvious how the poles move under variation of the coupling constant $\ell(\ell+1)$ of the repulsive centrifugal barrier.
For the interaction $V = V_C + V_S$ (cf. Eqs. 6.3-4) the T matrix can be written as

$$T = T_C + T_{CS}\ , \quad T_{CS} = -\frac{|g^C\rangle\langle g^C|}{\lambda^{-1} + \langle g|G_C|g\rangle}\ , \tag{6.5}$$

where g^C is the Coulomb-modified form factor, and G_C the Coulomb resolvent. The poles of T correspond to vanishing denominator, see van Haeringen et al. (1977), who proved that for Coulomb attraction the number of bound states is always infinite for $s > 0$, regardless the value of λ, λ real. It is 0 or 1 for Coulomb repulsion.
When λ is varied from $-\infty$ to $+\infty$ all Coulomb poles move relatively little, except for that corresponding to $n = \ell+1$, which can move to infinity via the imaginary axis. All other poles are confined to one small interval. This fact is intimately connected to the rank-1 character of V, for a rank-r potential the poles can traverse r intervals. In particular for Eq. (6.1) and Coulomb repulsion there is 0 or 1 bound state. Kok (1980) has extended the analysis to complex values of λ.

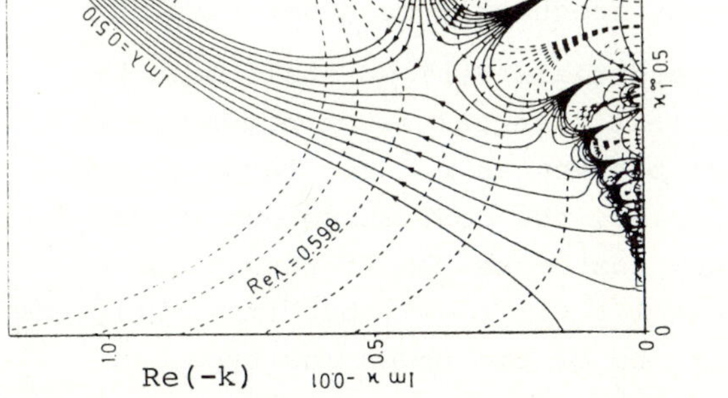

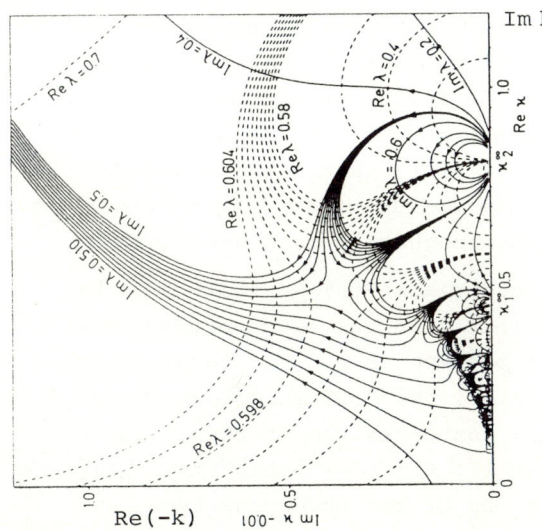

Fig. 6.3. Pole trajectories for Im λ fixed, Re λ varying (full lines), and for Re λ fixed, Im λ varying (broken lines). The arrows indicate the direction of increasing Re λ. Because k = iϰ the vertical axis is the positive imaginary k axis. Pole positions close to this axis on the left are "decaying bound states", with a finite left time determined by the value of Im k.

6.2. The number of two-body bound states

The number of bound two-body states depends on the asymptotic tail of the interaction. The attractive Coulomb tail gives rise to denumerable infinite bound states, regardless the character of the short-range part of the interaction. Let us consider now

$$V(r) = c\, r^{-\alpha}, \quad \alpha > 0, \quad r > R, \tag{6.6}$$

where V(r) may be arbitrary for $0 \leq r \leq R$ provided $r|V(r)|$ is integrable. For $c < 0$ the potential has an attractive tail. Then it supports an infinite number of bound states when $\alpha < 2$, whereas for $\alpha > 2$ this number is finite (Simon 1970, 1976). In the borderline case $c = 2$, i.e.,

$$V(r) = c\, r^{-2}, \quad r > R, \tag{6.7}$$

this number is infinite when $c < -\frac{1}{4}$ and finite when $c \geq -\frac{1}{4}$ (see van Haeringen 1985). In three-body equations singularities occur which are the same as those caused by an r^{-2} interaction. In fact, the Efimov effect [the fact that the number of bound states of a three-body system approaches infinity while purely short-range potentials, for example (6.1) are used, near selected strengths of the two-body interactions] is intimately connected to this picture.

7. SCATTERING IN TWO-BODY SYSTEMS AND EFFECTIVE-RANGE THEORY

7.1. Introduction

In the physical literature it is customary to regard local versus non-local, and short-range versus long-range potentials. The adjective "short-range" is mostly used for local potentials. However, it can mean that $|V(\vec{r})|$ for $r \to \infty$ decreases (i) faster than some exponential function $\exp(-\mu r)$, $\mu > 0$ (we call this exponentially bounded (e.b.)), or (ii) faster than any power of r, or (iii) faster than some specific power $r^{-\alpha}$, $\alpha > 0$ (mostly $\alpha = 1$), or (iv) sufficiently fast in order that $\int^{\infty} r^{\beta} |V(\vec{r})| d\vec{r} < \infty$, to mention the most frequently used definitions.

For local, rotationally invariant potentials V(r) we shall mean by short range that $\int^{\infty} |V(r)| dr < \infty$; if this integral is divergent we call V a long-range potential. When V(r) behaves as $r^{-\alpha}$ for large r, the dividing line is at $\alpha = 1$. In particular, the Coulomb potential has a long range in every context. In many textbooks "short-range" is described as "decreasing faster than 1/r for $r \to \infty$", but implicitly the potential is assumed (presumably) to be of the power-law type, at

least at infinity.
In Table 7.1 we give some typical properties of e.b., short-range, and long-range potentials. In Table 7.2 we give examples of nonlocal potentials that play an important role in models.

Behavior of $V(r),\ r\to\infty$	*Range*	*On-shell and half-shell T matrix existing*	*Effective-range function meromorphic near k=0*
$\sim e^{-\mu r},\ \mu>0$	*e.b.*	*yes*	*yes*
$\sim r^{-\alpha},\ \alpha>1$	*short*	*yes*	*no*
$\sim r^{-\alpha},\ 0<\alpha\leq 1$	*long*	*no*	*no*
($\alpha=1$: *Coulomb*)			

TABLE 7.1. Properties of some local central potentials.

Nonlocal potentials	→	*Separable (rank N)*	→	*Rational separable*	→	*Simple separable (rank 1)*	→	*Yamaguchi (rank 1,* $\ell = 0$*)*

TABLE 7.2. Specialization of nonlocal potentials, "→" stands for "contains as a special case".

Remarks

(1) The "range" of a local potential is determined only by its tail for large r.

(2) The classes of the local and the separable potentials are disjoint. The delta-shell potential can be regarded as a limiting case of local potentials, but also as a limiting case of nonlocal rank-one separable potentials.

(3) Existence of the on-shell and half-shell T matrix: When the half-shell T matrix is nonexistent it is convenient to define a <u>modified</u> half-shell T matrix with the help of so-called asymptotic states, $|\vec{k}\infty\pm\rangle_\alpha$ and $|k\ell\infty\pm\rangle_\alpha$, which can be associated with the local potential $V(r) = cr^{-\alpha}$, $r > R > 0$.

(4) Meromorphicity at k=0 of the effective-range function (ERF)

$$K_\ell(k^2) := k^{2\ell+1} \cot\delta_\ell(k). \tag{7.1}$$

When this function fails to be meromorphic at k=0 it is convenient to introduce a <u>modified</u> effective-range function (MERF) that <u>is</u> meromorphic at k=0. Van Haeringen and Kok (1982) introduced a simple and elegant formula for such a MERF; it involves the Jost solution and phase shift associated with a suitable long-range component of the potential, and the phase shift associated with the potential.

7.2. Separable potentials

Simple separable potentials have been given in Eqs. (6.1-3). Their charm lies in the fact that virtually all scattering quantities for interactions V_s and $V_c + V_s$ can be calculated in closed form, see van Haeringen (1985) Chaps. S and CS, respectively. Also the δ-shell potential, given essentially by $\langle r|g\rangle = \delta(r-R)$, $R > 0$, is very useful in this respect (see de Maag et al. 1984).

7.3. Power-law potentials

Let us consider some "long-range phenomena" of the potential (6.6). In Sec. 6.2 the number of bound states was discussed. Another interesting "long-range phenomenon" concerns the singularity of the scattering amplitude in the forward scattering direction (Kvitsinskii et al., 1983). Let us evaluate the Born approximation f_{BA} of the scattering amplitude associated with the potential $V(r) = cr^{-\alpha}$. This is essentially equal to the potential ir the momentum representation with p=p'. Setting for convenience $q = |\vec{p}-\vec{p}\,'|$, $q^2 = 2p^2(1-x)$, we get [Gradshteyn and Ryznik (1980) 3.944.5]

$$\begin{aligned} f_{BA}(x) &= -2\pi^2 \langle\vec{p}|V|\vec{p}\,'\rangle = -q^{-1} \lim_{\varepsilon\downarrow 0} \int_0^\infty e^{-\varepsilon r}\, V(r)\, \sin(qr)dr \\ &= -\tfrac{1}{2}\pi\, cq^{-1} \quad \text{if } \alpha = 2 \\ &= -c\, \Gamma(2-\alpha)\sin(\tfrac{1}{2}\pi\alpha)q^{\alpha-3} \quad \text{if } 0<\alpha<3,\ \alpha \neq 2. \end{aligned} \tag{7.2}$$

[The restriction $\alpha < 3$ originates from the behavior of V(r) at r = 0 and is therefore not really important.] Clearly $f_{BA}(x)$ is integrable with respect to x for $-1 \leq x \leq 1$ only if $\alpha > 1$. The Born approximation presumably contains the most singular parts of the T matrix in this respect. Further

$$\begin{aligned} \int_{-1}^{1} |f_{BA}(x)|^2\, dx &< \infty \quad \text{when } 2 < \alpha < 3 \\ &= \infty \quad \text{when } \alpha \leq 2. \end{aligned} \tag{7.3}$$

7.4. Positive eigenvalues

A somewhat unexpected "long-range phenomenon" is the existence of "bound states in the continuum". In 1929 von Neumann and Wigner already gave an example of a Hamiltonian possessing a positive eigenvalue with an eigenfunction that has a finite norm. For further details see the book of van Haeringen (1985).

7.5. Two-body scattering for potentials with Coulomb tail

The stationary scattering formalism is well known for $V = V_s + V_c$ with V_s a short-range potential. If $V_{s\ell}$ is separable simple formulas result, in particular for form factors as in (6.1-2), or linear combinations or parameter derivatives thereof, like the realistic Graz p-p interaction (Schweiger et al., 1983). Usually the parameters in such potentials are fitted to effective-range parameters determined from experiment.

We refer to the existing literature for the standard formulas (Taylor 1972, Newton 1982, van Haeringen 1985).

7.6. Effective-range functions and parameters

For exponentially bounded potentials the ERF (7.8) is meromorphic at k=0. For power-law-tailed potentials it is not. Hence only for e.b. potentials the expansion

$$K_\ell(k^2) = -a^{-1} + \tfrac{1}{2} r_0 k^2 + \ldots \tag{7.4}$$

makes sense: a is the scattering length and r_0 the effective range, both follow easily from the behavior of $\delta_\ell(k)$, and Eq. (7.1). The basic point of the usefulness of this expansion is that the detailed behavior of the short-range potential plays an insignificant part in the scattering at low energies, because the wavelength of the incident particle is so much larger than the actual range of the potential. Power-law tails make that K_ℓ is not meromorphic. The larger the power α, the less severe the singularity, however. Burke (1977) discusses qualitatively a number of cases from the literature. In particular, for the polarization potential

$$V = -\tfrac{1}{2}\, \alpha\, r^{-4} \tag{7.5}$$

he quotes O'Malley et al.'s result (1961),

$$k \cot\delta_0 = -\frac{1}{a} + \frac{\pi\alpha}{3a^2} k + \frac{2\alpha}{3a} k^2 \ln(\alpha k^2/16) + O(k^2), \tag{7.6}$$

which reveals, compared to (7.4), terms in k and $k^2 \ln k$. For $\ell \geq 1$

$$k^2 \cot\delta_\ell = (2\ell+3)(2\ell+1)2\ell-1)/\pi\alpha + \ldots \tag{7.7}$$

instead of

$$k^{2\ell+1} \cot\delta_\ell = -1/a_\ell + \tfrac{1}{2} r_{0\ell} k^2 + \ldots . \tag{7.8}$$

For interactions $V = V_s + V_c$, and V_s is e.b., also the ERF has to be modified. It may be taken as

$$K_{cs,\ell}(k^2) := c_{\ell\gamma}^{-1}\, k^{2\ell+1}[2\gamma\, H(\gamma) + C_0^2\{\cot\delta_{cs\ell}(k) - i\}], \tag{7.9}$$

where $C_0^2 = 2\pi\gamma/(e^{2\pi\gamma}-1)$,

$$c_{\ell\gamma}^{-1} := \binom{\ell+i\gamma}{\ell}\binom{\ell-i\gamma}{\ell} = \prod_{n=1}^{\ell} (1 + \gamma^2/n^2),$$

$$H(\gamma) := \psi(i\gamma) + (2i\gamma)^{-1} - \ell n(-i\gamma\ \mathrm{sgn}(s)) \tag{7.10}$$

and

$$\delta_{cs\ell}(k) = \delta_\ell(k) - \sigma_\ell(k), \tag{7.11}$$

where δ_ℓ is the phase shift associated with $V = V_c + V_s$; see Cornille and Martin (1962), Hamilton et al. (1973). Essentially the same formula is valid for Coulomb plus separable potentials. When V_s is a rank-one potential a simple formula for $K_{cs\ell}(k^2)$ results, so that closed expressions can be found for the so-called Coulomb-modified ER parameters defined by

$$K_{cs\ell}(k^2) = -\,1/a_{cs\ell} + \tfrac{1}{2}\, r_{cso}\, k^2 + \ldots, \tag{7.12}$$

see van Haeringen and Kok (1981) and de Maag et al. (1984).
The above formulas imply a completely different behavior of $\cot\delta_{cs\ell}$ for $k \to 0$ in the case of Coulomb repulsion and attraction respectively:

$$\text{Repulsion } (\gamma\to+\infty):\ \lim_{k\downarrow 0} \exp(-2\pi\gamma)\cot\delta_{cs\ell} = -\,\frac{a_B^{2\ell+1}(\ell!)^2}{2\pi a_{cs\ell}} \tag{7.13}$$

$$\text{Attraction } (\gamma\to-\infty):\ \lim_{k\downarrow 0} \cot\delta_{cs\ell} = \cot\delta_{cs\ell}(0) = -\,\frac{a_B^{2\ell+1}(\ell!)^2}{2\pi a_{cs\ell}}\,. \tag{7.14}$$

In the former case $\delta_{cs\ell}(k)$ vanishes extremely fast, $\propto \exp(-2\pi/ka_B)$, whereas in the latter $\delta_{cs\ell}$ is smoothly going to a nonvanishing finite value, cf. Fig. 7.2.
Bound states and resonances of V correspond to poles of the T matrix in the k plane. At these poles

$$\cot\delta_{cs\ell} = i,\ \delta_{cs\ell} = -i\infty \tag{7.15}$$

so that pole positions k obey (a_B is the Bohr radius)

$$\tfrac{1}{2}\, a_B\, K_{cs\ell}(k^2) = c_{\ell,-1/ka_B}^{-1}\, k^{2\ell}\, H(-1/ka_B). \tag{7.16}$$

Given the Coulomb-modified ER parameters this relation determines

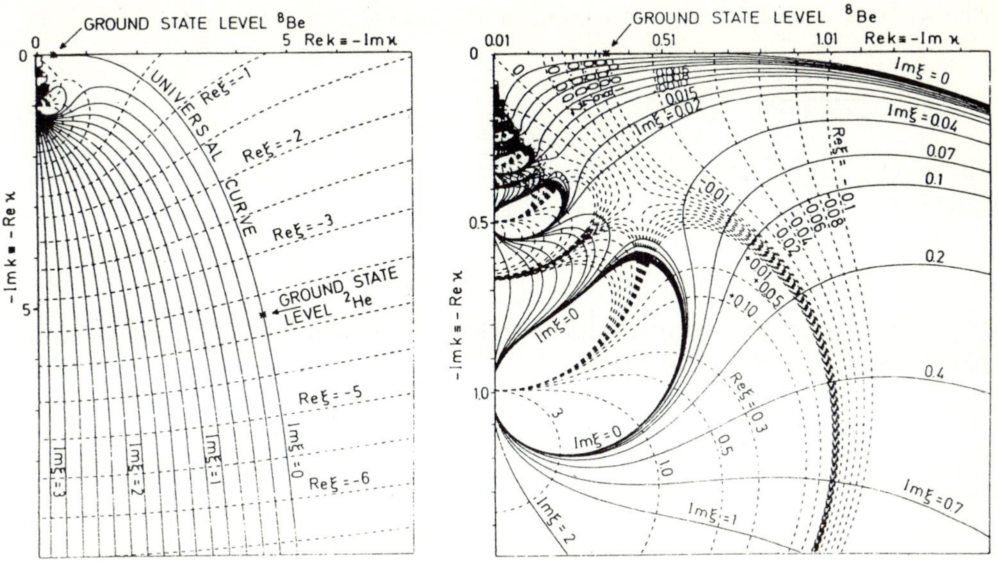

Fig. 7.1. Universal contour plot in fourth quadrant of the k plane. Scales are in units a_B^{-1}. Given k (experimenentally determined), one reads off the complex ξ. Given ξ, all corresponding complex pole positions can be found; ξ is the l.h.s. of (7.16) or the right detail near origin. The nonlinearity is obvious.

immediately the position of all poles in the complex k plane. This technique was used to very accurately determine the ^{8}Be ground state at E = -140 - i467 keV (Kok 1980), cf. Fig. 7.1. The restriction that V_s is e.b. is essential for the meromorphic character of both members of (7.9). In practice in composite-charged particle scattering the effective two-fragment potential does not obey this condition. This has led recently to an interesting discussion, which we recall in Sec. 9.1.

For Coulomb attraction the same relation can be solved in the upper half k plane. Given the (possible complex) scattering lengths the energy shifts of (possibly unstable) exotic atoms can all accurately be predicted. In fact, this gives the direct connection to the quantum-defect theory, initiated by Seaton (1955, 1958). The quantum defect of bound states can be introduced by $E_n = -a_B^{-2}/(n-\mu)^2$, $n = \ell+1, \ell+2, \ldots$.

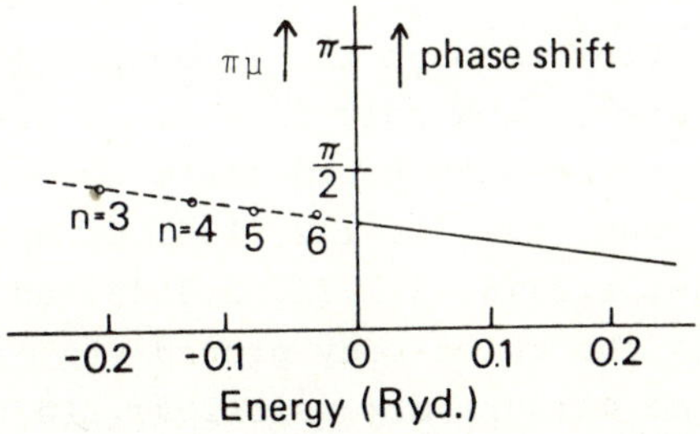

Fig. 7.2. ^{3}P quantum defect for Be and the phase shift for e-Be$^+$ scattering.

Here μ is a slowly varying function of energy. One can easily derive

$$\frac{\cot\delta_{cs\ell}(k^2)}{1 - \exp(2\pi\gamma)} \approx \cot\pi\mu$$

and in the limit $k^2 \to 0$ $\delta_{cs\ell}(0) = \pi\mu$. Bound-state energies can often be obtained from spectroscopic data. In Fig. 7.2. the extrapolation from positive to negative energy observables is done for e-Be^+ scattering in the triplet P state.

Modern multichannel quantum-defect theory (Seaton 1983, Kostelecký and Nieto 1985) is extremely useful in the description of atomic systems. Long-range modified scattering parameters are discussed in much more detail and in broader context by Badalyan et al. (1982).

8. THREE-BODY BOUND STATES

Operator equations for three-body scattering have been discussed at this meeting. In the coordinate representation they lead to integral-differential equations (Noyes 1970, Laverne and Cignoux 1973, Payne et al. 1980) that are convenient when local two-body potentials are used. In the momentum representation they lead to integral equations that are particularly convenient when separable potentials, or separable expansions of local or nonlocal potentials are as two-body interaction (Mitra 1962, Alt et al. 1967).
The homogeneous version of these equations describe the three-body bound states. In Sec. 8.1. we give results of numerical computations for simple three-boson systems. In Sec. 8.2. the number of three-body bound states for this system is discussed. In Sec. 8.3. the inclusion of the repulsive Coulomb potential between two of the particles is discussed. Different methods to handle this problem nonperturbatively are shown to lead to identical results (Kok et al. 1979, 1981, 1982, Lehman et al. 1984).

8.1. The three-boson bound states

Consider three identical bosons interacting through the Yamaguchi (ℓ=0) potential (6.1). There is one bound state if and only if $\lambda \geqq \lambda_s$, where $\lambda_s := -\langle g|G_0(0)|g\rangle^{-1} = 2\beta^3$. The first three-body bound state occurs already for $\lambda \gtrsim 0.8115\ \lambda_s$. For λ/λ_s = 0.85, 0.9, 1, 1.2, 1.6, 2, and 3 one finds $3\kappa/\beta$ = 0.2529, 0.4358, 0.7466, 1.2778, 2.1778, 2.9549, and 4.5906, respectively. Here κ determines the three-body ground-state energy $E = -\kappa^2$, the factor $3/\beta$ makes the numbers dimensionless. In the

following section we shall use two-body parameters that fit to some degree strength and range of the effective nucleon-nucleon interaction in the triton system. Details are given by Kok et al. (1981) and Lehman et al. (1984).
The parameter $\beta = 1.082\ fm^{-1}$, and λ/λ_s is taken 1.2 so that $\lambda = 2.4\ (1.082)^3\ fm^{-3}$. The number $\lambda = 0.1540\ fm^{-3}$ quoted by Lehman et al. is correct, because they use a different normalization. With $\hbar^2/2\mu = 41.47$ $MeV.fm^2$ one finds the three-boson binding energy 8.807580 MeV.
For the same parameter β Lehman et al. find 8.80985 MeV with $\lambda = 0.1540$. Presumably their λ is differently normalized by a factor $2\pi^2$ so that here $\lambda/\lambda_s = 1.19988$. Their value $\hbar^2/2\mu$ is quoted as 41.5016 MeV fm^2. Hence for $\beta = 1.082\ fm^{-1}$ the two sources give $E_3 = 0.21238\ fm^{-2}$ for $\lambda/\lambda_s = 1.20000$ (Kok et al.) and $E_3 = -0.21228\ fm^{-2}$ for $\lambda/\lambda_s = 1.19988$ (Lehman et al.).

8.2. The number of three-boson states

The three-boson system of the previous section develops with increasing λ a bound state at $\lambda/\lambda_s \approx 0.8115$. The first excited state occurs when this strength ≈ 0.9865. Then a number of Efinov states appear and disappear: The first Efinov state appears at 0.99937 and disappears at 1.013. The second Efinov state exists on the interval [0.99995,1.00075]. An ever increasing number of such states exists for strength intervals even closer around the value one. The original explanation of these states is due to Efinov (1970), who notes that the effective interaction with a r^{-2} tail develops near strength one. Amado et al. (1971, 1972) discuss the singularities in the kernel of the integral equation for that strength. Numerically the effect was investigated by Stelbovics and Dodd (1972), Baltar et al. (1976), Møller (1977, 1978), and Venema (1980). The last two sources study three-body pole trajectories, the numbers quoted in this section are taken form Venema (1980).
For very large strength parameters again a third bound state (second excited state) develops, after that a fourth, etc..
For a more general discussion of the number of bound states of an N-body system we refer to Simon (1970, 1976) and Klaus and Simon (1980).

8.3. Inclusion of the Coulomb potential

Inclusion of the Coulomb interaction in a nuclear three-body problem is considered difficult even when only two of the particles are charged. This is the reason that only very few exact (numerically) momentum-space, bound-state, three-body calculations with two charged particles have been performed (Kok et al. 1979, 1981, 1982, Lehman et al. 1984), although the formalism was ready more than a decade earlier (Veselova

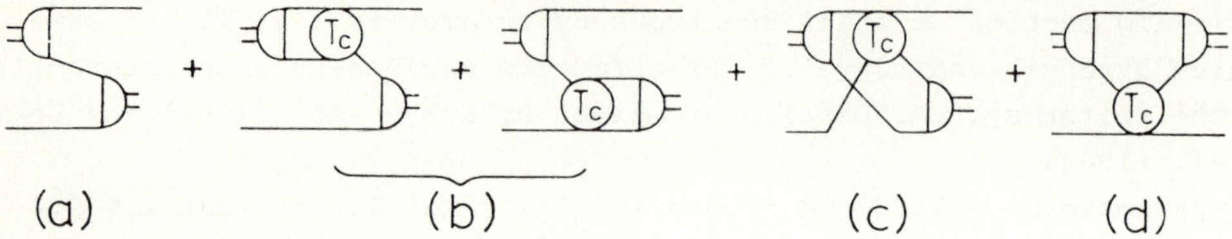

Fig. 8.1. The driving terms

1970, Alt et al. 1978).
In Veselova's system of integral equations the Coulomb T matrix appears linearly in some of the kernels, and one has to deal with T_c fully off-shell. Nevertheless, after p.w. projection equations in a single variable result because of the assumed separability of V_s. This is in contrast with the approach of Lehman et al., in which the problem is formulated in terms of wave-function components, and does not involve T_c, but only V_c. However, p.w. projection in this case leads to two-variable integral equations. The equivalence of the two formulations has been proven by Lehman et al. (1984).
In a rather symbolic notation the integral equations can be written as

$$X = Z + Z\tau X, \tag{8.1}$$

where the driving term Z consists of several components depicted in Fig. 8.1.
Furthermore, τ is a two-body propagator, possibly Coulomb modified, and X is the unknown. Explicit forms of Z and τ are given in Kok et al. 1981, before and after p.w. projection, with all normalization factors properly included. Each diagram and its singularities is discussed in detail. The most difficult one is diagram (d). By using the splitting $T_c = V_c + (T_c - V_c)$ various contributions result, which can be computed separately. In many terms part of the integrations can be performed analytically. Separating out all singularities and the integration of the remaining smooth functions is discussed in detail in the aforementioned source, which is available upon request.
In addition to a full calculation (leading to a binding energy of the model ^{3}He of 7.937952 MeV as compared to the binding of the model ^{3}H of 8.807580 MeV, cf. Sec. 8.1) many approximations have been investigated, viz.

- take only S wave part of the Coulomb potential [negligible effects in this case]
- contributions of each of the diagrams separately [diagram (d) contributes 21%, diagram (c) 19%, diagram (b) 43%, and the occurrence of the

Coulomb two-body propagator τ_c 17% to the Coulomb energy 869.628 keV]
- different numerical methods and mesh sizes
- take only first Born term V_c of T_c throughout the equations (this saves one or two orders of magnitude in computer time).

Higher-order-Born contributions tend to lessen the Coulomb energy: In first-oder Born approximation one finds 888.081 keV instead of 869.628 keV, a difference of approximately 2%.
These results are in very good agreement with those of Lehman et al.

9. SCATTERING IN THREE-BODY SYSTEMS AND EFFECTIVE-RANGE THEORY

9.1. Introduction

One lesson we learn from few-body conferences is the importance of correctly treating singularities of scattering amplitudes for slowly decreasing potentials. For example, above the elastic p-d threshold, and at the breakup threshold (Veselova and Faddeev 1981) there are singularities which are not purely Coulombic. The analysis of such singularities often is far from simple. For example, the driving terms of the type depicted in Fig. 8.1 have been analyzed by Kharchenko and Shadchin (1983). They derive the asymptotic behavior of the distorting ('polarization') potential in pd scattering, and separate out explicitly an r^{-2} (attractive) and r^{-3} behavior. In fact, expressions for the corresponding effective dipole and quadrupole moments of the deuteron induced in pd scattering due to the Coulomb rescattering of the protons are given. Their result is wrong (see Kok 1985). This illustrates that very delicate cancellations between different contributions may occur. The leading power of r that makes a nonvanishing contribution to the tail after the 1/r potential is the r^{-4} term. This is the well-known polarization potential, very familiar in atomic physics. For example Burke (1977) recalls a well-known result in second-order perturbation theory in the adiabatic approximation, for scattering of elections off neutral atoms.
At the Few-Body meeting at Tbilisi in 1984 Kvitsinskii and Merkuriev (1985) reported (in Russian) numerical results which showed that the standard definition of a Coulomb-modified scattering length in the presence of a polarization potential gives minus infinity! The first source discussing this phenomenon is presumably the paper of Oppenheim Berger and Spruch (1965). Once the problem was appreciated many authors could explain its solution (Kvitsinsky and Merkuriev 1985; Kvitsinsky 1985; Kuperin et al. 1985; Gibson 1985; Bencze et al. 1985; Bencze and

Chandler 1985; Kuznichev and Zepelova 1985).

9.2. p-d scattering at very low energies

From (7.13) it is clear that for $\ell=0$

$$a_{cs} = -a_B \lim_{k \to 0} \delta_{cs} \exp(2\pi/ka_B)/2\pi \tag{9.1}$$

for e.b. potentials. Oppenheim Berger and Spruch used a Born approximation to show that if in addition an r^{-4} potential is present, convergence of δ_{cs} is $O(k^5)$ as $k \to 0$, instead of exponential, so that $a_{cs} = -\infty$. Bencze et al. proved that with proper reinterpretation of δ_{cs} in (9.1) [i.e., substitute for δ_{cs} the phase shift relative to the total long-range phase shift, of Coulomb + polarization] the relation (9.1) still can be used. Moreover, calculations in a screening approach presumably compute the correct experimental number [which is obtained by extrapolation down from positive energy values]. A computational method using extrapolation in energy (and realistic interactions) also should give the same value as found by the experimentalists.

The net result of all explanatory efforts the last few years is that confusion persists: For the same realistic model potentials different groups find conflicting answers for the doublet p-d scattering length, see Kok (1985).

9.3. Scattering calculations at higher energies

At higher energies p-d phase-shift calculations of different groups seem to agree more or less below breakup.

A very important result is reported by Alt, Sandhas and Ziegelmann (1985), and Aguiar et al. (1986). They demonstrate that all conventional approximate methods to include Coulomb effects give unsatisfactory results! You have to do the full complicated calculation. The only approximation these authors did make in the course of their calculation is to replace in the various diagrams the two-body Coulomb amplitude by its Born term. Much insight into this approximation is provided by the work of Kok and van Haeringen (1980 etc., see for a list of sources van Haeringen 1985).

9.4. Concluding remarks

The material presented so far consists of merely introductory remarks to the full subject of this paper. Only very elementary examples have been covered (in the spirit of Part 5 of Feynman 1985). Many pieces of valid work, ideas, results, references, and names have not received the attention they deserve. That reflects another great problem of the Coulomb interaction in few-body physics: the multitude and broad range of its products, the broad-range Coulomb problem.

REFERENCES

Aguiar C.E.M., Brinati J.R., and Martins M.H.P. (1986) Elastic scattering of deuteron in a three-body system with Coulomb interaction. Rio de Janeiro preprint.
Alt E.O. (1986) The Coulomb force in few body reactions. In "Few-body methods", Ed. by T.K. Lim et al. World Scientific, Singapore.
Alt E.O. (1986) Calculation of proton-deuteron scattering observables in the screeening approach, and comparison with approximate treatments. Mainz Prepr. MZ-TH/86-15, Sept.
Alt E.O., Grassberger P., and Sandhas W. (1967) Nucl. Phys. B 2:167.
Alt E.O., Sandhas W., and Ziegelmann H. (1978) Coulomb effects in three-body reactions with two charged particles. Phys. Rev. C 17:1981-2005.
Alt E.O., Sandhas W., and Ziegelmann H. (1985) Calculation of proton-deuteron phase parameters including the Coulomb force. Nucl. Phys. A 445:429-461.
Amado R.D. and Adhikari S.K. (1972) Phys. Lett. B 40:11; (1972) Phys. Rev. C 6:1484.
Amado R.D. and Noble J.V. (1971) Phys. Lett. B 35:25; (1972) Phys. Rev. D 5:1992.
Badalyan A.M., Kok L.P., Polikarpov M.I., and Simonov Yu.A. (1982) Resonances in coupled channels in nuclear and particle physics. Phys. Rep. 82:31-177.
Baltar V.L., Ferreira E.M., and Antunes A.C. (1976) Nucl. Phys. A 265:365.
Balmer (1885) Wied. Ann. 25.
Bencze G. and Chandler C. (1985) Coulomb polarization effects in low energy p-d elastic scattering. Albuquerque preprint.
Bencze G., Chandler C, Friar J.L., Gibson A.G, and Payne G.L. (1986) Low energy scattering theory for Coulomb plus long-range potentials. Albuquerque preprint.
Burke P.G. (1977) Potential scattering in atomic physics. Plenum Press, New York.
Cajori F. (1946) Newton's Principia. Motte's translation revisited. University of California Press, Berkeley.
Chandler C. (1986) The Coulomb problem in nonrelativistic scattering theory. In Proceedings Few-Body XI, Tokyo-Sendai.
Cornille H. and Martin A. (1962) Nuovo Cim. 26:298-327.
Coulomb C.A. (1785,1787) Memoires de l'Academie Royale des Sciences de l'Institut Imperiale de France. Paris, 1788, pp. 569-577,578-611.
De Alfaro V. and Regge T. (1965) Potential Scattering. Wiley, New York.
de Maag J.W. et al. (1984) Coulomb-modified scattering parameters for Coulomb-plus-separable potentials for all l. J. Math. Phys. 25:684-692.
Dollard J.D. (1964) J. Math. Phys. 5:729 and (1966) 7:802.
Dollard J.D. (1971) Rocky Mtn. J. Math. 1:5 and (1972) 2:317.
Dusek J. (1982) Czech. J. Phys. B 32:1325-1348 and (1983) J. Math. Phys. 24:2471-2480.
Efimov V. (1970) Phys. Lett. B 33:563.
Enns V. (1979) Comm. Math. Phys. 65:151.
Faddeev L.D. (1969) In Three body problem. Ed. by J.S.C. McKee et al. NHPC, Amsterdam.
Feynman R.P. (1985) "Surely You're Joking, Mr. Feynman!". Norton & Co., New York.
Gibson A.G. (1985) On low energy scattering theory with Coulomb potentials. Budapest preprint KFKI-1985-79.
Gillmor C.S. (1971) Coulomb and the evolution of physics and engineering in eighteenth-century France. Princeton University Press, Princeton. See part. pp. 182, 193, 232.
Goodmanson D.M. and Taylor J.R. (1980)Coulomb scattering as the limit of scattering off smoothly screened Coulomb potentials. J. Math. Phys. 21:2202-2207.
Gorshkov V.G. (1961) Zh. Eksp. Teor. Fiz. 40:1481-1491.
Gradshteyn I.S. and Ryzhik I.M. (1980) Table of integrals, series, and products, corrected and enlarged edition. Academic Press, New York.
Hamilton J. et al. (1973) Nucl. Phys. B60:443.
Hormander L. (1976) The existence of wave operators in scattering theory. Math. Z. 146:69-91.
Isozaki H. and Kitada H. (1985) J. Fac. Sci. Univ. Tokyo, Sec. IA, Math 32:77.
Kharchenko V.F. and Shadchin S.A. (1983) Kiev, preprint ITP-83-101E.
Klaus M. and Simon B. (1980) Coupling constant thresholds in nonrelativistic quantum mechanics I. Short-range two-body case. Ann. Phys. (N.Y.) 130:251-281. (1980) II. Two-cluster thresholds in n-body systems. Commun. Math. Phys. 78:153-168.

Koch J.H. et al. (1971) Phys.Rev. Lett. 26:1465.
Kok L.P. (1980) Accurate determination of the ground-state level of the 2He nucleus. Phys. Rev. Lett. 45:427-430.
Kok L.P. (1980) Coulomb level shifts by a complex Yamaguchi potential. Phys. Rev. C 22:2404-2408.
L.P. Kok (1985) Few-body problem with Coulomb interactions. In Few-Body problems in Physics. Ed. by L.D. Faddeev et al. World Scientific, Singapore.
Kok L.P., Struik D.J., and van Haeringen H. (1979) On the exact solution of three-particle equations with Coulomb interaction I. The driving terms. Report 151, University of Groningen, 40 p. Available upon request.
Kok L.P., Struik D.J., Holwerda J.E., and van Haeringen H. (1981) On the exact solution of three-particle equations with Coulomb interaction II. Bound states. Report 170, University of Groningen, 48 p. Available upon request.
Kok L.P. and van Haeringen H. (1980) Off-shell Coulomb T matrix in connection with the exact solution of three-particle equations with Coulomb interaction. Phys. Rev. C 21:512-517.
Kok L.P. and van Haeringen H. (1982) Bound state solution in momentum space of three-particle equations with Coulomb interaction. Czech. J. Phys. B 32:311-315.
Kostelecky V.A. and Nieto M.M. (1985) Analytical wave functions for atomic quantum-defect theory. Phys. Rev. A 32: 3243-3246.
Krell, M. (1971) Phys. Rev. Lett. 26:584.
Kroeger H., Nachabe A.M., and Slobodrian R.J. (1986) Phys. Rev. C 33:1208.
Kroeger H. and Slobodrian R.J. (1984) Phys. Rev. C 30:1390.
Kuperin Yu.A., Kvitsinsky A.A., and Merkuriev S.P. (1985) Low energy scattering of three charged particles. Barcelona preprint FT-FP-5-85.
Kuzmichev V.E. and Zepelova M.L. (1985) in Proc. Eur. Few-Body X (CRIP, Budapest) 91-93.
Kvitsinskii A.A., Komarov I.V., and Merkuriev S.P. (1983) Singularities of the scattering amplitude for slowly decreasing potentials. Yad. Fiz. 38:101-114.
Kvitsinskii A.A. and Merkuriev S.P. (1985) In Few-Body problems in Physics. Ed. by L.D. Faddeev et al. World Scientific, Singapore.
Kvitsinsky A.A. and Merkuriev S.P. (1985) Polarization potential and low-energy characteristics of pd scattering. Yad. Fiz. 41:647-654.
Kvitsinsky A.A. (1985) Low energy scattering for potentials including power-type terms in addition to the Coulomb interaction. Teor. Mat. Fiz. 65:226-237.
Laverne A. and Cignoux C. (1973) Nucl. Phys. A 203:597.
Lehman D.R., Eskandarian A., Gibson B.F., and Maximon L.C. (1984) Momentum-space solution of a bound-state nuclear three-body problem with two charged particles. Phys. Rev. C 29:1450-1460.
McVoy K.W. (1967) Collision theory. In Fundamentals in nuclear theory. Ed. by A. De-Shalit and C. Villi. IAEA, Vienna.
Merkuriev S.P. (1980) Ann. Phys. (N.Y.) 130:395.
Merkuriev S.P. (1981) Three-body Coulomb scattering. Acta Phys. Austriaca Suppl. 23:65-110.
Mitra A.N. (1962) Nucl. Phys. 32:529.
Moeller K. (1977,1978) Z.f.K. Dresden, Internal reports 327,351,357,377.
Motte A. (1729) Sir Isaac Newton's Mathematical Principles of Natural Philosophy and his system of the world. [Translation from Latin into English of the third edition (1726) of Newton (1687).]
Mulherin D. and Zinnes I.I. (1970) J. Math. Phys. 11:1402.
Newton I.S. (1687) Philosophiae Naturalis Principia Mathematica. Printed by J. Streater, for the Royal Society, London.
Newton R.G. (1982) Scattering Theory of Waves and Particles. 2nd Ed. Springer, New York.
Noyes H.P. (1970) In Three body problem. Ed. by J.S.C. McKee et al. NHPC, Amsterdam.
Nussenzweig H.M. (1959) The poles of the S-matrix of a rectangular potential well or barrier. Nucl. Phys. 11, 499-521.
Okubo S. and Feldman D. (1960) Phys. Rev. 117:279-291;292-306.
O'Malley T.F. et al. (1961) Modification of effective-range theory in the presence of a long-range (r-4) potential. J. Math. Phys. 2:491-498.
Oppenheim Berger R. and Spruch L. (1965) Phys. Rev. 138:B1106-B1115.
Pauli W. (1926) Z. Phys. 36:336-363.
Payne G.L. et al. (1980) Phys. Rev. C 22:823-831;832-841.
Reed M. and Simon B. (1972) Methods of Modern Mathematical Physics. Academic Press, New York. Vol. I: Functional Analysis. (1975) Vol. II: Fourier Analysis, Self-Adjoint-

ness. (1978) Vol. IV: Analysis of Operators. (1979) Vol. III: Scattering Theory.
Schroedinger E. (1926) Ann. Phys. (Leipzig) 79:361-376;489-527;734-756, 80:437-490, 81:109-139.
Schweiger W. et al. (1983) Phys. Rev. C 27:515-522; 28:1414-1416.
Schwinger J. (1964) Coulomb Green's function. J. Math, Phys. 5:1606-1608.
Seaton M.J. (1955) Comp. Rend. 240:1317; (1958) Mon. Not., R. Astron. Soc. 118:504.
Seaton M.J. (1983) Rep. Progr. Phys. 46:167.
Simon B. (1970) On the infinitude or finiteness of the number of bound states of an N-body quantum system, I. Helv. Phys. Acta 43:607-630.
Simon B. (1976) On the number of bound states of two-body Schroedinger operators - a review. In Studies in Mathematical Physics. Ed. by E.H. Lieb et al. Princeton University Press, Princeton, pp. 305-326.
Stelbovics A.T. and Dodd L.R. (1972) Phys. Lett. 39:450.
Taluktar B. et al. (1984) J. Math. Phys. 24:683-686.
Taylor J.R. (1972) Scattering theory. Wiley, New York.
Taylor J.R. (1974) A new rigorous approach to Coulomb scattering. Nuovo Cim. B23:313-34.
van Haeringen, H. (1978) J. Math. Phys. 19:1379-1380.
van Haeringen, H. (1985) Charged-Particle Interactions. Theory and Formulas. Coulomb Press Leyden, Leiden.
van Haeringen H. and Kok L.P. (1981) Exact phase shifts and scattering parameters for Coulomb plus simple separable potentials for all l. Phys. Lett. A 82 (1981) 317-320.
van Haeringen H. and Kok L.P. (1982) Modified effective-range function. Phys Rev. A 26:1218-1225.
van Haeringen H., van der Mee C.V.M., and van Wageningen, R. (1977) The number of bound states for the Coulomb plus Yamaguchi potential. J. Math. Phys. 18:941-943.
van Haeringen, H. and van Wageningen, R. (1975) J. Math. Phys. 16:1441-1452.
Venema M. (1980) Het Efimov effekt in Amado's model (in Dutch). Report 163, University of Groningen, 79 p. Available upon request.
Veselova A.M. (1970) Separation of two-particle Coulomb singularities in a system of three charged particles. Teor. Mat. Fiz. 3:542-546.
Veselova A.M. (1973) Preprint ITF-73-106P. Kiev.
Veselova A.M. (1978) Teor. Mat. Fiz. 35:395.
Veselova A.M. and Faddeev L.D. (1981) Vestnik LGU Ser. Fiz. 22:42-46.
Wentzel G. (1926) Zwei Bemerkungen ueber die Zerstreuung korpuskularer Sthralen als Beugungserscheinung. Z. Phys. 40:590-593.
Yamaguchi Y. (1954) Phys. Rev. 95:1628-1634.

CALCULATION OF ELECTROMAGNETIC OBSERVABLES IN FEW-BODY SYSTEMS

B.F. Gibson

School of Physics
University of Melbourne
Parkville Victoria 3052
Australia

and

Theoretical Division*
Los Alamos National Laboratory
Los Alamos, New Mexico, 87545
U.S.A.

An introduction to the calculation of electromagnetic observables in few-body systems is given by studying two examples in the trinucleon system: 1) the elastic electron scattering charge form factor in configuration space and momentum space and 2) the two-body photodisintegration of ^{3}H leading to a neutron-deuteron final state in a separable potential formalism. In the discussion of charge form factor calculations, a number of related topics are touched upon: the relation of structure in Ψ to the properties of simple NN forces, the Faddeev and Schrödinger solution to the harmonic oscillator problem, the Rosenbluth formula for electron scattering from a spin-1/2 nuclear target (e.g., the proton or ^{3}He), and the charge density operator. Formulae for ^{3}He and ^{3}H charge form factors in a central force approximation are given in configuration and momentum space. The physics of these form factors is discussed in light of results from realistic nucleon-nucleon potential model calculations, including the effects of two-pion-exchange three-body force models. Topics covered are the rms charge radii, characterization of the charge form factors, properties of the charge densities, and the Coulomb energy of ^{3}He. In the discussion of the ^{3}H photodisintegration, the Siegert form of the electric dipole operator (in the long wave length limit) is derived as are the separable potential equations which describe the off-shell transition amplitudes which connect nucleon-plus-corrected-pair states. Expressions for the Born amplitudes required to complete the two-body photodisintegration amplitude calculation are given. Numerical results for a model central force problem are discussed and compared with an approximate calculation. Comparisons with ^{3}H(γ,n)d and ^{3}He(γ,p)d data are made, and the significant feaures of the exact theoretical calculation are outlined.

*Permanent address

LECTURE I. TRINUCLEON FORM FACTORS FROM ELASTIC ELECTRON SCATTERING

I. Introduction

Now that we have learned how to generate few-body wave functions in numerous ways, let us explore how one relates these ideas to the physical world of experimental observables. Most of us approach the everyday world around us by means of five senses. Of these, one of the most important is sight. How do we see? Photons from a source (natural or artificial) are scattered from objects and detected by our eyes. Experience has taught our super computer brains how to process the electronic signals generated by the optic nerve into meaningful images. Photon scattering is one of our oldest analytical tools.

Photon scattering is not the only such process that provides us with meaningful information about the world around us. Each of you is familiar with the story of Rutherford scattering and the discovery of the nucleus. In that case, alpha particle scattering revealed that the atom was not a uniform charge distribution. That example contains an important lesson: new physics is discovered when experimental observation differs from the result of a model calculation that is as complete as possible in terms of the known, relevant physics - not when undetermined parameters are varied (or chosen) to put a curve through the data. The pupose of calculations performed within the context of an exact theory is to explain honest differences with otherwise successful approximate theoretical prescriptions and to elucidate or discover novel aspects of physics.

The simplest picture of the nucleus is obtained in much the same manner as we use light to see, although we employ the "virtual photon" exchanged between an electron and the nucleus in the electron scattering process to define our image. Using elastic electron scattering, one can explore the charge and magnetic moment densities of the nucleus. The relativistic theory was first written down in the early 1930's by Mott.[1] The fact that nuclear charge distributions and sizes could be extracted from such experiments was laid out clearly by Rose[2] in the late 1940's. It was in the 1950's that elastic electron scattering became a feasible experimental tool[3] and Hofstadter won the Nobel Prize in 1961 for his work in electron scattering investigations of the structure of the nucleus. It is from this background that we ask such questions as whether we can provide a quantitative model of the charge and magnetic moment distributions of the few-nucleon systems - those nuclei whose wave functions we can generate from nonrelativistic Hamiltonians incorporating realistic representations of the nucleon-nucleon interaction.

II. Qualitative Aspects of the Relation between Ψ and V

Because one is investigating a density or $|\Psi|^2$ function in studying the differential cross section measured in unpolarized elastic electron scattering, the historical approach to analysis of the data has been to postulate a plausible analytic form for the density and to ask how well the data can be described by varying the model parameters. Such a procedure should be utilized with caution, because it is clear from the Schrödinger equation

$$[T + V]\,\Psi = E\,\Psi \qquad (1)$$

that important features of the nucleon-nucleon interaction will be reflected in the wave function Ψ and therefore in the density $|\Psi|^2$. Assuming a given form for Ψ makes a definite statement about the underlying Hamiltonian which has been implicitly assumed.

Let us examine the spatially symmetric S-state components that result from using three different central potential forms to generate the trinucleon wave function: smooth Gaussian (G), one term Yukawa (MT IV model of Malfliet and Tjon[4]), and a two term Yukawa with short-range repulsion (MT V model of Malfliet and Tjon[4]). Figure 1 depicts the Schrödinger wave function for the G model where $\vec{x}$ and $\vec{y}$ are the Jacobi coordinates of the ^{3}H system and θ is the angle between the two vectors.[5] (See Appendix A.) As expected, Ψ for this smooth potential shows no sharp structure. This wave function is reminiscent of that which results from solving the harmonic oscillator (HO) Hamiltonian. However, it differs in that the wave function falls off exponentially. (We shall return to the HO problem below.) The point to retain is that a smooth ad hoc wave function assumption implies an assumption that the underlying two-body potential is reasonably smooth. This point is emphasized by comparing Fig.1 with Fig.2, in which we plot Ψ for the MT IV potential model.[5] The Ψ for the Yukawa potential exhibits a definite kink, or ridge structure, due to the 1/x singualrity of the Yukawa form. When $y \simeq x/2$, one pair of interacting nucleons lie close together and the 1/x singularity leads to a discontinuity in the first derivative of Ψ along that line. This occurs for θ=0, a collinear configuration, but not for θ=90^0, an isosceles triangle configuration, in which overlapping pairs cannot occur for $x \neq 0 \neq y$. The ridge structure is even more apparent in the small S'-state components of Ψ, when the singlet and triplet potentials differ, as shown in Fig.3, where one of the S'-state components is plotted for the MT II-IV model. Thus, even a purely attractive potential having a simple Yukawa singularity leads to structure in the Schrödinger wave function.

Let us turn to the MT V model with short-range repulsion, whose wave function is plotted[5] in Fig.4. One anticipates that Ψ will be small whenever any two of the nucleons are close together. Clearly this has led to the "death valley" in Ψ along $y \simeq x/2$ and along $x=0$.

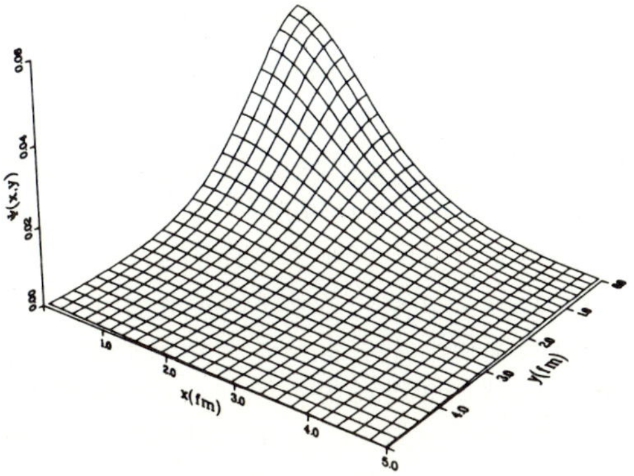

Fig.1 The Schrödinger wave function at fixed angle $\theta=0^0$ for the Gaussian potential model.

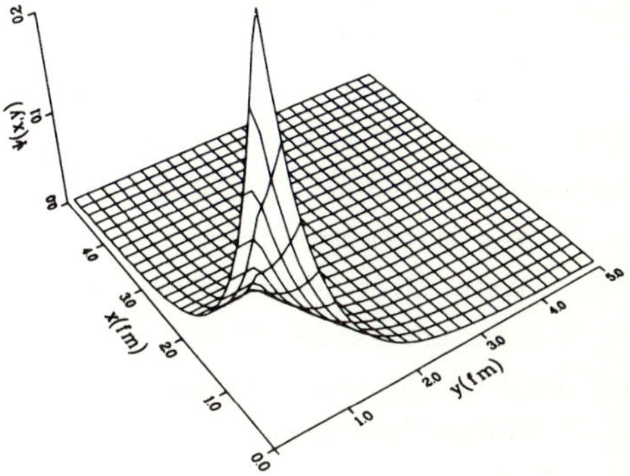

Fig.2 The Schrödinger wave function at fixed angle $\theta=0^0$ for the Yukawa potential (MT IV).

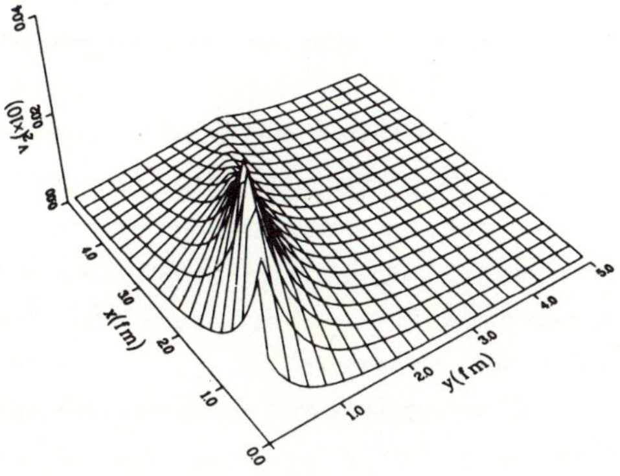

Fig.3 The Schrödinger wave function component $\sqrt{Z}$ of the S'-state at fixed angle $\theta=0^0$ for the MT II-IV potential model.

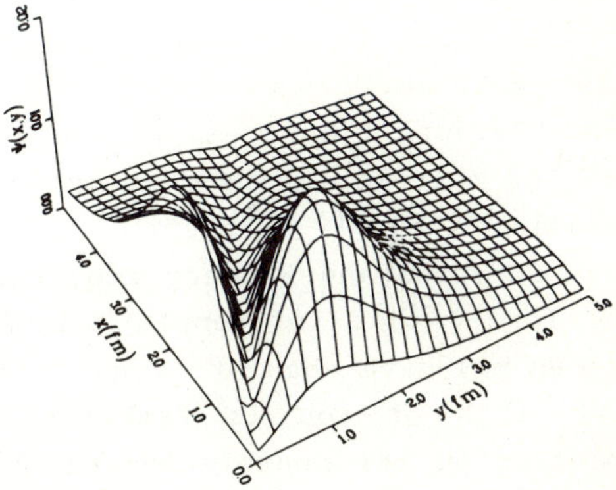

Fig.4 The Schrödinger wave function at fixed angle $\theta=0^0$ for the MT V potential model.

The two peaks in Ψ correspond to configurations in which the three nucleons are separated by about 1 fm, the distance at which the MT V potential attains its greatest depth (maximum attraction). For $\theta=90^0$ (not shown) these two peaks merge into one corresponding to an equilateral triangle configuration with internucleon separations of about 1 fm. Note that Ψ for the MT V model is suppressed near the origin, in contrast to Ψ for purely attractive potentials which has a maximum at the origin. In the neighborhood of x=y=0, all three particles are close together and the short-range repulsion of the potential has its maximum effect.

We have seen that as the potential model becomes more sophisticated, the complexity of the corresponding trinucleon wave function increases. Yukawa forms introduce kinks or ridges. Short-range repulsion produces peaks and valleys as the nucleons localize at positions corresponding to maximum attraction in the potential. It is also clear that simple (smooth) ad hoc model wave function hypotheses imply hidden assumptions about the smoothness (lack of repulsion) of the underlying potentials. "Forewarned is forearmed".

These effects in Ψ are reflected in calculated observables; e.g., the electron scattering cross sections. To illustrate this point, we show in Fig.5 Born approximation model results for scattering off gold using charge distributions of the uniform [$\rho(r) = \rho_o$, $r < r_o$] and exponential [$\rho(r) = \rho_o e^{-r/a}$] forms.[3,6] As the surface of the distribution becomes sharper, nodes develop; the spacing of the zeros is determined effectively by the size of the system. Note that a Gaussian charge distribution leads to a straight line on such a plot. As we shall see later, the trinucleon data (and alpha particle data as well) do not support the choice of such a simple, smooth model wave function.

Because I have emphasized structure and because we have heard so much of calculating Faddeev amplitudes, let us return to the harmonic oscillator model to ensure that the distinction between Schrödinger wave function and Faddeev amplitude is clear. It is the structure in the former that influences calculated observables. Structure in the latter may have absolutely no effect upon the observables.

Each of you is familiar with the Schrödinger wave function solution to the nonrelativistic harmonic oscillator problem for three spinless particles: (here we redefine $\vec{y}$ to be $\sqrt{3}/2\ \vec{y}$)

$$[E + \frac{1}{M}(\vec{\nabla}_x^2 + \vec{\nabla}_y^2)]\Psi = \frac{\beta^2}{M}(\vec{x}^2 + \vec{y}^2)\Psi . \qquad (2a)$$

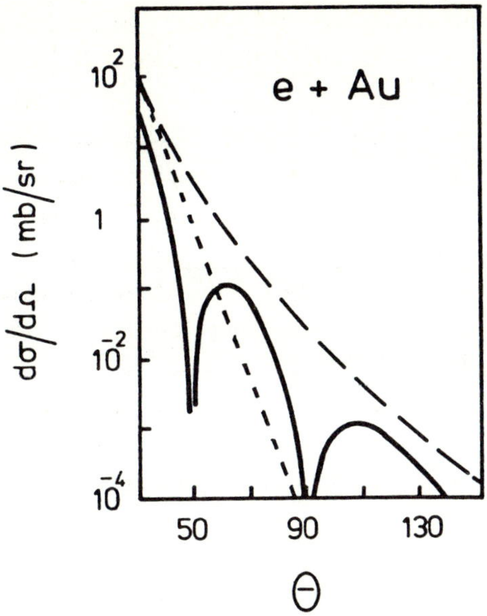

Fig.5 Cross section angular distributions for electron scattering from gold for a uniform charge distribution (solid line) an exponential charge distribution (dashed line) and a Gaussian charge distribution (dotted line).

The three-body solution is easily verified to be the familiar Gaussian function

$$\Psi = 3e^{-\beta(\vec{x}^2 + \vec{y}^2)/2} , \tag{2b}$$

where $E = 2\beta N/M$. (See Fig.6) One can generate this solution by solving the corresponding Faddeev equation[7]

$$[E + \frac{1}{M}(\vec{\nabla}_x^2 + \vec{\nabla}_y^2)]\psi = V\,\Psi = \frac{2\beta^2}{3M}\, x^2\, \Psi \tag{3a}$$

for the Faddeev amplitude $\psi(\vec{x},\vec{y})$ and then constructing the Schrödinger wave function from

$$\Psi(\vec{x},\vec{y}) = \psi(\vec{x},\vec{y}) + \psi(\tfrac{1}{2}\,\vec{x}+\vec{y},\ -\tfrac{3}{4}\,\vec{x} - \tfrac{1}{2}\,\vec{y})$$
$$+ \psi(-\tfrac{1}{2}\,\vec{x}-\vec{y},\ -\tfrac{3}{4}\,\vec{x} - \tfrac{1}{2}\,\vec{y}) . \tag{3b}$$
$$= \psi_1 + \psi_2 + \psi_3 .$$

This has been done, but one can also obtain a solution for ψ if, as in this case, Ψ is already known by solving the equation

$$[2\beta N + \vec{\nabla}_x^2 + \vec{\nabla}_y^2]\psi = 2\beta^2\, x^2\, e^{-\beta(\vec{x}^2+\vec{y}^2)/2} . \tag{4a}$$

The solution is of the form

$$\psi = \Psi + \xi$$

$$= e^{-\beta\rho^2/2} + \frac{\beta(\vec{x}^2-\vec{y}^2)}{2\rho^{N+1}} \, [\lambda \, J_{N+1}(\sqrt{2N\beta}\,\rho)$$

$$+ \pi \, Y_{N+1}(\sqrt{2N\beta}\,\rho) \int_0^\rho dt \, t^{N+2} \, e^{-t^2/2} \, J_{N+1}(\sqrt{2N\beta}\,t)$$

$$- \pi \, J_{N+1}(\sqrt{2n\beta}\,\rho) \int_0^\rho dt \, t^{N+2} \, e^{-t^2/2} \, Y_{N+1}(\sqrt{2N\beta}\,t)] , \quad (4b)$$

where $\rho^2 = \vec{x}^2 + \vec{y}^2$ and λ is an arbitrary constant. Obviously one has

$$\xi_1 + \xi_2 + \xi_3 = 0 \quad (5)$$

because of the $(\vec{x}^2-\vec{y}^2)$ factor in ξ_1 ; i.e., it does not contribute to Ψ in anyway. The ξ component of ψ results because the operator in the Faddeev equation Eq.(3a) is not L^2. Nonetheless, it illustrates

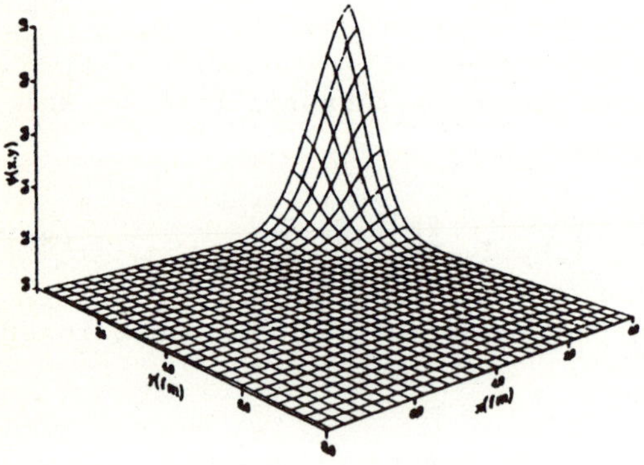

Fig.6 The Schrödinger wave function for the three-body harmonic oscillator problem.

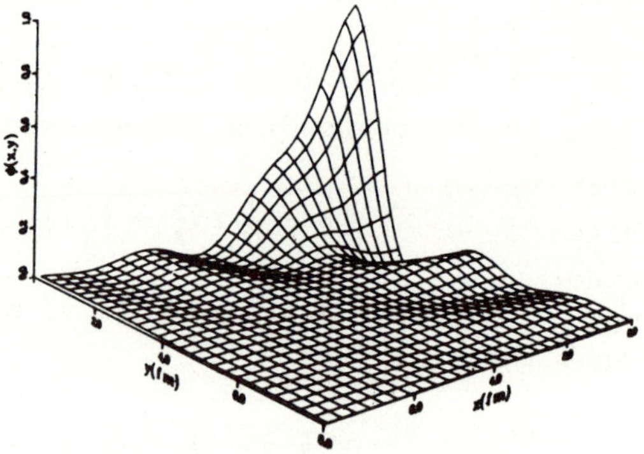

Fig.7 The Faddeev amplitude without the arbitrary component (λ=0) for the three-body harmonic oscillator problem.

the point that structure in the Faddeev amplitudes (see Fig.7) may cancel in the sums needed to generate the Schrödinger wave function. (The arbitrary component of ψ has been removed by setting $\lambda \equiv 0$.) Compare Figures 6 and 7. In particular, the ground-state wave function for the A=3 system must be positive definite, but the Faddeev amplitude for a potential with strong short-range repulsion will not be positive definite.[5,8] It is the cancellation between positive and negative parts of the Faddeev amplitudes in Eq.(3b) that leads to the deep valley seen in Fig.4. Thus, one must be careful about trying to reach conclusions concerning experimental observables based on an examination of the Faddeev amplitudes for a given model. It is the full Schrödinger wave function that defines the features of the observables.

III. Summary of Elastic Electron Scattering Formulae

Because a complete treatment of electron-proton scattering would require an hour unto itself and anything less would leave you wondering about the slight-of-hand performance, I refer you to the book by Bjorken and Drell[9] if the subject is not already familiar. I will quote here only the salient points and remark that they are not the subject of this lecture.

Consider the diagram shown in Fig.8 which depicts lowest order (single photon exchange) scattering of an electron from a proton. The electron with 4-momentum $k^\mu = (\varepsilon, \vec{k})$ scatters into the final state 4-momentum $k'^\mu = (\varepsilon', \vec{k}')$, while the proton recoils from initial state with 4-momentum $P^\mu = (E, \vec{p})$ to the final state with $P'^\mu = (E', \vec{p}')$. A virtual photon with 4-momentum transfer $q^\mu = k^\mu - k'^\mu = P'^\mu - P^\mu = (\omega, \vec{q})$ is exchanged in the process. We note that q^2 $(= q_\mu q^\mu) = \omega^2 - \vec{q}^2 \leq 0$, which is space-like. Hence, electron scattering is restricted to the kinematic region for which $\vec{q}^2 \geq \omega^2$. Figure 1 describes only the exchange of a single photon, but because higher order diagrams involve higher powers of the fine structure constant $\alpha(\simeq 1/137)$, the lowest order diagram should account for most of the scattering amplitude. The differential cross section for scattering an electron from a physical proton is

$$\frac{d\sigma}{d\Omega} = \sigma_M \left[F_1^2 - \frac{\kappa_p^2 q^2}{4M^2} F_2^2 - \frac{q^2}{2M^2} (F_1 + \kappa_p F_2)^2 \tan^2 \frac{\theta}{2} \right] , \tag{6}$$

where

$$\sigma_M = \left[\frac{\alpha \cos \frac{\theta}{2}}{2\varepsilon \sin^2 \frac{\theta}{2}} \right]^2 \tag{7}$$

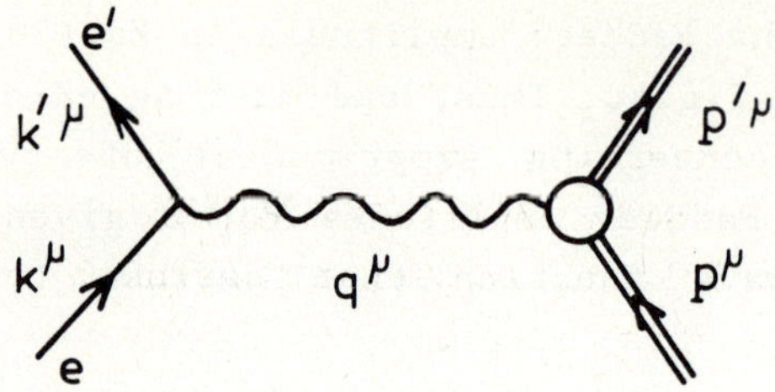

Fig.8. Lowest order diagram for electron scattering from a proton.

is the Mott cross section which describes the scattering from a point charge in the absence of any internal structure of the proton. Equation (6) is the famous Rosenbluth formula. This is the relativistic analog of the famous Rutherford cross section (with the charge of the nucleus set to 1). The F_1 and F_2 are structure functions associated with the charge and magnetic moment densities of the proton due to such vertex corrections as depicted in Fig.9a and 9b, and κ is the anomalous magnetic moment (κ_p = 1.79 while κ_n = -1.91). We note that the Dirac and Pauli structure form factors F_1 and F_2 can be separated experimentally by measuring the cross section as a function of θ while holding the momentum transfer q constant.

The combinations of the form factors given by

$$G_E = F_1 + \frac{\kappa q^2}{4M^2} F_2 \tag{8a}$$

$$G_M = F_1 + \kappa F_2 \tag{8b}$$

have a more direct geometrical interpretation, and one usually finds Eq.(6) rewritten as

$$\frac{d\sigma}{d\Omega} = \frac{\sigma_M}{(1-q^2/4M^2)} \left\{ G_E^2 - \frac{q^2}{4M^2} G_M^2 \left[1 + 2 \tan^2 \frac{\theta}{2} \left(1 - \frac{q^2}{4M^2}\right)\right] \right\} . \tag{9}$$

These are the nucleon form factors that we have used in our ^{3}H and ^{3}He form factor studies. Numerical values were taken from the 8.2 fit of Höhler et al.[10]

The charge form factor $F(q^2)$ is proportional to a matrix element of the form[9]

$$M \sim \int d^3p_i \; u(\vec{p}_f) \; u(\vec{p}_i) \tag{10a}$$

where $\vec{p}_i$ and $\vec{p}_f$ are related by $\vec{p}_f = \vec{p}_i - \vec{q}$. Thus, the form factor

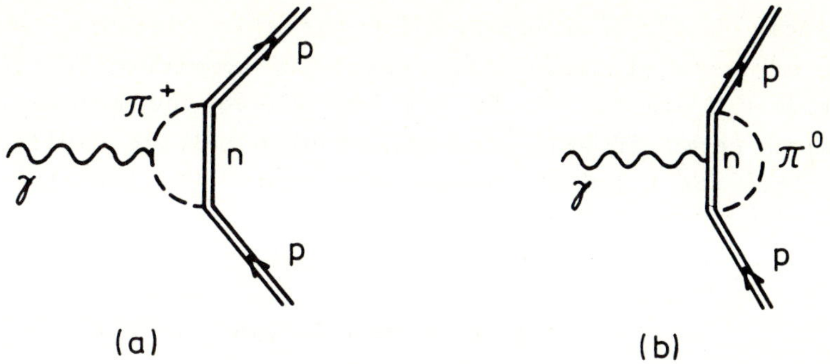

Fig. 9 Electromagnetic vertex corrections for the proton

structure function is a folding of the momentum space wave functions at the interaction vertex:

$$M \sim \int d^3p\ u(\vec{p} - \vec{q})\ u(\vec{p})\ . \tag{10b}$$

For calculations utilizing configuration space wave functions, it is more convenient to work with the Fourier transform of this expression:

$$M \sim \int d^3r\ e^{i\vec{q}\cdot\vec{r}}\ u^2(\vec{r}). \tag{10c}$$

The $u^2(\vec{r})$ can be identified with the charge density in the impulse approximation when one discusses nuclear systems.

The result for elastic electron scattering from the trinucleons is exactly that given by Eq.(9), because the nuclei have $J^\pi = \frac{1}{2}^+$. The formula for scattering from ^4He is even simpler, because that spin-0 object has no magnetic moment. Thus one need only set $G_M^2 \equiv 0$. The deuteron, being a spin-1 object, is more complex; there exists an electric quadrupole density in addition to a monopole charge density, as was pointed out by Schiff[11] and others.[3] Because of the increased complexity of the formulae for the spin-1 system, they will be left as a literature search exercise for those interested.

Restricting our attention to the trinucleons, we would like to determine four form factors: a charge and a magnetic moment form factor for each of ^3H and ^3He. To orient ourselves, let us consider the case in which we assume that the nucleon-nucleon force is purely central. (With no tensor force, there are no D-state components in the A=3 wave function.) We will also neglect differences between ^3He and ^3H due to the Coulomb force acting between the two protons in ^3He. In this model, the trinucleon wave function has two components (see Appendix B for a discussion of the spin, isospin, and spatial structure of these states.): the spatially symmetric S-state and the mixed-symmetry S'-state.[12] If one has solved the Faddeev amplitude equations in terms of spin-singlet and spin-triplet amplitudes $\psi^{(s)}$ and $\psi^{(t)}$, then the S and S' Faddeev amplitudes are given by[4,5]

$$\psi^{S} = [\psi^{(s)} - \psi^{(t)}]/\sqrt{2} \;, \tag{11a}$$

$$\psi^{S'} = [\psi^{(s)} + \psi^{(t)}]/\sqrt{2} \;. \tag{11b}$$

Because $\psi^{(s)}$ and $\psi^{(t)}$ have opposite signs, ψ^{S} is the dominant amplitude. [If the nucleon interaction were spin independent ($V^{(s)} \equiv V^{(t)}$), then $\psi^{S'} \equiv 0$.] The symmetrized combination of S-state amplitudes which define the S-state component of the Schrödinger wave function is then

$$u = \psi_1^S + \psi_2^S + \psi_3^S \;. \tag{12a}$$

The S'-state amplitude combinations are

$$v_1 = \frac{1}{\sqrt{6}} [\psi_2^{S'} + \psi_3^{S'} - 2\,\psi_1^{S'}] \;, \tag{12b}$$

$$v_2 = \frac{1}{\sqrt{2}} [\psi_3^{S'} - \psi_2^{S'}] \;. \tag{12c}$$

The Schrödinger wave function for ^{3}He $\Psi = u\phi_a + (v_2\phi_1 - v_1\phi_2)$ can then be used to evaluate matrix elements of the charge density operator, which we give here in impulse approximation:

$$\rho_c = \sum_{i=1}^{3} [\tfrac{1}{2}(1 + \tau_{iz})\, \rho_c^p\, (\vec{r}-\vec{r}_i) + \tfrac{1}{2}(1 - \tau_{iz})\, \rho_c^n\, (\vec{r}-\vec{r}_i)] . \tag{13}$$

The τ_{iz} are the unit isospin operators that act on the η's in the ϕ's, as defined in Appendix B. We limit our consideration to the nonrelativistic, impulse approximation form of the charge density operator for reasons outlined in Appendix C. It suffices to note here that exchange current contributions to the charge operator are essentially relativistic corrections and they are ambiguous.[13] Such is not the case for the isovector exchange current contributions to the magnetic moment density, but time limitation prevents us from considering that problem here.

Because ^{3}He has a charge of 2, we write its form factor as

$$2\,F_c({}^3\mathrm{He}) = \langle \Psi \,|e^{i\vec{q}\cdot\vec{r}}\,\rho_c|\,\Psi\rangle \qquad (14)$$

and evaluate Eq.(13) using the wave function Ψ defined in terms of the u and v functions of Eqs. (11 and 12). By changing variables from $\vec{r}$ to $\vec{r}-\vec{r}_i$, one can factor the nucleon form factors G^p and G^n from the expression. The expectation values of the τ_{iz} operators are easily evaluated leading to the simple relation[12,14]

$$2F_c({}^3\mathrm{He}) = (2G^p + G^n)\,F_1 - 2(G^p - G^n)\,F_2\,, \qquad (15)$$

where we have defined the body form factors F_1 and F_2 to be

$$F_1 = \iint d^3x\, d^3y\, e^{i\frac{2}{3}\vec{q}\cdot\vec{y}}\,[u^2(\vec{x},\vec{y}) + v_1^2(\vec{x},\vec{y}) + v_2^2(\vec{x},\vec{y})] \qquad (16a)$$

and

$$F_2 = -\iint d^3x\, d^3y\, e^{i\frac{2}{3}\vec{q}\cdot\vec{y}}\,u(x,y)\,v_1(\vec{x},\vec{y})\,. \qquad (16b)$$

Using the corresponding wave function for ^{3}H, we obtain in that case

$$F_c({}^3\mathrm{H}) = (G^p + 2G^n)\,F_1 + 2(G^p - G^n)\,F_2\,. \qquad (17)$$

We have made several simplifying assumptions to obtain these expressions. Because the S' state is only a few % of the normalization, let us drop the $v_1^2 + v_2^2$ terms in F_1. (Note that at $\vec{q}=0$, F_2 vanishes because u is symmetric in x and v_1 is antisymmetric in $\vec{x}$; thus the expressions in Eqs (14 and 16) have the proper normalizations.) Similarly, the neutron charge form factor is small, and we shall neglect it. Then we find

$$F_c({}^3\mathrm{He}) = G^p(F_1 - F_2) \qquad (18a)$$

and

$$F_c({}^3\mathrm{H}) = G^p(F_1 + 2\,F_2)\,. \qquad (18b)$$

Because F_2 is defined in such a way as to be positive, we have $F_c({}^3\mathrm{He}) < F_c({}^3\mathrm{H})$ for small q^2 which implies that the radius of ^{3}He must be greater than that for ^{3}H.[14] We shall discuss the physics of this later, but for the moment we emphasize that it has nothing to do with Coulomb repulsion in ^{3}He. There are no Coulomb effects in the present analysis.

The expressions for F_1 and F_2 that correspond to Eq.(18) are explicitly

$$F_1(q^2) = \iint d^3x\, d^3y\; e^{i\frac{2}{3}\vec{q}\cdot\vec{y}}\; u^2(\vec{x},\vec{y}) \tag{19a}$$

and

$$F_2(q^2) = -\iint d^3x\, d^3y\; e^{i\frac{2}{3}\vec{q}\cdot\vec{y}}\; u(\vec{x},\vec{y})\; v_1(\vec{x},\vec{y}) \ , \tag{19b}$$

in terms of configuration space wave functions. In momentum space, they are

$$F_1(q^2) = \iint d^3\vec{k}\; d^3\vec{p}\; u(\vec{k},\vec{p})\; u(\vec{k},\vec{p} - \tfrac{2}{3}\,\vec{q}) \tag{20a}$$

and

$$F_2(q) = -\iint d^3\vec{k}\; d^3\vec{p}\; u(\vec{k},\vec{p})\; v_1(\vec{k},\vec{p} - \tfrac{2}{3}\,\vec{q}) \ . \tag{20b}$$

These latter momentum space expressions were first used to actually calculate ^3H and ^3He charge and magnetic moment form factors for realistic nucleon-nucleon force wave functions solutions of the Faddeev equations.[15]

IV. Physics of the Trinucleon Form Factors

The form factor is a function of q^2 and not q even though the exponential argument in Eqs.(16) and (19) is linear in $\vec{q}$. Odd powers vanish in the transform because of the isotropic dependence of the charge density upon the direction of the external vector $\vec{q}$. For a general density $\rho(r)/4\pi$, one obtains

$$F = \frac{1}{4\pi}\int d^3r\; \rho(r)\; e^{i\vec{q}\cdot\vec{r}} \tag{21a}$$

$$= \int_0^\infty dr\; r^2\rho(r)\; j_o(qr) \tag{21b}$$

which is clearly even in powers of q. Expanding Eq.(21b) for small values of q, one obtains the usual expansion of the form factor in terms of the rms radius $\langle r^2\rangle$ of the system:

$$F(q^2) \simeq 1 - \frac{1}{6}\langle r^2\rangle\, q^2 \tag{22}$$

where

$$\langle r^2 \rangle = \int_0^\infty dr\ r^4 \rho(r) \qquad (23)$$

and

$$1 = \int_0^\infty dr\ r^2 \rho(r)\ . \qquad (24)$$

It is clear from Eq.(22) that the larger the size of the system, the faster the form factor falls away from $F(q^2=0) = 1$.

Let us apply this result to an analysis of the trinucleons. Martino recently reported values for the rms radii of ^3He and ^3H of 1.93(3) fm and 1.81(5) fm respectively.[16] The ^3He charge density has a radius some 6% larger than that of ^3H. If all the nucleon-nucleon forces were equal, the two radii would be the same. (See Fig.10a.) They are not. The neutron-proton spin-triplet force is stronger, binding the deuteron, whereas the neutron-proton singlet force, the proton-proton force, and the neutron-neutron force all lead to just unbound singlet states. The charge in ^3He is carried by the like pair of nucleons; the charge in ^3H is carried by the odd nucleon. This is illustrated in Figs. 10b and 10c. Because the spin-triplet and spin-singlet forces are not equivalent, the like pair of nucleons is distributed differently from the odd nucleon. In particular, the interaction between like nucleons is weaker than the average neutron-proton interaction, such that the like nucleons lie farther from the centre-of-mass than does the odd nucleon - the like nucleon pair distribution is more extended in space.[12,17] Because the charge radius is the average distance between the protons and the centre-of-mass, the ^3He charge radius is greater than the ^3H charge radius. This manifests itself in the wave function through the appearance of the S'-state wave function component. This S'-state of mixed spatial symmetry is a spin(isospin)-space correlation. It is responsible for introducing the F_2 body form factor in Eqs.(15),(17), and (18). It breaks the isoscalar symmetry in the trinucleon charge form factors, introducing an isovector component. Please understand that the difference between $F_c(^3He)$ and $F_c(^3H)$ discussed here has absolutely nothing to do with Coulomb repulsion between the two protons in ^3He. It arises strictly from the spin dependence of the nucleon-nucleon force.

How do Faddeev calculations of the trinucleon wave functions fare with respect to the measured radii? The answer to the question is not simple, because the radius is sensitive to the outer parts of

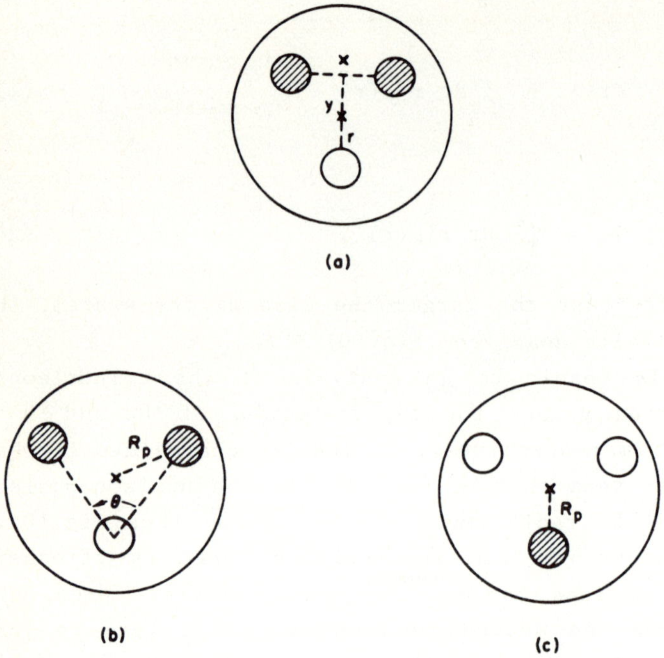

Fig. 10 Schematic model of ^{3}He with with identical forces between protons (shaded) and neutrons in (a). ^{3}He and ^{3}H are shown in (b) and (c) when the nn and pp forces are weaker than the average np force. R_p is the "charge radius".

the wave function which are in turn sensitive to the binding energy E_B of the system. (The binding energy depends upon the force model selected and the number of channels included in the calculation.) The asymptotic form of the S-state wave function component is proportional to $\exp(-\kappa\rho)/\rho^{5/2}$ where $\kappa = (m\ E_B)^{1/2}$ and ρ is the usual hyperspherical coordinate ($\rho^2 = x^2 + y^2$). If one assumes that this form is valid over all space, then one obtains[17]

$$\langle r^2 \rangle^{1/2} = 1/(2\kappa) \sim E_B^{-1/2} . \tag{25}$$

This is the binding energy dependence of the isoscalar or mass radius of the F_1 body form factor. In Fig.11 are shown the results from many calculations of the Los Alamos-Iowa Faddeev group. The symbols refer only to the number of channels and do not indicate whether a two-body or a two-body plus three-body potential model was used. The scaling behavior is clear, although for the ^{3}He and ^{3}H charge densities it is more nearly E_B^{-1} than the $E_B^{-1/2}$ which holds for the isosclar radius.[17] Clearly, a model which produces the correct binding energy for the trinucleons will give essentially the correct radii.

How do Faddeev calculations fare for the full form factors, which test more than the asymptotic properties of Ψ? Except for the

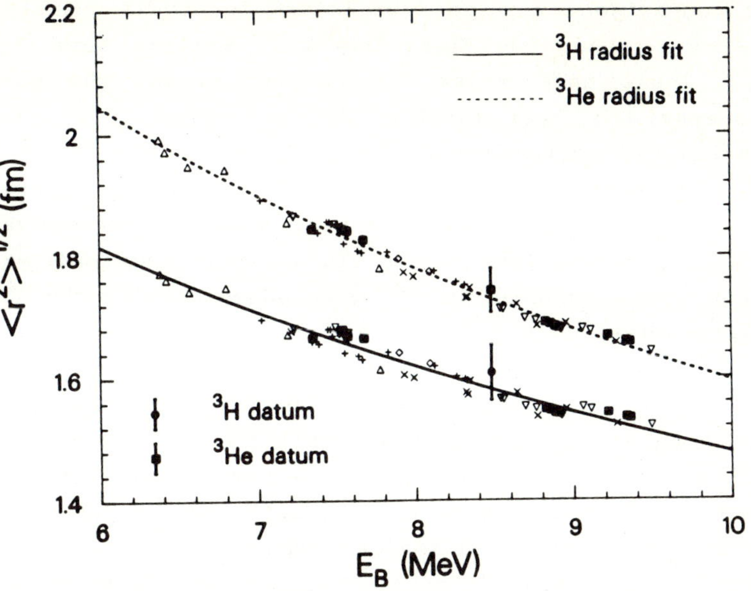

Fig. 11. The ^{3}He and ^{3}H rms charge radii plotted versus the triton bending energy E_B for various model Hamiltonians. The curves are fits to the theoretical values shown.

very-low-q^2 region, which is determined essentially by the rms radius of the system, the charge form factor for each of ^{3}He and ^{3}H is determined by three numbers in the traditional nuclear physics regime ($q^2 < 30\ fm^{-2}$).[20] These are the positions of the first diffraction minimum and the secondary diffraction maximum and the value of the form factor at that latter value of q^2. The most recent Saclay fits[21] to the world's trinucleon form factor data are characterized by

^{3}He:

$$q^2_{min} = 11.0 \pm 0.7\ fm^{-2}$$

$$q^2_{max} \simeq 15.65\ fm^{-2}$$

$$F(q^2_{max}) = -(5.9 \pm 0.3) \times 10^{-3}$$

^{3}H:

$$q^2_{min} = 12.6 \pm 0.5\ fm^{-2}$$

$$q^2_{max} \simeq 17.25\ fm^{-2}$$

$$F(q^2_{max}) = -(3.95 \pm 0.4) \times 10^{-3}\ .$$

For comparison, results from Ref.20 for these quantities in the case of ^3H are shown in Figs. 12-14. Nucleon form factors are included. These observables are plotted versus the corresponding binding energy for each model. The triangles, x's, circles, and inverted triangles correspond to the Reid soft core (RSC),[22] Argonne V_{14} (AV15),[23] super soft core (C) (SSCC),[24] and de Tourreil-Rouben-Sprung (B) (TRSB)[25] two-body potential models, respectively. Two-pion-exchange three-nucleon forces [Tucson-Melbourne (TM),[26] Brazilian (BR),[27] and Urbana-Argonne (UA)[28]] were added only to the RSC and AV14 models. All points with $E_B > 7.7$ MeV contain a three-body force. In each figure there is a band trending upward with increasing binding energy. [Points with small E_B which lie far off the band correspond primarily to three-channel calculations; they have severely truncated tensor forces and cannot be said to be particularly realistic.] In each case the AV14 model tends to produce larger values of q^2_{min} or q^2_{max} than the RSC model, and smaller values of $|F(q^2_{max})|$. A plot for the position of the first diffraction minima in ^3He is shown in Fig.15 for comparison. The results of these impulse approximation calculations may be summarized as follows: our minima and maxima are at too large a value of q^2 while the values of the maxima are too small, compared with experiment.

This is depicted most clearly in Figs.16 and 17, which compare our RSC 34-channel form factor curves corresponding to three different three-body force models with the experimental data.[21,29-32] The various three-body forces increase the magnitude of the form factor in the region of the secondary maximum but not enough to agree with the data. Moreover, there is a serious problem at more moderate momentum transfers which stems from the fact that the diffraction minima occur at the wrong locations. In the model presented, the fit to the low-q^2 data is best without the inclusion of a three-body force.

The ad hoc addition to either form factor of a component which vanishes at $q^2=0$ and is negative in the region of the diffraction minimum and secondary maximum would alleviate the problems.[20] Such a negative component would shift the form factor minimum and maximum to smaller values of q^2 and would increase the size of the form factor maximum. This simple structural behaviour accounts for the helpfulness of meson-exchange currents. However, we reiterate that there is no fundamental difference between certain pion-exchange contributions to the charge operator and the inclusion of relativistic corrections in the two-nucleon and three-nucleon Hamiltonians (ΔH). The matrix elements of the charge operator have a strength which can be dialed

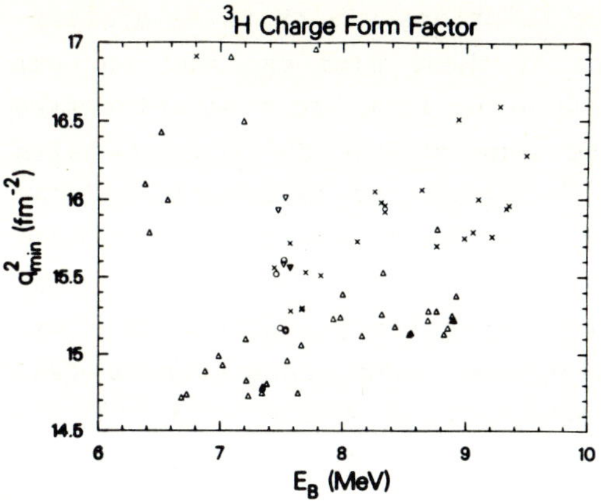

Fig. 12. Position of the first diffraction minimum of the ^3H charge form factor plotted versus the binding energy E_B for different combinations of the two-body and three-body force models. The triangles, X's, circles, and inverted triangles correspond to the RSC, AV14, SSCC, and TRSB two-body force models.

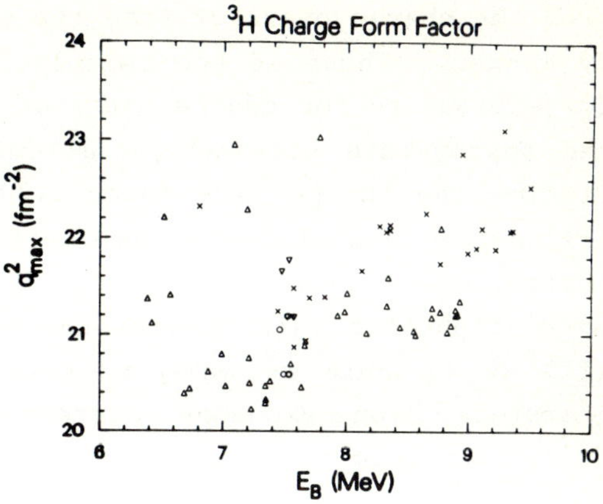

Fig. 13. Position of the secondary diffraction maximum of the ^3H charge form factor plotted versus the binding energy E_B. The data set and symbols are the same as in Fig. 12.

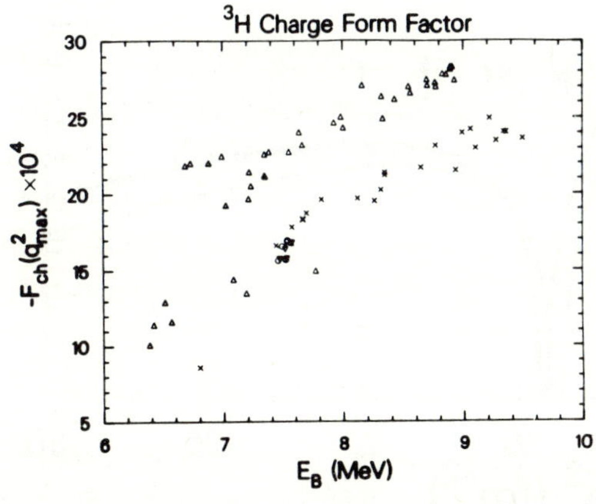

Fig. 14. Magnitude of the secondary diffraction maximum of the ^3H charge form factor plotted versus the binding energy E_B. The data set and symbols are the same as in Fig. 12.

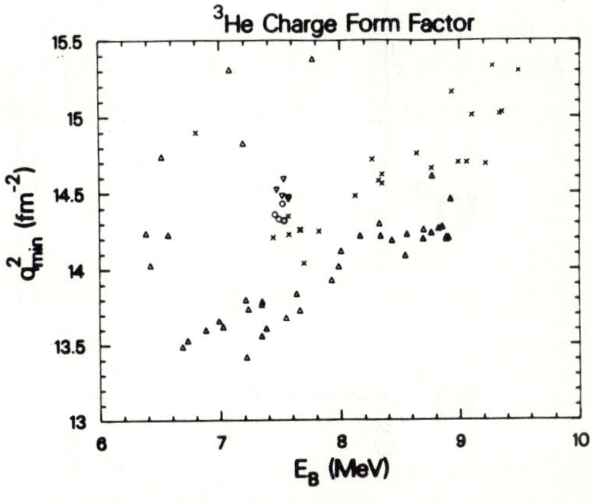

Fig. 15. The ^3He case as in Fig.12.

from the charge operator into the wave functions via ΔH in an arbitrary manner. Those ad hoc calculations of these pion exchange current corrections to the charge operator have heretofore had a negative sign and appropriate strength to alleviate some of the difficulties with fitting the charge form factors.[33] Therefore, it is imperative that trinucleon calculations be performed which include relativistic corrections. [One would prefer a model calculation with the minimal correct physics which avoids the $(v/c)^2$ expansion.] Only in this way will we be able to make a clear statement about relativistic (and therefore pion exchange current) effects and their role in the trinucleon form factors.

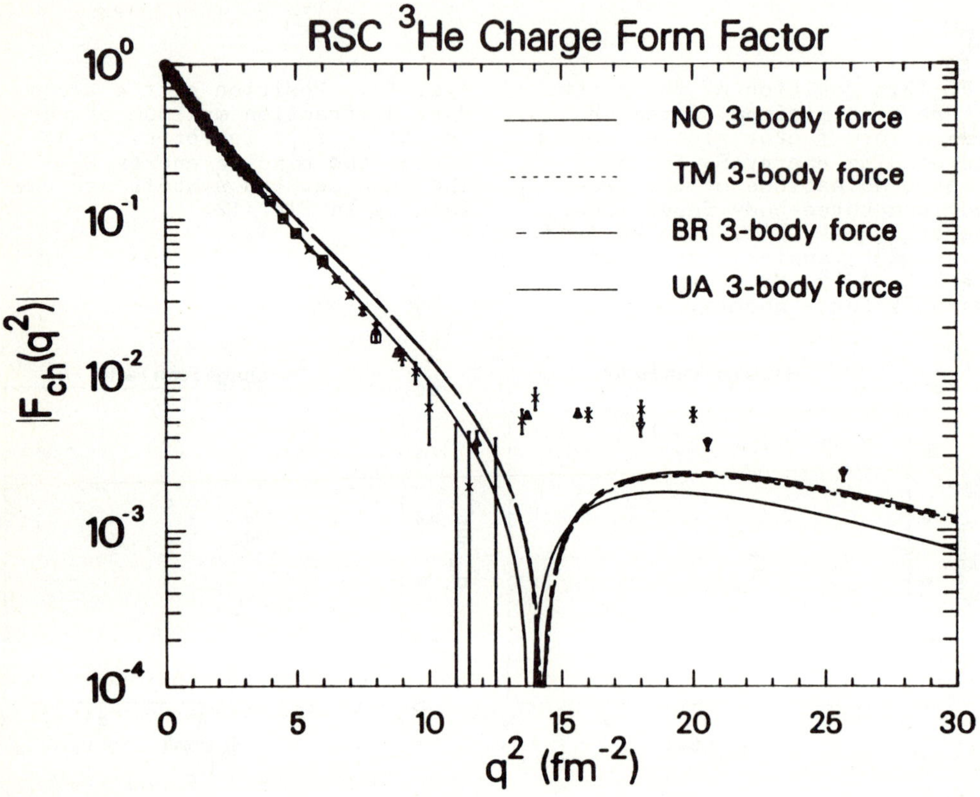

Fig. 16. The magnitude of the RSC ^{3}He charge form factor in impulse approximation for several three-body force models plotted versus q^2, together with the experimental data

What does all of this say about the charge densities of ^{3}He and ^{3}H? We plot the point-nucleon impulse approximation charge densities[20] in Figs. 18 and 19. The ^{3}He charge density has a maximum at the origin when no three-body force is included. This is modified to be a slight minimum (except for the TM model) when a three-body force is included. The size of this depression is much smaller than Sick obtained when he Fourier transformed the ^{3}He form factor data.[34] This

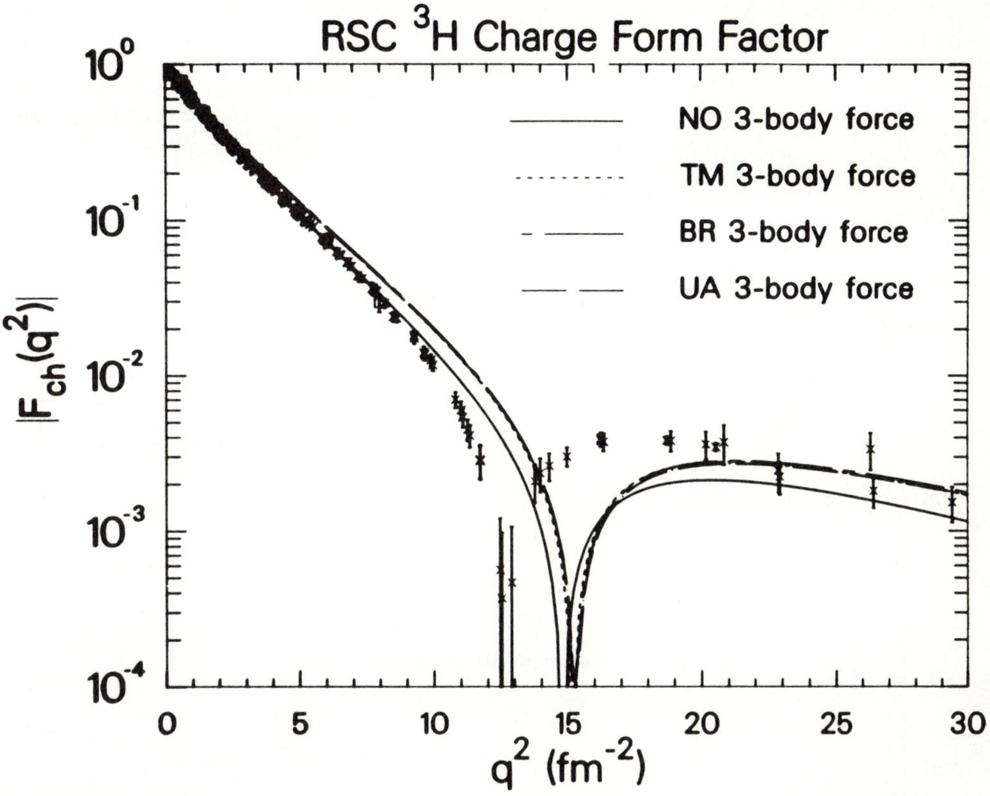

Fig. 17. The magnitude of the RSC ^{3}H charge form factor in impulse approximation for several three-body force models plotted versus q^2, together with the experimental data.

is a reflection of the smaller secondary maximum in the calculated form factors. It is important to realize that the Fourier transform of the experimental form factor is not necessarily properly interpreted as a "charge density." Furthermore, the size of the hole corresponds to less than 1% of the total charge of ^{3}He, which is the order of magnitude of relativistic corrections. The difference between the TM curve near the origin and the other three-body force curves reflects the form factor differences at much larger values of q^2 than those shown.

The ^{3}H charge density has a small hole for each model. This is caused by the L=2 (D-wave) component of the wave function. This component has a completely symmetric spin-quartet (S=3/2) wave func-

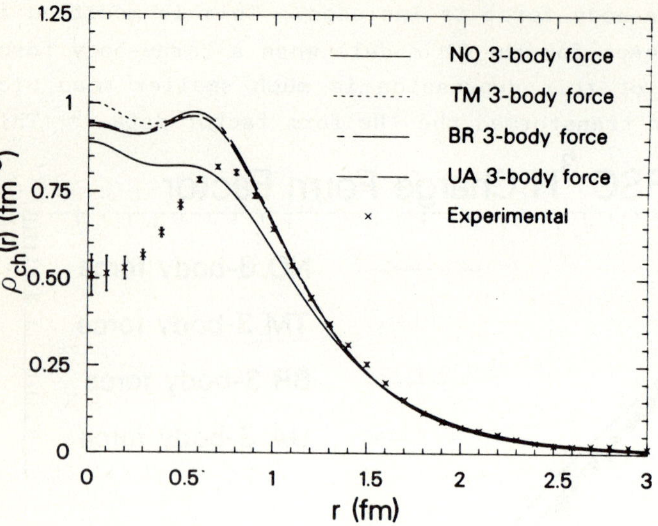

Fig.18. The RSC ^{3}He charge density in the impulse approximation for several three-body force models plotted versus r.

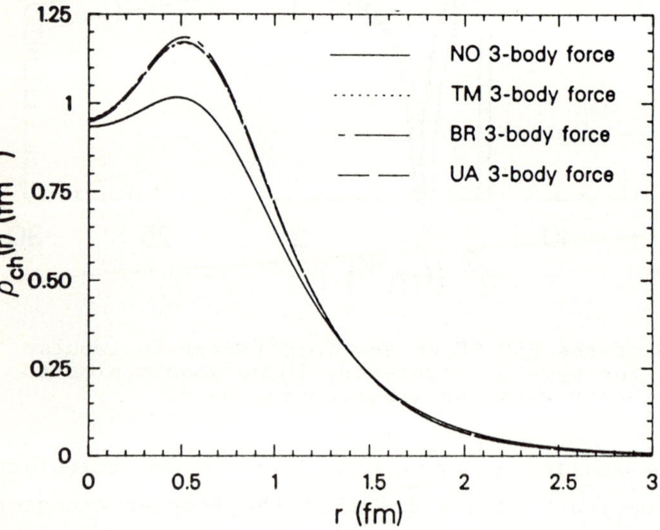

Fig. 19. The RSC ^{3}H charge density in the impulse approximation for several three-body force models plotted versus r.

tion. Consequently, the two neutrons in the L=2 component of the triton wave function must be in a relative odd-parity state (to satisfy the Pauli principle) as must be the remaining proton. Therefore, the charge density contribution from this wave function component must vanish at the origin. There is no similar restriction in ^{3}He, because there are two protons and singling out one of them leaves the remaining neutron-proton pair in any orbital state.

Finally, although it is difficult to observe the effect in Fig. 18 and 19, the increased binding due to the inclusion of a three-body force draws in the charge density toward the origin at large r. All curves correspond to a normalization of unity.

As a closing note, let us consider the relation between the charge form factors and the Coulomb energy E_C due to the Coulomb force acting between the two protons in ^{3}He. Clearly, E_C depends upon the size of ^{3}He and therefore the binding energy. Friar[35] and Fabre de la Ripelle[36] independently proposed exploiting the hyperspherical approximation, which leads to an estimate (E_C^H) for the Coulomb energy in terms of the charge density (and therefore the charge form factor). If one considers the geometrical picture depicted in Fig.10a, then for an equilateral triangle corresponding to the dominant S-state component of Ψ the distance x between the two protons is $\sqrt{3}$ r. Consequently, for a smooth operator such as 1/x, one obtains as a reasonable approximation

$$E_C^H \simeq E_C \, , \tag{26}$$

where

$$E_C = \langle \Psi | \frac{\alpha}{x} | \Psi \rangle \tag{27a}$$

and

$$E_C^H = \frac{1}{\sqrt{3}} \langle \Psi | \frac{\alpha}{r} | \Psi \rangle \, . \tag{27b}$$

The idea is to replace the two-body correlation function (required to calculate $\langle \frac{1}{x} \rangle$ by the one-body charge density (needed to evaluate $\langle \frac{1}{y} \rangle$). There is no a priori reason that this must work. If $\frac{1}{x}$ were instead $\delta(x)$ and one tried to replace it by $\delta(y)$, the approximation would obviously fail badly, as $\langle \Psi | \delta(x) | \Psi \rangle \simeq 0$ whereas one can see from Figs. 18 and 19 that $\langle \Psi | \delta(y) | \Psi \rangle \neq 0$. Nevertheless, we have shown by actual calculation that it works remarkably well for the Coulomb energy.[17] This can be seen in Fig.20, where E_C is plotted versus E_C^H . The E_C^H approximation is less than 1% larger than E_C for all models. The difference arises because the correlation function is suppressed more than the charge density for small values of their argument when there is short-range repulsion.[37]

The approximation is quite useful because we have available experimental charge form factor data which can be used to calculate E_C^H. One obtains[38] 638 ± 10 keV for the Coulomb energy of ^{3}He. (This is smaller than the 650 keV one would obtain for $E_B \simeq 8.5$ MeV in model calculations, because the experimental form factors are of larger

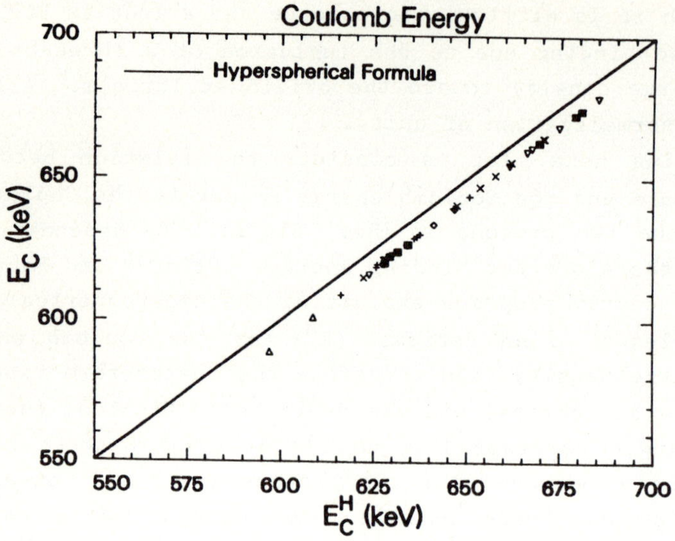

Fig. 20. The ^3He Coulomb energy E_C plotted versus the hyperspherical approximation.

magnitude in the region of the secondary maximum where they are negative.) The experimental binding energy difference between ^3H and ^3He is 764 keV. The fact that this binding energy difference is larger than the Coulomb energy is a clear indication of the presence of charge symmetry breaking forces in the nuclear Hamiltonian. That is, the neutron-neutron and proton-proton strong interactions are not identical. The known mixing of the ρ and ω and the π^0 and η would lead one to predict at least a small charge asymmetry in the nucleon-nucleon force. However, the size of the effect seen here is not fully understood.

LECTURE II. Two-Body Photodisintegration of the Triton

I. Introduction

The photon makes an ideal probe of the nucleus. The interaction operator is reasonably well understood. Thus, one may ask questions of the nuclear system independent of the interaction mechanism. We have seen how the virtual photon of electron scattering can be used to study the charge density of the trinucleons. Let us now look at how the real photon can be used to investigate the principal physics of the A=3 continuum.

Before turning to that problem, I would like to enumerate a few of the interesting aspects of low-energy photonuclear physics, lest you think that trinucleon photodisintegration is the only story. It was only a little more than 50 years ago that the first photonuclear experiment took place:[41] $^2H + \gamma \rightarrow n + p$. The inverse of that reaction (thermal capture of neutrons by hydrogen), with a cross section (of 330 mb) some 10% larger than theoretical models could account for, produced the first incontrovertible evidence for meson exchange current effects in nuclei.[42] The threshold $n+d \rightarrow {}^3H+\gamma$ reaction has a cross section (of 0.52 mb) some 600 times smaller,[43] and meson exchange current effects are enhanced (to 50%) relative to the standard nucleon current transition.[44] By exploring such processes in which normally dominant reaction mechanisms are suppressed, one can investigate details of nuclear physics which would otherwise be difficult to see. Another example is the forward (0°) photodisintegration of deuterium.[45] Because the normally dominant E1

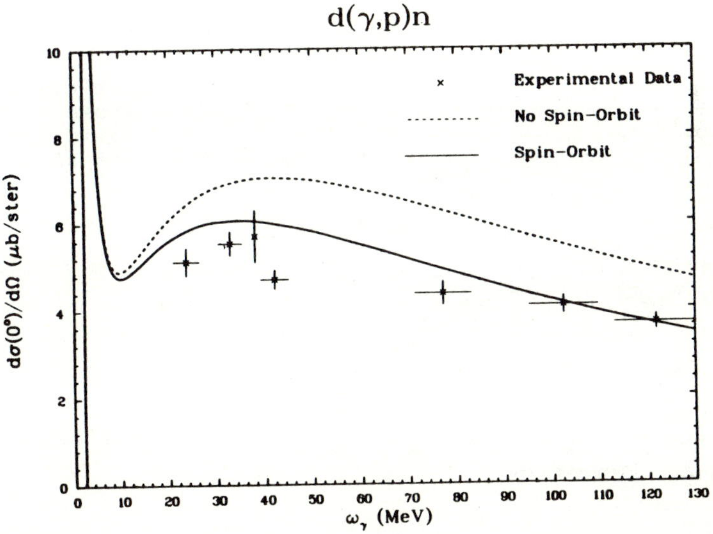

Fig. 21. Model calculations of the $^2H(\gamma,p)n$ reaction at 0° with and without the relativistic spin-orbit contribution to the E1 operator. The data are from Ref. 45.

transition from the L=0 component of the initial state to the L=1 partial wave of the final state vanishes in this geometry, one can clearly observe noncentral force effects.[46] In particular, the spin orbit terms provide a 20% enhancement of $d\sigma/d\Omega$ for $\theta=0^0$ as can be seen in Fig. 21. I will not have time to discuss the 85%-90% suppression of the T=1/2 channel in the three-body photodisintegration of ^{3}H and ^{3}He, compared to the T=3/2 channel.[47-50] It is, however, intimately related by three-body unitarity to the two-body breakup channel. Finally, one of the long standing puzzles in nuclear physics is the deviation from 1 of the ratio of photoneutrons and photoprotons from an alpha particle; i.e., experimentally one finds (see Fig. 22)

$$\sigma[^4He(\gamma,p)^3H]/\sigma[^4He(\gamma,n)^3He] \neq 1$$

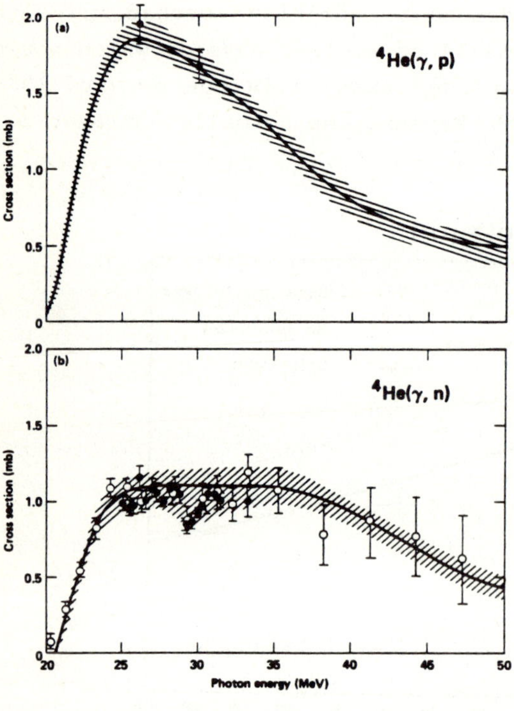

Fig. 22. The ^{4}He$(\gamma,p)^3$H and ^{4}He$(\gamma,n)^3$He cross section evaluation (solid line); data in (a) are from Ref. 51

for $E_\gamma < 30$ MeV, in contradiction to that which one would expect on the basis of simple charge symmetry arguments.[52] The list goes on.

In addition to the fact that few-body photonuclear reactions yield interesting physics, there are purely theoretical reasons for attempting to model these reactions.[50] At low energy only a few multipoles are important. That is not to say that higher multipoles can necessarily be neglected, but they can be treated adequately in Born approximation. As a consequence, large partial wave sums do not hide the physics. One can methodically examine a reaction and come to an understanding of the physics; e.g. the 0° photodisintegration of deuterium. Because only a few multipoles and partial waves dominate the problem, "exact" model calculations are not impossibly long. Thus, photonuclear reactions are an appropriate field in which to study the application of exact equation techniques in nuclear physics.

We shall examine below, as our example, the electric dipole (E1) two-body photodisintegration of tritium:

$$^{3}H + \gamma \rightarrow n + d \ .$$

We employ a separable potential formalism and restrict our consideration to central forces. This model has been investigated thoroughly by Barbour and Phillips[48] and independently by Gibson and Lehman[53] using a description of the nd off-shell scattering that resembles closely the formalism of Alt, Grassberger, and Sandhas.[54] The E2 contribution was shown to be negligible for photon energies of less than 40 MeV by Barbour and Hendry.[55] The physics points of interest are (1) the cross section calculation does depend on whether the initial and final states are eigenfunctions of the same Hamiltonian, (2) the full calculation is 50% larger than the plane-wave Born approximation result at the peak of the 90° differential cross section, and (3) the enhancement is due to using exact equations which incorporate proper three-body unitarity and therefore couple the two-body (nd) breakup channel to the T=1/2 three-body breakup (nnp) channel.[53]

II. The E1 operator

For those not familiar with photonuclear reactions, the matrix elements of the multipole series are known to go like $EL \sim (kR)^{L}$ at low energy, where k is the photon momentum (energy) in fm^{-1} and R is the size of the nuclear system of interest. We have seen that the trinucleon radius is less than 2 fm. A photon of 20 MeV energy has $k \simeq 0.1\ fm^{-1}$. Thus the ratio of E2 to E1 matrix elements is $\simeq 0.1$, and the E1 operator dominates the low-energy photodisintegration reaction.

Recall that the charge density operator relevant to the calculation of charge form factors is of the form

$$\rho(x_i) = e\delta(\vec{x} - \vec{x}_i)\,\frac{1}{2}\,(1 + \tau_{iz}) \tag{28}$$

for a point charge. Meson exchange current contributions to the charge density operator are $(v/c)^2$ relativistic corrections. What do we find for the E1 photodisintegration operator?

Formally, we wish to investigate the Hamiltonian

$$H_{Total} = H + H' , \tag{29}$$

where H' is the interaction which will be treated perturbatively and H is the nuclear Hamiltonian composed of kinetic energy and pair interactions. That is, we assume

$$H = H_o + V \tag{30}$$

where the initial- and final-state eigenfunctions satisfy

$$H|\Psi_i\rangle = E_i|\Psi_i\rangle \tag{31a}$$

and

$$H|\Psi_f\rangle = E_f|\Psi_f\rangle . \tag{31b}$$

The matrix element M_{if} which determines the transition from Ψ_i to Ψ_f is then given by

$$M_{if} = \langle\Psi_f|H'|\Psi_i\rangle . \tag{32}$$

The cross section for the reaction is

$$\sigma = \int \overline{\sum_{if}}\, |\, M_{if}\, |^2\, df, \tag{33}$$

where $\overline{\sum_{if}}$ is the sum over final spin states and average over initial spin states and $\int df$ is the required phase space integral.

The transition matrix element involves the interaction of the photon field $\vec{\varepsilon}\exp(i\vec{k}.\vec{r})$, where $\vec{\varepsilon}$ is the photon polarization and $\vec{k}$ its momentum vector ($\vec{\varepsilon}.\vec{k} = 0$), with the nuclear current $\vec{J}$. The current $\vec{J}$ consists of a nucleon component and a meson exchange component:

$$\vec{J} = \vec{J}_N + \vec{J}_{MEC} . \tag{34}$$

The nucleon current has the expected form

$$\vec{J}_N(\vec{x}_i) = \frac{e}{2m} \{ \vec{p}_i , \delta(\vec{x} - \vec{x}_i) \} . \quad (35)$$

That is, it is porportional to $\vec{p}/m$, the velocity of the nucleon. Note that it is of order (v/c), whereas the charge density is $(v/c)^0$ and its exchange current corrections are $(v/c)^2$. Unfortunately, meson exchange current corrections to $\vec{J}$ are also of order (v/c) and cannot, therefore, be neglected. However, it was pointed out by Siegert[56] in the late 1930's that one can include the principal effects of meson exchange currents in the long wave length limit (i.e. the low-energy region in which we are interested) by making use of the charge continuity equation

$$\vec{\nabla}.\vec{J} = - \frac{d}{dt} \rho = -i[\rho,H]. \quad (36)$$

Long wave length limit means to lowest order in $\vec{k}$. In a nucleons-only regime, where $J_{MEC} = 0$, the electric dipole current is

$$J \sim \vec{p}/m = \dot{\vec{r}} = i[\vec{r},H] . \quad (37)$$

In that case, the transition matrix element becomes

$$M_{if} = i\langle\Psi_f| \vec{\varepsilon}.[\vec{r},H]|\Psi_i\rangle \quad (38a)$$

$$= i(E_f - E_i) \langle\Psi_f|\vec{\varepsilon}.\vec{r}|\Psi_i\rangle . \quad (38b)$$

Thus, the long wave length limit form of the electric dipole operator is $\vec{\varepsilon}.\vec{r}$. The same result is achieved when the general current $\vec{J}$ is used by writing

$$\vec{\varepsilon}\, e^{i\vec{k}.\vec{r}} = \int_0^1 ds \{\vec{\nabla} [\vec{\varepsilon}.\vec{r}\, e^{is\vec{k}.\vec{r}}] - isr \times [\vec{k} \times \vec{\varepsilon}]e^{is\vec{k}.\vec{r}}\} . \quad (39)$$

The second term generates magnetic multipoles, so that we concentrate on the first. Thus, we consider

$$M_{if} = \langle\Psi_f|\int_0^1 ds\, \vec{J} . \vec{\nabla}[\vec{\varepsilon}.\vec{r}\, e^{is\vec{k}.\vec{r}}]|\Psi_i\rangle . \quad (40)$$

Performing an integration by parts and utilizing the current conservation relation to replace the $\vec{\nabla}.\vec{J}$ operation with the commutator of a point nucleon charge density and H, we obtain

$$M_{if} = i\langle\Psi_f|[\int_0^1 ds\, \vec{\varepsilon}.\vec{r}\, e^{is\vec{k}.\vec{r}}, H]|\Psi_i\rangle . \quad (41)$$

Again, because the initial and final states are eigenfunctions of H, we can evaluate the commutator and perform the $\int ds$ to obtain

$$M_{if} = (E_f - E_i) \sum_L (i)^L \langle\Psi_f|\vec{\varepsilon}.\vec{r}\, \frac{(\vec{k}.\vec{r})^{L-1}}{L!}|\Psi_i\rangle , \quad (42)$$

the leading term of which is

$$M_{if} = i\omega\, \langle\Psi_f|\vec{\varepsilon}.\vec{r}|\Psi_i\rangle , \quad (43)$$

where $\omega = E_f - E_i = |\vec{k}|$. Thus, the Siegert (long wave length) limit of the electric dipole operator which includes the meson exchange current as well as the nucleon current ($\vec{p}/m$) is $\vec{\varepsilon}.\vec{r}$.

Let me conclude this discussion by pointing out that there is an additional reason for using the Siegert form of the transition operator when one is forced to use approximate solutions to the nuclear Hamiltonian. Because the Siegert operator is related to the charge density, we are able to enforce some physical intuition in normalizing the bound-state spatial density; the normalization for the current density of the nucleus is unknown. Furthermore, care should be exercised to avoid the temptation to use the current form of the operator even when exact eigenstates of the nuclear Hamiltonian exist, as they do for the separable potential model. The requirement of gauge invariance introduces a gauge transformation in all nonlocal, momentum-dependent potentials when $\vec{\varepsilon}.\vec{p}$ is used. This complexity (see Yamaguchi[57] for a discussion of ^2H photodisintegration) is avoided by use of the Siegert forms of the E1 and E2 operators $\vec{\varepsilon}.\vec{r}$ and $\frac{1}{2}(\vec{\varepsilon}.\vec{r})(\vec{k}.\vec{r})$; the appropriate meson exchange currents are properly included.

III. Separable Potential Formalism

The nuclear Hamiltonian is assumed to be of the form given in Eq.(30), when the potential operator is

$$V = \sum_{\alpha=1}^{3} V_\alpha \equiv V_{23} + V_{31} + V_{12} . \quad (44)$$

The eigenstates of H are assumed to be those corresponding to a three-body bound state, a scattering state comprised of a nucleon plus a bound pair, and a scattering state of three unbound nucleons. Thus we have

$$H|\Psi_B\rangle = -E_B|\Psi_B\rangle \ , \ E_B > 0 \ ; \tag{45}$$

$$H|\Psi_{\alpha n\vec{p}}\rangle = E^{(2)}_{\alpha n}|\Psi_{\alpha n\vec{p}}\rangle \ , \ E^{(2)}_{\alpha n} = \frac{p^2}{2m_\alpha} - \varepsilon_{\alpha n} \ , \ \varepsilon_{\alpha n} > 0 \ ; \tag{46}$$

$$H|\Psi_{\alpha n\vec{p}\vec{k}}\rangle = E^{(3)}_{\alpha n}|\Psi_{\alpha n\vec{p}\vec{k}}\rangle \ , \ E^{(3)}_{\alpha n} = \frac{p^2}{2m_\alpha} + \frac{k^2}{2\mu_\alpha} \ . \tag{47}$$

The reduced masses are $m_\alpha = M_\alpha(M_\beta + M_\gamma)/\Sigma M_\alpha$ and $\mu_\alpha = M_\beta M_\gamma/(M_\beta + M_\gamma)$, where M_α is the mass of particle α. For three equal mass nucleons, these reduce to 2M/3 and M/2, respectively. The subscripts in Eq.(47) mean that nucleon α moves relative to the center-of-mass of the pair $\beta\gamma$ with momentum $\vec{p}$, while β and γ move relative to each other with momentum $\vec{k}$. The subscript n in Eqs.(46) and (47) denotes the remaining quantum numbers such as spin and isospin.

We are concerned here only with two-body photodisintegration, which is described by the transition amplitude matrix element

$$A_2(\alpha,n,\vec{p}) = \langle\Psi^{(-)}_{\alpha n\vec{p}}|H'|\Psi_B\rangle \ , \tag{48}$$

where the superscript (-) denotes the outgoing state which corresponds asymptotically to an incoming wave boundary condition. The two-body scattering state is a solution of the equivalent equations ($\eta > 0$; $E = p^2/2m_\alpha - \varepsilon_{\alpha n}$)

$$|\Psi^{(-)}_{\alpha n\vec{p}}\rangle = |\Phi_{\alpha n\vec{p}}\rangle - G_\alpha(E-i\eta)\sum_{\beta\neq\alpha} V_\beta|\Psi^{(-)}_{\alpha n\vec{p}}\rangle \tag{49}$$

and

$$|\Psi^{(-)}_{\alpha n\vec{p}}\rangle = |\Phi_{\alpha n\vec{p}}\rangle - G(E-i\eta)\sum_{\beta\neq\alpha} V_\beta|\Phi_{\alpha n\vec{p}}\rangle \ , \tag{50}$$

with the resolvent operators defined as

$$G_\alpha(z) = [H_0+V_\alpha - z]^{-1} \tag{51}$$

and

$$G(z) = [H-z]^{-1} \ . \tag{52}$$

The $|\Phi_{\alpha n\vec{p}}\rangle$ denotes the asymptotic scattering state comprised of a nucleon α moving freely with respect to the $\beta\gamma$ bound pair. If Eq.(50)

is written in terms of the distortion operator

$$\bar{\Omega}_{\alpha}^{(-)} \equiv \bar{\Omega}_{\alpha}(E-i\eta) = 1 - G(E-i\eta) \sum_{\beta\neq\alpha} V_{\beta} \tag{53}$$

and substituted into Eq.(48), we obtain

$$A^{2}(\alpha,n,\vec{p}) = \langle\Phi_{\alpha n\vec{p}}|\Omega_{\alpha}^{(+)}H'|\Psi_{B}\rangle , \tag{54}$$

where $\Omega_{\alpha}^{(+)} = (\bar{\Omega}_{\alpha}^{(-)})^{+}$. The crux of this operator manipulation is that a Faddeev-like equation can be written for $\Omega_{\alpha}^{(+)}$:

$$\Omega_{\alpha}^{(+)} = (\bar{\Omega}_{\alpha}^{(-)})^{\dagger} = 1 - \sum_{\beta\neq\alpha} V_{\beta}\, G(E+i\eta) = G_{\alpha}^{-1}(E+i\eta)G(E+i\eta) \tag{55a}$$

$$= 1 - \sum_{\infty\neq\alpha} V_{\beta}G_{\beta}G_{\beta}^{-1}G \tag{55b}$$

$$= 1 - \sum_{\beta\neq\alpha} V_{\beta}G_{\beta}\Omega_{\beta}^{(+)} , \tag{55c}$$

Equation (55c) can then be reexpressed in terms of the two-body t-matrix operator as

$$\Omega_{\alpha}^{(+)} = 1 - \sum_{\beta=1}^{3} \tilde{\delta}_{\alpha\beta}T_{\beta}G_{0}\Omega_{\beta}^{(+)} , \tag{56}$$

because

$$T_{\beta}(z)\, G_{0}(z) = V_{\beta}G_{\beta}(z) , \tag{57}$$

where $\tilde{\delta}_{\alpha\beta} = 1 - \delta_{\alpha\beta}$ and $G_{0}(z) = (H_{0}-z)^{-1}$. If one then iterates Eq.(56) to obtain

$$\Omega_{\alpha}^{(+)} = 1 - \sum_{\beta=1}^{3} \tilde{\delta}_{\alpha\beta}T_{\beta}G_{0} + \sum_{\beta=1}^{3}\sum_{\gamma=1}^{3} \tilde{\delta}_{\alpha\beta}T_{\beta}G_{0}\tilde{\delta}_{\beta\gamma}T_{\gamma}G_{0} - + \cdots \tag{58}$$

and regroups terms as

$$\Omega_{\alpha}^{(+)} = 1 - \sum_{\gamma=1}^{3} (\tilde{\delta}_{\alpha\gamma} - \sum_{\beta=1}^{3} \delta_{\alpha\beta}T_{\beta}G_{0}\delta_{\beta\gamma} + - \cdots)T_{\gamma}G_{0} , \tag{59}$$

then it is possible to recognize the expression in parenthesis in Eq.(59) as $G_{0}^{-1}X_{\alpha\gamma}$, where $X_{\alpha\gamma}$ is the transition operator that connects particle-plus-correlated-pair states. Therefore, one can write

$$\Omega_{\alpha}^{(+)} = 1 - \sum_{\gamma=1}^{3} G_{0}^{-1}X_{\alpha\gamma}T_{\gamma}G_{0} , \tag{60}$$

where

$$X_{\alpha\beta}(z) = G_0(z)\bar{\delta}_{\alpha\beta} - \sum_{\gamma} X_{\alpha\gamma}(z)T_{\gamma}(z)\bar{\delta}_{\gamma\beta}G_0(z) \tag{61a}$$

or

$$X_{\alpha\beta}(z) = G_0(z)\bar{\delta}_{\alpha\beta} - G_0(z)\sum_{\gamma} \bar{\delta}_{\alpha\gamma}T_{\gamma}(z)X_{\gamma\beta}(z) \; . \tag{61b}$$

The three-body dynamics of the continuum state now reside in the transition operator $X_{\alpha\gamma}$. The two-body photodisintegration amplitude can be written as

$$A^2(\alpha,n,\vec{p}) = \langle\Phi_{\alpha n\vec{p}}|H'|\Psi_B\rangle$$

$$- \sum_{\gamma=1}^{3} \langle\Phi_{\alpha n\vec{p}}|G_0^{-1}(z)X_{\alpha\gamma}(z)T_{\gamma}(z)G_0(z)H'|\Psi_B\rangle \; , \tag{62}$$

where $z = p^2/2m\alpha - \varepsilon_{\alpha n} + i\eta$, and the three-body dynamics and the photodisintegration operator have been separated.

The application of Eq.(62) to the photodisintegration of ^3H requires knowledge of both $T_\alpha(z)$ and H'. For this illustration, we assume that $T_\alpha(z)$ results from an attractive, central-force, spin-dependent interaction[57]

$$v_n(k,k') = -\frac{\lambda_n}{2\mu}\, g_n(k)g_n(k') \; , \tag{63}$$

where the λ_n are the strengths of the interactions and

$$g_n(k) = \langle\vec{k}|g_n\rangle$$

are the momentum dependent form factors which determine the ranges of the interactions. This separable form of the nucleon-nucleon interaction can support a bound state, and we will assume that the spin-triplet (t) potential does but that the spin-singlet (s) potential does not. The deuteron bound-state wave function is

$$\langle\chi_t| = N_2 \langle g_n|G_0^{(2)}(-\varepsilon_t) \; ; \; \varepsilon_t>0 \; , \tag{64}$$

where N_2 is the normalization constant chosen so that $\langle\chi_t|\chi_t\rangle = 1$ and $G_0^{(2)}(z)$ is the free-particle resolvent for two nucleons [in contrast to $G_0(z)$ which is the free-particle resolvent for three nucleons]. In this picture we have

$$T_\alpha(z) \quad = \quad - \sum_{n=s}^{t} |g_{\alpha n}\rangle \tau_{\alpha n}(z) \langle g_{\alpha n}| (|SI\rangle\langle SI|)_{\alpha n} ,$$

where

$$\tau_{\alpha n}(z) \quad = \quad \frac{\lambda_n}{2\mu_\alpha} \left(1 + \frac{\lambda_n}{2\mu_\alpha} \int d^3k \, \frac{g_n^2(k)}{z - k^2/2\mu_\alpha} \right)^{-1} . \tag{66}$$

The upper case S(I) denote the total spin (isospin) wave functions for three nucleons. (See Appendix B for details.) Our asymptotic continuum state becomes

$$\langle \Phi_{\alpha n \vec{p}} | \; = \; N_2 \, \langle g_{\alpha n} \, \vec{p} | G_0 \left(\frac{p^2}{2m_\alpha} - \varepsilon_{\alpha n} \right) \tag{67}$$

with $\langle g_{\alpha n} \, \vec{p}| = \langle g_{\alpha n}|\langle \vec{p}|$, and we can write Eq.(62) as

$$A^2(\alpha, n, \vec{p}) \; = \; N_2 \, \{ \langle g_{\alpha n} \vec{p} | G_0(z) H' | \Psi_B \rangle$$

$$+ \sum_{\beta=1}^{3} \sum_{n'=s}^{t} \int d^3p' \, \langle g_{\alpha n} \, \vec{p} | X_{\alpha\beta}(z) | g_{\beta n'} \, \vec{p}' \rangle \, \tau_{\beta n'} \left(z - \frac{p'^2}{2m_\beta} \right) \tag{68}$$

$$\times \langle g_{\beta n'} \, \vec{p}' | G_0(z) H' | \Psi_B \rangle \}$$

where we have suppressed the spin-isospin projection operator in the second term of the expression and used the identity $\int d^3p' |p'\rangle\langle p'| = 1$.

For three identical nucleons, one must symmetrize the amplitude:

$$M_2^n(z, \vec{p}) \quad = \quad \sqrt{\frac{T}{3}} \sum_{\alpha=1}^{3} A_2(\alpha, n, \vec{p}) \; . \tag{69}$$

The resulting symmetrized expression can be written as

$$M_2^n(z, \vec{p}) \quad = \quad B_n(z, \vec{p}) \tag{70}$$

$$+ \sum_{n'=s}^{t} \int d^3p' \, \langle \vec{p} | X_{nn'}(z) | p' \rangle \, \tau_{n'} \left(z - \frac{3p'^2}{4M} \right) B_{n'}(z, \vec{p}') ,$$

where $z = 3p^2/4M - \varepsilon_d + i\eta$ and the deuteron binding energy $\varepsilon_d \equiv \gamma^2/M = 2.225$ MeV. The amplitudes appearing in Eq.(70), written in off-shell form (z not equal to $3p^3/4M - \varepsilon_d + i\eta$ and $|\vec{p}|$ not the same as $|\vec{p}'|$), are

$$B_n(z,\vec{p}) = N_2 \sqrt{\tfrac{T}{3}} \sum_{\alpha=1}^{3} \langle g_{\alpha n}\vec{p}|G_0(z)H'|\Psi_B\rangle \tag{71}$$

and

$$\langle\vec{p}|X_{nn'}(z)|\vec{p}'\rangle = \frac{1}{3}\sum_{\alpha=1}^{3}\sum_{\beta=1}^{3} \langle g_{\alpha n}\vec{p}|X_{\alpha\beta}(z)|g_{\beta n'}\vec{p}'\rangle . \tag{72}$$

The off-shell three-particle transition amplitude $\langle\vec{p}|X_{nn'}(z)|\vec{p}'\rangle$ satisfies the integral equation

$$\langle\vec{p}|X_{nn'}(z)|\vec{p}'\rangle = \langle\vec{p}|Z_{nn'}(z)|\vec{p}'\rangle + \sum_{m=s}^{t} \int d^3p'' \langle\vec{p}|X_{nm}(z)|\vec{p}''\rangle \tau_m\left(z - \frac{3p''^2}{4M}\right)\langle\vec{p}''|Z_{mn'}(z)|\vec{p}'\rangle , \tag{73}$$

where

$$\langle\vec{p}|Z_{nn'}(z)|\vec{p}'\rangle = \frac{1}{3}\sum_{\alpha=1}^{3}\sum_{\beta=1}^{3} \tilde{\delta}_{\alpha\beta} \langle g_{\alpha n}\vec{p}|G_0(z)|g_{\beta n'}\vec{p}'\rangle . \tag{74}$$

The calculational method to be used to obtain the two-body disintegration matrix element is now clear on the basis of Eqs.(70) and (73). For those who prefer a graphical representation, a vivid description is given in Fig.23. The $X_{nn'}$ amplitudes are obtained by solving the coupled integral equations driven by the one-nucleon exchange term $Z_{nn'}$. The matrix describing the two-body photodisintegration is obtained as an integral relation involving these off-shell amplitudes and the Born terms for the disintegration process. It should be noted that one can treat any weak process by this method, since the perturbative operator H' has not yet been specified.

Taking for our ansatz the electric dipole operator, we have

$$H' = \frac{1}{2} e \sum_i \vec{\varepsilon}\cdot\vec{r}_i \, \tau_{iz} \tag{75}$$

where r_i are the nucleon center-of-mass coordinates and τ_{iz} is the z-component isospin Pauli operator for nucleon i. We include only the dominant S-state component of the triton ground state for this example (see Appendix B):

$$|\Psi_B\rangle = u\,\bar{\phi}_a , \tag{76}$$

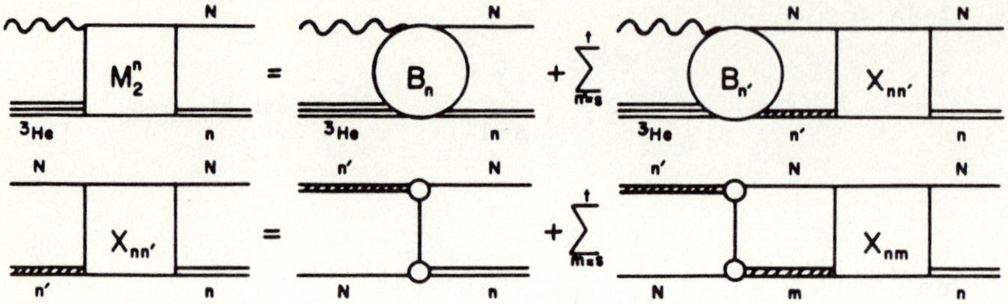

Fig. 23. A graphical representation of the equations used to generate the two-body photodisintegration transition amplitude. The wavy line represents the photon, the double line (n) the deuteron, and N is a neutron in the case of ^{3}H photodisintegration. The cross-hatched double line indicates that a particular correlated pair plus nuleon are off shell.

where the bar in $\bar{\phi}_a$ denotes ^{3}H instead of ^{3}He. Acting with the operator H' on Ψ_B, we obtain

$$H'|\Psi_B\rangle = \frac{e}{2\sqrt{3}} \vec{\varepsilon} \cdot [\frac{2}{\sqrt{3}} \vec{y}\, \bar{\phi}_2 + \vec{x}\, \bar{\phi}_1]\, u(\vec{k},\vec{p}) \; . \tag{77}$$

The final-state spin-isospin projection in Eq.(71) involves only $\chi_1\bar{n}_2$ and, because the deuteron wave function is S-wave, we get

$$B_t(z,\vec{p}) = \frac{eMN_2}{\sqrt{6}} \int d^3k \; \frac{g_t(k)[\vec{\varepsilon}\cdot\vec{y}\; u(\vec{k},\vec{p})]}{3p^2/4 + k^2 - Mz} \tag{78}$$

where $\vec{y} = -i\vec{\nabla}_p$. In the model in which we are working, the electric-dipole operator connects the $^2S_{1/2}$ ground state to the $^2P_{1/2}$ continuum state.

Let us specialize the equations to a calculable form by making partial-wave decompositions:

$$B_n(z,\vec{p}) = \vec{\varepsilon}\cdot p\, \mathscr{B}_n(z,p) \; , \tag{79a}$$

$$\langle\vec{p}|X_{nn'}|\vec{p}\,'\rangle = \sum_L (2L+1)\; X^L_{nn'}(p,p';z)\; P_L(\hat{p}\cdot\hat{p}') \; , \tag{79b}$$

$$\langle\vec{p}|Z_{nn'}|\vec{p}\,'\rangle = \sum_L (2L+1)\; Z^L_{nn'}(p,p';z)\; P_L(\hat{p}\cdot\hat{p}') \; , \tag{79c}$$

where $P_L(\cos\theta)$ is the Legendre function for angular momentum L. After some algebra, we obtain for M_2^t the expression

$$M_2^t(z,\vec{p}) = \vec{\varepsilon}.p\ \{\mathscr{B}_t(z,p)$$

$$+\ 4\pi \sum_{n=s}^{t} \int_0^\infty p'^2dp'X^1_{tn}(p,p';z)\tau_n\left(z - \frac{3p'^2}{4M}\right) \mathscr{B}_n(z,p')\} \qquad (80a)$$

$$\equiv \vec{\varepsilon}.p\ \mathscr{M}_2^t(z,p)\ . \qquad (80b)$$

The $\mathscr{B}_s(z,p)$ in Eq.(80) can be obtained in the same manner as $\mathscr{B}_t(z,p)$ using the projection $\chi_2\bar{\eta}_1$; the result is the same except that $g_s(k)$ replaces $g_t(k)$. The X^1_{tt} and X^1_{ts} amplitudes are generated by solving the coupled integral equations

$$X^1_{nn'}(p,p';z) = Z^1_{nn'}(p,p';z) +$$

$$4\pi \sum_{m=s}^{t} \int_0^\infty p''^2dp''Z^1_{n'm}(p',p'';z)\tau_m\left(z - \frac{3p''^2}{4M}\right)X^1_{nm}(p,p'';z)\ , \qquad (81)$$

where the driving terms are given by

$$Z^1_{nn'}(p,p';z) = C_{nn'} \int_{-1}^{1} dx\ \frac{P_1(x)g_n(q^2)g_{n'}(q'^2)}{p^2 + p'^2 + pp'x - Mz} \qquad (82)$$

with the coordinate definitions

$$q^2 = \frac{1}{4}p^2 + p'^2 + pp'x\ , \qquad (83a)$$

$$q'^2 = p^2 + \frac{1}{4}p'^2 + pp'x\ , \qquad (83b)$$

$$x = \vec{p}.\vec{p}'/pp'\ . \qquad (83c)$$

Note that we made use of the relation

$$Z^L_{nn'}(p,p';z) = Z^L_{n'n}(p',p;z)\ . \qquad (84)$$

The spin-isospin coefficient matrix is

$$[C_{nn'}] = \begin{bmatrix} C_{tt} & C_{ts} \\ C_{st} & C_{ss} \end{bmatrix} = \begin{bmatrix} \frac{1}{4} & -\frac{3}{4} \\ -\frac{3}{4} & \frac{1}{4} \end{bmatrix}\ . \qquad (85)$$

Once $M_2^t(z,\vec{p})$ is obtained, the differential cross section is constructed in the standard way:

$$d\sigma = 2\pi^2 E_\gamma |\mathcal{M}_2^t(\frac{3p^2}{4M} - \frac{\gamma^2}{M}, p)|^2 \sin^2\theta \rho_f , \tag{86}$$

where E_γ is the photon energy $(=|\vec{k}|)$, θ is the centre-of-mass angle of the ejected nucleon with respect to the photon direction $\hat{k}$, and ρ_f is the density of final states.

To summarize, we must solve the coupled integral equations given in Eq.(81) for the X amplitudes. The inhomogeneous terms for these equations are defined by Eq.(82). The resulting X_{tt}^1 and X_{ts}^1 must be combined with the Born amplitudes defined by Eq. (78) and (79a) as indicated in Eqs (80) to obtain $\mathcal{M}_2^t$ which is required by the cross section expression in Eq.(86).

IV. Numerical Methods

How are these equations solved in practice. One method[53] is to solve the coupled integral equations, Eq. (81), for the half-off-shell nucleon-plus-correlated-pair X amplitudes using standard contour rotation techniques. The variables p' and p" are rotated from the real axis into the fourth quadrant: $p' \to p' e^{i\phi}$ and $p'' \to p'' e^{-i\phi}$. The rotation angle ϕ is limited by the singularity in the inhomogeneous term, $Z_{nn'}^1(p,p';\frac{3p^2}{4M} - \frac{\gamma^2}{M})$, coming from the energy denominator $p^2 + p'^2 + pp'x - MZ = 0$. To avoid this singularity, the rotation angle must be chosen such that

$$\phi < \tan^{-1}(\frac{2\gamma}{p}) . \tag{87}$$

In practice this places a stringent limit on the energy for which contour rotation can be used to solve the separable potential equations.

Having obtained the amplitudes $X_{tn'}^1(p,p'e^{-i\phi};3p^2/4M-\gamma^2/M)$, the amplitude $M_2^t(3p^2/4M - \gamma^2/M,\vec{p})$ is computed by rotating the p' integration in the second term on the right-hand-side of Eq. (79a). This is helpful because the bound-state pole of τ_t is avoided. However, this rotation is possible only if no singularities of τ_n or $\mathcal{B}_n$ interfere. It is easy to show that this is the case for τ_n; that fact was used in solving Eq. (81). However, the $\mathcal{B}n$ are more complicated. Using the fact that the spectator function, i.e. the integral equation generated component of the bound state wave function to be discussed below, can be fitted very accurately with analytic forms of the type

$$u(p) = (1 + \bar{\alpha}p^2 + \bar{\beta}p^4 + \bar{\gamma}p^6 + \bar{\delta}p^8)^{-1} , \qquad (88)$$

one can break $\mathscr{B}_n$ into a sum of two types of terms: those that require only a single k integration ($k = |\vec{k}|$) and those that require both a k integration and an angular integration. Assuming $p' \to p'e^{-i\phi}$, we found that if the k integration in those terms that do not involve the angular integration are rotated 45° ($k \to ke^{-i\pi/4}$), then no singularities are encountered. Singularities in the angular-integration terms are avoided by rotating k the same as p', i.e., $k \to ke^{-i\phi}$. Throughout, it is assumed that for the p' rotation there is no contribution from the circular arc at infinity. For the integral in Eq. (79a), this can be shown to be true.

The Faddeev amplitude ψ_1 which makes up the S-state component of the wave function has the simple form

$$\psi_1 = N_3 \, [g_t(k)u_t(p) - g_s(k)u_s(p)]/(k^2 + \tfrac{3}{4}\,p^2 + ME_B) \qquad (89)$$

where the $u_n(p)$ are the singlet and triplet spectator functions obtained by means of a homogeneous set of coupled integral equations[59] analogous in form to Eq.(81). These bound-state equations are well known. However, in order to conveniently evaluate $\vec{\nabla}_p\Psi$ as required in the Born terms of Eq.(78), the spectator functions were fitted to the analytic form given in Eq.(88). Barbour and Phillips[48] chose another method for this part of the calculation. Instead of solving for the bound-state wave function using the Hamiltonian that generated the continuum wave function, they assumed a form like that generated by S-wave separable interactions, set the binding energy to the experimental value, and used the rms radius to fix the remaining parameter defining the spectator function. Such a phenomenological approach overemphasizes the asymptotic region. When combined with the E1 operator, this leads to a significant overestimate of the cross section near the peak. We shall return to this point in the next section.

V. Sample Numerical Results

The most important feature of the two-body photodisintegration cross section is the enhancement in the peak region of the full calculation over the plane wave Born approximation (PWBA), as shown in the 90° differential cross sections in Fig.24. The peak cross section for the full calculation is 40-50% larger. This type of effect was first reported in Ref. 48. However, the fascinating reason for this enhancement was not clear until the publication of Ref. 53. There it was shown that the on-shell distorted wave Born approximation result (DWBA) was actually smaller than the PWBA. On-shell neutron-deuteron final-state rescattering reduces the cross section ~10%, not increases it. Furthermore, retaining the off-shell rescattering in the triplet neutron-deuteron amplitude does not account for the large enhancement. The enhancement in the full amplitude comes from the off-shell scattering in the singlet correlated-pair-plus-nucleon intermediate state that leads to an on-shell neutron-plus-deuteron final state.

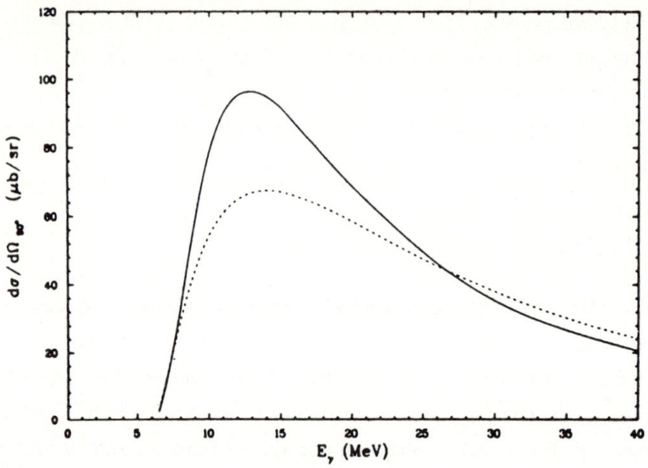

Fig. 24. Comparison of the $^3H(\gamma,d)n$ 90° differential cross section calculted with the complete solution of the separable potential equation (solid curve) with the plane wave Born approximation for the same model.

That is, the enhancement comes from an N+d* rescattering which takes the d* to a physical d. The two-body breakup channel is absorbing strength from the three-body breakup channel. This is possible only in a formalism that properly includes three-body unitarity. The two-body and T=1/2 three-body breakup channels are not independent. Their intimate connection cannot be ignored. Exact equation approaches were needed to understand the physics.

Data for the $^3H(\gamma,n)d$ reaction total cross section are shown Fig.25. In the electric dipole approximation, the total cross section is $8\pi/3$ times the 90° differential cross section. Thus the model calculations are qualitatively correct, which is all one can hope for in the simple model we have constructed. The comparison of the model with the data is better seen in a study of $^3He(\gamma,p)d$ shown in

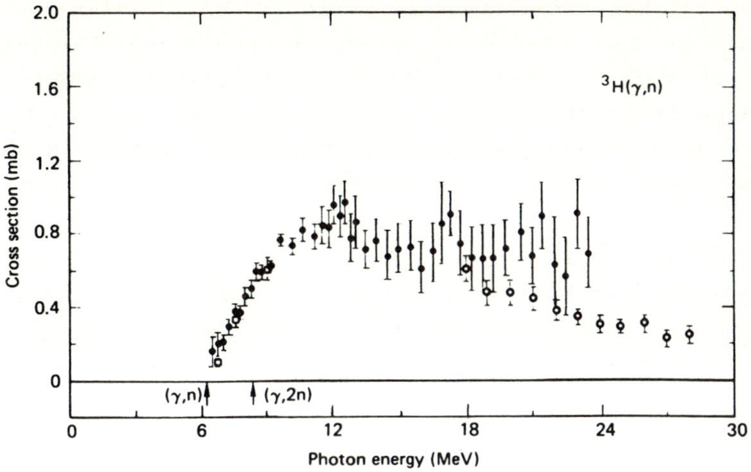

Fig. 25. The total cross section for the $^3H(\gamma,n)d$ reaction as reported in Ref. 60.

Fig. 26. Here the solid curve gives quite a reasonable representation of the data. The dashed curve is a calculation performed within the context of this model but using the Ref. 48 prescription for constructing a phenomenological ground-state wave function. Recall that the correct analytic form was used; the binding energy was

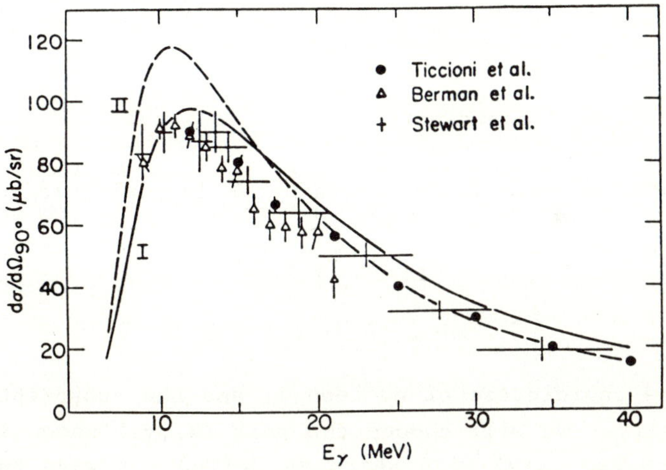

Fig. 26. Comparison of selected $^3He(\gamma,d)p$ 90° differential cross section data (Ref.61) with the calculations of Ref.53 (solid curve) and Ref. 48.

chosen to be the experimental value; the remaining parameter determining the spectator function was fitted to the rms radius. On the basis of such a curve, Barbour and Phillips concluded that the photodisintegration data were incompatible with the rms radius of ^{3}He. A better conclusion from Fig.26 is that one should use ground-state and continuum wave functions generated by the same Hamiltonian. Approximations in physics can be tricky and are often difficult to justify a priori.

In summary, we have examined a very simple model calculation of the two-body photodisintegration of ^{3}H but one which encompasses much of the important physics. We have seen how to produce a calculation from an abstract theory. Finally, we have examined the solution to part of what was once a real puzzle in photonuclear physics:[47,50] Why was the two-body cross section so large (compared to the three-body cross section)?

Acknowledgements

The work of the author is performed under the auspices of the U.S. Department of Energy. The author would like to thank his collaborators, J.L. Friar, G.L. Payne, and D. R. Lehman, for their assistance in compiling these lectures. Finally, the author wishes to thank the School of Physics at Melbourne University for its hospitality during the preparation of the manuscript and in particular Mrs. E. Smart for her technical assistance in that task.

Appendix A. Jacobi Coordinates

The centre-of-mass Jacobi coordinates in configuration space are defined by

$$\vec{x}_i = \vec{r}_j - \vec{r}_k \tag{A1a}$$

and

$$\vec{y}_i = \frac{1}{2}(\vec{r}_j + \vec{r}_k) - \vec{r}_i \,. \tag{A1b}$$

The r_i are the coordinates of nucleon i, and the subscripts are to be taken cyclically. We will choose the pair (x_1, y_1) shown in Fig.A1 to be the coordinates (x,y) with which we define our wave functions and amplitudes. The other two pairs can then be expressed in terms of $\vec{x}$ and $\vec{y}$ by the relations

$$\vec{x}_2 = \frac{1}{2}\vec{x} + \vec{y} \tag{A2a}$$

$$y_2 = -\frac{3}{4}\vec{x} - \frac{1}{2}\vec{y} \qquad \text{(A2b)}$$

and

$$\vec{x}_3 = -\frac{1}{2}\vec{x} - \vec{y} \qquad \text{(A3a)}$$

$$\vec{y}_3 = \frac{3}{4}\vec{x} - \frac{1}{2}\vec{y} \qquad \text{(A3b)}$$

The configuration space Schrödinger wave function Ψ describing three spinless bosons can then be expressed in terms of the single Faddeev amplitude ψ for that same system as

$$\Psi(\vec{x}_1,\vec{y}_1) = \psi(\vec{x}_1,\vec{y}_1) + \psi(\vec{x}_2,\vec{y}_2) + \psi(\vec{x}_3,\vec{y}_3) \qquad \text{(A4a)}$$

$$= \psi_1 + \psi_2 + \psi_3 \quad . \qquad \text{(A4b)}$$

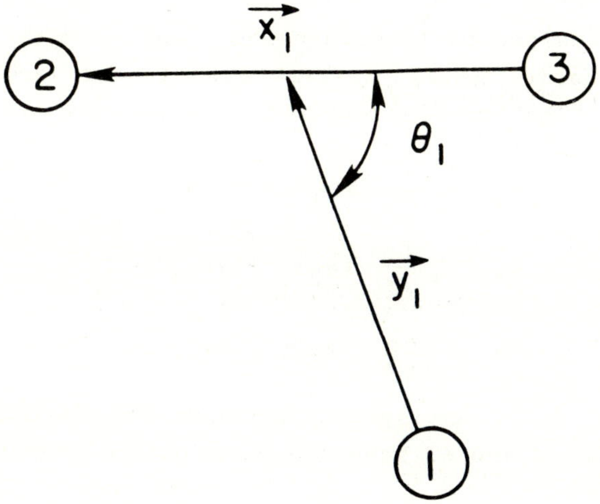

Fig. A1. The Jacobi coordinates in configuration space for the three-body problem.

Appendix B. Spin-Isospin Formalism

The doublet spin states for three nucleons have the form

$$\chi_1 = |[1, \tfrac{1}{2}]\tfrac{1}{2}> \tag{B1a}$$

$$= \frac{1}{\sqrt{6}} [(++-) + (+-+) - 2 (-++)] \tag{B1b}$$

$$\chi_2 = |[0, \tfrac{1}{2}]\tfrac{1}{2}> \tag{B1c}$$

$$= \frac{1}{\sqrt{2}} [(++-) - (+-+)] \tag{B1d}$$

Here a + (or -) means that the nucleon corresponding to that position has spin up (or down). Similarly, the isospin functions η_1 and η_2 can be defined to have the forms in Eqs.(B1) where a + (or -) means that the nucleon is a proton (or neutron); such functions describe ^{3}He and ^{3}H functions are obtained by interchanging + and - in the η's.

The spin-isospin basis states are linear combinations of these χ's and η's. The combinations which we require are

$$\phi_a = \frac{1}{\sqrt{2}} (\chi_1\eta_2 - \chi_2\eta_1) \tag{B2a}$$

$$\phi_1 = \frac{1}{\sqrt{2}} (\chi_2\eta_2 - \chi_1\eta_1) \tag{B2b}$$

$$\phi_2 = \frac{1}{\sqrt{2}} (\chi_2\eta_1 + \chi_1\eta_2) \tag{B2c}$$

The function ϕ_a is fully antisymmetric under the interchange of any two nucleons, while ϕ_1 and ϕ_2 have the same mixed symmetry properties as do the χ's and η's.

The Pauli principle requires that the overall wave function be fully antisymmetric in the interchange of all coordinates of any pair of nucleons. The function ϕ_a u satisfies this requirement, where u is spatially symmetric. This is the $^2S_{1/2}$ component of Ψ. which is denoted by S. Combinations of ϕ_1 and ϕ_2 with spatial functions of mixed symmetry that describe the trinucleons must also possess the ϕ_a antisymmetry property. There is one: $\phi_1 v_2 - \phi_2 v_1$. This is the $^2S_{1/2}$ state of mixed symmetry which is denoted by S'. The full wave function is then of the form

$$\Psi = \phi_a u + (\phi_1 v_2 - \phi_2 v_1) . \tag{B3}$$

Appendix C. Exchange Currents

Over 10 years ago pion-exchange current contributions to the charge density operator were calculated and then later applied in the case of ^3He. (See Ref.33.) Effects were found which were of the right sign and of sufficient magnitude to resolve the disagreement between theory and experiment concerning the size and shape of the charge form factor. Friar,[13] however, showed about the same time that these corrections to the charge density operator (1) are relativistic corrections [i.e., $(v/c)^2$ compared to 1], (2) contain important momentum-dependent terms which have never been included in ^3He calculations, (3) are model dependent, reflecting the physical difference between pseudoscalar and pseudovector couplings of pions and nucleons,[39] and (4) are ambiguous, reflecting a unitary ambiguity which arises in different methods of calculating these operators.

The fact that these isoscalar exchange currents are of relativistic origin means that one must do a relativistic calculation (including refitting the nucleon-nucleon interactions to the two-body data) to include them in a meaningful wave. The pseudoscalar versus pseudovector coupling model dependence is unavoidable. The unitary ambiguity is vexing because it would vanish if the wave functions calculated with a given potential were used with a commensurate form of the charge operator - such matrix elements could be free of any ambiguity. But realistic potentials to date have the wrong form to correspond to any of the allowed unitary representations of the charge density operator.

Exchange currents of the isovector ilk are not relativistic corrections compared to the nuclear current, $\vec{p}/M$. They contribute in a non-negligible manner to the magnetic density form factors of the trinucleons. Friar has discussed these in great detail.[13] See also the exchange current references in the magnetic moment investigation of Ref.40.

References

1. N.F. Mott, Proc. Roy. Soc. (London) A124 (1929) 426; A135 (1932) 429.
2. M.E. Rose, Phys. Rev. 73 (1949) 279.
3. R. Hofstadter, Rev. Mod. Phys. 28 (1956) 214; Ann. Rev. Nucl. Sci. 7 (1957) 231.
4. R.A. Malfliet and J.A. Tjon, Nucl. Phys. A127 (1969) 161.
5. J.L. Friar, B.F. Gibson, and G.L. Payne, Z. Phys. A301 (1981) 309.
6. D.R. Yennie, D.G. Ravenhall, and R.N. Wilson, Phys. Rev. 95 (1954) 500.

7. J.L. Friar, B.F. Gibson, and G.L. Payne, Phys. Rev. C22 (1980) 284.
8. J.L. Friar, B.F. Gibson, and G.L. Payne, Z. Phys. A312 (1983) 169.
9. J.D. Bjorken and S.D. Drell, Relativistic Quantum Mechanics, (McGraw-Hill, New York, 1964) ch.7.
10. G. Hoehler et al., Nucl. Phys. B114 (1976) 505.
11. L.I. Schiff, Phys. Rev. 92 (1953) 988.
12. L.I. Schiff, Phys. Rev. 133 (1964) B802.
13. J.L. Friar, Ann. Phys. (N.Y.) 104 (1977) 380; Phys. Lett. 59B (1975) 145; in New Vistas in Electronuclear Physics (Plenum, New York, 1986).
14. B.F. Gibson, Proceedings of the International Conference on Photonuclear Reactions and Applications, ed. by B.L. Berman (Lawrence Livermore Laboratory, CONF-730301, 1973) p 373.
15. J.A. Tjon, B.F. Gibson, and J.S. O'Connell, Phys. Rev. Lett. 25 (1970) 540.
16. J. Martino, Proceedings of the International Symposium on the Three-Body Force in the Three-Body System, ed. by B.L. Berman and B.F. Gibson, in Lecture Notes in Physics (Springer-Verlag, Heidelberg, 1986) Vol.260.
17. J.L. Friar, B.F. Gibson, C.R. Chen, and G.L. Payne, Phys. Lett. 161B (1985) 241.
18. C.R. Chen, G.L. Payne, J.L. Friar, and B.F. Gibson, Phys. Rev. C 31 (1985) 2266.
19. C.R. Chen, G.L. Payne, J.L. Friar, and B.F. Gibson, Phys. Rev. C 33 (1986) 1740.
20. J.L. Friar, B.F. Gibson, G.L. Payne, and C.R. Chen, Phys. Rev. C 34 (1986).
21. F.-P. Juster et al., Phys. Rev. Lett. 55 (1985) 2261; S. Platchkov and B. Frois, private communication.
22. R.V. Reid, Ann. Phys. (N.Y.) 50 (1968) 411; B.D. Day, Phys. Rev. C 24 (1981) 1203, provides the higher partial waves.
23. R.B. Wiringa, R.A. Smith, and T.L. Ainsworth, Phys. Rev. C 29 (1984) 1207.
24. R. de Tourreil and D.W.L. Sprung, Nucl. Phys. A201 (1973) 193.
25. R. de Tourreil, B. Rouben, and D.W.L. Sprung, Nucl. Phys. A242 (1975) 445.
26. S.A. Coon, M.D. Scadron, P.C. McNamee, B.R. Barrett, D.W.E. Blatt, and B.H.J. McKellar, Nucl. Phys. A317 (1979) 242.
27. H.T. Coelho, T.K. Das, and M.R. Robilotta, Phys. Rev. C 28 (1983) 1812.
28. J. Carlson, V.R. Pandharipande, and R.B. Wiringa, Nucl. Phys. A401 (1983) 59; R.B. Wiringa, ibid A401 (1983) 86.
29. H. Collard et al., Phys. Rev. 138 (1965) B57.
30. J.S. McCarthy, I. Sick, R.R. Whitney, and M.R. Yearian, Phys. Rev. Lett. 25 (1970) 884; J.S. McCarthy, I. Sick, and R.R. Whitney, Phys. Rev. C 15 (1977) 1396; R.G. Arnold et al., Phys. Rev. Lett. 40 (1978) 1429.
31. D.H. Beck, J. Asai, and D.M. Skopik, Phys. Rev. C 25 (1982) 1152; D.H. Beck, et al., ibid 30 (1984) 1403; D.H. Beck, private communication.
32. J.M. Cavedon et al., Phys. Rev. Lett. 49 (1982) 986; P.C. Dunn et al., Phys. Rev. C 27 (1983) 71; G.A. Retzlaff and D.M. Skopik, ibid 29 (1984) 1194; C.R. Otterman et al., Nucl. Phys. A436 (1985) 688.
33. W.M. Kloet and J.A. Tjon, Phys. Lett. 49B (1974) 419; E. Hadjimichael, R. Bornais, and B. Goulard, Phys. Rev. Lett. 48 (1982) 583; Phys. Rev. C 27 (1983) 831; T. Saskawa, A. Fukunaga, and S. Ishikawa, Czechoslovak Journal of Physics B36 (1986) 312.
34. I. Sick, in Lecture Notes in Physics (Springer-Verlag, Berlin, 1978) Vol. 87, p236.

35. J.L. Friar, Nucl. Phys. A156 (1970) 43.
36. M. Fabre de la Ripelle, Fizika 4 (1972) 1.
37. M. Fabre de la Ripelle, Prog. Theor. Phys. 40 (1968) 1454; Rev. Bras. Fis. (Suppl.) (1980) 219.
38. J.L. Friar and B.F. Gibson, Phys. Rev. C 15 (1977) 1779.
40. E.L. Tomusiak, M. Kimura, J.L. Friar, B.F. Gibson, G.L. Payne, and J. Dubach, Phys. Rev. C 32 (1985) 2075.
41. C. Chadwick and M. Goldhaber, Nature 134 ()1935) 237.
42. G.E. Brown and D.O. Riska, Phys. Lett. 38B (1972) 193; M. Gari and A.H. Hoffman, Phys. Rev. C 1 (1973) 994.
43. J.S. Merritt, J.G.V. Taylor, and A.W. Boyd, Nucl. Sci. Eng. 34 (1968) 195; E.T. Jurney et al., Phys. Rev. C 25 (1982) 2810.
44. L.I. Schiff, Phys. Rev. 52 (1937) 149; A.C. Phillips, Phys. Rev. 170 (1968) 952; Nucl. Phys. A184 (1972) 337.
45. R.J. Hughes et al., Nucl. Phys. A267 (1976) 329; J.F. Gilot et al., Phys. Lett. 47 (1981) 304; H.O. Meyer et al., Phys. Rev. Lett. 52 (1984) 1759.
46. A. Cambi, B. Mosconi, and P. Ricci, Phys. Rev. Lett. 48 (1982) 462; J. Phys. G10 (1984) L11; J.L. Friar, B.F. Gibson, and G.L. Payne, Phys. Rev. C 30 (1984) 441.
47. G. Barton, Nucl. Phys. A104 (1967) 189.
48. I.M. Barbour and A.C. Phillips, Phys. Rev. C 1 (1970) 165.
49. B.F. Gibson and D.R. Lehman, Phys. Rev. C 13 (1976) 477.
50. B.F. Gibson, Nucl. Phys. A353 (1981) 85c.
51. J.R. Calarco et al., Phys. Rev. C 27 (1983) 1866; B.L. Berman et al., Phys. Rev. C 22 (1980) 2273; L. Ward et al.,. Phys. Rev. C 24 (1981) 317.
52. B.F. Gibson, Nucl. Phys. A195 (1972) 449.
53. B.F. Gibson and D.R. Lehman, Phys. Rev. C 11 (1975) 29.
54. E.O. Alt, P. Grassberger, and W. Sandhas, Nucl. Phys. B2 (1967) 167.
55. I.M. Barbour and J.A. Hendry, Phys. Lett. 38B (1972) 151.
56. A.J.F. Siegert, Phys. Rev. 52 (1937) 787.
57. Y. Yamaguchi, Phys. Rev. 95 (1955) 1628.
58. J.H. Hetherington and L.M. Schick, Phys. Rev. 137 (1965) B935; 156 (1967) 1647.
59. B.F. Gibson and G.J. Stephenson, Jr., Phys. Rev. C 8 (1973) 1222; A.G. Sitenko and V.F. Kharchenko, Nucl. Phys. 49 (1963) 15.
60. D.D. Faul et al., Phys. Rev. Lett. 44 (1980) 129; Phys. Rev. C 24 (1981) 849.
61. J.R. Stewart, R.C. Morrison, and J.S. O"Connell, Phys. Rev. 138 (1965) B372; B.L. Berman, L.J. Keoster, and J.H. Smith, Phys. Rev. 133 (1964) B117; G. Ticcioni et al., Phys. Lett. 46B (1973) 369.

PART VI

OTHER APPLICATIONS OF SCATTERING THEORY METHODS

SCATTERING THEORY METHODS IN REACTING PLASMAS

D. Bollé*
Instituut voor Theoretische Fysica
Universiteit Leuven
B-3030 Leuven, Belgium

1. INTRODUCTION

One of the first results in deriving a connection between the statistical behavior of a system and the scattering processes of the constituent particles has been obtained by Beth and Uhlenbeck [1] and Gropper [2]. They showed that the second virial coefficient can be expressed in closed form in terms of the two-body bound-state energies and the scattering phase shifts. Many works have appeared on a generalization of this result to higher virial coefficients. For a list of references we refer to [3]. A complete S-matrix formulation of statistical mechanics via the virial expansion was advocated by Dashen, Ma and Bernstein [4].

Quite recently some reviews have been written on the status of this approach and its applications [3], [5], [6]. In this contribution we mainly discuss the recent use of these scattering theory methods in the study of the statistical behavior of charged particle systems, in particular strongly coupled, partially ionized plasmas. We especially look at the formation of bound states and the structure of the Planck-Larkin partition function, the lowering of the plasma continuum (Mott-effect) as a result of this formation and the corresponding structure of the many-fugacities expansion for the equation of state.

2. S-MATRIX APPROACH TO STATISTICAL MECHANICS

The thermodynamic properties of a system can be obtained from the knowledge of the grand canonical partition function Ξ defined by [7]

---

* Onderzoeksleider N.F.W.O., Belgium.

$$\Xi(z,V,T) = \sum_{N=0}^{\infty} z^N \frac{1}{N!} \text{Tr } e^{-\beta H_N}, \tag{1}$$

where z is the fugacity, $z = \exp(\beta\mu)$, with μ the chemical potential, where $\beta = 1/kT$ with k Boltzmann's constant, where T denotes the temperature and H_N the Hamiltonian for the N-particle system. V is the volume of the system and Tr denotes the trace in the N-particle space. The pressure is then given by

$$\beta P = V^{-1} \ln \Xi \tag{2}$$

and the average particle number reads

$$N = z \frac{\partial}{\partial z} \ln \Xi . \tag{3}$$

The partition function Ξ can be expanded in powers of the fugacity using the Ursell and Mayer cluster expansion [7]

$$\Xi(z,V,T) = \exp \{ V \sum_{n=1}^{\infty} b_n z^n \}, \tag{4}$$

where the coefficients b_n are the cluster integrals. Assuming the latter exist in the thermodynamic limit, they read

$$b_1 = \lambda^{-3} , \tag{5}$$

$$b_2 = \frac{2^{3/2}}{2!\,\lambda^3} \text{Tr } [e^{-\beta H_2} - e^{-\beta H_{2,0}}] , \tag{6}$$

$$b_3 = \frac{3^{3/2}}{3!\,\lambda^3} \text{Tr } [e^{-\beta H_3} - e^{-\beta H_{3,0}} - \sum_{\alpha=1}^{3} (e^{-\beta H_{3,\alpha}} - e^{-\beta H_{3,0}})], \tag{7}$$

Here λ is the thermal wavelength

$$\lambda = (2\pi\hbar^2 \beta /m)^{1/2}, \tag{8}$$

and H_2, $H_{2,0}$ are the total and free two-particle Hamiltonians, H_3, $H_{3,\alpha}$ and $H_{3,0}$ are the total, α-channel and free three-particle Hamiltonians. The factors in front of the trace come from the center-of-mass motion. We have assumed Boltzmann statistics. We remark that exchange effects do not introduce anything fundamentally new but working out their details may be highly non-trivial. The operator structure inside the trace of (6) and (7) is of the same type as that in the two-and three-body connected resolvent difference. The subtraction terms eliminate disconnected processes. The nth cluster coefficient only involves n- and fewer-particle effects.

Once we have an explicit form for the b_n, then the grand canonical partition function and all other thermodynamic properties of the system are determined. THe equation of state e.g. can be obtained as a series in the density of particles $\rho = N/V$ by elimination of the fugacity z using eqs. (2)-(4). The result is

$$\beta P = \sum_{n=1}^{\infty} a_n \rho^n , \tag{9}$$

where the coefficients a_n in this expansion, which are the virial coefficients, can be completely written down in terms of the cluster coefficients, viz.

$$\begin{aligned} a_1 &= 1, \qquad a_2 = - b_2 b_1^{-2} , \\ a_3 &= 4a_2^2 - 2\, b_3 b_1^{-3} , \ldots \end{aligned} \tag{10}$$

Our task is now to evaluate these cluster integrals in terms of scattering quantities. The standard method is to use the Watson transform [8], which connects the statistical operator, exp $(-\beta H)$, with the resolvent of H, $R(z) = (H-z)^{-1}$, z the complex energy, i.e.

$$e^{-\beta H} = - \frac{1}{2\pi i} \oint_C dz\, e^{-\beta z} R(z) , \tag{11}$$

where C is a contour around the spectrum of H in the complex energy plane. E.g. for H_2 we integrate along the following path: around the positive real axis from $\Gamma - i\eta$ to $\Gamma + i\eta$, $\Gamma, \eta > 0$ avoiding the origin by a small circle, along the circle $\{\Gamma \exp(i\theta) \}\ \theta \in [\arcsin(\eta,\Gamma), 2\pi - \arcsin(\eta,\Gamma)]\}$ and finally encircling all negative-energy bound-state positions, $E = -x_j^2$, $j = 1,2,\ldots N_b$, clockwise, i.e. along $\{\varepsilon_j \exp(i\theta_j) \}\ \theta_j \in [0,2\pi]$, ε_j sufficiently small $\}$. For simplicity, we assume that there are no zero-energy resonances and/or no zero-energy bound states.

Using then the relation between the connected resolvent difference and the on-energy-shell S-matrix, S(E), viz. [4], [9]

$$2 \,\mathrm{Im}\, \mathrm{Tr}\, [R_2(E+io) - R_{2,0}(E+io)] = -i\, \mathrm{Tr}\, S^*(E) \frac{d}{dE} S(E), \tag{12}$$

we obtain for b_2(eq.(6))

$$b_2(\beta)= \sqrt{2}\,\lambda^{-3}\{\sum_{j=1}^{N_b} e^{\beta\chi_j^2} +(2\pi)^{-1}\int_0^\infty dE\, e^{-\beta E}\mathrm{Tr}[-iS^*(E)\frac{d}{dE}S(E)]\}. \qquad (13)$$

Eq. (13) is a generalization of the Beth-Uhlenbeck result to non-spherically symmetric interactions. Indeed, in the case of spherical symmetry we know that the logarithmic derivative of the S-matrix is given by the sum of the energy derivatives of the partial wave phase shifts.

A similar, complete treatment for b_3 has been given by Buslaev and Merkuriev [10] . They find that an analogous expression to (13), only involving the connected three-particle S-matrix difference, that Dashen et al. [4] and Smith [11] believe to be valid, is not correct (for E>0). Due to strong singularities of [S(E)-1] on the diagonal, originating from the rescattering singularities in the three-to-three S-matrix, they were forced to add counterterms. The latter can be expressed in terms of the two-particle S-matrices but the final formula looks very complicated and one would like to have an independent confirmation of this results in order to resolve the controversy. In this context we mention the work of Servadio [12] during the last years, where he investigates what constraint unitarity imposes on the "truly three-body" scattering amplitude. This unitarity calculation is now completed and the author claims one could set up a calculation for b_3, using similar techniques and in analogy with his work on b_2 [13]. He already finds that three-body on-shell properties are sufficient to determine the third virial coefficient and he is confident he should be able to produce at least an alternative check on some of these counterterms [12].

For a review of calculations of b_2 and the higher cluster integrals in specific models, in different approximations ... we refer to [5].

Returning to expression (13) we see that the bound states have an exponential temperature dependence. Therefore we can say that they enter the cluster expansion like new particles. The equation of state (9) however does not reflect this fact because it is of a single density form. The way to improve upon this situation was suggested by Hill [14] who argued (in the context of classical mechanics) that one must introduce two fugacitites, two densities, one for the free particles and one for the stable clusters.

So we start from the grand canoncial partition function written as

$$\Xi(z_1,z_2,V,T) = \exp\{V \sum_{i,j} b_{ij}\, z_1^i z_2^j\}, \tag{14}$$

where z_1, z_2 are the fugacities of the free particles, respectively the two-particle cluster and the b_{ij} are the cluster coefficients for i free particles and j clusters. The average particle numbers are

$$N_i = z_i \frac{\partial}{\partial z_i} \ln \Xi \quad , \; i=1,2. \tag{15}$$

They both depend on β whereas the total particle number $N=N_1+2N_2$ does not. The equation of state can then be written in terms of the corresponding densities as

$$\beta P = \sum_{i,j} a_{ij}\, \rho_1^i\, \rho_2^j , \tag{16}$$

with a_{ij} the virial coefficients for i free particles and j clusters. These can be expressed in terms of the b_{ij} by using (14)-(16) and eliminating the fugacities. The result is (see e.g. [3], [15], [16] and references therein)

$$\begin{aligned}
a_{01} &= 1 , \quad a_{10} = 1 \\
a_{02} &= -b_{02}b_{01}^{-2} , \; a_{11} = -\, b_{11}b_{10}^{-1}b_{01}^{-1}, \; a_{20} = -b_{20}b_{10}^{-2}, \\
a_{03} &= -2b_{03}b_{01}^{-3} + 4b_{02}^{2}b_{01}^{-4} , \; a_{30} = -2b_{30}b_{10}^{-3} + 4b_{20}^{2}b_{10}^{-4} , \\
a_{21} &= -2b_{21}b_{10}^{-2}b_{01}^{-1} + 4b_{11}b_{20}\, b_{10}^{-3}b_{01}^{-1} + b_{11}^{2}b_{10}^{-2}b_{01}^{-2} , \ldots
\end{aligned} \tag{17}$$

We remark that we have taken a two-component system because the relations become unwieldy for the general case. The b_{ij} can again be expressed in terms of scattering parameters. (We only evaluate Boltzmann-like contricutions). Two b_{ij} do not involve scattering at all. Their values follow immediately from their definitions i.e.

$$b_{10} = \lambda_1^{-3} , \quad b_{01} = \lambda_2^{-3}\; e^{\beta\chi^2} , \tag{18}$$

where $\lambda_1 \lambda_2$ are the thermal wavelengths for the free particles, respectively the bound cluster with mass 2m. For the other b_{ij} we find (see e.g. [15])

$$b_{20}=\sqrt{2}\,\lambda_1^{-3}(2\pi)^{-1}\int_0^\infty dE\; e^{-\beta E}\mathrm{Tr}\,[-i\,S^*(E)\frac{d}{dE}\,S(E)],$$

$$b_{11}=3^{3/2}\,\lambda_1^{-3}e^{\beta\chi^2}(2\pi)^{-1}\int_0^\infty dE\; e^{-\beta E}\sum_{\alpha=1}^{3}\mathrm{Tr}_\alpha[-i\sum_{\gamma=0}^{3}S^*_{\alpha\gamma}(E-\chi^2)\frac{d}{dE}S_{\gamma\alpha}(E-\chi^2)],$$

$$\ldots. \qquad (19)$$

where S is the two-body on-shell S-matrix and $S_{\alpha\beta}$ are the three-body on-shell channel S-matrices.

If one now constructs the virials according to (17) the exponential χ^2 dependence cancels out. Basically, the effect of this factor in (19) is to control the relative amount of free particles and clusters. This is seen again in the expression for the chemical mass-action law. When chemical equilibrium is established in the system, the chemical potential for the free particles, μ_1, and that for the bound cluster, μ_2, satisfy $\mu_2=2\mu_1$. In terms of the fugacitites we then have $z_1^2=z_2$. Together with eqs. (14)-(15) this leads to (see e.g. [15] ,[17])

$$\frac{\rho_1^2}{\rho_2}=\frac{e^{-\beta\chi^2}}{2^{3/2}\lambda_1^3}\,[1+(\frac{4b_{20}}{b_{10}^2}-\frac{b_{11}}{b_{10}b_{01}})\,\rho_1+(\frac{2b_{11}}{b_{10}b_{01}}-\frac{2b_{02}}{b_{01}^2})\,\rho_2+\ldots] \qquad (20)$$

The first term in this chemical mass-action law is the so-called Saha equation and represents, together with the ideal gas law, i.e. $\beta P=\rho_1+\rho_2$ (cfr. eq. (16)), a simple equation of state for the dimerization of the system. The density dependent terms exhibit the deviation from this ideal behavior due to collisions. We remark that the coefficients of these terms are again independent of χ^2.

For systems with Coulomb interactions (e.g. plasmas) , there are some difficulties due to the long-range nature of that interaction. E.g. the fact that there is an infinite number of bound states accumulating at E=0 already causes the sum over these bound states in eq. (13) to diverge. In the next section we show how to overcome this difficulty.

3. THE PLANCK-LARKIN PARTITION FUNCTION

We first recall some known results from scattering theory. For spherically symmetric scattering , there exists the well-known Levinson theorem [18] for the partial-wave phase shift, δ_ℓ , viz.

$$\delta_\ell(\infty) - \delta_\ell(0) = -\pi N_{b,\ell}, \quad N_{b,\ell}: \text{bound states} \tag{21}$$

This result can be generalized to obtain a whole set of sum rules connecting the S-matrix and bound-state energies, that are valid for non spherically symmetric scattering [6]. We present, without derivation, the following ones

$$\int_0^\infty dE \{ i \, \mathrm{Tr}[S^*(E)\frac{d}{dE} S(E)] - \frac{1}{4\pi\sqrt{E}} \int d^3x V(\underline{x}) \} = 2\pi (N_b + N_o) + \pi q, \tag{22}$$

$$\int_0^\infty dE \, E\{i \, \mathrm{Tr}[S^*(E)\frac{d}{dE} S(E)] - \frac{1}{4\pi\sqrt{E}} \int d^3x V(\underline{x}) - \frac{1}{16\pi E^{3/2}} \int d^3x V^2(\underline{x}) \}$$
$$= 2\pi \sum_{j=1}^{N_b} (-\chi_j^2), \tag{23}$$

where we assume, for convenience, in the rest of this paper that there are no zero-energy bound-states, i.e. $N_o=0$, and no zero-energy resonances , i.e. q=0. These rules (22) and (23) are valid for shortrange interactions including the Yukawa potential.

We now consider again eq. (13) for $b_2(\beta)$. Writing the S-matrix part as a total differential in E, doing a partial integration and using Levinson theorem (22), we arrive at [6], [19]

$$b_2(\beta) = \sqrt{2}\,\lambda^{-3} \{ \sum_{j=1}^{N_b} e^{\beta\chi_j^2} - N_b - (16\pi\beta)^{-1/2} \int d^3x \, V(\underline{x})$$
$$-(2\pi^2)^{-1/2}\lambda^{-3}\beta \int_0^\infty dE \, e^{-\beta E} \int_E^\infty dE_1 \{ \mathrm{Tr}[-iS^*(E_1)\frac{d}{dE_1} S(E_1)] -$$
$$- \frac{1}{4\pi\sqrt{E_1}} \int d^3x \, V(\underline{x}) \} \tag{24}$$

Following a similar procedure, now using the sum rule (23) , we get

$$b_2(\beta) = \sqrt{2}\,\lambda^{-3}\{ \sum_{j=1}^{N_b} e^{\beta\chi_j^2} - N_b - \beta \sum_j \chi_j^2 - (16\pi\beta)^{-1/2} \int d^3x \, V(\underline{x})$$
$$-(8\pi)^{-1}\beta^{1/2} \int d^3x \, V^2(\underline{x}) + O(\beta^2) \} \tag{25}$$

Going on in the same way, using higher-order sum rules, we would obtain the well-known Wigner-Kirkwood expansion [5] ,[7] for the second virial coefficient. For more details we refer to [6],[19]. What is special about this derivation is that we explicitly see a cancellation between bound-state and scattering contributions. This cancellation is rigorously valid on a fully quantum-mechanical level. It also holds for higher cluster coefficients as can be easily shown e.g. for b_3 using the form in terms of the connected resolvent difference [5] and the three-body sum rules derived in [20]. It stays even valid when there is an infinite number of bound states as has been verified explicitly in a model calculation for the third cluster coefficient of binary mixtures of light and heavy particles allowing for the Efimov effect [21].

The sum of the first three terms of (25) in partial wave form, i.e.

$$b_2^{PL}(\beta) \equiv \sqrt{2}\,\lambda^{-3} \sum_{\ell} (2\ell+1) \sum_{j} (e^{\beta\chi_{j,\ell}^2} - 1 - \beta\chi_{j,\ell}^2) \tag{26}$$

can be written, in the case of scattering by the Coulomb potential, as

$$b_2^{PL}(\beta) = \sqrt{2}\,\lambda^{-3} \sum_{n=1}^{\infty} n^2 (e^{-\beta E_n} - 1 + \beta E_n) \tag{27}$$

with

$$E_n = -\frac{m e^4}{2\hbar^2 n^2} = -\frac{e^2}{2a_B n^2}, \quad n=1,2,3... \tag{28}$$

where a_B is the Bohr radius. It is clear that also in this case, the modified bound-state sum (27) is finite.

The expression (26) (or (27)) is known as the Planck-Larkin partition function in plasma physics (See e.g. [17] ,[22]). We have presented here a (rigorous) derivation of its underlying structure on the basis of higher-order scattering sum rules.

For low-density plasmas, the effective Hamiltonian can be replaced by the Hamiltonian of an isolated two-particle system. For hydrogen plasma e.g., this system has an infinite number of bound levels. In a first, crude approximation to chemical equilibrium calculations, to obtain e.g. the degree of ionization, one considers only the deepest level, E_1, as a bound state and neglects all excited states. In more refined calculations, one also takes into account some lower excited states. The higher ones, near to the continuum edge, have quite extended wave functions and low stability. They are considered as quasi-free and they are treated on the same footing as the scattering contributions. An important question is then: what is the appropriate border between bound and quasi-free discrete states? An answer is then given by eqs. (26), (27). Indeed, in first instance we can say that states with high principal quantum number such that their energy is below the mean thermal energy, i.e. $(-E_n\beta) \ll 1$, are suppressed in (26),(27). For refinements of the value of this border we refer to the next section.

For dense plasmas the Coulomb interaction is dynamically screened. This is one of the effects of the collective behavior of the charged particles. This effect can be taken into account approximately by considering static screening realized by replacing the Coulomb potential by a Debye (Yukawa) potential

$$V(r) = -e^2/r \quad \exp(-r/r_D), \tag{29}$$

where the Debye length r_D is a function of the temperature and the density of the protons, ρ_p, viz.

$$r_D^{-2} = 8\pi e^2 \rho_p \beta \ . \tag{30}$$

The non-modified bound-state sum is now finite, since the potential (29) has a finite number of levels. However, since these levels are dependent upon the screening length r_D,and hence functions of the temperature and density, the following happens. As r_D decreases the upper levels move into the continuum and the bound-state sum changes discontinuously. The (-1) subtraction in (26), due to Levinson's theorem, resolves this unphysical difficulty. The second subtraction ensures a finite $\beta \to 0$ limit for $b_2^{PL}(\beta)$.

Here it is interesting to remark that at a certain point all bound levels disappear in the continuum. This is called the Mott-transition. It happens at a finite density. To give an idea of the numbers we note that e.g. for hydrogen plasmas , in the region were more than 50% of the particles are charged, an approximation in the framework of the confined atom model (atoms are assumed to be enclosed into spheres of radii which are half of the mean distance between two protons in the system) gives that this transition is realized for proton densities of about 5.10^{22} cm^{-3} . This roughly corresponds to an $r_D \cong a_B$. For more details see [23] , chapter 6.5.

We can conclude this section with the statement that, in any case, the Planck-Larkin method is the appropriate way to separate the bound levels into a part associated with the real clusters ("composites") and a part which is treated as being delocalized. It has been used for exactly that purpose in the determination of the equation of state for plasmas, as we will discuss now.

4. EQUATION OF STATE FOR REACTING PLASMAS

The equation of state for plasmas has been an active research area for many years. It is not our intention to review the subject here (see [23] ,[24] and references therein), but we just want to give an idea where and how (many-body) scattering theory can be useful.

From the foregoing section,it is clear that for a given temperature, we have to distinguish three regions in a reacting plasma

(i) a low-density region: here real bound states and discrete quasi-free states exist; the latter have to be treated together with the continuum

(ii) an intermediate density region: here we find only real bound states

(iii) a high-density region:there exist only continuum states of the energy of internal motion.

We have already indicated in section 3 how the Planck-Larkin method determines the borderline ($-E_n\beta \cong 1$) between discrete bound levels and discrete quasi-free states. Some improvements upon this can be made in the following way. First, the value of the energy levels E_n should of course, take into account the effects of the surrounding plasma. This is a difficult, many-body problem. It can be attacked in principle by a systematic quantum statistical approach like the finite temperature Green's function method, starting from the Bethe-Salpeter equation [23]. Calculations for the bound levels in the screened latter approximation and including the interaction with bound states of the plasma in Born approximation show that the correction for the levels of two oppositely charged particles in a Z=1 plasma is of order of the density of the particles, while for $Z \neq 1$ plasmas it might be of order r_D^{-1}. (see [23] , chapter 4.4 and references therein). Secondly, the shift of the continuum Δ due to the "merging" of the energy levels has to be determined more precisely. Indeed, a simple approximation of the single-particle self-energy corrections yields, after averaging ([23] ,chapter 4.4)

$$\Delta = - e^2/r_D + O(\rho) \ . \tag{31}$$

which is nothing else than the chemical potential correction in the Debye approximation (29),(30). This roughly corresponds to the Mott criterium $r_D \cong a_B$.

To give an idea of these approximations we present the results [25] of their application to the surface of the sun (solar photosphere) having a T=5785 K and a density $\rho \cong 10^{-7}$ g cm^{-3}. Using the Saha equation

$$\frac{\rho}{\rho_e^2} = \lambda^3 \, \tilde{b}_2^{PL}(\beta)\, e^{\beta\Delta} \tag{32}$$

$$\tilde{b}_2^{PL}(\beta) = \sum_{E_{n\ell} < \Delta} (2\ell+1)\left(e^{-\beta E_{n\ell}} - 1 + \beta E_{n\ell}\right) \tag{33}$$

with ρ_e the free electron density, Δ given by (31) and $E_{n\ell}$ the Coulomb levels (28), one expects the hydrogen levels to merge between n=100 and n = 1000. However, the highest hydrogen line experimentally observed in the solar photosphere corresponds to n=17 [26] .

One should try to overcome the shortcomings of the present theory by developing better approximations to Δ, including e.g. electromagnetic radiation and nuclear reactions with or between the neutral particles [16] , [23]. So, Δ in (32)-(33) has then to be replaced by $\mu_{ex}(\rho,T)$, the exchange and interaction part of the chemical potential for the Coulomb contribution. Furthermore, eq. (32) should now be modified to include virial coefficient terms for the short-range interactions (see eq. (20), second and third term). Clearly, a lot of work is needed in this direction.

There has been some recent work on a direct generalization of the many-fugacity virial expansion for the equation of state (see section 1) to high Z plasmas. Since the details of the formal derivation are rather long and involved, and since a useful discussion of some of the methods and approximations employed would bring us too far aside, we refer to [24] ,[27] .

REFERENCES

1. G.E. Uhlenbeck and E. Beth, Physica 3, 729 (1936); E. Beth and G.E. Uhlenbeck, Physica 4, 915 (1937)
2. L. Gropper, Phys. Rev. 50, 963 (1936)
3. D. Bollé, Nucl.Phys. A353 ,377 (1981)

4. R. Dashen, S. Ma and H. Bernstein, Phys. Rev.187,345 (1969); R. Dashen and S. Ma, J. Math. Phys.11, 1136(1970);12,689 (1971)
5. W.G. Gibson, in "Few-Body Methods. Principles and Applications" ed. by T.K. Lim, C.G. Bao, D.P. Hou, S. Huber (World Scientific Singapore 1986)
6. D. Bollé, in Mathematics + Physics: Lectures on Recent Results, ed. L. Streit (World Scientific, Singapore 1986), Vol.2,p.84.
7. K. Huang, Statistical Mechanics (Wiley, N.Y. , 1963)
8. K.M. Watson, Phys. Rev. 103,489 (1956)
9. M.L. Goldberger and K.M. Watson, Collision Theory (Wiley, N.Y.1964)
10. V.S. Buslaev and S.P. Merkuriev, Proc. Steklov Inst. Math. 110, 28 (1970); Theor. Math. Phys. 5,1216 (1970); Sov. Phys. Dokl.14, 1055 (1970); S.P. Merkuriev, Zap. Nauchn. Semin.LOMI 95(1976)
11. F.T. Smith, J. Chem. Phys.38,1304 (1963); Phys. Rev. 131,2803(1963)
12. S. Servadio, "Truly-three-body" scattering and unitarity I & II, University of Pisa preprints IFUP TH 13-14/86 and the references cited therein.
13. S. Servadio, Phys. Rev. A4, 1256 (1971)
14. T.L. Hill, J. Chem. Phys. 23,617 (1955); "Statistical Mechanics", (McGraw Hill, N.Y. 1956), Chap.5.
15. T.A. Osborn, Phys. Rev.A16, 334 (1977)
16. C. Ray Smith, Ramarao Inguva and K.D. Lain, Phys. Rev. A23,3285 (1981)
17. W. Ebeling, Physica 73,573 (1974)
18. R.G. Newton, Scattering Theory of Waves and Particles (Springer, N.Y., 1982) 2nd ed.
19. D. Bollé, Ann. Phys. 121,131 (1979)
20. D. Bollé and T.A. Osborn, Phys. Rev.A26,3062 (1982)
21. W. Hoogeveen and J. A. Tjon, Physica 108A, 77 (1981)
22. F.J. Rogers, Phys. Rev. A19, 375 (1979)
23. W.D. Kraeft, D. Kremp, W. Ebeling and G. Röpke, Quantum Statistics of Charged Particle Systems (Plenum, N.Y. ,1986)
24. F.J. Rogers, Phys. Rev. A24, 1531 (1981)
25. W. Ebeling, W.D. Kraeft, D. Kremp and G. Röpke, Ap.J. 290, 24 (1985)
26. C.A. Rouse, Ap. J. 272, 377 (1983)
27. F.J. Rogers, Phys. Rev. A29, 868 (1984)

ON STATIONARY TWO-BODY SCATTERING THEORY IN TWO DIMENSIONS

F. Gesztesy
Institute for Theoretical Physics
University of Graz
A-8010 Graz, Austria

1. Introduction

The aim of this contribution is to review stationary scattering theory in two dimensions as developed in [8,9,11,14]. In order to explain the problems involved we quickly recall some of the basic principles of scattering in $\mathbb{R}^2$. Formally, the Hamiltonian H in $L^2(\mathbb{R}^2)$ is given by

$$H = -\Delta + V \tag{1.1}$$

where $-\Delta$ represents the kinetic energy term and $V(\underline{x})$ denotes the potential (cf.Eq.(3.1)) (we choose natural units $\hbar=2m=1$). The Møller operators in $L^2(\mathbb{R}^2)$ are then defined as strong limits

$$\Omega_{\pm} = \operatorname*{s-lim}_{t\to\pm\infty} e^{iHt}e^{-i(-\Delta)t} \tag{1.2}$$

and the unitary scattering operator in $L^2(\mathbb{R}^2)$ then reads

$$S = \Omega_+^* \Omega_- . \tag{1.3}$$

Due to energy conservation, i.e. due to the fact that S commutes with $-\Delta$, the scattering operator S can be decomposed w.r. to the spectral representation of $-\Delta$ and hence one obtains the direct integral decomposition

$$S \cong \int_{[0,\infty)}^{\oplus} dk\, S(k), \quad L^2(\mathbb{R}^2) \cong \int_{[0,\infty)}^{\oplus} dk\, L^2(S^1) \tag{1.4}$$

where S(k) denotes the on-shell scattering operator in $L^2(S^1)$ and S^1 is the unit circle in $\mathbb{R}^2$

$$S^1 = \{\underline{\omega} \in \mathbb{R}^2 \mid |\underline{\omega}| = 1\} \tag{1.5}$$

(and $\cong$ abbreviates unitary equivalence). Explicitly the structure of S(k) looks like

$$(S(k)\phi)(\underline{\omega}) = \phi(\underline{\omega}) + e^{i\pi/4}(k/2\pi)^{1/2}\int_{S^1} d\omega' f(k,\underline{\omega},\underline{\omega}')\phi(\underline{\omega}'), k>0, \phi \in L^2(S^2) \tag{1.6}$$

where $f(k,\underline{\omega},\underline{\omega}')$ denotes the on-shell scattering amplitude. Our main concern in this contribution will be an analysis of S(k) and $f(k,\underline{\omega},\underline{\omega}')$. In particular we shall be interested in their threshold behavior as $k\to 0_+$ and in a derivation of Levinson's theorem in Sect.3. In Sect.2 we shall study the special case where the potential V is spherically symmetric. In that case angular momentum conservation yields

$$S(k) = \bigoplus_{\ell=0}^{\infty} S_\ell(k), \quad S_\ell(k) = \exp[2i\delta_\ell(k)], \quad k>0 \tag{1.7}$$

where $\delta_\ell(k)$ denote the partial wave scattering phase shifts. After deriving the effective range formalism, we again study threshold properties of $S_\ell(k)$ and Levinson's theorem. For physical motivations behind these investigations one might consult the corresponding references listed up in [8,9,11,14]. Finally, in Sect.4 we consider a two-dimensional supersymmetric magnetic field system and show how to calculate Witten's (regularized) index and the axial anomaly in terms of Krein's spectral shift function [15,17,23].

2. Spherically symmetric interactions

As described in the introduction, the main purpose of this section is to analyze the partial wave on-shell scattering matrix $S_\ell(k) = \exp[2i\delta_\ell(k)]$, $k>0$, $\ell\in \mathbb{N}_0$ in some detail.

2.1. Preliminaries

Let V be a real-valued short-range potential of the type

$$\int_0^R dr\, r|\ell n(r)|^2|V(r)| + \int_R^\infty dr\, e^{2ar}|V(r)|<\infty \tag{2.1}$$

for some 0<R<1 and some a>0. Next we introduce regular and irregular solutions $F_\ell(r)$, $G_\ell(r)$ of the zero-energy Schrödinger equation ($\hbar=2m=1$)

$$-\psi_\ell''(r)+[(\ell^2-4^{-1})r^{-2}+V(r)]\psi_\ell(r)=0, \quad r>0, \quad \ell\in \mathbb{N}_0 \tag{2.2}$$

uniquely defined by [8,9]

$$F_\ell(r)=F_\ell^{(o)}(r)-\int_0^r dr' g_\ell^{(o)}(r,r')V(r')F_\ell(r'), \tag{2.3}$$

$$G_\ell(r)=G_\ell^{(o)}(r)+\int_r^\infty dr' g_\ell^{(o)}(r,r')V(r')G_\ell(r'), \quad r>0 \tag{2.4}$$

where

$$g_\ell^{(o)}(r,r')=G_\ell^{(o)}(r)F_\ell^{(o)}(r')-G_\ell^{(o)}(r')F_\ell^{(o)}(r), \quad r,r'>0 \tag{2.5}$$

denotes the free zero-energy Volterra kernel and $F_\ell^{(o)}(r)$, $G_\ell^{(o)}(r)$ are appropriate solutions of the free Schrödinger equation (i.e. of Eq.(2.2) with V=0)

$$F_\ell^{(o)}(r)=r^{(2\ell+1)/2}, \quad \ell\geq 0, \quad r\geq 0, \tag{2.6}$$

$$G_0^{(o)}(r)=-r^{1/2}\ell n(r), \quad G_\ell^{(o)}(r)=(2\ell)^{-1}r^{(1-2\ell)/2}, \quad \ell\geq 1, \quad r>0. \tag{2.7}$$

In addition we also need (ir)regular solutions $F_\ell(k,r)$, $G_\ell(k,r)$, $k>0$ of the Schrödinger equation corresponding to nonzero energy $k^2>0$

$$-\psi_\ell''(k,r)+[(\ell^2-4^{-1})r^{-2}+V(r)-k^2]\psi_\ell(k,r)=0,\quad k,r>0,\quad \ell\in\mathbb{N}_0 \tag{2.8}$$

uniquely defined by [8,9]

$$F_\ell(k,r)=F_\ell^{(o)}(k,r)-\int_0^r dr' g_\ell^{(o)}(k,r,r')V(r')F_\ell(k,r'), \tag{2.9}$$

$$G_\ell(k,r)=G_\ell^{(o)}(k,r)+\int_r^\infty dr' g_\ell^{(o)}(k,r,r')V(r')G_\ell(k,r'),\quad k,r>0, \tag{2.10}$$

where now

$$g_\ell^{(o)}(k,r,r')=G_\ell^{(o)}(k,r)F_\ell^{(o)}(k,r')-G_\ell^{(o)}(k,r')F_\ell^{(o)}(k,r),\quad k,r,r'>0, \tag{2.11}$$

$$F_\ell^{(o)}(k,r)=(k/2)^{-\ell}\Gamma(\ell+1)r^{1/2}J_\ell(kr), \tag{2.12}$$

$$G_\ell^{(o)}(k,r)=-i(\pi/2)(k/2)^{\ell}\Gamma(\ell+1)^{-1}r^{1/2}H_\ell^{(2)}(kr),\quad k,r>0 \tag{2.13}$$

(cf. [1]). Moreover the Jost function $\mathcal{F}_\ell(k)$ is given by

$$\mathcal{F}_\ell(k)=W(G_\ell(k),F_\ell(k))=1+\int_0^\infty dr\, G_\ell^{(o)}(k,r)V(r)F_\ell(k,r),\quad k>0 \tag{2.14}$$

(here W denotes the Wronskian). We also mention that $F_\ell^{(o)}(k,r)$ is real for $k\geq 0$ and entire with respect to k^2, implying

$$F_\ell^{(o)}(k,r)=F_\ell^{(o)}(-k,r)=\overline{F_\ell^{(o)}(k,r)},\quad k,r\geq 0. \tag{2.15}$$

Moreover analytic continuation in $G_\ell^{(o)}(k,r)$ gives

$$G_\ell^{(o)}(e^{-i\pi}k,r) = \overline{G_\ell^{(o)}(k,r)} + i(2k)^{2\ell}\Gamma(2\ell+1)^{-2}\Gamma((2\ell+1)/2)^2F_\ell^{(o)}(k,r),\quad k,r>0 \tag{2.16}$$

implying the fact that $g_\ell^{(o)}(k,r,r')$ as well as $F_\ell(k,r)$ are real for $k\geq 0$ and entire with respect to k^2, i.e.

$$g_\ell^{(o)}(k,r,r')=g_\ell^{(o)}(-k,r,r')=\overline{g_\ell^{(o)}(k,r,r')},\quad k\geq 0,\ r,r'>0, \tag{2.17}$$

$$F_\ell(k,r)=F_\ell(-k,r)=\overline{F_\ell(k,r)},\quad k,r\geq 0. \tag{2.18}$$

Our basic object, the on-shell partial wave scattering matrix is now given by

$$S_\ell(k)=\exp[2i\delta_\ell(k)]=\mathcal{F}_\ell(k)/\mathcal{F}_\ell(e^{-i\pi}k),\quad k>0. \tag{2.19}$$

We choose $\delta_\ell(\infty)=0$ to get uniqueness of the phase shifts. For later purposes we also note that

$$\cot[\delta_\ell(k)]=i[\mathcal{F}_\ell(k)+\mathcal{F}_\ell(e^{-i\pi}k)]/[\mathcal{F}_\ell(k)-\mathcal{F}_\ell(e^{-i\pi}k)]$$

$$=\int_0^\infty dr\ \mathrm{Re}\left[G_\ell^{(o)}(k,r)\right]V(r)F_\ell(k,r)/\int_0^\infty dr\ \mathrm{Im}\left[G_\ell^{(o)}(k,r)\right]V(r)F_\ell(k,r),\ k>0. \tag{2.20}$$

2.2. Scattering lengths and threshold states

Using assumption (2.1) (cf. [9]) for improvements) and Eq.(2.3) one derives the asymptotic relation

$$F_\ell(r) \underset{r\to\infty}{=} f_\ell F_\ell^{(o)}(r)-\left[\int_0^\infty dr' F_\ell^{(o)}(r')V(r')F_\ell(r')\right]G_\ell^{(o)}(r)+o\left(\min\left(F_\ell^{(o)}(r),G_\ell^{(o)}(r)\right)\right) \tag{2.21}$$

where we abbreviate

$$f_\ell=W(G_\ell,F_\ell)=1+\int_0^\infty dr G_\ell^{(o)}(r)V(r)F_\ell(r). \tag{2.22}$$

Eqs.(2.6) and (2.7) show that for all $\ell\geq 1$, $F_\ell^{(o)}(r)$ dominates $G_\ell^{(o)}(r)$ whereas for $\ell=0$, $G_o^{(o)}(r)$ dominates $F_o^{(o)}(r)$ as $r\to\infty$. Next let

$$h_\ell = -(d^2/dr^2)+\left[\ell^2-4^{-1}\right]r^{-2} \dotplus V \tag{2.23}$$

denote the form sum of the kinetic energy operator and the potential V in $L^2((0,\infty);dr)$. Concerning zero-energy (threshold) properties of h_ℓ we have the following possibilities:

Case A. h_ℓ, $\ell\geq 0$ has no zero-energy resonance respectively no zero-energy bound state.

This turns out to be the generic case.

Case B. h_o has a zero-energy (s-wave) resonance iff

$$\int_0^\infty dr\ F_o^{(o)}(r)V(r)F_o(r) = 0. \tag{2.24}$$

Case C. h_1 has a zero-energy (p-wave) resonance iff

$$f_1 = 0. \tag{2.25}$$

Case D. h_ℓ, $\ell\geq 2$ has a zero-energy bound state iff

$$f_\ell = 0. \tag{2.26}$$

In all cases the corresponding zero-energy resonance ($\ell=0,1$) resp. bound state function ($\ell\geq 2$) is given by $F_\ell(r)$. Its asymptotic behavior w.r. to r reads if

$\ell = 0$ and $\int_0^\infty dr\ F_o^{(o)}(r)V(r)F_o(r) = 0$:

$$F_o(r) \underset{r\to 0_+}{=} 0(r^{1/2}),\ F_o(r) \underset{r\to\infty}{=} 0(r^{1/2}) \tag{2.27}$$

if $\ell\geq 1$ and $f_\ell=0$:

$$F_\ell(r) \underset{r\to 0_+}{=} 0(r^{(2\ell+1)/2}),\ F_\ell(r) \underset{r\to\infty}{=} 0(r^{(1-2\ell)/2}). \tag{2.28}$$

In addition one can show that in the case $\ell=0$

$$\int_0^\infty dr\, F_0^{(o)}(r)V(r)F_0(r) = 0 \quad \text{implies } f_0 \neq 0 \tag{2.29}$$

and for $\ell\geq 1$

$$f_\ell = 0 \quad \text{implies} \quad \int_0^\infty dr\, F_\ell^{(o)}(r)V(r)F_\ell(r) \neq 0. \tag{2.30}$$

Let us indicate the proof of Eqs. (2.29) and (2.30). For that purpose one introduces another zero-energy solution $H_\ell(r)$ of Eq.(2.2) by

$$H_\ell(r)=F_\ell^{(o)}(r)+\int_r^\infty dr' g_\ell^{(o)}(r,r')V(r')H_\ell(r'). \tag{2.31}$$

Similar to Eq.(2.21), one now investigates the asymptotic behavior of $F_\ell'(r)$, $G_\ell(r)$, $G_\ell'(r)$, $H_\ell(r)$, $H_\ell'(r)$ as $r\to 0_+$ and as $r\to\infty$. Calculating Wronskians at $r=0_+$ and at $r\to\infty$ then yields

$$W(G_\ell,H_\ell)=f_\ell\left[1-\int_0^\infty drG_\ell^{(o)}(r)V(r)H_\ell(r)\right]$$

$$+\int_0^\infty drF_\ell^{(o)}(r)V(r)H_\ell(r)\int_0^\infty dr' G_\ell^{(o)}(r')V(r')G_\ell(r')=1,$$

$$W(H_\ell,F_\ell)=\int_0^\infty drF_\ell^{(o)}(r)V(r)H_\ell(r)=\int_0^\infty drF_\ell^{(o)}(r)V(r)F_\ell(r)$$

and hence

$$1=\int_0^\infty drF_\ell^{(o)}(r)V(r)F_\ell(r)\int_0^\infty dr' G_\ell^{(o)}(r')V(r')G_\ell(r')$$

$$+f_\ell\left[1-\int_0^\infty drG_\ell^{(o)}(r)V(r)H_\ell(r)\right]. \tag{2.32}$$

Thus Eqs.(2.29) and (2.30) follow. The above argument is taken from [24]. For a different proof cf. [35].

Finally we introduce the scattering length a_ℓ. The asymptotic relation (2.21) then leads to the definition

$$a_\ell=f_\ell^{-1}\int_0^\infty drF_\ell^{(o)}(r)V(r)F_\ell(r), \quad f_\ell\neq 0. \tag{2.33}$$

We emphasize that the dimension of a_ℓ equals $|\text{length}|^{2\ell}$ following the standard terminology in three dimensions we still call a_ℓ the "scattering length"). We also note the curious fact that a_0 vanishes in the presence of a zero-energy resonance whereas a_ℓ diverges in the presence of a threshold state for $\ell\geq 1$ (as is familiar from three dimensions).

2.3. The effective range expansion

The crucial step in deriving the effective range expansion now consists in the splitting of $\text{Re}\left[G_\ell^{(o)}(k,r)\right]$ into an entire function $\tilde{G}_\ell^{(o)}(k,r)$

w.r. to k^2 and a remainder term:

$$\mathrm{Re}[G_\ell^{(o)}(k,r)] = \tilde{G}_\ell^{(o)}(k,r)$$

$$+\{-i+(2/\pi)[\ln(2i/k)-C]\}(\pi/2)(k/2)^\ell \Gamma(\ell+1)^{-1} r^{1/2} J_\ell(kr),\ k,r>0 \qquad (2.34)$$

(here C denotes Euler's constant) and [1]

$$\tilde{G}_\ell^{(o)}(k,r) = -\Gamma(\ell+1)^{-1}(k/2)^\ell [\ln(r)-C] r^{1/2} J_\ell(kr)$$

$$+2^{-(2\ell+1)}\Gamma(\ell+1)^{-1}k^{2\ell}r^{(2\ell+1)/2}\sum_{m=0}^{\infty}[\Psi(m+1)+\Psi(\ell+m+1)](-1)^m(kr/2)^{2m}/[m!(\ell+m)!]$$

$$+2^{-1}\Gamma(\ell+1)^{-1}r^{(-2\ell+1)/2}\sum_{p=0}^{\ell-1}[(\ell-p-1)!/p!](kr/2)^{2p} \qquad (2.35)$$

where $\Psi(z)=\Gamma'(z)/\Gamma(z)$ denotes the psi function [1] (if $\ell=0$, the last term on the r.h.s. of Eq. (2.35) is interpreted to be zero). By inspection we get

$$\tilde{G}_\ell^{(o)}(k,r) = \tilde{G}_\ell^{(o)}(-k,r) = \overline{\tilde{G}_\ell^{(o)}(k,r)},\ k\geq 0,\ r>0 \qquad (2.36)$$

and

$$\tilde{G}_\ell^{(o)}(k,r) \underset{k\to 0}{=} G_\ell^{(o)}(r) + O(k^2),\ r>0. \qquad (2.37)$$

Similarly one infers

$$F_\ell^{(o)}(k,r) \underset{k\to 0}{=} F_\ell^{(o)}(r) + O(k^2),\ r\geq 0, \qquad (2.38)$$

$$F_\ell(k,r) \underset{k\to 0}{=} F_\ell(r) + O(k^2),\ r\geq 0. \qquad (2.39)$$

Inserting the above results into Eq.(2.20) then yields the effective range expansion [8,9]

$$\Gamma(\ell+1)^{-2}(k/2)^{2\ell}[(\pi/2)\cot[\delta_\ell(k)]-\ln(k/2)-C] \underset{k\to 0}{=} -a_\ell^{-1}+O(k^2). \qquad (2.40)$$

Here a_ℓ is precisely the scattering length introduced in Eq.(2.33) and moreover, due to to assumption (2.1), the r.h.s. of Eq. (2.40) is analytic w.r. to k^2 around the threshold $k^2=0$ [8,9]. At this point we would like to emphasize that actually a much more general result has been derived in [8,9]: First of all all results derived so far are proved under the additional influence of a Coulomb potential γ/r, $\gamma\in\mathbb{R}$ and secondly all results were generalized to dimensions $n\geq 2$. (In fact practically all results of Sects. 2.2 and 2.3 generalize to n-dimensions after replacing ℓ by $\ell+[(n-2)/2]$. For the low-energy behavior of the ($n\geq 2$ dimensional) cross section and applications to the Ramsauer-Townsend effect we also refer to [8,9].

2.4. Threshold properties of Jost functions, Levinson's theorem

We first analyze $\mathcal{F}_\ell(k)$ near the threshold $k=0$. Inserting Eq. (2.39) and (cf. [1])

$$G_\ell^{(o)}(k,r) \underset{k\to 0_+}{=} \begin{cases} -F_o^{(o)}(r)[\ell n(k)+(i\pi/2)+C-\ell n(2)]+G_o^{(o)}(r)+0(k^2\ell n(k)), \ell=0 \\ G_1^{(o)}(r)-4^{-1}F_1^{(o)}(r)k^2\ell n(k)+0(k^2), \ \ell=1 \\ G_\ell^{(o)}(r)+[8\ell(\ell-1)]^{-1}r^{(5-2\ell)/2}k^2+0(k^4\ell n(k)), \ \ell\geq 2 \end{cases} \tag{2.41}$$

into Eq. (2.14) we obtain

$$\mathcal{F}_\ell(k) \underset{k\to 0_+}{=} \begin{cases} A_o\ell n(k)+B_o+0(k^2\ell n(k)), \ \ell=0 \\ A_1+B_1k^2\ell n(k)+0(k^2), \ \ell=1 \\ A_\ell+B_\ell k^2+0(k^4\ell n(k)), \ \ell\geq 2. \end{cases} \tag{2.42}$$

Here

$$A_o=-\int_o^\infty dr\ F_o^{(o)}(r)V(r)F_o(r), \quad A_\ell=f_\ell, \quad \ell\geq 1,$$

$$B_o=f_o-[(i\pi/2)+C-\ell n(2)]\int_o^\infty dr\ F_o^{(o)}(r)V(r)F_o(r), \tag{2.43}$$

$$B_1=-4^{-1}\int_o^\infty dr\ F_1^{(o)}(r)V(r)F_1(r).$$

In particular

$$A_\ell=0 \quad \text{implies} \quad B_\ell\neq 0, \quad \ell\geq 0. \tag{2.44}$$

In order to prove assertion (2.44) we note that for $\ell=0,1$, Eq. (2.44) is equivalent to Eqs. (2.29) and (2.30). For $\ell\geq 2$ (i.e. in those angular momentum sectors where $A_\ell=0$ implies a threshold bound state of h_ℓ) one can follow the arguments in [33].

As an immediate application of the result (2.42) we obtain Levinson's theorem [11,14]

$$\delta_\ell(0) = \pi[N_\ell + D_\ell], \ \ell\geq 0 \tag{2.45}$$

where N_ℓ denotes the number of negative bound states of h_ℓ (which are all simple) and

$$D_o = 0, \qquad D_\ell = \begin{cases} 0, & f_\ell \neq 0 \\ 1, & f_\ell = 0, \quad \ell\geq 1. \end{cases} \tag{2.46}$$

The result (2.45) (and its generalization to nonspherically symmetric potentials) has first been derived in [11,14]. At first sight it contains unexpected features when compared to the well known three-dimensional analog (in obvious notation)

$$\delta_\ell^{(3)}(0) = \pi\left[N_\ell^{(3)}+D_\ell^{(3)}\right],$$

$$D_0^{(3)} = \begin{cases} 0, & f_0^{(3)} \neq 0 \\ 1/2, & f_0^{(3)} = 0 \end{cases} \qquad D_\ell^{(3)} = \begin{cases} 0, & f_\ell^{(3)} \neq 0 \\ 1, & f_\ell^{(3)} = 0, \quad \ell \geq 1. \end{cases} \tag{2.47}$$

In fact due to Eq.(2.46), a possible zero-energy s-wave resonance has no influence at all in Eq.(2.45), whereas a zero-energy p-wave resonance contributes like a threshold bound state.
We finally remark that in complete analogy to Eq. (2.47), the corresponding two-dimensional result (2.45) follows by a contour integration of $(d/dk)\ln[\mathcal{F}_\ell(k)]$ in the lower complex plane. An extension of Eq. (2.45) to certain spherically symmetric long-range interactions appeared recently in [35]. Eq. (2.45) was also recently considered in [25]. We emphasize that Eq.(2.45) immediately generalizes to $n\geq 2$ dimensions [11,14]. In particular the case $n=4$, and $\ell=0$ exhibits the same phenomena as $n=2$ and $\ell=1$.
Finally, inserting Eq.(2.42) into Eq.(2.19) we obtain the threshold behavior for the scattering matrix

$$S_0(k) = \begin{cases} 1+[i\pi/\ln(k)]+0((\ln(k))^{-2}), & \int_0^\infty dr\, F_0^{(0)}(r)V(r)F_0(r)\neq 0 \\ 1+0(k^2), & \int_0^\infty dr\, F_0^{(0)}(r)V(r)F_0(r)=0, \end{cases} \tag{2.48}$$

$$S_1(k) = \begin{cases} 1+0(k^2), & f_1\neq 0 \\ 1+0((\ln(k))^{-1}), & f_1=0, \end{cases} \tag{2.49}$$

$$S_\ell(k) = \begin{cases} 1+0(k^4), & f_\ell\neq 0 \\ 1+0(k^2), & f_\ell=0, \quad \ell\geq 2. \end{cases} \tag{2.50}$$

3. Nonspherically symmetric interactions

The aim of this section is to discuss the on-shell scattering operator $S(k)$ in $L^2(S^1)$ and to derive Levinson's theorem.

3.1. Preliminaries

Let V be real-valued and satisfying

$$\int_{\mathbb{R}^2} d^2x\,|V(\underline{x})|^{1+\delta}<\infty, \quad \int_{\mathbb{R}^2} d^2x V(\underline{x})\neq 0, \quad \int_{\mathbb{R}^2} d^2x\, e^{2a|\underline{x}|}\,|V(\underline{x})|<\infty \tag{3.1}$$

for some $\delta, a>0$. The Hamiltonian H in $L^2(\mathbb{R}^2)$ is then defined as the form sum

$$H = -\Delta \dotplus V \tag{3.2}$$

with $-\Delta$ the kinetic energy operator. Next we introduce the splitting

$$V = u.v, \quad v(\underline{x}) = |V(\underline{x})|^{1/2}, \; u(\underline{x}) = v(\underline{x})\,\mathrm{sgn}(V(\underline{x})) \tag{3.3}$$

and define the transition operator $T(k)$ in $L^2(\mathbb{R}^2)$ as

$$T(k) = [1 + uR_o(k)v]^{-1}, \; \mathrm{Im}k \geq 0, \; k \neq 0, \; k^2 \notin \sigma_p(H) \tag{3.4}$$

where $\sigma_p(H)$ denotes the point spectrum of H and $R_o(k)$ is the free resolvent

$$R_o(k) = (-\Delta - k^2)^{-1}, \; \mathrm{Im}k > 0 \tag{3.5}$$

with integral kernel (cf. [1])

$$R_o(k,\underline{x},\underline{y}) = (i/4)H_o^{(1)}(k|\underline{x}-\underline{y}|), \quad \underline{x} \neq \underline{y}. \tag{3.6}$$

In order to exhibit the singularity of $R_o(k)$ as $k \to 0$ we decompose [38]

$$uR_o(k)v = (2\pi)^{-1}[-\ell n(k) + (i\pi/2) + \ell n(2) + \Psi(1)](v,.)u + M(k), \; \mathrm{Im}k > -a, k \neq 0 \tag{3.7}$$

where $M(k)$ is Hilbert-Schmidt for all $\mathrm{Im}k > -a$. In particular the integral kernel of $M(0) \equiv M_{o,o}$ reads

$$M_{o,o}(\underline{x},\underline{y}) = -(2\pi)^{-1}u(\underline{x})\ell n|\underline{x}-\underline{y}|v(\underline{y}), \; \underline{x} \neq \underline{y}. \tag{3.8}$$

Next we define

$$P = (v,u)^{-1}(v,.)u, \; Q = 1 - P. \tag{3.9}$$

We remark that at this point our assumption $(v,u) \neq 0$ enters since it distinguishes the cases whether $(v,.)u$ is nilpotent or not. A complete treatment of the exceptional case $(v,u)=0$ for one-dimensional systems appeared in [12]. On the basis of these results one expects practically all results in the following to go through in the case $(v,u)=0$ without significant changes. Given definition (3.9) we obtain

$$T(k) = \{1 + (2\pi)^{-1}(v,u)[-\ell n(k) + (i\pi/2) + \ell n(2) + \Psi(1)]P + M(k)\}^{-1},$$
$$\mathrm{Im}k > -a, \; k \neq 0, \; k^2 \notin \sigma_p(H). \tag{3.10}$$

Before we can analyze Eq. (3.10) in more detail we need the following technical result [30]

$$(z + \sigma P + M_{o,o})^{-1} = Q(z + QM_{o,o}Q)^{-1}Q + O(\sigma^{-1}),$$
$$z \in \mathbb{C}\backslash\mathbb{R}, \; |\sigma| \text{ large enough} \tag{3.11}$$

where the r.h.s. of Eq. (3.11) turns out to be norm analytic with respect to σ^{-1} around $\sigma^{-1}=0$. Identifying σ in Eq.(3.11) with $(2\pi)^{-1}(v,u)\left[-\ell n(k)+\ldots\right]$ in Eq.(3.10), we infer that the low-energy behavior of $T(k)$ is intimately connected to the problem whether $1+QM_{00}Q$ is invertible or not (and hence to the eigenvalue problem $QM_{00}Q\phi=-\phi$, $\phi\in L^2(\mathbb{R}^2)$). Moreover the threshold behavior of $T(k)$ as $k\to 0$ also crucially depends on the zero-energy properties of H as shown below. First we recall the following result: Assume that -1 is an eigenvalue of $QM_{00}Q$. Let

$$\mathcal{V}=\{\phi\in L^2(\mathbb{R}^2)\,|\,QM_{00}Q\phi=-\phi\}, \tag{3.12}$$

$$\mathcal{W}=\{\chi\in \mathcal{V}\,|\,(v,M_{00}\chi)=0\}. \tag{3.13}$$

Then

(i) $\dim\mathcal{W}=\dim\mathcal{V}$ or $\dim\mathcal{W}=\dim\mathcal{V}-1$. (3.14)

(ii) $M_{00}\chi=-\chi$ for all $\chi\in\mathcal{W}$. (3.15)

(iii) If $\phi_0\in\mathcal{V}\setminus\mathcal{W}$ then $M_{00}\phi_0=-\phi_0+(v,u)^{-1}(v,M_{00}\phi_0)u$. (3.16)

The above is a slightly improved version of Lemma 7.3 in [30] and has been proved in [13]. Concerning properties of zero-energy solutions of Schrödinger's equation we have the following results [11,14] (cf. also [32]):

Assume that $QM_{00}Q\phi=-\phi$ for some $\phi\in L^2(\mathbb{R}^2)$ and define

$$\psi(\underline{x})=-(v,u)^{-1}(v,M_{00}\phi)-(2\pi)^{-1}\int_{\mathbb{R}^2} d^2y\,\ell n|\underline{x}-\underline{y}|\,v(\underline{y})\phi(\underline{y}). \tag{3.17}$$

Then

(i) $\psi\in L^2_{loc}(\mathbb{R}^2)$, $\nabla\psi\in L^2(\mathbb{R}^2)$ and $H\psi=0$ in the sense of distributions. (3.18)

(ii) $u(\underline{x})\psi(\underline{x})=-\phi(\underline{x})$ a.e. (3.19)

(iii) $\psi+(v,u)^{-1}(v,M_{00}\phi)-(2\pi)^{-1}|\underline{x}|^{-2}\underline{x}(\underline{y}v,\phi)\in L^2(\mathbb{R}^2)$ (3.20)

in particular

$\psi\in L^2(\mathbb{R}^2)$ is equivalent to $(v,M_{00}\phi)=(\underline{y}v,\phi)=0$. (3.21)

Given the above results we are able to introduce the following case distinctions concerning zero-energy properties of H:

<u>Case I.</u> -1 is not an eigenvalue of $QM_{00}Q$.

<u>Case II.</u> -1 is an eigenvalue of $QM_{00}Q$ of multiplicity $M\leq 3$, $QM_{00}Q\phi_j=-\phi_j$, $\phi_j\in L^2(\mathbb{R}^2)$, $0\leq j\leq 2$ and

a) $M=1$, $c_1^{(0)}\neq 0$

or

b) $M\leq 2$, $c_1^{(j)}=0$, $\underline{c}_2^{(j)}\neq 0$, $1\leq j\leq 2$ and if $M=2$ then $\underline{c}_2^{(1)}$, $\underline{c}_2^{(2)}$ are linearly independent

or

c) $2\leq M\leq 3$, $c_1^{(o)}\neq 0$, $c_1^{(j)}=0$, $\underline{c}_2^{(j)}\neq 0$, $1\leq j\leq 2$ and if $M=3$ then $\underline{c}_2^{(1)}$, $\underline{c}_2^{(2)}$ are linearly independent.

Case III. -1 is an eigenvalue of $QM_{oo}Q$ of multiplicity $N\in\mathbb{N}$, $QM_{oo}Q\phi_j=-\phi_j$, $\phi_j\in L^2(\mathbb{R}^2)$, $3\leq j\leq 2+N$ and $c_1^{(j)}=\underline{c}_2^{(j)}=0$, $3\leq j\leq 2+N$.

Case IV. -1 is an eigenvalue of $QM_{oo}Q$ of multiplicity $M+N$, $1\leq M\leq 3$, $N\in\mathbb{N}$, $QM_{oo}Q\phi_j=-\phi_j$, $\phi_j\in L^2(\mathbb{R}^2)$, $0\leq j\leq 2+N$, and

a) $c_1^{(o)}\neq 0$, $c_1^{(j)}=\underline{c}_2^{(j)}=0$, $3\leq j\leq 2+N$, $M=1$

or

b) $c_1^{(j)}=0$, $\underline{c}_2^{(j)}\neq 0$, $1\leq j\leq 2$, $c_1^{(j)}=\underline{c}_2^{(j)}=0$, $3\leq j\leq 2+N$, $1\leq M\leq 2$

or

c) $c_1^{(o)}\neq 0$, $c_1^{(j)}=0$, $\underline{c}_2^{(j)}\neq 0$, $1\leq j\leq 2$, $c_1^{(j)}=\underline{c}_2^{(j)}=0$, $3\leq j\leq 2+N$, $2\leq M\leq 3$.

Here the coefficients $c_1^{(j)}$, $\underline{c}_2^{(j)}$ are defined by

$$c_1^{(j)}=(v,u)^{-1}(v,M_{oo}\phi_j),\ \underline{c}_2^{(j)}=(2\pi)^{-1}(\underline{y}v,\phi_j). \tag{3.22}$$

Clearly case I describes the absence of zero-energy resonances and zero-energy bound states of H and hence represents the generic case (cf. case A in Sect. 2.2). Cases IIa)-c) denote the various possibilities of zero-energy resonances of H. E.g. if V is spherically symmetric then case IIa) corresponds to an s-wave resonance (i.e. to case B) whereas case IIb) corresonds to p-wave resonances (cf. case C). Case IIc) represents a possible mixture of both s-wave and p-wave resonances. Case III denotes the threshold bound state case (and hence coincides with case D). Finally cases IVa)-c) describe possible admixtures of cases II-III. By the results (3.14)-(3.16) and (3.17)-(3.21) the above list of cases is complete. We also remark that up to now the assumptions on V could be relaxed considerably [11].

3.2. Threshold properties, Levinson's theorem

Given assumptions (3.1) (cf. [11] for considerable improvements), the on-shell scattering amplitude $f(k,\underline{\omega},\underline{\omega}')$ in two dimensions is given by

$$f(k,\underline{\omega},\underline{\omega}')=(8\pi k)^{-1/2}e^{-3\pi i/4}(ve^{ik\underline{\omega}\cdot\underline{x}},T(k)ue^{ik\underline{\omega}'\cdot\underline{y}}),$$

$$k>0,\ k^2\notin\mathcal{E},\ \underline{\omega},\underline{\omega}'\in S^1. \tag{3.23}$$

Here T(k) has been defined in Eq. (3.4), (.,.) on the r.h.s. of Eq. (3.23) denotes the scalar product in $L^2(\mathbb{R}^2)$ and $\underline{x},\underline{y}$ are integration variables in obvious notation. Moreover the exceptional set $\mathcal{E}$ reads

$$\mathcal{E}=\{k^2>0 \mid uR_o(k)vg=-g \text{ for some } g\in L^2(\mathbb{R}^2), k>0\}. \tag{3.24}$$

Under assumptions (3.1), $\mathcal{E}$ is known to be a discrete set (which consists precisely of positive, embedded eigenvalues of H). In order to establish the low-energy behavior of $f(k,\underline{\omega},\underline{\omega}')$ in all cases I-IV, we evidently first need the corresponding behavior of T(k). For that purpose we first note that M(k) is of the type

$$M(k)=\sum_{n=0}^{\infty} k^{2n}\tilde{M}_{n,o}+\ell n(k)\sum_{n=0}^{\infty} k^{2n}\tilde{M}_{o,n}$$

$$\equiv \sum_{m,n=0}^{\infty} [1/\ell n(k)]^m [k^2 \ell n(k)]^n M_{m,n} \tag{3.25}$$

where all expansions in Eq.(3.25) are valid in Hilbert-Schmidt norm. Because of Eq.(3.6), the integral kernels of $M_{m,n}$ are given by (cf. also Eq.(3.8))

$$M_{m,m}(\underline{x},\underline{y})=(-1)^m[4\pi 4^m (m!)^2]^{-1}\{[i\pi+\ell n(4)+2\Psi(m+1)]u(\underline{x})|\underline{x}-\underline{y}|^{2m}v(\underline{y})$$

$$-2u(\underline{x})[\ell n|\underline{x}-\underline{y}|]|\underline{x}-\underline{y}|^{2m}v(\underline{y}),\ m\geq 1,\ \underline{x}\neq\underline{y},$$

$$M_{m-1,m}(\underline{x},\underline{y})=(-1)^{m-1}[2\pi 4^m (m!)^2]^{-1}u(\underline{x})|\underline{x}-\underline{y}|^{2m}v(\underline{y}),\ m\geq 1,$$

$$M_{m,n}(\underline{x},\underline{y})=0 \text{ if } m\leq n-2 \text{ or } m\geq n+1. \tag{3.26}$$

Motivated by Eqs.(3.10), (3.11) and (3.25) we finally mention another technical result, viz.

$$(1+QM_{o,o}Q+\varepsilon)^{-1}=\varepsilon^{-1}P_o+\sum_{n=0}^{\infty}(-\varepsilon)^n T_o^{\,n+1},\ |\varepsilon| \text{ small enough} \tag{3.27}$$

where P_o denotes the projection onto the eigenspace of $QM_{o,o}Q$ to the eigenvalue -1 and $T_o=n\text{-}\lim_{\varepsilon\to 0}(1+QM_{o,o}Q+\varepsilon)^{-1}[1-P_o]$ represents the corresponding reduced resolvent [27]. Expansion (3.27) is a norm convergent one. In view of Eq.(3.25), ε will be identified with quantities like $1/\ell n(k)$, $k^2\ell n(k)$ or k^2 depending on the cases involved. Putting together all these results, some rather tedious calculations finally yield the following norm convergent Laurent expansion for T(k) around the threshold

$$T(k)=\sum_{m=-\infty,n=-1}^{\infty}[1/\ell n(k)]^m[k^2\ell n(k)]^n t_{m,n},\ |k| \text{ small enough}. \tag{3.28}$$

Since the detailed computations of $t_{m,n}$ turn out to be extremely involved [11] we only indicate the leading order terms in expansion

(3.28) in the two simplest cases:

Case I. $T(k) \underset{k\to 0}{=} (1+QM_{00}Q)^{-1}Q+O(1/\ell n(k))$.

Case IIa). $T(k) \underset{k\to 0}{=} [2\pi |c_1^{(o)}|^2]^{-1} |\phi_o><\mathrm{sgn}(V)\phi_o| \ell n(k)+O(1)$.

Similarly one has $T(k) \underset{k\to 0}{=} O((k^2 \ell n(k))^{-1})$ in cases IIb),c) and $T(k) \underset{k\to 0}{=} O(k^{-2})$ in cases III and IV. These expansions for T(k) now immediately yield corresponding expansions for the on-shell scattering amplitude and the scattering operator by inserting Eq.(3.28) into Eqs. (3.23) and (1.6). In particular we get

$$S(k) \underset{k\to 0_+}{=} 1 + o(1) \tag{3.29}$$

in all cases I-IV in contrast to three dimensions (cf.[3,4,7,10,16,26,34]). Finally, by noting

$$2\mathrm{Im}\{\mathrm{Tr}[(H-k^2)^{-1}-R_o(k)]\}=(2ik)^{-1}\mathrm{Tr}[(d/dk)\ell n S(k)] \tag{3.30}$$

and

$$\mathrm{Tr}[(H-k^2)^{-1}-R_o(k)]=-\mathrm{Tr}[R_o(k)vT(k)uR_o(k)] \tag{3.31}$$

a standard contour integration of the function

$$F(k^2)=\mathrm{Tr}[(H-k^2)^{-1}-R_o(k)]+Dk^{-2}-k^{-2}\int_{\mathbb{R}^2} d^2x V(\underline{x}) \tag{3.32}$$

yields Levinson's theorem in the form [11,14]

$$(i/2)\int_o^\infty dk \mathrm{Tr}[(d/dk)\ell n S(k)]=\pi N_b+\pi D+4^{-1}\int_{\mathbb{R}^2} d^2x V(\underline{x}) \tag{3.33}$$

(assuming $\mathcal{E} = \emptyset$). Here N_b denotes the total number of negative bound states of H (counting multiplicity) and

$$D^{I}=D^{IIa)}=0,\ D^{IIb)}=M,\ D^{IIc)}=M-1,\ D^{III}=D^{IVa)}=N,$$

$$D^{IVb)}=M+N,\ D^{IVc)}=M+N-1. \tag{3.34}$$

In the generic case I, Eq. (3.33) has first been derived in [19] (cf. also [37]).

As a final comment we emphasize that relaxing the exponential falloff to a power falloff (i.e. assuming only

$$\int_{\mathbb{R}^2} d^2x(1+|\underline{x}|)^m |V(\underline{x})|<\infty$$

for appropriate m>2) all analytic expansions in Sects. 2 and 3 turn into asymptotic expansions (the order of which depends on m).

4. A two-dimensional supersymmetric system

The purpose of this section is to discuss the "canonical" example of a two-dimensional supersymmetric magnetic field system.

4.1. Preliminaries

We briefly establish some of the basic relations in supersymmetric quantum mechanics for one degree of freedom. For physical motivations we refer to [18,36,39,40] and the references therein. Let $\mathcal{H}$ be a complex, separable Hilbert space and A be a closed, densely defined operator in $\mathcal{H}$. Define self-adjoint, nonnegative operators H_j, j=1,2 by

$$H_1 = A^* A, \quad H_2 = AA^* \tag{4.1}$$

and the corresponding supercharge Q and supersymmetric Hamiltonian H in $\mathcal{H} \oplus \mathcal{H}$ by

$$Q = \begin{pmatrix} 0 & A^* \\ A & 0 \end{pmatrix}, \quad H=Q^2= \begin{pmatrix} H_1 & 0 \\ 0 & H_2 \end{pmatrix}. \tag{4.2}$$

Assuming $(H_1-z_o)^{-1}-(H_2-z_o)^{-1}$, $z_o \in \mathbb{C} \setminus [0,\infty)$ to be a trace class operator, Witten's regularized index $\Delta(z)$ is defined by [18]

$$\Delta(z)=-z\mathrm{Tr}[(H_1-z)^{-1}-(H_2-z)^{-1}], \quad z \in \mathbb{C}\setminus[0,\infty) \tag{4.3}$$

and Witten's index Δ is then given by [40]

$$\Delta = \lim_{\substack{z\to 0 \\ |\mathrm{Re}z| \le C_o |\mathrm{Im}z|}} \Delta(z) \tag{4.4}$$

(for some $C_o>0$) whenever the limit exists. The relevance of these definitions becomes clear in the case where A is a Fredholm operator i.e., where A is a closed operator with a closed range such that dimKer(A) and dimKer(A^*) are finite (here Ker(A) abbreviates the null space of A i.e. Ker(A)= $\{f \in \mathcal{H} | Af=0\}$ and "dim" denotes the dimension). In that case one can prove that

$$i(A) = \Delta \tag{4.5}$$

where i(A) denotes the Fredholmindex of A i.e.

$$i(A) = \mathrm{dimKer}(A) - \mathrm{dimKer}(A^*). \tag{4.6}$$

The result (4.5) easily follows after analyzing the Laurent expansions of $(H_j-z)^{-1}$ near z=0 [17,23]. Since actually dimKer(A)=dimKer(A^*A) we get in the case where A is Fredholm that

$$i(A) = \mathrm{dimKer}(H_1) - \mathrm{dimKer}(H_2) \tag{4.7}$$

and hence i(A) measures precisely the difference of zero-energy states of H_1 and H_2. In this context $\Delta(z)$ represents a regularized index of A. What happens if A is no longer a Fredholm operator? If we simply extend the definition (4.7) of i(A) to this case (whenever dimKer(A), dimKer(A^*) are finite) equality (4.5) does not continue to hold in general. In fact then Δ can take on every real number [15,17] as shown in the next section. Before we analyze this case in more detail, we first need some additional preparatory material. First of all we recall that A is Fredholm iff $\inf \sigma_{ess}(A^*A)>0$ (σ_{ess}-the essential spectrum). We also note that $H_1=A^*A$ and $H_2=AA^*$ are essentially isospectral [20] i.e. that

$$\sigma(H_1)\setminus\{0\} = \sigma(H_2)\setminus\{0\} \tag{4.8}$$

and

$$\begin{aligned} &H_1f=Ef,\ E\neq 0 \text{ implies } H_2(Af)=E(Af),\ f\in \mathrm{Dom}(H_1),\\ &H_2g=E'g,\ E'\neq 0 \text{ implies } H_1(A^*g)=E'(A^*g),\ g\in \mathrm{Dom}(H_2) \end{aligned} \tag{4.9}$$

with multiplicities preserved. Moreover our trace class hypothesis assures the existence of a real-valued, measurable function ξ_{12} on $\mathbb{R}$ (Krein's spectral shift function [6,31]) such that

$$(1+\lambda^2)^{-1}\xi_{12}\in L^2(\mathbb{R};d\lambda),\ \xi_{12}(\lambda)=0,\quad \lambda<0 \text{ and}$$

$$\mathrm{Tr}\left[(H_1-z)^{-1}-(H_2-z)^{-1}\right]=-\int_0^\infty d\lambda\,\xi_{12}(\lambda)(\lambda-z)^{-2},\ z\in \sigma(H_1)\cap\sigma(H_2). \tag{4.10}$$

In addition if $S_{12}(k)$, $k=\lambda^{1/2}$ denotes the on-shell scattering operator for the pair (H_1,H_2) then [6,31]

$$\det S_{12}(k)=\exp\left[-2\pi i\xi_{12}(k^2)\right] \text{ for a.e. } k^2\in\sigma_{ac}(H_2) \tag{4.11}$$

(σ_{ac} the absolutely continuous spectrum). Combining Eqs.(4.3), (4.4) and (4.10) we infer under the additional assumption that ξ_{12} is bounded on $\mathbb{R}$ and piecewise continuous in $(-\varepsilon,\varepsilon)$ for some $\varepsilon>0$ that [15, 17,22,23]

$$\Delta = -\xi_{12}(0_+). \tag{4.12}$$

Similarly, defining the axial anomaly A by [18]

$$A = -\lim_{\substack{z\to\infty\\ |\mathrm{Re}z|\leq C_1|\mathrm{Im}z|}} \Delta(z) \tag{4.13}$$

(C_1 a positive constant) we obtain [15,17,22,23]

$$A = \xi_{12}(\infty). \tag{4.14}$$

Here we needed in addition that ξ_{12} is bounded on $\mathbb{R}$ and that $\lim_{\lambda\to\infty} \xi_{12}(\lambda)=\xi_{12}(\infty)$ exists. We emphasize that the results (4.12) and (4.14) are completely independent of the fact whether A is Fredholm or not!

Finally we turn to invariance results against small perturbations of the quantities defined above. Let B be another closed, densely defined operator in $\mathcal{H}$ which is relatively compact w.r. to A and define on Dom(A) (the domain of A)

$$A_\beta = A + \beta B, \quad \beta \in \mathbb{R}. \tag{4.15}$$

If A is Fredholm, then the invariance of the Fredholm index, i.e.

$$i(A + \beta B) = i(A), \qquad \beta \in \mathbb{R} \tag{4.16}$$

is a well known fact [27] due to the relative compactness assumption. Here we shall mention a result which drops the assumption that A is Fredholm but assumes much stronger hypothesis on B than just relative compactness: Denote by $H_{j,\beta}$, $j=1,2$, $\xi_{12,\beta}$, $\Delta(\beta,z)$, $\Delta(\beta)$, $A(\beta)$ the quantities which result after replacing throughout A by A_β. Moreover suppose that B satisfies a relative trace class condition w.r. to $|A|$ (cf. [17,23] for more details) then

$$\Delta(\beta,z) = \Delta(z), \quad z \in \mathbb{C}\setminus[0,\infty), \qquad \beta \in \mathbb{R}. \tag{4.17}$$

The strategy to prove Eq.(4.17) simply consists in showing $(\partial/\partial\beta)F(\beta,z)=0$

$$F(\beta,z) \equiv \mathrm{Tr}\left[(H_{1,\beta}-z)^{-1}-(H_{2,\beta}-z)^{-1}\right], \quad z \in \mathbb{C}\setminus[0,\infty) \tag{4.18}$$

using commutation formulas of the type [20]

$$(A_\beta^* A_\beta - z)^{-1} A_\beta^* \subseteq A_\beta^*(A_\beta A_\beta^* - z)^{-1} \quad \text{etc.} \tag{4.19}$$

Hence Eq.(4.17) also yields the invariance of $\Delta(\beta)$ and $A(\beta)$. In particular Eq.(4.17) (under appropriate assumptions on the relative interaction H_1-H_2) implies the invariance of Krein's spectral shift function itself

$$\xi_{12,\beta}(\lambda)=\xi_{12}(\lambda), \quad \lambda \in \sigma_{ac}(H_2), \ \beta \in \mathbb{R} \tag{4.20}$$

assuming $\xi_{12,\beta}$ to be bounded and piecewise continuous on $\mathbb{R}$. The above supersymmetric formalism allows an interesting generalization if one replaces Q, H in Eq. (4.2) by

$$Q_m = \begin{pmatrix} m & A^* \\ A & -m \end{pmatrix}, \quad H_m = Q_m^2 = \begin{pmatrix} H_1 + m^2 & 0 \\ 0 & H_2 + m^2 \end{pmatrix}, \quad m \in \mathbb{R}\setminus\{0\}. \tag{4.21}$$

In that case one can define the spectral asymmetry η_m [5,21]

$$\eta_m = \lim_{t\to 0_+} \eta_m(t), \tag{4.22}$$

$$\eta_m(t) = \mathrm{Tr}\left[Q_m H_m^{-1/2} e^{-tH_m}\right], \quad m \in \mathbb{R}\setminus\{0\}. \tag{4.23}$$

A simple generalization of Eq. (4.10) then yields [15,17,23]

$$\eta_m(t) = m\int_0^\infty d\lambda\, \xi_{12}(\lambda)(d/d\lambda)\left[(\lambda+m^2)^{-1/2} e^{-t(\lambda+m^2)}\right] \tag{4.24}$$

implying

$$\eta_m = -(m/2)\int_0^\infty d\lambda\, \xi_{12}(\lambda)(\lambda+m^2)^{-3/2}, \quad m \in \mathbb{R}\setminus\{0\}. \tag{4.25}$$

The quantity η_m plays an important role in connection with fractional charge quantum numbers in external quantum field problems as extensively discussed in [36]. Clearly the invariance (4.20) of ξ_{12} implies that of $\eta_m(t), \eta_m$.

4.2. A supersymmetric magnetic field system in two dimensions

Let $\mathcal{H} = L^2(\mathbb{R}^2)$ and define

$$A = \overline{\left[(-i\partial_1 - a_1) + i(i\partial_2 + a_2)\right] \Big|_{C_0^\infty(\mathbb{R}^2)}} \tag{4.26}$$

where the bar denotes the closure operation and where

$$\underline{a} = (\partial_2\phi, -\partial_1\phi), \quad \partial_j \equiv \partial/\partial x_j, \quad j = 1,2 \tag{4.27}$$

and ϕ satisfies the requirements

$\phi \in C^2(\mathbb{R}^2)$ is real-valued,

$$\phi(\underline{x}) \underset{|\underline{x}|\to\infty}{=} -F\ln(|\underline{x}|) + C + O(|\underline{x}|^{-\varepsilon}), \tag{4.28}$$

$$(\nabla\phi)(\underline{x}) \underset{|\underline{x}|\to\infty}{=} -F|\underline{x}|^{-2}\underline{x} + O(|\underline{x}|^{-1-\varepsilon}), \quad C,F \in \mathbb{R}, \varepsilon>0,$$

$(\Delta\phi)^{1+\delta}, (1+|\underline{x}|^{\delta})(\Delta\phi) \in L^1(\mathbb{R}^2)$ for some $\delta>0$.

Then

$$H_j = \left[(-i\nabla - \underline{a})^2 - (-1)^j b\right] \Big|_{H^{2,2}(\mathbb{R}^2)}, \quad j=1,2 \tag{4.29}$$

($H^{2,2}(\mathbb{R}^2)$ the standard Sobolev space) where the magnetic field b is given by

$$b(\underline{x}) = (\partial_1 a_2 - \partial_2 a_1)(\underline{x}) = -(\Delta\phi)(\underline{x}) \tag{4.30}$$

and

$$F = (2\pi)^{-1} \int_{\mathbb{R}^2} d^2x \; b(\underline{x}) \tag{4.31}$$

represents the magnetic flux. Since $\sigma_{ess}(H_j) = [0,\infty)$, A is never a Fredholm operator in this example. We have the following results [15,17,23]:

$$\Delta(z) = \Delta = -F, \quad z \in \mathbb{C}\backslash[0,\infty), \tag{4.32}$$

$$\mathcal{A} = F, \tag{4.33}$$

$$\xi_{12}(\lambda) = F\theta(\lambda), \; \lambda \in \mathbb{R}. \tag{4.34}$$

(Here $\theta(x)=1$, $x \geq 0$, $\theta(x)=0$, $x<0$). Moreover one can prove [2] that

$$i(A)\mathrm{sgn}(F) = \theta(-F)\dim\mathrm{Ker}(A) - \theta(F)\dim\mathrm{Ker}(A^*)$$

$$= \begin{cases} -N \text{ if } |F| = N+\varepsilon \\ -(N-1) \text{ if } |F| = N \\ 0 \text{ if } F=0, \quad 0<\varepsilon<1, \; N \in \mathbb{N}. \end{cases} \tag{4.35}$$

(Here $\mathrm{sgn}(x) = \pm 1$ for $x \gtrless 0$ and $\mathrm{sgn}(0)=0$). In order to indicate a proof of Eqs.(4.32)-(4.34) we first study a special example (cf.[29]). Let

$$\phi(R,r) = \begin{cases} -(Fr^2/2R^2), & r \leq R \\ -(F/2)[1+\ln(r^2/R^2)], & r \geq R, \; R>0 \end{cases} \tag{4.36}$$

and denote the corresponding Hamiltonians in Eq.(4.29) by $H_j(R)$, $j=1,2$. Next let U_ε, $\varepsilon>0$ be the unitary group of dilations in $L^2(\mathbb{R}^2)$, viz.

$$(U_\varepsilon g)(\underline{x}) = \varepsilon^{-1} g(\underline{x}/\varepsilon), \quad \varepsilon>0, \quad g \in L^2(\mathbb{R}^2). \tag{4.37}$$

Then a straightforward calculation yields

$$U_\varepsilon H_j(R) U_\varepsilon^{-1} = \varepsilon^2 H_j(\varepsilon R), \quad \varepsilon, R>0, \quad j=1,2. \tag{4.38}$$

Denoting by $S_{12}(R)$ the scattering operator in $L^2(\mathbb{R}^2)$ associated with the pair $(H_1(R), H_2(R))$, then $S_{12}(R)$ is decomposable w.r. to the spectral representation of $H_2(R)P_{ac}(H_2(R))$ (P_{ac} the projection onto the absolutely continuous spectral subspace). Let $S_{12}(k,R)$ in $L^2(S^1)$ denote the fibers of $S_{12}(R)$ (i.e. the on-shell scattering operator associated with $(H_1(R), H_2(R)))$. Then Eq. (4.38) implies

$$S_{12}(k,R) = S_{12}(\varepsilon k,R/\varepsilon), \tag{4.39}$$

$$\xi_{12}(k^2,R) = \xi_{12}(\varepsilon^2 k^2,R/\varepsilon), \quad \varepsilon,k>0. \tag{4.40}$$

Applying now the invariance result (4.22), we infer that $\xi_{12}(\lambda)$ cannot depend on R as long as the flux F is kept fixed in Eq.(4.36). Thus Eq.(4.40) implies $\xi_{12}(\lambda)=\xi_{12}(\varepsilon^2\lambda)$, $\varepsilon,\lambda>0$ which in turn implies that ξ_{12} is in fact energy independent. Thus $\Delta(z)$ is independent of z and hence we can calculate its constant value in the high-energy limit $z\to\infty$. Iterating the resolvent equation for H_1 and using certain Bessel function estimates one then proves that [17,23]

$$\Delta(z) \underset{|z|\to\infty}{=} z[\mathrm{Tr}\ (H_2-z)^{-1}2b(H_2-z)^{-1}]+o(1)$$

$$\underset{|z|\to\infty}{=} z\ \mathrm{Tr}[R_o(\sqrt{z})2bR_o(\sqrt{z})]+o(1). \tag{4.41}$$

But

$$z\ \mathrm{Tr}[R_o(\sqrt{z})2bR_o(\sqrt{z})]=-(2\pi)^{-1}\int_{\mathbb{R}^2} d^2x\, b(\underline{x})=-F \tag{4.42}$$

then yields Eq. (4.32). Eqs. (4.33) and (4.34) now trivially follow.

We remark that the result (4.32) has been shown in [28] by using certain approximations in a path integral approach. The above treatment of [17,23] seems to be the first rigorous and nonperturbative one in the case where the flux F is arbitrary (i.e. non quantized).

Finally we mention that the generalized situation described at the end of Sect. 4.1 can also be analyzed in detail. E.g. for the regularized spectral asymmetry $\eta_m(t)$ Eqs. (4.25) and (4.34) immediately yield

$$\eta_m(t)=\mathrm{sgn}(m)Fe^{-tm^2}, \quad m\in\mathbb{R}\backslash\{0\}, \quad t>0 \tag{4.43}$$

implying the known result for the spectral asymmetry as $t\to 0_+$ [36]

$$\eta_m = \mathrm{sgn}(m)F, \quad m\in\mathbb{R}\backslash\{0\}. \tag{4.44}$$

Acknowledgements

I am particularly indebted to D. Bollé, C. Danneels, H. Grosse, W. Schweiger and B. Simon for all joint collaborations which led to the results presented above.

It is a great pleasure to thank L.S.Ferreira, A.C.Fonseca and L.Streit for their kind invitation to a most stimulating conference.

References

1. M.Abramowitz, I.A.Stegun, "Handbook of Mathematical Functions", Dover, New York, 1972.
2. Y.Aharonov, A.Casher, Phys.Rev. A19 (1979), 2461.
3. S.Albeverio, F.Gesztesy, R.Høegh-Krohn, Ann.Inst.H.Poincaré A37 (1982), 1.
4. S.Albeverio, D.Bollé, F.Gesztesy, R.Høegh-Krohn, L.Streit, Ann. Phys. 148 (1983), 308.
5. M.Atiyah, V.Patodi, I.Singer, Proc.Cambridge Phil.Soc. 77 (1975), 42; 78 (1975), 405; 79 (1976), 71.
6. M.S.Birman, M.G.Krein, Sov.Math.Dokl. 3 (1962), 740.
7. D.Bollé, Sum rules in scattering theory and applications to statistical mechanics, in "Mathematics + Physics Vol.2", L.Streit (ed.), World Scientific, 1986.
8. D.Bollé, F.Gesztesy, Phys.Rev.Lett. 52 (1984), 1469.
9. D.Bollé, F.Gesztesy, Phys.Rev.A30 (1984), 1279 and Phys.Rev.A33 (1986), 3517.
10. D.Bollé, S.F.J.Wilk, J.Math.Phys. 24 (1983), 1555.
11. D.Bollé, F.Gesztesy, C.Danneels, in preparation.
12. D.Bollé, F.Gesztesy, M.Klaus, J.Math.Anal.Appl. (in print).
13. D.Bollé, F.Gesztesy, S.F.J.Wilk, J.Operator Theory 13 (1985), 3.
14. D.Bollé, F.Gesztesy, C.Danneels, S.F.J.Wilk, Phys.Rev.Lett. 56 (1986), 900.
15. D.Bollé, F.Gesztesy, H.Grosse, B.Simon, Krein's spectral shift function and Fredholm determinants as efficient methods to study supersymmetric quantum mechanics, Lett. Math. Phys. (in print).
16. D.Bollé, F.Gesztesy, C.Nessmann, L.Streit, Rep. Math. Phys. (in print).
17. D.Bollé, F.Gesztesy, H.Grosse, W.Schweiger, B.Simon, Witten index, axial anomaly and Krein's spectral shift function in supersymmetric quantum mechanics, preprint, 1986.
18. C.Callias, Commun. Math. Phys. 62 (1978), 213.
19. M.Cheney, J.Math.Phys. 25 (1984), 1449.
20. P.A.Deift, Duke Math.J. 45 (1978), 267.
21. T.Eguchi, P.B.Gilkey, A.J.Hanson, Phys. Rep. 66 (1980), 213.
22. F.Gesztesy, Scattering theory for one-dimensional systems with non-trivial spatial asymptotics, in "Recent developments in the theory of Schrödinger operators", Springer Lecture Notes in Math., ed. by E. Balslev.

23. F.Gesztesy, B.Simon, in preparation.
24. F.Gesztesy, G.Karner, L.Streit, J. Math. Phys. 27 (1986), 249.
25. W.G.Gibson, Phys.Lett. 117A (1986), 107.
26. A.Jensen, T.Kato, Duke Math. J. 46 (1979), 583.
27. T.Kato, "Perturbation Theory for Linear Operators", Springer, Berlin, 1980.
28. A.Kihlberg, P.Salomonson, B.S.Skagerstam, Z.Phys. C28 (1985), 203.
29. J.Kiskis, Phys.Rev.D15 (1977), 2329.
30. M.Klaus, B.Simon, Ann. Phys. 130 (1980), 251.
31. M.G.Krein, Sov.Math.Dokl. 3 (1962), 707.
32. M.Murata, J.Func. Anal. 49 (1982), 10.
33. R.G.Newton, J.Math.Phys. 1 (1960), 319.
34. R.G.Newton, J.Math.Phys. 18 (1977), 1348.
35. R.G.Newton, Low-energy scattering for medium-range potentials, preprint, 1986.
36. A.J.Niemi, G.W.Semenoff, Phys.Rep. 135 (1986), 100.
37. T.Osborn, K.B.Sinha, D.Bollé, C.Danneels, J.Math.Phys. 26 (1985), 2796.
38. B.Simon, Ann.Phys. 97 (1976), 279.
39. M.Stone, Ann. Phys. 155 (1984), 56.
40. E.Witten, Nucl.Phys. B202 (1982), 253.

Dilation Analytic Methods

HEINZ K. H. SIEDENTOP

Institut für Mathematische Physik
Technische Universität Carolo-Wilhelmina
Mendelssohnstraße 3
3300 Braunschweig
Germany

ABSTRACT

Complex scaling and some of its variants are reviewed. Bounds on resonances (energy and lifetimes) are derived by combining the complex scaling methods with a variational principle for the multiplicity of eigenvalues and a generalization of Rouché's theorem for meromorphic functions with values in some trace ideal. - The method is illustrated with a particular simple example, a particle in a well.

1. Introduction

The method of dilation analyticity was developed by Aguilar and Combes [1] and Balslev and Combes [2] for defining and localizing resonances of one particle and multi particle hamiltonians. Independently van Winter [3, 4] developed the subject.

Many theoretical interesting results concerning e. g. the absence of singular continuous spectrum, the absence of positive bound states, where derived with these methods. A review of these results can be found in Reed and Simon [5]. Since the dilation methods require certain analyticity properties, the most simple quantum mechanical systems like a particle in a box cannot be treated. Simon [6] circumvented this difficulty by introducing the exterior complex dilation where the co-ordinates are left unchanged within a sphere. Avron and Herbst expanded the method in such a way that also particles in constant electric and magnetic fields became treatable. (For a review see Herbst [7].) Graffi and Yajima [8] treated time dependendent hamiltonians with exterior complex scaling. Combes et al. [9] revisited the exterior complex scaling correcting a technical point in [6, 8] concerning the analyticity of the corresponding analytic families of dilated hamiltonians. Hislop and Sigal [10] introduced the "soft" exterior scaling where the region of unscaled co-ordinates does not jump on a sphere to the region of scaled co-ordinates. Balslev [11], Cycon [12], Hunziker [13] and Sigal [14] introduced other variants of complex scaling.

The method of complex dilations has been applied extensively to calculate resonances. For reviews we refer the reader to Reinhardt [15], Junker [16], and Ho [17]. However, since it leads to non normal operators, and the resonances are described by the eigenvalues of these non normal operators, the question arises, how to bound these eigenvalues rigorously. Some methods to obtain bounds applying for self-adjoint operators (e. g. Weinstein's bounds (Yosida [18]) have generalizations to normal operators but do not generalize to the case of non normal operators. This problem was addressed by Moiseyev [19] who excluded resonances in circles of the complex plane, by Engdahl and Brändas [20] using a Cauchy integral method showing the existence of resonances in certain regions which are surrounded by a set of circles, and by the present author (Siedentop [21–24]) using a variational principle for the multiplicity of non normal operators and a generalization of Rouché's theorem for operators.

2. Complex Scaling

According to Weisskopf and Wigner [25] a resonance is a pole of the scattering amplitude continued across the essential spectrum of a given hamiltonian H. The modern understanding of resonances as poles of the S-matrix (see e. g. Taylor [26]) developed from this starting point. Schwinger [27] and Lovelace [28] modified the point of view and defined resonances to be poles of the analytically continued resolvent matrix elements. Later, Hagedorn [29, 30] and others showed the equivalence of both definitions under certain hypotheses. We shall forget about these assumptions and pretend the equivalence of the definitions:

Definition 1: *Let $H = H_0 + V$ be a hamiltonian in a Hilbert space $\mathfrak{H}$ and D a dense set of vectors in $\mathfrak{H}$, such that for $\psi \in D$*

$$f_\psi(z) = (\psi, (z - H)^{-1}\psi)$$

has a meromorphic coninuation across $\sigma_{ess}(H)$ from above and $(\psi, (z - H_0)^{-1}\psi)$ has an analytic continuation across $\sigma_{ess}(H)$. Then the poles of f_ψ are called resonances.

The problem is to find the poles of f_ψ. One way to circumvent the continuation problem is to find an operator H' who has its eigenvalues exactly at the location of the poles of f_ψ. Helffer and Sjöstrand [31] introduced such operators in their quasi-classical treatment of resonances. - Complex dilation is another way for reaching the same goal. In the following we describe the dilation method and some of its variants. We proceed in four steps: Firstly we construct a unitary group by co-ordinate transformation, secondly we unitarily transform the hamiltonians and assume a complex continuation in the group parameter exists, thirdly we analyze the spectrum of the dilated hamiltonian, and finally we identify an operator H' having the desired property. We first treat the one particle case.

<u>1. Step:</u> Let $F_\vartheta : \mathbb{R}^d \to \mathbb{R}^d$ be a flow. Then

$$[u_F(\vartheta)\varphi](x) = \sqrt{\det F'_\vartheta(x)}(\varphi \circ F_\vartheta)(x)$$

generates a unitary group on $L^2(\mathbb{R}^d)$ which is easily shown by change of variables.

Example 1: *Let ϑ be a real parameter.*

i) Dilations: Choose $F(\vartheta)$ as

$$F(\vartheta)(x) = e^{\vartheta} x.$$

ii) Boosts: Let $b \in \mathbb{R}^d$ and choose $F(\vartheta)$ as

$$F(\vartheta)(x) = b\vartheta + x.$$

iii) Exterior dilations: Inside a sphere of radius R the co-ordinates are left unchanged, outside they are dilated. Choose $F(\vartheta)$ as

$$F(\vartheta)(x) = \begin{cases} x & |x| \leq R \\ [R + e^{\vartheta}(|x| - R)]\frac{x}{|x|} & |x| > R. \end{cases}$$

2. Step: With the unitary group $u_F(\vartheta)$ we may unitarily transform the free hamiltonian, the potential, and the hamiltonian. Denote the corresponding operators by $H_0(\vartheta)$, $V(\vartheta)$, and $H(\vartheta)$. Suppose that unitary transformation leaves the quadratic form domain of these operators invariant. In the following we shall make some assumption about the continuation of the operators to complex ϑ. Assume:

i) $H_0(\vartheta)$ has a known analytic continuation into some domain $D \subset \mathbb{C}$ as a bounded operator valued function $H_0(\vartheta) : S_+ \to S_-$ with suitable scale spaces S_+ and S_-.

ii) $V : S_+ \to S_-$ compact.

iii) $V(\vartheta)$ has an analytic continuation as bounded operator from $S_+ \to S_-$.

Then the potential V is called analytic in the domain D with respect to F.

Example 2: *Consider the dilations of the previous example. In this case the scale space S_+ and S_- are the Sobolev space $H_1(\mathbb{R}^d)$ and $H_{-1}(\mathbb{R}^d)$. The dilated free hamiltonian H_0 is $H_0(\vartheta) = -e^{-2\vartheta}\Delta$, and if the potential is local, it is the multiplication operator $V(e^{\vartheta}x)$. If the domain D is $D = \{q \in \mathbb{C} \,||\, \operatorname{Im}\vartheta| < \alpha\}$, then the potential V is said to belong to $\mathcal{F}_\alpha$. If the continuation (iii) extends even to the boundary of this domain, V is said to be in $\overline{\mathcal{F}_\alpha}$. The union of all $\mathcal{F}_\alpha$ is called the set of all dilation analytic potentials. - This set is not empty. The Coulomb potential belongs to $\mathcal{F}_\infty$, the Yukawa potential belongs to $\overline{\mathcal{F}_{\frac{\pi}{2}}}$.*

3. Step: In order to identify the desired operator H' we need to investigate the spectrum of the dilated hamiltonian. The location of the essential spectrum depends on the flow we choose. We thus specialize to one particular example, the dilations. However, if possible, we keep the notation general, so that the generalization to other flows becomes apparent.

i) Essential spectrum:

$$\sigma_{ess}(H(\vartheta)) = e^{-2\operatorname{Im}\vartheta}[0, \infty). \tag{1}$$

The proof of this formula uses the relative form compactness of $V(\vartheta)$ which follows from the relative H_0 form compactness of V: Since $\sigma_{ess}(H_0) = \sigma(H_0) = [0, \infty)$, we obtain $\sigma_{ess}(H_0(\vartheta)) = e^{-2\vartheta}[0, \infty) = e^{-2\operatorname{Im}\vartheta}[0, \infty)$. Applying Weyl's theorem gives formula (1).

ii) Independence of the spectrum of $H(\vartheta)$ of $\operatorname{Re}\vartheta$: $\sigma(H(\vartheta)) = \sigma(H(\vartheta+\varphi))$, $\varphi \in \mathbb{R}$.

Since $u_F(\vartheta+\varphi) = u_F(\vartheta)u_F(\varphi)$ for real ϑ and φ, we have the relation $u_F(\varphi)H(\vartheta)u_F(\varphi)^{-1} = H(\vartheta+\varphi)$. Thus $H(\vartheta)$ and $H(\vartheta+\varphi)$ are unitarily equivalent, even if ϑ is in D.

iii) Eigenvalues of $H(\vartheta)$ are independent of ϑ, if ϑ is away from the essential spectrum: Let $0 < \operatorname{Im}\vartheta' < \operatorname{Im}\vartheta < \frac{\pi}{2}$. Then $\sigma_d(H(\vartheta')) \subset \sigma_d(H(\vartheta))$ and $\mathbb{R} \cap \sigma_d(H(\vartheta)) = \sigma_{pp}(H) \setminus \{0\}$.

To prove the first statement one observes that the eigenvalues $E_i(\vartheta)$ of $H(\vartheta)$ are analytic functions of ϑ as long as they are away from the essential spectrum of $H(\vartheta)$. Now by the above, $E_i(\vartheta+\varphi) = E_i(\vartheta)$ for real φ. Thus the E_i are constant, at least locally.

For a detailed proof of the above statements we refer the reader to the monograph by Reed and Simon [5]. The following figure illustrates the situation.

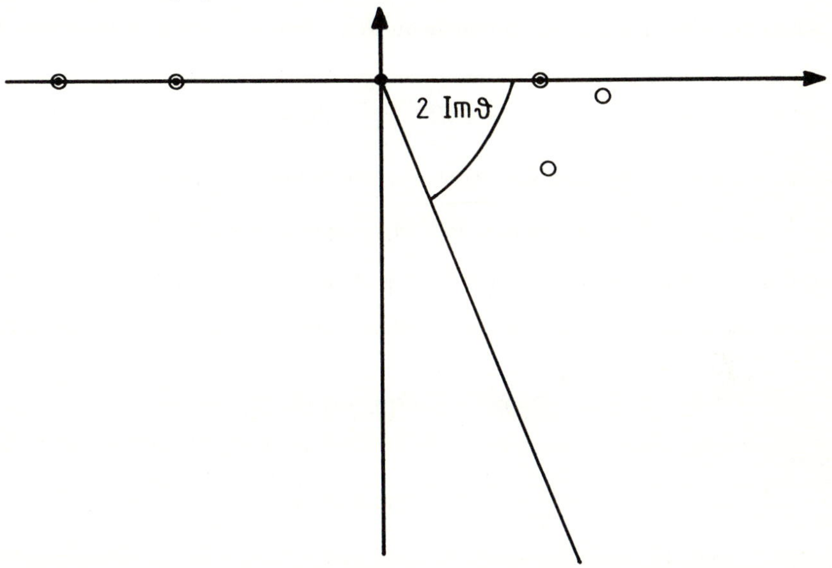

Fig. 1: The dots denote eigenvalues of H, and circles the discrete spectrum of $H(\vartheta)$. The essential spectrum of $H(\vartheta)$ (full line) is rotated by an angle $-2\operatorname{Im}\vartheta$ from the positive real axis.

4. Step: We shall now connect the discrete spectrum of the dilated operator $H(\vartheta)$ with the poles of the resolvent matrix elements. Let $V \in \mathcal{F}_\alpha$ and

$$N_D = \{\psi \in L^2(\mathbb{R}^d) | u_F(\vartheta)\psi \text{ has an analytic continuation from } \mathbb{R} \text{ to } D\}.$$

N_D is dense. The analytic vectors of the generator of $u_F(\vartheta)$ provide such a set. Then define $f(z,\vartheta) = (\psi(\overline{\vartheta}), (H(\vartheta)-z)^{-1}\psi(\vartheta))$. $f(z,\vartheta)$ is analytic in ϑ in the region $\{\vartheta \in \mathbb{C} | -\min\{\alpha, \frac{1}{2}\arg z\} < \operatorname{Im}\vartheta < \min\{\alpha, \frac{\pi}{2}\}\}$. Since for real φ

$$\begin{aligned} f(z,\varphi) =& (u_F(\varphi)\psi, (H(\varphi)-z)^{-1}u_F(\varphi)\psi) = (u_F(\varphi)\psi, u_F(\varphi)(H-z)^{-1}\psi) \\ =& (\psi, (H-z)\psi) = f(z,0), \end{aligned}$$

$f(z,\vartheta)$ provides an analytic continuation of $f(z,0)$ from $\mathbb{C}\setminus\sigma(H(0))$ to $\mathbb{C}\setminus\sigma(H(\vartheta))$. Thus the poles of the matrixelements coincide with the poles of the resolvent of $H(\vartheta)$. We therefore set $H' = H(\vartheta)$. This concludes the dilation analytic methods for one particle Schrödinger operators. In order to treat multiparticle Schrödinger operators we first reduce on the center of mass system, i. e. given the hamiltonian with pair interaction V_{ij}

$$\tilde{H} = \sum_{i=1}^{N} -\frac{1}{m_i}\Delta_i + \sum_{\substack{i,j=1\\ i<j}}^{N} V_{ij}$$

in $L^2(\mathbb{R}^{dN})$ we introduce the scalar product

$$(x,y) = \sum_{i=1}^{N} m_i (x_i, y_i)_i \tag{2}$$

on $\mathbb{R}^{dN}$ where $(.,.)_i$ are given scalar products on $\mathbb{R}^d$. Then the configuration space X is defined as

$$X = \{x \in \mathbb{R}^{dN} | \sum_{i=1}^{N} m_i x_i = 0\}.$$

The orthogonal space X_c describes the center of mass motion. We have

$$X_c = X^{\perp} = \{x \in \mathbb{R}^{dN} | x_1 = x_2 = \dots = x_N = 0\}.$$

As it should be, the dimension of X is $d(N-1)$, and the one of X_c is d.

Corresponding to this orthogonal decomposition of $\mathbb{R}^{dN}$ the Hilbert space splits into a tensor product

$$L^2(\mathbb{R}^{dN}) = L^2(X_c) \otimes L^2(X).$$

Furthermore

$$\tilde{H} = -\Delta_c + H$$

where

$$H = -\Delta + \sum_{\substack{i,j=1\\ i<j}}^{N} V_{ij}, \tag{3}$$

and Δ_c and Δ are the Laplace operators with respect to the metric introduced by (2).

The bound states and poles are defined as the poles of the (meromorphically continued) matrix elements of the operator H.

We now assume the pair interactions V_{ij} to be in $\mathcal{F}_\alpha$ for a given positive α. Because of the rich kinematical structure we have to introduce some more notation. A cluster decomposition $C = \{C_1, \dots, C_k\}$ of $\{1, 2, \dots, N\}$ is a set of non empty disjoint subsets, the clusters, of $\{1, 2, \dots, N\}$ such that $C_1 \cup \dots \cup C_k = \{1, \dots, N\}$.

To each cluster C_l there corresponds a cluster hamiltonian $H_{C_l} = -\Delta + \sum_{\substack{i,j\in C_l\\ i<j}} V_{ij}$ defined on the space $L^2(X^{C_l})$, $X^{C_l} = \{x \in \mathbb{R}^{d\#(C_l)} | \sum_{i\in C_l} m_i x_i = 0\}$, obtained by separating out the

center of mass motion of the Cluster C_l. - Given a cluster decomposition $C = \{C_1, ..., C_k\}$ let $\Sigma_C := \sigma_d(H_{C_1}(\vartheta)) + ... + \sigma_d(H_{C_k}(\vartheta))$, and $\Sigma(\vartheta) = \bigcup_{\#(C)\geq 2} \Sigma_C(\vartheta)$. Finally, assuming $\Sigma(0) \neq \emptyset$, set $\Sigma_{\text{min}} = \min \Sigma(0)$.

With these preliminaries we may formulate a result similar to the one-body case:

Let H have dilation analytic two-body interactions $V_{ij} \in \mathcal{F}_\alpha$. Then:

i) $\sigma(H(\vartheta))$ and $\Sigma(\vartheta)$ depend only on the imaginary part of ϑ.

ii) $\sigma_{ess}(H(\vartheta)) = \Sigma(\vartheta) + e^{-2\vartheta}[0, \infty)$.

iii) For $0 < \operatorname{Im} \vartheta < \min\{\alpha, \frac{\pi}{2}\}$

$$\sigma_d(H(0)), \Sigma(\vartheta) \subset I\!R \cup (\Sigma_{\text{min}} + \{\mu | -2 \operatorname{Im} \vartheta < \arg \mu < 0\})$$

and

$$I\!R \cap \Sigma(\vartheta) = \Sigma(0), \quad I\!R \cap \sigma_d(H(\vartheta)) = \sigma_{pp}(H) \setminus \Sigma(0)$$

Moreover for $0 < \operatorname{Im} \vartheta' < \operatorname{Im} \vartheta < \frac{1}{2}\pi$

$$\Sigma(\vartheta') \subset \Sigma(\vartheta) \text{ and } \sigma_d(H(\vartheta')) \subset \sigma_d(H(\vartheta)).$$

Thus we obtain the following picture.

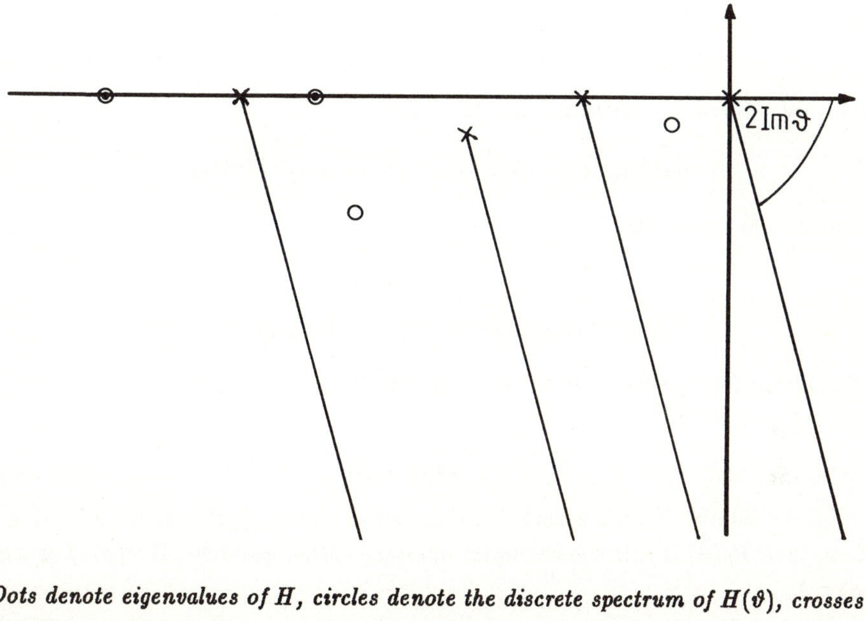

Fig. 2: Dots denote eigenvalues of H, circles denote the discrete spectrum of $H(\vartheta)$, crosses denote thresholds. The essential spectrum starts at the thresholds as halflines rotated by $-2 \operatorname{Im} \vartheta$.

In particular the multisheet structure of the functions $(\psi, (z - H)\psi)$ becomes apparent.

To prove the above statements one proceeds analogously to the one-body case except for localizing the essential spectrum of $H(\vartheta)$. In the one-body case we made use of Weyl's theorem because of the relative form compactness of the potential. In the multiparticle case there are directions in configuration space where the potential does not decay at all. This property is reflected in its non relatively compactness. To circumvent this problem one may use the Weinberg-van Winter machinery proceeding as in the proof of the *HVZ* theorem. For details we refer to Simon [32].

Since we required nothing but $V_{ij} \in \mathcal{F}_\alpha$, examples of potentials which may be treated by the dilation method are Coulomb and Yukawa interactions. It is however important that we restrict ourselves to pair interactions. It is not possible to treat molecules in the static approximation (one-body external Coulomb potential). One may bypass this difficulty by using exterior dilation (Simon [6]) or Hunziker's method (Hunziker [13]).

3. Integral equations for resonance states

For bound states there is a host of integral equations with well behaved kernels. See e. g. the notes of Gesztesy, Glöckle, and Fonseca in this proceedings. Despite the fact that the resonance states are decaying and thus are certainly no bound states the dilation analytic formalism describes them as belonging to the discrete spectrum of certain operators. One may thus expect that in a similar way integral equations for resonance states may be obtained.

In the one particle case we have the following result: Let $d \leq 3$, $\alpha > 0$, $V \in \mathcal{F}_\alpha$ and $0 < \operatorname{Im} \vartheta < \alpha$. Then for $\operatorname{Im} k < 0$ and $0 > \arg k^2 > -2 \operatorname{Im} \vartheta$ the point k^2 is a resonance, if and only if

$$R_k(\vartheta)\varphi = \varphi$$

where $R_k(\vartheta)$, the dilated Rollnik kernel, has the integral kernel

$$[R_k(\vartheta)](x,y) = [|V|^{\frac{1}{2}}(\vartheta)](x) G_0(\vartheta)(x,y) V^{\frac{1}{2}}(\vartheta)(y).$$

Here, assuming local potentials,

$$|V|^{\frac{1}{2}}(\vartheta)(x) = |V(\vartheta)(x)|^{\frac{1}{2}},$$
$$V^{\frac{1}{2}}(\vartheta)(x) = e^{i \arg[V(\vartheta)](x)} |V|^{\frac{1}{2}}(\vartheta)(x).$$

$G_0(\vartheta)$ is the dilated free resolvent. In three dimensions it has the kernel

$$G_0(\vartheta)(x,y) = -\frac{1}{4\pi} \frac{e^{2\vartheta}}{|x-y|} e^{ie^{\vartheta} k |x-y|}.$$

For $V_{ij} \in \mathcal{F}_\alpha$, the dilated Rollnik kernel is a compact operator. Furthermore if $V(\vartheta) \in R$, the Rollnik class, then $R_k(\vartheta)$ is a Hilbert-Schmidt operator. More generally, if $V(\vartheta) \in R + L^p(\mathbb{R}^d)$ then $R_k(\vartheta) \in \mathcal{I}_p$.

The first claim follows on a formal level by algebra. For a proof one proceeds as in the bound state case (see e. g. Simon [33]).

To prove the Hilbert-Schmidt property one estimates as follows

$$\operatorname{tr} |R_k(\vartheta)|^2 = \int d^d x d^d y |[V(\vartheta)](x)[V(\vartheta)](y)|^2 |G_0(\vartheta)(x,y)|^2 \leq c \int d^d x d^d y \frac{[V(\vartheta)](x)[V(\vartheta)](y)}{|x-y|^2} < \infty.$$

If the potential decreases fast enough, so that it cancels a possible increase at infinity of the kernel of the free dilated Greens function $G_0(\vartheta)$ in the expression for $R_k(\vartheta)(x,y)$, we may observe that the undilated Rollnik kernel $R_k(0)$ has already a continuation into the lower half plane. Resonances are then also described by eigenvalues one of these operators (Grossmann and Wu [34,35], Siedentop [24]).

If the potential has discontinuities in a finite region only but is analytic outside we may also use the exterior dilated Rollnik kernel $R_k(R,\vartheta)$ instead of $R_k(\vartheta)$.

In the multi particle case the situation is similar, but due to the kinematical richness, the notation becomes more complicated. One possibility is the use of the dilated symmetrized Weinberg-van Winter kernel $I_s(\vartheta,k^2)$. For $V_{ij}\in\mathcal{F}_\alpha$ and $0<\operatorname{Im}\vartheta<\alpha$ this kernel is a compact operator. The following result holds. If k^2 is a resonance on a part of a sheet of the energy surface which is "uncovered" by the dilation, then $I_s(\vartheta,k^2)$ has eigenvalue one.

Moreover, if $V_{ij}(\vartheta)\in R+L^{2p}(\mathbb{R}^d)$ for some $p\in\mathbb{N}$, then $I_s(\vartheta,k^2)$ is in $\mathcal{I}_{2p}$. (See Simon [32], Siedentop [36])

Contrary to the one particle case the existence of a resonance is only sufficient for the existence of an eigenvalue one of the Weinberg-van Winter kernel. Even for bound states there are counter examples for the reverse implication (Federbush [37], Newton [38]). To side step this difficulty one may use instead the dilated Faddeev equations (Balslev and Skibsted [39]).

We shall not prove the above statements here. We shall however give a definition of the dilated symmetrized Weinberg-van Winter kernel. It is, apart from the symmetrization, which allows for more singular potentials ($r^{-\alpha}$ with $0<\alpha<2$), the sum of the connected graphs of the expansion of the dilated resolvent.

First, we define a string: A string S is a family of cluster decompositions $\{C_N,...,C_k\}$ such that C_{j+1} is obtained from C_j by dividing a cluster of C_j, and such that C_j has exactly j clusters. k is called the index $i(S)$ of S.

By iCj with $i,j\in\{1,...,N\}$ we denote the fact that i,j are indices of a common cluster of C. By $\sim iCj$ we denote the contrary.

Then, given a cluster decomposition C,

$$R_C(\vartheta,E)=(1+[(H_0-E)(\vartheta)]^{-\frac{1}{2}}\sum_{\substack{iCj\\ i<j}}V_{ij}[(H_0-E)(\vartheta)]^{-\frac{1}{2}})^{-1}$$

is the reduced dilated resolvent. Furthermore, given two cluster decompositions C and C', let

$$I_{CC'}=(H_0-E)(\vartheta)^{-\frac{1}{2}}(-\sum_{\substack{iC'j\\ \sim iCj\\ i<j}}V_{ij}(\vartheta))(H_0-E)(\vartheta)^{-\frac{1}{2}}$$

Given a string $S_0=\{C_N,C_{N-1},...,C_k\}$ define

$$R_{S_0}(\vartheta,E)=R_{C_N}(\vartheta,E)I_{C_NC_{N-1}}R_{C_{N-1}}(\vartheta,E)I_{C_{N-1}C_{N-2}}\cdots I_{C_{k+1}C_k}R_{C_k}(E).$$

With these notations we finally may write down the dilated symmetrized Weinberg-van Winter kernel

$$I_S(\vartheta, E) = - \sum_{S=\{C_N,\ldots,C_2\}} R_S(\vartheta, E)[\sum_{\substack{\sim iC_2 j\\ i<j}} (H_0 - E)(\vartheta)^{-\frac{1}{2}} V_{ij}(\vartheta)(H_0 - E)(\vartheta)^{-\frac{1}{2}}].$$

4. Bounds on the energy and lifetime of resonances

In the two preceeding chapters we have reduced the calculation of the resonances to a spectral problem for the dilated Schrödinger operator or an associated integral equation. However, we paid a price. These operators cannot be self-adjoint since they have also non-real spectrum. Even worse, they are in general not even normal operators.

Most methods for bracketing eigenvalues of self-adjoint operators do not apply in this case. There are some methods like Weinstein's bounds (Yosida [18]) which generalize to normal operators. However, general non normal operators are unaccessible by it. - In the following we shall introduce two methods for obtaining upper and lower bounds on real and imaginary parts of eigenvalues of non normal operators. They are in some sense complementary. The first one (Siedentop [40]), a variational principle for the multiplicity of eigenvalues which goes back to an idea of Müller [41], excludes resonances in some subsets of the complex plane. The second one, a non commutative version of Rouché's theorem yields detailed information on the number of eigenvalues (Siedentop [22]).

Lemma 1 (Variational principle for the multiplicity): *Let $\mathfrak{H}$ be a separable complex Hilbert space, $1 \leq p \leq \infty$, and $A \in I_{2p}$. Then for any $B \in I_{2p}$*

$$s(B) := \operatorname{tr} |(A-1)B + A|^{2p} \geq \begin{cases} d & \text{if 1 is an eigenvalue of } A \text{ of multiplicity } d \\ 0 & \text{if 1 is no eigenvalue of } A. \end{cases} \tag{4}$$

Moreover

$$B_{\min} = \begin{cases} A(1-A)|_{(\mathrm{Kern}(1-A))^{\perp}}{}^{-1} & \text{on } (1-A)(\mathfrak{H}) \\ 0 & \text{on } ((1-A)(\mathfrak{H}))^{\perp} \end{cases}$$

yields equality in (4), i. e. $B_{\min}$ is a minimizing element.

Proof: First, we remark that $s(B)$ is well defined because of the ideal property of I_{2p}. For the same reason $B_{\min}$ is in I_{2p}.

Introduce $C = B + 1$ and $F = [(A-1)B + A]^p = [(A-1)C + 1]^p$. Furthermore decompose every $\mathfrak{x} \in \mathfrak{H}$ into a part $\mathfrak{x}_{\|}$ parallel to $\mathrm{Ker}(1 - A^*)$ and its orthogonal part $\mathfrak{x}_{\perp}$. Then

$$\| F\mathfrak{x} \|^2 = (F^*F\mathfrak{x}, \mathfrak{x}) = (F^*[(F\mathfrak{x})_{\|} + (F\mathfrak{x})_{\perp}], \mathfrak{x}) = ((F\mathfrak{x})_{\|} + F^*(F\mathfrak{x})_{\perp}, \mathfrak{x})$$

$$= ((F\mathfrak{x})_{\|}, \mathfrak{x}) + \| (F\mathfrak{x})_{\perp} \|^2 \geq (F\mathfrak{x}, \mathfrak{x}_{\|}) = (\mathfrak{x}, F\mathfrak{x}_{\|}) = \| \mathfrak{x}_{\|} \|^2 .$$

Letting $\mathfrak{x}$ run through an orthonormal basis $\mathfrak{e}_1, \mathfrak{e}_2, \mathfrak{e}_3, \ldots$, summing over the index i, and remarking that $\dim \mathrm{Ker}(A^* - 1) = \dim \mathrm{Ker}(A - 1)$ yields the desired inequality (4).

The second claim follows by inserting $B_{\min}$. We may now combine the results of the previous chapter with the above lemma.∎

Let $W_E(\vartheta)$ be the dilated Rollnik kernel as function of the energy.

Theorem 1: *Suppose $d \leq 3$, $V \in \mathcal{F}_\alpha$, $0 < \operatorname{Im}\vartheta < \alpha$, $p \in \mathbb{N} \cup \{\infty\}$, and $V(\vartheta) \in R + L^{2p}(\mathbb{R}^d)$. Then $E \in \{z | 0 > \arg z > -2\operatorname{Im}\vartheta\}$ cannot be a resonance, if there exists a $B_E \in \mathcal{I}_{2p}$ such that*

$$s_E(B_E) = \operatorname{tr}|(W_E(\vartheta) - 1)B_E + W_E(\vartheta)|^{2p} < 1. \tag{5}$$

Furthermore

$$B_{E,\min} = \begin{cases} W_E(\vartheta)(1 - W_E(\vartheta)|(\mathrm{Kern}(1 - W_E(\vartheta)))^{\perp}) & \text{on } (1 - W_E(\vartheta))(L^2(\mathbb{R}^d)) \\ 0 & \text{on } ((1 - W_E(\vartheta))(L^2(\mathbb{R}^d)))^{\perp} \end{cases} \tag{6}$$

minimizes s_E. In particular $s_E(B_E)$ can be made smaller than one by a particular E-dependent choice of trial operator, if E is not a resonance.

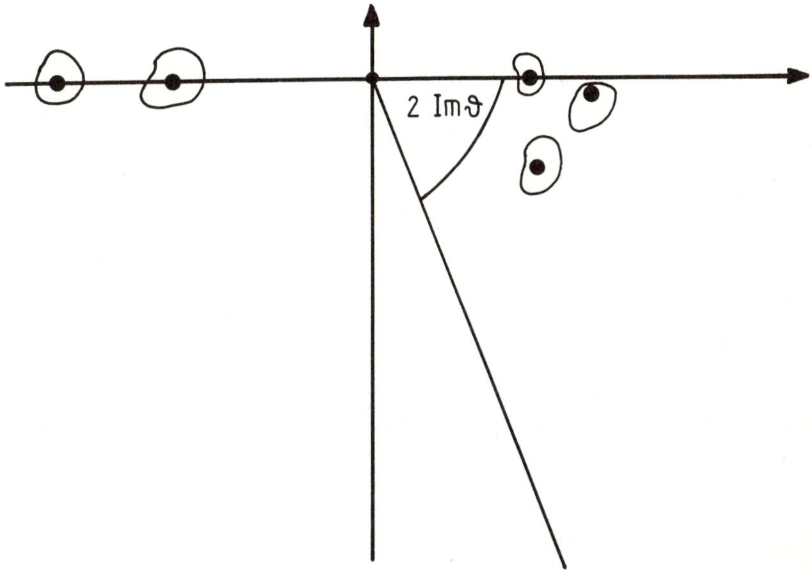

Fig. 3: Contourline $s_E(B_E) = 1$. The contourline encloses resonances and bound states. Bounds on the energy and lifetime of the resonances are obtained by projecting the enclosed sets onto the real and imaginary axis.

In order to evaluate (5) numerically we may choose an approximation to the minimizing operator (6). One possibility is to set

$$B = \sum_{\mu,\nu=1}^{n} \beta_{\mu,\nu} |\varphi_\mu >< \varphi_\nu|$$

where $\varphi_1, ..., \varphi_n$ are n linear independent functions in $L^2(\mathbb{R}^d)$. Inserting B into the variational principle (5) yields a function of the $\beta_{\mu,\nu}$ which may be minimized. A straight forward calculation yields for the minimizing matrix $\{\beta_{\mu,\nu}\}$

$$\beta = -W_G^{-1} \circ W_K \circ H_0^{-1}$$

where $(W_G)_{\mu,\nu} = (\varphi_\mu, |W_E(\vartheta) - 1|^2 \varphi_\nu)$, $(W_K)_{\mu,\nu} = (\varphi_\mu, (W_E(\vartheta)^* - 1)W_E(\vartheta)\varphi_\nu)$, and $(H_0)_{\mu,\nu} = (\varphi_\mu, \varphi_\nu)$.

We treat the following simple example to demonstrate the method: Let

$$V(x) = \begin{cases} -a & \text{for } |x| \leq \frac{b}{2} \\ 0 & \text{for } |x| > \frac{b}{2} \end{cases}$$

the one dimensional well potential of width b and depth a.

Instead of the dilated Rollnik kernel we may use the undilated Rollnik kernel directly in this case, since the potential is not only decreasing rapidly but is zero outside the interval $[-\frac{b}{2}, \frac{b}{2}]$ (see remark in chapter 3). As trial function we choose $n = 2m$ piecewise constant functions

$$\varphi_\mu(x) = \begin{cases} \sqrt{\frac{n}{2b}} & \text{for } (\mu - 1)\frac{b}{2n} \leq |x| \leq \mu\frac{b}{n} \\ 0 & \text{elsewhere} \end{cases}$$

for $1 \leq \mu \leq \frac{n}{2}$ and $\varphi_\mu = \text{sgn}\,(\varphi_{\mu-\frac{n}{2}})\varphi_{\mu-\frac{n}{2}}$ for $\frac{n}{2} + 1 \leq \mu \leq n$. Carrying through the calculations yields the following picture.

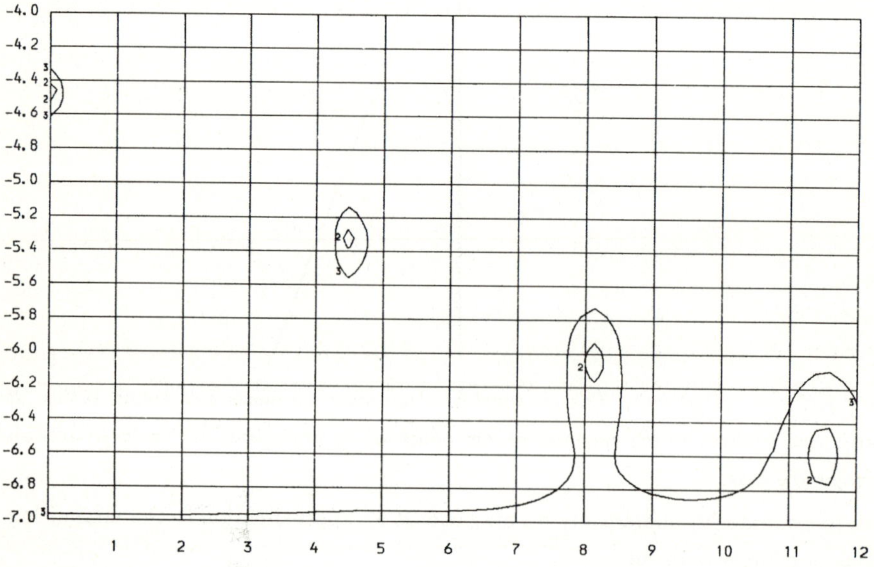

Fig. 4: Upper bound s as function of k. 1 denotes the contourline $s_k = 1$, 2 the contourline $s_k = 0.5$, and 3 the contourline $s_k = 0.1$. The picture is obtained by calculating 60 times 60 grid points and interpolation with $n = 5$,and $a = b = 1$.

Resonances lie in the regions encircled by contourlines. For the resonances on the left hand side - the contourline $s_k = 1$ is no longer visible, since the bounds are so narrow that they fall through mesh points. The following magnification shows this for the first resonance.

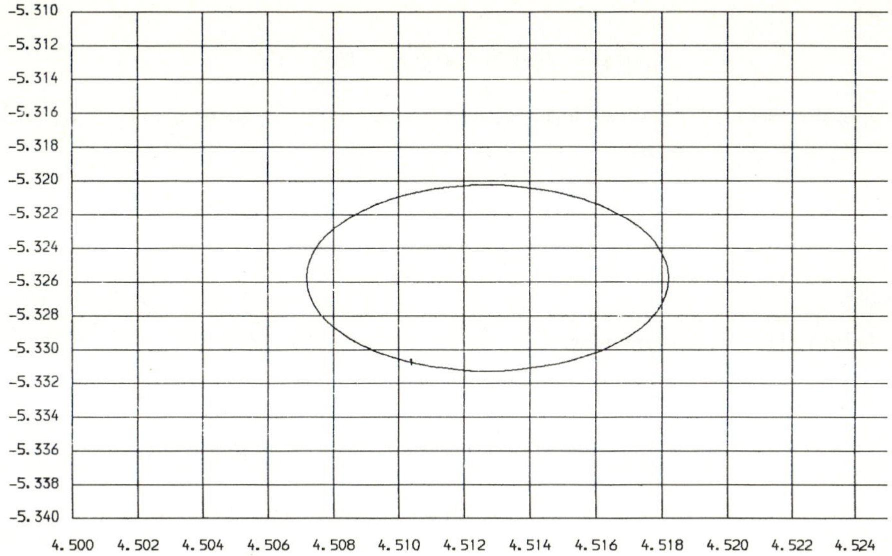

Fig. 5: Magnified part of figure 4.

Lemma 2 (non-commutative Rouché): *Let $\Gamma \subset \mathbb{C}$ be a simply connected domain enclosed by the smooth curve γ ($\partial\Gamma = \gamma$). Assume $1 \leq p \leq \infty$ and $f(z) = 1 + F(z)$, $g(z) = 1 + G(z)$ with $\mathcal{I}_p$-valued functions F and G, meromorphic in Γ and analytic on γ. Let the main part of Laurent series about any $z_0 \in \Gamma$ of F and G have only operators of finite rank as coefficients and assume*

$$\max_{z\in\gamma} \|f(z)g(z) - 1\|_p < 1. \tag{7}$$

Then

$$-\frac{1}{2\pi i}\text{tr} \oint_\gamma f'(z)f(z)^{-1}dz = \frac{1}{2\pi i}\text{tr} \oint_\gamma g'(z)g(z)^{-1}dz \tag{8}$$

Proof: Because of inequality (7) both f and g are invertible on γ. Thus, by Ribaric's and Vidav's theorem [42] $f(z)^{-1}$ and $g(z)^{-1}$ exist everywhere except for a discrete set, where the coefficients of the main part of the Laurent expansion are operators of finite rank. Thus the residues of $f'(z)f(z)^{-1}$ and $g'(z)g(z)^{-1}$ are finite rank operators and thus the right and left hand side of (8) exist.

Next we show that $-\frac{1}{2\pi i}\text{tr} \oint_\gamma h'(z)h(z)^{-1}dz$ is an integer if h has the same analyticity and trace ideal properties as f. To this end we approximate $h(z)$ by a sequence $h_n(z)$ of analytic functions of operators of finite rank which converges uniformly on γ. Then by the argument principle

$$\begin{aligned}&\frac{1}{2\pi i}\oint_\gamma \frac{d}{dz}\det(1 + h_n(z))dz \\ &= \frac{1}{2\pi i}\oint_\gamma \frac{d}{dz}\log e^{\text{tr}\ \log(1+h_n(z))}dz \\ &= \frac{1}{2\pi i}\text{tr} \oint_\gamma h_n'(z)h_n(z)^{-1}dz\end{aligned}$$

is a sequence of integers converging toward $\frac{1}{2\pi i}\text{tr} \oint_\gamma h'(z)h(z)^{-1}dz$ thus being an integer itself.

Now set

$$h_\kappa(z) = f(z)^{-1} + \kappa(g(z) - f(z)^{-1}).$$

h_κ is an operator $h(z)$ of the above form, since for every $A \in I_p$, there is a $B \in I_p$, such that $(1+A)^{-1} = 1 + B$, if $-1 \notin \sigma(A)$.

Thus,

$$\rho(\kappa) = \frac{1}{2\pi i} \oint h'_\kappa(z) h_\kappa(z)^{-1} dz$$

is an integer.

Now, $h_0(z) = f(z)^{-1}$, and therefore

$$\begin{aligned} \operatorname{tr} \oint h'_0(z) h_0(z)^{-1} dz &= \operatorname{tr} \oint (-1) f(z)^{-1} f'(z) f(z)^{-1} f(z) dz = -\lim_{n\to\infty} \operatorname{tr} \oint f_n(z)^{-1} f'_n(z) dz \\ &= -\lim_{n\to\infty} \oint \operatorname{tr} f'_n(z) f_n(z)^{-1} dz = -\operatorname{tr} \oint f'(z) f(z)^{-1} dz, \end{aligned}$$

where f_n converges uniformly to f on γ. Furthermore, $h_1(z) = g(z)$. Thus, for $\kappa = 0$

$$\frac{1}{2\pi i} \operatorname{tr} \oint_\gamma h'_\kappa(z) h_\kappa(z)^{-1} dz$$

yields the left hand side, for $\kappa = 1$ the right hand side of (8).

Finally, we may expand the inverse of $h_\kappa(z)$

$$\begin{aligned} h_\kappa(z)^{-1} &= \sum_{\nu=0}^{\infty} \left\{ f(z)[g(z) - f(z)^{-1}] \right\}^\nu f(z) \kappa^\nu \\ &= \sum_{\nu=0}^{\infty} [f(z) g(z) - 1]^\nu f(z) \kappa^\nu. \end{aligned}$$

By assumption this series converges in the I_p-norm. In particular ρ is continuous which proves the theorem, since $\rho(\kappa)$ is an integer for $0 \le \kappa \le 1$. ∎

We remark:

i) Condition (7) may be interpreted as condition on the approximation of the inverse of $f(z)$ by $g(z)$ on γ. If $g(z) = f(z)^{-1}$, then the left hand side of (7) would vanish.

ii) If we set $F(k) = -R_k(\vartheta)$, $p < \infty$, then the left hand side of (8) is the number of resonances N_Γ in Γ, if $\{k^2 | k \in \Gamma\}$ is in the sector $\{\mu \in \mathbb{C} \,|\, 0 > \arg \mu > -2 \operatorname{Im} \vartheta\}$: (We omit the argument ϑ in the following lines.)

$$\begin{aligned} N_\Gamma &= \frac{1}{2\pi i} \oint_\gamma \frac{d}{dk} \log \det_p (1 - R_k) dk = \frac{1}{2\pi i} \oint_\gamma \frac{d}{dk} \log \det(1 + \mathcal{R}_p(-R_k)) dk \\ &= \frac{1}{2\pi i} \operatorname{tr} \oint_\gamma \frac{d}{dk} \log(1 + \mathcal{R}_p(-R_k)) dk, \end{aligned}$$

where $\mathcal{R}_p(-R_k) = (1 - R_k) \exp[\sum_{j=1}^{p-1} \frac{R_k^j}{j}] - 1$. Introducing this expression yields

$$N_\Gamma = \frac{1}{2\pi i} \operatorname{tr} \oint_\gamma \left\{ [\frac{d}{dk}(1 + \mathcal{R}_p(-R_k))][1 + \mathcal{R}_p(-R_k)]^{-1} \right\} dk$$

$$= \frac{1}{2\pi i} \mathrm{tr} \oint_\gamma \left\{ \frac{d}{dk} [(1 - R_k) \exp[\sum_{j=1}^{p-1} \frac{R_k^j}{j}]] [(1 - R_k) \exp[\sum_{j=1}^{p-1} \frac{R_k^j}{j}]]^{-1} \right\} dk,$$

$$= \frac{1}{2\pi i} \mathrm{tr} \oint_\gamma \left\{ -R_k' \circ \exp[\sum_{j=1}^{p-1} \frac{R_k^j}{j}] \circ (\exp[\sum_{j=1}^{p-1} \frac{R_k^j}{j}])^{-1} (1 - R_k)^{-1} \right.$$

$$\left. +(1 - R_k) \frac{d}{dk} (\exp[\sum_{j=1}^{p-1} \frac{R_k^j}{j}]) \exp[\sum_{j=1}^{p-1} \frac{R_k^j}{j}]^{-1} (1 - R_k)^{-1} \right\} dk$$

$$= \frac{1}{2\pi i} \mathrm{tr} \oint_\gamma R_k' (1 - R_k)^{-1} dk$$

$$+ \frac{1}{2\pi i} \mathrm{tr} \oint_\gamma (1 - R_k) \frac{d}{dk} (\exp[\sum_{j=1}^{p-1} \frac{R_k^j}{j}]) \exp[\sum_{j=1}^{p-1} \frac{R_k^j}{j}]^{-1} (1 - R_k)^{-1} dk$$

$$= \frac{1}{2\pi i} \mathrm{tr} \oint_\gamma R_k' (1 - R_k)^{-1} dk.$$

(For the definition of the normalized Fredholm determinant see Simon [43].)

iii) The function g is arbitrary apart from the hypothesis of the theorem. Thus we may choose it in such a way that the right hand side of (8) becomes calculable.

One possible choice would be to set

$$g(z) = (1 + \sum_{\mu,\nu=1}^{n} \alpha_{\mu,\nu}(z) |\varphi_\nu >< \varphi_\mu|)^{-1}$$

where the coefficients $\alpha_{\mu,\nu}$ are obtained by a Galerkin method, i. e. by projecting $R_k(\vartheta)$ onto the space generated by $\varphi_1, ..., \varphi_n$.

Another choice would be to use the renormalized Fredholm series (see also Simon [43]) to approximate the inverse of $f(z)$.

Both methods yield meromorphic approximations of the inverse.

REFERENCES

[1] J. Aguilar, J. M. Combes: A class of analytic pertubations for one-body Schrödinger operators. Commun. Math. Phys. **22** (1971) 269-279

[2] E. Balslev, J. M. Combes: Spectral properties of many-body Schrödinger operators with dilation analytic interactions. Commun. Math. Phys. **22** (1971) 280-294

[3] C. van Winter: Complex dynamical variables for multiparticle systems with analytic interaction I. J. Math. Anal. Appl. **47** (1974) 633-670

[4] C. van Winter: Complex dynamical variables for multiparticle systems with analytic interaction II. J. Math. Anal. Appl. **48** (1974) 368-399

[5] M. Reed, B. Simon: Methods of modern mathematical physics IV. Analysis of operators. Academic press, New York 1978

[6] B. Simon: The definition of molecular resonance curves by the method of exterior complex scaling. Phys. Lett. **71 A** (1979) 211-214

[7] I. W. Herbst: Schrödinger operators with external homogenous electric and magnetic fields. In: G. Velo, A. S. Wightman (eds.). Rigorous atomic and molecular physics. Plenum Press, New York 1981

[8] V. S.. Graffi, K. Yajima: Exterior complex scaling and the AC-Stark effect in a Coulomb field. Commun. Math. Phys. **89** (1983) 277-301

[9] J. M. Combes, P. Duclos, M. Klein, R. Seiler: The shape resonance. To appear in Anal. Inst. H. Poincaré. Preprint 1986

[10] P. D. Hislop, J. M. Sigal: Shape resonances in quantum mechanics. For the Proceedings of the Int. Conf. on Diff. Equ. and Math. Physics, Birmingham, Alabama 1986. Preprint 1986

[11] E. Balslev: Analytic scattering theory of two-body Schrödinger operators. J. Funct. Analysis **29** , (1978) 375-396

[12] H. L. Cycon: Resonances defined by modified dilations. Helv. Phys. Acta **58** (1985) 969-981

[13] W. Hunziker: Distortion analyticity and molecular resonance curves. Preprint 1986

[14] I. M. Sigal: Complex transformation method and resonances in one-body quantum systems. Ann. Inst. Henri Poincaré. Phys. Théor. **41** (1984) 103-114

[15] W. P. Reinhardt: Complex coordinates in the theory of atomic and molecular structure and dynamics. Ann. Rev. Phys. Chem. **33** (1982) 223-255

[16] B. R. Junker: Recent computational developments in the use of complex scaling in resonance phenomena. Advances in atomic and molecular physics **18** (1982) 207-263

[17] J. K. Ho: The method of complex coordinate rotation and its applications to atomic collision processes. Physics Reports **99** (1983) 1-68

[18] K. Yosida: Functional analysis. 6th edition, Springer-Verlag, Berlin 1980

[19] N. Moiseyev: Resonances by the complex coordinate method with hermitian hamiltonian. Chem. Phys. Lett. **99** (1983) 364

[20] E. Engdahl, E. Brändas: Resonance regions determined by projection operator formulation. Preprint 1986

[21] H. K. H. Siedentop: Bound on resonance eigenvalues of Schrödinger operators. Phys. Rev. Lett. **99A** (1983) 65-68

[22] H. K. H. Siedentop: On the width of resonances. Z. Phys. A **316** (1984) 367-369

[23] H. K. H. Siedentop: On a generalization of Rouché's theorem for trace ideals with applications for resonances of Schrödinger operators. To appear, J. Math. Analysis Applic.

[24] H. K. H. Siedentop: On the localization of resonances. To appear in Int. Journ. Quantum Chemistry

[25] V. Weisskopf, E. P. Wigner: Berechnung der natürlichen Linienbreite auf Grund der Diracschen Lichttheorie, Z. Phys. **63** (1930) 54-73

[26] J. R. Taylor: Scattering Theory: The quantum theory of nonrelativistic collisions. John Wiley & Sons, Inc. New York 1972

[27] J. Schwinger: Field theory of unstable particles, Ann. Phys. **9** (1960) 169-193

[28] C. Lovelace: Scottish Universities' Summer School (R. C. Moorhouse, ed.) , Oliver and Boyd, Edinburgh 1963

[29] G. A. Hagedorn: Asymptotic completness for a class of four particle Schrödinger operators. Bull. Am. Math. Soc. **84** (1978) 155-156

[30] G. A. Hagedorn: A link between scattering resonances and dilation analytic resonances in few body quantum mechanics. Commun. Math. Phys. **65** (1979) 181-188

[31] B. Helffer, J. Sjöstrand: Resonances en limite semiclassique. To appear in Bull. de la Soc. Math. Fran.

[32] B. Simon: Quadratic form techniques and the Balslev-Combes theorem. Commun. Math. Phys. **27**, (1972) 1-9

[33] B. Simon: Quantum mechanics for hamiltonians defined as quadratic forms. Princeton University Press, Princeton 1971

[34] A. Grossmann, T. T. Wu: Schrödinger scattering amplitude. I. Journ. Math. Phys. **2** (1961) 710-713

[35] A. Grossmann, T. T. Wu: Schrödinger scattering amplitude. III. Math. Phys. **3** (1962) 684-689

[36] H. K. H. Siedentop: Localization of discrete spectrum of multiparticle Schrödinger operators. Z. Naturforsch. **40a** (1985) 1052-1058

[37] P. Federbush: Existence of spurious solutions to many body Bethe-Salpeter equations. Phys. Rev. **148** (1966) 1551-1552

[38] R. Newton: Spurious solutions of three particle equations. Phys. Rev. **153** (1967) 1502

[39] E. Balslev, E. Skibsted: Boundedness of two and three-body resonances. Ann. Inst. Henri Poincaré **43** (1985) 369-397

[40] H. K. H. Siedentop: Dimension of eigenspaces of Schrödinger operators - local Birman-Schwinger bound. Rep. Math. Phys. **21** (1985) 383-389

[41] A. M. K. Müller: Variation principle for probability amplitudes. Phys. Lett. **11** (1964) 238-239

[42] M. Ribaric, I. Vidav: Analytic properties of the inverse $A(z)^{-1}$ of an analytic linear operator-valued function $A(z)$. Arch. Rational Mech. Anal. **32** (1969) 298-310

[43] B. Simon: Trace ideals and their applications. London Mathematical Society. Lecture Notes 35. Cambridge University Press. Cambridge 1979

SEMICLASSICAL METHODS IN FEW-BODY SYSTEMS

H.J. Korsch and R. Möhlenkamp
Fachbereich Physik, University of Kaiserslautern
D-6750 Kaiserslautern, Federal Republic of Germany

1. Introduction: Semiclassical Mechanics

In the present article we give a short review of semiclassical methods in few body systems. The long history of these techniques - the first important contribution dates back to 1817 /1/ - and the enormous growth of this field necessarily leads to a very selective and subjective choice of the presented material. The reader is therefore strongly urged to take a look at the excellent existing reviews /2-5/, textbooks /6-9/ and conference reports /10-11/. Here we will concentrate on the semiclassical methods applied to scattering theory. Examples are mainly taken from atomic or molecular dynamics, e.g. the inelastic collision of an atom with a diatomic molecule at collision energies of a few eV. The discussed methods, however, have a much wider range of applications, covering low energy electron-atom collisions, chemical reactions and heavy-ion collisions at several hundred MeV. For the sake of simplicity we use in the following the word 'molecule' to describe bound few-body systems.

Basically there are two different types of approximations which are called 'semiclassical': The first - which we prefer to call 'quasiclassical' - is a combination of purely classical treatments of some degrees of freedom with a quantum treatment of the remaining ones, e.g. time dependent classical trajectory approximations (for a short review see /4,12/). Here we only discuss the second category of approximations, which construct the rigorous limit of quantum mechanics for $\hbar\to 0$. We confine ourselves to the so-called first-order semiclassical approximation-for higher order methods see /13/.

The main advantage of the semiclassical approximations is the fact that they describe directly the 'physics' of the process under investigation, which is more or less hidden in the quantum description. The semiclassical methods are therefore very efficient interpretative tools in order to 'understand' a phenomenon. This is directly achieved by relating the quantum wavefunction to an underlying family of classical trajectories. The price we have to pay is the complexity of the pattern of classical paths, their not easily recognizable topology and - possibly most important - the fact that we have approximated a linear theory (quantum mechanics) by a nonlinear one (classical mechanics), with all types of dynamical consequences of nonlinearity. In numerous applications the semiclassical method has proved to be invaluably helpful in analysing quantum processes - examples will be given below - here we only want to mention the semiclassical theory of inversion of experimental data, e.g. the construction of interaction potentials from spectroscopic or scattering data /14/.

The most widely known semiclassical result is the WKB-wavefunction for a particle moving in a potential V(R)

$$\psi(R) = \begin{cases} \dfrac{C}{2\sqrt{|P(R)|}}\, e^{\frac{1}{\hbar}\int_{R_0}^{R} |P(R')|\, dR'} & R < R_0 \\ \dfrac{C}{\sqrt{P(R)}} \sin\left[\dfrac{1}{\hbar}\displaystyle\int_{R_0}^{R} P(R')\, dR' + \dfrac{\pi}{4}\right], & R > R_0 \end{cases} \tag{1}$$

where $P(R) = (2m(E-V(R)))^{1/2}$ is the classical momentum and R_0 is the classical turning point, i.e. $P(R_0) = 0$. The simple WKB-wavefunction (1) can be used to accentuate several typical features of semiclassical approximations: R_0 separates the classical accessible region, which is assumed to be $R > R_0$, from the classically forbidden region. Classically the particle is reflected at R_0. In the region $R > R_0$ each point can be reacted by two trajectories, a direct ($\nu = 1$) and a reflected ($\nu = 2$) one. Each trajectory ν produces a contribution $A_\nu \exp(iS_\nu/\hbar)$ to the wavefunction:

$$\psi(R) = C' \sum_\nu A_\nu^{1/2}\, e^{\frac{i}{\hbar}(S_\nu + \delta_\nu)}, \tag{2}$$

where

$$S_\nu = \int_{\tilde{R}}^{R} P(R)\, dR \tag{3}$$

is the action integral along the path ν with an arbitrary reference point $\tilde{R}$. A_ν is the classical probability for finding the particle at distance R: $A = (\partial S/\partial R)^{-1} = P^{-1}(R)$, i.e. A is proportional to the time interval that the particle spends at point R. δ_ν is a phase shift of the wave due to reflections at the turning point: here we have $\delta_1 = 0$ (no reflections) and $\delta_2 = -\pi/2$. At the classical turning point R_0 the WKB-wavefunction diverges and the forbidden region $R < R_0$ can be reached via complex continuation, giving an exponential decay. The connection of the classical allowed and the classical forbidden wave across the turning point where the wavefunction breaks down has been extensively examined (see e.g. /7/). The breakdown in the vicinity of R_0 can be overcome by uniformization techniques (see chapter 4). In contrast to the uniformized semiclassical results the nonuniformized ones are called 'primitive' semiclassical.

If the particle is bound in a one-dimensional potential well we get a superposition of semiclassical waves reflected on the left (R_1) and right (R_2) turning point in (2), leading to destructive interference, with exceptions at those energies, where the action integral

$$N(E) = \frac{1}{2\pi} \oint P(R)\, dR \tag{4}$$

over a complete oscillation equals a half integer multiple of $\hbar$:

$$N(E_n) = (n + 1/2)\,\hbar \qquad n = 0, 1, \ldots \quad , \tag{5}$$

which is the famous WKB quantization condition. The 1/2-term in (5) is due to phase jumps caused by reflections at the turning points.

In the case of d-dimensional bound systems the quantization condition (5) can be generalized, provided that the system is integrable. (A system is called integrable, if there exist d independent constants of motion, which are 'in involution', i.e. their pairwise poisson brackets vanish). In this case the classical motion is restricted to lie on an d-torus in phase space and d independent action integrals can be defined by

$$N_k(E) = \frac{1}{2\pi} \oint_k P(R)\, dR \qquad k = 1, \ldots, d \tag{6}$$

where the integration path is taken along one of the independent closed paths on the torus. The semiclassical quantization condition now reads

$$N_k(E) = (n_k + \alpha_k \pi/4)\,\hbar \qquad \begin{array}{l} k = 1, \ldots, d \\ n_k = 0, 1, \ldots \end{array} \tag{7}$$

i.e. the fluxes on the torus are quantized. The integers α_k are the Maslov indices, which are determined by the number of reflections at caustics /15/. The canonically conjugate variables to the actions N_k are the phase variables w_k . The Hamiltonian in terms of these action-angle variables does only depend on the N_k and

$$H(N_1, \dots, N_d) = E \tag{8}$$

determines together with (7) the energy eigenvalues $E(n_1, n_2, \dots, n_d)$. This treatment can be approximately extended to almost integrable systems. The semiclassical quantization of a non-integrable system - which is the typical case, i.e. 'most' systems are nonintegrable - turns out to be a highly nontrivial problem. For a discussion of semiclassical quantization see /15,16/.

In the following chapter we will turn to scattering situations starting first with an instructive review of semiclassical elastic potential scattering and a discussion of basic semiclassical techniques. Chapter 3 will introduce the semiclassical S-matrix for inelastic collisions. For simplicity we consider first vibrational excitation in collinear collisions. Chapter 4 generalizes to higher dimensional cases with emphasis on the rainbow catastrophe structure, with a short introduction into the theory of uniformization. As an application rotational excitation of a diatomic molecule is discussed in chapter 5. We conclude with a few general remarks in chapter 6.

2. Elastic Scattering and Basic Semiclassical Techniques

We consider the case of elastic potential scattering. In this case the differential cross section is the absolute square of the scattering amplitude

$$f(\vartheta) = \frac{-i\hbar}{(2mE)^{1/2}} \sum_{\ell=0}^{\infty} (\ell + \tfrac{1}{2}) \{\exp(2i\eta_\ell) - 1\} P_\ell(\cos\vartheta) \tag{9}$$

in the angular momentum expansion. The semiclassical analysis of elastic scattering is one of the best - however not fully - unterstood applications of semiclassical methods in scattering. For reviews see /2,9,17/. The semiclassical scattering amplitude can be obtained from (9) by executing a typical series of approximation steps:

(1) Sums over quantum numbers are converted into integrals ($\Delta\ell$ is 'small' if $\hbar \to 0$). Here we can use the exact (!) Poisson summation formula ($L = (\ell + 1/2)\hbar$)

$$\sum_{\ell=0}^{\infty} g(\ell) = \frac{1}{\hbar} \sum_{M=-\infty}^{+\infty} e^{-iM\pi} \int_0^{\infty} dL \, g(L/\hbar - 1/2) \, e^{2\pi i M L/\hbar} \tag{10}$$

This is actually a sum over integrals and we will see below that different values of M determine topologically different classes of trajectories.

(2) All quantities in the integrals are replaced by their semiclassical (asymptotic) expressions: The scattering phase shift η_ℓ defined by the asymptotic form of the radial wavefunction

$$\psi_\ell(R) \xrightarrow[R\to\infty]{} A \sin\left\{P_\ell R/\hbar - \ell\pi/2 + \eta_\ell\right\} \tag{11}$$

is replaced by the semiclassical phase shift (compare with eq. (1) for $R \to \infty$)

$$\eta_\ell = \eta(L)/\hbar = \lim_{R\to\infty} \frac{1}{\hbar}\left\{\int_{R_{0L}}^{R} P_L(R')\,dR' - PR\right\} + L\pi/2\hbar, \tag{12}$$

where

$$P_L(R) = \left[2m\left(E - V(R) - L^2/2mR^2\right)\right]^{1/2} \tag{13}$$

is the radial momentum and L is the classical angular momentum. $\eta(L)$ can be rewritten as

$$\eta(L) = -\int_0^P R\,dP_L + L\pi/2. \tag{12'}$$

In case of more than one turning point R_{0L} must be chosen as the outermost of these. We further approximate the Legendre polynomial $P_\ell(\cos\vartheta)$ by the asymptotic form

$$P_\ell(\cos\vartheta) = (\pi L \sin\vartheta/2\hbar)^{-1/2} \cos(L\vartheta/\hbar - \pi/4), \tag{14}$$

valid for ϑ not too close to 0 or π. The resulting expression for the scattering amplitude is

$$f(\vartheta) = -\frac{i}{2}(\pi\hbar m E \sin\vartheta)^{-1/2} \sum_{M=-\infty}^{+\infty} e^{-iM\pi}\left(e^{-i\pi/4} I_M^{+} + e^{i\pi/4} I_M^{-}\right) \tag{15}$$

with

$$I_M^{\pm} = \int_0^{\infty} dL\, L^{1/2}\, e^{\frac{i}{\hbar}\{2\eta(L) + L(\pm\vartheta + 2\pi M)\}} \quad (16)$$

(3) In the limit $\hbar \to 0$ the exponent in (16) is rapidly varying and therefore only those L-regions give nonvanishing contributions, where the phase is stationary: We evaluate $I_M^{\pm}$ by means of stationary phase approximation. This method approximates the exponent to second order in the vicinity of the points where the derivative vanishes. The resulting Gaussian integral can be evaluated analytically.

$$\int g(x)\, e^{\frac{i}{\hbar}f(x)}\, dx \approx \sum_{\nu;\, f'(x_\nu)=0} g(x_\nu) \left(\frac{2\pi\hbar}{f''(x_\nu)}\right)^{1/2} e^{i\{f(x_\nu)/\hbar + \pi/4\}} \quad (17)$$

The stationary phase condition gives

$$2\,\frac{d\eta(L)}{dL} = \mp\vartheta - 2\pi M \quad (18a)$$

and we furthermore note that $2\,d\eta/dL$ is identical with the classical deflection function $\Theta(L)$:

$$2\,\frac{d\eta(L)}{dL} = \Theta(L) = \pi - 2L \int_{R_0(L)}^{\infty} \frac{dR}{R^2 P_L(R)} \; . \quad (18b)$$

The final primitive semiclassical result is (for classically allowed processes)

$$f(\vartheta) = (2mE\sin\vartheta)^{-1/2} \sum_{\nu} \left| L(\Theta)\Big/\frac{d\Theta}{dL}\right|^{1/2} e^{i\{S_\nu/\hbar + \delta_\nu \pi/4\}} \; . \quad (19)$$

The summation extends over all paths satisfying the stationarity condition (18a,b). The phase δ_ν is determined by the topology of the classical path (reflections at caustics, number of circulations around the centre):

$$\delta_\nu = -\,\mathrm{sign}\,\frac{d\Theta}{dL}\Big|_\nu \mp 1 - 4M \; , \quad (20)$$

where "$\mp$" is the "$\mp$"-sign in (18a) and M counts the number of windings around the centre. S_ν is the classical collision action

$$S_\nu = S(L_\nu, \Theta) = 2\eta(L_\nu) - L_\nu \Theta$$

with $L_\nu = L(\Theta_\nu)$. (21)

Neglecting the interference oscillations in $|f(\vartheta)|^2$ we recover the classical differential cross section:

$$\frac{d\sigma}{d\Omega}(\vartheta) = \frac{1}{2mE\sin\vartheta} \sum_{\nu} L_\nu \Big/ \left|\frac{d\Theta}{dL}\right|_\nu \tag{22}$$

or - in the initial-value representation -:

$$\frac{d\sigma}{d\Omega}(\vartheta) = \frac{1}{2mE\sin\vartheta} \int_0^\infty dL\, L \sum_M \left\{ \delta(\Theta(L) - \vartheta + 2\pi M) + \delta(\Theta(L) + \vartheta + 2\pi M) \right\}, \tag{23}$$

where we sum over all initial conditions, i.e. all orbital angular momenta (impact parameter) of the incoming particle. The δ-functions pick up all trajectories with desired scattering angle ϑ.

Classically forbidden processes, i.e. scattering into the shadow region, diffraction effects for purely repulsive potentials, resonances from tunnelling through potential barriers etc., can be very conveniently and successfully described in the semiclassical equation (19) by complex-valued classical trajectories satisfying the prescribed boundary conditions (see, e.g. /2-5,17/ or the discussion of diffraction effects in /18/).

(4) The last step is the elimination of the divergencies at the rainbows by means of uniformization techniques. We defer a discussion of this point to chapter 4 below.

The inclusion of contributions with $M \neq 0$ is only important for diffraction effects close to the forward or backward direction /2,18/ or for orbiting scattering at very low energies /2,19/, where the energy is close to the barrier of the effective potential. In this region also tunnelling resonances become important, for a semiclassical treatment see /20/. In the following we restrict ourselves to M=o.

A typical example is shown in the upper part of Fig. 1 (taken from /21/). The interaction potential possesses an attractive minimum, which leads to a classical deflection function Θ(L) with an attractive minimum at a negative deflection angle $\Theta_R = \Theta(L_R)$, the rainbow angle. For $\vartheta < |\Theta_R|$ we have three contributing classical trajectories, at $\vartheta = |\Theta_R|$ two of them coalesce and disappear into the complex plane. For $\vartheta > |\Theta_R|$ a classically allowed and a forbidden trajectory remain.

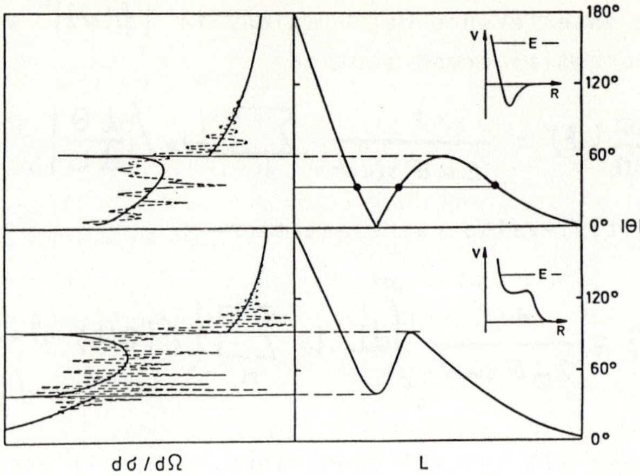

Fig.1. Elastic rainbow scattering for two typical interactions V(R) (taken from /21/)

The classical cross section diverges at the rainbow due to the vanishing of the denominator $d\Theta/dL$. The quantum cross sections show a characteristic interference pattern of fast and slow oscillations, which agrees almost exactly with the three-trajectory semiclassical formula (19), with the exception of a small region close to Θ_R. The lower part of Fig. 1 shows a different situation, where the potential (in this case a He^+-Ne interaction) is positive with a characteristic 'hump', which produces a deflection function with a 'pocket' and a double-sided rainbow /21/.

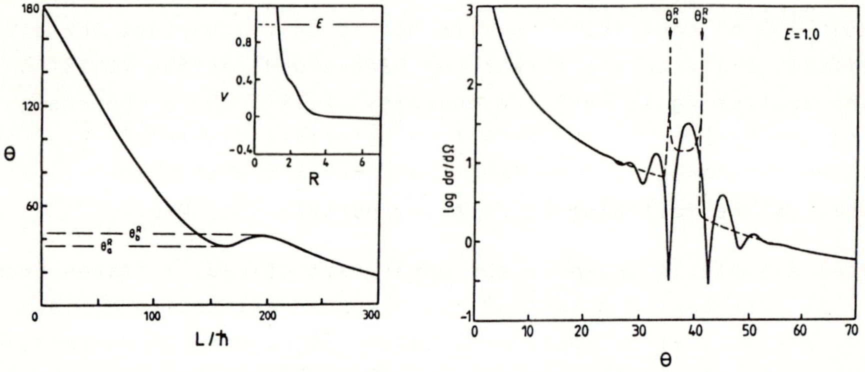

Fig. 2 Rainbow scattering from a purely repulsive potential /21/

It may be interesting to note, that such double-sided rainbows occur also for purely repulsive potentials ($dV/dR < 0$ for all R), an example is given in Fig. 2 (taken from /21/).

The semiclassical techniques described above have found numerous applications in different contexts, e.g. evaluation of matrix-elements (Franck-Condon factors, Clebsch-Gondon coefficients,...), in quantum statistical mechanics (e.g. a semiclassical representation of the partition function and related quantities /22/)... Last not least we would like to mention a semiclassical approximation of the off-shell scattering T-matrix /23/.

3. The Semiclassical S-Matrix: Collinear Vibrational Excitation

As a characteristic example of an inelastic scattering event we discuss a collinear-three particle system, where a particle 3 collides with a bound pair (1,2). For convenience we assume pair interactions between 1 and 2 (e.g. harmonic oscillator described by the Hamiltonian H_0) as well as 2 and 3 (described by an interaction potential V, e.g. exponential repulsion).

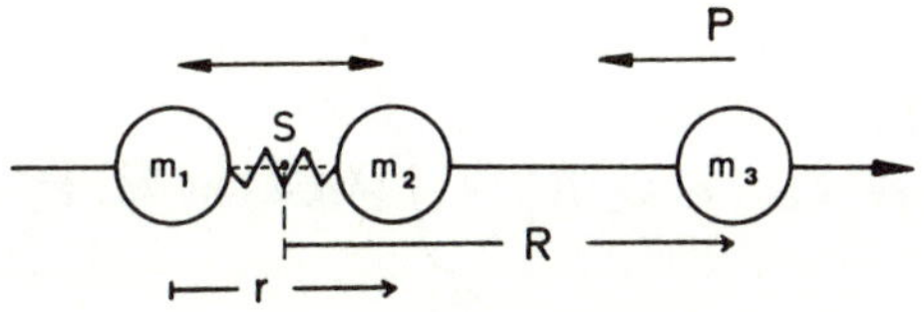

Fig. 3 Collinear vibrational excitation

We choose coordinates r (distance between 1 and 2) and relative coordinate R (distance of 3 from the (1,2) - center of mass) in the center of mass system. We are interested in the S-matrix element $S_{n_2 n_1}$ for transition between initial vibrational state n_1 and final state n_2. The transition probability is

$$P_{n_2 n_1} = | S_{n_2 n_1} |^2 . \tag{24}$$

There are several possibilities to derive the desired expression for the semiclassical S-matrix. Here we follow the derivation given by Child /4,24,25/ and Marcus /26/. First we describe the vibrational motion of the oscillator in terms of action angle variables N and w (see chapter 1). The Schrödinger equation now reads

$$\left\{ H_0\left(\hbar\left(-i\frac{\partial}{\partial w}+\frac{1}{2}\right)\right) - \frac{\hbar^2}{2m}\frac{\partial^2}{\partial R^2} + V(w,R)\right\}\Psi = E\Psi . \quad (25)$$

The asymptotic (V=o) solution of (25) is

$$\Psi^{\circ}(w,R) = P_1^{-1/2}\, e^{\frac{i}{\hbar}\{(N_1-\hbar/2)w + P_1 R\}} \quad (26)$$

where $P_1 = m\dot{R}$ is the asymptotic momentum ($m = m_3(m_1+m_2)/(m_1+m_2+m_3)$). The initial action N_1 is related to the initial vibrational quantum number n_1 by $N_1 = (n_1+1/2)\hbar$ (eq. (6)). The usual semiclassical ansatz for the solution

$$\Psi(w,R) = A(w,R)\, e^{\frac{i}{\hbar}\{W(w,R) - \hbar w/2\}} \quad (27)$$

and expansion of A, W in terms of $\hbar$ shows that W(w,R) is a solution of the Hamilton-Jacobi equation

$$H_0\left(\frac{\partial W}{\partial w}\right) + \frac{1}{2m}\left(\frac{\partial W}{\partial w}\right)^2 + V(w,R) = E , \quad (28)$$

which is solved by the line integral

$$W(w,R) = \int_{w_1}^{w} N\,dw + w_1 N_1 + \int_{R_1}^{R} P\,dR + R_1 P_1 - \pi\hbar/2 \quad (29).$$

evaluated along the classical trajectory with energy E and initial action N starting at (w_1,R_1). These starting values depend, of course, on (w,R) and the $w_1N_1+R_1P_1$ term removes the (w_1,R_1) dependence of $\partial W/\partial w$ and $\partial W/\partial R$. The constant $-\pi\hbar/2$ is only added for the outgoing solution and takes care of the phase jump at the turning point in the R-motion (see the discussion following eq. (2)). The normalization constant A(w,R) in (27) is obtained as

$$A(w,R) = \left(\frac{1}{P(R)}\frac{\partial w_1}{\partial w}\right)^{1/2} . \quad (30)$$

The angle variable w has in the asymptotic region a simple time dependence

$$w(t) = \omega t + \bar{w} , \quad (31)$$

where ω is the classical oscillation frequency $\omega = \partial H_0/\partial N_1$. This time dependence can be easily removed by transforming from variables (w,R) to ($\overline{w}$,t) with

$$t = mR/P. \tag{32}$$

When this transformation is carried out the S-matrix elements S_{n_2,n_1} can be computed by projecting the asymptotic semiclassical wavefunction (27) onto the asymptotic states with wavefunctions of type (26). The final result is (ref. /4/, eq. (178))

$$S_{n_2 n_1} = \frac{1}{2\pi}\int_0^{2\pi}\left(\frac{\partial\overline{w}}{\partial\overline{w}_1}\right)^{1/2} e^{\frac{i}{\hbar}\Delta(\overline{w}_1)}\, d\overline{w}_1 \tag{33}$$

with

$$\Delta(\overline{w}_1) = \left[N(N_1,\overline{w}_1) - N_2\right]\overline{w}(N_1,\overline{w}_1) - \int_{N_1}^{N(N_1,\overline{w}_1)} \overline{w}\,dN - \int_{P_1}^{P(N_1,\overline{w}_1)} R\,dP. \tag{34}$$

The S-matrix element is again a diffraction integral taken over all initial values of the angle variable $\overline{w}_1$.

As a next approximation step the integral (33) is evaluated by the stationary phase approximation. Stationarity of the phase leads to the condition

$$N_2 = N(N_1,\overline{w}_1), \tag{35}$$

i.e. only those trajectories ν contribute, which solve equation (35). The resulting primitive semiclassical S-matrix is

$$S_{n_2 n_1} = \sum_\nu \left[\frac{2\pi}{\hbar}\frac{\partial N}{\partial\overline{w}_1}\right]^{-1/2} e^{\frac{i}{\hbar}\Delta_\nu}. \tag{36}$$

It is illustrative to look at a simple consequence of the periodicity of the phase $\Delta(\overline{w}_1)$. In writing (33) or (36) we have implicitely assumed that n_1 and n_2 have integer values. Let us drop this restriction for a moment and integrate over <u>all</u> initial phases from $-\infty$ to $+\infty$. In this case we have additional roots of eq. (35) at $\overline{w}_{\nu\mu} = \overline{w}_\nu + 2\pi\mu$ $\mu = 0,\pm 1,\pm 2,\pm\ldots$ and

$$\Delta_{\nu\mu} = \Delta_\nu + (N_1 - N_2)2\pi\mu \tag{37}$$

and hence

$$S_{n_2 n_1} \sim \sum_{\mu=-\infty}^{+\infty} e^{\frac{i}{\hbar}(N_1 - N_2) 2\pi\mu} \sim \sum_{\tilde{n}=-\infty}^{+\infty} \delta(n_1 - n_2 - \tilde{n}) \quad (38)$$

i.e. we obtain the selection rule, that n_1-n_2 must be an integer. Eqs. (33) or (36) must usually be solved numerically by integrating the classical equations of motion. Typical transition probabilities are shown in Fig. (4).

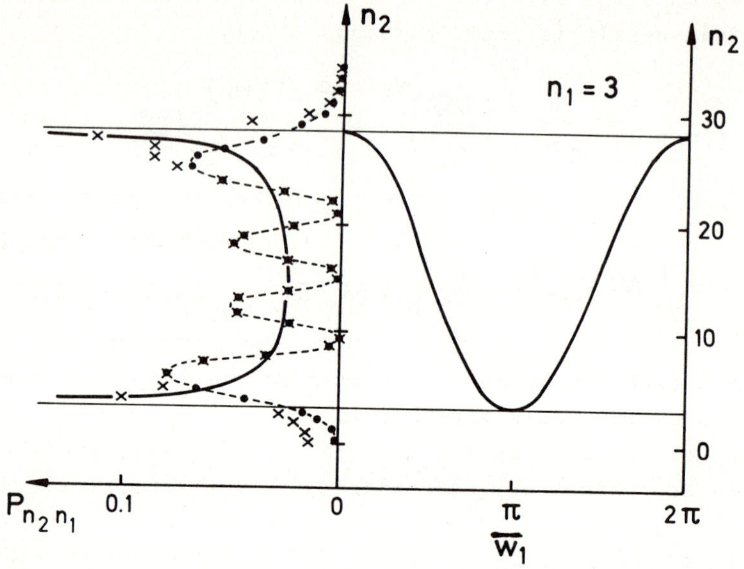

Fig. 4 Vibrational excitation function n_2 $(\overline{w}_1)$ and primitive semiclassical (x), quantum (•) as well as classical transition probabilities

The results are actually shown for harmonic oscillator and hard-sphere interaction V, where quantum /27/ and semiclassical /28/ results can be obtained in closed form. The parameters in Fig. 4 are $E_{kin}=48\hbar\omega$ and m=1/9. The quantum results (•) are compared with classical (—) and primitive semiclassical (x) computations. Two classical trajectories contribute in the classical allowed region, in the classical forbidden region complex classical trajectories must be used. At the boundaries of the classically allowed region the classical and primitive semiclassical results show the typical divergence which may be called 'vibrational rainbow'. The initial value representation (33) avoids this divergence at the rainbow and has been shown to be in good agreement with quantum results.

In the case of the following extremely simplified model the analysis can be carried out in closed form /24,29/. Let us assume that the oscillator is harmonic and the collision with the atom 3 is modelled by a time dependent force $f(t)$. In this case a straightforward derivation yields ($\hbar = \omega = 1$)

$$\Delta(\bar{w}_1) = N_1 \bar{w}_1 + \frac{\alpha}{2}\sqrt{2N_1}\cos\bar{w}_1 - N_2\,\bar{w}(\bar{w}_1) \tag{39}$$

with

$$\alpha = \int_{-\infty}^{+\infty} f(t)\cos t\,dt \tag{40}$$

and $\bar{w}(\bar{w}_1)$ is given by

$$\tan\bar{w} = \tan\bar{w}_1 - \alpha/\sqrt{2N_1}\cos\bar{w}_1 . \tag{41}$$

The phase $\Delta(\bar{w}_1)$ is stationary at $\bar{w}_1$ values satisfying the stationarity condition

$$N_2 = N_1 - \alpha\sqrt{2N_1}\sin\bar{w}_1 + \alpha^2/2 , \tag{42}$$

with the same qualitative behaviour as shown in fig. 4. The primitive semiclassical transition probabilities are

$$P_{n_2 n_1} = A_{n_2 n_1}\sin^2(\Delta_{n_2 n_1} + \pi/4) , \tag{43}$$

where $\Delta_{n_2 n_1}$ is the stationary value of $\Delta(\bar{w}_1)$ and the amplitude A is given by

$$A_{n_2 n_1} = 4\left[2\pi\left\{(N_1+N_2)\alpha^2 - (N_2-N_1)^2 - \alpha^4/4\right\}\right]^{-1/2} . \tag{44}$$

The theory outlined above can be easily extended to the treatment of three-dimensional collision of vibrating and rotating 'diatomic molecule' or even more complicated situations, e.g. polyatomic molecular reactions .... . (Note, however, the concluding remarks in chapter 6!).

In the following we will discuss semiclassical techniques from a more general point of view within the framework of catastrophe theory.

4. Rainbow Catastrophes

In the preceding sections we have discussed the semiclassical relationship between the quantum scattering amplitude and the classical trajectory field for simplified scattering problems. The results can be generalized to higher-dimensional situations. As an example rotational excitation is discussed in detail in chapter 5. Here we will discuss semiclassical scattering from a more general point of view. Typically the semiclassical scattering amplitude is given in terms of oscillatory integrals of the from

$$f(\vec{c}) = \int g(\vec{c},\vec{s})\, e^{\frac{i}{\hbar}\Delta(\vec{c},\vec{s})}\, d\vec{s} \tag{45}$$

with

$$\Delta(\vec{c},\vec{s}) = \vec{c}\cdot\vec{s} - \Gamma(\vec{s}) \tag{46}$$

in the initial value representation, which are of the same structure as those found in sections 2 and 3 (compare eqs. (16,33)). The variables $\vec{c}=(c_1,\ldots,c_d)$ denote the final values of the observables in the scattering event (e.g. scattering angle, internal quantum numbers or action variables) and $\vec{s}=(s_1,\ldots,s_d)$ are initial variables for the classical trajectories (e.g. orbital angular momentum and angle variables). In order to keep the discussion general we will adopt the language of catastrophe theory (see below) and call the variables $\vec{s}$ 'state variables' and the $\vec{c}$ 'control variables'. The stationary points of the integral (45)

$$\Delta_{\vec{s}}(\vec{c},\vec{s}) = 0 \quad \Longleftrightarrow \quad \vec{c} = \vec{\nabla}_{\vec{s}}\,\Gamma(\vec{s}) \tag{47}$$

induce a mapping (a Legendre transformation) between the variables $\vec{s}$ and $\vec{c}$, i.e. the smooth function $\Gamma(\vec{s})$ acts as a generator of this mapping. The stationary phase approximation to (45) is then

$$f(\vec{c}) = (2\pi\hbar)^{d/2} \sum_{\nu} \frac{g(\vec{c},\vec{s}_\nu)}{|\det D|_\nu^{1/2}}\, e^{i\{\Delta_\nu/\hbar + \frac{\pi}{4}\operatorname{sig} D_\nu\}} \tag{48}$$

(sig D is the signature of D) with the Hessian matrix

$$D = (D_{ik}) = \left(\frac{\partial^2\Delta}{\partial s_i\,\partial s_k}\right) \tag{49}$$

where the sum runs over the solution of (47). (Care must be taken, however, of the fact, that not <u>all</u> stationary points - in particular not all complex ones - contribute to (48). The proper choice depends on complex integration path deformations.) The determinant of D is the Jacobian of the mapping (47)

$$\det D = \left| \frac{\partial(c_1,\dots,c_d)}{\partial(s_1,\dots,s_d)} \right| . \tag{50}$$

The variation of the control parameters $\vec{c}$ causes the roots $\vec{s}$ of (50) to move in state-space. At certain values two or more of these roots coalesce and the mapping becomes singular, i.e. the Jacobian (50) will vanish and the primitive semiclassical result (48) will diverge. We call these critical values in control-space a rainbow catastrophe in analogy with the rainbow in potential scattering. The rainbow set in state space separates regions in which the mapping (47) is on-to-one.

The setup described by the gradient map (47) is precisely the object of interest in catastrophe theory /30-33/. Thom's theorem states, that the catastrophes are not at all arbitrary. In all generic cases the catastrophes (let us confine ourselves to the case of not more than five control parameters) are locally equivalent to a set of seven elementary catastrophes /30-33/, which are most conveniently characterized in terms of their 'relevant' number of state-variables (the 'dimension') and control-variables (the 'codimension'). The 'dimension' is restricted to the values one and two and the codimension is less or equal to five. For dimension one there are cuspoids (fold, cusp, swallowtail, ... with codimension increasing from one to five and for dimension two we have umbilics (elliptic-, hyperbolic-, parabolic- and symbolic-umbilic and their dual forms with codimensions three, four and five). Each of these elementary catastrophes can be described by a generating polynomial $P(C_1, \dots, C_m, S_1, \dots, S_n)$ with dimension $n \leq 2$ and codimension $m \leq 5$.

For the case of three-dimensonal atom-diatom scattering we have as control variables the scattering angle ϑ, the final rotational momentum J, it's projection onto a given axis M and the final vibrational action N, i.e. four control variables. We will therefore never encounter higher catastrophes as those on the elementary list, even if we vary the collision energy as parameter number five.

Along with the canonical catastrophe polynomials P(C,S) a set of canonical oscillatory integrals can be defined:

$$I(\vec{C}) = \int d\vec{S}\, e^{iP(\vec{C},\vec{S})} . \tag{51}$$

The most prominent member of this set of canonical integrals is the one related to the simplest catastrophe, the fold catastrophe:

$$Ai(C) = \frac{1}{2\pi}\int_{-\infty}^{+\infty} dS\, e^{i(S^3/3 + CS)}, \tag{52}$$

which is the Airy function. We will demonstrate some important aspects for the cusp-catastrophe integral: the Pearcey function /34/

$$Pe(C_1, C_2) = \int_{-\infty}^{+\infty} dS\, e^{i(S^4/4 + C_1 S^2/2 + C_2 S)} \tag{53}$$

This integral has three stationary points, two of them coalesce at the cusped rainbow curve $4C_1^3/27+C_2^2=0$ in control space, see Fig. 5a. Inside the cusp we have three real stationary points, outside only one together with two complex conjugate saddle points.

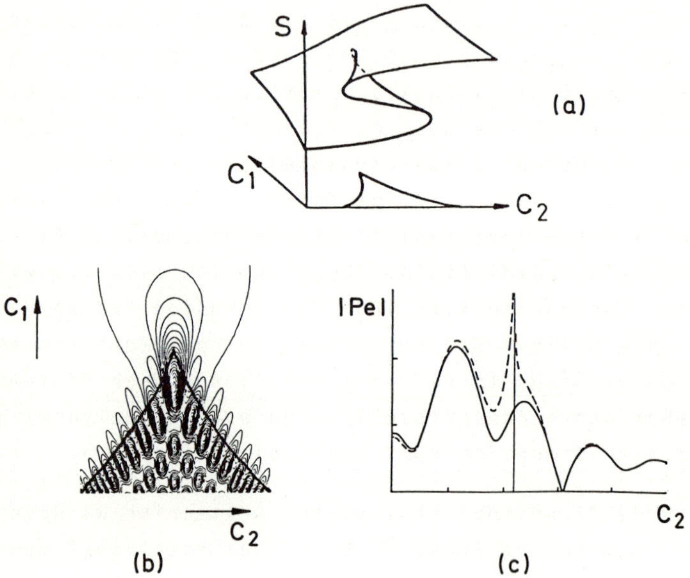

Fig. 5 Cusp catastrophe and canonical cusp oscillatory integral (Pearcey function)

The behaviour of the real stationary points as a function of (C_1, C_2) can be easily visualized in (C_1, C_2, S)-space, where they are localized on a (smooth!) manifold, whose projection gives the cusped rainbow in control space. Fig. 5b also shows the oscillatory pattern of the Pearcey function, which decorates the cusp /35/. Fig. 5c gives a cut through the cusp and shows a comparison of the Pearcey function with the stationary phase approximation /36/ (the oscillations outside the cusp are due to interference with a complex valued stationary point).

The canonical integrals (51) can be used in order to remove the divergencies from the primitive semiclassical results by means of a uniform mapping of the phase function of the diffraction integral (45) onto the phase of the canonical integral with the same catastrophe structure. For details see ref. /36-39/. We would like to note, however, that some additional considerations are necessary if the state space is periodic in some variables. For some uniformization results using Bessel functions see /40/.

5. Rotational Excitation

To illustrate the semiclassical method in more general applications we discuss the rotational excitation of a diatomic molecule by collision with an atom. The interaction is described by a potential $V(R,\gamma)$, where R is the distance of the atom from the molecular centre of mass and γ is the orientation of the molecular- with respect to the scattering-axis. For simplicity we assume a homonuclear molecule $V(R,\gamma)=V(R,\pi-\gamma)$ which is initially in the rotational ground state.

We treat the dynamics approximately by assuming a sudden collision, more precisely we use the so-called infinite-order-sudden (IOS) approximation, where the angle γ of the molecule is frozen during the collision. An exact quantum computation for the systems of interest here is still out of range at present times. It is assumed that the collision is energetically elastic, i.e. the molecule can only absorb rotational momentum and no energy (large rotational constant). This approximation has proved to be excellent for many realistic systems.

For the case of purely repulsive potentials the semiclassical scattering amplitude for rotational transitions $o\to j$ has been shown to be /41/

$$f(j\leftarrow 0,\vartheta) = \frac{1}{2\pi}(-1)^j(\hbar m E \sin\vartheta)^{-1/2} \cdot \int_0^\infty dL \int_0^{\pi/2} d\gamma\, (L\sin\gamma)^{1/2}\, e^{\frac{i}{\hbar}\{2\eta(L,\gamma)-\vartheta L-\mathcal{J}\gamma\}} \qquad (54)$$

In eq. (54) the symmetry of the potential has already been incorporated $(o\le\gamma\le\pi/2)$ and j must be an even integer. We have used the notation

$$\mathcal{J} = (j+1/2)\hbar \qquad (55)$$

and $\eta(L,\gamma)$ is the WKB-phase shift (12) for a γ-dependent potential $V(R,\gamma)$.

Evaluating the integral by two-dimensional stationary phase gives the primitive semiclassical result in the classically allowed region /41/.

$$f(j\leftarrow 0, \vartheta) = (-1)^j (\hbar m E \sin\vartheta)^{-1/2} \cdot \sum_\nu \left[\frac{L \sin\gamma}{|\det D|}\right]_\nu^{1/2} e^{i\{\Delta_\nu/\hbar + \frac{\pi}{4} \mathrm{sig}\, D_\nu\}} \tag{56}$$

with phase

$$\Delta = 2\eta(L,\gamma) - \vartheta L - J\gamma . \tag{57}$$

D is the Hessian matrix

$$D = \begin{pmatrix} 2\frac{\partial^2\eta}{\partial L^2} & 2\frac{\partial^2\eta}{\partial L\partial\gamma} \\ 2\frac{\partial^2\eta}{\partial L\partial\gamma} & 2\frac{\partial^2\eta}{\partial\gamma^2} \end{pmatrix} \tag{58}$$

and sig D is the signature of D.
The stationary phase conditions are

$$\vartheta = \Theta(L,\gamma) \quad , \quad J = \mathcal{J}(L,\gamma) \tag{59}$$

where

$$\Theta(L,\gamma) = 2\,\frac{\partial\eta(L,\gamma)}{\partial L}$$

is the classical deflection functions (18b) and

$$\mathcal{J}(L,\gamma) = 2\,\frac{\partial\eta(L,\gamma)}{\partial\gamma} \tag{60}$$

the classical excitation function, i.e. the classical rotational momentum transfered to the molecule. Eq. (59) must be solved for the root-trajectories (L_ν, γ_ν) and we find two such roots in the classically allowed region, i.e. two molecular orientations leading to a prescribed scattering result (ϑ, J). The amplitude in (56) can be rewritten as

$$\left[\frac{L \sin\gamma}{|\det D|}\right]^{1/2} = \left[L \sin\gamma \Big/ \left|\frac{\partial(\Theta, \mathcal{J})}{\partial(L,\gamma)}\right|\right]^{1/2} , \tag{61}$$

where the Jacobian of the transformation $(L,\gamma) \leftrightarrow (\vartheta, j)$ appears in the denominator. Classically forbidden processes can be treated by complex continuation of classical mechanics.

Dropping the interference phases in (56) we obtain the fully classical cross section

$$\frac{d\sigma}{d\Omega}(j\leftarrow 0,\vartheta) = \frac{\hbar}{mE}\sum_{\nu=1}^{2}\left[L\sin\gamma \Big/ \left|\frac{\partial(\Theta,\mathcal{J})}{\partial(L,\gamma)}\right|\right]_{\nu} \tag{62}$$

$$= \frac{\hbar}{mE}\iint dL\, d\gamma\; L\sin\gamma\; \delta(\vartheta - \Theta(L,\gamma))\,\delta(j - \mathcal{J}(L,\gamma)).$$

As a realistic application He-Na_2 scattering at 0.1eV collision energy /41/ is shown in figs. 6,7. The potential is an analytic fit to computed ab initio values. Fig. 6 shows the deflection- and excitation-functions $\Theta(L,\gamma)$ and $\mathcal{J}(L,\gamma)$ as contour lines on the (L,γ)-plane.

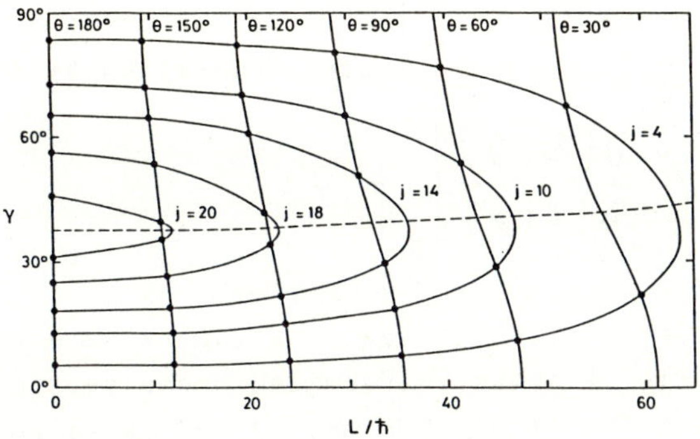

Fig. 6 Contour lines of the classical deflection function $\Theta(L,\gamma)$ and the classical rotational excitation function $\mathcal{J}(L,\gamma)$ for He-Na_2 collisions at 0.1eV collision energy. The dashed line is the rainbow curve.

The two contributing trajectories (L_ν, γ_ν) for various $(\Theta,\mathcal{J})$-values are marked by dots. At the dashed curve (the 'rainbow' curve) these trajectories coincide. The rainbow line $(L,\gamma)_R$ maps onto the rainbow line $\mathcal{J}_R(\Theta)$ in $(\Theta,\mathcal{J})$-space, (see Fig. 7;/41/) which is the boundary of the classically allowed region.

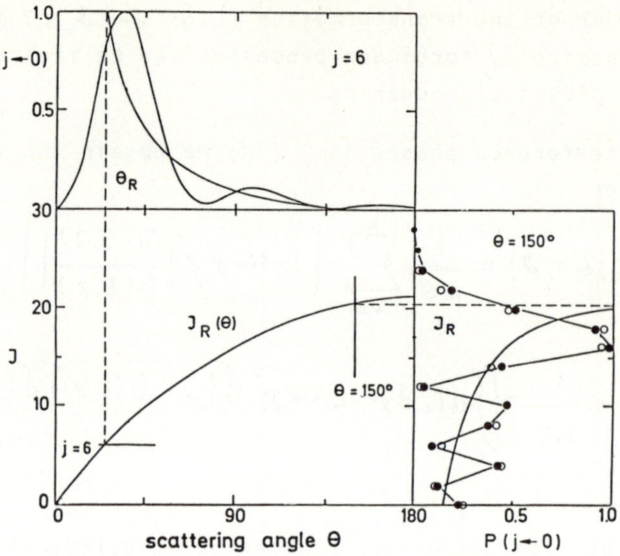

Fig. 7 Classical rotational rainbow curve $J_R(\theta)$ as well as classical, semiclassical and quantal rotational transition probabilities

The resulting probabilities for rotational excitation as a function of θ or J are also shown in Fig. 7, which illustrates the so-called rotational rainbow scattering /41,42/: The classical cross sections diverges at the classical rainbow curve $J_R(\theta)$ determined by the vanishing of the Jacobian

$$\left.\frac{\partial(\theta, J)}{\partial(L, \gamma)}\right|_R = 0 \; . \tag{63}$$

The primitive semiclassical cross sections (56) also diverge at the classical rainbow. Fig. 7 shows uniformized cross sections (o) (compare chapter 4), which are finite in this region and agree very well with quantum computations (•) as well as with experimental measurements /43/.

Very similar effects have been found in different fields of physics. Fig. 8 shows as an example rotational transition probabilities for Coulomb excitation of nuclei in heavy ion collisions ($^{84}Kr + ^{238}U$ at a collision energy of 385 MeV /44/, for similar results in nuclear scattering see /45/). The figure shows the classical rotational excitation probability $P(j \leftarrow o)$ at fixed scattering angle as a histogram. The dots are quantum results /44/. The rotational rainbow structure for this nuclear system is identical with the molecular ones. Very similar structures have been found in recent measurements of rotational excitation in electron molecule collisions (e-Na_2; backward scattering at 300eV collision energy) /46/.

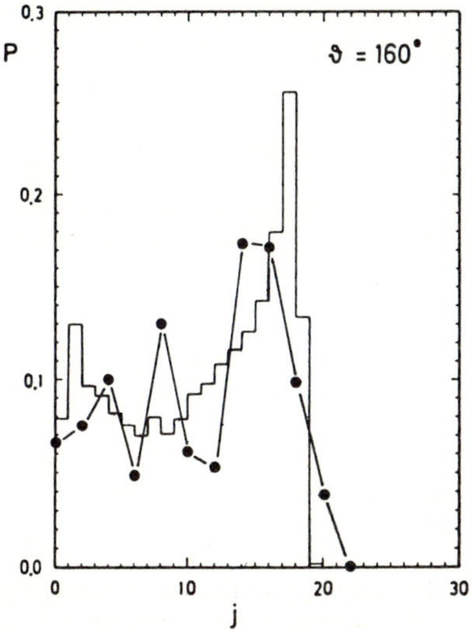

Fig. 8 Classical and (semi)quantal cross sections for rotational excitation of ^{238}U nuclei by collisions with ^{84}Kr at 385 MeV collision energy

It might be interesting to take a closer look at the classical results shown in Fig. 8, which show a characteristic deviation from the behaviour of the classical IOS cross sections shown in Fig. 7: The square-root singularity at the rotational rainbow has disappeared and two spikes for low and high j-values are observed. This is not an artefact of the statistics in the classical Monte-Carlo integration. It is due to deviations from the IOS-approximations and a closer analysis reveales, that the IOS square-root singularity is changed into a step at the maximum value of j and two logarithmic singularities at lower j values /47/.

The rainbow patterns discussed above have a very simple structure. This is due to the simple topology of the potential surface. Let us now examine more complicated situations, e.g. rotationally inelastic atom-diatom collisions, where the potential is anisotropic and possesses a potential minimum. Here the 'ordinary' potential well rainbows found for elastic scattering (chapter 2) and the rotational rainbows will coexist. They are, however, not independent and interact in a characteristic manner. First we assume, that the potential well is spherically symmetric and the anisotropic part remains purely repulsive. Fig. 9 shows the rainbow structure for such a system /20,36,48/. First we note, that negative deflection angles appear, which have been folded to positive scattering angles in Fig. 9. We find two rainbow structures: a fold type catastrophe (α) and a cusped structure in the low J, low ϑ region (β). Both rainbows can be viewed as part of a hyperbolic umbilic catastrophe.

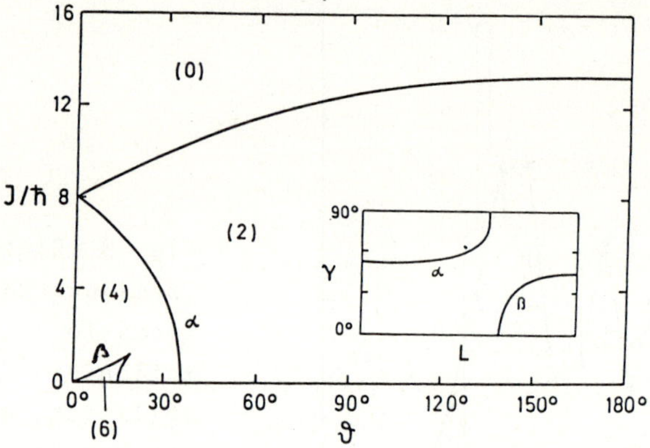

Fig. 9 Characteristic rotational rainbow patterns for anisotropic potentials with a spherically symmetric potential well

The number of contributing classical trajectories ranges from (o) to (6) in the different regions. The semiclassical interference pattern is complicated. As an example Fig. 10 shows the differential cross section for the 0→2 transition. In this case the cusp region is already classically inaccessible.

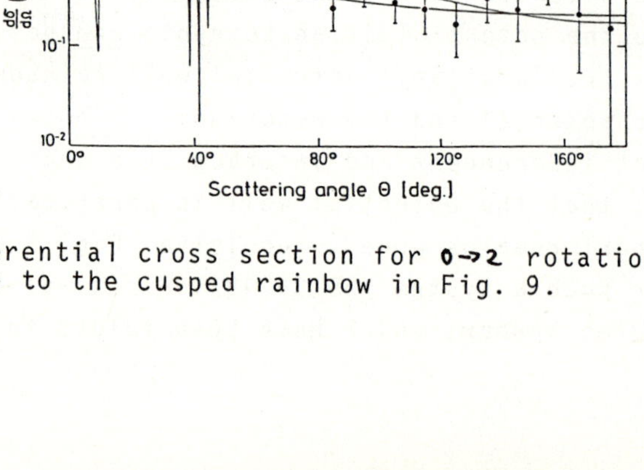

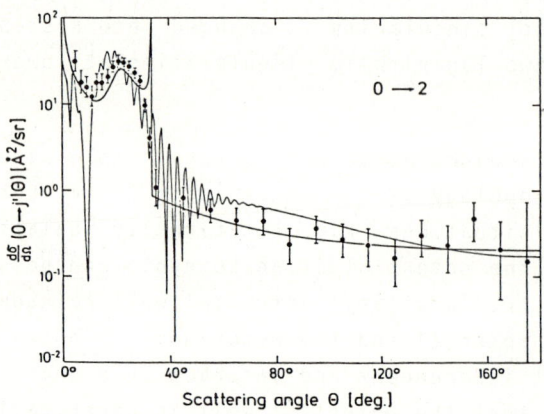

Fig. 10 Differential cross section for 0→2 rotational transitions close to the cusped rainbow in Fig. 9.

The strong singularity due to the coalescence of three trajectories at the cusp point (see chapter 4) produces a pronounced hump in the cross sections, which has been reproduced in exact classical trajectory computations (full dots in Fig. 10) /48/. The semiclassical uniformized cross sections have been studied in /36/, good agreement with quantum results has been obtained.

6. Concluding Remarks and a Final Warning

In the preceding chapters we have discussed some aspects of the semiclassical limit of quantum scattering. We have learned that the quantum S-matrix can be constructed from the classical trajectories, or more precisely from the classical trajectory field, whose local structure is determined by the classical catastrophes, classified by Thom's famous theorem. The global topology of the rainbow catastrophes is characteristic for a particular class of collision systems and can be analyzed once and for all. Knowing the rainbow structure one can directly deduce the root trajectory pattern and from this the quantum interference oscillations - in pictorial words phrased as 'sewing the wave flesh on the classical bones' /2/, which are organized by the skeleton of rainbow catastrophes. This allows an understanding of the features of quantum mechanics on different levels of abstraction.

Before closing this short review we find it inevitable to give credit to Keller's 'geometric theory of diffraction' /49/, who developed many aspects of semiclassical mechanics in the framework of wave optics/geometrical optics (see also Berry's beautiful articles on the 'semiclassical' theory of wave optics /37,50/ as well as Nye's fascinating experiments on the catastrophe optics of small water droplet diffraction (see /51/ and references therein).

We have described the general machinery of semiclassical dynamics and we have demonstrated its usefulnes in practical applications. A final warning seems to be appropriate, however, in order to avoid unwarranted generalizations. Semiclassical mechanics - as discussed above - is intimately linked to classical integrability (compare chapter 1), in particular the asymptotic states in scattering systems must be integrable. But even if this is the case, the collision systems can show features of nonintegrability, i.e. chaos in collisions, linked to classically chaotic 'sticking' trajectories. For recent discussions see /52/. The semiclassical limit of quantum mechanics for nonintegrable systems (espacially for those, who are not 'close' to integrable ones) is still far from being understood and therefore a challange for future research.

References:

/1/ N. Fröman and P.O. Fröman: "On the History of the so-called WKB-Method from 1817 to 1926", in /11/, pp.1-7

/2/ M.V. Berry and K.E. Mount, Rept. Progr. Phys. 35 (1972), 315

/3/ W.H. Miller, Adv. Chem. Phys. 25 (1974) 65; Adv. Chem. Phys. 30 (1975) 77

/4/ M.S. Child, in "Dynamics of Molecular Collisions", Part B, W.H. Miller Ed. (Plenum, N.Y., 1976) pp. 171-216

/5/ M.S. Child: "Semiclassical Reactive Scattering", in: "Theory of Chemical Reaction Dynamics", M. Baer, Ed. (CRC Press, 1985) Vol. III, p.247

/6/ R.P. Feynman and A.R. Hibbs: "Quantum Mechanics and Path Integrals" (McGraw-Hill, N.Y., 1965)

/7/ N. Fröman and P.O. Fröman: "JWKB Approximation" (North-Holland, Amsterdam, 1965)

/8/ V.P. Maslov and M.V. Fedoriuk: "Semiclassical Approximation in Quantum Mechanics" (Reidel, Dordrecht, 1981)

/9/ D.M. Brink: "Semiclassical Methods for Nucleus-Nucleus Scattering" (Cambridge Univ. Press 1985)

/10/ "Semiclassical Methods in Molecular Scattering and Spectroscopy" M.S. Child, Ed. (Reidel, Dordrecht, 1980)

/11/ "Semiclassical Descriptions of Atomic and Nuclear Collisions", Proceedings of the Niels Bohr Centennial Conf. Copenhagen, J. Bang and J. DeBoer, Eds. (North-Holland, Amsterdam, 1985)

/12/ F. Wolf, R.J. Allan and H.J. Korsch, Comm. At. Mol. Phys. 18 (1986) 107

/13/ N. Fröman: "Semiclassical and Higher-Order Approximations: Properties, Solution of Connection Problems", in /10/, pp. 1-44

/14/ U. Buck, Rev. Mod. Phys. 46 (1974) 369; R.J. LeRoy, in /10/, pp. 109-126; M.S. Child and R.B. Gerber, Mol. Phys. 38 (1979) 421; R. Schinke, J. Chem. Phys. 73 (1980) 6117

/15/ I.C. Percival, Adv. Chem. Phys. 36 (1977) 1

/16/ M.V. Berry: "Regular and Irregular Motion", Amer. Inst. Phys. Conf. Proc. Nr. 46, S. Jorna, Ed. (1978) 16-120

/17/ J.N.L. Connor: "Semiclassical Theory of Elastic Scattering", in /10/, pp. 45-108

/18/ H.J. Korsch and D. Leissing, J. Phys. B9 (1976) 1857

/19/ H.J. Korsch and K.E. Thylwe, J. Phys. B16 (1983) 793

/20/ H.J. Korsch: "Semiclassical Theory of Resonances", in: 'Resonances - Models and Phenomena', Lecture Notes in Physics (Springer), S. Albeverio et al. Eds. (1984), pp. 217-234; H.J. Korsch and K.E. Thylwe, J. Phys. B19 (1986) 2139; H.J. Korsch, R. Möhlenkamp and K.E. Thylwe, J. Phys. B19 (1986) 2151; R. Möhlenkamp and H.J. Korsch, Phys. Rev. A, to be published

/21/ H.J. Korsch and F. Wolf, Comm. At. Mol. Phys. 15 (1984) 139; R. Möhlenkamp, H.J. Korsch and R.J. Allan, J. Phys. B17 (1984) L 673

/22/ H.J. Korsch, J. Phys. A12 (1979)811; A12 (1979) 1521

/23/ H.J. Korsch, Phys. Rev. A14 (1976) 1645; H.J. Korsch and R. Möhlenkamp, J. Phys. B10 (1977) 3451; B15 (1982) 2187

/24/ M.S. Child: "The Classical S-Matrix", in /10/, pp.155-177

/25/ M.S. Child: "Molecular Collision Theory" (Academic Press, London, 1974)

/26/ R.A. Marcus, J. Chem. Phys. 54 (1971) 3965

/27/ P. Eckelt and H.J. Korsch, Chem. Phys. Lett. 11 (1971) 313

/28/ H.J. Korsch, unpublished

/29/ P. Pechukas and M.S. Child, Mol. Phys. 31 (1976) 973

/30/ R. Thom: "Structural Stability and Morphogenesis" (Benjamin Reading, 1975)

/31/ T. Poston and I. Stewart: "Catastrophe Theory and its Applications" (Pitman, London, 1978)

/32/ R. Gilmore: "Catastrophe Theory for Scientists and Engineers" (John Wiley, N.Y., 1981)

/33/ V.I. Arnold: "Singularity Theory" (Cambridge Univ. Press, 1981)

/34/ T. Pearcey, Philos. Mag. 37 (1946) 311

/35/ F.J. Wright, Thesis Bristol University 1977; M.V. Berry, J. Phys. A8 (1975) 566

/36/ F. Wolf and H.J. Korsch, J. Chem. Phys. 81 (1984) 3127

/37/ M.V. Berry, Adv. Phys. 25 (1976) 1

/38/ J.N.L. Connor, Discuss. Faraday Soc. 55 (1973) 51; Mol. Phys. 31 (1976) 33; J.N.L. Connor and D. Farrelly, J. Chem. Phys. 75 (1981) 2831

/39/ T. Uzer and M.S. Child, Mol. Phys. 46 (1982) 1371

/40/ J.R. Stine and R.A. Marcus, J. Chem. Phys. 59 (1973) 5145; K. Kreek, R.L. Ellis and R.A. Marcus, J. Chem. Phys. 61 (1974) 4540

/41/ H.J. Korsch and R. Schinke, J. Chem. Phys. 73 (1980) 1222; 75 (1981) 3850

/42/ R. Schinke and J.M. Bowman, in "Molecular Collision Dynamics", J.M. Bowman, Ed. (Springer, Heidelberg, 1983)

/43/ P.L. Jones, U. Hefter, A. Mattheus, J. Witt, K. Bergmann, W. Müller, W. Meyer and R. Schinke, Phys. Rev. A26 (1982) 1283

/44/ S. Landowne and A. Venturi: "Inelastic Scattering-Nuclear" in "Treatise on Heavy-Ion Science", Vol. 1, D.A. Bromley, Ed., pp. 355-460, especially § 4

/45/ S. Levit: "Semiclassical Functional-Integral Methods for Few and Many-Body Systems", in /11/, pp. 119-133

/46/ G. Ziegler, M. Rädle, O. Pütz, K. Jung, H. Ehrhardt and K. Bergmann, to be published

/47/ H.J. Korsch and D. Richards, J. Phys. B14 (1981) 1973; H.J. Korsch and D. Poppe, Chem. Phys. 69 (1982) 99; H.J. Korsch, Z.V. Lewis and D. Poppe, Z. Phys. A312 (1983) 277

/48/ R. Schinke, H.J. Korsch and D. Poppe, J. Chem. Phys. 77 (1982) 6005

/49/ J.B. Keller, Ann. Phys. N.Y. 4 (1958) 180; Proc. Symp. Appl. Math. 8 (1958) 27; J. Opt. Soc. Am. 52 (1962) 116

/50/ M.V. Berry, J. Phys. A8 (1975) 566

/51/ J.F. Nye, Proc. R. Soc. Lond. A403 (1986) 1

/52/ S.D. Bosanac, Phys. Rev. A32 (1985) 871; Ch. Schlier, preprint 1985; Ch. Jung, preprint 1986

LIST OF PARTICIPANTS

Y. *Akaishi*, Hokkaido University, Sapporo, JAPAN

A. *Arriaga*, Centro de Física Nuclear, Lisboa, PORTUGAL

A. *Barbosa*, Centro Física Nuclear, Lisboa, PORTUGAL

M.*Batinic*, Ruder Boskovic Institut, Zagreb, YOGOSLÁVIA

P. *Bicudo*, Centro de Física da Matéria Condensada, Lisboa, PORTUGAL

D. *Bollé*, Universiteit Leuven, Heverlee, BELGIQUE

E. *Brandas*, University of Uppsala, Uppsala, SWEDEN

J. *Buesco*, Faculdade Ciências de Lisboa, Lisboa, PORTUGAL

B. *Cabral*, Faculdade Ciências de Lisboa, Lisboa, PORTUGAL

S. *Coon*, University of Azizona, Tucson, USA

E. *Cravo*, Centro de Física Nuclear, Lisboa, PORTUGAL

M. *Crespo*, Centro de Física Nuclear, Lisboa, PORTUGAL

A. *Eiró*, Centro de Física Nuclear, Lisboa, PORTUGAL

M. *Fabre de la Ripelle*, Institut de Physique Nucléaire, Orsay, FRANCE

L. *Ferreira*, Universidade de Coimbra, Coimbra, PORTUGAL

J. *Fleisher*, Universitaet Bielefeld, Bielefeld, WEST GERMANY

A. *Fonseca*, Centro de Física Nuclear, Lisboa, PORTUGAL

J. *Friar*, Los Alamos National Laboratory, Los Alamos, USA

J. *Frohlich*, Osteirreichisches Forschungszentrum Seibersdorf, Seibersdorf, AUSTRIA

U. *Gastaldi*, CERN Geneve, SWITZERLAND

F. *Gesztesy*, Institut fur Theoretische Physik, Graz, AUSTRIA

B. *Gibson*, Los Alamos National Laboratory, Los Alamos, USA

W. *Glockle*, Ruhr-Universitaet Bochum, Bochum, WEST GERMANY

G. *Goulard*, Collège Militaire Royal de St.Jean, Quebec, CANADA

D. *Hennequin*, Collège Militaire Royal de St.Jean, Quebec, CANADA

H. *Hofmann*, Universitaet Erlangen, Nurnberg, WEST GERMANY

A.J. *Huizing*, Vrye Universiteit, Amsterdam, THE NETHERLANDS

O. *Kalush*, Ruhr - Universitaet Bochum, Bochum, WEST GERMANY

L. *Kok*, University of Groningen, Groningen, THE NETHERLANDS

J. *Korsch*, Universitaet Kaiserslautern, Kaiserslautern, WEST GERMANY

T. Lim, Drexel University, Philadelphia, USA

S. Liuti, Instituto Superiore di Sanita, Roma, ITALY

M. Mahlke, Ruhr - Universitaet Bochum, Bochum, WEST GERMANY

S. Oryu, Science University of Tokyo, Tokyo, JAPAN

G. Payne, University of Iowa, Iowa City, USA

M. Pena, Centro de Física Nuclear, Lisboa, PORTUGAL

W.Plessas, Institut fur Theorethische Physik, Graz, AUSTRIA

G. Rupp, Instituto Superior Técnico, Lisbon, PORTUGAL

F.D. Santos, Centro de Física Nuclear, Lisbon, PORTUGAL

T. Sasakawa, Tohoku University, Sendai, JAPAN

M. Sawicki, University of Warsaw, Warsaw, POLAND

H. Siedentop, Universitaet Carolo-Wilhelmina, Braunschweig, WEST GERMANY

K. Schmidt, New York University, New York, USA

I. Slaus, Ruder Boskovic Institute, Zagreb, YUGOSLÁVIA

E. Steinmetz, Institut fur Physik, Univertaet Mainz, WEST GERMANY

L. Streit, Universitaet Bielefeld, Bielefeld, WEST GERMANY

J. Tjon, Univ. of Utrecht, Utrecht, THE NETHERLANDS

V. Vento, University of Valencia, Valencia, SPAIN

J. Yun, Faculdade de Ciências de Lisboa, Lisboa, PORTUGAL

H. Witala, Ruhr - Universitaet Bochum, Bochum, WEST GERMANY

M. Wolker, Ruhr - Universitaet Bochum, Bochum WEST GERMANY

Lecture Notes in Physics

Vol. 237: Nearby Molecular Clouds. Proceedings, 1984. Edited by G. Serra. IX, 242 pages. 1985.

Vol. 238: The Free-Lagrange Method. Proceedings, 1985. Edited by M.J. Fritts, W.P. Crowley and H. Trease. IX, 313 pages. 1985.

Vol. 239: Geometrics Aspects of the Einstein Equations and Integrable Systems. Proceedings, 1984. Edited by R. Martini. V, 344 pages. 1985.

Vol. 240: Monte-Carlo Methods and Applications in Neutronics, Photonics and Statistical Physics. Proceedings, 1985. Edited by R. Alcouffe, R. Dautray, A. Forster, G. Ledanois and B. Mercier. VIII, 483 pages. 1985.

Vol. 241: Numerical Simulation of Combustion Phenomena. Proceedings, 1985. Edited by R. Glowinski, B. Larrouturou and R. Temam. IX, 404 pages. 1985.

Vol. 242: Exactly Solvable Problems in Condensed Matter and Relativistic Field Theory. Proceedings, 1985. Edited by B.S. Shastry, S.S. Jha and V. Singh. V, 318 pages. 1985.

Vol. 243: Medium Energy Nucleon and Antinucleon Scattering. Proceedings, 1985. Edited by H.V. von Geramb. IX, 576 pages. 1985.

Vol. 244: W. Dittrich, M. Reuter, Selected Topics in Gauge Theories. V, 315 pages. 1986.

Vol. 245: R.Kh. Zeytounian, Les Modèles Asymptotiques de la Mécanique des Fluides I. IX, 260 pages. 1986.

Vol. 246: Field Theory, Quantum Gravity and Strings. Proceedings, 1984/85. Edited by H.J. de Vega and N. Sánchez. VI, 381 pages. 1986.

Vol. 247: Nonlinear Dynamics Aspects of Particle Accelerators. Proceedings, 1985. Edited by J.M. Jowett, M. Month and S. Turner. VIII, 583 pages. 1986.

Vol. 248: Quarks and Leptons. Proceedings, 1985. Edited by C.A. Engelbrecht. X, 417 pages. 1986.

Vol. 249: Trends in Applications of Pure Mathematics to Mechanics. Proceedings, 1985. Edited by E. Kröner and K. Kirchgässner. VIII, 523 pages. 1986.

Vol. 250: Lie Methods in Optics. Proceedings 1985. Edited by J. Sánchez Mondragón and K.B. Wolf. XIV, 249 pages. 1986.

Vol. 251: R. Liebmann, Statistical Mechanics of Periodic Frustrated Ising Systems. VII, 142 pages. 1986.

Vol. 252: Local and Global Methods of Nonlinear Dynamics. Proceedings, 1984. Edited by A.W. Sáenz, W.W. Zachary and R. Cawley. VII, 263 pages. 1986.

Vol. 253: Recent Developments in Nonequilibrium Thermodynamics Fluids and Related Topics. Proceedings, 1985. Edited by J. Casas-Vázquez, D. Jou and J.M. Rubí. X, 392 pages. 1986.

Vol. 254: Cool Stars, Stellar Systems, and the Sun. Proceedings, 1985. Edited by M. Zeilik and D.M. Gibson. XI, 501 pages. 1986.

Vol. 255: Radiation Hydrodynamics in Stars and Compact Objects. Proceedings, 1985. Edited by D. Mihalas and K.-H. A. Winkler. VI, 454 pages. 1986.

Vol. 256: Dynamics of Wave Packets in Molecular and Nuclear Physics. Proceedings, 1985. Edited by J. Broeckhove, L. Lathouwers and P. van Leuven. VIII, 187 pages. 1986.

Vol. 257: Statistical Mechanics and Field Theory: Mathematical Aspects. Proceedings, 1985. Edited by T.C. Dorlas, N.M. Hugenholtz and M. Winnink. VII, 328 pages. 1986.

Vol. 258: Wm. G. Hoover, Molecular Dynamics. VI, 138 pages. 1986.

Vol. 259: R.F. Alvarez-Estrada, F. Fernández, J.L. Sánchez-Gómez, V. Vento, Models of Hadron Structure Based on Quantum Chromodynamics. VI, 294 pages. 1986.

Vol. 260: The Three-Body Force in the Three-Nucleon System. Proceedings, 1986. Edited by B.L. Berman and B.F. Gibson. XI, 530 pages. 1986.

Vol. 261: Conformal Groups and Related Symmetries – Physical Results and Mathematical Background. Proceedings, 1985. Edited by A.O. Barut and H.-D. Doebner. VI, 443 pages. 1986.

Vol. 262: Stochastic Processes in Classical and Quantum Systems. Proceedings, 1985. Edited by S. Albeverio, G. Casati and D. Merlini. XI, 551 pages. 1986.

Vol. 263: Quantum Chaos and Statistical Nuclear Physics. Proceedings, 1986. Edited by T.H. Seligman and H. Nishioka. IX, 382 pages. 1986.

Vol. 264: Tenth International Conference on Numerical Methods in Fluid Dynamics. Proceedings, 1986. Edited by F.G. Zhuang and Y.L. Zhu. XII, 724 pages. 1986.

Vol. 265: N. Straumann, Thermodymamik. VI, 140 Seiten. 1986.

Vol. 266: The Physics of Accretion onto Compact Objects. Proceedings, 1986. Edited by K.O. Mason, M.G. Watson and N.E. White. VIII, 421 pages. 1986.

Vol. 267: The Use of Supercomputers in Stellar Dynamics. Proceedings, 1986. Edited by P. Hut and S. McMillan. VI, 240 pages. 1986.

Vol. 268: Fluctuations and Stochastic Phenomena in Condensed Matter. Proceedings, 1986. Edited by L. Garrido. VIII, 413 pages. 1987.

Vol. 269: PDMS and Clusters. Proceedings, 1986. Edited by E.R. Hilf, F. Kammer and K. Wien. VIII, 261 pages. 1987.

Vol. 270: B.G. Konopelchenko, Nonlinear Integrable Equations. VIII, 361 pages. 1987.

Vol. 271: Nonlinear Hydrodynamic Modeling: A Mathematical Introduction. Edited by Hampton N. Shirer. XVI, 546 pages. 1987.

Vol. 272: Homogenization Techniques for Composite Media. Proceedings, 1985. Edited by E. Sanchez-Palencia and A. Zaoui. IX, 397 pages. 1987.

Vol. 273: Models and Methods in Few-Body Physics. Proceedings, 1986. Edited by L.S. Ferreira, A.C. Fonseca and L. Streit. XIX, 674 pages. 1987.